third edition

ORGANIC CHEMISTRY
A Brief Course

Robert C. Atkins
James Madison University

Francis A. Carey
University of Virginia

McGraw Hill

Boston   Burr Ridge, IL   Dubuque, IA   Madison, WI   New York   San Francisco   St. Louis
Bangkok   Bogotá   Caracas   Kuala Lumpur   Lisbon   London   Madrid   Mexico City
Milan   Montreal   New Delhi   Santiago   Seoul   Singapore   Sydney   Taipei   Toronto

McGraw-Hill Higher Education ⚬

A Division of The **McGraw-Hill** *Companies*

ORGANIC CHEMISTRY: A BRIEF COURSE, THIRD EDITION

Published by McGraw-Hill, a business unit of The McGraw-Hill Companies, Inc., 1221
Avenue of the Americas, New York, NY 10020. Copyright © 2002, 1997, 1990 by The
McGraw-Hill Companies, Inc. All rights reserved. No part of this publication may be reproduced
or distributed in any form or by any means, or stored in a database or retrieval system, without the
prior written consent of The McGraw-Hill Companies, Inc., including, but not limited to, in any
network or other electronic storage or transmission, or broadcast for distance learning.

Some ancillaries, including electronic and print components, may not be available to customers
outside the United States.

This book is printed on acid-free paper.

International 1 2 3 4 5 6 7 8 9 0 QPD/QPD 0 9 8 7 6 5 4 3 2 1
Domestic 1 2 3 4 5 6 7 8 9 0 QPD/QPD 0 9 8 7 6 5 4 3 2 1

ISBN 0–07–231944–5
ISBN 0–07–112162–5 (ISE)

Executive editor: *Kent A. Peterson*
Developmental editor: *Shirley R. Oberbroeckling*
Marketing manager: *Thomas D. Timp*
Lead project manager: *Peggy J. Selle*
Lead production supervisor: *Sandra Hahn*
Designer: *K. Wayne Harms*
Cover designer: *Kristy Goddard*
Senior supplement producer: *David A. Welsh*
Media technology producer: *Lori A. Welsh*
Compositor: *GTS Graphics, Inc.*
Typeface: *10/12 Times Roman*
Printer: *Quebecor World Dubuque, IA*

Library of Congress Cataloging-in-Publication Data

Atkins, Robert C. (Robert Charles), 1944–
    Organic chemistry : a brief course / Robert C. Atkins, Francis A. Carey. — 3rd ed.
        p. cm.
    Includes bibliographical references and index.
    ISBN 0–07–231944–5 — ISBN 0–07–112162–5 (ISE)
    1. Chemistry, Organic. I. Carey, Francis A., 1937– . II. Title.

QD253.2 .A74 2002
547—dc21
                                      2001030808
                                      CIP

INTERNATIONAL EDITION ISBN 0–07–112162–5
Copyright © 2002. Exclusive rights by The McGraw-Hill Companies, Inc., for manufacture
and export. This book cannot be re-exported from the country to which it is sold by McGraw-Hill.
The International Edition is not available in North America.

www.mhhe.com

ABOUT THE AUTHORS

Robert C. Atkins was born in Massachusetts and educated in New Jersey. He received an S.B. in chemistry from the Massachusetts Institute of Technology in 1966 and a Ph.D. in organic chemistry from the University of Wisconsin at Madison in 1970. Following a year of postdoctoral study at Columbia University he was appointed to the faculty of James Madison University, where he holds the rank of professor of chemistry. He is the author of *Study Guide to Accompany Chemistry; Principles and Applications* (McGraw-Hill, 1979) and the coauthor (with Professor Carey) of *Study Guide and Solutions Manual to Accompany Organic Chemistry* (McGraw-Hill, 2000).

Professor Atkins has done research in the area of pyrolysis reaction mechanisms. He also has an active interest in laboratory safety and has served for more than 25 years as his department's safety coordinator. Since 1985 Professor Atkins has also volunteered as a technical advisor to both a local and a regional hazardous materials team.

Bob and his wife Mary, a family nurse practitioner, have two grown children: Maureen, an event planner in Denver, CO; and David, a Buddhist monk living in South Korea. Bob enjoys spending his leisure time swimming, surf fishing, and attending operas.

Francis A. Carey is a native of Pennsylvania, educated in the public schools of Philadelphia, at Drexel University (B.S. in chemistry, 1959), and at Penn State (Ph.D. 1963). Following postdoctoral work at Harvard and military service, he was appointed to the chemistry faculty of the University of Virginia in 1966. Prior to retiring in 2000, he regularly taught the two-semester lecture courses in general chemistry and organic chemistry.

With his students, Professor Carey has published over forty research papers in synthetic and mechanistic organic chemistry. In addition to this text, he has written a text for the two-semester course entitled *Organic Chemistry,* and is coauthor (with Richard J. Sundberg) of *Advanced Organic Chemistry,* a two-volume treatment designed for graduate students and advanced undergraduates. He was a member of the Committee of Examiners of the Graduate Record Examination in Chemistry from 1993–2000.

Frank and his wife Jill, who is a teacher and director of a preschool as well as a church organist, are the parents of three grown sons and the grandparents of Riyad and Ava.

A B O U T   T H E   C O V E R

The cover image shows buckminsterfullerene (C_{60}) and its unusual electrostatic potential map as calculated using Spartan molecular modeling software. It is best understood by first thinking about a sheet of fused benzene rings, such as occur in graphite. In such a sheet, negative electrostatic potential is equally distributed above and below the plane of the atoms. Now curve the sheet, tending toward the ball. As the electrons inside the ball are brought closer together, they escape toward the outer surface where they are farther apart. The inside becomes a region of high positive potential (blue) and the outer surface a region of high negative potential (red) as shown in the lower right-hand corner of the cover.

This edition is dedicated to our families
for their love, support, and understanding.

BRIEF CONTENTS

CONTENTS

CHAPTER 5
ALKENES AND ALKYNES II: REACTIONS   120

CHAPTER 6
AROMATIC COMPOUNDS   147

CHAPTER 7
STEREOCHEMISTRY   182

CHAPTER 10
ALCOHOLS, ETHERS, AND PHENOLS   242

CHAPTER 11
ALDEHYDES AND KETONES   274

CHAPTER 19
SPECTROSCOPY   477

PREFACE

The students who take this course are a diverse and ambitious group and come from a variety of programs, including biology, nutrition, engineering, agricultural sciences, environmental sciences, and the allied health sciences. They share a common need to learn about organic compounds, their structure, properties, nomenclature, and applications, although not at the level of detail typical of the year-long organic chemistry course. These students will need to master the same reasoning processes as those who are enrolled in the more traditional two-semester sequence, and they need just as much guidance. By selectively revealing the logic of organic chemistry through our organization, pedagogy, problem solving, and illustrations, we provide that guidance.

Our goal in writing *Organic Chemistry: A Brief Course* is to prepare a text that is both modern in outlook and selective in coverage. This revision was undertaken in the same spirit. To foster this goal, you will see this edition is strengthened by:

- Strong **organization** with reaction mechanisms stressed early and often within a functional group approach to help a student learn concepts, not memorize them.

- Our including more material designed to help the student learn because we were **selective in the topic coverage.** Fewer topics equal more help and focused learning.

- Highly developed **pedagogy,** allowing ease of learning.

- **Numerous problems** within the text, approximately one third of which are multipart and include a detailed solution for part (a) for guiding the student through problem solving in organic chemistry. In addition, we provide numerous end-of-chapter problems. All problems have complete solutions described in the accompanying *Solutions Manual* prepared by the authors.

- New full-color **illustrations** using Spartan models and a new design. These enhance the student's understanding of concepts.

NEW TO THE THIRD EDITION

Full four-color design allows for color-coding of atoms in chemical equations and draws the student's attention to the sites of reactivity in organic molecules. It also provides electrostatic potential maps that enable the student to see the charge distribution in the molecule and connect structure to reactivity.

Spartan-generated molecular models accurately reflect molecular structure and electrostatic potential maps.

New figures were added and existing figures revised to take advantage of the new look to enhance learning.

Spectra from earlier editions using 60-MHz NMR have been replaced by spectra recorded at 200 MHz.

Molecules of biological importance have been generated from crystallographic coordinates in the Protein Data Bank and replace artist's drawing to provide the most accurate information available.

ORGANIZATION

As the Table of Contents indicates, this text is organized according to the families of organic compounds and their functional groups, but not rigidly so. Certainly alkanes and cycloalkanes, described in Chapter 2, need to be discussed early because they provide the simplest introduction to principles of organic structure, bonding, and nomenclature. Alcohols and alkyl halides (Chapter 3), however, follow immediately after alkanes and appear before alkenes and alkynes (Chapters 4 and 5) because alkenes are prepared from alcohols and alkyl halides by reactions that are useful in practice and instructive with respect to mechanism. Reactions in this text are rarely an end in themselves; instead, they are used as examples of the connectedness of organic chemistry and to develop mechanistic ideas.

Chapter 6, Aromatic Compounds, completes the coverage of the major classes of hydrocarbons. The focus then shifts to a structural chapter (Stereochemistry, Chapter 7) and a mechanistic chapter (Nucleophilic Substitution, Chapter 8). Chapter 9, Free Radicals, presents an integrated treatment of material separating free-radical intermediates from polar ones, emphasizing the differences between them.

Chapters 10–13 describe the major families of oxygen-containing functional groups, especially carbonyl compounds. Functional-group chemistry concludes with amines in Chapter 14.

Chapters 15–18 cover those organic substances that lie at the heart of biological chemistry (carbohydrates, lipids, amino acids, peptides and proteins, and nucleic acids). The coverage emphasizes structure and the similarity of biological processes to the fundamental transformations of organic chemistry.

Chapter 19 covers spectroscopy. The 60-MHz NMR spectra of earlier editions have been replaced by spectra recorded at 200 MHz. Instructors who use spectroscopy to reinforce concepts of structure and bonding may want to cover parts of Chapter 19 immediately after Chapter 3 or Chapter 6. Others may want to teach it after Chapter 14 by which time all of the families of organic compounds have been introduced. In many courses, spectroscopy is taught from a "hands-on" perspective in the laboratory. Chapter 19 can serve as a primary reading reference for such an approach.

PEDAGOGY

Chapter outline—This quick overview presents the main concepts of the chapter in outline form. The outline will help a student focus on the concepts of the chapters and manage their time in study and review of the concepts.

Summary tables—These annotated summary tables are much more than a list of reactions. They provide comments to remind the students about important features of each reaction.

Learning objectives—Our experience with lists of learning objectives has been that students welcome the guidance they provide, but find them to be most useful after they have been exposed to the material rather than as an introduction to it. Accordingly, we list the learning objectives at the end of each chapter.

End-of-chapter summaries—The summaries review the important topics of each chapter and frequently use accompanying tables for easy reference.

Boxed essays—The essays scattered throughout the text offer additional insights into the contemporary relevance of organic chemistry. They relate organic chemistry to areas such as bioorganic research, industry use, and environmental issues.

PROBLEM SOLVING

In the in-text problem-solving examples students are led through the reasoning process and drilled in its application. Many of the problems have multiple parts with a step-by-step solution provided for one of the questions as a guide to solving the others. Answers to all the in-chapter problems are included at the back of the book.

End-of-chapter problems are a comprehensive bank of problems that give students liberal opportunity to master skills by working the problems.

ILLUSTRATIONS AND DESIGN

All figures were redrawn to convey visual concepts clearly and forcefully. In addition, the authors created a number of new images using Spartan molecular modeling software. Now students can view electrostatic potential maps to see the charge distribution of a molecule in vivid color. These striking images afford the instructor a powerful means to lead students to a better understanding of organic molecules.

The new design emphasizes clarity. The color is carefully used to heighten interest and to create visual cues for important information.

SUPPLEMENTS

For the Student

Solutions Manual—Written by the authors with detailed solutions to all of the problems within the text.

Website—For each chapter students will find a summary, concise overview of the concepts in the chapter, and a short quiz to test their knowledge.

For the Instructor

- Test Bank—A collection of hundreds of multiple-choice questions available in print, Macintosh, or Windows format.
- Transparencies—A set of figures chosen by the authors to represent the major teaching points.
- Website—Containing PowerPoint slides of all the figures in the text, electronic Test Bank, and directions to the McGraw-Hill Companies PageOut course management tool.

ACKNOWLEDGMENTS

Special thanks go to the editorial staff at McGraw-Hill, Kent Peterson, Shirley Oberbroeckling, and Peggy Selle. Our copy editor, Linda Davoli, has been a valuable colleague and the source of many useful suggestions.

This text has benefited from the comments offered by numerous reviewers who offered valuable advice at various stages during this and the previous edition. We appreciate their help. Those who reviewed the manuscript for this edition include:

Jeff Albert, *South Dakota State University*
Ardeshir Azadina, *Michigan State University*
William F. Berkowitz, *City University of New York—Queens*

Richard Blatchly, *Keene State College*
Lance Crist, *Georgetown University*
Alvan Hengge, *Utah State University*
Robert H. Higgins, *Fayetteville State College*
Steven Holmgren, *Montana State University*
Richard P. Johnson, *University of New Hampshire*
Brenda Kesler, *San Jose State University*
Thomas Lectka, *Johns Hopkins University*
Rita S. Majerle, *South Dakota State University*
William A. Meena, *Rock Valley College*
Nicholas Natale, *University of Idaho*
Jung Oh, *Kansas State University—Salina*
Claire R. Olander, *Appalachian State University*
Robert H. Paine, *Rochester Institute of Technology*
Dilip K. Paul, *Pittsburg State University*
Michael Rathke, *Michigan State University*
Carey S. Reed, *Penn State—Altoona*
Michael Sady, *Western Nevada Community College*
Ralph Shaw, *Southeastern Louisiana University*
Cynthia Somers, *Red Rocks Community College*
Denise Tridle, *Highland Community College*

Comments, suggestions, and questions are welcome. Our e-mail addresses are

Robert C. Atkins (atkinsrc@jmu.edu)

Francis A. Carey (fac6q@virginia.edu.)

A GUIDE TO USING THIS TEXT

INTEGRATED TEXT AND VISUALS

The authors created a new full-color art program using Spartan software. Because visualization is so important to understanding, these illustrations work hand in hand with the text to convey information.

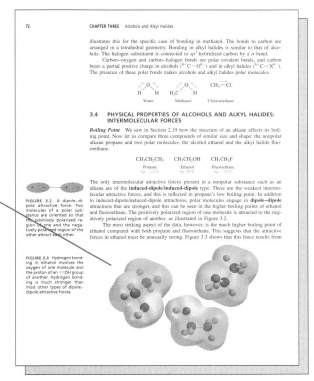

illustrates this for the specific case of bonding in methanol. The bonds to carbon are arranged in a tetrahedral geometry. Bonding in alkyl halides is similar to that of alcohols. The halogen substituent is connected to sp^3 hybridized carbon by a σ bond.

Carbon–oxygen and carbon–halogen bonds are polar covalent bonds, and carbon bears a partial positive charge in alcohols ($^{\delta+}C$—$O^{\delta-}$) and in alkyl halides ($^{\delta+}C$—$X^{\delta-}$). The presence of these polar bonds makes alcohols and alkyl halides polar molecules.

Water   Methanol   Chloromethane

3.4 PHYSICAL PROPERTIES OF ALCOHOLS AND ALKYL HALIDES: INTERMOLECULAR FORCES

Boiling Point We saw in Section 2.19 how the structure of an alkane affects its boiling point. Now let us compare three compounds of similar size and shape: the nonpolar alkane propane and two polar molecules, the alcohol ethanol and the alkyl halide fluoroethane.

CH₃CH₂CH₃   CH₃CH₂OH   CH₃CH₂F
Propane   Ethanol   Fluoroethane

The only intermolecular attractive forces present in a nonpolar substance such as an alkane are of the **induced-dipole/induced-dipole** type. These are the weakest intermolecular attractive forces, and this is reflected in propane's low boiling point. In addition to induced-dipole/induced-dipole attractions, polar molecules engage in **dipole–dipole** attractions that are stronger, and this can be seen in the higher boiling points of ethanol and fluoroethane. The positively polarized region of one molecule is attracted to the negatively polarized region of another, as illustrated in Figure 3.2.

The most striking aspect of the data, however, is the much higher boiling point of ethanol compared with both propane and fluoroethane. This suggests that the attractive forces in ethanol must be unusually strong. Figure 3.3 shows that this force results from

FIGURE 3.2 A dipole–dipole attractive force. Two molecules of a polar substance are oriented so that the positively polarized region of one and the negatively polarized region of the other attract each other.

FIGURE 3.3 Hydrogen bonding in ethanol involves the oxygen of one molecule and the proton of an —OH group of another. Hydrogen bonding is much stronger than most other types of dipole-dipole attractive forces.

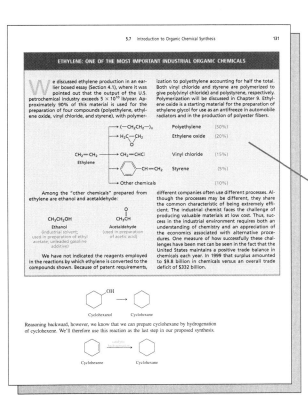

ETHYLENE: ONE OF THE MOST IMPORTANT INDUSTRIAL ORGANIC CHEMICALS

We discussed ethylene production in an earlier boxed essay (Section 4.1), where it was pointed out that the output of the U.S. petrochemical industry exceeds 5×10^{10} lb/year. Approximately 90% of this material is used for the preparation of four compounds (polyethylene, ethylene oxide, vinyl chloride, and styrene), with polymerization to polyethylene accounting for half the total. Both vinyl chloride and styrene are polymerized to give poly(vinyl chloride) and polystyrene, respectively. Polymerization will be discussed in Chapter 9. Ethylene oxide is a starting material for the preparation of ethylene glycol for use as an antifreeze in automobile radiators and in the production of polyester fibers.

CH₂=CH₂ → (—CH₂CH₂—)ₙ   Polyethylene   (50%)
→ H₂C—CH₂   Ethylene oxide   (20%)
CH₂=CH₂ → CH₂=CHCl   Vinyl chloride   (15%)
Ethylene
→ CH=CH₂   Styrene   (5%)
→ Other chemicals   (10%)

Among the "other chemicals" prepared from ethylene are ethanol and acetaldehyde:

CH₃CH₂OH   CH₃CH
Ethanol   Acetaldehyde
(industrial solvent; used in preparation of ethyl acetate; unleaded gasoline additive)   (used in preparation of acetic acid)

We have not indicated the reagents employed in the reactions by which ethylene is converted to the compounds shown. Because of patent requirements, different companies often use different processes. Although the processes may be different, they share the common characteristic of being extremely efficient. The industrial chemist faces the challenge of producing valuable materials at low cost. Thus, success in the industrial environment requires both an understanding of chemistry and an appreciation of the economics associated with alternative procedures. One measure of how successfully these challenges have been met can be seen in the fact that the United States maintains a positive trade balance in chemicals each year. In 1999 that surplus amounted to $9.8 billion in chemicals versus an overall trade deficit of $332 billion.

OH
Cyclohexanol → Cyclohexane

Reasoning backward, however, we know that we can prepare cyclohexane by hydrogenation of cyclohexene. We'll therefore use this reaction as the last step in our proposed synthesis.

Cyclohexene → Cyclohexane

INSTRUCTIVE BOXED ESSAYS

The essays in the book are designed to help students think and learn by relating concepts to biological, environmental, and other real-world applications. Examples include *Methane and the Biosphere, Natural and "Designed" Enediyne Antibiotics,* and *Ethylene: One of the Most Important Industrial Organic Chemicals.*

PROBLEM SOLVING—BY EXAMPLE

Problem-solving strategies and skills are emphasized throughout. Understanding of topics is continually reinforced by problems that appear within topic sections. For many problems, sample solutions are given.

4.7   DEHYDRATION OF ALCOHOLS

In the dehydration of alcohols, the H and OH are lost from adjacent carbons. An acid catalyst is necessary.

$$H-\overset{|}{\underset{|}{C}}-\overset{|}{\underset{|}{C}}-OH \xrightarrow{H^+} \ \ >C=C< \ + H_2O$$

Alcohol   Alkene   Water

Before dehydrogenation of ethane became the dominant method, ethylene was prepared by heating ethyl alcohol with sulfuric acid.

$$CH_3CH_2OH \xrightarrow[160°C]{H_2SO_4} CH_2=CH_2 + H_2O$$

Ethyl alcohol   Ethylene   Water

Concentrated sulfuric acid (H_2SO_4) and phosphoric acid (H_3PO_4) are the acids most frequently used in alcohol dehydrations. Other alcohols dehydrate in a manner similar to ethyl alcohol.

Cyclohexanol   Cyclohexene   Water
(79–87%)

2-Methyl-2-propanol   2-Methylpropene   Water
(82%)

PROBLEM 4.10   Identify the alkene obtained on dehydration of each of the following alcohols:

(a) 3-Ethyl-3-pentanol   (c) 2-Propanol
(b) 1-Propanol   (d) 2,3,3-Trimethyl-2-butanol

Sample Solution   (a) The hydrogen and the hydroxyl are lost from adjacent carbons in the dehydration of 3-ethyl-3-pentanol.

3-Ethyl-3-pentanol   3-Ethyl-2-pentene   Water

The hydroxyl group is lost from a carbon that bears three equivalent ethyl substituents. Elimination can occur in any one of three equivalent directions to give the same alkene, 3-ethyl-2-pentene.

In the preceding examples, including those of Problem 4.10, only a single alkene could be formed from each alcohol. When the alcohol is capable of yielding two or more different alkenes,

---

Hydrocarbons that contain a carbon–carbon triple bond are classified as **alkynes.** Alkynes are named in the IUPAC system in a manner similar to alkenes, however the suffix *-yne* replaces *-ane* (Section 4.12).

The carbon–carbon triple bond in alkynes is composed of a σ and two π components (Section 4.13). The σ component contains two electrons in an orbital generated by the overlap of *sp*-hybridized orbitals on adjacent atoms. Each of these carbons also has two 2*p* orbitals, which overlap in pairs so as to give two π orbitals. Alkynes have a linear arrangement of their bonds in the —C≡C— unit.

Alkynes are prepared from geminal or vicinal dihalides by double dehydrohalogenation reactions (Section 4.14).

ADDITIONAL PROBLEMS

Alkene Nomenclature and Structure

4.18   Write structural formulas for each of the following:

(a) 1-Heptene
(b) 3-Ethyl-2-pentene
(c) *cis*-3-Octene
(d) *trans*-1,4-Dichloro-2-butene
(e) (*Z*)-3-Methyl-2-hexene
(f) (*E*)-3-Chloro-2-hexene
(g) 1-Bromo-3-methylcyclohexene
(h) 1-Bromo-6-methylcyclohexene
(i) 4-Methyl-4-penten-2-ol

4.19   Write a structural formula or build a molecular model and give a correct IUPAC name for each alkene of molecular formula C_7H_{14} that has a *tetrasubstituted* double bond.

4.20   Give the IUPAC names for each of the following compounds:

(a) $(CH_3CH_2)_2C=CHCH_3$
(b) $(CH_3CH_2)_2C=C(CH_2CH_3)_2$
(c) $(CH_3)_3CCH=CCl_2$

(d)

4.21   Each of the following compounds is found in nature. Write a structural formula or build a molecular model of each one, clearly showing the stereochemistry of each double bond.

(a) The sex attractant of the Mediterranean fruit fly, (*E*)-6-nonen-1-ol.
(b) Geraniol, a naturally occurring substance present in the fragrant oil of many plants that has a pleasing, rose-like odor. Geraniol is the *E* isomer of

$$(CH_3)_2C=CHCH_2CH_2C=CHCH_2OH$$
$$\overset{|}{CH_3}$$

(c) Nerol, a naturally occurring substance, that is a stereoisomer of geraniol.
(d) The sex attractant of the codling moth, the 2*Z*, 6*E* stereoisomer of

$$CH_3CH_2C=CHCH_2CH_2C=CHCH_2OH$$
$$\overset{|}{CH_3} \qquad \overset{|}{CH_2CH_3}$$

4.22   Specify the hybridization of each carbon in the molecules shown. How many carbon–carbon σ bonds and carbon–carbon π bonds are present in each molecule?

(a) $(CH_3)_2C=C(CH_3)_2$
(b) $CH_2=CHCH_2CH=CH_2$
(c)

4.23   For which of the following alkenes are stereoisomeric forms possible?

(a) 1-Chloropropene   (b) 2-Chloropropene   (c) 1,2-Dichloropropene

---

. . . AND MORE PROBLEMS

Every chapter ends with a comprehensive bank of problems that give students liberal opportunity to master skills by working problems.

THE SUMMARY

Summaries ending each chapter are crafted to allow students to check their knowledge and revisit chapter content in a study-friendly format. Learning is reinforced through concise narrative and through summary tables that students find valuable.

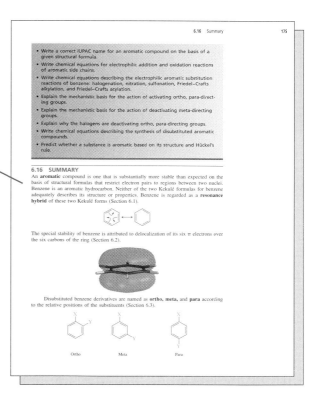

WEBSITE

The website at www.mhhe.com/atkins-carey is a resource that provides additional support for the third edition of *Organic Chemistry: A Brief Course*. The tutorial materials provide a short overview of the chapter content using Chime images to enhance and facilitate learning.

For the instructor, all of the figures within the text are available in PowerPoint. Also available is PageOut, a course management system.

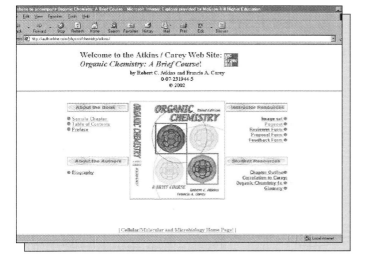

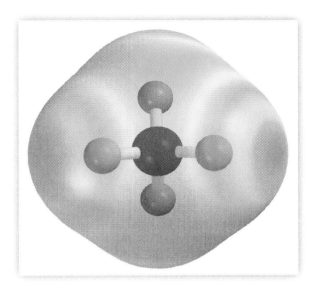

CHAPTER 1
CHEMICAL BONDING

Organic chemistry is the study of carbon compounds. It is a vast field, one that encompasses a great deal of information and that is inherently useful. Research in organic chemistry continues to provide new materials that enrich our lives, new drugs that extend them, and new knowledge that describes the chemical basis of life itself. An understanding of organic chemistry must begin with an understanding of molecular structure. **Structure** is the key to everything in chemistry. The properties of a substance depend on the atoms it contains and the way they are connected. What is less obvious, but very powerful, is the idea that someone who is trained in chemistry can look at a structural formula of a substance and tell you a lot about its properties. This chapter begins your training toward understanding the relationship between structure and properties in organic compounds. It reviews some fundamental principles of molecular structure and chemical **bonding.** By applying these principles you will learn to recognize which structural patterns are more stable than others and develop skills in communicating chemical information by way of structural formulas that will be used throughout your study of organic chemistry.

> A glossary of important terms may be found immediately before the index at the back of the book.

1.1 ATOMS, ELECTRONS, AND ORBITALS

Before discussing bonding principles, let's first review some fundamental relationships between atoms and electrons. Each element is characterized by a unique **atomic number Z,** which is equal to the number of protons in its nucleus. A neutral atom has equal numbers of protons, which are positively charged, and electrons, which are negatively charged. The electrons spend 90–95% of their time near the nucleus in regions of space called **orbitals.**

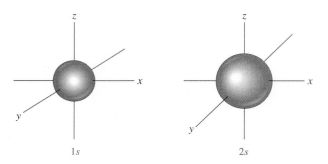

FIGURE 1.1 Boundary surfaces of a 1s orbital and a 2s orbital. The boundary surfaces enclose the volume where the probability of finding an electron is 90–95%.

Orbitals are described by specifying their size, shape, and directional properties. Spherically symmetrical ones such as shown in Figure 1.1 are called *s* **orbitals.** The letter *s* is preceded by the **principal quantum number** n (n = 1, 2, 3, etc.) which specifies the **shell** and is related to the energy of the orbital. An electron in a 1s orbital is likely to be found closer to the nucleus, is lower in energy, and is more strongly held than an electron in a 2s orbital.

A hydrogen atom (Z = 1) has one electron; a helium atom (Z = 2) has two. The single electron of hydrogen occupies a 1s orbital, as do the two electrons of helium. The respective electron configurations are described as:

$$\text{Hydrogen: } 1s^1 \qquad \text{Helium: } 1s^2$$

In addition to being negatively charged, electrons possess the property of **spin.** The **spin quantum number** of an electron can have a value of either $+\frac{1}{2}$ or $-\frac{1}{2}$. According to the **Pauli exclusion principle,** two electrons may occupy the same orbital only when they have opposite, or "paired," spins. For this reason, no orbital can contain more than two electrons. Because two electrons fill the 1s orbital, the third electron in lithium (Z = 3) must occupy an orbital of higher energy. After 1s, the next higher energy orbital is 2s. The third electron in lithium therefore occupies the 2s orbital, and the electron configuration of lithium is

$$\text{Lithium: } 1s^2 2s^1$$

A complete periodic table of the elements is presented on the inside back cover.

The **period** (or **row**) of the periodic table in which an element appears corresponds to the principal quantum number of the highest numbered occupied orbital (n = 1 in the case of hydrogen and helium). Hydrogen and helium are first-row elements; lithium (n = 2) is a second-row element.

With beryllium (Z = 4), the 2s level becomes filled, and the next orbitals to be occupied in the remaining second-row elements are the 2p_x, 2p_y, and 2p_z orbitals. These orbitals, portrayed in Figure 1.2, have a boundary surface that is usually described as "dumbbell-shaped." Each orbital consists of two "lobes," that is, slightly flattened spheres that touch each other along a nodal plane passing through the nucleus. The 2p_x, 2p_y, and 2p_z orbitals are equal in energy and mutually perpendicular.

The electron configurations of the first 12 elements, hydrogen through magnesium, are given in Table 1.1. In filling the 2p orbitals, notice that each is singly occupied before any one is doubly occupied. This is a general principle for orbitals of equal energy known as **Hund's rule.**

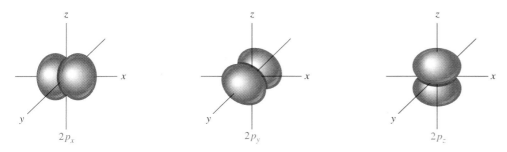

FIGURE 1.2   Boundary surfaces of the 2p orbitals.

Of particular importance in Table 1.1 are hydrogen, carbon, nitrogen, and oxygen.

Countless organic compounds contain nitrogen, oxygen, or both in addition to carbon, the essential element of organic chemistry. Most of them also contain hydrogen.

It is often convenient to speak of the **valence electrons** of an atom. These are the outermost electrons, the ones most likely to be involved in chemical bonding and reactions. For second-row elements these are the $2s$ and $2p$ electrons. Because four orbitals ($2s$, $2p_x$, $2p_y$, $2p_z$) are involved, the maximum number of electrons in the **valence shell** of any second-row element is 8. Neon, with all its $2s$ and $2p$ orbitals doubly occupied, has eight valence electrons and completes the second row of the periodic table.

> **PROBLEM 1.1**   How many valence electrons does carbon have?

Once the $2s$ and $2p$ orbitals are filled, the next level is the $3s$, followed by the $3p_x$, $3p_y$, and $3p_z$ orbitals. Electrons in these orbitals are farther from the nucleus than those in the $2s$ and $2p$ orbitals and are of higher energy.

> **PROBLEM 1.2**   Referring to the periodic table as needed, write electron configurations for all the elements in the third period.

Answers to all problems that appear within the body of a chapter are found in an Appendix. A brief discussion of the problem and advice on how to do problems of the same type are offered in the Solutions Manual.

TABLE 1.1	Electron Configurations of the First Twelve Elements of the Periodic Table						
		Number of Electrons in Indicated Orbital					
Element	Atomic Number Z	1s	2s	$2p_x$	$2p_y$	$2p_z$	3s
Hydrogen	1	1					
Helium	2	2					
Lithium	3	2	1				
Beryllium	4	2	2				
Boron	5	2	2	1			
Carbon	6	2	2	1	1		
Nitrogen	7	2	2	1	1	1	
Oxygen	8	2	2	2	1	1	
Fluorine	9	2	2	2	2	1	
Neon	10	2	2	2	2	2	
Sodium	11	2	2	2	2	2	1
Magnesium	12	2	2	2	2	2	2

Sample Solution The third period begins with sodium and ends with argon. The atomic number Z of sodium is 11, and so a sodium atom has 11 electrons. The maximum number of electrons in the 1s, 2s, and 2p orbitals is ten, and so the eleventh electron of sodium occupies a 3s orbital. The electron configuration of sodium is $1s^2 2s^2 2p_x^2 2p_y^2 2p_z^2 3s^1$.

Neon, in the second period, and argon, in the third, possess eight electrons in their valence shell; they are said to have a complete **octet** of electrons. Helium, neon, and argon belong to the class of elements known as **noble gases** or **rare gases.** The noble gases are characterized by an extremely stable "closed-shell" electron configuration and are very unreactive.

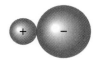

FIGURE 1.3 An ionic bond is the force of electrostatic attraction between oppositely charged ions, illustrated in this case by Na$^+$ (red) and Cl$^-$ (green). In solid sodium chloride, each sodium ion is surrounded by six chloride ions and vice versa in a crystal lattice.

1.2   IONIC BONDS

Atoms combine with one another to give **compounds** having properties different from the atoms they contain. The attractive force between atoms in a compound is a **chemical bond.** One type of chemical bond, called an **ionic bond,** is the force of attraction between oppositely charged species (**ions**) (Figure 1.3). Ions that are positively charged are referred to as **cations;** those that are negatively charged are **anions.**

Whether an element is the source of the cation or anion in an ionic bond depends on several factors, for which the periodic table can serve as a guide. In forming ionic compounds, elements at the left of the periodic table typically lose electrons, forming a cation that has the same electron configuration as the nearest noble gas. Loss of an electron from sodium, for example, gives the species Na$^+$, which has the same electron configuration as neon.

$$\text{Na·} \longrightarrow \text{Na}^+ + e^-$$

Sodium atom   Sodium ion   Electron
$1s^2 2s^2 2p^6 3s^1$ $1s^2 2s^2 2p^6$

Elements at the right of the periodic table tend to gain electrons to reach the electron configuration of the next higher noble gas. Adding an electron to chlorine, for example, gives the anion Cl$^-$, which has the same closed-shell electron configuration as the noble gas argon.

$$:\ddot{\text{Cl}}· \quad + \quad e^- \quad \longrightarrow \quad :\ddot{\text{Cl}}:^-$$

Chlorine atom   Electron   Chloride ion
$1s^2 2s^2 2p^6 3s^2 3p^5$ $1s^2 2s^2 2p^6 3s^2 3p^6$

PROBLEM 1.3 Which of the following ions possess a noble gas electron configuration?

(a) K$^+$ (b) H$^-$ (c) O$^-$ (d) F$^-$ (e) Ca^{2+}

Sample Solution (a) Potassium has atomic number 19, and so a potassium atom has 19 electrons. The ion K$^+$, therefore, has 18 electrons, the same as the noble gas argon. The electron configurations of K$^+$ and Ar are the same: $1s^2 2s^2 2p^6 3s^2 3p^6$.

Transfer of an electron from a sodium atom to a chlorine atom yields a sodium cation and a chloride anion, both of which have a noble gas electron configuration:

$$Na\cdot \quad + \quad :\overset{..}{\underset{..}{Cl}}\cdot \quad \longrightarrow \quad Na^+[:\overset{..}{\underset{..}{Cl}}:]^-$$

Sodium atom   Chlorine atom   Sodium chloride

Attractive forces between oppositely charged particles are what we mean by an **ionic bond** between two atoms.

Ionic bonds are very common in **inorganic** compounds, but rare in **organic** ones. What kinds of bonds, then, link carbon to other elements in millions of organic compounds? Instead of losing or gaining electrons, carbon *shares* electrons with other elements (including other carbon atoms) to give what are called covalent bonds.

1.3 COVALENT BONDS

The **covalent,** or **shared electron pair,** model of chemical bonding was first suggested by G. N. Lewis of the University of California in 1916. Lewis proposed that a sharing of two electrons by two hydrogen atoms permits each one to have a stable closed-shell electron configuration analogous to helium.

Gilbert Newton Lewis (born Weymouth, Massachusetts, 1875; died Berkeley, California, 1946) has been called the greatest American chemist.

Two hydrogen atoms,   Hydrogen molecule:
each with a single   covalent bonding by way of
electron   a shared electron pair

Structural formulas of this type in which electrons are represented as dots are called **Lewis structures.**

Covalent bonding in F_2 gives each fluorine 8 electrons in its valence shell and a stable electron configuration equivalent to that of the noble gas neon:

Two fluorine atoms, each   Fluorine molecule:
with seven electrons in   covalent bonding by way of
its valence shell   a shared electron pair

> **PROBLEM 1.4** Hydrogen is bonded to fluorine in hydrogen fluoride by a covalent bond. Write a Lewis formula for hydrogen fluoride.

The Lewis model limits second-row elements (Li, Be, B, C, N, O, F, Ne) to a total of 8 electrons (shared plus unshared) in their valence shells. Hydrogen is limited to 2. Most of the elements that we'll encounter in this text obey the **octet rule:**

> *In forming compounds they gain, lose, or share electrons to give a stable electron configuration characterized by 8 valence electrons as in the noble gases.*

When the octet rule is satisfied for carbon, nitrogen, oxygen, and fluorine, they have an electron configuration analogous to the noble gas neon.

Now let's apply the Lewis model to the organic compounds methane and carbon tetrafluoride.

Combine $\cdot\overset{.}{\underset{.}{C}}\cdot$ and four $H\cdot$ to write a H
                                  Lewis structure $H:\overset{..}{C}:H$
                                  for methane $\overset{..}{H}$

Combine $\cdot\ddot{\text{C}}\cdot$ and four $\cdot\ddot{\text{F}}:$ to write a Lewis structure for carbon tetrafluoride

$$:\!\ddot{\underset{\displaystyle\ddot{F}}{\overset{\displaystyle\ddot{F}}{\ddot{F}\!:\!C\!:\!\ddot{F}}}}\!:$$

Carbon has 8 electrons in its valence shell in both methane and carbon tetrafluoride. By forming covalent bonds to four other atoms, carbon achieves a stable electron configuration analogous to neon. Each covalent bond in methane and carbon tetrafluoride is quite strong—comparable to the bond between hydrogens in H_2 in bond dissociation energy.

> **PROBLEM 1.5**   Given the information that it has a carbon–carbon bond, write a satisfactory Lewis structure for C_2H_6 (ethane).

Representing a 2-electron covalent bond by a dash (—), the Lewis structures for hydrogen fluoride, fluorine, methane, and carbon tetrafluoride become:

$$\text{H}\!-\!\ddot{\text{F}}\!: \qquad :\!\ddot{\text{F}}\!-\!\ddot{\text{F}}\!: \qquad \text{H}\!-\!\overset{\displaystyle\text{H}}{\underset{\displaystyle\text{H}}{\text{C}}}\!-\!\text{H} \qquad :\!\ddot{\text{F}}\!-\!\overset{\displaystyle:\ddot{\text{F}}:}{\underset{\displaystyle:\ddot{\text{F}}:}{\text{C}}}\!-\!\ddot{\text{F}}\!:$$

Hydrogen   Fluorine   Methane   Carbon tetrafluoride
fluoride

1.4   DOUBLE BONDS AND TRIPLE BONDS

Lewis's concept of shared electron pair bonds allows for 4-electron **double bonds** and 6-electron **triple bonds.** Carbon dioxide (CO_2) has two carbon–oxygen double bonds, and the octet rule is satisfied for both carbon and oxygen. Similarly, the most stable Lewis structure for hydrogen cyanide (HCN) has a carbon–nitrogen triple bond.

Carbon dioxide: $:\ddot{\text{O}}::\text{C}::\ddot{\text{O}}:$ or $:\ddot{\text{O}}\!=\!\text{C}\!=\!\ddot{\text{O}}:$

Hydrogen cyanide: $\text{H}:\text{C}:::\text{N}:$ or $\text{H}\!-\!\text{C}\!\equiv\!\text{N}:$

Multiple bonds are very common in organic chemistry. Ethylene (C_2H_4) contains a carbon–carbon double bond in its most stable Lewis structure, and each carbon has a completed octet. The most stable Lewis structure for acetylene (C_2H_2) contains a carbon–carbon triple bond. Here again, the octet rule is satisfied.

Ethylene: $\underset{\displaystyle\text{H}}{\overset{\displaystyle\text{H}}{}}\!\text{C}::\text{C}\!\underset{\displaystyle\text{H}}{\overset{\displaystyle\text{H}}{}}$ or $\underset{\text{H}}{\overset{\text{H}}{}}\!\!\diagdown\!\!\text{C}\!=\!\text{C}\!\!\diagup\!\!\underset{\text{H}}{\overset{\text{H}}{}}$

Acetylene: $\text{H}:\text{C}:::\text{C}:\text{H}$ or $\text{H}\!-\!\text{C}\!\equiv\!\text{C}\!-\!\text{H}$

> **PROBLEM 1.6**   Write the most stable Lewis structure for each of the following compounds:
>
> (a) Formaldehyde, CH_2O. Both hydrogens are bonded to carbon. (A solution of formaldehyde in water was at one time used to preserve biological specimens.)
>
> (b) Tetrafluoroethylene, C_2F_4. (The starting material for the preparation of Teflon.)

(c) Acrylonitrile, C_3H_3N. The atoms are connected in the order CCCN, and all hydrogens are bonded to carbon. (The starting material for the preparation of acrylic fibers such as Orlon and Acrilan.)

Sample Solution (a) Each hydrogen contributes 1 valence electron, carbon contributes 4, and oxygen 6 for a total of 12 valence electrons. We are told that both hydrogens are bonded to carbon. Because carbon forms four bonds in its stable compounds, join carbon and oxygen by a double bond. The partial structure so generated accounts for 8 of the 12 electrons. Add the remaining four electrons to oxygen as unshared pairs to complete the structure of formaldehyde.

Partial structure showing   Complete Lewis structure
covalent bonds   of formaldehyde

1.5 POLAR COVALENT BONDS AND ELECTRONEGATIVITY

Electrons in covalent bonds are not necessarily shared equally by the two atoms they connect. If one atom has a greater tendency to attract electrons toward itself than the other, we say the electron distribution is **polarized,** and the bond is referred to as a **polar covalent bond.** Hydrogen fluoride, for example, has a polar covalent bond. Because fluorine attracts electrons more strongly than hydrogen, the electrons in the H—F bond are pulled toward fluorine, giving it a partial negative charge, and away from hydrogen giving it a partial positive charge. This separation of charge is called a **dipole.** This polarization of electron density is represented in various ways.

$$^{\delta+}H—F^{\delta-} \qquad\qquad \overset{\longleftrightarrow}{H—F}$$

(The symbols $^{\delta+}$ and $^{\delta-}$ indicate   (The symbol $\longleftrightarrow$ represents
partial positive and partial   the direction of polarization
negative charge, respectively.)   of electrons in the H—F bond.)

The tendency of an atom to draw the electrons in a covalent bond toward itself is referred to as its **electronegativity.** An **electronegative** element attracts electrons; an **electropositive** one donates them. Electronegativity increases across a row in the periodic table. The most electronegative of the second-row elements is fluorine; the most electropositive is lithium. Electronegativity decreases in going down a column. Fluorine is more electronegative than chlorine. The most commonly cited electronegativity scale was devised by Linus Pauling and is presented in Table 1.2.

PROBLEM 1.7 Examples of carbon-containing compounds include methane (CH_4), chloromethane (CH_3Cl), and methyllithium (CH_3Li). In which one does carbon bear the greatest partial positive charge? The greatest partial negative charge?

1.6 FORMAL CHARGE

Lewis structures frequently contain atoms that bear a positive or negative charge. If the molecule as a whole is neutral, the sum of its positive charges must equal the sum of its negative charges. An example is nitric acid, HNO_3:

Linus Pauling. (*Photograph provided by Linus Pauling Institute of Science and Medicine, Palo Alto, CA.*)

TABLE 1.2	Selected Values from the Pauling Electronegativity Scale						
	Group Number						
Period	I	II	III	IV	V	VI	VII
1	H 2.1						
2	Li 1.0	Be 1.5	B 2.0	C 2.5	N 3.0	O 3.5	F 4.0
3	Na 0.9	Mg 1.2	Al 1.5	Si 1.8	P 2.1	S 2.5	Cl 3.0
4	K 0.8	Ca 1.0					Br 2.8
5							I 2.5

As written, the structural formula for nitric acid depicts different bonding patterns for its three oxygens. One oxygen is doubly bonded to nitrogen, another is singly bonded to both nitrogen and hydrogen, and the third has a single bond to nitrogen and a negative charge. Nitrogen is positively charged. The positive and negative charges are called **formal charges,** and the Lewis structure of nitric acid would be incomplete were they to be omitted.

We calculate formal charges by counting the number of electrons "owned" by each atom in a Lewis structure and comparing this **electron count** with that of a neutral atom. Figure 1.4 illustrates how electrons are counted for each atom in nitric acid. Counting electrons for the purpose of computing the formal charge differs from counting to see if the octet rule is satisfied. A second-row element has a filled valence shell if the sum of all the electrons, shared and unshared, is 8. Electrons that connect two atoms by a covalent bond count toward filling the valence shell of both atoms. When calculating the formal charge, however, only half the number of electrons in covalent bonds can be considered to be "owned" by an atom.

> The number of valence electrons in an atom of a main-group element such as nitrogen is equal to its group number. In the case of nitrogen this is 5.

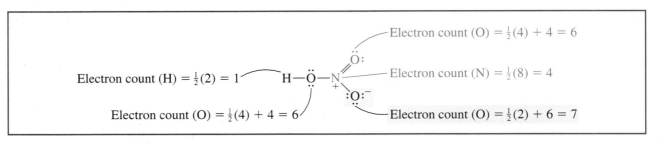

FIGURE 1.4   Counting electrons in nitric acid. The electron count of each atom is equal to half the number of electrons it shares in covalent bonds plus the number of electrons in its own unshared pairs.

To illustrate, let's start with the hydrogen of nitric acid. As shown in Figure 1.4, hydrogen is associated with only two electrons—those in its covalent bond to oxygen. It shares those two electrons with oxygen, and so we say that the electron count of each hydrogen is $\frac{1}{2}(2) = 1$. Because this is the same as the number of electrons in a neutral hydrogen atom, the hydrogen in nitric acid has no formal charge.

Moving now to nitrogen, we see that it has four covalent bonds (two single bonds + one double bond), and so its electron count is $\frac{1}{2}(8) = 4$. A neutral nitrogen has five electrons in its valence shell. The electron count for nitrogen in nitric acid is 1 less than that of a neutral nitrogen atom, so its formal charge is $+1$.

Electrons in covalent bonds are counted as if they are shared equally by the atoms they connect, but unshared electrons belong to a single atom. Thus, the oxygen that is doubly bonded to nitrogen has an electron count of 6 (four electrons as two unshared pairs + two electrons from the double bond). Because this is the same as a neutral oxygen atom, its formal charge is 0. Similarly, the OH oxygen has two bonds plus two unshared electron pairs, giving it an electron count of 6 and no formal charge.

The oxygen highlighted in yellow in Figure 1.4 owns three unshared pairs (6 electrons) and shares 2 electrons with nitrogen to give it an electron count of 7. This is 1 more than the number of electrons in the valence shell of an oxygen atom, and so its formal charge is -1.

The method described for calculating formal charge has been one of reasoning through a series of logical steps. It can be reduced to the following equation:

$$\text{Formal charge} = \frac{\text{group number in}}{\text{periodic table}} - \text{number of bonds} - \text{number of unshared electrons}$$

> It will always be true that a covalently bonded hydrogen has no formal charge (formal charge = 0).

> It will always be true that a nitrogen with four covalent bonds has a formal charge of $+1$. (A nitrogen with four co-valent bonds cannot have unshared pairs, because of the octet rule.)

> It will always be true that an oxygen with two covalent bonds and two unshared pairs has no formal charge.

> It will always be true that an oxygen with one covalent bond and three unshared pairs has a formal charge of -1.

PROBLEM 1.8   Determine the formal charge of all the atoms in each of the following species and the net charge on the species as a whole.

(a) $H-\overset{\overset{\displaystyle ..}{}}{\underset{\underset{\displaystyle H}{\displaystyle |}}{O}}-H$ (b) $H-\overset{\overset{\displaystyle ..}{}}{\underset{\underset{\displaystyle H}{\displaystyle |}}{C}}-H$ (c) $H-\overset{\overset{\displaystyle .}{}}{\underset{\underset{\displaystyle H}{\displaystyle |}}{C}}-H$ (d) $H-\underset{\underset{\displaystyle H}{\displaystyle |}}{C}-H$ (e) $:C\equiv N:$

Sample Solution   (a) Each hydrogen has a formal charge of 0, as is always the case when hydrogen is covalently bonded to one substituent. Oxygen has an electron count of 5.

$H-\overset{\overset{\displaystyle ..}{}}{\underset{\underset{\displaystyle H}{\displaystyle |}}{O}}-H$ Electron count of oxygen = $2 + \frac{1}{2}(6) = 5$

Unshared pair   Covalently bonded electrons

A neutral oxygen atom has 6 valence electrons; therefore, oxygen in this species has a formal charge of $+1$. The species as a whole has a unit positive charge. It is the hydronium ion H_3O^+.

Determining formal charges on individual atoms of Lewis structures is an important element in good "electron bookkeeping." So much of organic chemistry can be made more understandable by keeping track of electrons that it is worth taking some time at the beginning to become proficient at the seemingly simple task of counting electrons.

1.7   STRUCTURAL FORMULAS OF ORGANIC MOLECULES

Table 1.3 outlines a systematic procedure for writing Lewis structures. Notice that the process depends on knowing not only the molecular formula, but also the order in which the atoms are attached to one another. This order of attachment is called the **constitution,** or **connectivity,** of the molecule and is determined by experiment. Only rarely is it possible to deduce the constitution of a molecule from its molecular formula.

     Organic chemists have devised a number of shortcuts to speed the writing of structural formulas. Sometimes we leave out unshared electron pairs, but only when we are sure enough in our ability to count electrons to know when they are present and when they're not. We've already mentioned representing covalent bonds by dashes. In **condensed structural formulas** we leave out some, many, or all of the covalent bonds and use subscripts to indicate the number of identical groups attached to a particular atom. These successive levels of simplification are illustrated as shown for isopropyl alcohol ("rubbing alcohol").

$$
\begin{array}{ccc}
& \overset{\displaystyle H}{\underset{\displaystyle H}{|}}\ \ \overset{\displaystyle H}{\underset{\displaystyle \mathrel{:}\!O\!\mathrel{:}}{|}}\ \ \overset{\displaystyle H}{\underset{\displaystyle H}{|}} \\
H{-}C{-}C{-}C{-}H & & \\
\end{array}
$$

$$
\text{H—C—C—C—H} \qquad \text{written as} \qquad \text{CH}_3\text{CHCH}_3 \qquad \text{or condensed}
$$

H—C—C—C—H   written as CH_3CHCH_3 or condensed even further to $(\text{CH}_3)_2\text{CHOH}$ with OH below the central carbon.

PROBLEM 1.9   Expand the following condensed formulas so as to show all the bonds and unshared electron pairs.

(a) $\text{HOCH}_2\text{CH}_2\text{NH}_2$ (d) CH_3CHCl_2

(b) $(\text{CH}_3)_3\text{CH}$ (e) $\text{CH}_3\text{NHCH}_2\text{CH}_3$

(c) $\text{ClCH}_2\text{CH}_2\text{Cl}$ (f) $(\text{CH}_3)_2\text{CHCH}{=}\text{O}$

Sample Solution   (a) The molecule contains two carbon atoms, which are bonded to each other. Both carbons bear two hydrogens. One carbon bears the group HO—; the other is attached to —NH$_2$.

$$
\begin{array}{ccc}
H & H & H \\
| & | & | \\
H{-}O{-}C{-}C{-}N{-}H \\
| & | \\
H & H \\
\end{array}
$$

When writing the constitution of a molecule, it is not necessary to concern yourself with the spatial orientation of the atoms. There are many other correct ways to represent the constitution shown. What is important is to show the sequence OCCN (or its equivalent NCCO) and to have the correct number of hydrogens present on each atom.

     To locate unshared electron pairs, first count the total number of valence electrons brought to the molecule by its component atoms. Each hydrogen contributes 1, each carbon 4, nitrogen 5, and oxygen 6, for a total of 26. Ten bonds are shown, accounting for 20 electrons; therefore 6 electrons must be contained in unshared pairs. Add pairs of electrons to oxygen and nitrogen so that their octets are complete, two unshared pairs to oxygen and one to nitrogen.

$$
\begin{array}{ccc}
H & H & H \\
| & | & | \\
H{-}\ddot{\underset{..}{O}}{-}C{-}C{-}\underset{..}{N}{-}H \\
| & | \\
H & H \\
\end{array}
$$

| TABLE 1.3 | How to Write Lewis Structures |

Step	Illustration
1. The molecular formula and the connectivity are determined experimentally and are included among the information given in the statement of the problem.	Methyl nitrite has the molecular formula CH_3NO_2. All hydrogens are bonded to carbon, and the order of atomic connections is CONO.
2. Count the number of valence electrons available. For a neutral molecule this is equal to the sum of the valence electrons of the constituent atoms.	Each hydrogen contributes 1 valence electron, carbon contributes 4, nitrogen contributes 5, and each oxygen contributes 6 for a total of 24 in CH_3NO_2.
3. Connect bonded atoms by a shared electron pair bond represented by a dash (—).	For methyl nitrite we write the partial structure $$H-\underset{\underset{H}{\vert}}{\overset{\overset{H}{\vert}}{C}}-O-N-O$$
4. Count the number of electrons in shared electron pair bonds (twice the number of bonds), and subtract this from the total number of electrons to give the number of electrons to be added to complete the structure.	The partial structure in step 3 contains six bonds equivalent to 12 electrons. Because CH_3NO_2 contains 24 electrons, 12 more electrons need to be added.
5. Add electrons in pairs so that as many atoms as possible have 8 electrons. (Hydrogen is limited to 2 electrons.) When the number of electrons is insufficient to provide an octet for all atoms, assign electrons to atoms in order of decreasing electronegativity.	With four bonds, carbon already has 8 electrons. The remaining 12 electrons are added as indicated. Both oxygens have 8 electrons, but nitrogen (less electronegative than oxygen) has only 6. $$H-\underset{\underset{H}{\vert}}{\overset{\overset{H}{\vert}}{C}}-\ddot{\ddot{O}}-\ddot{N}-\ddot{\ddot{O}}:$$
6. If one or more atoms have fewer than 8 electrons, use unshared pairs on an adjacent atom to form a double (or triple) bond to complete the octet.	An electron pair on the terminal oxygen is shared with nitrogen to give a double bond. $$H-\underset{\underset{H}{\vert}}{\overset{\overset{H}{\vert}}{C}}-\ddot{\ddot{O}}-\ddot{N}=\ddot{\ddot{O}}:$$ The structure shown is the best (most stable) Lewis structure for methyl nitrite. All atoms except hydrogen have 8 electrons (shared + unshared) in their valence shell.
7. Calculate formal charges.	None of the atoms in the Lewis structure shown in step 6 possesses a formal charge. An alternative Lewis structure for methyl nitrite, $$H-\underset{\underset{H}{\vert}}{\overset{\overset{H}{\vert}}{C}}-\overset{+}{\ddot{O}}=\ddot{N}-\ddot{\ddot{O}}:^{-}$$ although it satisfies the octet rule, is less stable than the one shown in step 6 because positive charge is separated from negative charge.

> As you practice, you will begin to remember patterns of electron distribution. A neutral oxygen with two bonds has two unshared electron pairs. A neutral nitrogen with three bonds has one unshared pair.

With practice, writing structural formulas for organic molecules soon becomes routine and can be simplified even more. For example, a chain of carbon atoms can be represented by drawing all of the C—C bonds while omitting individual carbons. The resulting structural drawings can be simplified still more by stripping away the hydrogens.

$$CH_3CH_2CH_2CH_3 \qquad becomes \qquad \qquad simplified\ to$$

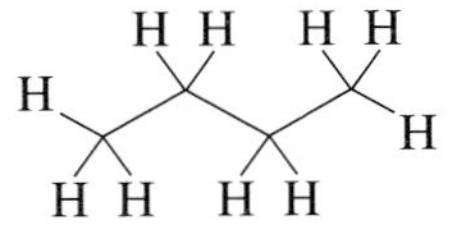

In these simplified representations, called **bond-line formulas** or **carbon skeleton diagrams,** the only atoms specifically written in are those that are neither carbon nor hydrogen bound to carbon. Hydrogens bound to these **heteroatoms** are shown, however.

$$CH_3CH_2CH_2CH_2OH \qquad becomes \qquad \qquad OH$$

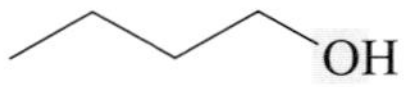

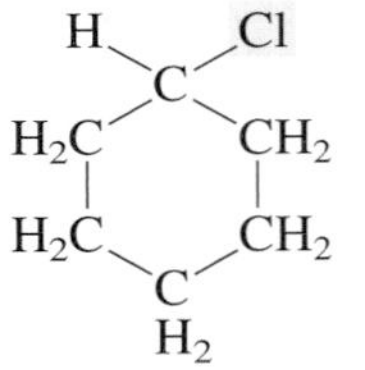

becomes

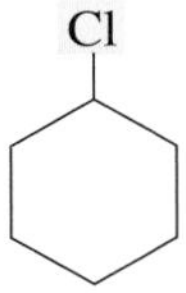

PROBLEM 1.10 Expand the following bond-line representations to show all the atoms including carbon and hydrogen.

(a)

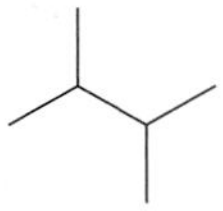

(c) 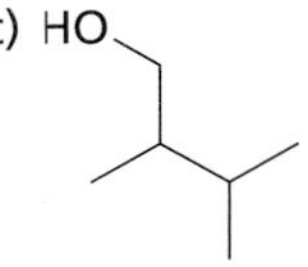 HO

(b)

(d)

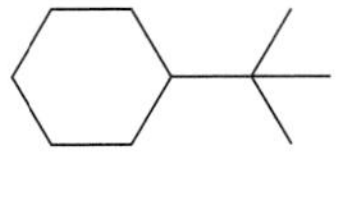

Sample Solution (a) A carbon appears at each bend in the chain and at the ends of the chain. Each of the six carbon atoms bears the appropriate number of hydrogen substituents so that it has four bonds.

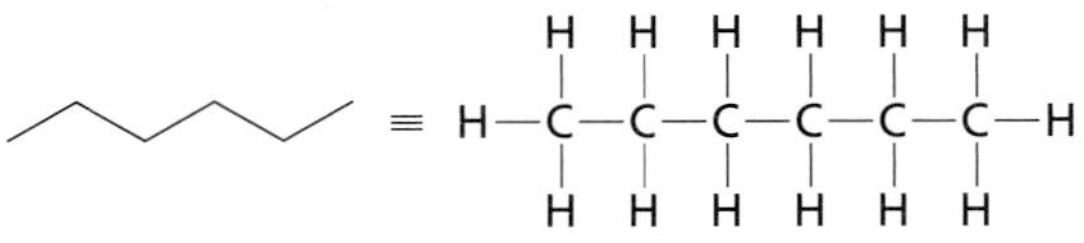

$$\equiv \quad H-\overset{\displaystyle H}{\underset{\displaystyle H}{C}}-\overset{\displaystyle H}{\underset{\displaystyle H}{C}}-\overset{\displaystyle H}{\underset{\displaystyle H}{C}}-\overset{\displaystyle H}{\underset{\displaystyle H}{C}}-\overset{\displaystyle H}{\underset{\displaystyle H}{C}}-\overset{\displaystyle H}{\underset{\displaystyle H}{C}}-H$$

Alternatively, the structure could be written as $CH_3CH_2CH_2CH_2CH_2CH_3$ or in condensed form as $CH_3(CH_2)_4CH_3$.

1.8   ISOMERS AND ISOMERISM

You may have noticed that the compounds in Problems 1.10a and 1.10b have the same molecular formula (C_6H_{14}). Different compounds that have the same molecular formula are called **isomers.** As you will see in this and later chapters, isomerism is very common in organic chemistry.

We can illustrate isomerism by referring to two different compounds, *nitromethane* and *methyl nitrite*, both of which have the molecular formula CH_3NO_2. Nitromethane,

The suffix -*mer* in the word *isomer* is derived from the Greek word *meros,* meaning "part," "share," or "portion." The prefix *iso-* is also from Greek (*isos,* meaning "the same"). Thus isomers are different molecules that have the same parts (elemental composition).

$$\begin{array}{cc}
\text{Nitromethane} & \text{Methyl nitrite}
\end{array}$$

used to power race cars, is a liquid with a boiling point of 101°C. Methyl nitrite is a gas boiling at −12°C, which when inhaled causes dilation of blood vessels. Isomers that differ in the order in which their atoms are bonded are often referred to as **structural isomers.** A more modern term is **constitutional isomer.** As noted in the previous section, the order of atomic connections that defines a molecule is termed its **constitution,** and we say that two compounds are **constitutional isomers** if they have the same molecular formula but differ in the order in which their atoms are connected.

PROBLEM 1.11 Write structural formulas for all the constitutionally isomeric compounds having the given molecular formula.

(a) C_2H_6O (b) C_3H_8O (c) $C_4H_{10}O$

Sample Solution   (a) Begin by considering the ways in which two carbons and one oxygen may be bonded. There are two possibilities: C—C—O and C—O—C. Add the six hydrogens so that each carbon has four bonds and each oxygen two. There are two constitutional isomers: ethyl alcohol and dimethyl ether.

$$\begin{array}{cc}
\text{Ethyl alcohol} & \text{Dimethyl ether}
\end{array}$$

The phenomenon of isomerism is closely linked to the development of organic chemistry as a science. At one time it was believed that organic compounds could arise only through the action of some "vital force." Organic chemistry, according to this view, was truly the chemistry of living systems. Inorganic chemistry, on the other hand, existed in a separate domain—the world of water, metals, minerals, and the like. The laboratory synthesis of organic compounds from inorganic ones was believed to be impossible unless some vital force was present to forge the link between living and nonliving matter.

The experiment that began the downfall of this doctrine of "vitalism" was the discovery by Friedrich Wöhler in 1828 that crystals of urea were formed when a solution of ammonium cyanate in water was evaporated.

$$NH_4^{+\,-}OCN \longrightarrow O{=}C(NH_2)_2$$

Ammonium cyanate   Urea
(an inorganic compound)   (an organic compound)

The reaction that Wöhler observed was the conversion of a compound to its isomer; both ammonium cyanate and urea have the molecular formula CH_4N_2O. What made the transformation noteworthy was that ammonium cyanate was classified as inorganic, but urea was accepted as an organic substance because it had been isolated earlier from urine. Without the aid of some vital force, an inorganic substance had been transformed into an organic one. Wöhler could not describe the nature of the transformation in structural terms because chemists had not yet begun to think of substances as having defined structures at that time. His work demonstrated a fundamental flaw in the doctrine of vitalism, however, and over the next 30 years organic chemistry outgrew vitalism.

Beginning approximately in 1858, what we now know as the **structural theory of organic chemistry** was independently proposed by August Kekulé (Germany), Archibald Couper (Scotland), and Alexander Butlerov (Russia). Its fundamental tenets are that carbon has four bonds in its stable compounds and has the capacity to bond to other carbon atoms so as to form long chains. Constitutional isomers are possible because a particular elemental composition can accommodate more than one pattern of atoms and bonds.

1.9   RESONANCE

When writing a Lewis structure, we restrict a molecule's electrons to certain well-defined locations, either linking two atoms by a covalent bond or as unshared electrons on a single atom. Sometimes more than one Lewis structure can be written for a molecule, especially those that contain multiple bonds. An example often cited in introductory chemistry courses is ozone (O_3). Ozone occurs naturally in large quantities in the upper atmosphere, where it screens the surface of the earth from much of the sun's ultraviolet rays. Were it not for this ozone layer, most forms of surface life on earth would be damaged or even destroyed by the rays of the sun. The following Lewis structure for ozone satisfies the octet rule; all three oxygens have 8 electrons in their valence shell.

This Lewis structure, however, doesn't accurately portray the bonding in ozone, because the two terminal oxygens are bonded differently to the central oxygen. The central oxygen is depicted as doubly bonded to one and singly bonded to the other. Because it is generally true that double bonds are shorter than single bonds, we would expect ozone to exhibit two different O—O bond lengths, one of them characteristic of the O—O single bond distance (147 pm in hydrogen peroxide, H—O—O—H) and the other one characteristic of the O=O double bond distance (121 pm in O_2). Such is not the case. Both bond distances in ozone are exactly the same (128 pm)—somewhat shorter than the single-bond distance and somewhat longer than the double-bond distance.

> *The structure of ozone requires that the central oxygen must be* identically bonded *to both terminal oxygens.*

To deal with circumstances such as the bonding in ozone, the notion of **resonance** between Lewis structures was developed. According to the resonance concept, when more than one Lewis structure may be written for a molecule, a single structure is not sufficient to describe it. Rather, the true structure has an electron distribution that is a "hybrid" of all the possible Lewis structures that can be written for the molecule. In the case of ozone, two equivalent Lewis structures may be written. We use a double-headed arrow to represent resonance between these two Lewis structures.

Bond distances in organic compounds are usually 1–2 Å (1 Å = 10^{-10} m). Because the angstrom (Å) is not an SI unit, we will express bond distances in picometers (1 pm = 10^{-12} m). Thus, 128 pm = 1.28 Å.

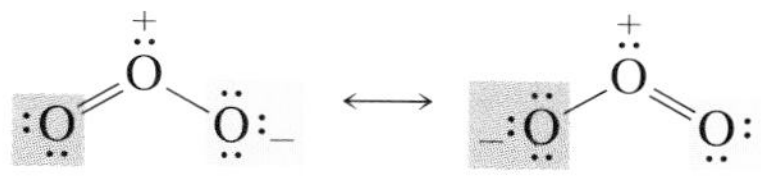

It is important to remember that the double-headed resonance arrow does not indicate a *process* in which the two Lewis structures interconvert. Ozone, for example, has a *single* structure; it does not oscillate back and forth between two Lewis structures, rather its true structure is not adequately represented by any single Lewis structure.

Resonance attempts to correct a fundamental defect in Lewis formulas. Lewis formulas show electrons as being **localized;** they either are shared between two atoms in a covalent bond or are unshared electrons belonging to a single atom. In reality, electrons distribute themselves in the way that leads to their most stable arrangement. This sometimes means that a pair of electrons is **delocalized,** or shared by several nuclei. What we try to show by the resonance description of ozone is the delocalization of the lone-pair electrons of one oxygen and the electrons in the double bond over the three atoms of the molecule. Organic chemists often use curved arrows to show this electron delocalization. Alternatively, an average of two Lewis structures is sometimes drawn using a dashed line to represent a "partial" bond. In the dashed-line notation the central oxygen is linked to the other two by bonds that are halfway between a single bond and a double bond, and the terminal oxygens each bear one half of a unit negative charge.

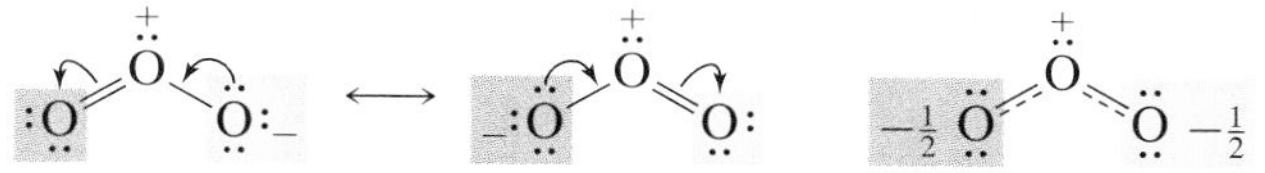

Curved arrow notation   Dashed-line notation
(electron delocalization in ozone)

The rules to be followed when writing resonance structures are summarized in Table 1.4.

PROBLEM 1.12 Electron delocalization can be important in ions as well as in neutral molecules. Using curved arrows, show how an equally stable resonance structure can be generated for each of the following anions:

(a)

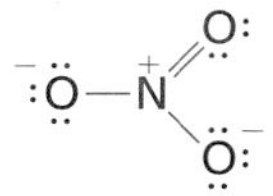

(c)

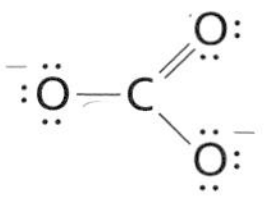

(b)

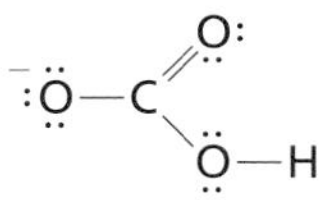

(d)

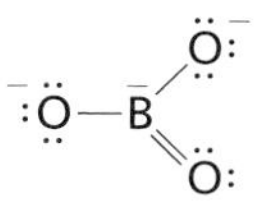

Sample Solution (a) When using curved arrows to represent the reorganization of electrons, begin at a site of high electron density, preferably an atom that is negatively charged. Move electron pairs until a proper Lewis structure results. For nitrate ion, this can be accomplished in two ways:

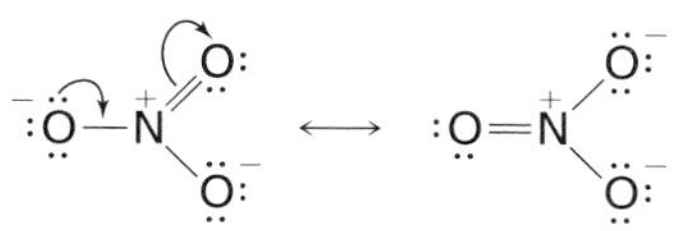

TABLE 1.4	Introduction to the Rules of Resonance*

Rule	Illustration
1. Atomic positions (connectivity) must be the same in all resonance structures; only the electron positions may vary among the various contributing structures.	The structural formulas  $CH_3-\overset{+}{N}\!\!\begin{smallmatrix}\ddot{O}:\\ \\ \ddot{O}:^-\end{smallmatrix}$ and $CH_3-\ddot{O}-\ddot{N}=\ddot{O}:$ A B represent different compounds, not different resonance forms of the same compound. A is a Lewis structure for *nitromethane;* B is *methyl nitrite.*
2. Lewis structures in which second-row elements own or share more than 8 valence electrons are especially unstable and make no contribution to the true structure. (The octet rule may be exceeded for elements beyond the second row.)	Structural formula C, 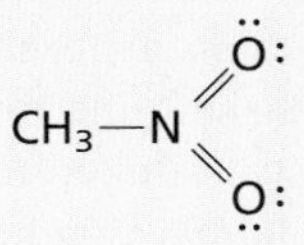 $CH_3-N\begin{smallmatrix}\ddot{O}:\\ \\ \ddot{O}:\end{smallmatrix}$ C has 10 electrons around nitrogen. It is not a permissible Lewis structure for nitromethane and so cannot be a valid resonance form.
3. When two or more structures satisfy the octet rule, the most stable one is the one with the smallest separation of oppositely charged atoms.	The two Lewis structures D and E of methyl nitrite satisfy the octet rule: 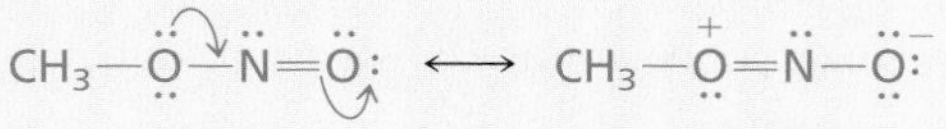 $CH_3-\ddot{O}-\ddot{N}=\ddot{O}:$ ⟷ $CH_3-\overset{+}{\ddot{O}}=\ddot{N}-\ddot{O}:^-$ D E Structure D has no separation of charge and is more stable than E, which does. The true structure of methyl nitrite is more like D than E.

(Continued)

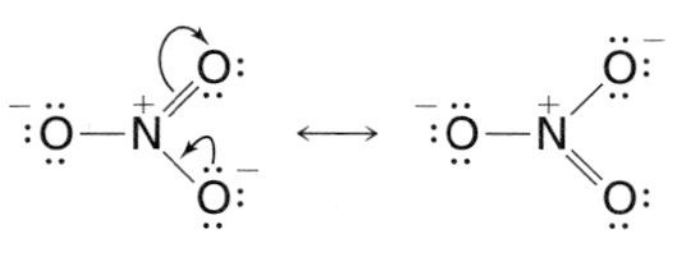

$:\ddot{O}-\overset{+}{N}\begin{smallmatrix}\ddot{O}:\\ \\ \ddot{O}:^-\end{smallmatrix}$ ⟷ $:\ddot{O}-\overset{+}{N}\begin{smallmatrix}\ddot{O}:^-\\ \\ \ddot{O}:\end{smallmatrix}$

Three equally stable Lewis structures are possible for nitrate ion. The negative charge in nitrate is shared equally by all three oxygens.

It is good chemical practice to represent molecules by their most stable Lewis structure. The ability to write alternative resonance forms and to compare their relative stabilities, however, can provide insight into both molecular structure and chemical behavior. This will become particularly apparent in the last two thirds of this text, where the resonance concept will be used regularly.

TABLE 1.4	Introduction to the Rules of Resonance* *(Continued)*

Rule	Illustration
4. Among resonance forms in which a negative charge is shared by two or more atoms, the most stable resonance form is the one in which negative charge resides on the most electronegative atom.	The most stable Lewis structure for cyanate ion is F because the negative charge is on its oxygen. $:N \equiv C - \overset{..}{\underset{..}{O}}:^{-}$ $\longleftrightarrow$ $^{-}:\overset{..}{N} = C = \overset{..}{\underset{..}{O}}:$ F G In G the negative charge is on nitrogen. Oxygen is more electronegative than nitrogen and can better support a negative charge.
5. Each contributing Lewis structure must have the same number of electrons and the same *net* charge, although the formal charges of individual atoms may vary among the various Lewis structures.	The Lewis structures $CH_3 - \overset{+}{N} \overset{\overset{\displaystyle \overset{..}{O}:}{\diagup}}{\underset{\underset{\displaystyle :\overset{..}{\underset{..}{O}}:^{-}}{}}{}}$ and $CH_3 - \overset{\overset{\displaystyle :\overset{..}{\underset{..}{O}}^{-}}{\diagup}}{\underset{\underset{\displaystyle \overset{..}{\underset{..}{O}}:^{-}}{}}{N:}}$ H I are *not* resonance forms of one another. Structure H has 24 valence electrons and a net charge of 0; I has 26 valence electrons and a net charge of -2.
6. Electron delocalization stabilizes a molecule. A molecule in which electrons are delocalized is more stable than implied by any of the individual Lewis structures that may be written for it. The degree of stabilization is greatest when the contributing Lewis structures are of equal stability.	Nitromethane is stabilized by electron delocalization more than methyl nitrite is. The two most stable resonance forms of nitromethane are *equivalent* to each other. $CH_3 - \overset{+}{N} \overset{\overset{\displaystyle \overset{..}{O}:}{\diagup}}{\underset{\underset{\displaystyle :\overset{..}{\underset{..}{O}}:^{-}}{}}{}}$ $\longleftrightarrow$ $CH_3 - \overset{\overset{\displaystyle :\overset{..}{\underset{..}{O}}^{-}}{\diagup}}{\underset{\underset{\displaystyle \overset{..}{O}:}{}}{\overset{+}{N}}}$ The two most stable resonance forms of methyl nitrite are *not* equivalent. $CH_3 - \overset{..}{\underset{..}{O}} - \overset{..}{N} = \overset{..}{\underset{..}{O}}: \longleftrightarrow CH_3 - \overset{..}{\underset{..}{O}} = \overset{+}{N} - \overset{..}{\underset{..}{O}}:^{-}$

* These are the most important rules to be concerned with at present. Additional aspects of electron delocalization, as well as additional rules for its depiction by way of resonance structures, will be developed as needed in subsequent chapters.

1.10 THE SHAPES OF SOME SIMPLE MOLECULES

So far our concern has emphasized "electron bookkeeping." We now turn our attention to the shapes of molecules.

Methane, for example, is described as a tetrahedral molecule because its four hydrogens occupy the corners of a tetrahedron with carbon at its center as the various methane models in Figure 1.5 illustrate. We often show three-dimensionality in structural formulas by using a solid wedge (▬◤) to depict a bond projecting from the paper toward the reader and a dashed wedge (⠤⠤⠤) to depict one receding from the paper. A simple line (—) represents a bond that lies in the plane of the paper (Figure 1.6).

MOLECULAR MODELS

As early as the nineteenth century many chemists built scale models to better understand molecular structure. We can gain a clearer idea about the features that affect structure and reactivity when we examine the three-dimensional shape of a molecule. Several types of molecular models are shown for methane in Figure 1.5. Probably the most familiar are ball-and-stick models (Figure 1.5b), which direct approximately equal attention to the atoms and the bonds that connect them. Framework models (Figure 1.5a) and space-filling models (Figure 1.5c) represent opposite extremes. Framework models emphasize the pattern of bonds of a molecule while ignoring the sizes of the atoms. Space-filling models emphasize the volume occupied by individual atoms at the cost of a clear depiction of the bonds; they are most useful in cases in which one wishes to examine the overall molecular shape and to assess how closely two nonbonded atoms approach each other.

The earliest ball-and-stick models were exactly that: wooden balls in which holes were drilled to accommodate dowels that connected the atoms. Plastic versions, including relatively inexpensive student sets, became available in the 1960s and proved to be a valuable learning aid. Precisely scaled stainless steel framework and plastic space-filling models, although relatively expensive, were standard equipment in most research laboratories.

Computer graphics-based representations are rapidly replacing classical molecular models. Indeed, the term "molecular modeling" as now used in organic chemistry implies computer generation of models. The methane models shown in Figure 1.5 were all drawn on a personal computer using software that possesses the feature of displaying and printing the same molecule in framework, ball-and-stick, and space-filling formats. In addition to permitting models to be constructed rapidly, even the simplest software allows the model to be turned and viewed from a variety of perspectives.

More sophisticated programs not only draw molecular models, but also incorporate computational tools that provide useful insights into the electron distribution. Figure 1.5d illustrates this higher level approach to molecular modeling by using colors to display the electric charge distribution within the boundaries defined by the space-filling model. Figures such as 1.5d are called **electrostatic potential maps.** They show the transition from regions of highest to lowest electron density according to the colors of the rainbow. The most electron-rich regions are red; the most electron-poor are blue. For methane, the overall shape of the electrostatic potential map is similar to the volume occupied by the space-filling model. The most electron-rich regions are closer to carbon and the most electron-poor regions closer to the hydrogen atoms.

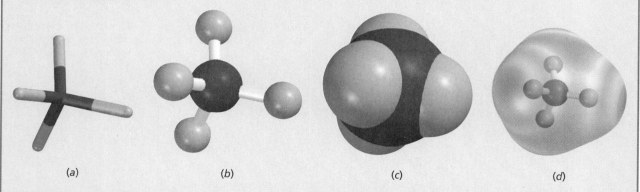

(a)   (b)   (c)   (d)

FIGURE 1.5   (a) A framework (tube) molecular model of methane (CH_4). A framework model shows the bonds connecting the atoms of a molecule, but not the atoms themselves. (b) A ball-and-stick (ball-and-spoke) model of methane. (c) A space-filling model of methane. (d) An electrostatic potential map superimposed on a ball-and-stick model of methane. The electrostatic potential map corresponds to the space-filling model, but with an added feature. The colors identify regions according to their electric charge, with red being the most negative and blue the most positive.

The tetrahedral geometry of methane is often explained in terms of the **valence shell electron-pair repulsion (VSEPR) model.** The VSEPR model rests on the idea that an electron pair, either a bonded pair or an unshared pair, associated with a particular atom will be as far away from the atom's other electron pairs as possible. Thus, a tetrahedral geometry permits the four bonds of methane to be maximally separated and is characterized by H—C—H angles of 109.5°, a value referred to as the **tetrahedral angle.**

Water, ammonia, and methane share the common feature of an approximately tetrahedral arrangement of four electron pairs. Because we describe the shape of a molecule according to the positions of its atoms rather than the disposition of its electron pairs, however, water is said to be **bent,** and ammonia is **trigonal pyramidal** (Figure 1.7). The H—O—H angle in water (105°) and the H—N—H angle in ammonia (107°) are slightly less than the tetrahedral angle.

Boron trifluoride (BF_3; Figure 1.8a) is a **trigonal planar** molecule. There are 6 electrons, 2 for each B—F bond, associated with the valence shell of boron. These three bonded pairs are farthest apart when they are coplanar, with F—B—F bond angles of 120°.

> **PROBLEM 1.13** The salt sodium borohydride, $NaBH_4$, has an ionic bond between Na^+ and the anion BH_4^-. What are the H—B—H angles in the borohydride anion?

Multiple bonds are treated as a single unit in the VSEPR model. Formaldehyde (Figure 1.8b) is a trigonal planar molecule in which the electrons of the double bond and those of the two single bonds are maximally separated. A linear arrangement of atoms in carbon dioxide (Figure 1.9) allows the electrons in one double bond to be as far away as possible from the electrons in the other double bond.

> **PROBLEM 1.14** Specify the shape of the following:
>
> (a) H—C≡N: Hydrogen cyanide (c) :N̈=N⁺=N̈:⁻ Azide ion
>
> (b) H_4N^+ Ammonium ion (d) CO_3^{2-} Carbonate ion

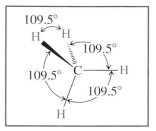

FIGURE 1.6 A wedge-and-dash drawing of the structure of methane. A solid wedge projects from the plane of the paper toward you; a dashed wedge projects away from you. A bond represented by a line drawn in the customary way lies in the plane of the paper.

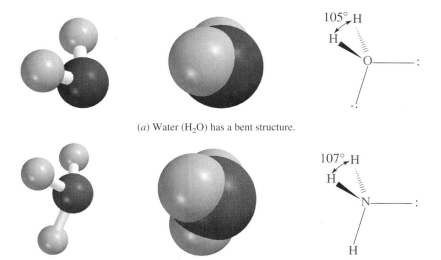

(a) Water (H_2O) has a bent structure.

(b) Ammonia (NH_3) has a trigonal pyramidal structure.

FIGURE 1.7 Ball-and-spoke and space-filling models and wedge-and-dash drawings of (a) water and (b) ammonia. The shape of a molecule is described in terms of its atoms. An approximately tetrahedral arrangement of electron pairs translates into a bent geometry for water and a trigonal pyramidal geometry for ammonia.

(a) BF₃   (b) H₂C=O

FIGURE 1.8 (a) The trigonal planar geometry of boron trifluoride (BF₃) can be readily seen. Six electrons are in the valence shell of boron, a pair for each covalent bond to fluorine. The three pairs of electrons are farthest apart when the F—B—F angle is 120°. (b) A model of formaldehyde (H₂C=O) showing the trigonal planar geometry of the bonds to carbon. Many molecular models, including those shown here, show only the connections between atoms without differentiating among single, double, or triple bonds.

FIGURE 1.9 Ball-and-spoke model showing the linear geometry of carbon dioxide (O=C=O).

Sample Solution (a) The structure shown accounts for all the electrons in hydrogen cyanide. No unshared electron pairs are associated with carbon, and so the structure is determined by maximizing the separation between its single bond to hydrogen and the triple bond to nitrogen. Hydrogen cyanide is a *linear* molecule.

1.11 MOLECULAR POLARITY

We can combine our knowledge of molecular geometry with a feel for the polarity of chemical bonds to predict whether a molecule is polar. Formaldehyde (H₂C=O, Figure 1.8b), for example, is polar. The small H—C dipoles point in the same direction as the larger C=O bond dipole. Carbon dioxide, on the other hand, is nonpolar. Even though polar bonds are present, the individual C=O bond dipoles cancel each other.

$$\underset{H}{\overset{H}{\diagdown}}C=\ddot{\underset{\cdot\cdot}{O}}:\qquad :\ddot{\underset{\cdot\cdot}{O}}=C=\ddot{\underset{\cdot\cdot}{O}}:$$

Formaldehyde   Carbon dioxide

Carbon tetrachloride, with four polar C—Cl bonds and a tetrahedral shape, is nonpolar because the four bond dipoles cancel one another, as shown in Figure 1.10. The C—Cl and C—H bond dipoles do not cancel in CH₂Cl₂, and dichloromethane is therefore polar.

PROBLEM 1.15 Which of the following compounds would you expect to have a dipole moment? If the molecule has a dipole moment, specify its direction.

(a) BF₃   (b) H₂O   (c) CH₄   (d) CH₃Cl   (e) HCN

Sample Solution (a) Boron trifluoride is planar with 120° bond angles. Although each boron–fluorine bond is polar, their combined effects cancel and the molecule has no dipole moment.

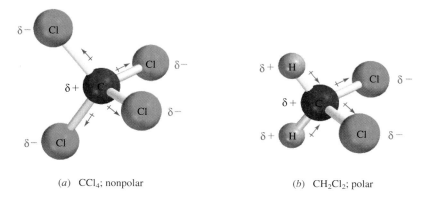

(*a*) CCl_4; nonpolar   (*b*) CH_2Cl_2; polar

FIGURE 1.10 (*a*) Carbon tetrachloride (CCl_4) is nonpolar because the individual bond dipoles cancel one another. (*b*) The H—C bond dipoles reinforce the C—Cl bond dipoles in dichloromethane (CH_2Cl_2). The molecule is polar.

1.12 *sp³* HYBRIDIZATION AND BONDING IN METHANE

There was a vexing puzzle in the early days of the development of theories of bonding in methane (CH_4). Because covalent bonding requires the overlap of half-filled orbitals of the connected atoms, carbon with an electron configuration of $1s^2 2s^2 2p_x{}^1 2p_y{}^1$ has only two half-filled orbitals (Figure 1.11*a*), so how can it have bonds to four hydrogens?

In the 1930s Linus Pauling offered an ingenious solution to the puzzle. His model began with a simple idea: "promoting" one of the $2s$ electrons to the empty $2p_z$ orbital gives four half-filled orbitals and allows for four C—H bonds (Figure 1.11*b*). The electron configuration that results ($1s^2 2s^1 2p_x{}^1 2p_y{}^1 2p_z{}^1$), however, is inconsistent with the fact that all of these bonds are equivalent and directed toward the corners of a tetrahedron. The second part of Pauling's idea was novel: mix together (**hybridize**) the four valence orbitals of carbon ($2s$, $2p_x$, $2p_y$, and $2p_z$) to give four half-filled orbitals of equal energy (Figure 1.11*c*). The four new orbitals in Pauling's scheme are called *sp³* **hybrid orbitals** because they come from one s orbital and three p orbitals.

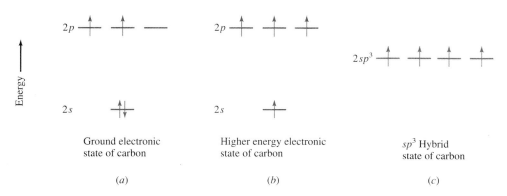

FIGURE 1.11 (*a*) Electron configuration of carbon in its most stable state. (*b*) An electron is "promoted" from the $2s$ orbital to the vacant $2p$ orbital. (*c*) The $2s$ orbital and the three $2p$ orbitals are combined to give a set of four equal-energy *sp³*-hybridized orbitals, each of which contains one electron.

FIGURE 1.12 The two lobes of each sp^3 hybrid orbital are of different size. More of the electron density is concentrated on one side of the nucleus than on the other.

Each sp^3 hybrid orbital has two lobes of unequal size (Figure 1.12), making the electron density greater on one side of the nucleus than the other. In a bond to hydrogen, it is the larger lobe of a carbon sp^3 orbital that overlaps with a hydrogen 1s orbital. The orbital overlaps corresponding to the four C—H bonds of methane are portrayed in Figure 1.13. Orbital overlap along the internuclear axis generates a bond with rotational symmetry that is called a **sigma (σ) bond.** In this case the bond is a C($2sp^3$)—H($1s$) σ bond.

A tetrahedral arrangement of four σ bonds is characteristic of sp³-*hybridized carbon.*

The peculiar shape of sp^3 hybrid orbitals turns out to have an important consequence. Because most of the electron density in an sp^3 hybrid orbital lies to one side of a carbon atom, overlap with a half-filled 1s orbital of hydrogen, for example, on that side produces a stronger bond than would result otherwise. If the electron probabilities were equal on both sides of the nucleus, as it would be in a p orbital, half of the time the electron would be remote from the region between the bonded atoms, and the bond would be weaker. Thus, not only does Pauling's orbital hybridization proposal account for carbon forming four bonds rather than two, these bonds are also stronger than they would be otherwise.

1.13 BONDING IN ETHANE

The orbital hybridization model of covalent bonding is readily extended to carbon–carbon bonds. As Figure 1.14 illustrates, ethane is described in terms of a carbon–carbon σ bond joining two CH_3 (**methyl**) groups. Each methyl group consists of an sp^3-hybridized carbon attached to three hydrogens by sp^3–1s σ bonds. Overlap of the remaining half-filled orbital of one carbon with that of the other generates a σ bond between them. Here is a third kind of σ bond, one that has as its basis the overlap of two sp^3-hybridized orbitals.

In general, you can expect that carbon will be sp³-*hybridized when it is directly bonded to four atoms.*

The orbital hybridization model of bonding is not limited to compounds in which all the bonds are single, but can be adapted to compounds with double and triple bonds, as described in the following two sections.

FIGURE 1.13 The sp^3 hybrid orbitals are arranged in a tetrahedral fashion around carbon. Each orbital contains one electron and can form a bond with a hydrogen atom to give a tetrahedral methane molecule. (*Note:* Only the major lobe of each sp^3 orbital is shown. As indicated in Figure 1.12, each orbital contains a smaller back lobe, which has been omitted for the sake of clarity.)

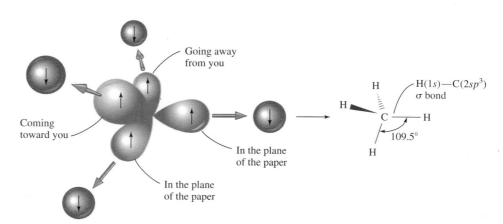

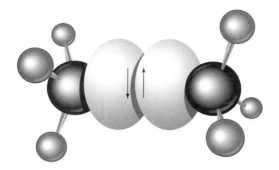

FIGURE 1.14 Orbital overlap description of the *sp³–sp³* σ bond between the two carbon atoms of ethane.

1.14   *sp²* HYBRIDIZATION AND BONDING IN ETHYLENE

Ethylene is a planar molecule, as the structural representations of Figure 1.15 indicate. Because *sp³* hybridization is associated with a tetrahedral geometry at carbon, it is not appropriate for ethylene, which has a trigonal planar geometry at both of its carbons. The hybridization scheme is determined by the number of atoms to which the carbon is directly attached. In ethane, four atoms are attached to carbon by σ bonds, and so four equivalent *sp³* hybrid orbitals are required. In ethylene, three atoms are attached to each carbon, so three equivalent hybrid orbitals are required. These three orbitals are generated by mixing the carbon 2*s* orbital with two of the 2*p* orbitals and are called ***sp²* hybrid orbitals.** One of the 2*p* orbitals is left unhybridized (Figure 1.16).

Each carbon of ethylene uses two of its *sp²* hybrid orbitals to form σ bonds to two hydrogen atoms, as illustrated in the first part of Figure 1.17. The remaining *sp²* orbitals, one on each carbon, overlap along the internuclear axis to give a σ bond connecting the two carbons.

As Figure 1.17 shows, each carbon atom still has, at this point, an unhybridized 2*p* orbital available for bonding. These two half-filled 2*p* orbitals have their axes perpendicular to the framework of σ bonds of the molecule and overlap in a side-by-side manner to give what is called a **pi (π) bond.** According to this analysis, the carbon–carbon double bond of ethylene is viewed as a combination of a σ bond plus a π bond. The additional

Another name for ethylene is **ethene.**

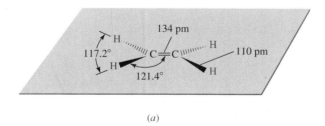

(*a*)

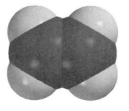

(*b*)

FIGURE 1.15 (*a*) All the atoms of ethylene lie in the same plane. All the bond angles are close to 120°, and the carbon–carbon bond distance is significantly shorter than that of ethane. (*b*) A space-filling model of ethylene.

FIGURE 1.16 (*a*) Electron configuration of carbon in its most stable state. (*b*) An electron is "promoted" from the 2*s* orbital to the vacant 2*p* orbital. (*c*) The 2*s* orbital and two of the three 2*p* orbitals are combined to give a set of three equal-energy *sp*²-hybridized orbitals. One of the 2*p* orbitals remains unchanged.

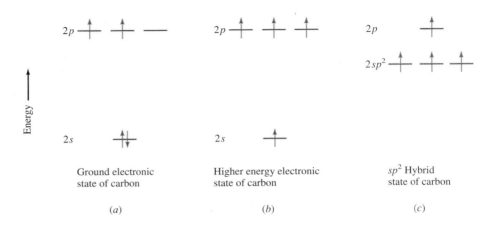

Ground electronic state of carbon

Higher energy electronic state of carbon

*sp*² Hybrid state of carbon

(*a*)

(*b*)

(*c*)

FIGURE 1.17 The carbon–carbon double bond in ethylene has a σ component and a π component. The σ component arises from overlap of *sp*²-hybridized orbitals along the internuclear axis. The π component results from a side-by-side overlap of 2*p* orbitals.

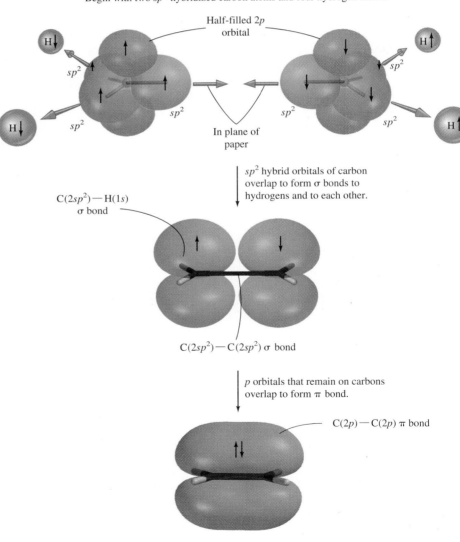

Begin with two *sp*²-hybridized carbon atoms and four hydrogen atoms:

Half-filled 2*p* orbital

In plane of paper

*sp*² hybrid orbitals of carbon overlap to form σ bonds to hydrogens and to each other.

C(2*sp*²)—H(1*s*) σ bond

C(2*sp*²)—C(2*sp*²) σ bond

p orbitals that remain on carbons overlap to form π bond.

C(2*p*)—C(2*p*) π bond

increment of bonding makes a carbon–carbon double bond both stronger and shorter than a carbon–carbon single bond.

Electrons in a π bond are called **π electrons.** The probability of finding a π electron is highest in the region above and below the plane of the molecule. The plane of the molecule corresponds to a nodal plane, where the probability of finding a π electron is zero.

In general, you can expect that carbon will be sp²-hybridized when it is directly bonded to three atoms.

1.15   *sp* HYBRIDIZATION AND BONDING IN ACETYLENE

One more hybridization scheme is important in organic chemistry. It is called **sp hybridization** and applies when carbon is directly bonded to two atoms, as it is in acetylene. The structure of acetylene is shown in Figure 1.18 along with its bond distances and bond angles.

Because each carbon in acetylene is bonded to two other atoms, the orbital hybridization model requires each carbon to have two equivalent orbitals available for the formation of σ bonds as outlined in Figure 1.19. According to this model the carbon $2s$ orbital and one of the $2p$ orbitals combine to generate a pair of two equivalent sp hybrid orbitals. These two sp orbitals share a common axis, but their major lobes are oriented at an angle of 180° to each other. Two of the original $2p$ orbitals remain unhybridized. Their axes are perpendicular to each other and to the common axis of the pair of sp hybrid orbitals.

As portrayed in Figure 1.20, the two carbons of acetylene are connected to each other by a $2sp–2sp$ σ bond, and each is attached to a hydrogen substituent by a $2sp–1s$ σ bond. The unhybridized $2p$ orbitals on one carbon overlap with their counterparts on the other to form two π bonds. The carbon–carbon triple bond in acetylene is viewed as a multiple bond of the $\sigma + \pi + \pi$ type.

In general, you can expect that carbon will be sp-hybridized when it is directly bonded to two atoms.

> **PROBLEM 1.16**   Give the hybridization state of each carbon in the following compounds:
>
> (a) Carbon dioxide (O=C=O)   (d) Propene (CH₃CH=CH₂)
> (b) Formaldehyde (H₂C=O)   (e) Acetone [(CH₃)₂C=O]
> (c) Ketene (H₂C=C=O)   (f) Acrylonitrile (CH₂=CHC≡N)
>
> **Sample Solution**   (a) Carbon in CO₂ is directly bonded to two other atoms. It is *sp*-hybridized.

> Another name for acetylene is **ethyne.**

(a)

(b)

FIGURE 1.18   Acetylene is a linear molecule as indicated in the (*a*) structural formula and a (*b*) space-filling model.

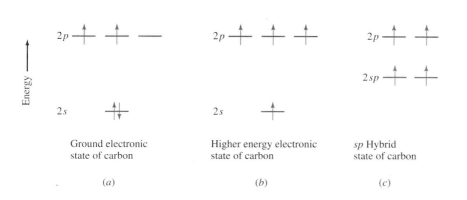

Ground electronic state of carbon

(*a*)

Higher energy electronic state of carbon

(*b*)

sp Hybrid state of carbon

(*c*)

FIGURE 1.19   (*a*) Electron configuration of carbon in its most stable state. (*b*) An electron is "promoted" from the 2*s* orbital to the vacant 2*p* orbital. (*c*) The 2*s* orbital and one of the three 2*p* orbitals are combined to give a set of two equal-energy *sp*-hybridized orbitals. Two of the 2*p* orbitals remain unchanged.

FIGURE 1.20 A description of bonding in acetylene based on *sp* hybridization of carbon. The carbon–carbon triple bond is viewed as consisting of one σ bond and two π bonds.

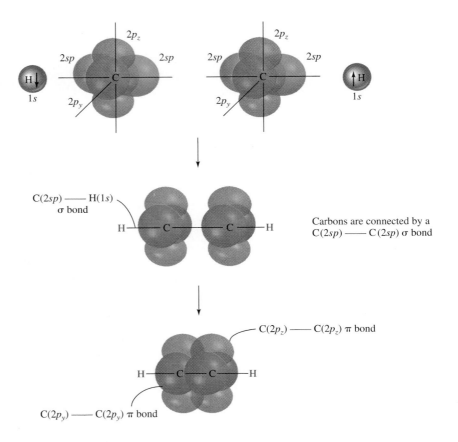

C(2*sp*) —— H(1*s*)
σ bond

Carbons are connected by a C(2*sp*) —— C(2*sp*) σ bond

C(2*p_z*) —— C(2*p_z*) π bond

C(2*p_y*) —— C(2*p_y*) π bond

LEARNING OBJECTIVES

This chapter has reviewed the principles of structure and bonding that will be useful as you learn about the chemistry of carbon compounds. It emphasized chemical bonds and electron "bookkeeping." Its main objective was to provide you with the skills essential in writing proper structural formulas. The skills you have learned in this chapter should enable you to:

- Write the electron configuration corresponding to a neutral atom or to an ion derived from it when given the atomic number of any element between hydrogen and argon in the periodic table.

- Describe the difference between ionic and covalent bonding.

- State the octet rule and discuss its significance.

- Determine the direction of polarization of a covalent bond on the basis of the difference in electronegativity of the atoms that it connects.

- Calculate the formal charges on atoms in Lewis structures.

- Write the structures of organic molecules using condensed structural and bond-line formulas.

- Write Lewis structures for constitutionally isomeric substances.

Continued

- Understand the difference between resonance and isomerism and use the resonance concept to describe electron delocalization in molecules and ions.
- Use the valence shell electron-pair repulsion model to predict the shapes of simple molecules.
- Predict the polarity of a molecule knowing its shape.
- Understand the sp^3, sp^2, sp orbital hybridization bonding models.

1.16   SUMMARY

The **electron configuration** of an atom describes the arrangement of electrons in regions of space called **orbitals** (Section 1.1). The outermost electrons, or **valence electrons,** are involved in bonding. Chemical bonds are classified as ionic or covalent. An **ionic bond** (Section 1.2) is the electrostatic attraction between two oppositely charged ions and occurs in substances such as sodium chloride. Chlorine, an electronegative element, gains an electron to form a negatively charged **anion.** Sodium, a metal, loses an electron to form a positively charged **cation.**

Carbon does not normally participate in ionic bonds; **covalent bonding** (Section 1.3) is observed instead. A covalent bond results from the sharing of a pair of electrons between two atoms. Double bonds correspond to the sharing of 4 electrons, and triple bonds to the sharing of 6 (Section 1.4). Lewis structures for covalently bonded molecules are written on the basis of **octet rule.** The most stable structures are those in which the atoms of second-row elements are associated with 8 electrons (shared plus unshared) in their valence shells. Unshared electron pairs are also known as **lone pairs.**

Covalent bonds between atoms of different electronegativities are **polarized,** meaning that the electrons in the bond are drawn closer to the more electronegative atom (Section 1.5).

$$\overset{\delta+}{C} \overset{\delta-}{X}$$

Polarization of C—X bond when X
is more electronegative than carbon

A particular atom in a Lewis structure may be neutral, positively charged, or negatively charged. We refer to the charge of an atom in a Lewis structure as its **formal charge,** and we can calculate the formal charge by comparing the electron count of an atom in a molecule with that of the neutral atom itself. The procedure is described in Section 1.6.

Structural formulas describe the **constitution** of a molecule. Table 1.3 (Section 1.7) outlines procedures for writing Lewis structures of organic compounds. **Isomers** (Section 1.8) are different compounds that have the same molecular formula. They are different compounds because their structures are different. Isomers that differ in the order of their atomic connections are described as **constitutional isomers.**

Resonance between Lewis structures is a device used to describe electron delocalization in molecules (Section 1.9). Many molecules are not adequately described on the basis of a single Lewis structure because the Lewis rules restrict electrons to the region between only two nuclei. In those cases, the true structure is better understood as a hybrid of all possible structures that have the same atomic positions but different electron distribution. The most fundamental rules for resonance are summarized in Table 1.4.

The **valence shell electron-pair repulsion (VSEPR)** method (Section 1.10) predicts molecular geometries on the basis of repulsive interactions between the pairs of electrons that surround a central atom. A tetrahedral arrangement provides for the maximum separation of four electron pairs, a trigonal planar geometry is optimal for three electron pairs, and a linear arrangement is best for two electron pairs.

Knowledge of a molecule's shape and the polarity of its individual bonds can be used to determine whether it is **polar** or **nonpolar** (Section 1.11).

Bonding in organic compounds is often described according to an **orbital hybridization model** (Section 1.12). The sp^3 **hybridization** state of carbon is derived by mixing its $2s$ and the three $2p$ orbitals to give a set of four equivalent orbitals that have their axes directed toward the corners of a tetrahedron. The four bonds in methane are C—H σ bonds generated by overlap of carbon sp^3 orbitals with hydrogen $1s$ orbitals.

The carbon–carbon bond in ethane (CH_3CH_3) is a σ bond generated by overlap of an sp^3 orbital of one carbon with the sp^3 orbital of the other (Section 1.13).

Carbon is sp^2**-hybridized** in ethylene, and the double bond can be viewed as having a σ component and a π component. The sp^2 hybridization state is derived by combining the $2s$ and two of the three $2p$ orbitals of carbon. Three equivalent sp^2 orbitals result and the axes of these orbitals are coplanar. Overlap of an sp^2 orbital of one carbon with an sp^2 orbital of another produces a σ bond between them. Each carbon still has one unhybridized p orbital available for bonding, and "side-by-side" overlap of the p orbitals of adjacent carbons gives a π bond between them (Section 1.14).

The π bond in ethylene generated by
overlap of p orbitals of adjacent carbons

Carbon is sp**-hybridized** in acetylene, and the triple bond is of the $\sigma + \pi + \pi$ type. The $2s$ orbital and one of the $2p$ orbitals are combined to give two equivalent sp orbitals, which have their axes collinear. A σ bond between two carbons is supplemented by two π bonds formed by overlap of pairs of unhybridized p orbitals (Section 1.15).

The triple bond of acetylene has a σ bond component and two π bonds;
the two π bonds are shown here and are perpendicular to each other.

ADDITIONAL PROBLEMS

Electronic Configuration, Lewis Structures, Formal Charge, and Resonance

1.17 Write the electron configuration for each of the following ions. Which of these ions possesses a noble gas electron configuration?

    (a) Li^+ (b) Mg^+ (c) Mg^{2+} (d) S^{2-}

1.18 The atomic number Z of an element and the electron configuration corresponding to an ion derived from that element are shown here. Identify the ion in each case.

    (a) $(Z = 9)$: $1s^2 2s^2 2p^6$ (d) $(Z = 13)$: $1s^2 2s^2 2p^6 3s^1$
    (b) $(Z = 12)$: $1s^2 2s^2 2p^6 3s^1$ (e) $(Z = 13)$: $1s^2 2s^2 2p^6$
    (c) $(Z = 12)$: $1s^2 2s^2 2p^6$ (f) $(Z = 16)$: $1s^2 2s^2 2p^6 3s^2 3p^6$

1.19 In each of the following groups, identify the compound that is most likely to have an ionic bond.

    (a) CO, NO, O_2, CaO (c) CCl_4, $MgCl_2$, Cl_2O, Cl_2
    (b) LiF, BF_3, CF_4, F_2 (d) PBr_3, $AlBr_3$, KBr, $BrCl$

1.20 Each of the following species will be encountered at some point in this text. They all have the same number of electrons binding the same number of atoms and the same arrangement of bonds; they are *isoelectronic*. Specify which atoms, if any, bear a formal charge in the Lewis structure given and the net charge for each species.

    (a) :N≡N: (d) :N≡O:
    (b) :C≡N: (e) :C≡O:
    (c) :C≡C:

1.21 You will meet all the following isoelectronic species in this text. Repeat the previous problem for these three structures.

    (a) :Ö=C=Ö: (b) :N̈=N=N̈:   (c) :Ö=N=O:

1.22 Consider structural formulas A, B, C, and D:

    H—C̈=N=O̤:   H—C≡N—Ö: H—C≡N=O: H—C̈=N̈—Ö:
        A   B   C   D

    (a) Which structures contain a positively charged carbon?
    (b) Which structures contain a positively charged nitrogen?
    (c) Which structures contain a positively charged oxygen?
    (d) Which structures contain a negatively charged carbon?
    (e) Which structures contain a negatively charged nitrogen?
    (f) Which structures contain a negatively charged oxygen?
    (g) Which structures are electrically neutral (contain equal numbers of positive and negative charges)? Are any of them cations? Anions?
    (h) Which structure is the most stable?
    (i) Which structure is the least stable?

1.23 In each of the following pairs, determine whether the two represent resonance forms of a single species or depict different substances. If two structures are not resonance forms, explain why.

    (a) :N̈—N≡N: and   :N=N=N:
    (b) :N̈—N≡N: and   :N̈—N=N̈:
    (c) :N̈—N≡N: and   :N̈—N̈—N̈:

1.24 Among the following four structures, one is *not* a permissible resonance form. Identify the wrong structure. Why is it incorrect?

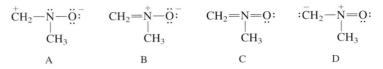

A　　　　　　B　　　　　　C　　　　　　D

1.25 Keeping the same atomic connections and moving only electrons, write a more stable Lewis structure for each of the following. Be sure to specify formal charges, if any, in the new structure.

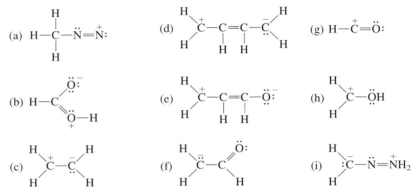

1.26 Determine the formal charge at all the atoms in each of the following species and the net charge on the species as a whole.

(a) $H-\ddot{O}-H$
　　　|
　　　H

(b) $H-\ddot{C}-H$
　　　|
　　　H

(c) $H-\dot{C}-H$
　　　|
　　　H

(d) $H-C-H$
　　　|
　　　H

(e) $H-\ddot{C}-H$

1.27 What is the formal charge of oxygen in each of the following Lewis structures?

(a) $CH_3\ddot{O}:$　　　　(b) $(CH_3)_2\ddot{O}:$　　　　(c) $(CH_3)_3O:$

Molecular Structure, Isomers, and Polarity

1.28 For each of the following molecules that contain polar covalent bonds, indicate the positive and negative ends of the dipole, using the symbol $\longleftrightarrow$. Refer to Table 1.2 as needed.

(a) HCl　　　　(c) HI　　　　(e) HOCl
(b) ICl　　　　(d) H_2O

1.29 Write Lewis formulas of the following molecules or ions, showing all electron pairs.

(a) PH_3
(b) AlH_4^-
(c) $COCl_2$　(all atoms bonded to carbon)
(d) HCO_3^-　(hydrogen is bonded to oxygen)
(e) $^+NO_2$　(order of atoms is ONO)
(f) NO_2^-　(order of atoms is ONO)

1.30 Using the VSEPR approach to molecular geometry, predict the shape of each of the species in Problem 1.29.

1.31 Predict whether either of the following molecules is polar.
(a) $(CH_3)_2O$ (b) CS_2

1.32 Write a Lewis structure for each of the following organic molecules:
(a) C_2H_5Cl (ethyl chloride: sprayed from aerosol cans onto skin to relieve pain)
(b) C_2H_3Cl [vinyl chloride: starting material for the preparation of poly(vinyl chloride), or PVC, plastics]
(c) $C_2HBrClF_3$ (halothane: a nonflammable inhalation anesthetic; all three fluorines are bonded to the same carbon)
(d) $C_2Cl_2F_4$ (Freon 114: formerly used as a refrigerant and as an aerosol propellant; each carbon bears one chlorine)

1.33 Each of the following molecular formulas represents two constitutionally isomeric substances. Write Lewis structures for the two isomers in each case.
(a) C_4H_{10} (b) C_3H_7Cl (c) C_2H_6O (d) C_2H_7N

1.34 Write structural formulas for all the constitutional isomers of molecular formula C_3H_6O that contain
(a) Only single bonds   (b) One double bond

1.35 Expand the following structural representations so as to more clearly show all the atoms and any unshared electron pairs.

(a) 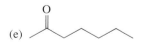   A component of high-octane gasoline

(b) 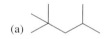   Occurs in bay and verbena oil

(c) 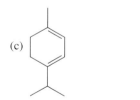   Pleasant-smelling substance found in marjoram oil

(d) 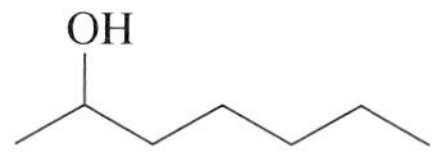   Present in oil of cloves

(e)   Found in Roquefort cheese

(f)     Benzene: parent compound of a large family of organic substances

(g) 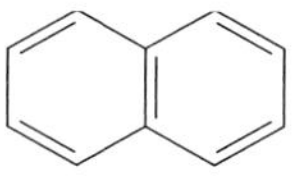   Naphthalene: sometimes used as a moth repellent

(h) 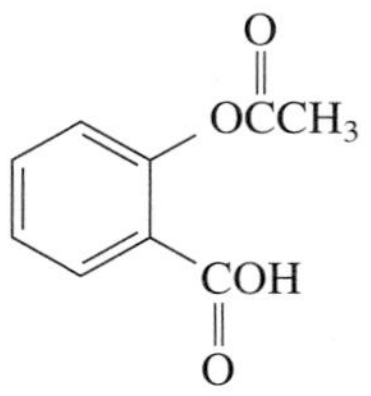   Aspirin

(i) 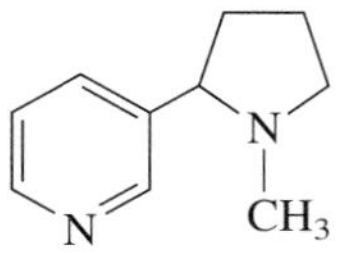   Nicotine: a toxic substance present in tobacco

1.36 Molecular formulas of organic compounds are customarily presented in the fashion $C_2H_5BrO_2$. The number of carbon and hydrogen atoms is presented first, followed by the other atoms in alphabetical order. Give the molecular formulas corresponding to each of the compounds in Problem 1.35. Are any of them isomers?

1.37 Select the compounds in Problem 1.35 in which all the carbons are
    (a) sp^3-hybridized   (b) sp^2-hybridized

Do any of the compounds in Problem 1.35 contain an sp-hybridized carbon?

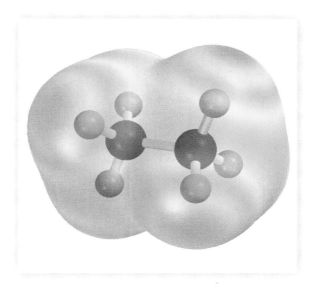

CHAPTER 2
ALKANES AND CYCLOALKANES

N ow that we've reviewed the various bonding models, we are ready to examine organic compounds in respect to their structure, reactions, properties, and applications. Were we to list the physical and chemical properties of each of the more than 8 million organic compounds separately, it would tax the capacity of even a powerful computer. Yet someone who is trained in organic chemistry can simply look at the structure of a substance and make reasonably confident predictions about its properties, including how it will behave in a chemical reaction.

Organic chemists associate particular structural units, called **functional groups,** with characteristic patterns of reactivity; they look at large molecules as collections of functional groups attached to nonreactive frameworks. Not only does this "functional group approach" have predictive power, but time and experience have shown that it organizes the material in a way that makes learning organic chemistry easier for most students.

We'll begin the chapter with a brief survey of various kinds of hydrocarbons—compounds that contain only carbon and hydrogen—introduce some functional groups, then return to hydrocarbons to discuss alkanes in some detail. The names of alkanes may seem strange at first, but they form the foundation for the most widely accepted system of organic nomenclature. The fundamentals of this nomenclature system, the **IUPAC rules,** constitute one of the main topics of this chapter.

2.1 CLASSES OF HYDROCARBONS

Hydrocarbons are compounds that contain only carbon and hydrogen and are divided into two main classes: **aliphatic** hydrocarbons and **aromatic** hydrocarbons. This classification dates from the nineteenth century, when organic chemistry was almost

33

exclusively devoted to the study of materials from natural sources, and terms were coined that reflected a substance's origin. Two sources were fats and oils, and the word *aliphatic* was derived from the Greek word *aleiphar* ("fat"). Aromatic hydrocarbons, irrespective of their own odor, were typically obtained by chemical treatment of pleasant-smelling plant extracts.

Aliphatic hydrocarbons include three major groups: alkanes, alkenes, and alkynes. **Alkanes** are hydrocarbons in which all the bonds are single bonds, **alkenes** contain one or more carbon–carbon double bonds, and **alkynes** contain one or more carbon–carbon triple bonds. Examples of the three classes of aliphatic hydrocarbons are the two-carbon compounds *ethane*, *ethylene*, and *acetylene*.

> Bonding in ethane, ethylene, and acetylene was discussed in Sections 1.13–1.15.

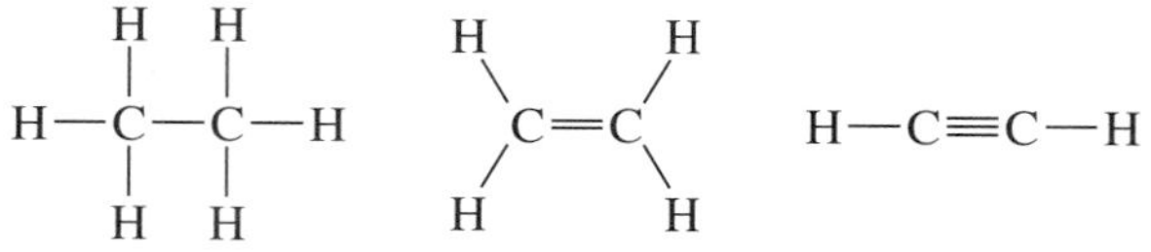

Ethane
(alkane)

Ethylene
(alkene)

Acetylene
(alkyne)

Another name for aromatic hydrocarbons is **arenes.** Arenes have properties that are much different from alkanes, alkenes, and alkynes. The most important aromatic hydrocarbon is *benzene*.

> Bonding in benzene will be discussed in Section 6.1.

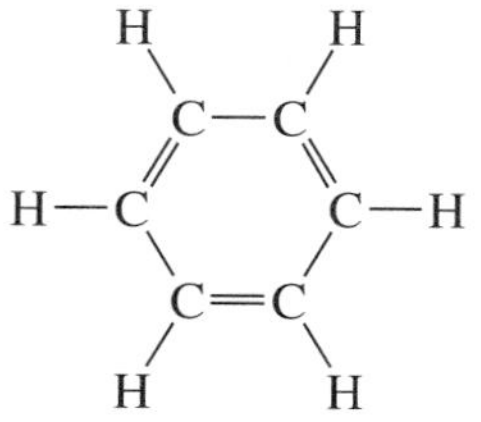

Benzene
(arene)

Many of the principles of organic chemistry can be developed by examining the series of hydrocarbons in the order: alkanes, alkenes, alkynes, and arenes. Alkanes are introduced in this chapter, alkenes and alkynes in Chapters 4 and 5, and arenes in Chapter 6.

2.2   REACTIVE SITES IN HYDROCARBONS

A **functional group** is the structural unit responsible for a given molecule's reactivity under a particular set of conditions. It can be as small as a single hydrogen atom, or it can encompass several atoms. The functional group of an alkane is any one of its hydrogens. A reaction that we shall discuss in Chapter 9 is one in which an alkane reacts with chlorine. For example:

$$CH_3CH_3 \; + \; Cl_2 \; \longrightarrow \; CH_3CH_2Cl \; + \; HCl$$

Ethane   Chlorine   Chloroethane   Hydrogen chloride

One of the hydrogen atoms of ethane is replaced by chlorine. This replacement of hydrogen by chlorine is a characteristic reaction of all alkanes and can be represented by the equation:

$$R-H + Cl_2 \longrightarrow R-Cl + HCl$$

Alkane   Chlorine   Alkyl chloride   Hydrogen chloride

In the general equation the functional group ($-H$) is shown explicitly, and the remainder of the alkane molecule is abbreviated as R. This is a commonly used notation that helps focus our attention on the functional group transformation without being distracted by the parts of the molecule that remain unaffected. A hydrogen atom in one alkane is very much like the hydrogen of any other alkane in its reactivity toward chlorine. Our ability to write general equations such as the one shown illustrates why the functional group approach is so useful in organic chemistry.

A hydrogen atom is a functional unit in alkenes and alkynes as well as in alkanes. These hydrocarbons, however, contain a second functional group as well. The carbon–carbon double bond is a functional group in alkenes, and the carbon–carbon triple bond is a functional group in alkynes.

A hydrogen atom is a functional group in arenes, and we represent arenes as ArH to reflect this. What will become apparent when we discuss the reactions of arenes, however, is that their chemistry is much richer than that of alkanes, and it is therefore more appropriate to consider the ring in its entirety as the functional group.

2.3 THE KEY FUNCTIONAL GROUPS

As a class, alkanes are not particularly reactive compounds, and the H in RH is not a particularly reactive functional group. Indeed, when a group other than hydrogen is present on an alkane framework, that group is almost always the functional group. Table 2.1 lists examples of some compounds of this type. All will be discussed in later chapters.

Some of the most important families of organic compounds, those that contain the carbonyl group ($C=O$), deserve separate mention and are listed in Table 2.2. Carbonyl-containing compounds rank among the most abundant and biologically significant classes of naturally occurring substances.

Carbonyl group chemistry is discussed in a block of three chapters (Chapters 11–13).

TABLE 2.1	Functional Groups in Some Important Classes of Organic Compounds		
Class	**Generalized Abbreviation**	**Representative Example**	**Name of Example***
Alcohol	ROH	CH_3CH_2OH	Ethanol
Alkyl halide	RCl	CH_3CH_2Cl	Chloroethane
Amine†	RNH_2	$CH_3CH_2NH_2$	Ethanamine
Epoxide	R_2C-CR_2	H_2C-CH_2	Oxirane
	O	O	
Ether	ROR	$CH_3CH_2OCH_2CH_3$	Diethyl ether
Nitrile	$RC\equiv N$	$CH_3CH_2C\equiv N$	Propanenitrile
Nitroalkane	RNO_2	$CH_3CH_2NO_2$	Nitroethane
Thiol	RSH	CH_3CH_2SH	Ethanethiol

* Most compounds have more than one acceptable name.

† The example given is a *primary* amine (RNH_2). *Secondary* amines have the general structure R_2NH; *tertiary* amines are R_3N.

TABLE 2.2	Classes of Compounds That Contain a Carbonyl Group		
Class	**Generalized Abbreviation**	**Representative Example**	**Name of Example**
Aldehyde	$\overset{\displaystyle O}{\overset{\displaystyle \|}{R\overset{}{C}H}}$	$\overset{\displaystyle O}{\overset{\displaystyle \|}{CH_3\overset{}{C}H}}$	Ethanal
Ketone	$\overset{\displaystyle O}{\overset{\displaystyle \|}{R\overset{}{C}R}}$	$\overset{\displaystyle O}{\overset{\displaystyle \|}{CH_3\overset{}{C}CH_3}}$	2-Propanone
Carboxylic acid	$\overset{\displaystyle O}{\overset{\displaystyle \|}{R\overset{}{C}OH}}$	$\overset{\displaystyle O}{\overset{\displaystyle \|}{CH_3\overset{}{C}OH}}$	Ethanoic acid
Carboxylic acid derivatives:			
Acyl halide	$\overset{\displaystyle O}{\overset{\displaystyle \|}{R\overset{}{C}X}}$	$\overset{\displaystyle O}{\overset{\displaystyle \|}{CH_3\overset{}{C}Cl}}$	Ethanoyl chloride
Acid anhydride	$\overset{\displaystyle O\ \ O}{\overset{\displaystyle \|\ \ \|}{R\overset{}{C}O\overset{}{C}R}}$	$\overset{\displaystyle O\ \ O}{\overset{\displaystyle \|\ \ \|}{CH_3\overset{}{C}O\overset{}{C}CH_3}}$	Ethanoic anhydride
Ester	$\overset{\displaystyle O}{\overset{\displaystyle \|}{R\overset{}{C}OR}}$	$\overset{\displaystyle O}{\overset{\displaystyle \|}{CH_3\overset{}{C}OCH_2CH_3}}$	Ethyl ethanoate
Amide	$\overset{\displaystyle O}{\overset{\displaystyle \|}{R\overset{}{C}NR_2}}$	$\overset{\displaystyle O}{\overset{\displaystyle \|}{CH_3\overset{}{C}NH_2}}$	Ethanamide

PROBLEM 2.1 Many compounds contain more than one functional group. The structure of prostaglandin E_1, a hormone that regulates the relaxation of smooth muscles, contains two different kinds of carbonyl groups. Classify each one (aldehyde, ketone, carboxylic acid, ester, amide, acyl chloride, or carboxylic acid anhydride).

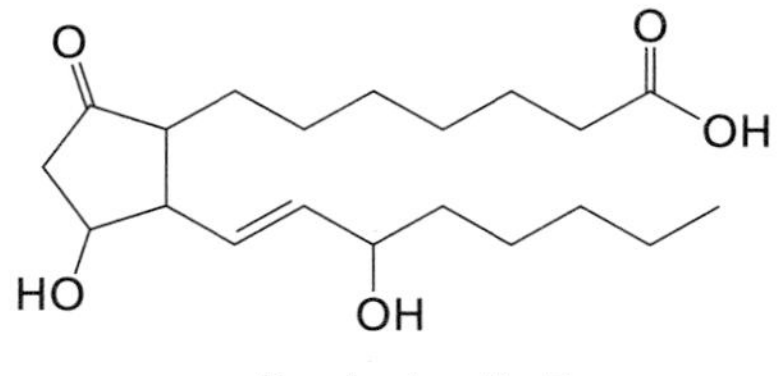

Prostaglandin E_1

The reactions of the carbonyl group feature prominently in *organic synthesis*—the branch of organic chemistry that plans and carries out the preparation of compounds of prescribed structure.

2.4   INTRODUCTION TO ALKANES: METHANE, ETHANE, AND PROPANE

Alkanes have the general molecular formula C_nH_{2n+2}. The simplest one, **methane** (CH_4), is also the most abundant. Large amounts are present in our atmosphere, in the ground, and in the oceans. Methane has been found on Jupiter, Saturn, Uranus, Neptune, and Pluto, and even on Halley's Comet.

Ethane (C_2H_6, CH_3CH_3) and **propane** (C_3H_8, $CH_3CH_2CH_3$) are second and third, respectively, to methane in many ways. Ethane is the alkane next to methane in structural simplicity, followed by propane. Ethane ($\approx 10\%$) is the second and propane ($\approx 5\%$) the third most abundant component of natural gas, which is $\approx 75\%$ methane. The characteristic odor of natural gas we use for heating our homes and cooking comes from trace amounts of unpleasant-smelling sulfur-containing compounds such as ethanethiol (see Table 2.1) that are deliberately added to it in order to warn us of potentially dangerous leaks. Natural gas is colorless and nearly odorless, as are methane, ethane, and propane.

Methane is the lowest boiling alkane, followed by ethane, then propane.

	CH_4	CH_3CH_3	$CH_3CH_2CH_3$
	Methane	Ethane	Propane
Boiling point:	$-160°C$	$-89°C$	$-42°C$

> See the boxed essay: *"Methane and the Biosphere"* that accompanies this section.

> Boiling points cited in this text are at 1 atm (760 mm of mercury) unless otherwise stated.

This will generally be true as we proceed to look at other alkanes; as the number of carbon atoms increases, so does the boiling point. All the alkanes with four carbons or less are gases at room temperature and atmospheric pressure. With the highest boiling point of the three, propane is the easiest one to liquefy. We are all familiar with "propane tanks." These steel containers maintain a propane-rich mixture of hydrocarbons called liquefied petroleum gas (LPG) in a liquid state under high pressure as a convenient clean-burning fuel.

The structural features of methane, ethane, and propane are summarized in Figure 2.1. All of the carbon atoms are sp^3-hybridized, all of the bonds are σ bonds, and the bond angles at carbon are close to tetrahedral.

> **PROBLEM 2.2** How many carbons are sp^3-hybridized in propane? How many σ bonds occur in this molecule? Identify the orbital overlaps that give rise to each σ bond.

2.5 CONFORMATIONS OF ETHANE AND PROPANE

In addition to the constitution of ethane, another aspect of its structure commands our attention: its **conformation.**

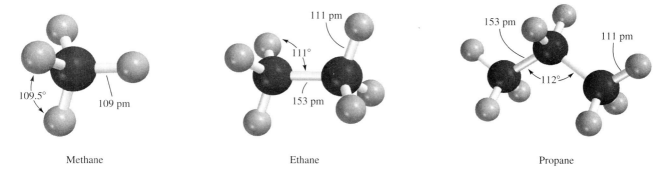

FIGURE 2.1 Structures of methane, ethane, and propane showing bond distances and bond angles.

METHANE AND THE BIOSPHERE*

One of the things that environmental scientists do is to keep track of important elements in the biosphere—in what form do these elements normally occur, to what are they transformed, and how are they returned to their normal state? Careful studies have given clear, although complicated, pictures of the "nitrogen cycle," the "sulfur cycle," and the "phosphorus cycle," for example. The "carbon cycle," begins and ends with atmospheric carbon dioxide. It can be represented in an abbreviated form as:

$$CO_2 + H_2O + energy \underset{respiration}{\overset{photosynthesis}{\rightleftarrows}} carbohydrates$$

$$\downarrow$$

naturally occurring substances of numerous types

Methane is one of literally millions of compounds in the carbon cycle, but one of the most abundant. It is formed when carbon-containing compounds decompose in the absence of air (**anaerobic** conditions). The organisms that bring this about are called **methanoarchaea.** Cells can be divided into three types: **archaea, bacteria,** and **eukarya.** Methanoarchaea are one kind of archaea and may rank among the oldest living things on earth. They can convert a number of carbon-containing compounds, including carbon dioxide and acetic acid, to methane.

Virtually anywhere water contacts organic matter in the absence of air is a suitable place for methanoarchaea to thrive—at the bottom of ponds, bogs, and rice fields, for example. Marsh gas (swamp gas) is mostly methane. Methanoarchaea live inside termites and grass-eating animals. One source quotes 20 L/day as the methane output of a large cow.

The scale on which methanoarchaea churn out methane, estimated to be 10^{11}–10^{12} lb/year, is enormous. About 10% of this amount makes its way into the atmosphere, but most of the rest simply ends up completing the carbon cycle. It exits the anaerobic environment where it was formed and enters the aerobic world where it is eventually converted to carbon dioxide by a variety of processes.

When we consider sources of methane we have to add "old" methane, methane that was formed millions of years ago but became trapped beneath the earth's surface, to the "new" methane just described. Firedamp, an explosion hazard to miners, occurs in layers of coal and is mostly methane. Petroleum deposits, formed by microbial decomposition of plant material under anaerobic conditions, are always accompanied by pockets of natural gas, which is mostly methane.

An interesting thing happens when trapped methane leaks from sites under the deep ocean floor. If the pressure is high enough (50 atm) and the water cold enough (4°C), the methane doesn't simply bubble to the surface. Individual methane molecules become trapped inside clusters of 6–18 water molecules forming **methane clathrates** or **methane hydrates.** Aggregates of these clathrates stay at the bottom of the ocean in what looks like a lump of dirty ice. Ice that burns. Far from being mere curiosities, methane clathrates are potential sources of energy on a scale greater than that of all known oil reserves combined. At present, it is not economically practical to extract the methane, however.

Methane clathrates have received recent attention from a different segment of the scientific community. While diving in the Gulf of Mexico in 1997, a research team of biologists and environmental scientists were surprised to find a new species of worm grazing on the mound of a methane clathrate. What were these worms feeding on? Methane? Bacteria that live on the methane? A host of questions having to do with deep-ocean ecosystems suddenly emerged. Stay tuned.

*The biosphere is the part of the earth where life is; it includes the surface, the oceans, and the lower atmosphere.

You will find it helpful at this point to construct a molecular model of ethane and view the two conformations from a variety of perspectives.

Conformations are different spatial arrangements of a molecule that are generated by rotation about single bonds.

Two of the many conformations of ethane, the **staggered conformation** and the **eclipsed conformation,** are depicted in Figure 2.2. The C—H bonds in the staggered conformation are arranged so that each one bisects the angle made by a pair of C—H bonds on

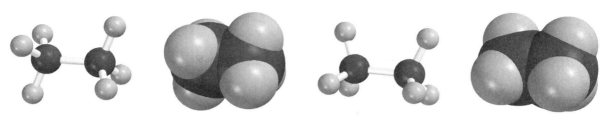

(*a*) Staggered conformation of ethane   (*b*) Eclipsed conformation of ethane

the adjacent carbon. In the eclipsed conformation each C—H bond is aligned with a C—H bond on the adjacent carbon. The staggered and eclipsed conformations interconvert by rotation around the carbon–carbon bond.

Among the various ways in which the staggered and eclipsed forms are portrayed, wedge-and-dash, sawhorse, and Newman projection drawings are especially useful. These are shown for the staggered conformation of ethane in Figure 2.3 and for the eclipsed conformation in Figure 2.4.

We used **wedge-and-dash** drawings earlier (Section 1.10) and so Figures 2.3*a* and 2.4*a* are familiar to us. A **sawhorse drawing** (Figures 2.3*b* and 2.4*b*) shows the conformation of a molecule without having to resort to different styles of bonds. In a **Newman projection** (Figures 2.3*c* and 2.4*c*), we sight down the C—C bond, and represent the front carbon by a point and the back carbon by a circle. Each carbon has three substituents that are placed symmetrically around it.

FIGURE 2.2 The staggered and eclipsed conformations of ethane shown as ball-and-spoke and space-filling models

Newman projections were devised by Professor Melvin S. Newman of Ohio State University in the 1950s.

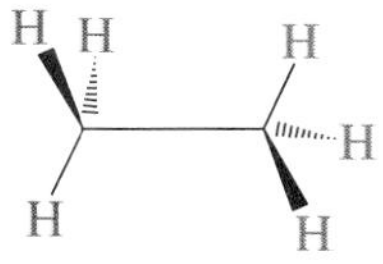

(*a*) Wedge-and-dash

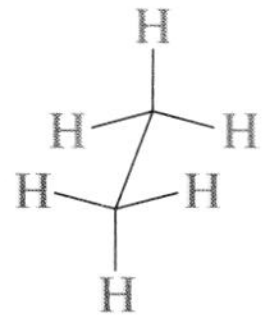

(*b*) Sawhorse

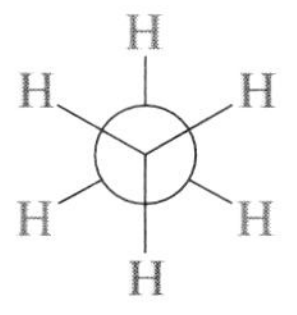

(*c*) Newman projection

FIGURE 2.3 Some commonly used representations of the staggered conformation of ethane.

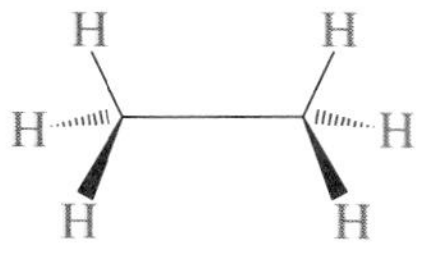

(*a*) Wedge-and-dash

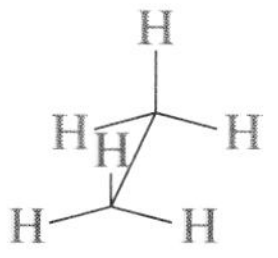

(*b*) Sawhorse

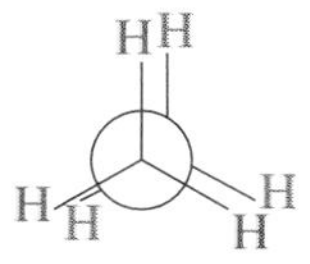

(*c*) Newman projection

FIGURE 2.4 Some commonly used representations of the eclipsed conformation of ethane.

You may have noticed that, in fact, an infinite number of conformations of ethane exist that differ by only tiny increments of rotation about the carbon–carbon bond. Are all conformations possible? How fast is the process of rotation about the carbon–carbon bond? Which conformation is the most stable? Which one is the least stable? Questions of this type arise with almost all chemical substances (not just organic compounds and not just alkanes), and their study is called **conformational analysis.** In the case of ethane, the *staggered conformation is the most stable,* and the eclipsed form is the least stable of all conformations. Rotation about the carbon–carbon bond is extremely fast (several million times per second at room temperature), and conformations interconvert rapidly. At any instant, most of the ethane molecules exist in the staggered conformation. The staggered conformation is most stable because it allows for the maximum separation of electron pairs on adjacent atoms. The VSEPR model (Section 1.10) predicts molecular shapes on the basis of maximum separation of electron pairs on a single atom. In ethane the electron pairs of the C—H bonds of one carbon are farthest away from the electron pairs of the C—H bonds of the adjacent carbon when the bonds are staggered.

Because it is less stable than the staggered conformation, we say the eclipsed conformation of ethane is **strained** and identify that strain as being due to the eclipsing of bonds on adjacent atoms. This type of strain is called **torsional strain.** In the following section we will see a second type of strain that, combined with torsional strain, is an important consideration in the conformational analysis of higher alkanes.

> **PROBLEM 2.3** Construct a molecular model of propane and use it to draw Newman projections of the staggered and eclipsed conformations of propane. How do they differ from the conformations of ethane?

2.6 ISOMERIC ALKANES: THE BUTANES

Methane is the only alkane of molecular formula CH_4, ethane the only one that is C_2H_6, and propane the only one that is C_3H_8. Beginning with C_4H_{10}, however, constitutional isomers (Section 1.8) are possible; two alkanes have this particular molecular formula. In one, called **n-butane,** four carbons are joined in a continuous chain. The *n* in *n*-butane stands for "normal" and means that the carbon chain is unbranched. The second isomer has a branched carbon chain and is called **isobutane.**

Make molecular models of the two isomers of C_4H_{10}.

$$CH_3CH_2CH_2CH_3 \qquad\qquad CH_3CHCH_3 \quad \text{or} \quad (CH_3)_3CH$$
$$\qquad\qquad\qquad\qquad\qquad\qquad\qquad\quad | $$
$$\qquad\qquad\qquad\qquad\qquad\qquad\qquad CH_3$$

	n-Butane	Isobutane
Boiling point:	−0.4°C	−10.2°C
Melting point:	−139°C	−160.9°C

As noted earlier (Section 1.13), CH_3 is called a *methyl* group. In addition to having methyl groups at both ends, *n*-butane contains two CH_2, or **methylene** groups. Isobutane contains three methyl groups bonded to a CH unit. The CH unit is called a **methine** group.

"Butane" lighters contain about 5% *n*-butane and 95% isobutane in a sealed container. The pressure produced by the two compounds (about 3 atm) is enough to keep them in the liquid state until opening a small valve emits a fine stream of the vaporized mixture across a spark which ignites it.

n-Butane and isobutane have the same molecular formula but differ in the order in which their atoms are connected. They are **constitutional isomers** of each other (Section 1.8). Because they are different in structure, they can have different properties. Both are gases at room temperature, but *n*-butane boils almost 10°C higher than isobutane and has a melting point that is over 20°C higher.

The bonding in *n*-butane and isobutane continues the theme begun with methane, ethane, and propane. All of the carbon atoms are sp^3-hybridized, all of the bonds are σ

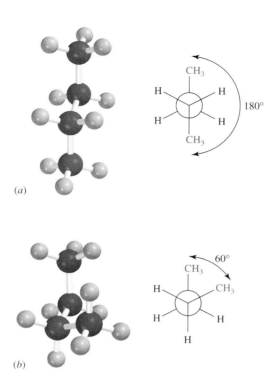

FIGURE 2.5 The (a) anti and (b) gauche conformations of butane shown as ball-and-spoke models (*left*) and as Newman projections (*right*). The gauche conformation is less stable than the anti because of the van der Waals strain between the methyl groups.

bonds, and the bond angles at carbon are close to tetrahedral. This generalization holds for all alkanes regardless of the number of carbons they have.

The most stable conformation of *n*-butane is shown in Figure 2.5(*a*). It has a zigzag arrangement of its carbon chain, and all the bonds are staggered. A Newman projection of this conformation, sighting down the C-2—C-3 bond, is shown at the right. As can be seen in the Newman projection, not only are the bonds to C-2 and C-3 staggered with respect to each other, but the angle between the bonds to the methyl groups is 180°. We call this the **anti** conformation.

Figure 2.5(*b*) depicts a second staggered conformation of *n*-butane called the **gauche** conformation. The methyl groups are much closer together in the gauche conformation, where the angle between them is only 60°, than they are in the anti, and the gauche conformation is slightly less stable than the anti. The destabilization of a molecule that results when two of its atoms are too close to each other is called **van der Waals strain** or **steric hindrance.** Because anti and gauche are both staggered conformations, they are free of torsional strain. They are rapidly interconverted at room temperature, and at any instant approximately 65% of the molecules of *n*-butane exist in the anti conformation and 35% in the gauche.

> **PROBLEM 2.4** Draw a Newman projection formula of the most stable conformation of isobutane.

2.7   HIGHER ALKANES

n-Alkanes are alkanes that have an unbranched carbon chain. ***n*-Pentane** and ***n*-hexane** are *n*-alkanes possessing five and six carbon atoms, respectively.

$$CH_3CH_2CH_2CH_2CH_3 \qquad CH_3CH_2CH_2CH_2CH_2CH_3$$

n-Pentane   *n*-Hexane

Their condensed structural formulas can be abbreviated even more by indicating within parentheses the number of methylene groups in the chain. Thus, *n*-pentane may be written as $CH_3(CH_2)_3CH_3$ and *n*-hexane as $CH_3(CH_2)_4CH_3$. This shortcut is especially convenient with longer chain alkanes. The laboratory synthesis of the "ultralong" alkane $CH_3(CH_2)_{388}CH_3$ was achieved in 1985; imagine trying to write a structural formula for this compound in anything other than an abbreviated way!

PROBLEM 2.5 An *n*-alkane of molecular formula $C_{28}H_{58}$ has been isolated from a certain fossil plant. Write a condensed structural formula for this alkane.

n-Alkanes have the general formula $CH_3(CH_2)_xCH_3$ and are said to belong to a **homologous series** of compounds, that is, one in which successive members differ by a $—CH_2—$ group.

Unbranched alkanes are sometimes referred to as "straight-chain alkanes," but, as we saw for *n*-butane in Section 2.5, their chains are not straight but instead tend to adopt the "zigzag" shape portrayed in the bond-line formulas introduced in Section 1.7.

Bond-line formula of *n*-pentane   Bond-line formula of *n*-hexane

PROBLEM 2.6 Much of the communication between insects involves chemical messengers called **pheromones**. A species of cockroach secretes a substance from its mandibular glands that alerts other cockroaches to its presence and causes them to congregate. One of the principal components of this **aggregation pheromone** is the alkane shown in the bond-line formula that follows. Give the molecular formula of this substance, and represent it by a condensed formula.

Three isomeric alkanes have the molecular formula C_5H_{12}. The unbranched isomer is, as we have seen, *n*-pentane. The isomer with a single methyl branch is called **isopentane.** The third isomer has a three-carbon chain with two methyl branches. It is called **neopentane.**

> Make molecular models of the three isomers of C_5H_{12}.

n-Pentane:	$CH_3CH_2CH_2CH_2CH_3$	or	$CH_3(CH_2)_3CH_3$	or	
Isopentane:	$CH_3CHCH_2CH_3$ $\quad\ \ \mid$ $\quad\ \ CH_3$	or	$(CH_3)_2CHCH_2CH_3$	or	
Neopentane:	$\quad\ \ CH_3$ $\quad\ \ \mid$ CH_3CCH_3 $\quad\ \ \mid$ $\quad\ \ CH_3$	or	$(CH_3)_4C$	or	

Do any additional C_5H_{12} isomers occur? Although answering that question is straightforward for an alkane with only five carbons (there are no more isomers), it is less so for alkanes with more carbons. As Table 2.3 dramatically shows, the number of isomers increases rapidly with an increasing number of carbon atoms.

TABLE 2.3	The Number of Constitutionally Isomeric Alkanes of Particular Molecular Formulas

Molecular Formula	Number of Constitutional Isomers
CH_4	1
C_2H_6	1
C_3H_8	1
C_4H_{10}	2
C_5H_{12}	3
C_6H_{14}	5
C_7H_{16}	9
C_8H_{18}	18
C_9H_{20}	35
$C_{10}H_{22}$	75
$C_{15}H_{32}$	4,347
$C_{20}H_{42}$	366,319
$C_{40}H_{82}$	62,491,178,805,831

The best way to ensure that you have written all the isomers of a particular molecular formula is to work systematically, beginning with the unbranched chain and then shortening it while adding branches one by one. It is essential that you be able to recognize when two different-looking structural formulas are actually the same molecule written in different ways. The key point is the *connectivity* of the carbon chain. For example, the following group of structural formulas do *not* represent different compounds; they are just a portion of the many ways we could write a structural formula for isopentane. Each one has a continuous chain of four carbons with a methyl branch located one carbon from the end of the chain.

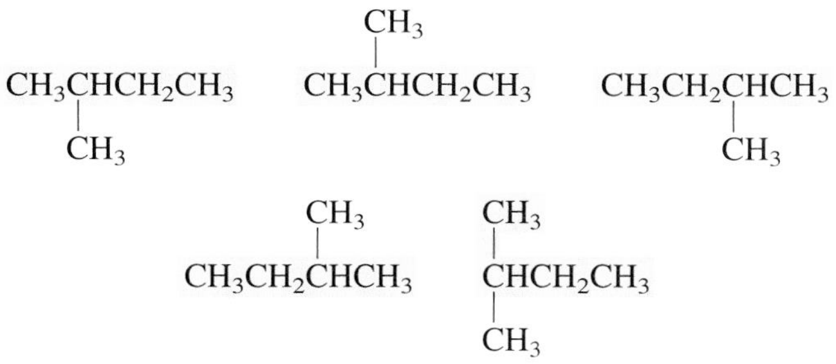

$$CH_3CHCH_2CH_3 \qquad \begin{array}{c} CH_3 \\ | \\ CH_3CHCH_2CH_3 \end{array} \qquad CH_3CH_2CHCH_3$$
$$\quad | \qquad\qquad\qquad\qquad\qquad\qquad\qquad\qquad\qquad | $$
$$\quad CH_3 \qquad\qquad\qquad\qquad\qquad\qquad\qquad\qquad\quad CH_3$$

$$\begin{array}{cc} CH_3 & CH_3 \\ | & | \\ CH_3CH_2CHCH_3 & CHCH_2CH_3 \\ & | \\ & CH_3 \end{array}$$

PROBLEM 2.7 Write condensed and bond-line formulas for the five isomeric C_6H_{14} alkanes.

Sample Solution When writing isomeric alkanes, it is best to begin with the unbranched isomer.

$$CH_3CH_2CH_2CH_2CH_2CH_3 \qquad \text{or}$$

Next, remove a carbon from the chain and use it as a one-carbon (methyl) branch at the carbon atom next to the end of the chain.

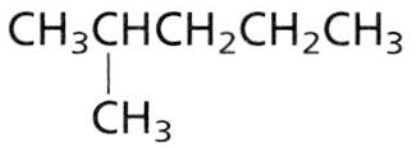

$$CH_3CHCH_2CH_2CH_3 \qquad \text{or}$$
$$\quad | $$
$$\quad CH_3$$

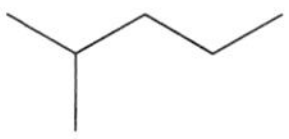

Now, write structural formulas for the remaining three isomers. Be sure that each one is a unique compound and not simply a different representation of one written previously.

A second question presents itself, namely, how can we identify alkanes so that each one has a unique name? Once again, the problem is not so difficult for C_5H_{12} with only three isomers. The same is not true for higher alkanes, however. As with writing structures, being able to name compounds in a systematic way becomes of paramount importance. By following a set of rules—presented in the following section—you will always get the same systematic name for given compound. Conversely, two different compounds will always have different names.

2.8   IUPAC NOMENCLATURE OF UNBRANCHED ALKANES

Nomenclature in organic chemistry is of two types: **common** (or "trivial") and **systematic.** Some common names existed long before organic chemistry became an organized branch of chemical science. Methane, ethane, propane, *n*-butane, isobutane, *n*-pentane, isopentane, and neopentane are common names. One simply memorizes the name that goes with a compound in just the same way that one matches names with faces. So long as the number of names and compounds are few, the task is manageable. But millions of organic compounds are already known, and the list continues to grow! A system built on common names is not adequate to the task of communicating structural information. Beginning in 1892, chemists developed a set of rules for naming organic compounds based on their structures, which we now call the **IUPAC rules.** *IUPAC* stands for the "International Union of Pure and Applied Chemistry."

The IUPAC rules assign names to unbranched alkanes as shown in Table 2.4. Methane, ethane, propane, and butane are retained for CH_4, CH_3CH_3, $CH_3CH_2CH_3$, and $CH_3CH_2CH_2CH_3$, respectively. Thereafter, the number of carbon atoms in the chain is specified by a Latin or Greek prefix preceding the suffix *-ane,* which identifies the compound as a member of the alkane family. Notice that the prefix *n-* is not part of the IUPAC system. The IUPAC name for $CH_3CH_2CH_2CH_3$ is butane, not *n*-butane.

TABLE 2.4	IUPAC Names of Unbranched Alkanes		
Number of Carbon Atoms	**Name**	**Number of Carbon Atoms**	**Name**
1	Methane	11	Undecane
2	Ethane	12	Dodecane
3	Propane	13	Tridecane
4	Butane	14	Tetradecane
5	Pentane	15	Pentadecane
6	Hexane	16	Hexadecane
7	Heptane	17	Heptadecane
8	Octane	18	Octadecane
9	Nonane	19	Nonadecane
10	Decane	20	Icosane*

* Spelled "eicosane" prior to 1979 version of IUPAC rules.

> **PROBLEM 2.8** What is the IUPAC name of the alkane described in Problem 2.6 as a component of the cockroach aggregation pheromone?

In Problem 2.7 you were asked to write structural formulas for the five isomeric alkanes of molecular formula C$_6$H$_{14}$. In the next section you will see how the IUPAC rules generate a unique name for each isomer.

2.9 APPLYING THE IUPAC RULES: THE NAMES OF THE C$_6$H$_{14}$ ISOMERS

We can present and illustrate the most important of the IUPAC rules for alkane nomenclature by naming the five C$_6$H$_{14}$ isomers. By definition (see Table 2.4), the unbranched C$_6$H$_{14}$ isomer is hexane.

CH$_3$CH$_2$CH$_2$CH$_2$CH$_2$CH$_3$

IUPAC name: **hexane**
(common name: *n*-hexane)

> You might find it helpful to make molecular models of all the C$_6$H$_{14}$ isomers.

The IUPAC rules name branched alkanes as *substituted derivatives* of the unbranched alkanes listed in Table 2.4. Consider the C$_6$H$_{14}$ isomer represented by the structure

CH$_3$CHCH$_2$CH$_2$CH$_3$
|
CH$_3$

Step 1

Pick out the *longest continuous carbon chain,* and find the IUPAC name in Table 2.4 that corresponds to the unbranched alkane having that number of carbons. This is the parent alkane from which the IUPAC name is to be derived.

In this case, the longest continuous chain has *five* carbon atoms; the compound is named as a derivative of pentane. The key word here is *continuous.* It does not matter whether the carbon skeleton is drawn in an extended straight-chain form or in one with many bends and turns. All that matters is the number of carbons linked together in an uninterrupted sequence.

Step 2

Identify the substituent groups attached to the parent chain.

The parent pentane chain bears a methyl (CH$_3$) group as a substituent.

Step 3

Number the longest continuous chain in the direction that gives the lowest number to the substituent at the first point of branching.

The numbering scheme

1 2 3 4 5
CH$_3$CHCH$_2$CH$_2$CH$_3$ is equivalent to   2 3 4 5
| CH$_3$CHCH$_2$CH$_2$CH$_3$
CH$_3$ | CH$_3$

Both schemes count five carbon atoms in their longest continuous chain and bear a methyl group as a substituent at the second carbon. An alternative numbering sequence that begins at the other end of the chain is incorrect:

$$\overset{5}{C}H_3\overset{4}{C}H\overset{3}{C}H_2\overset{2}{C}H_2\overset{1}{C}H_3 \qquad \text{(Methyl group attached to C-4)}$$
$$\underset{|}{\overset{|}{C}H_3}$$

Step 4

Write the name of the compound. The parent alkane is the last part of the name and is preceded by the names of the attached groups and their numerical locations (**locants**). Hyphens separate the locants from the words.

$$CH_3CHCH_2CH_2CH_3$$
$$\underset{CH_3}{|}$$

IUPAC name: **2-methylpentane**

The same sequence of four steps gives the IUPAC name for the isomer that has its methyl group attached to the middle carbon of the five-carbon chain.

$$CH_3CH_2CHCH_2CH_3$$
$$\underset{CH_3}{|}$$

IUPAC name: **3-methylpentane**

Both remaining C_6H_{14} isomers have two methyl groups as substituents on a four-carbon chain. Thus the parent chain is butane. When the same substituent appears more than once, use the multiplying prefixes *di-, tri-, tetra-,* and so on. A separate locant is used for each substituent, and the locants are separated from one another by commas and from the words by hyphens.

$$\overset{CH_3}{\underset{|}{\underset{CH_3}{|}}}$$
$$CH_3CCH_2CH_3$$

IUPAC name: **2,2-dimethylbutane**

$$\overset{CH_3}{\underset{|}{}}$$
$$CH_3CHCHCH_3$$
$$\underset{CH_3}{|}$$

IUPAC name: **2,3-dimethylbutane**

PROBLEM 2.9 Phytane is a naturally occurring alkane produced by the alga *Spirogyra* and is a constituent of petroleum. The IUPAC name for phytane is 2,6,10,14-tetramethylhexadecane. Write a structural formula for phytane.

PROBLEM 2.10 Derive the IUPAC names for

(a) The isomers of C_4H_{10}
(b) The isomers of C_5H_{12}

(c) $(CH_3)_3CCH_2CH(CH_3)_2$
(d) $(CH_3)_3CC(CH_3)_3$

Sample Solution (a) C_4H_{10} has two isomers. Butane (see Table 2.4) is the IUPAC name for the isomer that has an unbranched carbon chain. The other isomer has three carbons in its longest continuous chain with a methyl branch at the central carbon; its IUPAC name is 2-methylpropane.

$$CH_3CH_2CH_2CH_3$$

IUPAC name: **butane**
(common name: *n*-butane)

$$CH_3CHCH_3$$
$$\underset{CH_3}{|}$$

IUPAC name: **2-methylpropane**
(common name: isobutane)

So far, the only branched alkanes that we've named have methyl groups attached to the main chain. What about groups other than CH_3? What do we call these groups, and how do we name alkanes that contain them?

2.10 ALKYL GROUPS

An alkyl group lacks one of the hydrogens of an alkane. A methyl group (CH_3—) is an alkyl group derived from methane (CH_4). Unbranched alkyl groups in which the point of attachment is at the end of the chain are named in IUPAC nomenclature by replacing the *-ane* endings of Table 2.4 by *-yl*.

CH_3CH_2— $CH_3(CH_2)_5CH_2$— $CH_3(CH_2)_{16}CH_2$—

Ethyl group   **Heptyl** group   **Octadecyl** group

The dash at the end of the chain represents a potential point of attachment for some other atom or group.

Carbon atoms are classified according to their degree of substitution by other carbons. A **primary** carbon is *directly* attached to one other carbon. Similarly, a **secondary** carbon is directly attached to two other carbons, a **tertiary** carbon to three, and a **quaternary** carbon to four. Alkyl groups are designated as primary, secondary, or tertiary according to the degree of substitution of the carbon at the potential point of attachment.

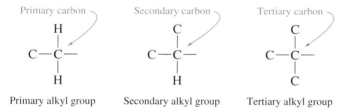

Primary alkyl group   Secondary alkyl group   Tertiary alkyl group

Ethyl (CH_3CH_2—), heptyl [$CH_3(CH_2)_5CH_2$—], and octadecyl [$CH_3(CH_2)_{16}CH_2$—] are examples of primary alkyl groups.

Branched alkyl groups are named by using the longest continuous chain that begins at the point of attachment as the base name. Thus, the systematic names of the two C_3H_7 alkyl groups are propyl and 1-methylethyl. Both are better known by their common names, *n*-propyl and isopropyl, respectively.

$$CH_3CH_2CH_2- \qquad \overset{\overset{\displaystyle CH_3}{\displaystyle |}}{\underset{2}{CH_3}\underset{1}{CH}}- \quad \text{or} \quad (CH_3)_2CH-$$

Propyl group   **1-Methylethyl** group
(common name: *n*-propyl)   (common name: isopropyl)

An isopropyl group is a secondary alkyl group. Its point of attachment is to a secondary carbon atom, one that is directly bonded to two other carbons.

The C_4H_9 alkyl groups may be derived either from the unbranched carbon skeleton of butane or from the branched carbon skeleton of isobutane. Those derived from butane are the butyl (*n*-butyl) group and the 1-methylpropyl (*sec*-butyl) group.

$$CH_3CH_2CH_2CH_2-$$

$$\overset{\displaystyle CH_3}{\overset{\displaystyle |}{CH_3CH_2\underset{3\quad 2\quad 1}{CH}-}}$$

Butyl group
(common name: *n*-butyl)

1-Methylpropyl group
(common name: *sec*-butyl)

Those derived from isobutane are the 2-methylpropyl (isobutyl) group and the 1,1-dimethylethyl (*tert*-butyl) group. Isobutyl is a primary alkyl group because its potential point of attachment is to a primary carbon. *tert*-Butyl is a tertiary alkyl group because its potential point of attachment is to a tertiary carbon.

$$\overset{\displaystyle CH_3}{\overset{\displaystyle |}{\underset{3\quad 2\quad 1}{CH_3CHCH_2}-}} \quad \text{or} \quad (CH_3)_2CHCH_2-$$

$$\overset{\displaystyle CH_3}{\overset{\displaystyle |}{\underset{2}{CH_3}\underset{|\,1}{C}-}} \quad \text{or} \quad (CH_3)_3C-$$
$$\underset{\displaystyle CH_3}{}$$

2-Methylpropyl group
(common name: isobutyl)

1,1-Dimethylethyl group
(common name: *tert*-butyl)

> The names and structures of the most frequently encountered alkyl groups are given on the inside back cover.

In addition to methyl and ethyl groups, *n*-propyl, isopropyl, *n*-butyl, *sec*-butyl, isobutyl, and *tert*-butyl groups will appear often throughout this text. Although these are common names, they have been integrated into the IUPAC system and are an acceptable adjunct to systematic nomenclature. You should be able to recognize these groups on sight and to give their structures when needed.

2.11 IUPAC NAMES OF HIGHLY BRANCHED ALKANES

By combining the basic principles of IUPAC notation with the names of the various alkyl groups, we can develop systematic names for highly branched alkanes. We'll start with the following alkane, name it, then increase its complexity by successively adding methyl groups at various positions.

$$\overset{\displaystyle CH_2CH_3}{\overset{\displaystyle |}{\underset{1\ \ 2\ \ 3\ \ 4\ \ 5\ \ 6\ \ 7\ \ 8}{CH_3CH_2CH_2CHCH_2CH_2CH_2CH_3}}}$$

As numbered on the structural formula, the longest continuous chain contains eight carbons, and so the compound is named as a derivative of octane. Numbering begins at the end nearest the branch, and so the ethyl substituent is located at C-4, and the name of the alkane is **4-ethyloctane.**

What happens to the IUPAC name when a methyl replaces one of the hydrogens at C-3?

$$\overset{\displaystyle CH_2CH_3}{\overset{\displaystyle |}{\underset{1\quad 2\quad 3}{CH_3CH_2CHCH}\underset{4\ \ 5\ \ 6\ \ 7\ \ 8}{CH_2CH_2CH_2CH_3}}}$$
$$\underset{\displaystyle CH_3}{}$$

The compound becomes an octane derivative that bears a C-3 methyl group and a C-4 ethyl group.

> *When two or more different substituents are present, they are listed in alphabetical order in the name.*

The IUPAC name for this compound is **4-ethyl-3-methyloctane.**

Replicating prefixes such as *di-*, *tri-*, and *tetra-* (see Section 2.9) are used as needed but are ignored when alphabetizing. Adding a second methyl group to the original structure, at C-5, for example, converts it to **4-ethyl-3,5-dimethyloctane.**

$$CH_2CH_3$$
$$|$$
$$\overset{1}{C}H_3\overset{2}{C}H_2\overset{3}{C}HCHCHCH_2CH_2CH_3$$
$$\underset{4}{|} \quad \underset{5}{|} \underset{6}{} \quad 7 \quad 8$$
$$CH_3 \quad CH_3$$

Italicized prefixes such as *sec-* and *tert-* are ignored when alphabetizing except when they are compared with each other. *tert*-Butyl precedes isobutyl, and *sec*-butyl precedes *tert*-butyl.

PROBLEM 2.11   Give an acceptable IUPAC name for each of the following alkanes:

(a)

$$CH_2CH_3$$
$$|$$
$$CH_3CH_2CHCHCHCH_2CHCH_3$$
$$| \quad | \quad |$$
$$CH_3 \quad CH_3 \quad CH_3$$

(c)

$$CH_3$$
$$|$$
$$CH_3CH_2CHCH_2CHCH_2CHCH(CH_3)_2$$
$$| \quad |$$
$$CH_2CH_3 \quad CH_2CH(CH_3)_2$$

(b) $(CH_3CH_2)_2CHCH_2CH(CH_3)_2$

Sample Solution   (a) This problem extends the preceding discussion by adding a third methyl group to 4-ethyl-3,5-dimethyloctane, the compound just described. It is, therefore, an **ethyltrimethyloctane**. Notice, however, that the numbering sequence needs to be changed to adhere to the rule of numbering from the end of the chain nearest the first branch. When numbered properly, this compound has a methyl group at C-2 as its first-appearing substituent.

$$CH_2CH_3$$
$$|$$
$$\overset{8}{C}H_3\overset{7}{C}H_2\overset{6}{C}HCHCHCH_2CHCH_3$$
$$\underset{5}{|} \quad | \quad |$$
$$CH_3 \quad CH_3 \quad CH_3$$

5-Ethyl-2,4,6-trimethyloctane

Finally, when equal locants are generated from two different numbering directions, choose the direction that gives the lower number to the substituent that appears first in the name. (Remember, substituents are listed alphabetically.)

The IUPAC nomenclature system is inherently logical and incorporates healthy elements of common sense into its rules. Granted, some long, funny-looking, hard-to-pronounce names are generated. Once you know the code (rules of grammar) though, it becomes a simple matter to convert those long names to unique structural formulas.

A tabular summary of the IUPAC rules for alkane nomenclature appears on page 62.

2.12   CYCLOALKANE NOMENCLATURE

Cycloalkanes are alkanes that contain a ring of three or more carbons. They are frequently encountered in organic chemistry and are characterized by the molecular formula C_nH_{2n}. Some examples include:

Cycloalkanes are one class of **alicyclic** (*aliphatic cyclic*) hydrocarbons.

$$H_2C{-\!\!\!-}CH_2$$
$$\diagdown \; \diagup$$
$$CH_2$$

usually represented as   ▽

Cyclopropane

If you make a molecular model of cyclohexane, you will find its shape to be very different from a planar hexagon. We'll discuss the reasons why beginning in Section 2.14.

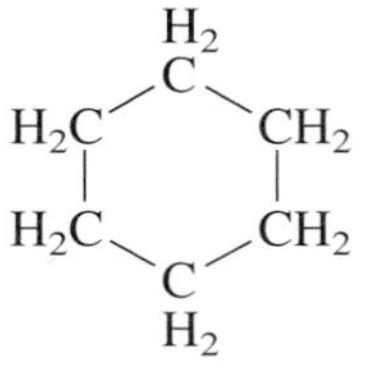

$$
\begin{array}{c}
\text{H}_2 \\
\text{C} \\
\text{H}_2\text{C} \quad\quad \text{CH}_2 \\
| \quad\quad\quad | \\
\text{H}_2\text{C} \quad\quad \text{CH}_2 \\
\text{C} \\
\text{H}_2
\end{array}
$$

usually represented as

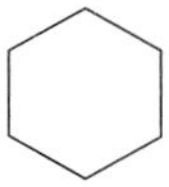

Cyclohexane

As you can see, cycloalkanes are named, under the IUPAC system, by adding the prefix *cyclo-* to the name of the unbranched alkane with the same number of carbons as the ring. Attached groups are identified in the usual way. Their positions are specified by numbering the carbon atoms of the ring in the direction that gives the lowest number to the substituted carbon at the first point of difference.

—CH$_2$CH$_3$

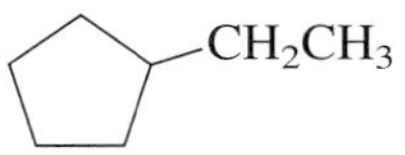

Ethylcyclopentane

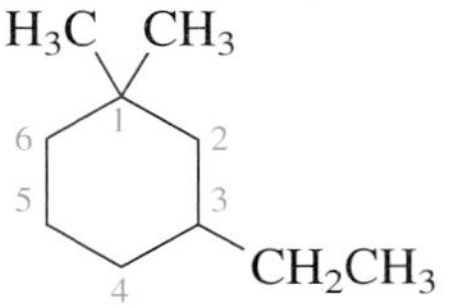

3-Ethyl-1,1-dimethylcyclohexane
(not 1-ethyl-3,3-dimethylcyclohexane, because first point of difference
rule requires 1,1,3 substitution pattern rather than 1,3,3)

When the ring contains fewer carbon atoms than an alkyl group attached to it, the compound is named as an alkane, and the ring is treated as a cycloalkyl substituent:

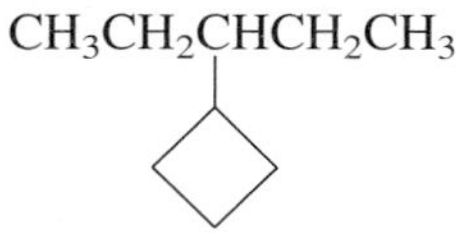

CH$_3$CH$_2$CHCH$_2$CH$_3$

3-Cyclobutylpentane

PROBLEM 2.12 Name each of the following compounds:

(a) —C(CH$_3$)$_3$

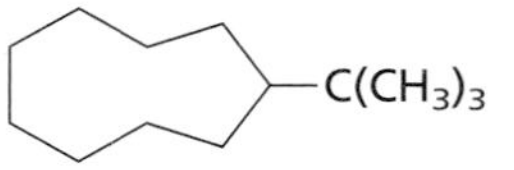

(c)

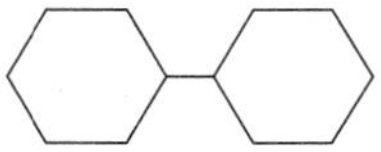

(b) (CH$_3$)$_2$CH   H$_3$C   CH$_3$

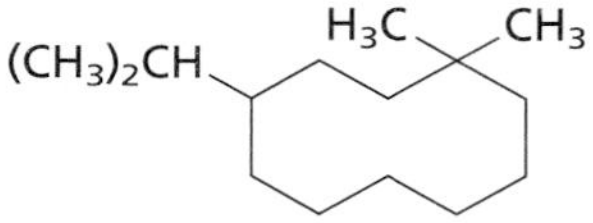

Sample Solution (a) The molecule has a *tert*-butyl group bonded to a nine-membered cycloalkane. It is *tert*-butylcyclononane. Alternatively, the *tert*-butyl group could be named systematically as a 1,1-dimethylethyl group, and the compound would then be named (1,1-dimethylethyl)cyclononane. (Parentheses are used when necessary to avoid ambiguity. In this case the parentheses alert the reader that the locants 1,1 refer to substituents on the alkyl group and not to ring positions.)

2.13   CONFORMATIONS OF CYCLOALKANES

Conformational analysis is far simpler in cyclopropane than in any other cycloalkane. Cyclopropane's three carbons are, of geometric necessity, coplanar.

All adjacent pairs of bonds
are eclipsed in cyclopropane

As can be seen in this depiction of cyclopropane, the three C—H bonds on the upper face of the ring are eclipsed, as are the three on the bottom face. Thus, cyclopropane incorporates an element of **torsional strain** (Section 2.5) into its structure. A more serious source of strain exists in cyclopropane, however. An equilateral triangle has angles of 60°, yet the bond angles at carbon when attached to four atoms or groups are ideally 109.5°. This distortion of the bond angles at carbon from the tetrahedral value is referred to as **angle strain** and makes cyclopropane less stable than other members of the alkane and cycloalkane family.

Cyclopropane is the only planar cycloalkane. Cyclobutane has less angle strain than cyclopropane and can reduce torsional strain by adopting the nonplanar "puckered" conformation shown in Figure 2.6*a*. Cyclopentane exists in a nonplanar conformation to relieve torsional strain; angle strain is relatively small because the 108° angles of a regular pentagon are close to the 109.5° angles of sp^3-hybridized carbon. One of the nonplanar conformations of cyclopentane, the envelope conformation, is shown in Figure 2.6*b*.

The torsional strain of planar
cyclopentane can be readily
seen with a molecular
model.

2.14   CONFORMATIONS OF CYCLOHEXANE

Six-membered rings occur more often than rings of any other size in organic compounds. Consequently, six-membered rings have been studied more extensively and their conformations are well understood. A planar conformation of cyclohexane would suffer from angle strain because the angles of a regular hexagon are 120°. It would also have a significant amount of torsional strain. The most stable conformation of cyclohexane is a nonplanar conformation known as the **chair conformation,** shown in Figure 2.7. A second, much less stable, nonplanar conformation is the **boat conformation,** shown in Figure 2.8.

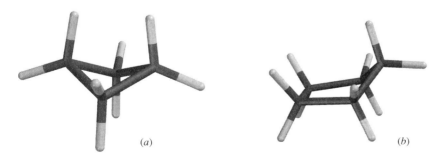

(*a*)   (*b*)

FIGURE 2.6 (*a*) Nonplanar ("puckered") conformation of cyclobutane. (*b*) Envelope conformation of cyclopentane

FIGURE 2.7 (*a*) A ball-and-spoke model and (*b*) a space-filling model of the chair conformation of cyclohexane.

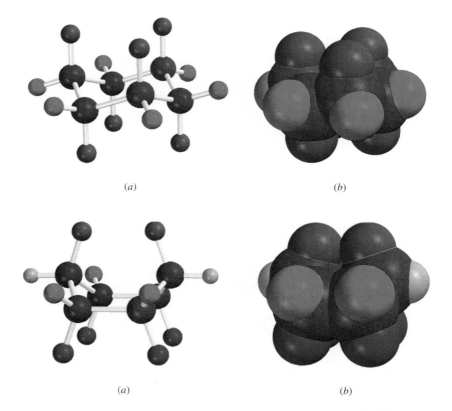

(*a*)   (*b*)

FIGURE 2.8 (*a*) A ball-and-spoke model and (*b*) a space-filling model of the boat conformation of cyclohexane. The close approach of the two uppermost hydrogens is clearly evident in the space-filling model.

(*a*)   (*b*)

While both are free of angle strain, the chair is free of torsional strain as well.

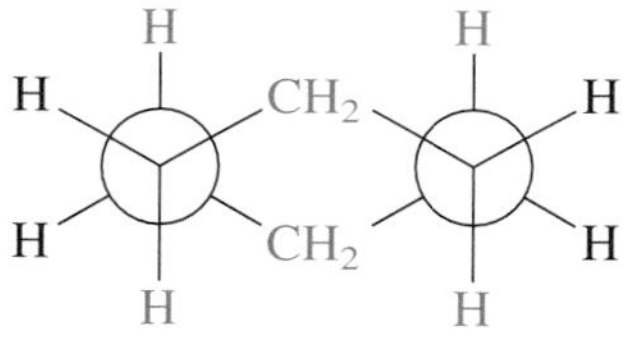

Make a molecular model of the chair conformation of cyclohexane, and turn it so that you can look down one of the C—C bonds.

Staggered arrangement of
bonds in chair conformation
of cyclohexane

Considering the chair conformation in detail reveals some surprising features. The 12 hydrogen atoms are not all identical but are divided into two groups, as shown in Figure 2.9. Six of the hydrogens, called **axial** hydrogens, have their bonds parallel to a vertical axis that passes through the ring's center. These axial bonds alternately are directed up and down on adjacent carbons. The second set of six hydrogens, called **equatorial** hydrogens, are located approximately along the equator of the molecule. Notice that the four bonds to each carbon are arranged tetrahedrally, consistent with an sp^3 hybridization of carbon.

The conformational features of six-membered rings are fundamental to organic chemistry, so it is essential that you have a clear understanding of the directional properties of axial and equatorial bonds and be able to represent them accurately. Figure 2.10 on page 54 offers some guidance on the drawing of chair cyclohexane rings. Be sure to study the figure before attempting the following problem.

PROBLEM 2.13 Given the chair conformations of cyclohexane shown here, draw the indicated carbon–hydrogen bonds.

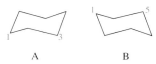

A   B

(a) Axial C—H on C-1 of A
(b) Equatorial C—H on C-3 of A
(c) Equatorial C—H on C-1 of B
(d) Axial C—H on C-5 of B

Sample Solution (a) Carbon in position 1 lies below its nearest neighbors; it is "down." Axial bonds point alternately straight up and straight down and take their direction from the carbon atom to which they are attached. Draw the axial bond to C-1 straight down.

H

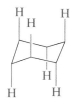

Axial C—H bonds

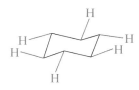

Equatorial C—H bonds

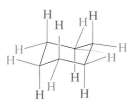

Axial and equatorial bonds together

FIGURE 2.9 Axial and equatorial bonds in cyclohexane.

2.15 CONFORMATIONAL INVERSION (RING FLIPPING) IN CYCLOHEXANE

We have seen that alkanes are not locked into a single conformation. Rotation around the central carbon–carbon bond in butane occurs rapidly, interconverting anti and gauche conformations. Cyclohexane, too, is conformationally mobile. Through a process known as **ring inversion, chair–chair interconversion,** or, more simply, **ring flipping,** one chair conformation is converted to another chair.

The most important result of ring inversion is that *any substituent that is axial in the original chair conformation becomes equatorial in the ring-flipped form and vice versa.*

X axial; Y equatorial   X equatorial; Y axial

The consequences of this point are developed for a number of monosubstituted cyclohexane derivatives in the following section, beginning with methylcyclohexane.

2.16 CONFORMATIONAL ANALYSIS OF MONOSUBSTITUTED CYCLOHEXANES

Ring inversion in methylcyclohexane differs from that of cyclohexane in that the two chair conformations are not equivalent. In one chair the methyl group is axial; in the other it is equatorial. At room temperature approximately 95% of the molecules of

(1) Begin with the chair conformation of cyclohexane.

(2) Draw the axial bonds before the equatorial ones, alternating their direction on adjacent atoms. Always start by placing an axial bond "up" on the uppermost carbon or "down" on the lowest carbon.

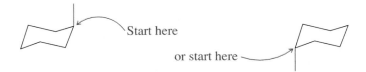

Start here

or start here

Then alternate to give

in which all the axial bonds are parallel to one another

(3) Place the equatorial bonds so as to approximate a tetrahedral arrangement of the bonds to each carbon. The equatorial bond of each carbon should be parallel to the ring bonds of its two nearest neighbor carbons.

Place equatorial bond at C-1 so that it is parallel to the bonds between C-2 and C-3 and between C-5 and C-6.

Following this pattern gives the complete set of equatorial bonds.

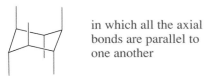

(4) Practice drawing cyclohexane chairs oriented in either direction.

and

FIGURE 2.10 A guide to representing the orientations of the bonds in the chair conformation of cyclohexane.

methylcyclohexane are in the chair conformation that has an equatorial methyl group, whereas only 5% of the molecules have an axial methyl group.

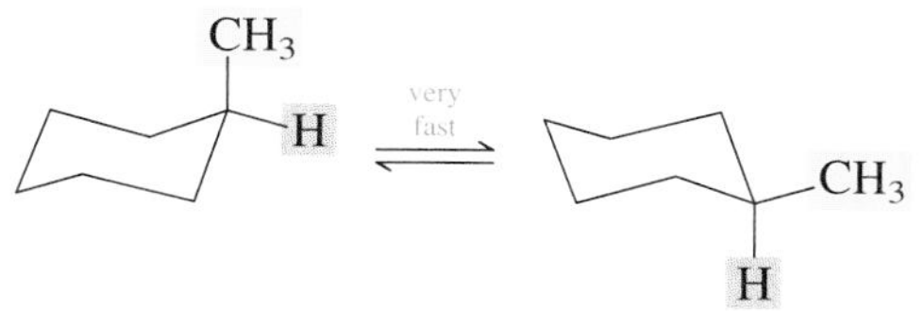

5%   95%

When two conformations of a molecule are in equilibrium with each other, the one with the lower free energy predominates. Why is equatorial methylcyclohexane more stable than axial methylcyclohexane?

A methyl group is less crowded when it is equatorial than when it is axial. One of the hydrogens of an axial methyl group is within 190–200 pm of the axial hydrogens at C-3 and C-5. This distance is less than the sum of the van der Waals radii of two hydrogens (240 pm) and causes van der Waals strain in the axial conformation. When the methyl group is equatorial, it experiences no significant crowding.

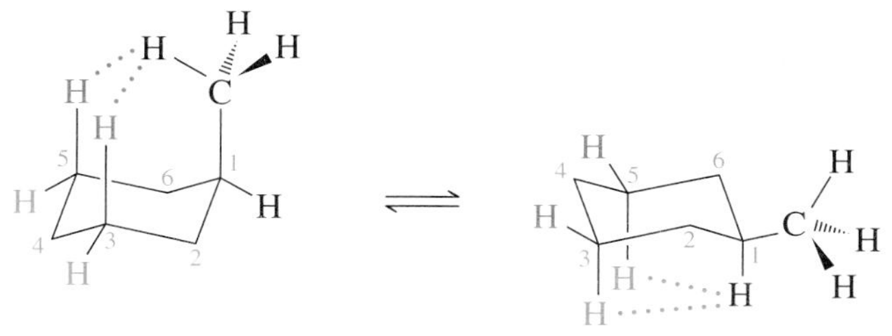

Van der Waals strain between hydrogen of axial CH₃ and axial hydrogens at C-3 and C-5

Smaller van der Waals strain between hydrogen at C-1 and axial hydrogens at C-3 and C-5

The greater stability of an equatorial methyl group, compared with an axial one, is another example of a **steric effect.** An axial substituent is said to be crowded because of **1,3-diaxial repulsions** between itself and the other two axial substituents located on the same side of the ring.

The same reasoning can explain the observed conformation of other substituted cyclohexanes. The larger group tends to be equatorial, and this tendency increases as the group becomes progressively "bulkier." Although the ratio of equatorial to axial methyl-cyclohexane conformations is 95:5, that ratio increases to greater than 999:1 for *tert*-butylcyclohexane.

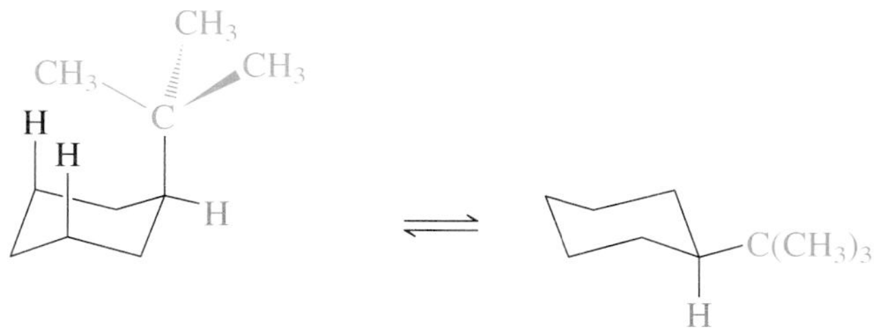

Less than 0.01%
(serious 1,3-diaxial repulsions involving *tert*-butyl group)

Greater than 99.99%
(decreased van der Waals strain)

> Highly branched groups such as *tert*-butyl are commonly described as "bulky."

> **PROBLEM 2.14** Draw the most stable conformation of 1-*tert*-butyl-1-methyl-cyclohexane.

In general, a group is "bulky" in organic chemistry if it is highly branched. A very long carbon chain is no bulkier than a shorter one, but a branched alkyl group is bulkier than an unbranched one.

2.17 DISUBSTITUTED CYCLOALKANES: STEREOISOMERS

When a cycloalkane bears two substituents on different carbons—methyl groups, for example—these substituents may be on the same or on opposite sides of the ring. When substituents are on the same side, we say they are **cis** to each other; if they are on opposite sides, they are **trans** to each other. Both terms come from the Latin, in which *cis* means "on this side" and *trans* means "across."

cis-1,2-Dimethylcyclopropane   trans-1,2-Dimethylcyclopropane

> **PROBLEM 2.15** Exclusive of compounds with double bonds, four hydrocarbons are constitutional isomers of *cis*- and *trans*-1,2-dimethylcyclopropane. Identify these compounds.

The prefix *stereo*- is derived from the Greek word *stereos,* meaning "solid." **Stereochemistry** is the term applied to the three-dimensional aspects of molecular structure and reactivity.

The cis and trans forms of 1,2-dimethylcyclopropane are stereoisomers. **Stereoisomers** are isomers that have their atoms bonded in the same order—that is, they have the same constitution, but they differ in the arrangement of atoms in space. Stereoisomers of the cis–trans type are sometimes referred to as **geometric isomers.**

Stereoisomerism in disubstituted cyclohexanes is somewhat more complicated than in cyclopropanes because, as we have seen, the cyclohexane ring is not planar. Let us first examine *cis*-1,2-dimethylcyclohexane. Both methyl groups in the cis stereoisomer are on the same face of the molecule. As the following structures show, both are "up," that is, above the hydrogen on the same carbon. Recalling that ring flipping interconverts axial and equatorial positions, we can see that the molecule can adopt either of two equivalent chair conformations. In each one, one methyl group is axial and the other equatorial.

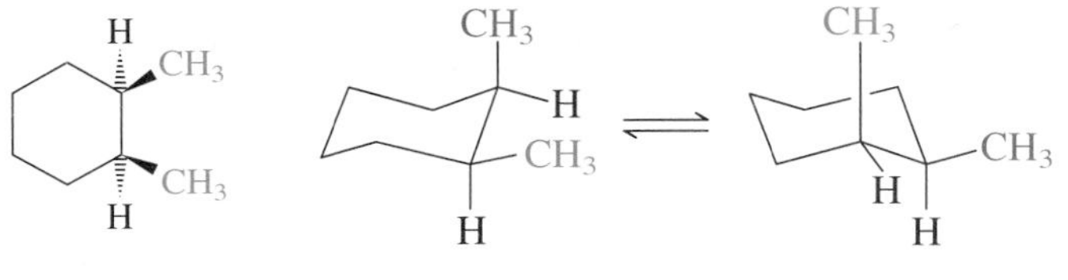

cis-1,2-Dimethylcyclohexane   cis-1,2-Dimethylcyclohexane

The situation is different for *trans*-1,2-dimethylcyclohexane, however. The two chair conformations are not equivalent. In one, both methyl groups are axial; in the other, both are equatorial.

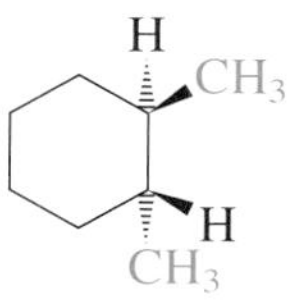

trans-1,2-Dimethylcyclohexane

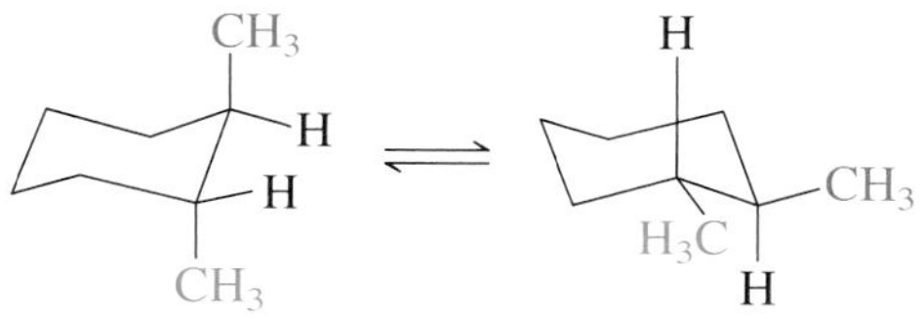

(both methyl groups
are axial: less stable
chair conformation)

(both methyl groups are
equatorial: more stable
chair conformation)

trans-1,2-Dimethylcyclohexane

The chair conformation in which both methyl groups are equatorial is more stable than the one in which both are axial and is the predominant one at equilibrium. We can understand why by recalling that equatorial substituents are less crowded than axial ones.

If two substituent groups are different, the preferred conformation will be the one in which the bulkier group is equatorial. Thus the most stable conformation of *cis*-1-*tert*-butyl-2-methylcyclohexane has an equatorial *tert*-butyl group and an axial methyl group.

PROBLEM 2.16   Draw the most stable conformation of *cis*- and *trans*-1-*tert*-butyl-2-methylcyclohexane.

All the properties of a molecule ultimately depend on its structure. Constitution or connectivity is an important element of molecular structure, but it is not the only one. The three-dimensional shape of a molecule—the arrangement of its atoms in space, or its **stereochemistry**—is also important. Many organic reactions and biochemical processes are known in which one stereoisomeric form of a substance reacts readily, but the other form is essentially inert under the same conditions.

2.18 POLYCYCLIC RING SYSTEMS

Organic compounds are not limited to a single ring, and many compounds contain two or more rings. Substances that contain two rings are referred to as **bicyclic,** those with three rings are called **tricyclic,** and so on. Camphene is a naturally occurring bicyclic hydrocarbon obtained from pine oil. Cortisone is a **steroid** that is formed in the outer layer of the adrenal gland and is commonly prescribed as an antiinflammatory drug. Steroids are a major class of tetracyclic lipids and are discussed in Chapter 16.

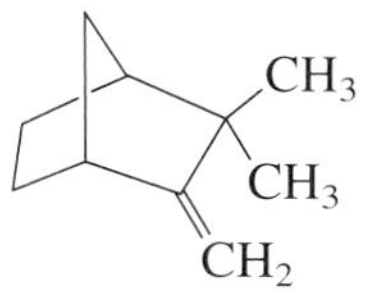

Camphene

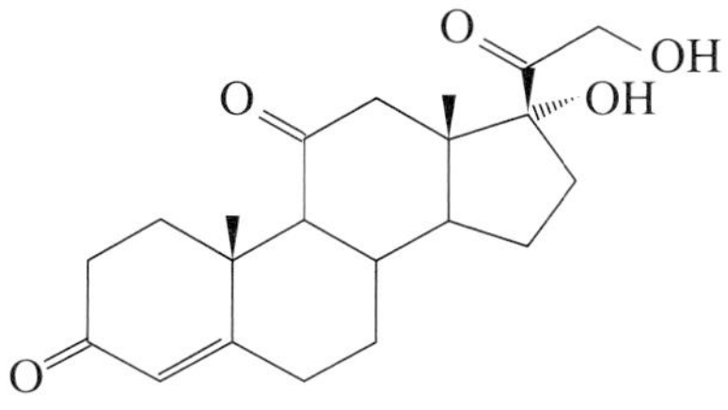

Cortisone

These are but two examples of the wide variety of compounds that contain more than one ring.

2.19   PHYSICAL PROPERTIES OF ALKANES AND CYCLOALKANES

Boiling Point   As we have seen earlier in this chapter, methane, ethane, propane, and butane are gases at room temperature. The unbranched alkanes, pentane (C_5H_{12}) through heptadecane ($C_{17}H_{36}$), are liquids, whereas higher homologs are solids. As shown in Figure 2.11, the boiling points of unbranched alkanes increase with the number of carbon atoms. Figure 2.11 also shows that the boiling points for 2-methyl-branched alkanes are lower than those of the unbranched isomer. This effect of chain branching can be clearly seen by comparing the three C_5H_{12} isomers.

$$CH_3CH_2CH_2CH_2CH_3 \qquad CH_3\underset{\underset{CH_3}{|}}{CH}CH_2CH_3 \qquad CH_3\underset{\underset{CH_3}{|}}{\overset{\overset{CH_3}{|}}{C}}CH_3$$

Pentane   2-Methylbutane   2,2-Dimethylpropane
(bp 36°C)   (bp 28°C)   (bp 9°C)

The most instructive way to consider the relation between boiling point and molecular structure is to ask yourself why any substance, pentane, for example, is a liquid rather than a gas. Pentane is a liquid at room temperature and atmospheric pressure because the attractive forces between molecules are greater in the liquid state than in the vapor. These **intermolecular attractive forces** must be overcome to vaporize pentane, or any other substance.

The strength of the intermolecular attractive forces is directly related to the surface area of the molecule. Branched isomers have lower boiling points than their unbranched counterparts because they are more compact and have a smaller surface area. The shapes of these isomers are clearly evident in the space-filling models depicted in Figure 2.12.

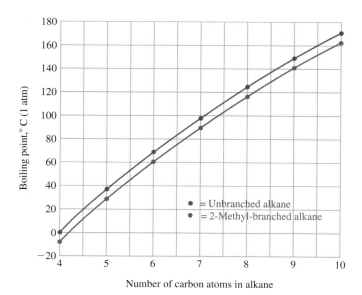

FIGURE 2.11   Boiling points of unbranched alkanes and their 2-methyl-branched isomers.

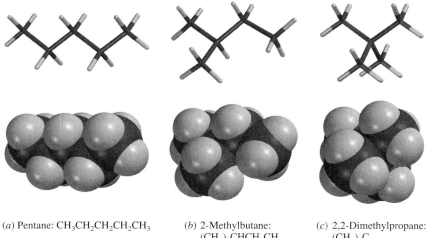

(a) Pentane: CH₃CH₂CH₂CH₂CH₃   (b) 2-Methylbutane: (c) 2,2-Dimethylpropane:
                                          (CH₃)₂CHCH₂CH₃   (CH₃)₄C

FIGURE 2.12 Space-filling models of (a) pentane, (b) 2-methylbutane, and (c) 2,2-dimethyl-propane. The most branched isomer, 2,2-dimethylpropane, has the most compact, most spherical, three-dimensional shape.

PROBLEM 2.17 Match the boiling points with the appropriate alkanes.
Alkanes: octane, 2-methylheptane, 2,2,3,3-tetramethylbutane
Boiling points (°C, 1 atm): 106, 116, 126

Although the boiling point trends noted above clearly indicate the presence of intermolecular forces in alkanes, it might seem that two nearby molecules A and B of a nonpolar substance such as an alkane would be unaffected by each other.

The electron clouds of neighboring molecules "feel" each other's presence, however, and a temporary distortion of the electron clouds results in an **induced-dipole/induced-dipole** attractive force.

The resulting weak attraction is one example of a **van der Waals force.** Van der Waals forces of the induced-dipole/induced-dipole type are the most important intermolecular forces present in the liquid state of an alkane. We will encounter additional attractive van der Waals forces in the next chapter.

Solubility in Water A familiar physical property of alkanes is contained in the adage "oil and water don't mix." Alkanes—indeed all hydrocarbons—are virtually insoluble in water. A more general statement about solubility is "like dissolves like." That is, polar solutes tend to dissolve in polar solvents, and nonpolar solutes tend to dissolve in nonpolar solvents.

Why is it that nonpolar substances (such as alkanes) are soluble in one another, but not soluble in water? The intermolecular attractive forces in a nonpolar solvent are the same as those in a nonpolar solute: induced-dipole/induced-dipole attractions. Thus hexane, for example, can readily dissolve in decane because the solute–solvent attractions in the solution are comparable to the forces present in each component.

For an alkane to dissolve in water, however, the attractive forces between the alkane and water would have to be strong enough to replace the dipole–dipole attractive forces between water molecules. They are not. Alkanes, being nonpolar, interact only weakly with water molecules. Alkanes are less dense than water, with densities in the 0.6–0.8 g/mL range, and thus float on water as the environmental damage from oil spills demonstrates. The exclusion of nonpolar molecules, such as alkanes, from water is called the **hydrophobic effect.**

2.20   CHEMICAL PROPERTIES: COMBUSTION OF ALKANES

As a group, alkanes are relatively unreactive, but they do burn in air. Their combination with oxygen is called **combustion.** On combustion in air, alkanes are converted to carbon dioxide and water.

$$CH_3CH_2CH_3 \ + \ 5O_2 \ \longrightarrow \ 3CO_2 \ + \ 4H_2O$$

   Propane   Oxygen   Carbon dioxide Water

The combustion of alkanes is **exothermic,** meaning that it gives off heat, and is a principal source of energy in our society. Natural gas is, as we have noted, predominantly methane accompanied by smaller amounts of ethane, propane, and butane. Petroleum, from the Latin words *petra* ("rock") and *oleum* ("oil"), is the source of many of the fuels we use every day. The complex mixture of materials present in petroleum (also called **crude oil**) can be separated into simpler mixtures by distillation. The fraction boiling in the range 30–150°C is called **straight-run gasoline** and contains, among other substances, alkanes with 5–10 carbon atoms. Kerosene is the petroleum fraction boiling at 175–325°C; it is principally C_8 through C_{14} hydrocarbons and is used as diesel fuel. Higher boiling fractions are used as lubricating oils, greases, and asphalt.

Petroleum is much more than a source of gasoline, and refineries do much more than make automobile fuel. Petroleum is far more valuable as a source of **petrochemicals** than as a source of gasoline. Petroleum fractions can be "cracked" to give ethylene and other hydrocarbons, and these compounds lead to a host of products that we use in our everyday lives. We will describe some materials derived from ethylene and other petrochemicals in later chapters.

> Straight-run gasoline is not a satisfactory fuel for automobile engines because its "octane rating" is too low. Premature ignition of the fuel gives rise to engine "knock" and robs high-compression engines of their power. A refinery process known as **reforming** is one way of enhancing the octane rating of the gasoline fraction.

LEARNING OBJECTIVES

This chapter used the family of hydrocarbons known as alkanes to introduce the concepts of structure and nomenclature in organic chemistry. The skills you have learned in this chapter should enable you to:

- Recognize the hydrocarbon families, functionally substituted derivatives of alkanes, and classes of compounds containing the carbonyl group.
- Give the IUPAC names of the unbranched alkanes having up to 20 carbon atoms.

Continued

- Given an alkane or cycloalkane, write its IUPAC name.
- Given the IUPAC name for an alkane or cycloalkane, write its structural formula.
- Recognize by common name and structure the alkyl groups that contain up to four carbon atoms.
- Recognize and represent conformations of particular molecules by wedge-and-dash, Newman projection, and sawhorse formulas.
- Draw a chair conformation for a cyclohexane derivative, clearly showing substituent(s) in axial or equatorial orientations as appropriate.
- Know the meaning of the terms *eclipsed conformation, staggered conformation, anti conformation,* and *gauche conformation.*
- Know the meaning of the terms *angle strain, torsional strain,* and *van der Waals strain.*
- Given the chair conformation for a cyclohexane derivative, draw a structural formula for its ring-flipped form.
- Understand the difference between constitutional isomers and stereoisomers.
- Understand how branching affects the boiling point of an alkane.

2.21   SUMMARY

The classes of hydrocarbons are **alkanes, alkenes, alkynes,** and **arenes** (Section 2.1). Alkanes are hydrocarbons in which all of the bonds are *single* bonds and are characterized by the molecular formula C_nH_{2n+2}.

Functional groups are the structural units responsible for the characteristic reactions of a molecule. The functional groups in an alkane are its hydrogens (Section 2.2). Other families of organic compounds, listed in Table 2.1, bear more reactive functional groups, and the hydrocarbon chain to which they are attached can often be viewed as a supporting framework for the reactive function (Section 2.3). Several classes of organic compounds contain the **carbonyl group,** $\diagdown C = O$ (Table 2.2).

The simplest alkane is **methane** CH_4 (Section 2.4); **ethane** is C_2H_6, and **propane** is C_3H_8. Constitutional isomers are possible for alkanes with four or more carbons (Section 2.6). A single alkane may have different names; a name may be a **common name** or it may be a **systematic name** developed by a well-defined set of rules. The system that is the most widely used in chemistry is **IUPAC nomenclature** (Sections 2.8 to 2.12). According to IUPAC nomenclature, alkanes are named as derivatives of unbranched parents. Substituents on the longest continuous chain are identified and their positions specified by number. The IUPAC rules for naming alkanes are summarized in Table 2.5.

Conformations are different spatial arrangements of a molecule that are generated by rotation about single bonds. The most stable and least stable conformations of ethane are the **staggered** and the **eclipsed,** respectively (Section 2.5)

TABLE 2.5	Summary of IUPAC Nomenclature of Alkanes

Rule	Example
1. Find the longest continuous chain of carbon atoms, and assign a basis name to the compound corresponding to the IUPAC name of the unbranched alkane having the same number of carbons.	The longest continuous chain in the alkane shown is six carbons. 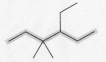 This alkane is named as a derivative of **hexane.**
2. List the substituents attached to the longest continuous chain in alphabetical order. Use the prefixes *di-, tri-, tetra-,* and so on, when the same substituent appears more than once. Ignore these prefixes when alphabetizing.	The alkane bears two methyl groups and an ethyl group. It is an **ethyldimethylhexane.**
3. Number the chain in the direction that gives the lower locant to a substituent at the first point of difference.	When numbering from left to right, the substituents appear at carbons 3, 3, and 4. When numbering from right to left the locants are 3, 4, and 4; therefore, number from left to right.
	The correct name is **4-ethyl-3,3-dimethylhexane.**
4. When two different numbering schemes give equivalent sets of locants, choose the direction that gives the lower locant to the group that appears first in the name.	In the following example, the substituents are located at carbons 3 and 4 regardless of the direction in which the chain is numbered.
	Ethyl precedes methyl in the name; therefore **3-ethyl-4-methylhexane** is correct.

Staggered conformation of ethane
(most stable conformation)

Eclipsed conformation of ethane
(least stable conformation)

Torsional strain is the destabilization that results from the eclipsing of bonds. Staggered conformations are more stable than eclipsed because they have no torsional strain.

The two staggered conformations of butane are not equivalent. The **anti** conformation is more stable than the **gauche** (Section 2.6).

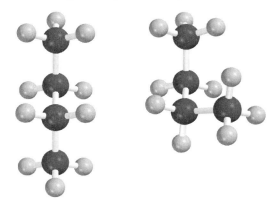

Anti conformation of butane   Gauche conformation of butane

Neither staggered conformation incorporates any torsional strain. The gauche conformation is less stable than the anti because of van der Waals strain involving the methyl groups.

Cyclopropane is planar and strained (angle strain and torsional strain). Cyclobutane is nonplanar and less strained than cyclopropane. Cyclopentane has two nonplanar conformations, one of which is the envelope (Section 2.13).

The chair is by far the most stable conformation for cyclohexane and its derivatives (Sections 2.14 to 2.17). The chair conformation is free of angle strain, torsional strain, and van der Waals strain. The C—H bonds in cyclohexane are not all equivalent but are divided into two sets of six each, called **axial** and **equatorial.** Cyclohexane undergoes a rapid conformational change referred to as **ring inversion,** or **ring flipping.** The process of ring inversion causes all axial bonds to become equatorial, and vice versa.

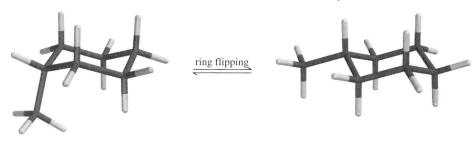

Methyl group axial (less stable)   Methyl group equatorial (more stable)

Substituents on a cyclohexane ring are more stable when they occupy equatorial sites than when they are axial. Branched substituents, especially *tert*-butyl, have a pronounced preference for the equatorial position. The relative stabilities of stereoisomeric disubstituted (and more highly substituted) cyclohexanes can be assessed by analyzing chair conformations for van der Waals strain involving axial substituents.

Alkanes and cycloalkanes are essentially nonpolar. The forces of attraction between molecules are relatively weak van der Waals forces. Because of their smaller surface area (Section 2.19), branched alkanes have lower boiling points than their unbranched isomers. Being nonpolar, alkanes are insoluble in water.

Alkanes and cycloalkanes burn in air to give carbon dioxide, water, and heat. This process is called **combustion** (Section 2.20).

ADDITIONAL PROBLEMS

Functional Groups

2.18 The structure of the antiinflammatory drug cortisone is shown in Section 2.18 (page 57). Identify the functional groups present in this substance.

2.19 (a) Complete the structure of the pain-relieving drug ibuprofen on the basis of the fact that ibuprofen is a carboxylic acid that has the molecular formula $C_{13}H_{18}O_2$, X is an isobutyl group, and Y is a methyl group.

$$X—\underset{}{\bigcirc}—\overset{\overset{\textstyle Y}{|}}{CH}—Z$$

(b) Mandelonitrile may be obtained from peach flowers. Derive its structure from the template in part (a) given that X is hydrogen, Y is the functional group that characterizes alcohols, and Z characterizes nitriles.

2.20 Isoamyl acetate is the common name of the substance most responsible for the characteristic odor of bananas. Write a structural formula for isoamyl acetate, given the information that it is an ester in which the carbonyl group bears a methyl substituent and there is a 3-methylbutyl group attached to one of the oxygens.

2.21 *n*-Butyl mercaptan is the common name of a foul-smelling substance obtained from skunk fluid. It is a thiol of the type RX, where R is an *n*-butyl group and X is the functional group that characterizes a thiol. Write a structural formula for this substance.

2.22 Some of the most important organic compounds in biochemistry are the α-*amino acids*, represented by the general formula shown.

$$\overset{\overset{\textstyle O}{\|}}{\underset{\underset{\textstyle {}^+NH_3}{|}}{RCHCO^-}}$$

Write structural formulas for the following a-amino acids.

(a) Alanine (R = methyl)
(b) Valine (R = isopropyl)
(c) Leucine (R = isobutyl)
(d) Isoleucine (R = *sec*-butyl)
(e) Serine (R = XCH_2, where X is the functional group that characterizes alcohols)
(f) Cysteine (R = XCH_2, where X is the functional group that characterizes thiols)
(g) Aspartic acid (R = XCH_2, where X is the functional group that characterizes carboxylic acids)

Structure and Nomenclature

2.23 Write the structures and give the IUPAC names for all the alkanes of molecular formula C_7H_{16} that:

(a) Are named as methyl-substituted derivatives of hexane.
(b) Are named as dimethyl derivatives of pentane.
(c) Are named as ethyl-substituted derivatives of pentane.

2.24 Give the molecular formula and the IUPAC name for each of the following compounds:

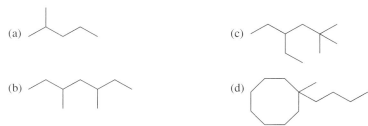

2.25 Rewrite the following condensed structural formulas as carbon-skeleton (bond-line) formulas and give the IUPAC name for each one.

(a) $(CH_3)_3CCH_2CH_2CH_3$

(b) $CH_3CH_2CH_2CH(CH_2CH_3)_2$

(c) $CH_3CH_2CHCH_2CHCH_2CH_2CH_3$
$\qquad\quad |\qquad\quad |$
$\qquad\quad CH_3\quad CH_2CH_3$

(d) $CH_3CH_2CHCH_2C(CH_3)_3$
$\qquad\quad |$
$\qquad\quad CH_2CH(CH_3)_2$

2.26 Write a structural formula for each of the following compounds:

(a) 3-Ethyloctane
(b) 6-Isopropyl-2,3-dimethylnonane
(c) 4-*tert*-Butyl-3-methylheptane
(d) 4-Isobutyl-1,1-dimethylcyclohexane
(e) *sec*-Butylcycloheptane
(f) Cyclobutylcyclopentane

2.27 Which of the compounds in each of the following groups are isomers?

(a) Butane, cyclobutane, isobutane, 2-methylbutane
(b) Cyclohexane, hexane, methylcyclopentane, 1,1,2-trimethylcyclopropane
(c) Ethylcyclopropane, 1,1-dimethylcyclopropane, 1-cyclopropylpropane, cyclopentane
(d) 4-Methyltetradecane, 2,3,4,5-tetramethyldecane, pentadecane, 4-cyclobutyldecane

2.28 Write the structural formula of a compound of molecular formula $C_4H_8Cl_2$ in which

(a) All the carbons belong to methylene groups
(b) None of the carbons belong to methylene groups

2.29 Female tiger moths signify their presence to male moths by giving off a sex attractant. The sex attractant (pheromone) has been isolated and found to be a 2-methyl-branched alkane having a molecular weight of 254. What is this material?

2.30 How many σ bonds are in pentane? In cyclopentane?

2.31 Hectane is the IUPAC name for the unbranched alkane which contains 100 carbon atoms.

(a) What is the molecular formula of hectane?
(b) Write the condensed molecular formula for hectane in the form $CH_3(CH_2)_nCH_3$.
(c) How many σ bonds are in hectane?
(d) How many alkanes have names of the type X-methylhectane? (Examples include 2-methylhectane, 3-methylhectane, etc.)
(e) How many alkanes have names of the type 2,X-dimethylhectane?

Conformations of Alkanes and Cycloalkanes

2.32   Which has more torsional strain, cyclopropane or the planar conformation of cyclopentane? Which has more angle strain?

2.33   Draw a Newman projection formula (looking down the C-1—C-2 bond) for the most stable conformation of 2,2-dimethylpropane.

2.34   Write Newman projection formulas for two different staggered conformations of 2,3-dimethylbutane (as viewed down the C-2—C-3 bond).

2.35   Draw Newman projection formulas for the three most stable conformations of 2-methylbutane (as viewed down the C-2—C-3 bond). One of these conformations is less stable than the other two. Which one? Why?

2.36   Determine whether the two structures in each of the following pairs represent **constitutional isomers**, different **conformations** of the same compound, or **stereoisomers** that cannot be interconverted by rotation about single bonds.

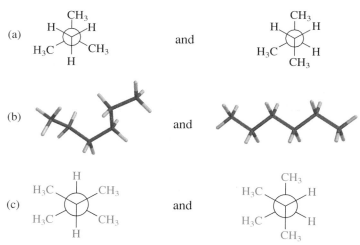

(d) *cis*-1,2-Dimethylcyclopentane and *trans*-1,3-dimethylcyclopentane

Cyclohexane Conformations and Stereochemistry

2.37   Draw a clear conformational depiction of the most stable conformation of:
   (a) 1,1,3-Trimethylcyclohexane   (b) 1,1,4-Trimethylcyclohexane

2.38   Draw chair conformations of 1,1,3-trimethylcyclohexane and 1,1,4-trimethylcyclohexane that are less stable than those in Problem 2.37.

2.39   Draw both possible chair conformations for each of the following compounds, clearly showing the orientation of each substituent (axial or equatorial). Indicate which conformation is the more stable one.

(a)

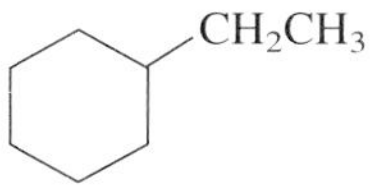

CH₂CH₃

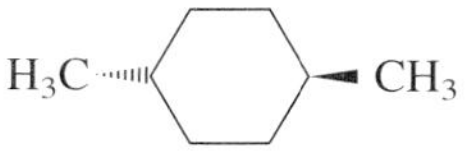

(c) 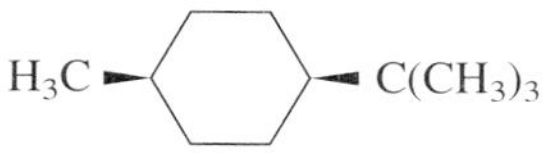 H₃C ➤⟨ ⟩◄ C(CH₃)₃

(b) H₃C ⁘⁘⁘⟨ ⟩◄ CH₃

(d) 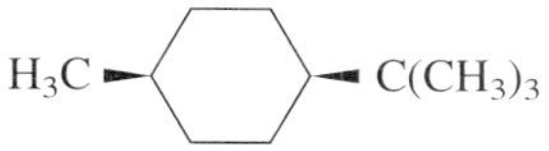 H₃C ➤⟨ ⟩◄ C(CH₃)₃

2.40 (a) Which stereoisomer of 1,3-dimethylcyclohexane exists in two equivalent chair conformations?

(b) Draw the chair conformation of this stereoisomer.

(c) Draw the 1,3-dimethylcyclohexane stereoisomer that has two nonequivalent chair conformations, and specify which conformation is more stable.

2.41 Write a structural formula for the most stable conformation of each of the following compounds:

(a) *cis*-1-Isopropyl-3-methylcyclohexane

(b) *trans*-1-Isopropyl-3-methylcyclohexane

(c) *cis*-1-*tert*-Butyl-4-ethylcyclohexane

2.42 Identify the more stable stereoisomer in each of the following pairs, and give the reason for your choice:

(a) *cis*- or *trans*-1-Isopropyl-2-methylcyclohexane

(b) *cis*- or *trans*-1-Isopropyl-3-methylcyclohexane

(c) *cis*- or *trans*-1-Isopropyl-4-methylcyclohexane

(d) 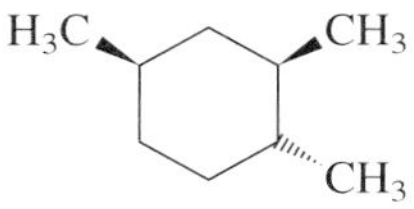 H₃C⟍ ⟍CH₃   H₃C⟍   ⟍CH₃
          ⟨   ⟩   or   ⟨   ⟩
              ⁘⁘CH₃   CH₃

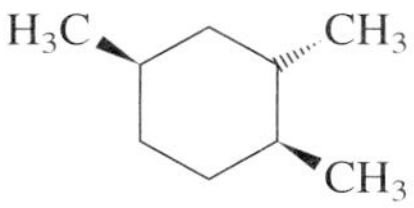

2.43 The following are representations of two forms of glucose. The six-membered ring is known to exist in a chair conformation in each form. Draw clear representations of the most stable conformation of each. Are they two different conformations of the same molecule, or are they stereoisomers? Which substituents (if any) occupy axial sites?

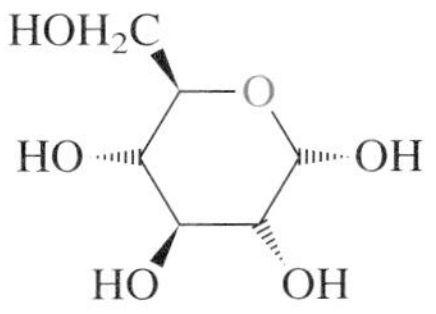

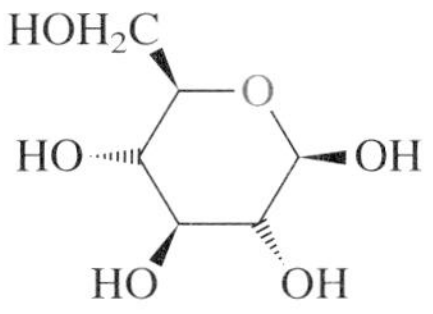

HOH₂C⟍   O   HOH₂C⟍   O
HO⁘⁘⟨   ⟩⁘⁘OH   HO⁘⁘⟨   ⟩◄OH
        HO   OH   HO   OH

2.44 A typical steroid skeleton is shown along with the numbering scheme used for this class of compounds. Specify in each case whether the designated substituent is axial or equatorial.

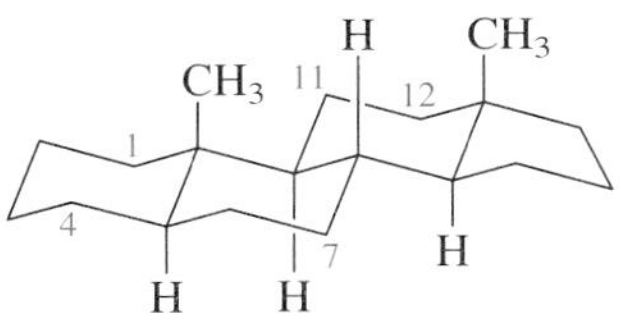

H CH₃
CH₃ 11 ⟍ ⁘ 12
1
4   7
H   H   H

(a) Substituent at C-1 cis to the methyl groups

(b) Substituent at C-4 cis to the methyl groups

(c) Substituent at C-7 trans to the methyl groups

(d) Substituent at C-11 trans to the methyl groups

(e) Substituent at C-12 cis to the methyl groups

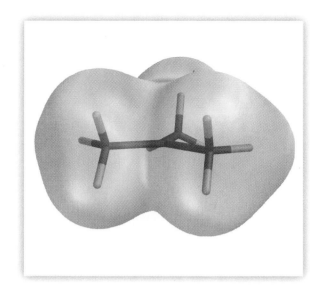

CHAPTER 3
ALCOHOLS AND ALKYL HALIDES

Our first two chapters established some fundamental principles concerning the *structure* of organic molecules. In this chapter we begin our discussion of organic chemical *reactions* by directing attention to **alcohols** and **alkyl halides.** These two rank among the most useful classes of organic compounds because they often serve as starting materials for the preparation of numerous other families.

One reaction leading to alkyl halides that will be described in this chapter illustrates functional group transformations. In this reaction the hydroxyl (—OH) group of an alcohol is replaced by halogen on treatment with a hydrogen halide.

$$R-\boxed{OH} \ + \quad H-X \quad \longrightarrow \quad R-X \ + H-\boxed{OH}$$

Alcohol   Hydrogen halide   Alkyl halide   Water

This reaction is classified as a **substitution,** a term that describes the relationship between reactants and products—one functional group replaces another. In this chapter we go beyond the relationship of reactants and products and consider the mechanism of each reaction. A **mechanism** attempts to show *how* starting materials are converted into products during a chemical reaction.

While developing these themes of reaction and mechanism, we will also use alcohols and alkyl halides as vehicles to extend the principles of IUPAC nomenclature, continue to develop concepts of structure and bonding, and see how structure affects properties. A review of **acids** and **bases** constitutes an important part of this chapter in which a qualitative approach to proton-transfer equilibria will be developed that will be used throughout the remainder of the text.

3.1 NOMENCLATURE OF ALCOHOLS AND ALKYL HALIDES

The IUPAC rules permit alcohols and alkyl halides to be named systematically in two ways. In one method, the alkyl group of the compound is named followed by a separate word specific for the functional group: fluoride, chloride, bromide, iodide, or alcohol. These names are illustrated in the following examples.

$$CH_3CH_2F \qquad (CH_3)_2CHCl \qquad (CH_3)_3CBr$$

Ethyl fluoride　　Isopropyl chloride　　*tert*-Butyl bromide

$$CH_3CHCH_2CH_2CH_3$$
$$|$$
$$I$$

1-Methylbutyl iodide　　　Cyclohexyl alcohol

The IUPAC rules permit certain common alkyl group names to be used. These include *n*-propyl, isopropyl, *n*-butyl, *sec*-butyl, isobutyl, and *tert*-butyl (see Section 2.10).

The second way of naming alkyl halides treats the halogen as a **halo- (fluoro-, chloro-, bromo-,** or **iodo-) substituent** on an alkane chain. The carbon chain is numbered in the direction that gives the substituted carbon the lower locant.

$$\overset{5}{C}H_3\overset{4}{C}H_2\overset{3}{C}H_2\overset{2}{C}H_2\overset{1}{C}H_2F \qquad \overset{1}{C}H_3\overset{2}{C}HCH_2CH_2CH_3 \qquad \overset{1}{C}H_3\overset{2}{C}H_2\overset{3}{C}HCH_2CH_3$$

1-Fluoropentane　　　　2-Bromopentane　　　　3-Iodopentane

When the carbon chain bears both a halogen and an alkyl substituent, the two substituents are considered of equal rank, and the chain is numbered so as to give the lower number to the substituent nearer the end of the chain.

$$\overset{1}{C}H_3\overset{2}{C}H\overset{3}{C}H_2\overset{4}{C}H_2\overset{5}{C}H\overset{6}{C}H_2\overset{7}{C}H_3 \qquad \overset{1}{C}H_3\overset{2}{C}H\overset{3}{C}H_2\overset{4}{C}H_2\overset{5}{C}H\overset{6}{C}H_2\overset{7}{C}H_3$$

5-Chloro-2-methylheptane　　　　2-Chloro-5-methylheptane

> **PROBLEM 3.1** Write structural formulas, and give two systematic names for each of the isomeric alkyl chlorides that have the molecular formula C_4H_9Cl.

The comparable names of alcohols are developed by identifying the longest chain that bears the hydroxyl group and replacing the *-e* ending of the corresponding alkane by the suffix *-ol*. The position of the hydroxyl group is indicated by number, choosing the sequence that assigns the lower locant to the carbon that bears the hydroxyl group.

$$CH_3CH_2OH \qquad CH_3CHCH_2CH_2CH_2CH_3 \qquad CH_3CCH_2CH_2CH_3$$

Ethanol　　　　　2-Hexanol　　　　2-Methyl-2-pentanol

Several alcohols are commonplace substances, well known by common names that reflect their origin (wood alcohol, grain alcohol) or use (rubbing alcohol). Wood alcohol is methanol (methyl alcohol, CH_3OH), grain alcohol is ethanol (ethyl alcohol, CH_3CH_2OH), and rubbing alcohol is 2-propanol [isopropyl alcohol, $(CH_3)_2CHOH$].

Hydroxyl groups take precedence over ("outrank") alkyl groups and halogen substituents in determining the direction in which a carbon chain is numbered.

$$\underset{7}{CH_3}\underset{6}{CH}\underset{5}{CH_2}\underset{4}{CH_2}\underset{3}{CH}\underset{2}{CH_2}\underset{1}{CH_3}$$

CH₃	OH

6-Methyl-3-heptanol
(not 2-methyl-5-heptanol)

trans-2-Methylcyclopentanol

$$\underset{3}{FCH_2}\underset{2}{CH_2}\underset{1}{CH_2OH}$$

3-Fluoro-1-propanol

PROBLEM 3.2 Write structural formulas, and give two systematic names for each of the isomeric alcohols that have the molecular formula $C_4H_{10}O$.

THE COMMON ALCOHOLS: METHYL, ETHYL, AND ISOPROPYL ALCOHOL

Until the 1920s, the major source of methyl alcohol (methanol) was its isolation as a byproduct in the production of charcoal from wood—hence its common name, **wood alcohol.** Now most of the 11 billion pounds of methanol produced annually in the United States is synthetic, prepared directly from carbon monoxide and hydrogen.

$$CO \quad + \quad 2H_2 \quad \xrightarrow[\text{400°C, pressure}]{\text{ZnO—Cr}_2\text{O}_3} \quad CH_3OH$$

Carbon monoxide   Hydrogen   Methanol

Almost one half of this methanol is converted to formaldehyde for incorporation into various resins and plastics. Methanol is also used as a solvent, as an antifreeze, and as a convenient clean-burning liquid fuel. Methanol is a colorless liquid that boils at 65°C and is miscible with water in all proportions. It is poisonous; drinking as little as 30 mL has been fatal. Ingestion of sublethal amounts can lead to blindness.

When vegetable matter ferments, its carbohydrates are converted to ethyl alcohol (ethanol) and carbon dioxide by enzymes present in yeast. Using glucose as a representative carbohydrate, the reaction may be written as

$$C_6H_{12}O_6 \quad \xrightarrow{\text{enzymes}} \quad 2CH_3CH_2OH \quad + \quad 2CO_2$$

Glucose   Ethanol   Carbon dioxide

Fermentation of barley produces beer; grapes give wine. The maximum ethanol content is about 15% because higher concentrations inactivate the enzymes, halting fermentation. Because ethanol boils at 78°C and water at 100°C, distillation of the fermentation broth gives "distilled spirits" of increased ethanol content. Whiskey is the aged distillate of fermented grain and contains slightly less than 50% ethanol.

Brandy and cognac are made by aging the distilled spirits from fermented grapes and other fruits and are about 70% ethanol. The characteristic flavors, odors, and colors of the various alcoholic beverages depend both on their origin and the way they are aged.

Synthetic ethanol is derived from petroleum via ethylene, $CH_2=CH_2$ (see boxed essay following Section 5.6). In the United States, some 1.2 billion pounds of synthetic ethanol is produced annually. It is relatively inexpensive and useful for industrial applications. To render it unfit for drinking, ethanol can be **denatured** by the deliberate addition of noxious materials, including methanol, thereby exempting it from the taxes imposed on ethanol used in beverages.

Our bodies are reasonably well equipped to metabolize ethanol, making it less dangerous than methanol. Alcohol abuse and alcoholism have been and remain, however, persistent problems in human societies.

Isopropyl alcohol [$(CH_3)_2CHOH$] is prepared commercially from petroleum via propene, $CH_3CH=CH_2$. With a boiling point of 82°C, isopropyl alcohol evaporates quickly from the skin, producing a cooling effect. Often containing dissolved oils and fragrances, isopropyl alcohol is the major component of rubbing alcohol. Isopropyl alcohol also possesses weak antibacterial properties and is used to maintain medical instruments in a sterile condition and to clean the skin before minor surgery.

Alcohols are among the most readily available organic compounds, both through isolation from natural sources and by synthesis. Because of this availability, alcohols are valuable starting materials for the preparation of a variety of other classes of compounds.

3.2 CLASSES OF ALCOHOLS AND ALKYL HALIDES

Alcohols and alkyl halides are classified as primary, secondary, or tertiary according to the classification of the carbon that bears the functional group (see Section 2.10). Thus, **primary alcohols** and **primary alkyl halides** are compounds of the type RCH_2G (where G is the functional group), **secondary alcohols** and **secondary alkyl halides** are compounds of the type R_2CHG, and **tertiary alcohols** and **tertiary alkyl halides** are compounds of the type R_3CG.

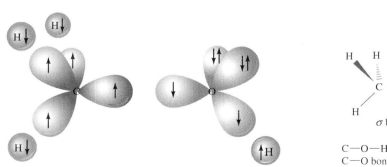

2,2-Dimethyl-1-propanol	2-Bromobutane	1-Methylcyclohexanol	2-Chloro-2-methylpentane
(a primary alcohol)	(a secondary alkyl halide)	(a tertiary alcohol)	(a tertiary alkyl halide)

> **PROBLEM 3.3** Classify the isomeric $C_4H_{10}O$ alcohols as being primary, secondary, or tertiary.

Many of the properties of alcohols and alkyl halides are affected by whether their functional groups are attached to primary, secondary, or tertiary carbons. We will see a number of cases in which a functional group attached to a primary carbon is more reactive than one attached to a secondary or tertiary carbon, as well as other cases in which the reverse is true.

> **PROBLEM 3.4** The European bark beetle is the insect most responsible for the spread of Dutch elm disease. The beetle bores into the bark of an elm tree and emits an aggregation pheromone which attracts other beetles to the site. The beetles carry with them a fungus which, if it becomes established and grows uncontrollably, can kill the tree. One component of the aggregation pheromone is 4-methyl-3-heptanol. Write a structural formula for this alcohol. Is it a primary, secondary, or tertiary alcohol?

3.3 BONDING IN ALCOHOLS AND ALKYL HALIDES

The carbon that bears the functional group is sp^3-hybridized in alcohols and alkyl halides. The hydroxyl group of an alcohol is attached to carbon by a σ bond generated by overlap of an sp^3 hybrid orbital of carbon with an sp^3 hybrid orbital of oxygen. Figure 3.1

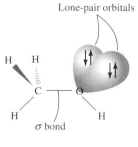

Lone-pair orbitals

C—O—H angle = 108.5°
C—O bond distance = 142 pm

σ bond

(a)   *(b)*

FIGURE 3.1 Orbital hybridization model of bonding in methanol. (*a*) The orbitals used in bonding are the 1*s* orbitals of hydrogen and *sp*³-hybridized orbitals of carbon and oxygen. (*b*) The bond angles at carbon and oxygen are close to tetrahedral, and the carbon–oxygen σ bond is about 10 pm shorter than a carbon–carbon single bond.

illustrates this for the specific case of bonding in methanol. The bonds to carbon are arranged in a tetrahedral geometry. Bonding in alkyl halides is similar to that of alcohols. The halogen substituent is connected to an sp^3-hybridized carbon by a σ bond.

Carbon–oxygen and carbon–halogen bonds are polar covalent bonds, and carbon bears a partial positive charge in alcohols ($^{\delta+}C{-}O^{\delta-}$) and in alkyl halides ($^{\delta+}C{-}X^{\delta-}$). The presence of these polar bonds makes alcohols and alkyl halides polar molecules.

$$\overset{\nwarrow O \nearrow}{\underset{H \qquad H}{}} \qquad \overset{\nwarrow O \nearrow}{\underset{H_3C \qquad H}{}} \qquad CH_3{-}\overset{\longmapsto}{Cl}$$

Water   Methanol   Chloromethane

3.4   PHYSICAL PROPERTIES OF ALCOHOLS AND ALKYL HALIDES: INTERMOLECULAR FORCES

Boiling Point   We saw in Section 2.19 how the structure of an alkane affects its boiling point. Now let us compare three compounds of similar size and shape: the nonpolar alkane propane and two polar molecules, the alcohol ethanol and the alkyl halide fluoroethane.

$CH_3CH_2CH_3$ CH_3CH_2OH CH_3CH_2F

Propane   Ethanol   Fluoroethane
bp: $-42°C$ bp: $78°C$ bp: $-32°C$

FIGURE 3.2 A dipole–dipole attractive force. Two molecules of a polar substance are oriented so that the positively polarized region of one and the negatively polarized region of the other attract each other.

The only intermolecular attractive forces present in a nonpolar substance such as an alkane are of the **induced-dipole/induced-dipole** type. These are the weakest intermolecular attractive forces, and this is reflected in propane's low boiling point. In addition to induced-dipole/induced-dipole attractions, polar molecules engage in **dipole–dipole** attractions that are stronger, and this can be seen in the higher boiling points of ethanol and fluoroethane. The positively polarized region of one molecule is attracted to the negatively polarized region of another, as illustrated in Figure 3.2.

The most striking aspect of the data, however, is the much higher boiling point of ethanol compared with both propane and fluoroethane. This suggests that the attractive forces in ethanol must be unusually strong. Figure 3.3 shows that this force results from

FIGURE 3.3 Hydrogen bonding in ethanol involves the oxygen of one molecule and the proton of an —OH group of another. Hydrogen bonding is much stronger than most other types of dipole–dipole attractive forces.

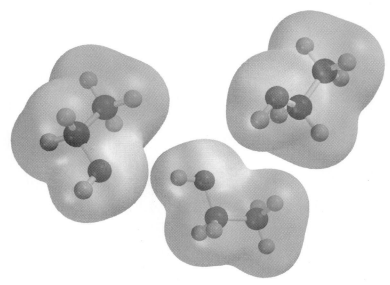

a dipole–dipole attraction between the positively polarized proton of the —OH group of one ethanol molecule and the negatively polarized oxygen of another. The term **hydrogen bonding** is used to describe dipole–dipole attractive forces of this type. The proton involved must be bonded to an electronegative element, usually oxygen or nitrogen. Protons in C—H bonds do not participate in hydrogen bonding. Thus fluoroethane, even though it is a polar molecule and engages in dipole–dipole attractions, does not form hydrogen bonds and, therefore, has a lower boiling point than ethanol.

Hydrogen bonding can be expected in molecules that have —OH or —NH groups. Individual hydrogen bonds are about 10–50 times weaker than typical covalent bonds, but their effects can be significant. More than other dipole–dipole attractive forces, intermolecular hydrogen bonds are strong enough to impose a relatively high degree of structural order on systems in which they are possible. As we will see in Chapters 17 and 18, the three-dimensional structures adopted by proteins and nucleic acids, the organic molecules of life, are dictated by patterns of hydrogen bonds.

> Hydrogen bonds between —OH groups are stronger than those between —NH groups, as a comparison of the boiling points of water (H_2O, 100°C) and ammonia (NH_3, −33°C) demonstrates.

> **PROBLEM 3.5** The constitutional isomer of ethanol, dimethyl ether (CH_3OCH_3), is a gas at room temperature. Suggest an explanation for this observation.

Table 3.1 lists the boiling points of some representative alkyl halides and alcohols. When comparing the boiling points of related compounds as a function of the *alkyl group*, we find that the boiling point increases with the number of carbon atoms, as it does with alkanes.

With respect to the *halogen* in a group of alkyl halides, the boiling point increases as one descends the periodic table; alkyl fluorides have the lowest boiling points, alkyl iodides the highest. This trend matches the order of increasing ease with which the electron distribution around an atom is distorted by a nearby electric field and is a significant factor in determining the strength of induced-dipole/induced-dipole and dipole/induced-dipole attractions. Forces that depend on induced dipoles are strongest when the halogen is iodine, and weakest when the halogen is fluorine.

The boiling points of the chlorinated derivatives of methane increase with the number of chlorine atoms because of an increase in the induced-dipole/induced-dipole attractive forces.

CH_3Cl	CH_2Cl_2	$CHCl_3$	CCl_4
Chloromethane (methyl chloride)	Dichloromethane (methylene dichloride)	Trichloromethane (chloroform)	Tetrachloromethane (carbon tetrachloride)

Boiling point: −24°C 40°C 61°C 77°C

| TABLE 3.1 | Boiling Points of Some Alkyl Halides and Alcohols |

Name of Alkyl Group	Formula	Functional Group X and Boiling Point, °C (1 atm)				
		X = F	X = Cl	X = Br	X = I	X = OH
Methyl	CH_3X	−78	−24	3	42	65
Ethyl	CH_3CH_2X	−32	12	38	72	78
Propyl	$CH_3CH_2CH_2X$	−3	47	71	103	97
Pentyl	$CH_3(CH_2)_3CH_2X$	65	108	129	157	138
Hexyl	$CH_3(CH_2)_4CH_2X$	92	134	155	180	157

Fluorine is unique among the halogens in that increasing the number of fluorines does not produce higher and higher boiling points.

	CH_3CH_2F	CH_3CHF_2	CH_3CF_3	CF_3CF_3
	Fluoroethane	1,1-Difluoroethane	1,1,1-Trifluoroethane	Hexafluoroethane
Boiling point:	$-32°C$	$-25°C$	$-47°C$	$-78°C$

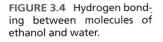

These boiling points illustrate why we should do away with the notion that boiling points always increase with increasing molecular weight.

Thus, although the difluoride CH_3CHF_2 boils at a higher temperature than CH_3CH_2F, the trifluoride CH_3CF_3 boils at a lower temperature than either of them. Even more striking is the observation that the hexafluoride CF_3CF_3 is the lowest boiling of any of the fluorinated derivatives of ethane. The boiling point of CF_3CF_3 is, in fact, only 11° higher than that of ethane itself. The reason for this behavior has to do with a decrease in induced-dipole/induced-dipole forces that accompanies the incorporation of fluorine substituents into a molecule. Their weak intermolecular attractive forces give fluorinated hydrocarbons (**fluorocarbons**) certain desirable physical properties such as that found in the "no stick" Teflon coating of frying pans. Teflon is a polymer (Section 9.8) made up of long chains of $—CF_2CF_2—$ units.

Solubility in Water Alkyl halides and alcohols differ markedly from one another in their solubility in water. All alkyl halides are insoluble in water, but low-molecular-weight alcohols (methyl, ethyl, *n*-propyl, and isopropyl) are soluble in water in all proportions. Their ability to participate in intermolecular hydrogen bonding not only affects the boiling points of alcohols, but also enhances their water solubility. Hydrogen-bonded networks of the type shown in Figure 3.4, in which alcohol and water molecules associate with one another, replace the alcohol–alcohol and water–water hydrogen-bonded networks present in the pure substances.

Higher alcohols become more hydrocarbon-like and less water-soluble. 1-Octanol, for example, dissolves to the extent of only 1 mL in 2000 mL of water. As the alkyl

FIGURE 3.4 Hydrogen bonding between molecules of ethanol and water.

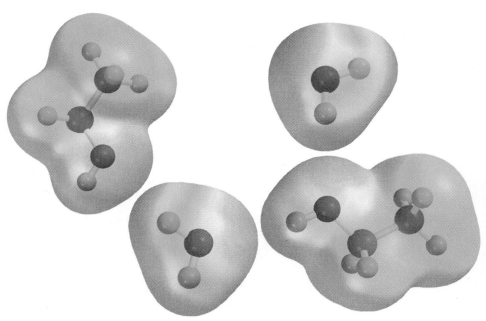

chain gets longer, the hydrophobic effect (see Section 2.19) becomes more important, to the point that it, more than hydrogen bonding, governs the solubility of alcohols.

Density Alkyl fluorides and chlorides are less dense, and alkyl bromides and iodides more dense, than water.

$$CH_3(CH_2)_6CH_2F \quad CH_3(CH_2)_6CH_2Cl \quad CH_3(CH_2)_6CH_2Br \quad CH_3(CH_2)_6CH_2I$$

Density (20°C):	0.80 g/mL	0.89 g/mL	1.12 g/mL	1.34 g/mL

Because alkyl halides are insoluble in water, a mixture of an alkyl halide and water separates into two layers. When the alkyl halide is a fluoride or chloride, it is the upper layer and water is the lower. The situation is reversed when the alkyl halide is a bromide or an iodide. In these cases the alkyl halide is the lower layer. Polyhalogenation increases the density. The compounds CH_2Cl_2, $CHCl_3$, and CCl_4, for example, are all more dense than water.

All liquid alcohols have densities of approximately 0.8 g/mL and are, therefore, less dense than water.

3.5 ACIDS AND BASES: GENERAL PRINCIPLES

A solid understanding of acid–base chemistry is a big help in understanding chemical reactivity. This section reviews some principles and properties of acids and bases.

According to the theory proposed by Svante Arrhenius, a Swedish chemist and winner of the 1903 Nobel Prize in chemistry, an acid ionizes in aqueous solution to liberate protons (H^+, hydrogen ions), whereas bases ionize to liberate hydroxide ions (HO^-). A more general theory of acids and bases was devised independently by Johannes Brønsted (Denmark) and Thomas M. Lowry (England) in 1923. In the Brønsted–Lowry approach, an acid is a **proton donor,** and a base is a **proton acceptor.**

$$B\overset{\frown}{:} + H\overset{\frown}{-}A \rightleftharpoons \overset{+}{B}-H + :A^-$$

Base	Acid	Conjugate acid	Conjugate base

The Brønsted–Lowry definitions of acids and bases are widely used in organic chemistry. As noted in the preceding equation, the **conjugate acid** of a substance is formed when it accepts a proton from a suitable donor. Conversely, the proton donor is converted to its **conjugate base.** A conjugate acid–base pair always differ by a single proton.

> **Curved arrow notation** is used to show the electron pair of the base abstracting a proton from the acid. The pair of electrons in the H—A bond becomes an unshared pair in the anion $^-$:A. Curved arrows track **electron movement,** not atomic movement.

> **PROBLEM 3.6** Write an equation for the reaction of ammonia ($:NH_3$) with hydrogen chloride (HCl). Use curved arrows to track electron movement, and identify the acid, base, conjugate acid, and conjugate base.

In aqueous solution, an acid transfers a proton to water. Water acts as a Brønsted base.

$$\underset{H}{\overset{H}{\diagdown}}\overset{\frown}{O:} + H\overset{\frown}{-}A \rightleftharpoons \underset{H}{\overset{H}{\diagdown}}\overset{+}{O}-H + :A^-$$

Water (base)	Acid	Conjugate acid of water	Conjugate base

The systematic name for the conjugate acid of water (H_3O^+) is **oxonium ion.** Its common name is **hydronium ion.**

The strength of an acid is measured by its **acid dissociation constant** or **ionization constant** K_a.

$$K_a = \frac{[H_3O^+][A^-]}{[HA]}$$

Table 3.2 lists a number of Brønsted acids and their acid dissociation constants. Strong acids are characterized by K_a values that are greater than that for hydronium ion (H_3O^+, $K_a = 55$). Essentially every molecule of a strong acid transfers a proton to water in dilute aqueous solution. Weak acids have K_a values less than that of H_3O^+; they are incompletely ionized in dilute aqueous solution.

A convenient way to express acid strength is through the use of pK_a, defined as follows:

$$pK_a = -\log_{10}K_a$$

Thus, water, with $K_a = 1.8 \times 10^{-16}$, has a pK_a of 15.7; ammonia, with $K_a \approx 10^{-36}$, has a pK_a of 36. The stronger the acid, the larger the value of its K_a and the smaller the value of pK_a. Water is a very weak acid, but is a far stronger acid than ammonia. Table 3.2 includes pK_a as well as K_a values for acids. Because both systems are widely used, you should practice converting K_a to pK_a and vice versa.

TABLE 3.2	Acid Dissociation Constants K_a and pK_a Values for Some Brønsted Acids*			
Acid	**Formula†**	**Dissociation constant, K_a**	**pK_a**	**Conjugate base**
Hydrogen iodide	HI	$\approx 10^{10}$	≈ -10	I^-
Hydrogen bromide	HBr	$\approx 10^9$	≈ -9	Br^-
Hydrogen chloride	HCl	$\approx 10^7$	≈ -7	Cl^-
Sulfuric acid	HOSO$_2$OH	1.6×10^5	-4.8	$HOSO_2O^-$
Hydronium ion	H—$\overset{+}{O}H_2$	55	-1.7	H_2O
Hydrogen fluoride	HF	3.5×10^{-4}	3.5	F^-
Acetic acid	CH$_3\overset{\overset{\displaystyle O}{\|\|}}{C}$OH	1.8×10^{-5}	4.7	$CH_3\overset{\overset{\displaystyle O}{\|\|}}{C}O^-$
Ammonium ion	H—$\overset{+}{N}H_3$	5.6×10^{-10}	9.2	NH_3
Water	HOH	$1.8 \times 10^{-16‡}$	15.7	HO^-
Methanol	CH$_3$OH	$\approx 10^{-16}$	≈ 16	CH_3O^-
Ethanol	CH$_3$CH$_2$OH	$\approx 10^{-16}$	≈ 16	$CH_3CH_2O^-$
Isopropyl alcohol	(CH$_3$)$_2$CHOH	$\approx 10^{-17}$	≈ 17	$(CH_3)_2CHO^-$
tert-Butyl alcohol	(CH$_3$)$_3$COH	$\approx 10^{-18}$	≈ 18	$(CH_3)_3CO^-$
Ammonia	H$_2$NH	$\approx 10^{-36}$	≈ 36	H_2N^-

* Acid strength decreases from top to bottom of the table. Strength of conjugate base increases from top to bottom of the table.

† The most acidic proton—the one that is lost on ionization—is highlighted.

‡ The "true" K_a for water is 1×10^{-14}. Dividing this value by 55.5 (the number of moles of water in 1 L of water) gives a K_a of 1.8×10^{-16} and puts water on the same concentration basis as the other substances in the table.

PROBLEM 3.7 Formic acid (present in the sting of ants) has a pK_a of 3.75. Oxalic acid (a poisonous substance found in rhubarb leaves) has a pK_a of 1.19. What is the value of K_a for each of these acids? Which acid is stronger?

An important part of the Brønsted–Lowry picture of acids and bases concerns the relative strengths of an acid and its conjugate base.

The stronger the acid, the weaker the conjugate base, and vice versa.

Ammonia (NH_3) is the weakest acid in Table 3.2. Its conjugate base, amide ion (H_2N^-), is therefore the strongest base. Hydroxide (HO^-) is a strong base, much stronger than the halide ions F^-, Cl^-, Br^-, and I^-, which are very weak bases. Fluoride is the strongest base of the halides but is 10^{12} times less basic than hydroxide ion.

PROBLEM 3.8 Comparing the conjugate bases of the acids in Problem 3.7, which is the stronger base, formate (the conjugate base of formic acid) or oxalate (the conjugate base of oxalic acid)?

In any proton-transfer process the position of equilibrium favors formation of the weaker acid and the weaker base.

$$\text{Stronger acid + stronger base} \underset{}{\overset{K > 1}{\rightleftharpoons}} \text{weaker acid + weaker base}$$

This is one of the most important equations in chemistry.

Table 3.2 is set up so that the strongest acid is at the top of the acid column, with the strongest base at the bottom of the conjugate base column. An acid will transfer a proton to the conjugate base of any acid that lies below it in the table, and the equilibrium constant for the reaction will be greater than one.

Table 3.2 contains both inorganic and organic compounds. Organic compounds are similar to inorganic ones when the functional groups responsible for their acid–base properties are the same. Thus, alcohols (ROH) are similar to water (HOH) in both their Brønsted acidity (ability to *donate* a proton from oxygen) and Brønsted basicity (ability to *accept* a proton on oxygen). Just as proton transfer to a water molecule gives oxonium ion (hydronium ion, H_3O^+), proton transfer to an alcohol gives an **alkyloxonium ion** (ROH_2^+).

| Alcohol | Acid | Alkyloxonium ion | Conjugate base |

We shall see that several important reactions of alcohols involve strong acids either as reagents or as catalysts to increase the rate of reaction. In all these reactions the first step is formation of an alkyloxonium ion by proton transfer from the acid to the oxygen of the alcohol.

PROBLEM 3.9 Write an equation for proton transfer from hydrogen chloride to *tert*-butyl alcohol. Use curved arrows to track electron movement, and identify the acid, base, conjugate acid, and conjugate base.

3.6   ACID–BASE REACTIONS: A MECHANISM FOR PROTON TRANSFER

Diagrams that show the change in potential energy as reactants proceed to products in a chemical reaction can help us understand more about reaction mechanisms. Molecules must become energized to undergo a chemical reaction. Kinetic (thermal) energy is absorbed by a molecule from collisions with other molecules and is transformed into potential energy. Every chemical reaction has a certain minimum increase in potential energy necessary for the process to occur, the **activation energy** (E_{act}). The point of maximum potential energy encountered by the reactants as they proceed to products is called the **transition state.**

> The structure that exists at the transition state is sometimes referred to as the **activated complex.**

Consider the transfer of a proton from hydrogen bromide to water:

$$:\ddot{Br}\!-\!H \; + \; :\ddot{O}\!\!\begin{array}{c}H\\ \,\\ H\end{array} \;\; \Longleftrightarrow \;\; :\ddot{Br}:^{-} \; + \; H\!-\!\overset{+}{\underset{H}{O}}\!\!\begin{array}{c}H\\ \,\\ \end{array}$$

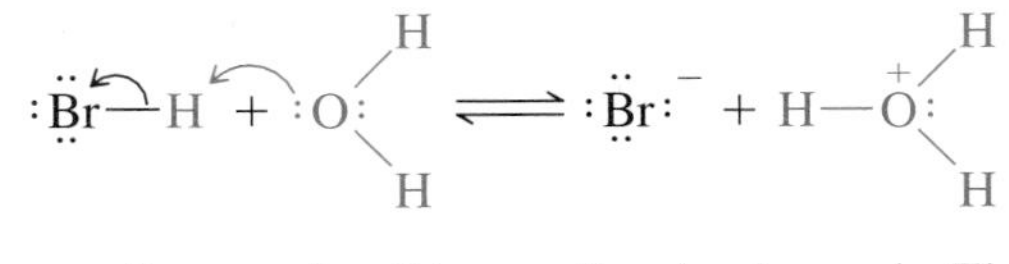

A potential energy diagram for this reaction is shown in Figure 3.5. Because the transfer of a proton from hydrogen bromide to water is exothermic, the products are placed lower in energy than the reactants. The diagram depicts the reaction as occurring in a single elementary step. An **elementary step** is one that involves only one transition state. A reaction can proceed by way of a single elementary step, in which case it is described as a **concerted** reaction, or by a series of elementary steps. In the case of proton transfer from hydrogen bromide to water, breaking of the H—Br bond and making of the $H_2\overset{+}{O}$—H bond occur "in concert" with each other. The species present at the transition state is not a stable structure and cannot be isolated or examined directly. Its structure is assumed to be one in which the proton being transferred is partially bonded to both bromine and oxygen simultaneously, although not necessarily to the same extent.

> Dashed lines in transition state structures represent **partial** bonds, that is, bonds in the process of being made or broken.

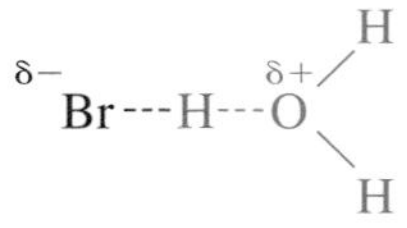

$$\overset{\delta-}{Br}\text{---}H\text{---}\overset{\delta+}{O}\!\!\begin{array}{c}H\\ \,\\ H\end{array}$$

The **molecularity** of an elementary step is given by the number of species that undergo a chemical change in that step. The elementary step

$$HBr + H_2O \Longleftrightarrow Br^{-} + H_3O^{+}$$

is **bimolecular** because it involves one molecule of hydrogen bromide and one molecule of water.

> **PROBLEM 3.10**   Represent the structure of the transition state for proton transfer from hydrogen chloride to *tert*-butyl alcohol.

Proton transfer from hydrogen bromide to water and alcohols ranks among the most rapid chemical processes and occurs almost as fast as the molecules collide with one another. Thus the height of the energy barrier separating reactants and products, the activation energy for proton transfer, must be quite low.

The concerted nature of proton transfer contributes to its rapid rate. The energy cost of breaking the H—Br bond is partially offset by the energy released in making the $H_2\overset{+}{O}$—H bond. Thus, the activation energy is far less than it would be for a hypothetical stepwise process involving an initial, unassisted ionization of the H—Br bond, followed by a combination of the resulting H^{+} with water.

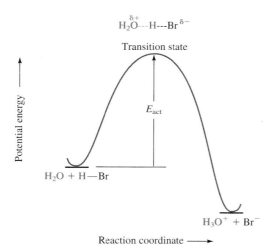

3.7 PREPARATION OF ALKYL HALIDES FROM ALCOHOLS AND HYDROGEN HALIDES

Much of what organic chemists do is directed toward practical goals. Chemists in the pharmaceutical industry synthesize new compounds as potential drugs for the treatment of disease. Agricultural chemicals designed to increase crop yields include synthetic organic compounds used for weed control, insecticides, and fungicides. Among the "building block" molecules used as starting materials to prepare new substances, alcohols and alkyl halides are especially valuable. In this and following sections you will see how alcohols can be used as starting materials for the preparation of alkyl halides. This method of alkyl halide preparation also serves as a focal point for developing some principles of reaction mechanisms. An additional method for the preparation alkyl halides, using alkanes as starting materials, is described in Chapter 9.

Alcohols react with hydrogen halides according to the general equation:

$$\underset{\text{Alcohol}}{\text{R}-\text{OH}} + \underset{\text{Hydrogen halide}}{\text{H}-\text{X}} \longrightarrow \underset{\text{Alkyl halide}}{\text{R}-\text{X}} + \underset{\text{Water}}{\text{H}-\text{OH}}$$

The order of reactivity of the hydrogen halides parallels their acidity: HI > HBr > HCl >> HF. Hydrogen iodide is used infrequently, however, and the reaction of alcohols with hydrogen fluoride is not a useful method for the preparation of alkyl fluorides.

Among the various classes of alcohols, tertiary alcohols are observed to be the most reactive and primary alcohols the least reactive.

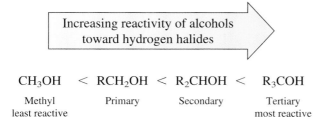

Increasing reactivity of alcohols
toward hydrogen halides

$$\underset{\substack{\text{Methyl}\\\text{least reactive}}}{\text{CH}_3\text{OH}} < \underset{\text{Primary}}{\text{RCH}_2\text{OH}} < \underset{\text{Secondary}}{\text{R}_2\text{CHOH}} < \underset{\substack{\text{Tertiary}\\\text{most reactive}}}{\text{R}_3\text{COH}}$$

Tertiary alcohols are converted to alkyl chlorides in high yield within minutes on reaction with hydrogen chloride at room temperature and below.

$$(CH_3)_3COH \quad + \quad HCl \quad \xrightarrow{25°C} \quad (CH_3)_3CCl \quad + \quad H_2O$$

2-Methyl-2-propanol   Hydrogen chloride   2-Chloro-2-methylpropane   Water
(*tert*-butyl alcohol)   (*tert*-butyl chloride) (78–88%)

The less reactive secondary and primary alcohols require heating with the more reactive hydrogen halide hydrogen bromide (HBr) for the corresponding alkyl halides to be formed successfully.

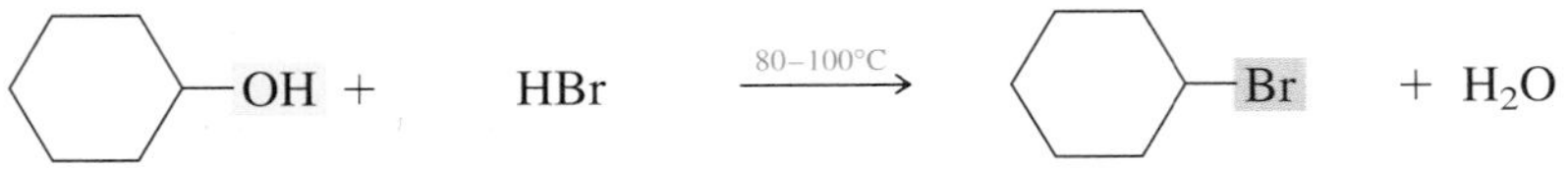

Cyclohexanol   Hydrogen bromide   Bromocyclohexane   Water
                                                    (73%)

$$CH_3(CH_2)_5CH_2OH \quad + \quad HBr \quad \xrightarrow{120°C} \quad CH_3(CH_2)_5CH_2Br \quad + \quad H_2O$$

1-Heptanol   Hydrogen   1-Bromoheptane   Water
                          bromide   (87–90%)

The same kind of transformation may be carried out by heating an alcohol with sodium bromide and sulfuric acid.

$$CH_3CH_2CH_2CH_2OH \quad \xrightarrow[\text{heat}]{NaBr, H_2SO_4} \quad CH_3CH_2CH_2CH_2Br$$

1-Butanol   1-Bromobutane (70–83%)
(*n*-butyl alcohol)   (*n*-butyl bromide)

We'll often find it convenient to write chemical equations in the abbreviated form shown here, in which reagents, especially inorganic ones, are not included in the body of the equation but instead are indicated over the arrow. Inorganic products—in this case, water—are usually omitted. These simplifications focus our attention on the organic reactant and its functional group transformation.

PROBLEM 3.11 Write chemical equations for the reaction that takes place between each of the following pairs of reactants:

(a) 2-Butanol and hydrogen bromide

(b) 3-Ethyl-3-pentanol and hydrogen chloride

(c) 1-Tetradecanol and hydrogen bromide

Sample Solution (a) An alcohol and a hydrogen halide react to form an alkyl halide and water. In this case 2-bromobutane was isolated in 73% yield.

2-Butanol   Hydrogen bromide   2-Bromobutane   Water

3.8   MECHANISM OF THE REACTION OF ALCOHOLS WITH HYDROGEN HALIDES

The reaction of an alcohol with a hydrogen halide is a **substitution.** A halogen, usually chlorine or bromine, replaces a hydroxyl group as a substituent on carbon. Calling the reaction a substitution tells us the relationship between the organic reactant and its product but does not reveal the mechanism. In developing a mechanistic picture for a particular reaction, we combine some basic principles of chemical reactivity with experimental observations to deduce the most likely sequence of elementary steps.

Consider the reaction of *tert*-butyl alcohol with hydrogen chloride:

$$(CH_3)_3COH \ + \ HCl \ \longrightarrow \ (CH_3)_3CCl \ + \ H_2O$$

| *tert*-Butyl alcohol | Hydrogen chloride | *tert*-Butyl chloride | Water |

The generally accepted mechanism for this reaction is presented as a series of three elementary steps in Figure 3.6. We say "generally accepted" because a reaction mechanism can never be proved to be correct. A mechanism is our best present assessment of how a reaction proceeds and must account for all experimental observations. If new experimental data appear that conflict with the mechanism, the mechanism must be modified to accommodate them. If the new data are consistent with the proposed mechanism, our confidence grows that it is likely to be correct.

We already know about step 1 of the mechanism outlined in Figure 3.6; it is an example of a Brønsted acid–base reaction of the type discussed in Section 3.5 and formed the basis of Problem 3.9.

Steps 2 and 3, however, are new to us. Step 2 involves dissociation of an alkyloxonium ion to a molecule of water and a **carbocation,** a species that contains a positively charged carbon. In step 3, this carbocation reacts with chloride ion to yield *tert*-butyl chloride. Both the alkyloxonium ion and the carbocation are **intermediates** in the reaction. They are not isolated, but are formed in one step and consumed in another

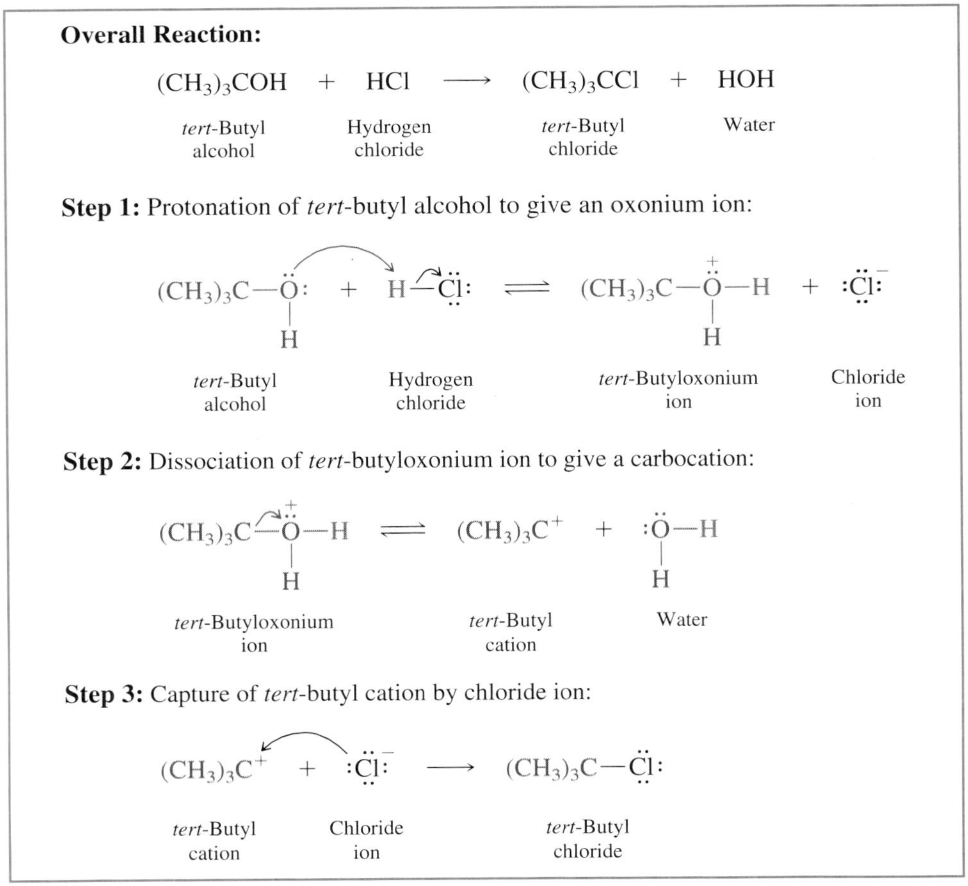

> If you have not already written out the solutions to Problem 3.9, you should do so now.

Overall Reaction:

$$(CH_3)_3COH \ + \ HCl \ \longrightarrow \ (CH_3)_3CCl \ + \ HOH$$

| *tert*-Butyl alcohol | Hydrogen chloride | *tert*-Butyl chloride | Water |

Step 1: Protonation of *tert*-butyl alcohol to give an oxonium ion:

$$(CH_3)_3C\overset{\cdot\cdot}{\underset{|}{O}}: \ + \ H\!-\!\overset{\cdot\cdot}{\underset{\cdot\cdot}{Cl}}: \ \rightleftharpoons \ (CH_3)_3C\overset{+}{\overset{\cdot\cdot}{\underset{|}{O}}}\!-\!H \ + \ :\overset{\cdot\cdot}{\underset{\cdot\cdot}{Cl}}:^-$$

| *tert*-Butyl alcohol | Hydrogen chloride | *tert*-Butyloxonium ion | Chloride ion |

Step 2: Dissociation of *tert*-butyloxonium ion to give a carbocation:

$$(CH_3)_3C\overset{+}{\overset{\cdot\cdot}{\underset{|}{O}}}\!-\!H \ \rightleftharpoons \ (CH_3)_3C^+ \ + \ :\overset{\cdot\cdot}{\underset{|}{O}}\!-\!H$$

| *tert*-Butyloxonium ion | *tert*-Butyl cation | Water |

Step 3: Capture of *tert*-butyl cation by chloride ion:

$$(CH_3)_3C^+ \ + \ :\overset{\cdot\cdot}{\underset{\cdot\cdot}{Cl}}:^- \ \longrightarrow \ (CH_3)_3C\!-\!\overset{\cdot\cdot}{\underset{\cdot\cdot}{Cl}}:$$

| *tert*-Butyl cation | Chloride ion | *tert*-Butyl chloride |

FIGURE 3.6 The mechanism of formation of *tert*-butyl chloride from *tert*-butyl alcohol and hydrogen chloride.

during the passage of reactants to products. If we add the equations for steps 1 through 3 together, the equation for the overall process results. A valid reaction mechanism must account for the consumption of all reactants and the formation of all products, be they organic or inorganic. So that we may better understand the chemistry expressed in steps 2 and 3, we need to examine carbocations in more detail.

3.9   STRUCTURE, BONDING, AND STABILITY OF CARBOCATIONS

A carbocation is a species that has a positively charged carbon. The structure of carbocations can be described by considering the simplest example, methyl cation (CH_3^+). The positively charged carbon contributes 3 valence electrons, and each hydrogen contributes 1 for a total of 6 electrons, which are used to form three C—H σ bonds. As we saw in Section 1.14, carbon is sp^2-hybridized when it is bonded to three atoms or groups. We therefore choose the sp^2-hybridization model for bonding shown in Figure 3.7. Carbon forms σ bonds to three hydrogens by overlap of its sp^2 orbitals with hydrogen $1s$ orbitals. The three σ bonds are coplanar. Remaining on carbon is an unhybridized $2p$ orbital that contains no electrons. The axis of this empty p orbital is perpendicular to the plane defined by the three σ bonds.

Carbocations are classified as primary, secondary, or tertiary according to the number of carbons that are directly attached to the positively charged carbon.

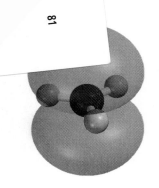

FIGURE 3.7 Structure of methyl cation CH_3^+. Carbon is sp^2-hybridized. Each hydrogen is attached to carbon by a σ bond formed by overlap of a hydrogen $1s$ orbital with an sp^2 hybrid orbital of carbon. All four atoms lie in the same plane. The unhybridized $2p$ orbital of carbon is unoccupied, and its axis is perpendicular to the plane of the atoms.

$$CH_3CH_2CH_2CH_2 \overset{\displaystyle H}{\underset{\displaystyle H}{-C^+}} \qquad CH_3CH_2CH_2 \overset{\displaystyle H}{\underset{\displaystyle CH_2CH_3}{-C^+}} \qquad CH_3 \!\!-\!\!\!\!+$$

primary carbocation   secondary carbocation   tertiary carbocation

> **PROBLEM 3.12**   Write structural formulas for all the isomeric carbocations of formula $C_4H_9^+$. Which ones are primary carbocations? Which are secondary? Which are tertiary?

Numerous studies have shown that the more stable a carbocation is, the faster it is formed. These studies also demonstrate that

Alkyl groups directly attached to the positively charged carbon stabilize a carbocation.

Thus, the observed order of carbocation stability is

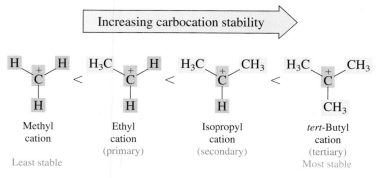

Increasing carbocation stability

Methyl cation	Ethyl cation (primary)	Isopropyl cation (secondary)	tert-Butyl cation (tertiary)
Least stable			Most stable

Carbocations are stabilized by substituents that release or donate electron density to the positively charged carbon. Alkyl groups release electron density better than hydrogen, so the more alkyl groups attached to the positively charged carbon, the more stable the carbocation. The electron-donating or electron-withdrawing effect of a group that is transmitted through σ bonds is called an **inductive effect.**

PROBLEM 3.13 Of the isomeric $C_5H_{11}{}^+$ carbocations, which one is the most stable?

When we say that tertiary carbocations are more stable than their secondary and primary counterparts, we are speaking in a relative sense. Even tertiary carbocations are highly reactive species, normally incapable of being isolated. The effect of carbocation stability is observed in the relative rates at which various alcohols react with hydrogen halides. Tertiary alcohols react faster than secondary alcohols because tertiary carbocations, being more stable than secondary carbocations, are formed faster and the rate of reaction is governed by the rate of carbocation formation.

The more stable an intermediate, the faster it is formed.

PROBLEM 3.14 Of the isomeric $C_5H_{12}O$ alcohols, which one reacts at the fastest rate with hydrogen chloride?

3.10 ELECTROPHILES AND NUCLEOPHILES

The positive charge on carbon and the vacant p orbital combine to make carbocations strongly **electrophilic** ("electron-loving," or "electron-seeking"). **Nucleophiles** are just the opposite. A nucleophile is "nucleus-seeking;" it has an unshared pair of electrons that it can use to form a covalent bond. Step 3 of the mechanism of the reaction of *tert*-butyl alcohol with hydrogen chloride is an example of a reaction between an electrophile and a nucleophile and is depicted from a structural perspective in Figure 3.8. The crucial electronic interaction is between an unshared electron pair of the nucleophilic chloride anion and the vacant $2p$ orbital of the electrophilic carbocation.

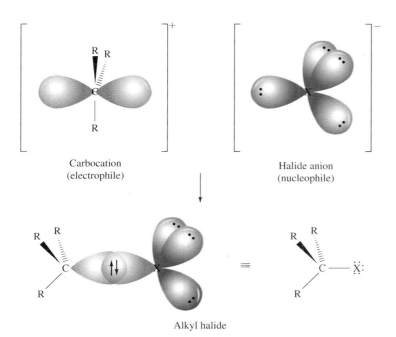

Carbocation
(electrophile)

Halide anion
(nucleophile)

Alkyl halide

FIGURE 3.8 Combination of a carbocation and a halide anion to give an alkyl halide.

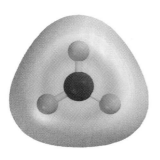

FIGURE 3.9 Electrostatic potential map of methyl cation (CH_3^+). The regions of lowest electron density are blue, are centered on carbon, and are located above and below the plane defined by the four atoms.

Figure 3.9 maps the electrostatic potential in methyl cation and shows that the region of positive charge coincides with where we expect the vacant $2p$ orbital to be—centered on carbon and above and below the plane of the atoms.

A number of years ago G. N. Lewis extended our understanding of acid–base behavior to include reactions other than proton transfers. According to Lewis:

An acid is an electron-pair acceptor and *a base is an electron-pair donor.*

Thus, carbocations are electron-pair acceptors and are **Lewis acids.** Halide anions are electron-pair donors and are **Lewis bases.** It is generally true that electrophiles are Lewis acids, and nucleophiles are Lewis bases.

3.11   REACTION OF PRIMARY ALCOHOLS WITH HYDROGEN HALIDES

Unlike tertiary and secondary carbocations, primary carbocations are too high in energy to be intermediates in chemical reactions. Because primary alcohols are converted, albeit rather slowly, to alkyl halides on treatment with hydrogen halides, they must follow some other mechanism that avoids carbocation intermediates. This alternative mechanism is believed to be one in which the carbon–halogen bond begins to form before the carbon–oxygen bond of the alkyloxonium ion is completely broken.

$$\overset{..}{\underset{..}{:}}\overset{-}{X}\overset{..}{:}\ +\ RCH_2{-}\overset{+}{\underset{..}{O}}H_2\ \longrightarrow\ \overset{\delta-}{\overset{..}{\underset{..}{:}}X}{-}{-}{-}CH_2{-}{-}\overset{\delta+}{\underset{..}{O}}H_2\ \longrightarrow\ \overset{..}{\underset{..}{:}}X{-}CH_2R\ +\ \overset{..}{\underset{..}{:}}H_2O\overset{..}{:}$$

| Halide ion | Primary alkyloxonium ion | Transition state | Primary alkyl halide | Water |

The halide nucleophile helps to "push off" a water molecule from the alkyloxonium ion. According to this mechanism, both the halide ion and the alkyloxonium ion are involved in the same bimolecular elementary step.

PROBLEM 3.15   1-Butanol and 2-butanol are converted to their corresponding bromides on being heated with hydrogen bromide. Write a suitable mechanism for each reaction.

LEARNING OBJECTIVES

This chapter introduced concepts of organic chemical reactivity by describing two classes of compounds: alkyl halides and alcohols. The skills you have learned in this chapter should enable you to:

- Write systematic names for alcohols and alkyl halides on the basis of a given structural formula.
- Write structural formulas for alcohols and alkyl halides given their systematic names.
- Explain how hydrogen bonding affects the boiling point and water solubility of alcohols.
- Give the Brønsted–Lowry definitions for acids and bases.

Continued

- Describe the relationship between acidity, K_a, and pK_a.
- Write a balanced chemical equation for the reaction of an alcohol with a hydrogen halide.
- Write chemical equations describing the mechanism of the reaction of an alcohol with a hydrogen halide.
- Explain the meaning of the terms *alkyloxonium ion, carbocation, nucleophile,* and *electrophile.*
- Describe the structure of a carbocation using the orbital hybridization bonding model.
- Compare the relative stabilities of a group of carbocations.
- Describe how the rate of reaction of alcohols with hydrogen halides depends on the structure of the alcohol.
- Give the definition of a Lewis acid and a Lewis base.

3.12 SUMMARY

Alcohols and alkyl halides can be named systematically as functional derivatives of alkanes, but in slightly different ways (Section 3.1). Alkyl halides are named as halo-substituted alkanes. Alcohols are named by replacing the *-e* ending of an alkane with *-ol.* In both types of compounds the hydrocarbon chain is numbered in the direction that gives the lower number to the carbon that bears the functional group.

$$CH_3CH_2CH_2CH_2CHCH_3 \qquad CH_3CH_2CH_2CH_2CHCH_3$$
$$\qquad\qquad\qquad | \qquad\qquad\qquad\qquad\qquad\qquad |$$
$$\qquad\qquad\qquad OH \qquad\qquad\qquad\qquad\qquad\qquad Br$$

2-Hexanol   2-Bromohexane

Alcohols and alkyl halides may also be named on the basis of the names of their alkyl groups. The alkyl group is named followed by the word *alcohol, fluoride, chloride, bromide,* or *iodide,* as appropriate.

$$CH_3CHCH_3 \qquad\qquad CH_3CHCH_3$$
$$\qquad | \qquad\qquad\qquad\qquad |$$
$$\qquad OH \qquad\qquad\qquad\qquad Br$$

Isopropyl alcohol   Isopropyl bromide

Alcohols and alkyl halides are classified as primary, secondary, or tertiary according to the degree of substitution at the carbon that bears the functional group (Section 3.2).

The halogens (especially fluorine and chlorine) and oxygen are more electronegative than carbon, and the carbon–halogen bond in alkyl halides and the carbon–oxygen bond in alcohols are polar covalent bonds. Carbon is positively polarized and the halogen or oxygen is negatively polarized (Section 3.3).

Dipole–dipole attractive forces give alcohols and alkyl halides higher boiling points than alkanes of similar molecular weight (Section 3.4). The attractive force between —OH groups in alcohols is called **hydrogen bonding.** Hydrogen bonding between the hydroxyl group and water molecules makes the water solubility of alcohols greater than that of hydrocarbons. Low-molecular-weight alcohols (methanol, ethanol, 1-propanol, 2-propanol) are soluble in water in all proportions. Alkyl halides are insoluble in water.

Brønsted acids are proton donors; **Brønsted bases** are proton acceptors (Section 3.5). In an acid–base reaction, the **conjugate acid** of a substance (the base) is formed when it accepts a proton from a suitable donor. Conversely, the proton donor (the acid) is converted to its **conjugate base** when it loses its proton.

$$B: \overset{\curvearrowright}{+} H\overset{\curvearrowleft}{-}A \;\rightleftharpoons\; \overset{+}{B}-H \;+\; :A^-$$

| Base | Acid | Conjugate acid | Conjugate base |

The **acid dissociation** constant is given by

$$K_a = \frac{[H_3O^+][A^-]}{[HA]}$$

and may be conveniently expressed as

$$pK_a = -\log_{10}K_a$$

The stronger the acid, the larger the value of K_a and hence the smaller the value of pK_a. The weaker the acid, the smaller the value of K_a and hence the larger the value of pK_a. An important corollary of the Brønsted view of acids and bases is that the stronger the acid, the weaker the conjugate base; the weaker the acid, the stronger the conjugate base.

Proton transfer from a Brønsted acid to the oxygen of water is a fast single-step, **bimolecular, concerted** process (Section 3.6).

Alcohols may be converted to alkyl halides by reaction with hydrogen halides (Section 3.7).

General equation:

$$ROH \;+\; HX \;\longrightarrow\; RX \;+\; H_2O$$

| Alcohol | Hydrogen halide | Alkyl halide | Water |

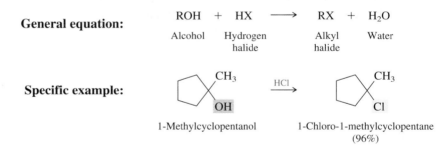

Specific example:

1-Methylcyclopentanol   1-Chloro-1-methylcyclopentane (96%)

Alcohol reactivity increases in the order methyl < primary < secondary < tertiary (Section 3.7). The reaction of an alcohol with a hydrogen halide proceeds by way of two positively charged intermediates, an **alkyloxonium ion** and a **carbocation.** The alkyloxonium ion is the **conjugate acid** of an alcohol. The carbocation is formed by dissociation of an alkyloxonium ion and is then captured by a halide ion to give an alkyl halide. Steps 1 to 3 describe the **mechanism** for the reaction of an alcohol with a hydrogen halide (Section 3.8).

$$(1) \quad ROH \;+\; HX \;\overset{fast}{\rightleftharpoons}\; R\overset{+}{O}H_2 \;+\; X^-$$

| Alcohol | Hydrogen halide | Alkyloxonium ion | Halide anion |

$$(2) \quad R\overset{+}{O}H_2 \;\overset{slow}{\longrightarrow}\; R^+ \;+\; H_2O$$

| Alkyloxonium ion | Carbocation | Water |

(3) R^+ + X^- $\xrightarrow{\text{fast}}$ RX

        Carbocation   Halide ion   Alkyl halide

Carbocations are characterized by the presence of a carbon atom that has only three atoms or groups bonded to it (Section 3.9). Such carbon atoms are sp^2-hybridized, and the three σ orbitals are coplanar. Carbocations are positively charged and have a vacant $2p$ orbital whose axis is perpendicular to the plane of the three σ orbitals.

Carbocations are stabilized by alkyl substituents attached directly to the positively charged carbon. Alkyl groups are **electron-releasing** substituents. Stability increases in the order:

        Least stable CH_3^+ < R_2CH^+ < R_2C^+ Most stable

Carbocations are strongly **electrophilic** (Lewis acids) and react with **nucleophiles** (Lewis bases) (Section 3.10).

    Primary alcohols do not react with hydrogen halides by way of carbocation intermediates. The nucleophilic species (Br^-) attacks the alkyloxonium ion and "pushes off" a water molecule from carbon in a bimolecular step (Section 3.11).

ADDITIONAL PROBLEMS

Structure and Nomenclature

3.16 Write structural formulas for each of the following alcohols and alkyl halides:
  (a) 1-Bromo-3-iodobutane
  (b) *sec*-Butyl alcohol
  (c) 3-Heptanol
  (d) *trans*-2-Chlorocyclopentanol
  (e) 2,6-Dichloro-4-methyl-4-octanol
  (f) *trans*-4-*tert*-Butylcyclohexanol
  (g) 1-Cyclopropylethanol
  (h) 2-Cyclopropylethanol

3.17 Give each of the following compounds a systematic name.
  (a) $(CH_3)_2CHCH_2CH_2CH_2Br$

  (b) $(CH_3)_2CHCH_2CH_2CH_2OH$

  (c) CF_3CH_2OH

  (d)

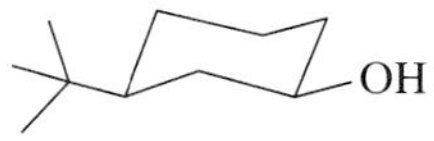

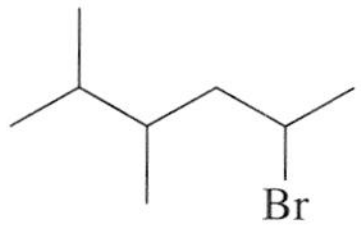

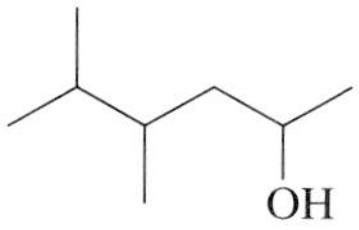

3.18 Write structural formulas or build molecular models for all the constitutionally isomeric alcohols of molecular formula $C_5H_{12}O$. Assign two systematic names to each one, and specify whether it is a primary, secondary, or tertiary alcohol.

3.19 A hydroxyl group is a somewhat "smaller" substituent on a six-membered ring than is a methyl group. That is, the preference of a hydroxyl group for the equatorial orientation is less pronounced than that of a methyl group. Given this information, write structural formulas or build molecular models for all the isomeric methylcyclohexanols, showing each one in its most stable conformation. Give an acceptable IUPAC name for each isomer.

3.20 (a) Menthol, used to flavor various foods and tobacco, is the most stable stereoisomer of 2-isopropyl-5-methylcyclohexanol. Draw or make a molecular model of its most stable conformation. Is the hydroxyl group cis or trans to the isopropyl group? To the methyl group?

(b) Neomenthol is a stereoisomer of menthol. That is, it has the same constitution but differs in the arrangement of its atoms in space. Neomenthol is the second most stable stereoisomer of 2-isopropyl-5-methylcyclohexanol; it is less stable than menthol but more stable than any other stereoisomer. Write the structure, or make a molecular model of neomenthol in its most stable conformation.

3.21 Provide a brief explanation of the difference in boiling point between $CH_3CH_2OCH_2CH_3$ (35°C) and $CH_3CH_2CH_2CH_2OH$ (117°C).

Acid–Base

3.22 Write the structure of the conjugate acid and the conjugate base of:

    (a) Water   (b) Methanol   (c) Ammonia

3.23 For each of the following pairs of ions, identify which is the stronger base:

    (a) HO^- or NH_2^-

    (b) F^- or Cl^-

    (c) $CH_3CH_2O^-$ or $CH_3\overset{\displaystyle O}{\overset{\|}{C}}O^-$

3.24 Each of the following pairs of compounds undergoes a Brønsted acid–base reaction whose equilibrium constant is greater than one. Give the products of each reaction and identify the acid, the base, the conjugate acid, and the conjugate base. Use the acid dissociation constants in Table 3.1 as a guide.

    (a) $HI + H_2O \rightleftharpoons$

    (b) $CH_3CH_2O^- + CH_3\overset{\displaystyle O}{\overset{\|}{C}}OH \rightleftharpoons$

    (c) $HF + H_2N^- \rightleftharpoons$

    (d) $CH_3\overset{\displaystyle O}{\overset{\|}{C}}O^- + HCl \rightleftharpoons$

3.25 For each of the following acid–base reactions, indicate whether the reaction favors products ($K > 1$) or reactants ($K < 1$). Refer to Table 3.1 for the relevant acid dissociation constants.

    (a) $(CH_3)_3CO^- + H_2O \rightleftharpoons$

    (b) $NH_3 + (CH_3)_2CHO^- \rightleftharpoons$

    (c) $HF + HO^- \rightleftharpoons$

3.26 (a) The acid dissociation constant of benzoic acid (a carboxylic acid) is 6.3×10^{-5}. What is the pK_a?

(b) The pK_a of chloroacetic acid is 2.9. Which acid is the stronger, chloroacetic or benzoic?

(c) The conjugate base of chloroacetic acid is the chloroacetate ion; that of benzoic acid is the benzoate ion. Which conjugate base is the stronger, chloroacetate or benzoate?

3.27 The pK_as of methanol (CH_3OH) and methanethiol (CH_3SH) are 16 and 11, respectively.

(a) Label the stronger and weaker acids and bases in the following equation:

$$CH_3OH + CH_3S^- \rightleftharpoons CH_3O^- + CH_3SH$$

(b) Is the value of the equilibrium constant K greater or less than 1?

3.28 Hydrogen cyanide (HCN) has a pK_a of 9.1. What is its K_a? Is cyanide ion (CN^-) a stronger or a weaker base than hydroxide ion (HO^-)?

Alcohol Reactions: Carbocations

3.29 (a) Write a chemical equation for the reaction of 2-butanol with hydrogen bromide.

(b) Write a stepwise mechanism for this reaction.

3.30 Repeat Problem 3.29 for the reaction of 1-methylcyclohexanol with hydrogen chloride.

3.31 Select the compound in each of the following pairs that will be converted to the corresponding alkyl bromide more rapidly on being treated with hydrogen bromide.

(a) 1-Butanol or 2-butanol
(b) 2-Methyl-2-butanol or 2-butanol
(c) 2-Methylbutane or 2-butanol
(d) 1-Methylcyclopentanol or cyclohexanol

3.32 Show the structure of the carbocation formed from the reaction of each of the following alcohols with hydrogen chloride. Rank the cations in order of decreasing stability.

$$(CH_3)_2CHCH_2CH_2OH \qquad (CH_3)_2CCH_2CH_3 \qquad (CH_3)_2CHCHCH_3$$
$$\qquad\qquad\qquad\qquad\qquad\qquad |\qquad\qquad\qquad\qquad\qquad |$$
$$\qquad\qquad\qquad\qquad\qquad\qquad OH\qquad\qquad\qquad\qquad OH$$

3.33 Which among the following carbocations is most stable? Which is least stable?

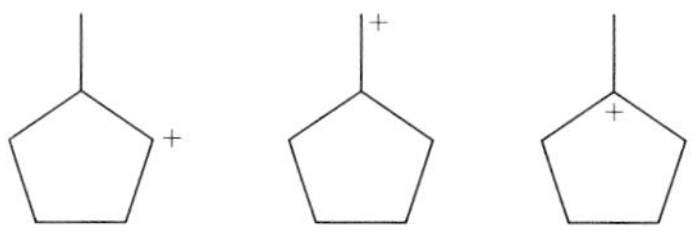

3.34 Experimental evidence indicates that the cyclopropyl cation is much less stable than the isopropyl cation. Why do you think this is so?

3.35 Two stereoisomers of 1-bromo-4-methylcyclohexane are formed when *trans*-4-methylcyclohexanol reacts with hydrogen bromide. Write structural formulas or make molecular models of:

(a) *trans*-4-Methylcylohexanol
(b) The carbocation intermediate in this reaction
(c) The two stereoisomers of 1-bromo-4-methylcyclohexane

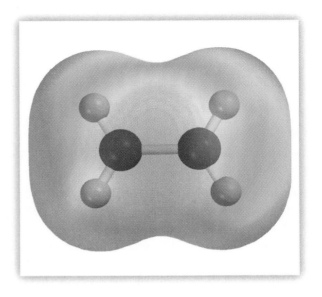

CHAPTER 4
ALKENES AND ALKYNES I: STRUCTURE AND PREPARATION

Alkenes are hydrocarbons that contain a carbon–carbon double bond, which is both an important structural unit and an important functional group in organic chemistry. The shape of an organic molecule is influenced by the presence of this bond, and the double bond is the site of most of the chemical reactions that alkenes undergo. Some representative alkenes include isobutylene (an industrial chemical), α-pinene (a fragrant liquid obtained from pine trees), and farnesene (a naturally occurring alkene with three double bonds).

$(CH_3)_2C{=}CH_2$

Isobutylene
(used in the production of synthetic rubber)

α-Pinene
(a major constituent of turpentine)

Farnesene
(present in the waxy coating found on apple skins)

Hydrocarbons characterized by the presence of a carbon–carbon triple bond are called **alkynes.** Examples of natural products that contain carbon–carbon triple bonds include tariric acid, from the seed of a Guatemalan plant, and cicutoxin, a poisonous substance isolated from water hemlock.

$$\underset{\text{Tariric acid}}{CH_3(CH_2)_{10}C\equiv C(CH_2)_4\overset{\displaystyle \overset{O}{\|}}{C}OH}$$

$$\underset{\text{Cicutoxin}}{HOCH_2CH_2CH_2C\equiv C-C\equiv CCH=CHCH=CHCH=CHCH\underset{\underset{\displaystyle OH}{|}}{CH}CH_2CH_2CH_3}$$

Diacetylene ($HC\equiv C-C\equiv CH$) has been identified as a compound of the hydrocarbon-rich atmospheres of Uranus, Neptune, and Pluto. It is also present in the atmospheres of Titan and Triton, satellites of Saturn and Neptune, respectively.

The present chapter describes the structure and preparation of alkenes and alkynes. Chapter 5 discusses their chemical reactions.

4.1 ALKENE NOMENCLATURE

We give alkenes IUPAC names by replacing the *-ane* ending of the corresponding alkane with *-ene*. The two simplest alkenes are **ethene** and **propene.** Both are also well known by their common names *ethylene* and *propylene.*

$$CH_2=CH_2 \qquad\qquad CH_3CH=CH_2$$

IUPAC name: **ethene**   IUPAC name: **propene**
(common name: ethylene)   (common name: propylene)

Ethylene is an acceptable synonym for ethene in the IUPAC system.

The longest continuous chain that includes the double bond forms the base name of the alkene, and the chain is numbered in the direction that gives the doubly bonded carbons their lower numbers. The locant (or numerical position) of only one of the doubly bonded carbons is specified in the name; it is understood that the other doubly bonded carbon must follow in sequence.

> Propylene, isobutylene, and other common names ending in *-ylene* are not acceptable IUPAC names.

$$\overset{1}{CH_2}=\overset{2}{CH}\overset{3}{CH_2}\overset{4}{CH_3} \qquad\qquad \overset{6}{CH_3}\overset{5}{CH_2}\overset{4}{CH_2}\overset{3}{CH}=\overset{2}{CH}\overset{1}{CH_3}$$

1-Butene   2-Hexene
(not 1,2-butene)   (not 4-hexene)

Carbon–carbon double bonds take precedence over alkyl groups and halogens in determining the main carbon chain and the direction in which it is numbered.

$$\overset{4}{CH_3}\underset{\underset{\displaystyle CH_3}{|}}{\overset{3}{CH}}\overset{2}{CH}=\overset{1}{CH_2} \qquad\qquad \overset{6}{BrCH_2}\overset{5}{CH_2}\overset{4}{CH_2}\underset{\underset{\displaystyle \overset{2}{CH}=\overset{1}{CH_2}}{|}}{\overset{3}{CH}}CH_2CH_2CH_3$$

3-Methyl-1-butene   6-Bromo-3-propyl-1-hexene
(not 2-methyl-3-butene)   (longest chain that contains double bond is six carbons)

Hydroxyl groups, however, outrank the double bond. Compounds that contain both a double bond and a hydroxyl group use the combined suffix *-en* + *-ol* to signify that both are present.

$$H \quad\quad\quad {}^6CH_3$$
$$\underset{1\quad2\quad3}{HOCH_2CH_2CH_2} {}^4C{=}C{}^5 \quad CH_3$$

5-Methyl-4-hexen-1-ol
(not 2-methyl-2-hexen-6-ol)

PROBLEM 4.1   Name each of the following using IUPAC nomenclature:

(a) $(CH_3)_2C{=}C(CH_3)_2$

(b) $(CH_3)_3CCH{=}CH_2$

(c) $(CH_3)_2C{=}CHCH_2CH_2CH_3$

(d) $CH_2{=}CHCH_2CHCH_3$
$\qquad\qquad\qquad |$
$\qquad\qquad\quad Cl$

(e) $CH_2{=}CHCH_2CHCH_3$
$\qquad\qquad\qquad |$
$\qquad\qquad\quad OH$

Sample Solution   (a) The longest continuous chain in this alkene contains four carbon atoms. The double bond is between C-2 and C-3, and so it is named as a derivative of 2-butene.

ETHYLENE

thylene was known to chemists in the eighteenth century and isolated in pure form in 1794. An early name for ethylene was *gaz oléfiant* (French for "oil-forming gas"), a term suggested to describe the fact that an oily liquid product is formed when two gases—ethylene and chlorine—react with each other.

$$CH_2{=}CH_2 + Cl_2 \longrightarrow ClCH_2CH_2Cl$$

Ethylene	Chlorine	1,2-Dichloroethane
(bp: −104°C)	(bp: −34°C)	(bp: 83°C)

The term *gaz oléfiant* was the forerunner of the general term *olefin,* formerly used as the name of the class of compounds we now call alkenes.

Ethylene occurs naturally in small amounts as a plant hormone. Hormones are substances that act as messengers and play regulatory roles in biological processes. Ethylene is involved in the ripening of many fruits, in which it is formed in a complex series of steps from a compound containing a cyclopropane ring:

$$\underset{CO_2^-}{\overset{\overset{+}{NH_3}}{\triangleright\!\!\triangleleft}} \xrightarrow[\text{steps}]{\text{several}} CH_2{=}CH_2 + \text{other products}$$

1-Amino-
cyclopropane-
carboxylic acid   Ethylene

Even minute amounts of ethylene can stimulate ripening, and the rate of ripening increases with the concentration of ethylene. This property is used to advantage, for example, in the marketing of bananas. Bananas are picked green in the tropics, kept green by being stored with adequate ventilation to limit the amount of ethylene present, and then induced to ripen at their destination by passing ethylene over the fruit.

Ethylene is the cornerstone of the world's mammoth petrochemical industry and is produced in vast quantities. In a typical year the amount of ethylene produced in the United States (5×10^{10} lb) exceeds the combined weight of all of its people. In one process, ethane from natural gas is heated to bring about its dissociation into ethylene and hydrogen:

$$CH_3CH_3 \xrightarrow{750°C} CH_2{=}CH_2 + H_2$$

Ethane	Ethylene	Hydrogen

This reaction is known as **dehydrogenation** and is simultaneously both a source of ethylene and one of the methods by which hydrogen is prepared on an industrial scale. Most of the hydrogen so generated is subsequently used to reduce nitrogen to ammonia for the preparation of fertilizer.

Some uses of ethylene will be seen in Chapter 5.

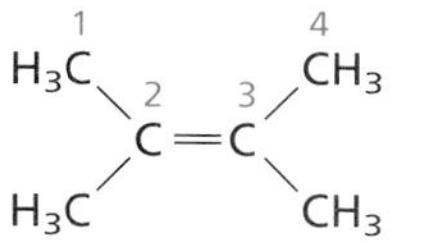

H₃C, CH₃ ... structure

$$H_3C \diagdown \underset{2}{C} = \underset{3}{C} \diagup CH_3$$

2,3-Dimethyl-2-butene

Identifying the alkene as a derivative of 2-butene leaves two methyl groups to be accounted for as substituents attached to the main chain. This alkene is 2,3-dimethyl-2-butene.

Cycloalkenes and their derivatives are named by adapting cycloalkane terminology to the principles of alkene nomenclature.

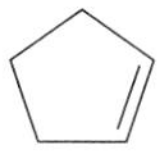

Cyclopentene 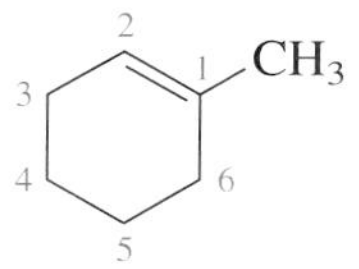   1-Methylcyclohexene 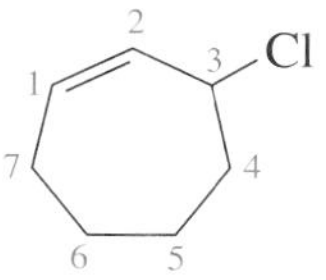   3-Chlorocycloheptene
(not 1-chloro-2-cycloheptene)

No locants are needed in the absence of substituents; it is understood that the double bond connects C-1 and C-2. Substituted cycloalkenes are numbered beginning with the double bond, proceeding through it, and continuing in sequence around the ring. The direction of numbering is chosen so as to give the lower of two possible locants to the substituent.

PROBLEM 4.2 Write structural formulas or build molecular models and give the IUPAC names of all the monochloro-substituted derivatives of cyclopentene.

Hydrocarbons that contain two carbon–carbon double bonds are called **alkadienes,** or more simply **dienes.** Dienes are **conjugated** when the two double bonds are joined by a single bond as in 1,3-pentadiene and **isolated** when the two double bonds are separated from each other by one or more sp^3-hybridized carbon atoms as in 1,4-pentadiene. A single carbon atom is common to two double bonds in **cumulated** dienes, as in 1,2-pentadiene.

Cumulated dienes are relatively rare.

$$CH_2{=}CH{-}CH{=}CHCH_3 \qquad CH_2{=}CHCH_2CH{=}CH_2 \qquad CH_2{=}C{=}CHCH_2CH_3$$

1,3-Pentadiene   1,4-Pentadiene   1,2-Pentadiene
(conjugated)   (isolated)   (cumulated)

4.2   STRUCTURE AND BONDING IN ALKENES

The structure of ethylene and the orbital hybridization model for the double bond were presented in Section 1.14. To review, Figure 4.1 depicts the planar structure of ethylene, its bond distances, and its bond angles. Each of the carbon atoms is sp^2-hybridized, and the double bond possesses a σ component and a π component. The σ component results when an sp^2 orbital of one carbon, oriented so that its axis lies along the internuclear axis, overlaps with a similarly disposed sp^2 orbital of the other carbon. Each sp^2 orbital contains one electron, and the resulting σ bond contains 2 of the 4 electrons of the double bond. The π bond contributes the other 2 electrons and is formed by a "side-by-side" overlap of singly occupied p orbitals of the two carbons.

Propene, $CH_3CH{=}CH_2$, has two different types of carbon–carbon bonds. The double bond is of the σ + π type, and the bond to the methyl group is a σ bond formed by sp^3–sp^2 overlap.

FIGURE 4.1 (a) The frame-work of σ bonds in ethylene showing bond distances in picometers and bond angles in degrees. All six atoms are coplanar. The carbon–carbon bond is a double bond made up of the σ component shown and the π component illustrated in (b). (b) The p or-bitals of two sp²-hybridized carbons overlap to produce a π bond. An electron pair in the π bond is shared by the two carbons.

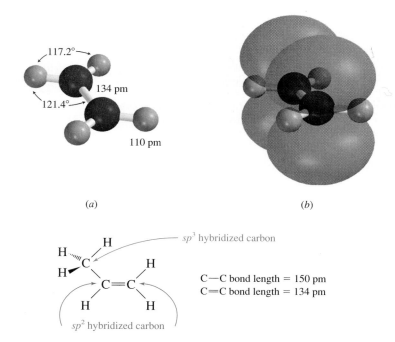

(a)   (b)

sp³ hybridized carbon

C—C bond length = 150 pm
C=C bond length = 134 pm

sp² hybridyzed carbon

PROBLEM 4.3 We can use bond-line formulas to represent alkenes in much the same way that we use them to represent alkanes. Consider the following alkene:

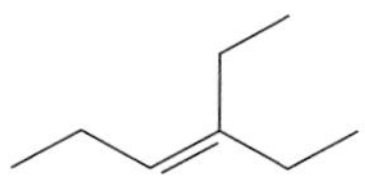

(a) What is the molecular formula of this alkene?

(b) What is its IUPAC name?

(c) How many carbon atoms are sp^2-hybridized in this alkene? How many are sp^3-hybridized?

(d) How many σ bonds are of the sp^2–sp^3 type? How many are of the sp^3–sp^3 type?

Sample Solution (a) Recall when writing bond-line formulas for hydrocarbons that a carbon occurs at each end and at each bend in a carbon chain. The appropriate number of hydrogens are attached so that each carbon has four bonds. Thus the compound shown is

$$CH_3CH_2CH{=}C(CH_2CH_3)_2$$

The general molecular formula for an alkene is C_nH_{2n}. Ethylene is C_2H_4; propene is C_3H_6. Counting the carbons and hydrogens of the compound shown (C_8H_{16}) reveals that it, too, corresponds to C_nH_{2n}.

With few exceptions, the properties of isolated dienes are like those of alkenes, and the double bonds in an isolated diene are best viewed as independent structural units. Conjugated dienes, on the other hand, exhibit properties that indicate a significant inter-action between the two double bonds. A conjugated diene, for example, is slightly more stable than an isomeric isolated diene. We will also see in Chapter 5 that conjugated dienes undergo certain reactions that isolated dienes do not.

(a) 1,4-Pentadiene
(isolated double bonds)

(b) 1,3-Pentadiene
(conjugated double bonds)

FIGURE 4.2 (a) Isolated double bonds are separated from each other by one or more sp^3-hybridized carbons and cannot overlap to give an extended π orbital. (b) In a conjugated diene, overlap of two π orbitals gives an extended π system encompassing four carbon atoms.

The factor most responsible for the increased stability of conjugated double bonds is the greater delocalization of their π electrons. As shown in Figure 4.2a, an isolated diene system is characterized by two separate π bonds. An sp^3-hybridized carbon insulates the two π bonds from each other. In a conjugated diene, however, mutual overlap of the two π systems, as shown in Figure 4.2b, generates an extended π system encompassing four adjacent carbons. The four π electrons are said to be **delocalized** over the four carbons. Whenever electrons are delocalized over several carbon atoms, the system is more stable than when these electrons feel the attractive force of fewer nuclei.

4.3   ISOMERISM IN ALKENES

Although ethylene is the only two-carbon alkene, and propene the only three-carbon alkene, there are *four* isomeric alkenes of molecular formula C_4H_8:

| 1-Butene | 2-Methylpropene | cis-2-Butene | trans-2-Butene |

Make molecular models of *cis*- and *trans*-2-butene to verify that they are different.

1-Butene has an unbranched carbon chain with a double bond between C-1 and C-2. It is a constitutional isomer of the other three. Similarly, 2-methylpropene, with a branched carbon chain, is a constitutional isomer of the other three.

The pair of isomers designated *cis*- and *trans*-2-butene have the same constitution; both have an unbranched carbon chain with a double bond connecting C-2 and C-3. They differ from each other, however, in that the cis isomer has both of its methyl groups on the same side of the double bond, but the methyl groups in the trans isomer are on opposite sides of the double bond. Recall from Section 2.17 that isomers that have the same constitution but differ in the arrangement of their atoms in space are classified as **stereoisomers.** *cis*-2-Butene and *trans*-2-butene are stereoisomers, and the terms *cis* and *trans* specify the *configuration* of the double bond.

Cis–trans stereoisomerism in alkenes is not possible when one of the doubly bonded carbons bears two identical substituents. Thus, neither 1-butene nor 2-methylpropene can have stereoisomers.

Stereoisomeric alkenes are sometimes referred to as *geometric isomers.*

1-Butene
(no stereoisomers possible)

2-Methylpropene
(no stereoisomers possible)

> **PROBLEM 4.4** How many alkenes have the molecular formula C_5H_{10}? Write their structures and give their IUPAC names. Specify the configuration of stereoisomers as cis or trans as appropriate.

In principle, *cis*-2-butene and *trans*-2-butene may be interconverted by rotation about the C-2=C-3 *double* bond. However, unlike rotation about the C-2—C-3 *single* bond in butane, which is quite fast, interconversion of the stereoisomeric 2-butenes does not occur under normal circumstances. It is sometimes said that rotation about a carbon–carbon double bond is *restricted,* but this is an understatement. Conventional laboratory sources of heat do not provide enough thermal energy for rotation about the double bond in alkenes to take place. As shown in Figure 4.3, rotation about a double bond requires the *p* orbitals of C-2 and C-3 to be twisted from their stable parallel alignment— in effect, the π component of the double bond must be broken at the transition state.

4.4   NAMING STEREOISOMERIC ALKENES BY THE *E–Z* NOTATIONAL SYSTEM

When one of the groups on either end of a double bond is the same as one on the other end, the configuration of the double bond can be described as cis or trans. Oleic acid,

FIGURE 4.3 Interconversion of *cis*- and *trans*-2-butene proceeds by cleavage of the π component of the double bond. The red balls represent the two methyl groups.

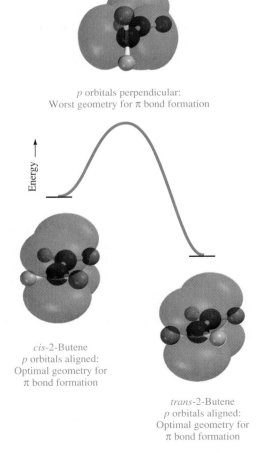

p orbitals perpendicular:
Worst geometry for π bond formation

Energy ⟶

cis-2-Butene
p orbitals aligned:
Optimal geometry for
π bond formation

trans-2-Butene
p orbitals aligned:
Optimal geometry for
π bond formation

for example, a material that can be obtained from olive oil, has a cis double bond. Cinnamaldehyde, responsible for the characteristic odor of cinnamon, has a trans double bond.

Oleic acid   Cinnamaldehyde

PROBLEM 4.5 Female gypsy moths attract males by sending a chemical signal known as a **pheromone**. The substance emitted by the female gypsy moth is an epoxide called disparlure (see Section 10.8). In the laboratory synthesis of disparlure the immediate precursor to the epoxide is *cis*-2-methyl-7-octadecene. Write a structural formula, including stereochemistry, for this compound.

The terms *cis* and *trans* are ambiguous, however, when it is not obvious which substituent on one carbon is "similar" or "analogous" to a reference substituent on the other. Fortunately, a completely unambiguous system for specifying double bond stereochemistry has been developed based on an *atomic number* criterion for ranking substituents on the doubly bonded carbons. When atoms of higher atomic number are on the *same* side of the double bond, we say that the double bond has the **Z** configuration, where Z stands for the German word *zusammen*, meaning "together." When atoms of higher atomic number are on *opposite* sides of the double bond, we say that the configuration is **E**. The symbol **E** stands for the German word *entgegen*, meaning "opposite."

Higher ⟶ Cl   Br ⟵ Higher   Higher ⟶ Cl   F ⟵ Lower

Lower ⟶ H   F ⟶ Lower   Lower ⟶ H   Br ⟶ Higher

Z configuration
Higher ranked substituents (Cl and Br)
are on same side of double bond

E configuration
Higher ranked substituents (Cl and Br)
are on opposite sides of double bond

The substituent groups on the double bonds of most alkenes are, of course, more complicated than in this example. The rules for ranking substituents, especially alkyl groups, are described in Table 4.1.

The priority rules were developed by R. S. Cahn and Sir Christopher Ingold (England) and Vladimir Prelog (Switzerland) in the context of a different aspect of organic stereochemistry; they will appear again in Chapter 7.

PROBLEM 4.6 Determine the configuration of each of the following alkenes as *Z* or *E* as appropriate:

(a) H_3C ⟍ ⟋ CH_2OH
  $C=C$
  H ⟋ ⟍ CH_3

(c) H_3C ⟍ ⟋ CH_2CH_2OH
  $C=C$
  H ⟋ ⟍ $C(CH_3)_3$

(b) H_3C ⟍ ⟋ CH_2CH_2F
  $C=C$
  H ⟋ ⟍ $CH_2CH_2CH_2CH_3$

(d) △ ⟍ ⟋ H
  $C=C$
  CH_3CH_2 ⟋ ⟍ CH_3

TABLE 4.1	Cahn–Ingold–Prelog Priority Rules

Rule	Example
1. Higher atomic number takes precedence over lower. Bromine (atomic number 35) outranks chlorine (atomic number 17). Methyl (C, atomic number 6) outranks hydrogen (atomic number 1).	The compound 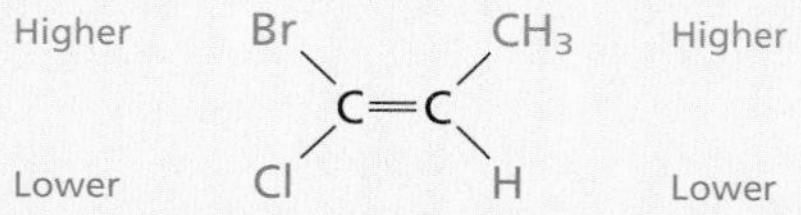 Higher Br CH_3 Higher C=C Lower Cl H Lower has the *Z* configuration. Higher ranked atoms (Br and C of CH_3) are on the same side of the double bond.

2. When two atoms directly attached to the double bond are identical, compare the atoms attached to these two on the basis of their atomic numbers. Precedence is determined at the first point of difference:
 	
Ethyl [—C(C,H,H)]   outranks   methyl [—C(H,H,H)]
 	
Similarly, *tert*-butyl outranks isopropyl, and isopropyl outranks ethyl:
 	
  —C(CH_3)_3 > —CH(CH_3)_2 > —CH_2CH_3
  —C(C,C,C) > —C(C,C,H) > —C(C,H,H)
	The compound 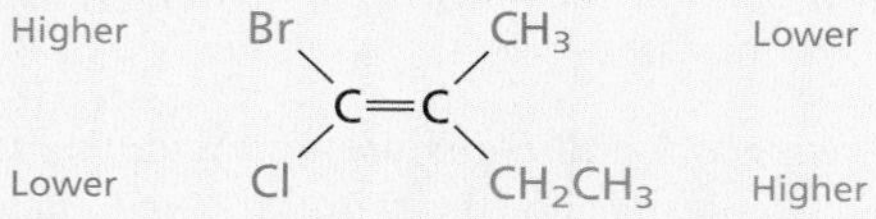
	
	Higher Br CH_3 Lower
	
	            C=C
	
	Lower Cl CH_2CH_3 Higher
	
	has the *E* configuration.

3. Work outward from the point of attachment, comparing all the atoms attached to a particular atom before proceeding further along the chain:
 	
—CH(CH_3)_2 [—C(C,C,H)]   outranks
 	
   —CH_2CH_2OH [—C(C,H,H)]
	The compound 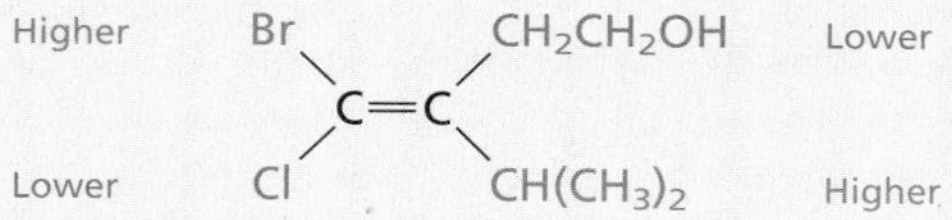
	
	Higher Br CH_2CH_2OH Lower
	
	            C=C
	
	Lower Cl $CH(CH_3)_2$ Higher
	
	has the *E* configuration.

4. When working outward from the point of attachment, always evaluate substituent atoms one by one, never as a group. Because oxygen has a higher atomic number than carbon,
 	
—CH_2OH [—C(O,H,H)]   outranks
 	
   —C(CH_3)_3 [—C(C,C,C)]
	The compound 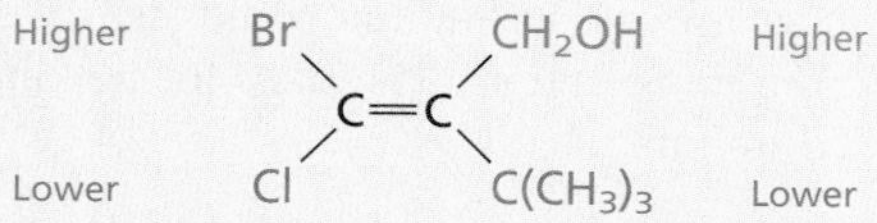
	
	Higher Br CH_2OH Higher
	
	            C=C
	
	Lower Cl $C(CH_3)_3$ Lower
	
	has the *Z* configuration.

5. An atom that is multiply bonded to another atom is considered to be replicated as a substituent on that atom:
 	
O
‖
—CH   is treated as if it were   —C(O,O,H)
 	
The group —CH=O [—C(O,O,H)]   outranks
 	
   —CH_2OH [—C(O,H,H)]
	The compound 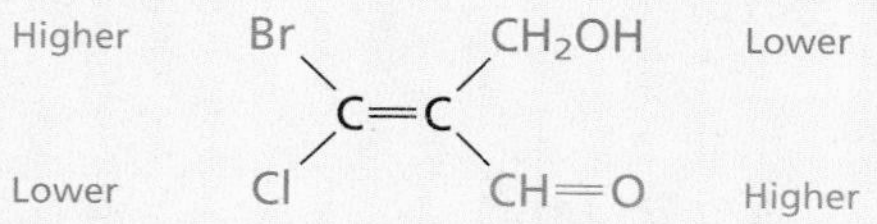
	
	Higher Br CH_2OH Lower
	
	            C=C
	
	Lower Cl CH=O Higher
	
	has the *E* configuration.

Sample Solution (a) One of the doubly bonded carbons bears a methyl group and a hydrogen. According to the rules of Table 4.1, methyl outranks hydrogen. The other carbon atom of the double bond bears a methyl and a —CH$_2$OH group. The —CH$_2$OH group is of higher priority than methyl.

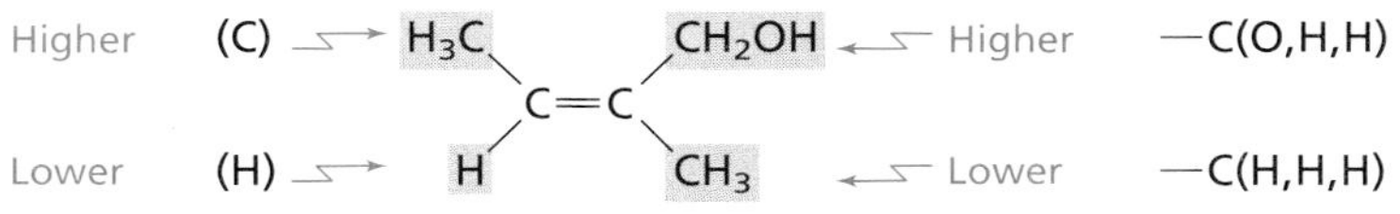

Higher   (C) → H$_3$C   CH$_2$OH ← Higher   —C(O,H,H)

Lower   (H) → H   CH$_3$ ← Lower   —C(H,H,H)

Higher ranked substituents are on the same side of the double bond; the configuration is Z.

A table on the inside of the back cover lists some of the more frequently encountered atoms and groups in order of increasing precedence. You should not attempt to memorize this table, but should be able to derive the relative placement of one group versus another.

4.5 RELATIVE STABILITIES OF ALKENES

The two most important factors governing alkene stability are:

1. Degree of substitution (alkyl substituents stabilize a double bond)
2. Van der Waals strain (destabilizing when alkyl groups are cis to each other)

Degree of Substitution. We classify double bonds as **monosubstituted, disubstituted, trisubstituted,** or **tetrasubstituted** according to the number of carbon atoms *directly* attached to the C=C structural unit.

Monosubstituted alkenes:

$$RCH=CH_2 \quad \text{as in} \quad CH_3CH_2CH=CH_2$$

1-Butene

Disubstituted alkenes:
(R and R′ may be the same or different)

$$RCH=CHR' \quad \text{as in} \quad CH_3CH=CHCH_3$$

cis- or *trans-*2-Butene

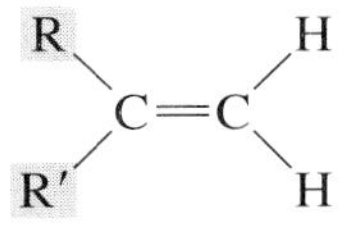

as in $(CH_3)_2C=CH_2$

2-Methylpropene

Trisubstituted alkenes:
(R, R′, and R″ may be the same or different)

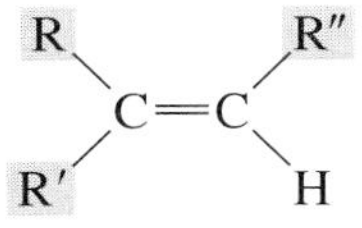

as in $(CH_3)_2C=CHCH_2CH_3$

2-Methyl-2-pentene

Tetrasubstituted alkenes:
(R, R′, R″, and R‴ may be the same or different)

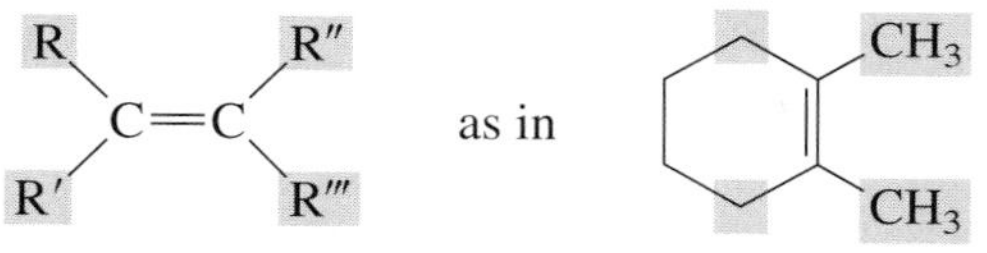

1,2-Dimethylcyclohexene

In the example shown, each of the highlighted ring carbons counts as a separate substituent on the double bond.

> **PROBLEM 4.7**   Write structural formulas or build molecular models and give the IUPAC names for all the alkenes of molecular formula C_6H_{12} that contain a trisubstituted double bond. (Don't forget to include stereoisomers.)

In general, alkenes with more highly substituted double bonds are more stable than isomers with less substituted double bonds.

> **PROBLEM 4.8**   Give the structure or make a molecular model of the most stable C_6H_{12} alkene.

Like the sp^2-hybridized carbons of carbocations, the sp^2-hybridized carbons of double bonds are electron attracting, and alkenes are stabilized by substituents that release electrons to these carbons. Alkyl groups are better electron-releasing substituents than hydrogen and are, therefore, better able to stabilize an alkene.

An effect that results when two or more atoms or groups interact so as to alter the electron distribution in a system is called an **electronic effect.** The greater stability of more highly substituted alkenes is an example of an electronic effect.

van der Waals Strain.

Alkenes are more stable when large substituents are trans to each other than when they are cis.

trans-2-Butene, for example, is 3 kJ/mol (0.7 kcal/mol) more stable than *cis*-2-butene. The source of this energy difference is illustrated in Figure 4.4, where it is seen that methyl groups approach each other very closely in *cis*-2-butene, but the trans isomer is free of strain. An effect that results when two or more atoms are close enough in space that a repulsion occurs between them is one type of **steric effect.** The greater stability of trans alkenes compared with their cis counterparts is an example of a steric effect.

> **PROBLEM 4.9**   Arrange the following alkenes in order of decreasing stability: 1-pentene; (*E*)-2-pentene; (*Z*)-2-pentene; 2-methyl-2-butene.

The difference in stability between stereoisomeric alkenes is even more pronounced with larger alkyl groups on the double bond. A particularly striking example compares *cis*- and *trans*-2,2,5,5-tetramethyl-3-hexene, in which the cis isomer is destabilized by the large van der Waals strain between the bulky *tert*-butyl groups on the same side of the double bond.

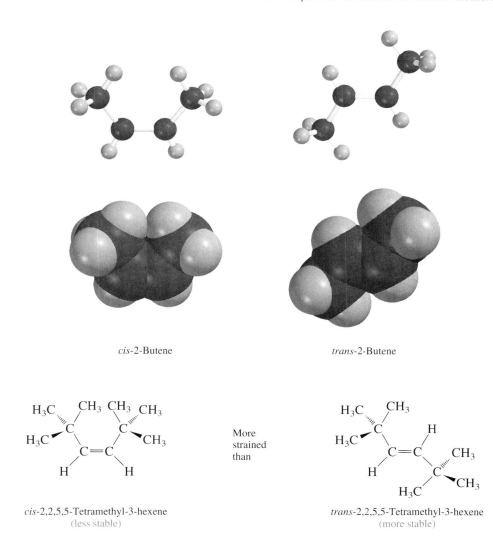

FIGURE 4.4 Ball-and-spoke and space-filling models of *cis*- and *trans*-2-butene. The space-filling model shows the serious van der Waals strain between two of the hydrogens in *cis*-2-butene. The molecule adjusts by expanding those bond angles that increase the separation between the crowded atoms. The combination of angle strain and van der Waals strain makes *cis*-2 butene less stable than *trans*-2-butene.

cis-2-Butene

trans-2-Butene

More
strained
than

cis-2,2,5,5-Tetramethyl-3-hexene
(less stable)

trans-2,2,5,5-Tetramethyl-3-hexene
(more stable)

4.6 PREPARATION OF ALKENES: ELIMINATION REACTIONS

The reaction used to prepare ethylene on an industrial scale (see boxed essay *Ethylene* on page 92) is not applicable to the laboratory preparation of alkenes. Alkenes are usually prepared in the laboratory by **elimination reactions,** reactions of the type:

$$X-\overset{|}{\underset{|}{C}}-\overset{|}{\underset{|}{C}}-Y \longrightarrow \overset{\diagdown}{\diagup}C=C\overset{\diagup}{\diagdown} + X-Y$$

> *Alkene formation requires that X and Y be substituents on adjacent carbon atoms.*

The principal methods for the preparation of alkenes are the **dehydration of alcohols** (X = H; Y = OH) and the **dehydrohalogenation of alkyl halides** (X = H; Y = Cl, Br, or I). Eliminations are one of the fundamental reactions of organic chemistry, therefore we will look at these two methods for preparing alkenes in some detail.

4.7   DEHYDRATION OF ALCOHOLS

In the dehydration of alcohols, the H and OH are lost from adjacent carbons. An acid catalyst is necessary.

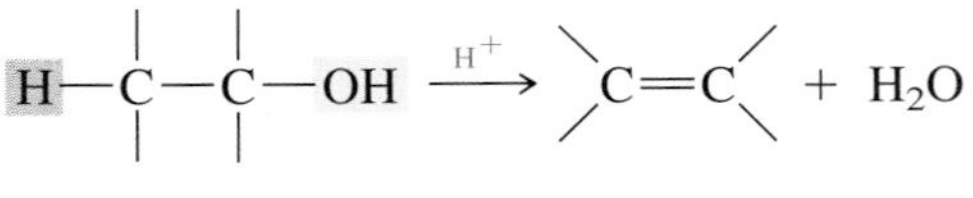

$$H-\overset{|}{\underset{|}{C}}-\overset{|}{\underset{|}{C}}-OH \xrightarrow{H^+} \underset{/}{\overset{\backslash}{C}}=\underset{\backslash}{\overset{/}{C}} + H_2O$$

Alcohol   Alkene   Water

Before dehydrogenation of ethane became the dominant method, ethylene was prepared by heating ethyl alcohol with sulfuric acid.

$$CH_3CH_2OH \xrightarrow[160°C]{H_2SO_4} CH_2{=}CH_2 + H_2O$$

Ethyl alcohol   Ethylene   Water

Concentrated sulfuric acid (H_2SO_4) and phosphoric acid (H_3PO_4) are the acids most frequently used in alcohol dehydrations. Other alcohols dehydrate in a manner similar to ethyl alcohol.

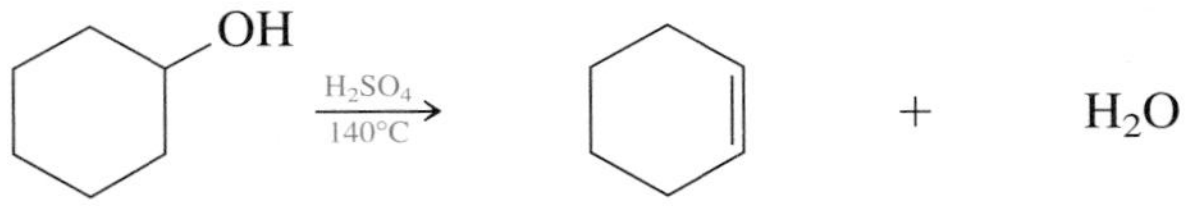

Cyclohexanol   Cyclohexene   Water
                      (79–87%)

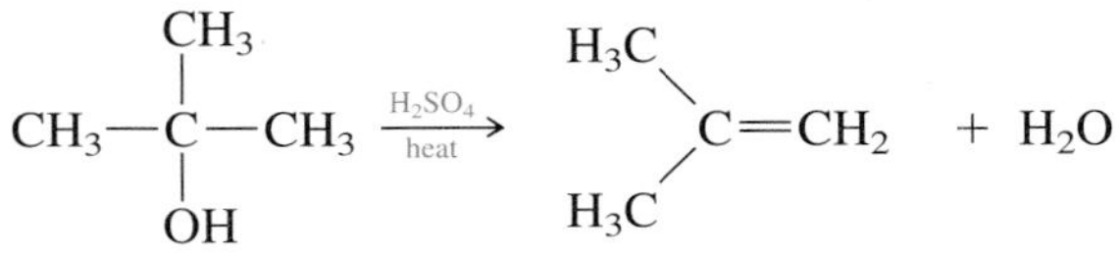

2-Methyl-2-propanol   2-Methylpropene   Water
                          (82%)

PROBLEM 4.10   Identify the alkene obtained on dehydration of each of the following alcohols:

(a) 3-Ethyl-3-pentanol   (c) 2-Propanol

(b) 1-Propanol   (d) 2,3,3-Trimethyl-2-butanol

Sample Solution   (a) The hydrogen and the hydroxyl are lost from adjacent carbons in the dehydration of 3-ethyl-3-pentanol.

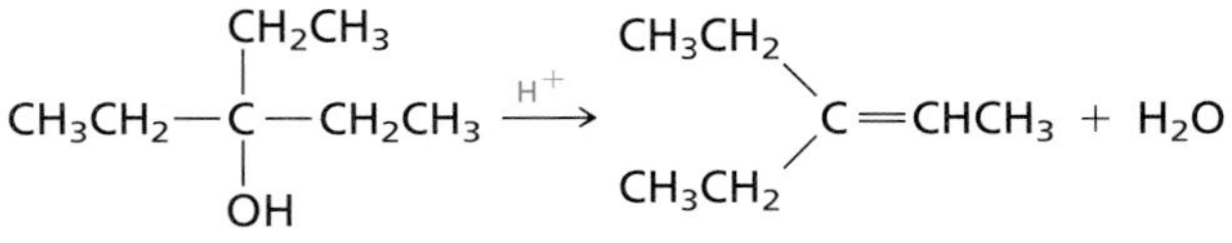

$$CH_3CH_2-\overset{\overset{\displaystyle CH_2CH_3}{|}}{\underset{\underset{\displaystyle OH}{|}}{C}}-CH_2CH_3 \xrightarrow{H^+} \underset{CH_3CH_2}{\overset{CH_3CH_2}{\diagdown}}C{=}CHCH_3 + H_2O$$

3-Ethyl-3-pentanol   3-Ethyl-2-pentene   Water

The hydroxyl group is lost from a carbon that bears three equivalent ethyl substituents. Elimination can occur in any one of three equivalent directions to give the same alkene, 3-ethyl-2-pentene.

In the preceding examples, including those of Problem 4.10, only a single alkene could be formed from each alcohol. When the alcohol is capable of yielding two or more different alkenes,

The alkene with the most highly substituted double bond predominates.

Dehydration of 2-methyl-2-butanol, for example, yields 2-methyl-2-butene, a trisubstituted alkene, as the major product. The isomeric disubstituted alkene 2-methyl-1-butene is the minor product.

2-Methyl-2-butanol   2-Methyl-1-butene (10%)   2-Methyl-2-butene (90%)

Dehydration of this alcohol is selective in respect to its *direction*. Elimination occurs in the direction that leads to the double bond between C-2 and C-3 more than between C-2 and C-1. Reactions that can proceed in more than one direction, but in which one direction is preferred, are said to be **regioselective.** Eliminations that are regioselective in the direction that yields the alkene with the most highly substituted double bond are said to obey the **Zaitsev rule.**

The Zaitsev rule is named after Alexander Zaitsev, a nineteenth-century Russian chemist credited with making the first general statement on this selectivity.

PROBLEM 4.11 Each of the following alcohols has been subjected to acid-catalyzed dehydration and yields a mixture of two isomeric alkenes. Identify the two alkenes in each case, and predict which one is the major product on the basis of the Zaitsev rule.

(a) $(CH_3)_2CCH(CH_3)_2$
     |
     OH

(b) H₃C OH

(c) OH

H

Sample Solution (a) Dehydration of 2,3-dimethyl-2-butanol can lead to either 2,3-dimethyl-1-butene by removal of a C-1 hydrogen or to 2,3-dimethyl-2-butene by removal of a C-3 hydrogen.

2,3-Dimethyl-2-butanol   2,3-Dimethyl-1-butene (minor product)   2,3-Dimethyl-2-butene (major product)

The major product is 2,3-dimethyl-2-butene. It has a tetrasubstituted double bond and is more stable than 2,3-dimethyl-1-butene, which has a disubstituted double bond. The major alkene arises by loss of a hydrogen from the adjacent carbon that has fewer attached hydrogens (C-3) rather than from the adjacent carbon that has the greater number of hydrogens (C-1).

In addition to being regioselective, alcohol dehydrations are stereoselective. A **stereoselective** reaction is one in which a single starting material can yield two or more stereoisomeric products, but gives one of them in greater amounts than any other. Alcohol dehydrations tend to produce the more stable stereoisomer of an alkene. Dehydration of 3-pentanol, for example, yields a mixture of *trans*-2-pentene and *cis*-2-pentene in which the more stable trans stereoisomer predominates.

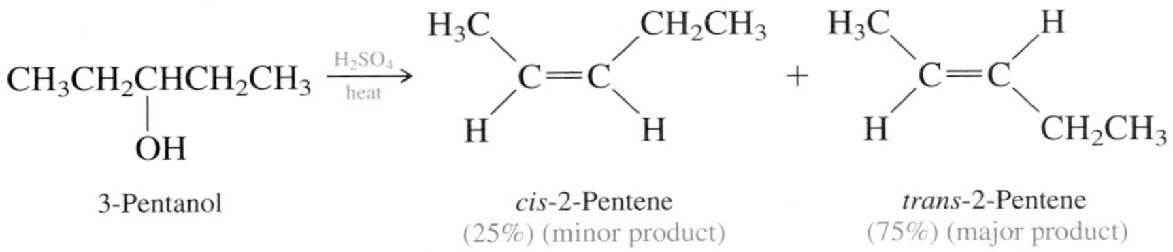

$$CH_3CH_2CHCH_2CH_3 \xrightarrow[\text{heat}]{H_2SO_4}$$

(structures shown)

3-Pentanol

cis-2-Pentene
(25%) (minor product)

trans-2-Pentene
(75%) (major product)

4.8   THE MECHANISM OF ACID-CATALYZED DEHYDRATION OF ALCOHOLS

The dehydration of alcohols and the conversion of alcohols to alkyl halides by treatment with hydrogen halides (Section 3.7) are similar in two important ways:

1. Both reactions are promoted by acids.

2. The relative reactivity of alcohols decreases in the order tertiary > secondary > primary.

These common features suggest that carbocations are key intermediates in alcohol dehydration, just as they are in the conversion of alcohols to alkyl halides. Figure 4.5 portrays a three-step mechanism for the sulfuric acid-catalyzed dehydration of tert-butyl alcohol. Steps 1 and 2 describe the generation of tert-butyl cation by a process similar to that which led to its formation as an intermediate in the reaction of tert-butyl alcohol

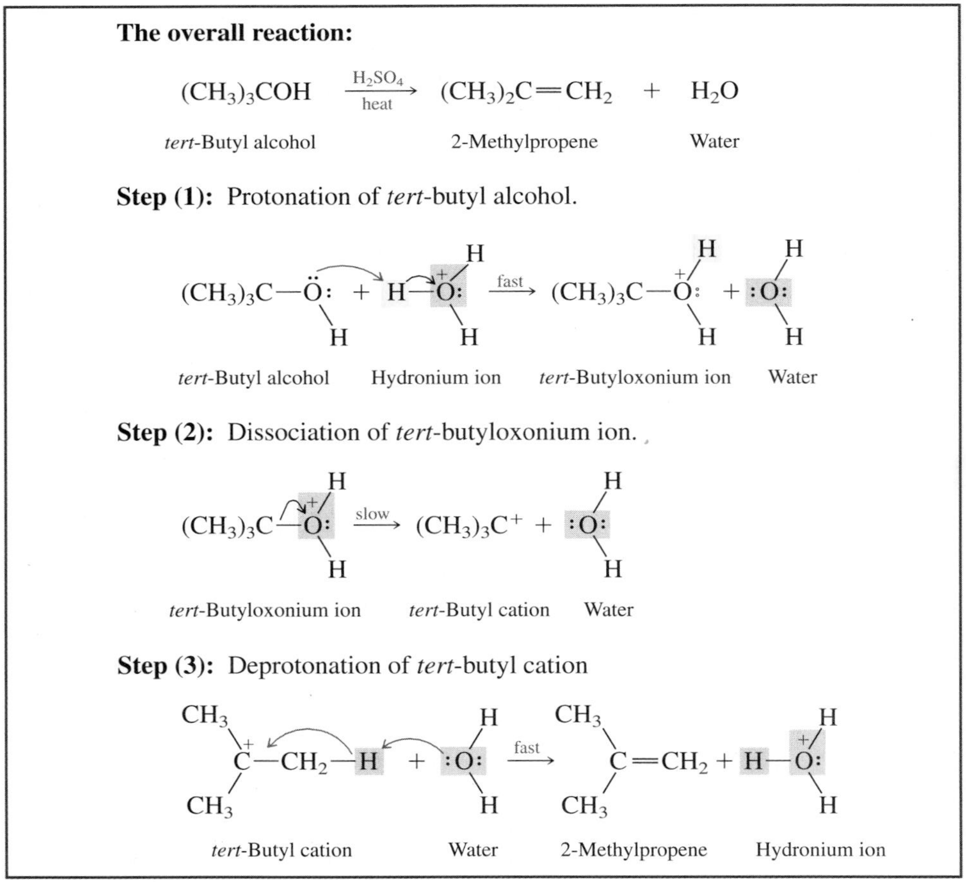

FIGURE 4.5 The mechanism for the acid-catalyzed dehydration of tert-butyl alcohol.

The overall reaction:

$$(CH_3)_3COH \xrightarrow[\text{heat}]{H_2SO_4} (CH_3)_2C=CH_2 + H_2O$$

tert-Butyl alcohol   2-Methylpropene   Water

Step (1): Protonation of tert-butyl alcohol.

$$(CH_3)_3C-\overset{..}{\underset{H}{O}}: + H-\overset{+}{\underset{H}{O}}: \xrightarrow{\text{fast}} (CH_3)_3C-\overset{+}{\underset{H}{O}}: + :\overset{..}{\underset{H}{O}}:$$

tert-Butyl alcohol   Hydronium ion   tert-Butyloxonium ion   Water

Step (2): Dissociation of tert-butyloxonium ion.

$$(CH_3)_3C-\overset{+}{\underset{H}{O}}: \xrightarrow{\text{slow}} (CH_3)_3C^+ + :\overset{..}{\underset{H}{O}}:$$

tert-Butyloxonium ion   tert-Butyl cation   Water

Step (3): Deprotonation of tert-butyl cation

$$\underset{CH_3}{\overset{CH_3}{C}}{}^{+}-CH_2-H + :\overset{..}{\underset{H}{O}}: \xrightarrow{\text{fast}} \underset{CH_3}{\overset{CH_3}{C}}=CH_2 + H-\overset{+}{\underset{H}{O}}:$$

tert-Butyl cation   Water   2-Methylpropene   Hydronium ion with hydrogen chloride. Step 3 in Figure 4.5, however, is new to us and is the step in which the double bond is formed.

Step 3 is an acid–base reaction in which the carbocation acts as a Brønsted acid, transferring a proton to a Brønsted base (water). This is the property of carbocations that is of the most significance to elimination reactions. Carbocations are strong acids; they are the conjugate acids of alkenes and readily lose a proton to form alkenes. Even weak bases such as water are sufficiently basic to abstract a proton from a carbocation.

> Step 3 in Figure 4.5 shows water as the base that abstracts a proton from the carbocation. Other Brønsted bases present in the reaction mixture that can function in the same way include *tert*-butyl alcohol and hydrogen sulfate ion.

PROBLEM 4.12   Write a structural formula for the carbocation intermediate formed in the dehydration of each of the alcohols in Problem 4.11 (Section 4.7). Using curved arrows, show how each carbocation is deprotonated by water to give a mixture of alkenes.

Sample Solution   (a) The carbon that bears the hydroxyl group in the starting alcohol is the one that becomes positively charged in the carbocation.

$$(CH_3)_2\overset{\underset{|}{OH}}{C}CH(CH_3)_2 \quad \xrightarrow[-H_2O]{H^+} \quad (CH_3)_2\overset{+}{C}CH(CH_3)_2$$

Water may remove a proton from either C-1 or C-3 of this carbocation. Loss of a proton from C-1 yields the minor product 2,3-dimethyl-1-butene. (This alkene has a disubstituted double bond.)

2,3-Dimethyl-1-butene

Loss of a proton from C-3 yields the major product 2,3-dimethyl-2-butene. (This alkene has a tetrasubstituted double bond.)

2,3-Dimethyl-2-butene

Like tertiary alcohols, secondary alcohols normally undergo dehydration by way of carbocation intermediates.

As noted earlier (Section 3.11) primary carbocations are too high in energy to be intermediates in most chemical reactions. If primary alcohols don't form primary carbocations, then how do they undergo elimination? A modification of our general mechanism for alcohol dehydration offers a reasonable explanation. For primary alcohols it is believed that a proton is lost from the alkyloxonium ion in the same step in which carbon–oxygen bond cleavage takes place. For example, the rate-determining step in the sulfuric acid-catalyzed dehydration of ethanol may be represented as:

| Water | Ethyloxonium ion | Hydronium ion | Ethylene | Water |

4.9   DEHYDROHALOGENATION OF ALKYL HALIDES

Dehydrohalogenation is the elimination of a hydrogen and a halogen from an alkyl halide. The reaction is carried out in the presence of a strong base, such as sodium ethoxide ($NaOCH_2CH_3$) in ethyl alcohol as solvent.

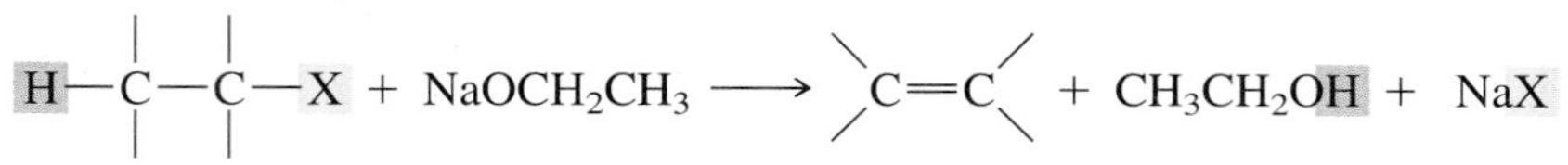

Alcohols (e.g., ethyl alcohol) are weak acids ($pK_a \approx 16$) and their conjugate bases (e.g., ethoxide ion) are strong bases (Section 3.5).

$$H-\overset{|}{\underset{|}{C}}-\overset{|}{\underset{|}{C}}-X \ + \ NaOCH_2CH_3 \ \longrightarrow \ \overset{\diagdown}{\diagup}C=C\overset{\diagup}{\diagdown} \ + \ CH_3CH_2OH \ + \ NaX$$

<table>
<tr><td>Alkyl
halide</td><td>Sodium
ethoxide</td><td>Alkene</td><td>Ethyl
alcohol</td><td>Sodium
halide</td></tr>
</table>

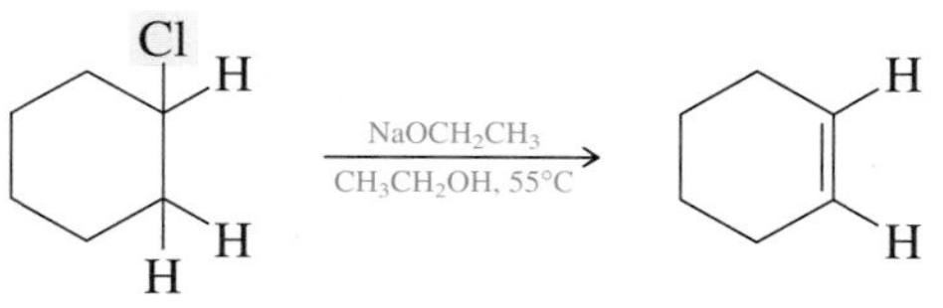

Cyclohexyl chloride

Cyclohexene
(100%)

The regioselectivity of dehydrohalogenation of alkyl halides follows the Zaitsev rule; elimination predominates in the direction that leads to the more highly substituted alkene.

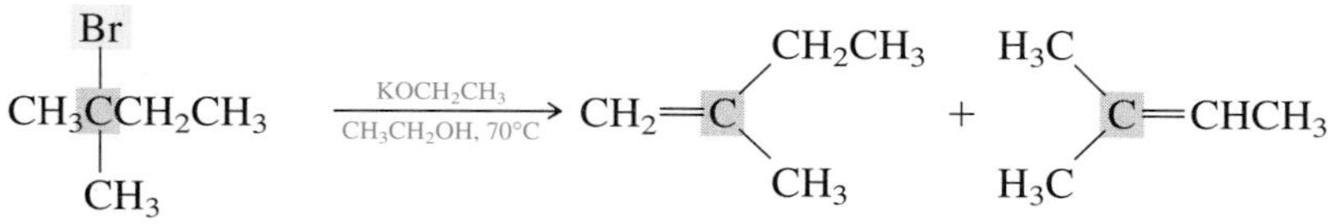

2-Bromo-2-methylbutane

2-Methyl-1-butene
(29%)

2-Methyl-2-butene
(71%)

PROBLEM 4.13   Write the structures of all the alkenes that can be formed by dehydrohalogenation of each of the following alkyl halides. Apply the Zaitsev rule to predict the alkene formed in greatest amount in each case.

(a) 2-Bromo-2,3-dimethylbutane   (c) 3-Bromo-3-ethylpentane

(b) *tert*-Butyl chloride   (d) 2-Bromo-3-methylbutane

Sample Solution   (a) First analyze the structure of 2-bromo-2,3-dimethylbutane with respect to the number of possible elimination pathways.

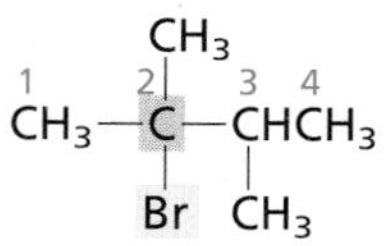

Bromine must be lost from C-2;
hydrogen may be lost from C-1 or from C-3

The two possible alkenes are

2,3-Dimethyl-1-butene
(minor product)

2,3-Dimethyl-2-butene
(major product)

The major product, predicted on the basis of Zaitsev's rule, is 2,3-dimethyl-2-butene. It has a tetrasubstituted double bond. The minor alkene has a disubstituted double bond.

In addition to being regioselective, dehydrohalogenation of alkyl halides is stereoselective and favors formation of the more stable stereoisomer. Usually, as in the case of 5-bromononane, the trans (or E) alkene is formed in greater amounts than its cis (or Z) stereoisomer.

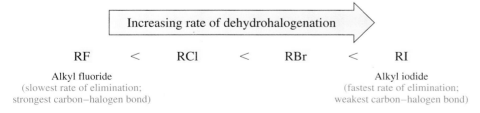

PROBLEM 4.14 Write structural formulas for all the alkenes that can be formed in the reaction of 2-bromobutane with potassium ethoxide.

4.10 THE E2 MECHANISM OF DEHYDROHALOGENATION

A mechanism proposed in the 1920s for dehydrohalogenation is still accepted as a valid description of how these reactions occur. A valid mechanism must account for these facts:

1. The reaction exhibits second-order kinetics; it is first order in alkyl halide and first order in base.

$$Rate = k[\text{alkyl halide}][\text{base}]$$

   Doubling the concentration of either the alkyl halide or the base doubles the reaction rate. Doubling the concentration of both reactants increases the rate by a factor of 4.

2. The rate of elimination depends on the halogen, the reactivity of alkyl halides increasing with decreasing strength of the carbon–halogen bond.

Increasing rate of dehydrohalogenation

RF   <   RCl   <   RBr   <   RI

Alkyl fluoride   Alkyl iodide
(slowest rate of elimination;   (fastest rate of elimination;
strongest carbon–halogen bond)   weakest carbon–halogen bond)

Iodide is the best **leaving group** in a dehydrohalogenation reaction, fluoride the poorest.

   What are the implications of these observations? Second-order kinetics suggest a bimolecular rate-determining step involving both a molecule of the alkyl halide and a molecule of base. Because the halogen with the weakest bond to carbon (iodine) reacts the fastest, we can conclude that the carbon–halogen bond breaks in the rate-determining step.

On the basis of these observations, a concerted (one-step) mechanism for dehydrohalogenation was proposed and given the mechanistic symbol **E2,** standing for **elimination bimolecular.**

Transition state for bimolecular elimination

In the E2 mechanism the three key elements

1. C—H bond breaking
2. C=C π bond formation
3. C—X bond breaking

are all taking place at the same transition state. The carbon–hydrogen and carbon–halogen bonds are in the process of being broken, the base is becoming bonded to the hydrogen, a π bond is being formed, and the hybridization of carbon is changing from sp^3 to sp^2. An energy diagram for the E2 mechanism is shown in Figure 4.6.

> **PROBLEM 4.15** Use curved arrows to track electron movement in the dehydrohalogenation of *tert*-butyl chloride by sodium methoxide by the E2 mechanism.

The regioselectivity of elimination is accommodated in the E2 mechanism by noting that a partial double bond develops at the transition state. Because alkyl groups stabilize double bonds, they also stabilize a partially formed π bond in the transition state. The more stable alkene therefore requires a lower energy of activation for its formation and predominates in the product mixture because it is formed faster than a less stable one.

4.11   A DIFFERENT MECHANISM FOR ALKYL HALIDE ELIMINATION: THE E1 MECHANISM

The E2 mechanism is a concerted process in which the carbon–hydrogen and carbon–halogen bonds both break in the same elementary step. What if these bonds break in separate steps?

One possibility is the two-step mechanism of Figure 4.7 (page 110), in which the carbon–halogen bond breaks first to give a carbocation intermediate, followed by deprotonation of the carbocation in a second step.

The alkyl halide, in this case 2-bromo-2-methylbutane, ionizes to a carbocation and a halide anion by a heterolytic cleavage of the carbon–halogen bond. Like the dissociation of an alkyloxonium ion to a carbocation, this step is rate-determining. Because the rate-determining step is unimolecular—that is, it involves only the alkyl halide and not the base—this mechanism is known by the symbol **E1,** standing for **elimination unimolecular.** It exhibits first-order kinetics.

$$\text{Rate} = k[\text{alkyl halide}]$$

The mechanism for proton transfer is a bimolecular, concerted process (Section 3.6).

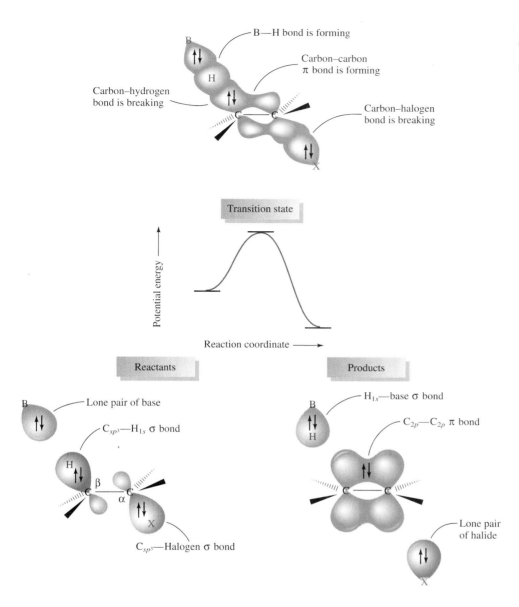

FIGURE 4.6 Potential energy diagram for concerted E2 elimination of an alkyl halide.

The reactivity order parallels the ease of carbocation formation, thus elimination by the E1 mechanism is observed only for tertiary and some secondary alkyl halides. The best examples of E1 eliminations are those carried out in the absence of added base. In the example cited in Figure 4.7, the base that abstracts the proton from the carbocation intermediate is a very weak one; it is a molecule of the solvent, ethyl alcohol. At even modest concentrations of strong base, elimination by the E2 mechanism is much faster than E1 elimination.

There is a strong similarity between the mechanism shown in Figure 4.7 and the one shown for alcohol dehydration in Figure 4.5. Indeed, we can describe the acid-catalyzed dehydration of alcohols as an E1 elimination of their conjugate acids. The main difference between the dehydration of 2-methyl-2-butanol and the dehydrohalogenation

The reaction:

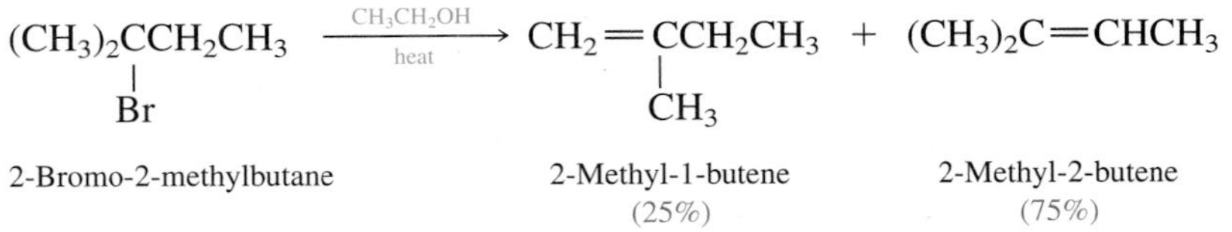

$$(CH_3)_2CCH_2CH_3 \xrightarrow[\text{heat}]{CH_3CH_2OH} CH_2=CCH_2CH_3 + (CH_3)_2C=CHCH_3$$
$$\overset{|}{Br} \qquad\qquad\qquad\qquad \overset{|}{CH_3}$$

2-Bromo-2-methylbutane   2-Methyl-1-butene   2-Methyl-2-butene
                                  (25%)   (75%)

The mechanism:

Step (1): Alkyl halide dissociates by heterolytic cleavage of carbon–halogen bond. (Ionization step)

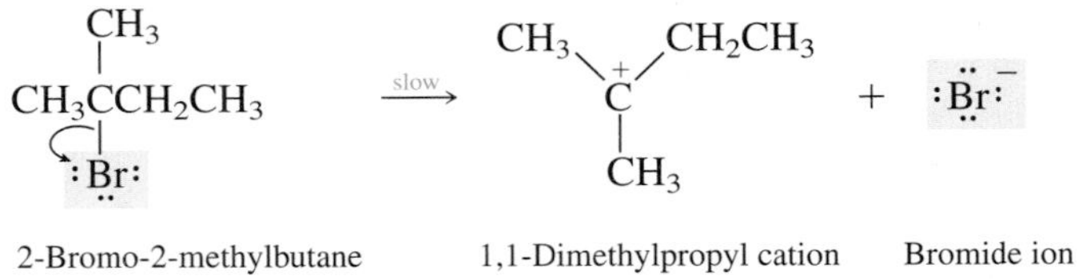

2-Bromo-2-methylbutane   1,1-Dimethylpropyl cation   Bromide ion

Step (2): Ethanol acts as a base to remove a proton from the carbocation to give the alkene products. (Deprotonation step)

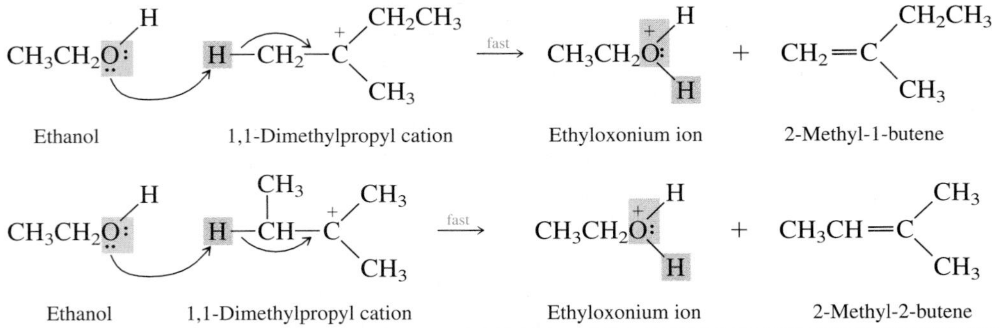

Ethanol   1,1-Dimethylpropyl cation   Ethyloxonium ion   2-Methyl-1-butene

Ethanol   1,1-Dimethylpropyl cation   Ethyloxonium ion   2-Methyl-2-butene

FIGURE 4.7   The E1 mechanism for the dehydrohalogenation of 2-bromo-2-methylbutane in ethanol.

of 2-bromo-2-methylbutane is the source of the carbocation. When the alcohol is the substrate, it is the corresponding alkyloxonium ion that dissociates to form the carbocation. The alkyl halide ionizes directly to the carbocation.

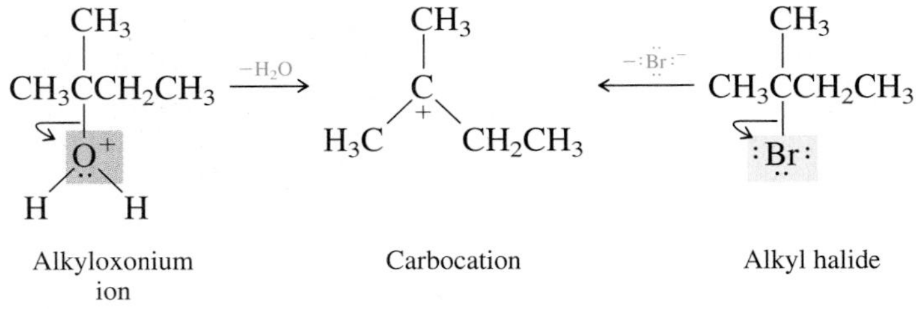

Alkyloxonium   Carbocation   Alkyl halide
ion

4.12   ALKYNE NOMENCLATURE

Hydrocarbons that contain a carbon–carbon triple bond are called **alkynes.** Noncyclic alkynes have the molecular formula C_nH_{2n-2}. Acetylene ($HC\equiv CH$) is the simplest alkyne. We call compounds that have their triple bond at the end of a carbon chain ($RC\equiv CH$) **monosubstituted,** or **terminal, alkynes.** Disubstituted alkynes ($RC\equiv CR'$) are said to have **internal** triple bonds.

In naming alkynes the usual IUPAC rules for hydrocarbons are followed, and the suffix *-ane* is replaced by *-yne*. Both acetylene and ethyne are acceptable IUPAC names for $HC\equiv CH$. The position of the triple bond along the chain is specified by number in a manner analogous to alkene nomenclature.

$$HC\equiv CCH_3 \qquad HC\equiv CCH_2CH_3 \qquad CH_3C\equiv CCH_3 \qquad (CH_3)_3CC\equiv CCH_3$$

Propyne   1-Butyne   2-Butyne   4,4-Dimethyl-2-pentyne

PROBLEM 4.16   Write structural formulas and give the IUPAC names for all the alkynes of molecular formula C_5H_8.

4.13   STRUCTURE AND BONDING IN ALKYNES: *sp* HYBRIDIZATION

An *sp* hybridization bonding model for the carbon–carbon triple bond was developed in Section 1.15 and is reviewed for acetylene in Figure 4.8. Linear geometries characterize the $-C\equiv C-C$ and $C-C\equiv C-C$ units of terminal and internal triple bonds, respectively. This linear geometry is responsible for the relatively small number of known cycloalkynes. Although few cycloalkynes occur naturally, they have gained recent attention when it was discovered that some of them hold promise as anticancer drugs (see the boxed essay *Natural and "Designed" Enediyne Antibiotics* on page 112).

4.14   PREPARATION OF ALKYNES BY ELIMINATION REACTIONS

Just as it is possible to prepare alkenes by dehydrohalogenation of alkyl halides, so may alkynes be prepared by a *double dehydrohalogenation* of dihaloalkanes. The dihalide may be a **geminal dihalide,** one in which both halogens are on the same carbon, or it may be a **vicinal dihalide,** one in which the halogens are on adjacent carbons.

Vicinal is derived from the Latin *vicinalis,* meaning "neighboring."

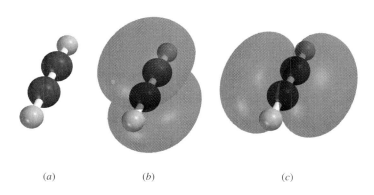

(a)   (b)   (c)

FIGURE 4.8 The carbon atoms of acetylene are connected by a $\sigma + \pi + \pi$ triple bond. Both carbon atoms are *sp*-hybridized, and each is bonded to a hydrogen by an *sp*–1*s* σ bond. The σ component of the triple bond arises by *sp*–*sp* overlap. Each carbon has two *p* orbitals, the axes of which are perpendicular to each other. One π bond is formed by overlap of the *p* orbitals shown in part (*b*), the other by overlap of the *p* orbitals shown in part (*c*). Each π bond contains two electrons.

Double dehydrohalogenation of a geminal dihalide:

$$(CH_3)_3CCH_2CHCl_2 \xrightarrow[NH_3]{3NaNH_2} (CH_3)_3CC\equiv CNa \xrightarrow{H_2O} (CH_3)_3CC\equiv CH$$

| 1,1-Dichloro-3,3-dimethylbutane | Sodium salt of alkyne product (not isolated) | 3,3-Dimethyl-1-butyne (56–60%) |

Double dehydrohalogenation of a vicinal dihalide:

$$CH_3(CH_2)_7CHCH_2Br \xrightarrow[NH_3]{3NaNH_2} CH_3(CH_2)_7C\equiv CNa \xrightarrow{H_2O} CH_3(CH_2)_7C\equiv CH$$
$$\overset{|}{Br}$$

| 1,2-Dibromodecane | Sodium salt of alkyne product (not isolated) | 1-Decyne (54%) |

NATURAL AND "DESIGNED" ENEDIYNE ANTIBIOTICS

Beginning in the 1980s, research directed toward the isolation of new drugs derived from natural sources identified a family of tumor-inhibitory antibiotic substances characterized by novel structures containing a $C\equiv C-C=C-C\equiv C$ unit as part of a 9- or 10-membered ring. With one double bond and two triple bonds (-*ene* + *di*- + -*yne*), these compounds soon became known as *enediyne* antibiotics. The simplest member of the class is *dynemicin A;* most of the other enediynes have even more complicated structures.

Enediynes hold substantial promise as anticancer drugs because of their potency and selectivity. Not only do they inhibit cell growth, they have a greater tendency to kill cancer cells than they do normal cells. The mechanism by which enediynes act involves novel chemistry unique to the $C\equiv C-C=C-C\equiv C$ unit, which leads to a species that cleaves DNA and halts tumor growth.

The history of drug development has long been based on naturally occurring substances. Often, however, compounds that might be effective drugs are produced by plants and microorganisms in such small amounts that their isolation from natural sources is not practical. If the structure is relatively simple, chemical synthesis provides an alternative source of the drug, making it available at a lower price. Equally important, chemical synthesis, modification, or both can improve the effectiveness of a drug. Building on the enediyne core of dynemicin A, for example, Professor Kyriacos C. Nicolaou and his associates at the Scripps Research Institute and the University of California at San Diego have prepared a simpler analog that is both more potent and more selective than dynemicin A. It is a "designed enediyne" in that its structure was conceived on the basis of chemical reasoning so as to carry out its biochemical task. The designed enediyne offers the additional advantage of being more amenable to large-scale synthesis.

Dynemicin A

"Designed" enediyne

> **PROBLEM 4.17** Give the structures of three isomeric dibromides that could be used as starting materials for the preparation of 3,3-dimethyl-1-butyne.

Double dehydrohalogenation of geminal or vicinal dihalides is most often used to prepare terminal alkynes. The methods used to prepare alkynes with internal triple bonds will be described in Chapter 5.

LEARNING OBJECTIVES

This chapter introduced the nomenclature, structure, and preparation of alkenes and alkynes. The skills you have learned in this chapter should enable you to:

- Write a structural formula for an alkene or alkyne on the basis of its systematic IUPAC name.
- Write a correct IUPAC name for an alkene or alkyne on the basis of a given structural formula.
- Recognize alkenes that can exist in stereoisomeric forms, and identify these as cis, trans, *E*, or *Z*, as appropriate.
- Compare isomeric alkenes with respect to their relative stability according to the degree of substitution and stereochemistry at the double bond.
- Explain the structural differences between conjugated, cumulated, and isolated double bonds in dienes.
- Write a chemical equation for the formation of an alkene by dehydration of an alcohol, and write a mechanism for this reaction.
- Write a chemical equation for the formation of an alkene by dehydrohalogenation of an alkyl halide.
- State and give an example of Zaitsev's rule.
- Describe the E2 mechanism of dehydrohalogenation, clearly identifying the role of the base, and show the bonding changes that occur at the transition state.
- Describe the E1 mechanism for dehydrohalogenation of alkyl halides.
- Write a chemical equation for the formation of an alkyne by the dehydrohalogenation of a geminal or vicinal dihalide.

4.15 SUMMARY

Alkenes and cycloalkenes contain carbon–carbon double bonds. According to **IUPAC nomenclature,** alkenes are named by substituting *-ene* for the *-ane* suffix of the alkane that has the same number of carbon atoms as the longest continuous chain that includes the double bond (Section 4.1). The chain is numbered in the direction that gives the lower number to the first-appearing carbon of the double bond. The double bond takes precedence over alkyl groups and halogens in dictating the direction of numbering, but is outranked by the hydroxyl group.

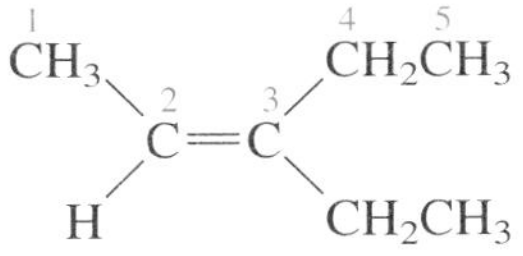

3-Ethyl-2-pentene 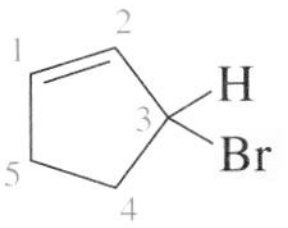   3-Bromocyclopentene   3-Buten-1-ol

Alkenes that contain two double bonds are called **alkadienes.** The double bonds in dienes are classified as **cumulated, conjugated,** or **isolated** (Section 4.1). Conjugated dienes are characterized by two double bonds that are joined by a single bond. In isolated dienes two double bonds are separated from each other by one or more sp^3 carbons.

Bonding in alkenes is described according to an sp^2 orbital hybridization model. The double bond unites two sp^2-hybridized carbon atoms and is made of a σ component and a π component (Section 4.2). The σ bond arises by overlap of an sp^2 hybrid orbital on each carbon. The π bond results from a side-by-side overlap of p orbitals.

The double bonds in isolated dienes are similar to double bonds in alkenes (Section 4.2). A conjugated diene is stabilized by **delocalization** of the π electrons over the four carbons of the $C=C-C=C$ unit.

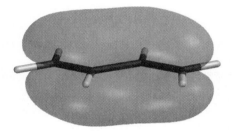

Rotation about the double bond of alkenes does not occur, and **stereoisomers** are possible (Section 4.3). The **configurations** of stereoisomeric alkenes are described according to two notational systems. One system adds the prefix *cis-* to the name of the alkene when identical substituents are on the same side of the double bond and the prefix *trans-* when they are on opposite sides. The other ranks substituents according to a system of rules based on atomic number (Section 4.4). The prefix Z is used for alkenes that have higher ranked substituents on the same side of the double bond; the prefix E is used when higher ranked substituents are on opposite sides.

$$H_3C \qquad CH_2CH_3 \qquad\qquad H_3C \qquad H$$
$$\diagdown C=C\diagup \qquad\qquad\qquad \diagdown C=C\diagup$$
$$H \qquad\qquad H \qquad\qquad\qquad H \qquad\qquad CH_2CH_3$$

cis-2-Pentene   *trans*-2-Pentene
[(Z)-2-pentene]   [(E)-2-pentene]

Comparing stereoisomeric alkenes, trans alkenes are generally more stable than cis. Alkyl substituents stabilize a double bond (Section 4.5). The general order of alkene stability is:

$$R_2C=CR_2 > R_2C=CHR > RCH=CHR > RCH=CH_2$$

Most stable   Least stable

Alkenes are prepared by **elimination** of alcohols and alkyl halides (Section 4.6). These reactions are summarized with examples in Table 4.2. In both cases, elimination proceeds in the direction that yields the more highly substituted double bond **(Zaitsev's rule).**

TABLE 4.2	Preparation of Alkenes by Elimination Reactions of Alcohols and Alkyl Halides

Reaction (section) and Comments	General Equation and Specific Example
Dehydration of alcohols (Sections 4.7 and 4.8) Dehydration requires an acid catalyst; the order of reactivity of alcohols is tertiary > secondary > primary. Elimination is regioselective and proceeds in the direction that produces the most highly substituted double bond. When stereoisomeric alkenes are possible, the more stable one is formed in greater amounts. A carbocation intermediate is involved.	$R_2CHCR'_2 \xrightarrow{H^+} R_2C{=}CR'_2 + H_2O$ $\quad\;$ OH Alcohol $\qquad$ Alkene $\quad$ Water 2-Methyl-2-hexanol H_2SO_4, 80°C 2-Methyl-1-hexene $\qquad$ + $\qquad$ 2-Methyl-2-hexene (19%) $\qquad\qquad\qquad\qquad$ (81%)
Dehydrohalogenation of alkyl halides (Sections 4.9 and 4.10) Strong bases cause a proton and a halide to be lost from adjacent carbons of an alkyl halide to yield an alkene. Regioselectivity is in accord with the Zaitsev rule. The order of halide reactivity is I > Br > Cl > F. A concerted E2 reaction pathway is followed.	$R_2CHCR'_2 + :B^- \longrightarrow R_2C{=}CR'_2 + H{-}B + X^-$ $\quad\;$ X Alkyl $\quad$ Base $\qquad$ Alkene $\qquad$ Conjugate $\quad$ Halide halide $\qquad\qquad\qquad\qquad\qquad$ acid of base 1-Chloro-1-methylcyclohexane $KOCH_2CH_3$, CH_3CH_2OH, 100°C Methylenecyclohexane $\quad$ + $\quad$ 1-Methylcyclohexene (6%) $\qquad\qquad\qquad\qquad\qquad$ (94%)

Carbocations are intermediates in the acid-catalyzed dehydration of alcohols, and are formed by dissociation of an alkyloxonium ion, ROH_2^+. Once formed, the carbocation loses a proton to form an alkene (Section 4.8).

When alkyl halides are treated with strong bases such as sodium ethoxide, dehydrohalogenation occurs by the **E2 (elimination bimolecular) mechanism** (Section 4.10). According to the E2 mechanism, the three elements of elimination, (1) removal of a proton from carbon by the base, (2) π bond formation, and (3) loss of halide from carbon, all occur in the same transition state.

In the absence of a strong base, alkyl halides eliminate by the **E1 (elimination unimolecular) mechanism** (Section 4.11). The E1 mechanism involves rate-determining ionization of the alkyl halide to a carbocation, followed by deprotonation of the carbocation.

Hydrocarbons that contain a carbon–carbon triple bond are classified as **alkynes.** Alkynes are named in the IUPAC system in a manner similar to alkenes, however the suffix *-yne* replaces *-ane* (Section 4.12).

The carbon–carbon triple bond in alkynes is composed of a σ and two π components (Section 4.13). The σ component contains two electrons in an orbital generated by the overlap of *sp*-hybridized orbitals on adjacent atoms. Each of these carbons also has two 2*p* orbitals, which overlap in pairs so as to give two π orbitals. Alkynes have a linear arrangement of their bonds in the —C≡C— unit.

Alkynes are prepared from **geminal** or **vicinal** dihalides by double dehydrohalogenation reactions (Section 4.14).

ADDITIONAL PROBLEMS

Alkene Nomenclature and Structure

4.18 Write structural formulas for each of the following:

(a) 1-Heptene

(b) 3-Ethyl-2-pentene

(c) *cis*-3-Octene

(d) *trans*-1,4-Dichloro-2-butene

(e) (*Z*)-3-Methyl-2-hexene

(f) (*E*)-3-Chloro-2-hexene

(g) 1-Bromo-3-methylcyclohexene

(h) 1-Bromo-6-methylcyclohexene

(i) 4-Methyl-4-penten-2-ol

4.19 Write a structural formula or build a molecular model and give a correct IUPAC name for each alkene of molecular formula C_7H_{14} that has a *tetrasubstituted* double bond.

4.20 Give the IUPAC names for each of the following compounds:

(a) $(CH_3CH_2)_2C\!=\!CHCH_3$

(b) $(CH_3CH_2)_2C\!=\!C(CH_2CH_3)_2$

(c) $(CH_3)_3CCH\!=\!CCl_2$

(d)

4.21 Each of the following compounds is found in nature. Write a structural formula or build a molecular model of each one, clearly showing the stereochemistry of each double bond.

(a) The sex attractant of the Mediterranean fruit fly, (*E*)-6-nonen-1-ol.

(b) Geraniol, a naturally occurring substance present in the fragrant oil of many plants that has a pleasing, rose-like odor. Geraniol is the *E* isomer of

$$(CH_3)_2C\!=\!CHCH_2CH_2C\!=\!CHCH_2OH$$
$$\qquad\qquad\qquad\qquad\quad |$$
$$\qquad\qquad\qquad\qquad\ CH_3$$

(c) Nerol, a naturally occurring substance that is a stereoisomer of geraniol.

(d) The sex attractant of the codling moth, the 2*Z*, 6*E* stereoisomer of

$$CH_3CH_2CH_2C\!=\!CHCH_2CH_2C\!=\!CHCH_2OH$$
$$\qquad\qquad\quad |\qquad\qquad\qquad\qquad |$$
$$\qquad\qquad\ CH_3\qquad\qquad\qquad CH_2CH_3$$

4.22 Specify the hybridization of each carbon in the molecules shown. How many carbon–carbon σ bonds and carbon–carbon π bonds are present in each molecule?

(a) $(CH_3)_2C\!=\!C(CH_3)_2$

(b) $CH_2\!=\!CHCH_2CH_2CH\!=\!CH_2$

(c)

4.23 For which of the following alkenes are stereoisomeric forms possible?

(a) 1-Chloropropene   (b) 2-Chloropropene   (c) 1,2-Dichloropropene

4.24 Arrange the following alkenes in order of decreasing stability: 1-pentene; *cis*-2-pentene; *trans*-2-pentene; 2-methyl-2-butene.

Preparation of Alkenes

4.25 What three alkenes may be formed from the acid-catalyzed dehydration of 2-pentanol? Which one will predominate?

4.26 (a) Write the structures or build molecular models of all the isomeric alcohols having the molecular formula $C_5H_{12}O$.
(b) Which one will undergo acid-catalyzed dehydration most readily?
(c) Write the structure of the most stable C_5H_{11} carbocation.
(d) Which alkenes may be derived from the carbocation in part (c)? Which of these will predominate?
(e) Which alcohol in part (a) will dehydrate to give only a pair of stereoisomers?

4.27 Write a sequence of steps describing the mechanism of the acid-catalyzed dehydration of 2-butanol. Use the curved-arrow notation to show electron movement in each step.

4.28 How many alkenes would you expect to be formed by dehydrohalogenation of each of the following alkyl bromides? Identify the alkenes in each case.
(a) 1-Bromohexane
(b) 2-Bromohexane
(c) 3-Bromo-3-methylpentane
(d) 3-Bromo-2,2-dimethylbutane

4.29 Choose the compound of molecular formula $C_7H_{13}Br$ that gives each alkene shown as the *exclusive* product of E2 elimination.

(a)

(b) $\bigcirc$=CH₂

(c) $\bigcirc$—CH₃

(d) ⬠ with CH₃/CH₃

4.30 Give the structures of two different alkyl bromides both of which yield the indicated alkene as the *exclusive* product of E2 elimination.
(a) $CH_3CH{=}CH_2$
(b) $(CH_3)_2C{=}CH_2$
(c) ◇ with CH₃/CH₃

4.31 (a) Write the structures or build molecular models of all the isomeric alkyl bromides having the molecular formula $C_5H_{11}Br$.
(b) Which one undergoes E1 elimination at the fastest rate?
(c) Which one is incapable of reacting by the E2 mechanism?
(d) Which ones can yield only a single alkene on E2 elimination?
(e) For which isomer does E2 elimination give two alkenes that are not constitutional isomers?
(f) Which one yields the most complex mixture of alkenes on E2 elimination?

4.32 The rate of the reaction

$$(CH_3)_3CCl + NaSCH_2CH_3 \longrightarrow (CH_3)_2C{=}CH_2 + CH_3CH_2SH + NaCl$$

is first order in $(CH_3)_3CCl$ and first order in $NaSCH_2CH_3$. Give the symbol (E1 or E2) for the most reasonable mechanism, and use curved-arrow notation to represent the flow of electrons.

4.33 Bimolecular elimination (the E2 mechanism) requires that the proton H and the halogen X, as shown in the general equation on page 108 (Section 4.10) be anti to one another. With that in mind, consider menthyl chloride and neomenthyl chloride having the structures shown. One of these stereoisomers undergoes elimination on treatment with sodium ethoxide in ethanol much more readily than the other. Which reacts faster, menthyl chloride or neomenthyl chloride? Why? (Molecular models will help here.)

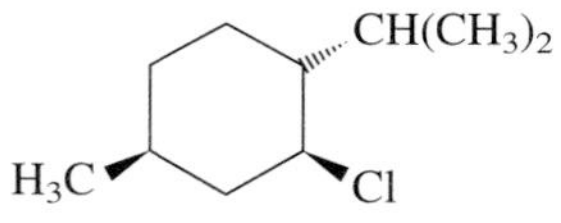
CH(CH3)2
H3C
Cl
Menthyl chloride

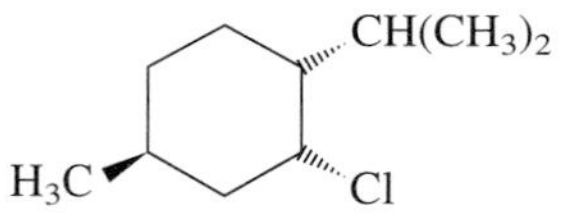
CH(CH3)2
H3C
Cl
Neomenthyl chloride

Alkadienes

4.34 Many naturally occurring substances contain several carbon–carbon double bonds: some isolated, some conjugated, and some cumulated. Identify the types of carbon–carbon double bonds found in each of the following substances:

(a) β-Springene (a scent substance from the dorsal gland of springboks)

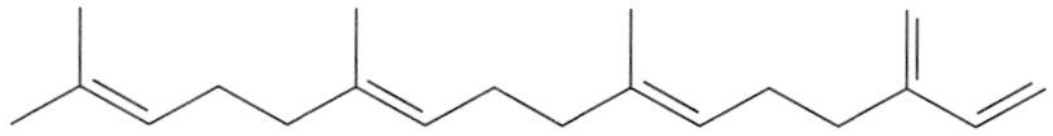

(b) Humulene (found in hops and oil of cloves)

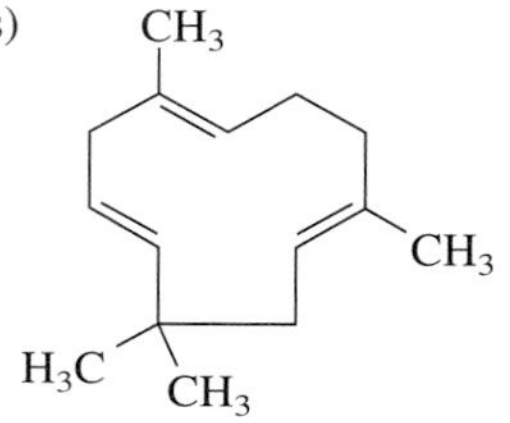
CH3
CH3
H3C
CH3

(c) Cembrene (occurs in pine resin)

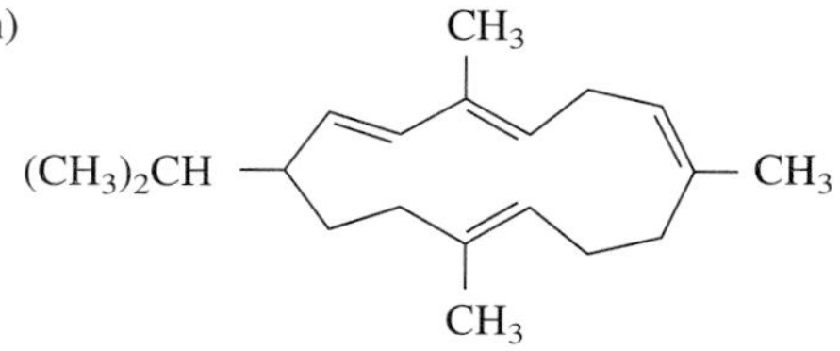
CH3
(CH3)2CH
CH3
CH3

(d) The sex attractant of the male dried-bean beetle

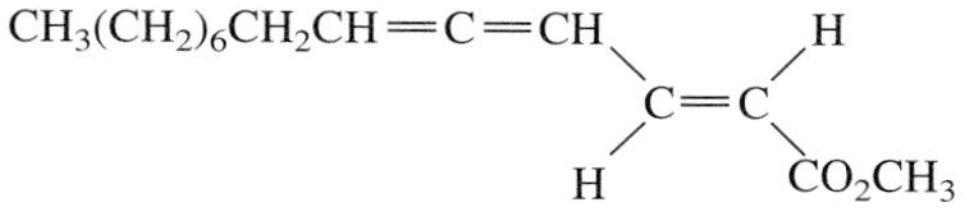
CH3(CH2)6CH2CH=C=CH
H
C=C
H
CO2CH3

4.35 A certain species of grasshopper secretes a substance containing a cumulated diene of molecular formula $C_{13}H_{20}O_3$ that acts as an ant repellent. The carbon skeleton and location of various substituents in this substance are indicated in the partial structure shown. Complete the structure, adding double bonds where appropriate.

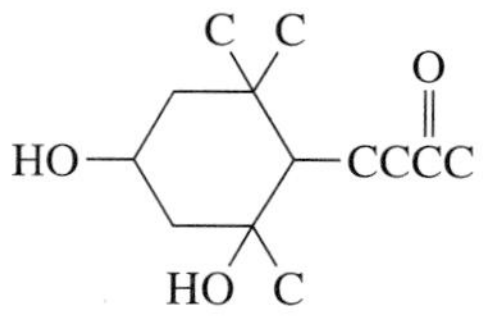
C C
O
HO
CCCC
HO C

4.36 Write structural formulas and give the IUPAC names for all the alkynes of molecular formula C_6H_{10}.

4.37 Provide the IUPAC name for each of the following alkynes:

    (a) $CH_3CH_2CH_2C\equiv CH$

    (b) $CH_3CH_2C\equiv CCH_3$

    (c) $CH_3C\equiv CCHCH(CH_3)_2$
$$\qquad\qquad\qquad\qquad |$$
$$\qquad\qquad\qquad\quad CH_3$$

4.38 Write a structural formula or build a molecular model of each of the following:

    (a) 1-Octyne

    (b) 2-Octyne

    (c) 3-Octyne

    (d) 4-Octyne

    (e) 2,5-Dimethyl-3-hexyne

    (f) 4-Ethyl-1-hexyne

4.39 Identify the compounds in Problem 4.38 as either internal or terminal alkynes.

4.40 By writing chemical equations, show how two isomeric substances having the formula $C_6H_{12}Cl_2$ could be converted exclusively to 1-hexyne.

Reactions

4.41 Give the structure of the organic product of each of the following reactions. If two or more isomeric alkenes are formed, show all of them and indicate which one is the major product.

(a)

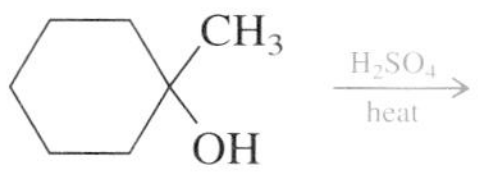

(b)

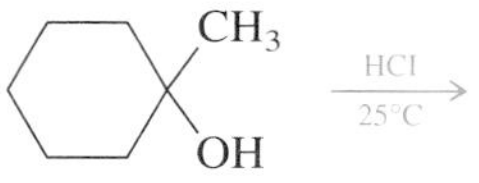

(c) Product of part (b) + $NaOCH_2CH_3$ $\xrightarrow{\text{warm}}$

(d)

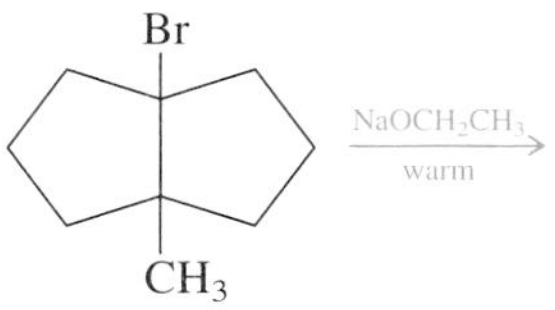

(e) $(CH_3)_3CCH_2CHCl$ $\xrightarrow[\text{2. } H_2O]{\text{1. } NaNH_2,\ NH_3}$
$$\qquad\qquad\qquad |$$
$$\qquad\qquad\quad Br$$

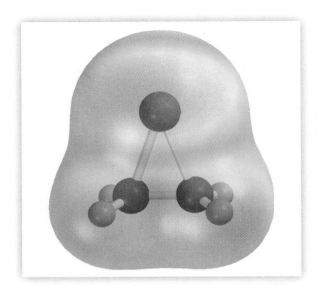

CHAPTER 5
ALKENES AND ALKYNES II: REACTIONS

Now that we know something of the structure and preparation of alkenes and alkynes, we will look at their chemical reactions. The characteristic reaction of alkenes and alkynes is **addition** to the double or triple bond. The general form of addition to an alkene may be represented as

$$A-B + \underset{\diagdown}{\overset{\diagup}{C}}=\underset{\diagup}{\overset{\diagdown}{C}} \longrightarrow A-\overset{|}{\underset{|}{C}}-\overset{|}{\underset{|}{C}}-B$$

The range of compounds represented as A—B in this equation is quite large, and their variety offers a wealth of opportunity for converting alkenes and alkynes to a number of other functional group types.

Alkenes and alkynes are commonly described as **unsaturated hydrocarbons** because they have the capacity to react with substances that add to them. Alkanes, on the other hand, are said to be **saturated** hydrocarbons and are incapable of undergoing addition reactions.

5.1 HYDROGENATION OF ALKENES

The relationship between reactants and products in addition reactions can be illustrated by the hydrogenation of alkenes to yield alkanes. **Hydrogenation** is the addition of H_2 to a multiple bond. An example is the reaction of hydrogen with ethylene to form ethane.

$$H_2C=CH_2 + H-H \xrightarrow{\text{Pt, Pd, Ni, or Rh}} H-CH_2-CH_2-H$$

Ethylene   Hydrogen   Ethane

The bonds in the product are stronger than the bonds in the reactants; two C—H σ bonds of an alkane are formed at the expense of the H—H σ bond and the π component of the alkene's double bond. Thus the overall reaction is exothermic.

The uncatalyzed addition of hydrogen to an alkene, although exothermic, is very slow. The rate of hydrogenation increases dramatically, however, in the presence of certain finely divided metal catalysts. Platinum is the hydrogenation catalyst most often used, although palladium, nickel, and rhodium are also effective. Metal-catalyzed addition of hydrogen is normally rapid at room temperature, and the alkane is produced in high yield, usually as the only product.

> The French chemist Paul Sabatier received the 1912 Nobel Prize in chemistry for his discovery that finely divided nickel is an effective hydrogenation catalyst.

$$(CH_3)_2C=CHCH_3 + H_2 \xrightarrow{\text{Pt}} (CH_3)_2CHCH_2CH_3$$

2-Methyl-2-butene   Hydrogen   2-Methylbutane (100%)

PROBLEM 5.1 What three alkenes yield 2-methylbutane on catalytic hydrogenation?

Catalytic hydrogenation of an alkene is believed to proceed by the series of steps shown in Figure 5.1. In this model for alkene hydrogenation, hydrogen atoms are transferred from the catalyst surface to the alkene. Although the two hydrogens are not transferred simultaneously, both add to the same face of the double bond, as shown in the following example:

$$\text{Dimethyl cyclohexene-1,2-dicarboxylate} + H_2 \xrightarrow{\text{Pt}} \text{Dimethyl cyclohexane-}cis\text{-1,2-dicarboxylate}$$

Dimethyl cyclohexene-1,2-dicarboxylate   Dimethyl cyclohexane-*cis*-1,2-dicarboxylate (100%)

The term **syn addition** describes the stereochemistry of reactions such as catalytic hydrogenation in which two atoms or groups add to the *same face* of a double bond. When atoms or groups add to *opposite faces* of the double bond, the process is called **anti addition.**

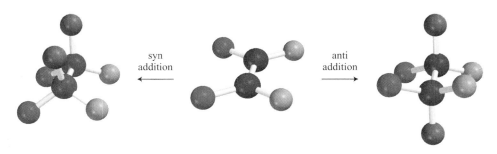

FIGURE 5.1 A mechanism for heterogeneous catalysis in the hydrogenation of alkenes.

Step 1: Hydrogen molecules react with metal atoms at the catalyst surface. The relatively strong hydrogen–hydrogen σ bond is broken and replaced by two weak metal–hydrogen bonds.

Step 2: The alkene reacts with the metal catalyst. The π component of the double bond between the two carbons is replaced by two relatively weak carbon–metal σ bonds.

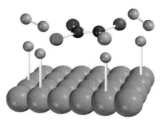

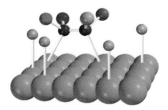

Step 3: A hydrogen atom is transferred from the catalyst surface to one of the carbons of the double bond.

Step 4: The second hydrogen atom is transferred, forming the alkane. The sites on the catalyst surface at which the reaction occurred are free to accept additional hydrogen and alkene molecules.

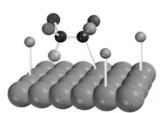

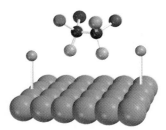

Commercial applications of the hydrogenation of carbon–carbon double bonds include the conversion of vegetable oils to margarine and will be discussed in Chapter 16.

5.2   ELECTROPHILIC ADDITION OF HYDROGEN HALIDES TO ALKENES

In many addition reactions the attacking reagent, unlike H_2, is a polar molecule. Hydrogen halides are among the simplest examples of polar substances that add to alkenes.

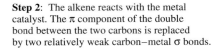

$$\diagdown\!\!\!\diagup C\!=\!C\diagup\!\!\!\diagdown \;\;+\;\; {}^{\delta+}H\!-\!X^{\delta-} \;\longrightarrow\; H\!-\!\underset{|}{\overset{|}{C}}\!-\!\underset{|}{\overset{|}{C}}\!-\!X$$

Alkene   Hydrogen halide   Alkyl halide

$$\underset{H}{\overset{CH_3CH_2}{\diagdown}}C\!=\!C\underset{H}{\overset{CH_2CH_3}{\diagup}} \;\;+\;\; HBr \;\longrightarrow\; CH_3CH_2CH_2\underset{Br}{CH}CH_2CH_3$$

cis-3-Hexene   Hydrogen bromide   3-Bromohexane
                                                                        (76%)

The mechanism we use to describe the addition of hydrogen halides to alkenes is called **electrophilic addition.** Recall from Section 3.10 that an electrophile is electron deficient, and is therefore "electron seeking." Positively charged species are electrophiles, as is the positively polarized hydrogen of a hydrogen halide.

Electrophiles react with substances that can act as a source of electrons. The π electrons of an alkene are more weakly held than the σ electrons and are capable of being attacked by electrophiles such as hydrogen halides, as illustrated in Figure 5.2.

We can use curved arrows to show how an alkene and hydrogen halide react. Recall from Section 4.8 that carbocations are the conjugate acids of alkenes. Acid–base reactions are reversible processes. An alkene, therefore, can accept a proton from a hydrogen halide to form a carbocation.

$$R_2C{=}CR_2 \;+\; H{-}\overset{..}{\underset{..}{X}}: \;\rightleftharpoons\; \overset{+}{R_2C}{-}\overset{\overset{H}{|}}{CR_2} \;+\; :\overset{..}{\underset{..}{X}}:^-$$

Alkene	Hydrogen halide	Carbocation	Anion
(base)	(acid)	(conjugate acid)	(conjugate base)

We've also seen (Section 3.10) that carbocations, when generated in the presence of halide anions, react with them to form alkyl halides.

$$\overset{+}{R_2C}{-}\overset{\overset{H}{|}}{CR_2} \;+\; :\overset{..}{\underset{..}{X}}:^- \;\longrightarrow\; R_2C{-}\overset{\overset{H}{|}}{\underset{\underset{:X:}{|}}{CR_2}}$$

Carbocation	Halide ion	Alkyl halide
(electrophile)	(nucleophile)	

FIGURE 5.2 Electrostatic potential maps of HCl and ethylene. When the two react, the interaction is between the electron-rich site (red) of ethylene and the electron-poor region (blue) of HCl. The electron-rich region of ethylene is associated with the π electrons of the double bond, while H is the electron-poor atom (blue) of HCl.

5.3   REGIOSELECTIVITY OF HYDROGEN HALIDE ADDITION: MARKOVNIKOV'S RULE

In principle, a hydrogen halide can add to an unsymmetrical alkene (an alkene in which the two carbons of the double bond are not equivalently substituted) in either of two directions. In practice, addition is so highly regioselective as to be considered regiospecific.

$$RCH{=}CH_2 \;+\; H{-}X \;\longrightarrow\; \underset{\underset{X}{|}}{RCH}{-}\underset{\underset{H}{|}}{CH_2} \quad \text{rather than} \quad \underset{\underset{H}{|}}{RCH}{-}\underset{\underset{X}{|}}{CH_2}$$

$$R_2C{=}CH_2 \;+\; H{-}X \;\longrightarrow\; \underset{\underset{X}{|}}{R_2C}{-}\underset{\underset{H}{|}}{CH_2} \quad \text{rather than} \quad \underset{\underset{H}{|}}{R_2C}{-}\underset{\underset{X}{|}}{CH_2}$$

$$R_2C{=}CHR \;+\; H{-}X \;\longrightarrow\; \underset{\underset{X}{|}}{R_2C}{-}\underset{\underset{H}{|}}{CHR} \quad \text{rather than} \quad \underset{\underset{H}{|}}{R_2C}{-}\underset{\underset{X}{|}}{CHR}$$

The regioselectivity of hydrogen halide addition to alkenes is stated as **Markovnikov's rule:**

When an unsymmetrically substituted alkene reacts with a hydrogen halide, the hydrogen adds to the carbon that has the greater number of hydrogens, and the halogen adds to the carbon having fewer hydrogens.

Markovnikov's rule is named after Vladimir Markovnikov, a nineteenth-century chemist who was a colleague of Alexander Zaitsev (Section 4.7) at the University of Kazan in Russia.

The following equations provide some examples of addition according to Markovnikov's rule.

$$CH_3CH_2CH{=}CH_2 \ + \qquad HBr \qquad \xrightarrow{\text{acetic acid}} \qquad CH_3CH_2CHCH_3$$
$$\underset{Br}{|}$$

| 1-Butene | Hydrogen bromide | 2-Bromobutane (80%) |

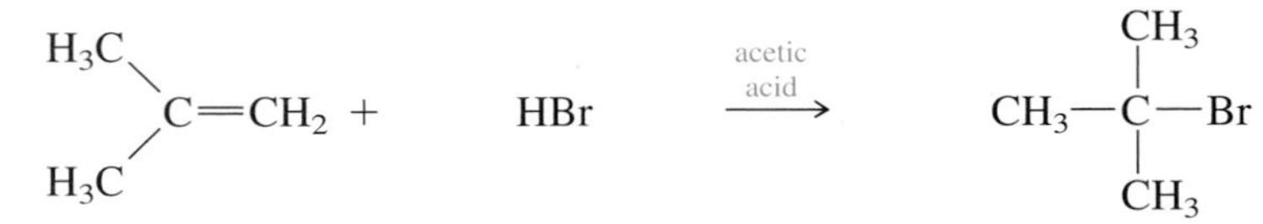

2-Methylpropene   Hydrogen bromide   2-Bromo-2-methylpropane (90%)

1-Methylcyclopentene   Hydrogen chloride   1-Chloro-1-methylcyclopentane (100%)

PROBLEM 5.2 Write the structure of the major organic product formed in the reaction of hydrogen chloride with each of the following:

(a) 2-Methyl-2-butene   (c) *cis*-2-Butene

(b) 2-Methyl-1-butene

(d) 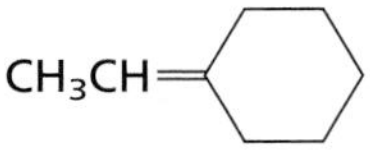 CH₃CH

Sample Solution (a) Hydrogen chloride adds to the double bond of 2-methyl-2-butene in accordance with Markovnikov's rule. The proton adds to the carbon that has one attached hydrogen, chlorine to the carbon that has none.

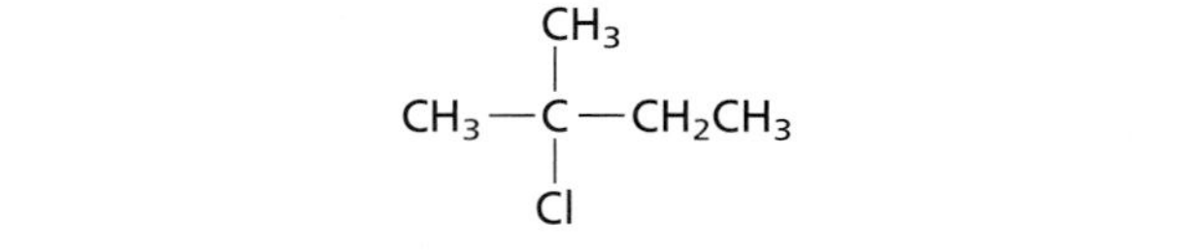

2-Methyl-2-butene

Chlorine becomes attached to this carbon   Hydrogen becomes attached to this carbon

$$CH_3{-}\underset{\underset{Cl}{|}}{\overset{\overset{CH_3}{|}}{C}}{-}CH_2CH_3$$

2-Chloro-2-methylbutane
(major product from Markovnikov addition of hydrogen chloride to 2-methyl-2-butene)

Markovnikov's rule, like Zaitsev's, organizes experimental observations in a form suitable for predicting the major product of a reaction. The reasons it works appear when we examine the mechanism of electrophilic addition in more detail.

5.4 MECHANISTIC BASIS FOR MARKOVNIKOV'S RULE

Let's compare the carbocation intermediates for addition of a hydrogen halide (HX) to an unsymmetrical alkene of the type $RCH{=}CH_2$ (a) according to Markovnikov's rule and (b) opposite to Markovnikov's rule.

(a) *Addition according to Markovnikov's rule:*

| Secondary carbocation | Halide ion | Observed product |

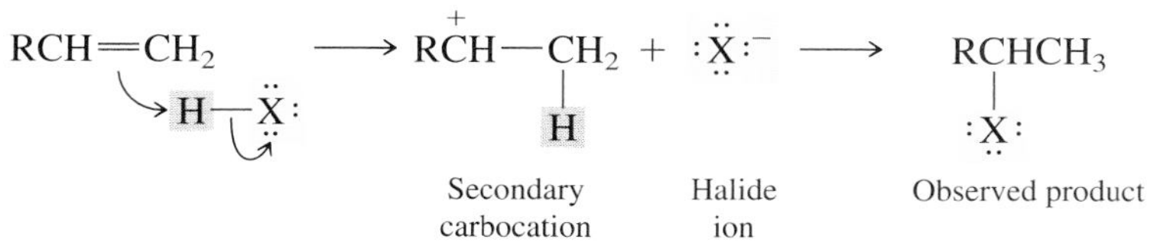

(b) *Addition opposite to Markovnikov's rule:*

| Primary carbocation | Halide ion | Not formed |

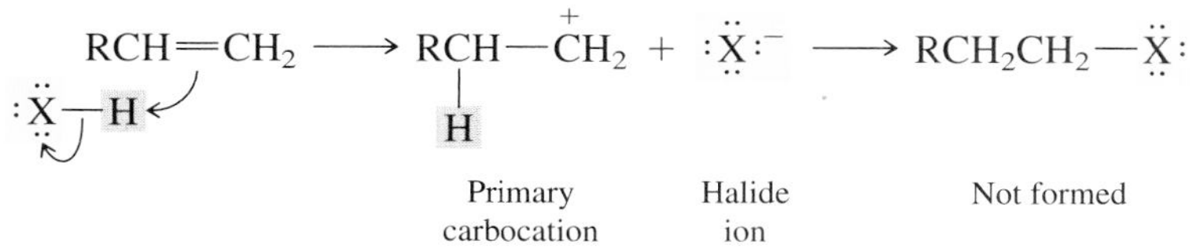

Recall from Section 3.9 that secondary carbocations are more stable than primary ones. The observed product of hydrogen halide addition is formed from the *more stable* of the two possible carbocation intermediates. The secondary carbocation is more stable and is formed faster than the less stable primary carbocation, and the observed product is the secondary alkyl halide. Thus, the regioselectivity expressed in Markovnikov's rule results because a proton adds to the carbon that has the greater number of hydrogens to give the more stable of two possible carbocations, and the product is formed via this intermediate.

PROBLEM 5.3 Give a structural formula for the carbocation intermediate that leads to the major product in each of the reactions of Problem 5.2 (Section 5.3).

Sample Solution (a) Protonation of the double bond of 2-methyl-2-butene can give a tertiary carbocation or a secondary carbocation.

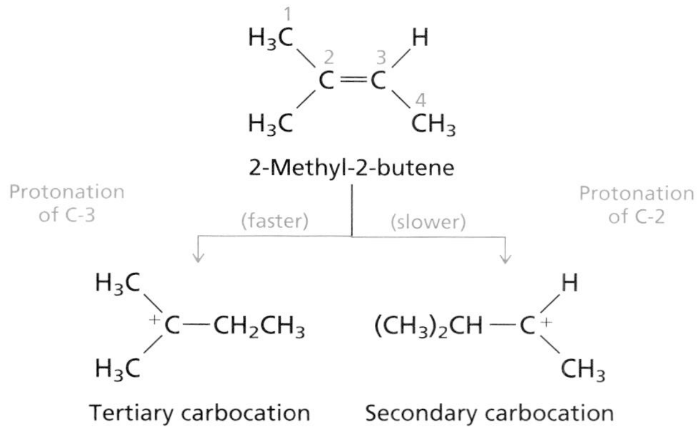

The product of the reaction is derived from the more stable carbocation—in this case, a tertiary carbocation is formed more rapidly than a secondary one.

5.5 ACID-CATALYZED HYDRATION OF ALKENES

Alkenes may be converted to alcohols by the addition of a molecule of water to the carbon–carbon double bond in the presence of an acid catalyst.

$$\underset{\text{Alkene}}{\diagup C = C \diagdown} + \underset{\text{Water}}{HOH} \xrightarrow{H^+} \underset{\text{Alcohol}}{H - \overset{|}{\underset{|}{C}} - \overset{|}{\underset{|}{C}} - OH}$$

Markovnikov's rule is followed:

$$\underset{\text{2-Methyl-2-butene}}{\overset{H_3C}{\underset{H_3C}{>}} C = C \overset{H}{\underset{CH_3}{<}}} \xrightarrow{50\% \ H_2SO_4/H_2O} \underset{\substack{\text{2-Methyl-2-butanol} \\ (90\%)}}{CH_3 - \overset{CH_3}{\underset{OH}{\overset{|}{\underset{|}{C}}}} - CH_2CH_3}$$

We can extend the general principles of electrophilic addition to acid-catalyzed hydration. In the first step of the mechanism shown in Figure 5.3, proton transfer to 2-methylpropene forms *tert*-butyl cation. This is followed in step 2 by reaction of the carbocation with a molecule of water acting as a nucleophile. The alkyloxonium ion formed in this step is simply the conjugate acid of *tert*-butyl alcohol. Deprotonation of the alkyloxonium ion in step 3 yields the alcohol and regenerates the acid catalyst.

> **PROBLEM 5.4** Instead of the three-step mechanism of Figure 5.3, the following two-step mechanism might be considered:
>
> 1. $(CH_3)_2C = CH_2 + H_3O^+ \xrightarrow{\text{slow}} (CH_3)_3C^+ + H_2O$
>
> 2. $(CH_3)_3C^+ + HO^- \xrightarrow{\text{fast}} (CH_3)_3COH$
>
> This mechanism cannot be correct! What is its fundamental flaw?

You may have noticed that the acid-catalyzed hydration of an alkene and the acid-catalyzed dehydration of an alcohol are the reverse of each other.

$$\underset{\text{Alkene}}{\diagup C = C \diagdown} + \underset{\text{Water}}{H_2O} \underset{}{\overset{H^+}{\rightleftharpoons}} \underset{\text{Alcohol}}{H - \overset{|}{\underset{|}{C}} - \overset{|}{\underset{|}{C}} - OH}$$

According to **Le Châtelier's principle,**

A system at equilibrium adjusts so as to minimize any stress applied to it.

When the concentration of water is increased, the system responds by consuming water. This means that proportionally more alkene is converted to alcohol; the position of equilibrium shifts to the right. Thus, when we wish to prepare an alcohol from an alkene, we employ a reaction medium in which the molar concentration of water is high—dilute sulfuric acid, for example.

On the other hand, alkene formation is favored when the concentration of water is kept low. The system responds to the absence of water by causing more alcohol

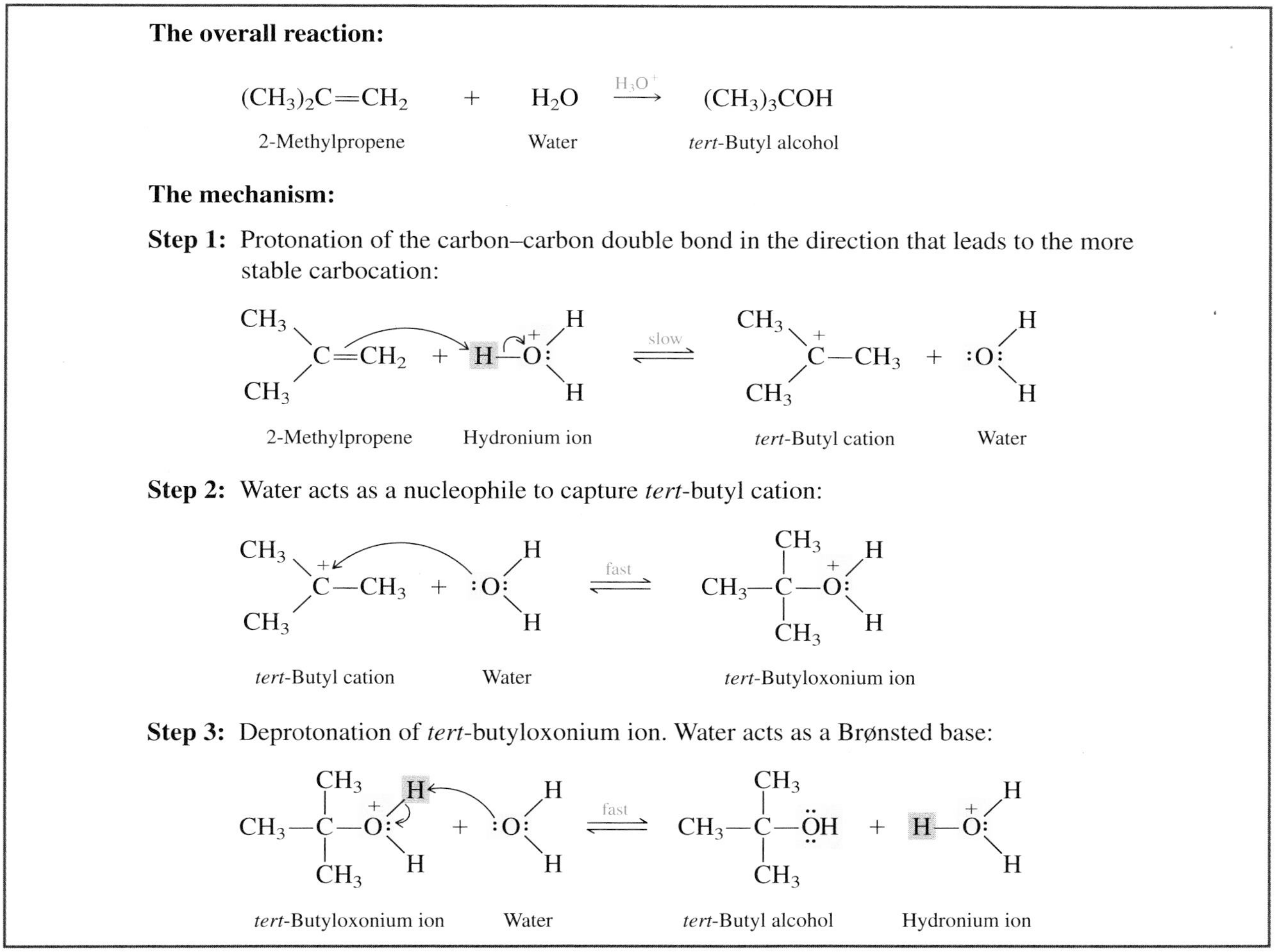

The overall reaction:

$$(CH_3)_2C{=}CH_2 \quad + \quad H_2O \xrightarrow{H_3O^+} \quad (CH_3)_3COH$$

2-Methylpropene   Water   *tert*-Butyl alcohol

The mechanism:

Step 1: Protonation of the carbon–carbon double bond in the direction that leads to the more stable carbocation:

2-Methylpropene   Hydronium ion   *tert*-Butyl cation   Water

Step 2: Water acts as a nucleophile to capture *tert*-butyl cation:

tert-Butyl cation   Water   *tert*-Butyloxonium ion

Step 3: Deprotonation of *tert*-butyloxonium ion. Water acts as a Brønsted base:

tert-Butyloxonium ion   Water   *tert*-Butyl alcohol   Hydronium ion

FIGURE 5.3   Mechanism of acid-catalyzed hydration of 2-methylpropene.

molecules to suffer dehydration, and when alcohol molecules dehydrate, they form more alkene. The amount of water in the reaction mixture is kept low by using concentrated strong acids as catalysts. Distilling the reaction mixture is an effective way of removing water as it is formed, causing the equilibrium to shift toward products. If the alkene is low-boiling, it too can be removed by distillation.

5.6   ADDITION OF HALOGENS TO ALKENES

Halogens react with alkenes by electrophilic addition to give **vicinal** dihalides.

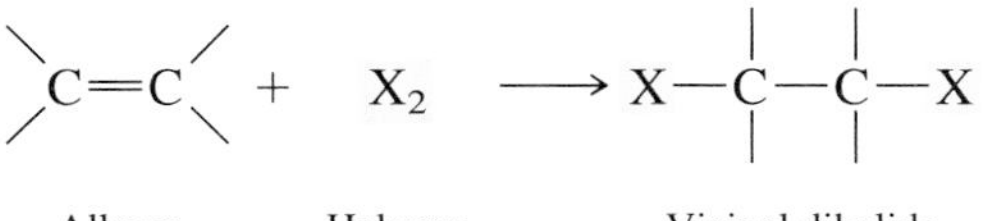

Vicinal dihalides were first seen in Section 4.14.

Alkene   Halogen   Vicinal dihalide

Two substituents, in this case the halogens, are vicinal if they are attached to adjacent carbons. The halogen is either chlorine (Cl_2) or bromine (Br_2), and the addition takes place rapidly at room temperature or below in a variety of solvents. For example,

$$CH_3CH=CHCH(CH_3)_2 \ + \ Br_2 \ \xrightarrow[0°C]{CHCl_3} \ CH_3CH-CHCH(CH_3)_2$$
$$\qquad\qquad\qquad\qquad\qquad\qquad\qquad\qquad\qquad\qquad | \quad\; |$$
$$\qquad\qquad\qquad\qquad\qquad\qquad\qquad\qquad\qquad\quad Br \quad Br$$

4-Methyl-2-pentene   Bromine   2,3-Dibromo-4-methylpentane
                                                             (100%)

> **PROBLEM 5.5**   Draw the structure of the product from the reaction of 2-methyl-propene with chlorine (Cl_2).

When we examine addition of chlorine and bromine to cycloalkenes, an important stereo-chemical feature of these reactions becomes apparent. The halogen substituents in the product are trans to each other.

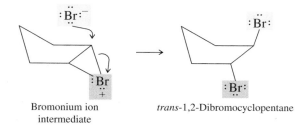

Cyclopentene   Bromine   *trans*-1,2-Dibromocyclopentane
                                               (80% yield; none of the cis
                                               isomer is formed)

The addition is exclusively anti; the two halogens have added to opposite faces of the double bond.

The generally accepted mechanism for halogen addition to alkenes is outlined in Figure 5.4 for the reaction of ethylene with bromine.

$$CH_2=CH_2 \ + \ Br_2 \ \longrightarrow \ BrCH_2CH_2Br$$

Ethylene   Bromine   1,2-Dibromoethane

The key intermediate formed in step 1 of this mechanism is a bridged **halonium ion** (in this case, a bridged **bromonium ion**). The bromine and both carbon atoms of the three-membered ring in the bromonium ion have octets of electrons. Because of this the bromonium ion is more stable than any carbocation intermediate capable of being formed in the reaction. Step 2 is the conversion of the bromonium ion to 1,2-dibro-moethane by reaction with bromide ion (Br^-) from the side opposite the carbon–bromine bond. Recalling that the vicinal dibromide formed from cyclopentene is exclusively the trans stereoisomer, we see that attack by Br^- occurs from the side opposite the C—Br bond of the bromonium ion intermediate.

Bromonium ion   *trans*-1,2-Dibromocyclopentane
intermediate

FIGURE 5.4 Mechanism of electrophilic addition of bromine to ethylene.

The overall reaction:

$$CH_2\!=\!CH_2 \quad + \quad Br_2 \quad \longrightarrow \quad BrCH_2CH_2Br$$

Ethylene   Bromine   1,2-Dibromoethane

The mechanism:

Step 1: Reaction of ethylene and bromine to form a bromonium ion intermediate:

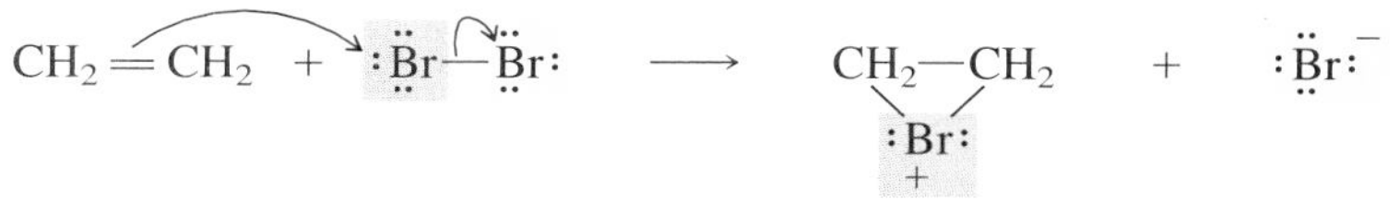

Ethylene   Bromine   Ethylenebromonium   Bromide
                                              ion   ion

Step 2: Nucleophilic attack of bromide anion on the bromonium ion:

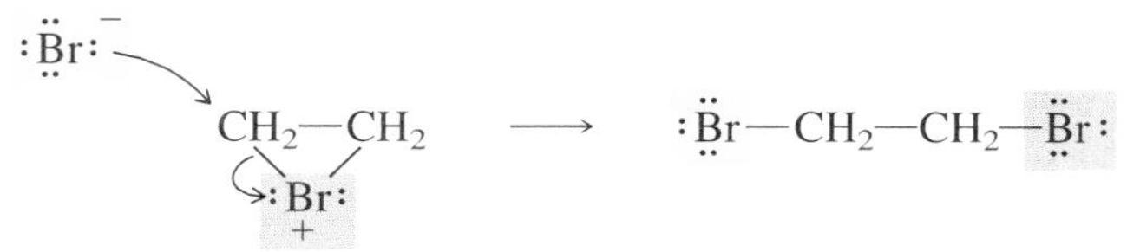

Bromide   Ethylenebromonium   1,2-Dibromoethane
ion   ion

In *aqueous* solution a different reaction occurs when chlorine and bromine react with alkenes. The product is a **vicinal halohydrin,** a compound that has a halogen and a hydroxyl group on adjacent carbons.

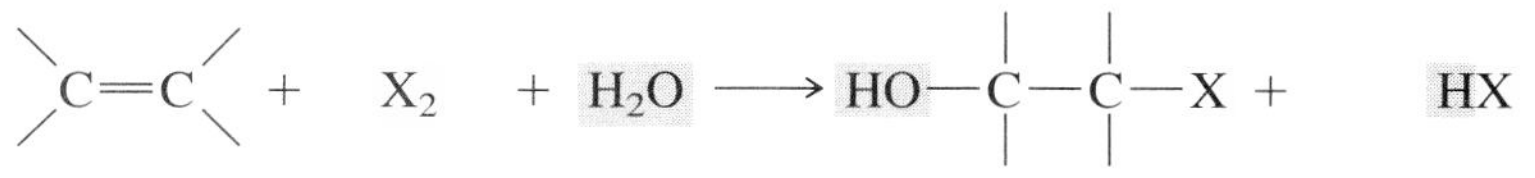

Alkene   Halogen   Water   Halohydrin   Hydrogen halide

$$CH_2\!=\!CH_2 + \quad Br_2 \quad \xrightarrow{H_2O} \quad HOCH_2CH_2Br$$

Ethylene   Bromine   2-Bromoethanol
                                            (70%)

Anti addition occurs. The halogen and the hydroxyl group add to opposite faces of the double bond.

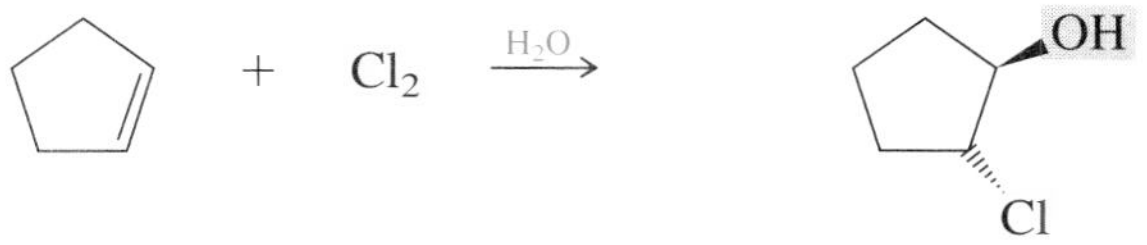

Cyclopentene   Chlorine   *trans*-2-Chlorocyclopentanol
                                    (52–56% yield; cis isomer not formed)

Halohydrin formation is mechanistically similar to halogen addition to alkenes. A halonium ion intermediate is formed, which is attacked by water in aqueous solution.

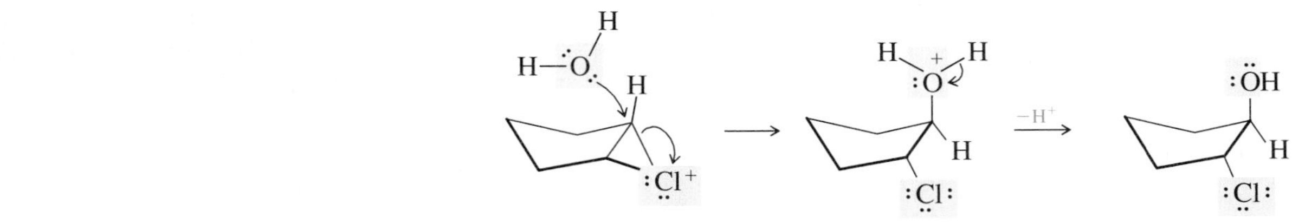

Chloronium ion
intermediate

trans-2-Chlorocyclopentanol

Halogen addition in the presence of water is regioselective; the addition occurs in accordance with Markovnikov's rule. That is, the electrophile (Cl or Br) adds to the less substituted end of the double bond, and the nucleophile (H_2O) adds to the more substituted end.

2-Methylpropene

1-Bromo-2-methyl-
2-propanol
(77%)

PROBLEM 5.6 Give the structure of the product formed when each of the following alkenes reacts with bromine in water:

(a) 2-Methyl-1-butene

(c) 3-Methyl-1-butene

(b) 2-Methyl-2-butene

(d) 1-Methylcyclopentene

Sample Solution (a) The hydroxyl group becomes bonded to the more highly substituted carbon of the double bond, and bromine bonds to the less highly substituted one.

2-Methyl-1-butene   Bromine   1-Bromo-2-methyl-2-butanol

5.7   INTRODUCTION TO ORGANIC CHEMICAL SYNTHESIS

An important concern to chemists is **synthesis,** the challenge of preparing a particular compound in an economical way and with confidence that the method chosen will lead to the desired structure. In this section we will introduce the topic of synthesis, emphasizing the need for systematic planning to decide what is the best sequence of steps to convert a specified starting material to a desired product (the **target molecule**).

A critical feature of synthetic planning is *to reason backward from the target to the starting material*. A second is *to always use reactions that you know will work*.

Let's begin with a simple example. Suppose you wanted to prepare cyclohexane, given cyclohexanol as the starting material. We haven't encountered any reactions so far that permit us to carry out this conversion in a single step.

ETHYLENE: ONE OF THE MOST IMPORTANT INDUSTRIAL ORGANIC CHEMICALS

We discussed ethylene production in an earlier boxed essay (Section 4.1), where it was pointed out that the output of the U.S. petrochemical industry exceeds 5×10^{10} lb/year. Approximately 90% of this material is used for the preparation of four compounds (polyethylene, ethylene oxide, vinyl chloride, and styrene), with polymerization to polyethylene accounting for half the total. Both vinyl chloride and styrene are polymerized to give poly(vinyl chloride) and polystyrene, respectively. Polymerization will be discussed in Chapter 9. Ethylene oxide is a starting material for the preparation of ethylene glycol for use as an antifreeze in automobile radiators and in the production of polyester fibers.

$(-CH_2CH_2-)_n$	Polyethylene	(50%)
H_2C-CH_2 with O bridge	Ethylene oxide	(20%)
$CH_2=CHCl$	Vinyl chloride	(15%)
$C_6H_5-CH=CH_2$	Styrene	(5%)
Other chemicals		(10%)

$CH_2=CH_2$ (Ethylene) →

Among the "other chemicals" prepared from ethylene are ethanol and acetaldehyde:

CH_3CH_2OH

Ethanol
(industrial solvent; used in preparation of ethyl acetate; unleaded gasoline additive)

$$CH_3CH \overset{O}{\underset{\|}{}}$$

Acetaldehyde
(used in preparation of acetic acid)

We have not indicated the reagents employed in the reactions by which ethylene is converted to the compounds shown. Because of patent requirements, different companies often use different processes. Although the processes may be different, they share the common characteristic of being extremely efficient. The industrial chemist faces the challenge of producing valuable materials at low cost. Thus, success in the industrial environment requires both an understanding of chemistry and an appreciation of the economics associated with alternative procedures. One measure of how successfully these challenges have been met can be seen in the fact that the United States maintains a positive trade balance in chemicals each year. In 1999 that surplus amounted to $9.8 billion in chemicals versus an overall trade deficit of $332 billion.

Cyclohexanol → Cyclohexane

Reasoning backward, however, we know that we can prepare cyclohexane by hydrogenation of cyclohexene. We'll therefore use this reaction as the last step in our proposed synthesis.

Cyclohexene $\xrightarrow{\text{catalytic hydrogenation}}$ Cyclohexane

Recognizing that cyclohexene may be prepared by dehydration of cyclohexanol, a practical synthesis of cyclohexane from cyclohexanol becomes apparent.

$$\text{Cyclohexanol} \xrightarrow[\text{heat}]{H_2SO_4} \text{Cyclohexene} \xrightarrow[\text{Pt}]{H_2} \text{Cyclohexane}$$

As a second example, consider the preparation of 1-bromo-2-methyl-2-propanol from *tert*-butyl alcohol.

$$(CH_3)_3COH \longrightarrow (CH_3)_2CCH_2Br$$
$$\underset{\text{OH}}{|}$$

tert-Butyl alcohol   1-Bromo-2-methyl-2-propanol

Begin by asking the question, "What kind of compound is the target molecule, and what methods can I use to prepare that kind of compound?" The desired product has a bromine and a hydroxyl on adjacent carbons; it is a vicinal bromohydrin. The only method we have learned so far for the preparation of vicinal bromohydrins involves the reaction of alkenes with Br_2 in water. Thus, a reasonable last step is

$$(CH_3)_2C{=}CH_2 \xrightarrow[H_2O]{Br_2} (CH_3)_2CCH_2Br$$
$$\underset{\text{OH}}{|}$$

2-Methylpropene   1-Bromo-2-methyl-2-propanol

We now have a new problem: Where does the necessary alkene come from? Alkenes are prepared from alcohols by acid-catalyzed dehydration (Section 4.7) or from alkyl halides by E2 elimination (Section 4.9). Because our designated starting material is *tert*-butyl alcohol, we can combine its dehydration with bromohydrin formation to give the correct sequence of steps:

$$(CH_3)_3COH \xrightarrow[\text{heat}]{H_2SO_4} (CH_3)_2C{=}CH_2 \xrightarrow[H_2O]{Br_2} (CH_3)_2CCH_2Br$$
$$\underset{\text{OH}}{|}$$

tert-Butyl alcohol   2-Methylpropene   1-Bromo-2-methyl-2-propanol

> **PROBLEM 5.7** Write a series of equations describing a synthesis of 1-bromo-2-methyl-2-propanol from *tert*-butyl bromide.

Often more than one synthetic route may be available to prepare a particular compound. Indeed, it is normal to find in the chemical literature that the same compound has been synthesized in a number of different ways. As we proceed through the text and develop a larger inventory of functional group transformations, our ability to evaluate alternative synthesis plans will increase. In most cases the best synthesis plan is the one with the fewest steps.

5.8   ELECTROPHILIC ADDITION REACTIONS OF CONJUGATED DIENES

Our discussion of chemical reactions of alkadienes will be limited to those of conjugated dienes. The reactions of isolated dienes are essentially the same as those of individual alkenes. The reactions of cumulated dienes are—like their preparation—so specialized that their treatment is better suited to an advanced course in organic chemistry.

Electrophilic addition is the characteristic chemical reaction of alkenes, and conjugated dienes undergo addition reactions with the same electrophiles that react with alkenes, and by similar mechanisms. As the reaction of 1,3-butadiene with hydrogen chloride illustrates, however, conjugated dienes exhibit a richer spectrum of reactivity than do simple alkenes.

$$CH_2{=}CHCH{=}CH_2 \xrightarrow{\text{HCl}} \underset{\underset{\displaystyle Cl}{|}}{CH_3CHCH{=}CH_2} + CH_3CH{=}CHCH_2Cl$$

1,3-Butadiene   3-Chloro-1-butene   1-Chloro-2-butene

Two products are formed. One of them, 3-chloro-1-butene, is said to be the product of **direct addition,** or **1,2 addition.** The other, 1-chloro-2-butene, is said to be the product of **conjugate addition,** or **1,4 addition.** Both correspond to proton addition to the end of the conjugated diene unit but differ in respect to the position of attachment of the halogen and the location of the double bond. The proton and the chlorine add to adjacent atoms in the formation of the 1,2-addition product; they add to the ends of the diene unit in forming the 1,4-addition product.

> Note that the numbers 1 and 2 in 1,2 addition and 1 and 4 in 1,4 addition do not refer to the locants in the IUPAC name of the compound but to the relative positions of carbons within the conjugated diene structural unit.

PROBLEM 5.8 Write structural formulas corresponding to the 1,2- and 1,4-addition products formed on reaction of each of the following alkadienes with hydrogen bromide.

(a) 2,4-Hexadiene   (b) 2,3-Dimethyl-1,3-butadiene   (c) 1,3-Cyclopentadiene

Sample Solution The hydrogen and bromine add to adjacent carbons in 1,2 addition and to the ends of the conjugated diene system in 1,4 addition.

1,2 Addition

$$\underset{1}{CH_3CH}{=}\underset{2}{CH}{-}CH{=}CHCH_3 \xrightarrow{\text{HBr}} CH_3CH_2{-}\underset{\underset{\displaystyle Br}{|}}{CH}{-}CH{=}CHCH_3$$

2,4-Hexadiene   4-Bromo-2-Hexene

1,4 Addition

$$\underset{1}{CH_3CH}{=}CH{-}CH{=}\underset{4}{CHCH_3} \xrightarrow{\text{HBr}} CH_3CH_2{-}CH{=}CH{-}\underset{\underset{\displaystyle Br}{|}}{CHCH_3}$$

2,4-Hexadiene   2-Bromo-3-Hexene

We can account for the formation of both products on the basis of the mechanism for hydrogen halide addition to alkenes once we recognize a unique feature of the carbocation intermediate that is formed by protonation of 1,3-butadiene.

$$\overset{+}{H} + CH_2{=}CHCH{=}CH_2 \longrightarrow CH_3\overset{+}{C}HCH{=}CH_2$$

This type of carbocation is called an **allylic carbocation.** The positive charge of an allylic carbocation is shared by two carbons. This delocalization of the positive charge can be described by the orbital overlap view presented in Figure 5.5 or with resonance structures:

FIGURE 5.5 Electron delocalization in allylic carbocation. (*a*) The π orbital of the double bond, and the vacant 2*p* orbital of the positively charged carbon. (*b*) Overlap of the π orbital and the 2*p* orbital gives an extended π orbital that encompasses all three carbons. The two electrons in the π bond are delocalized over two carbons in part (*a*) and over three carbons in part (*b*).

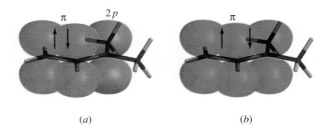

$$CH_3\overset{+}{C}H-CH=CH_2 \longleftrightarrow CH_3CH=CH-\overset{+}{C}H_2$$

A negatively charged chloride ion can react with the resonance-stabilized carbocation at either of the two carbons that share the positive charge. When chloride attacks the carbocation at its secondary carbon, 3-chloro-1-butene is formed; when chloride attacks the primary carbon, 1-chloro-2-butene is the product.

$$
\begin{array}{c}
CH_3\overset{+}{C}HCH=CH_2 \\
\updownarrow \\
CH_3CH=CH\overset{+}{C}H_2
\end{array}
\xrightarrow{Cl^-}
\underset{\underset{Cl}{|}}{CH_3CHCH=CH_2} + CH_3CH=CHCH_2Cl
$$

3-Chloro-1-butene   1-Chloro-2-butene

Allylic carbocations are more stable than simple alkyl carbocations and are formed more readily. Thus, reactions which proceed through carbocation intermediates occur relatively rapidly when the carbocation is allylic.

A mixture of 1,2 and 1,4 addition is observed as well in the addition of chlorine or bromine to conjugated dienes.

$$
CH_2=CHCH=CH_2 + Br_2 \xrightarrow{CHCl_3} \underset{\underset{Br}{|}}{BrCH_2CHCH=CH_2} + \underset{H}{\overset{BrCH_2}{\diagdown}}C=C\underset{CH_2Br}{\overset{H}{\diagup}}
$$

1,3-Butadiene   Bromine   3,4-Dibromo-1-butene   (*E*)-1,4-Dibromo-2-butene
                                              (37%)   (63%)

> **PROBLEM 5.9** Exclusive of stereoisomers, how many products are possible in the electrophilic addition of 1 mole of bromine to 2-methyl-1,3-butadiene?

5.9   ACIDITY OF ACETYLENE AND TERMINAL ALKYNES

Before describing addition reactions of alkynes, let us first discuss the property of terminal alkynes that most distinguishes them from other hydrocarbons—their acidity.

As measured by their acid-dissociation constants K_a (Section 3.5), hydrocarbons are exceedingly weak acids. A carbon–hydrogen bond has little tendency to ionize according to the equation.

$$R-H \rightleftharpoons R:^- + H^+$$

Hydrocarbon   Carbanion   Proton

Loss of a proton from a hydrocarbon leaves an anion in which the negative charge is borne by carbon; such a species is called a **carbanion.** Carbon is not very electronegative, and the formation of carbanions by ionization of C—H bonds is characterized by very small values of K_a. However, the *range* of K_a values exhibited by hydrocarbons varies enormously when comparing ethane, ethylene, and acetylene as a representative alkane, alkene, and alkyne, respectively. As measured by K_a acetylene is 10^{19} times more acidic than ethylene and 10^{36} times more acidic than ethane.

$$HC{\equiv}CH \quad > \quad CH_2{=}CH_2 \quad > \quad CH_3CH_3$$

Acetylene	Ethylene	Ethane
$K_a = 10^{-26}$	$\sim 10^{-45}$	$\sim 10^{-62}$
$pK_a = 26$	~ 45	~ 62
(strongest acid)		(weakest acid)

The greater acidity of acetylene is attributed to a hybridization effect. When carbon is *sp* hybridized, it more electronegative than when it is *sp²-* or *sp³*-hybridized. Thus, the carbanion that results from ionization of acetylene binds its electron pair more strongly than do the carbanions from ethylene and ethane, is more stable than they are, and its formation is associated with a larger equilibrium constant.

Terminal alkynes (RC≡CH) resemble acetylene in acidity.

$$(CH_3)_3CC{\equiv}CH \qquad K_a = 3 \times 10^{-26} \ (pK_a = 25.5)$$

3,3-Dimethyl-1-butyne

Hydrocarbons with internal triple bonds, compounds of the type RC≡CR′, lack protons bonded to *sp*-hybridized carbon and are more like alkanes in terms of their acid strength.

Although acetylene and terminal alkynes are far stronger acids than other hydrocarbons, we must remember that they are, nevertheless, very weak acids—much weaker than water and alcohols, for example. Hydroxide ion is too weak a base to convert acetylene to its anion in meaningful amounts. The position of the equilibrium described by the following equation lies overwhelmingly to the left:

$$H{-}C{\equiv}C{-}H + {}^{-}{:}\ddot{O}H \ \rightleftharpoons\ H{-}C{\equiv}C{:}^{-} + H{-}\ddot{O}H$$

Acetylene	Hydroxide ion	Acetylide ion	Water
(weaker acid)	(weaker base)	(stronger base)	(stronger acid)
$K_a = 10^{-26}$			$K_a = 1.8 \times 10^{-16}$
$pK_a = 26$			$pK_a = 15.7$

Because acetylene is a far weaker acid than water and alcohols, these substances are not suitable solvents for reactions involving acetylide ions. Acetylide is instantly converted to acetylene by proton transfer from compounds that contain hydroxyl groups.

Amide ion is a much stronger base than acetylide ion and converts acetylene to its conjugate base quantitatively.

$$H{-}C{\equiv}C{-}H + {}^{-}{:}\ddot{N}H_2 \ \rightleftharpoons\ H{-}C{\equiv}C{:}^{-} + H{-}\ddot{N}H_2$$

Acetylene	Amide ion	Acetylide ion	Ammonia
(stronger acid)	(stronger base)	(weaker base)	(weaker acid)
$K_a = 10^{-26}$			$K_a = 10^{-36}$
$pK_a = 26$			$pK_a = 36$

Solutions of sodium acetylide (HC≡CNa) may be prepared by adding sodium amide (NaNH₂) to acetylene in liquid ammonia as the solvent. Terminal alkynes react similarly to give species of the type RC≡CNa.

PROBLEM 5.10 Complete each of the following equations to show the conjugate acid and the conjugate base formed by proton transfer between the indicated species. Use curved arrows to show the flow of electrons, and specify whether the position of equilibrium lies to the side of reactants or products.

(a) $CH_3C\equiv CH + \ :\!\ddot{O}CH_3 \ \rightleftharpoons$

(b) $HC\equiv CH + H_2\ddot{C}CH_3 \ \rightleftharpoons$

(c) $CH_3C\equiv CCH_2OH + \ :\!\ddot{N}H_2 \ \rightleftharpoons$

Sample Solution (a) The equation representing the acid–base reaction between propyne and methoxide ion is

$$CH_3C\equiv C{-}H + \quad :\!\ddot{O}CH_3 \quad \rightleftharpoons \quad CH_3C\equiv C:^- \quad + \quad H{-}\ddot{O}CH_3$$

Propyne	Methoxide ion	Propynide ion	Methanol
(weaker acid)	(weaker base)	(stronger base)	(stronger acid)

Alcohols are stronger acids than acetylene, and so the position of equilibrium lies to the left. Methoxide ion is not a strong enough base to remove a proton from acetylene.

5.10 PREPARATION OF ALKYNES BY ALKYLATION

The reactions described so far for the preparation of alkynes involved introducing a triple bond into an existing carbon chain by double dehydrohalogenation of a geminal dihalide or a vicinal dihalide (Section 4.14). In this section we will see how to prepare alkynes while building longer carbon chains. By attaching alkyl groups to acetylene, more complex alkynes can be prepared.

$$H{-}C\equiv C{-}H \longrightarrow R{-}C\equiv C{-}H \longrightarrow R{-}C\equiv C{-}R'$$

Acetylene	Monosubstituted or terminal alkyne	Disubstituted derivative of acetylene

Reactions that attach alkyl groups to molecular fragments are called **alkylation** reactions. One way in which alkynes are prepared is by alkylation of acetylene.

Alkylation of acetylene involves a sequence of two separate operations. In the first one, acetylene is converted to its conjugate base by treatment with sodium amide.

$$HC\equiv CH + \quad NaNH_2 \quad \longrightarrow \quad HC\equiv CNa \quad + \quad NH_3$$

Acetylene	Sodium amide	Sodium acetylide	Ammonia

Next, an alkyl halide (the **alkylating agent**) is added to the solution of sodium acetylide. Acetylide ion acts as a nucleophile, displacing halide from carbon and forming a new carbon–carbon bond.

$$HC\equiv CNa + \quad RX \quad \longrightarrow HC\equiv CR + \quad NaX \qquad via \qquad HC\equiv C:^{\frown} R{-}X$$

Sodium acetylide	Alkyl halide	Alkyne	Sodium halide

This substitution reaction must be carried out with alkyl halides that are either methyl (CH_3X) or primary (RCH_2X). Acetylide ions are very basic, much more basic than hydroxide ion, for example, and react with secondary and tertiary alkyl halides by elimination. The reasons for this will be explored further in Chapter 8. The synthetic sequence is usually carried out in liquid ammonia as the solvent. Alternatively, diethyl ether or tetrahydrofuran may be used.

$$HC\equiv CNa \quad + \quad CH_3CH_2CH_2CH_2Br \xrightarrow{NH_3} CH_3CH_2CH_2CH_2C\equiv CH$$

Sodium acetylide   1-Bromobutane   1-Hexyne
                                                          (70–77%)

An analogous sequence using terminal alkynes as starting materials yields alkynes of the type $RC\equiv CR'$.

$$(CH_3)_2CHCH_2C\equiv CH \xrightarrow[NH_3]{NaNH_2} (CH_3)_2CHCH_2C\equiv CNa \xrightarrow{CH_3Br} (CH_3)_2CHCH_2C\equiv CCH_3$$

4-Methyl-1-pentyne   5-Methyl-2-hexyne
                                                          (81%)

Dialkylation of acetylene can be achieved by carrying out the sequence twice.

$$HC\equiv CH \xrightarrow[\substack{2.\ CH_3CH_2Br}]{1.\ NaNH_2,\ NH_3} HC\equiv CCH_2CH_3 \xrightarrow[\substack{2.\ CH_3Br}]{1.\ NaNH_2,\ NH_3} CH_3C\equiv CCH_2CH_3$$

Acetylene   1-Butyne   2-Pentyne
                                                          (81%)

PROBLEM 5.11   Outline efficient syntheses of each of the following alkynes from acetylene and any necessary organic or inorganic reagents:

(a) 1-Heptyne   (b) 2-Heptyne   (c) 3-Heptyne

Sample Solution   (a) An examination of the structural formula of 1-heptyne reveals it to have a pentyl group attached to an acetylene unit. Alkylation of acetylene, by way of its anion, with a pentyl halide is a suitable synthetic route to 1-heptyne.

$$HC\equiv CH \xrightarrow[NH_3]{NaNH_2} HC\equiv CNa \xrightarrow{CH_3CH_2CH_2CH_2CH_2Br} HC\equiv CCH_2CH_2CH_2CH_2CH_3$$

Acetylene   Sodium acetylide   1-Heptyne

5.11   ADDITION REACTIONS OF ALKYNES

We have just seen one important chemical property of alkynes, the acidity of acetylene and terminal alkynes. In this section we will explore several other reactions of alkynes. Like alkenes, alkynes undergo addition reactions.

Hydrogenation

The conditions for hydrogenation of alkynes are similar to those employed for alkenes. In the presence of finely divided platinum, palladium, nickel, or rhodium, two moles of hydrogen add to the triple bond of an alkyne to yield an alkane.

$$RC\equiv CR' + \quad 2H_2 \xrightarrow{Pt,\ Pd,\ Ni,\ or\ Rh} RCH_2CH_2R'$$

Alkyne   Hydrogen   Alkane

$$CH_3CH_2\underset{\underset{CH_3}{|}}{C}HCH_2C\equiv CH + \quad 2H_2 \xrightarrow{Ni} CH_3CH_2\underset{\underset{CH_3}{|}}{C}HCH_2CH_2CH_3$$

4-Methyl-1-hexyne   Hydrogen   3-Methylhexane
                                                          (77%)

Alkenes are intermediates in the hydrogenation of alkynes to alkanes, a fact which has led to the development of methods for preparation of alkenes from alkynes as starting materials. Using a specially prepared palladium catalyst known as **Lindlar palladium,** it is possible to halt the hydrogenation of an alkyne cleanly after 1 mole of H_2 has been consumed per mole of alkyne. Syn addition of hydrogen to the triple bond takes place at the surface of the metal, and the product is a cis (or Z) alkene.

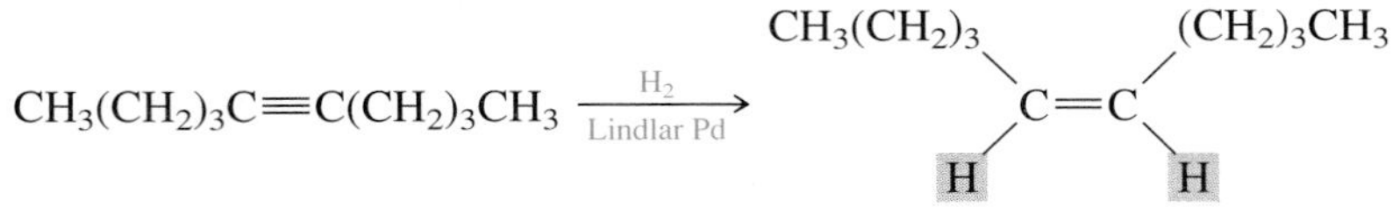

$$CH_3(CH_2)_3C{\equiv}C(CH_2)_3CH_3 \xrightarrow[\text{Lindlar Pd}]{H_2}$$

$$\begin{array}{c} CH_3(CH_2)_3 \qquad\qquad (CH_2)_3CH_3 \\ C{=}C \\ H \qquad\qquad\qquad H \end{array}$$

5-Decyne

(Z)-5-Decene
(87%)

> **PROBLEM 5.12** Write a series of equations showing how you could prepare *cis*-2-pentene from 1-butyne.

Lindlar palladium is said to be a "poisoned" catalyst. Treatment of palladium with a combination of lead acetate, barium sulfate, and a substance called quinoline degrades its catalytic activity so that, while it is still an effective catalyst for the hydrogenation of alkynes to alkenes, it is a poor catalyst for the hydrogenation of alkenes to alkanes.

Metal–Ammonia Reduction

A useful alternative to catalytic partial hydrogenation for converting alkynes to alkenes is reduction by a Group I metal (lithium, sodium, or potassium) in liquid ammonia. The unique feature of metal—ammonia reduction is that it converts alkynes to trans (or E) alkenes, whereas catalytic hydrogenation yields cis (or Z) alkenes. Thus, from the same alkyne one can prepare either a cis or a trans alkene by choosing the appropriate reaction conditions.

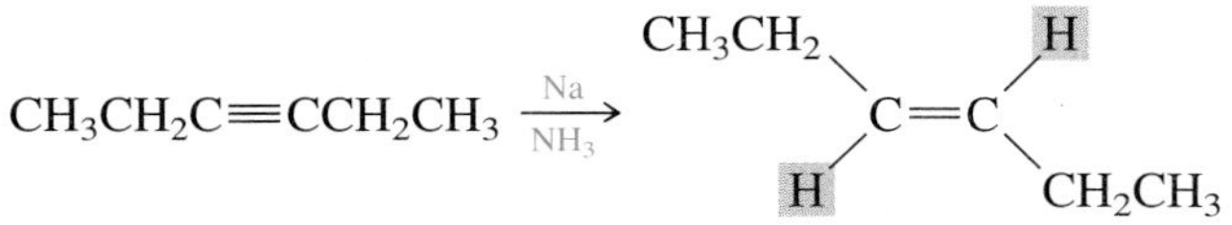

$$CH_3CH_2C{\equiv}CCH_2CH_3 \xrightarrow[\text{NH}_3]{Na}$$

$$\begin{array}{c} CH_3CH_2 \qquad\qquad H \\ C{=}C \\ H \qquad\qquad CH_2CH_3 \end{array}$$

3-Hexyne

(E)-3-Hexene
(82%)

> **PROBLEM 5.13** Suggest efficient syntheses of (*E*)- and (*Z*)-2-heptene from propyne and any necessary organic or inorganic reagents.

It is important to recognize the metal–ammonia reduction of alkynes does not involve hydrogen (H_2) in any way. It is a chemical reduction, not a hydrogenation. Sodium and ammonia are reactants, not catalysts.

Addition of Hydrogen Halides

Alkynes react with many of the same electrophilic reagents that add to the carbon–carbon double bond of alkenes. Hydrogen halides, for example, add to alkynes to form alkenyl halides.

$$RC \equiv CR' + \quad HX \quad \longrightarrow \quad RCH = CR'$$
$$\underset{X}{|}$$

Alkyne   Hydrogen halide   Alkenyl halide

The regioselectivity of addition follows Markovnikov's rule. A proton adds to the carbon that has the greater number of hydrogens, and halide adds to the carbon with the fewer hydrogens.

$$CH_3CH_2CH_2CH_2C \equiv CH \ + \quad HBr \quad \longrightarrow \quad CH_3CH_2CH_2CH_2C = CH_2$$
$$\underset{Br}{|}$$

1-Hexyne   Hydrogen bromide   2-Bromo-1-hexene
(60%)

In the presence of excess hydrogen halide, geminal dihalides are formed by sequential addition of two molecules of hydrogen halide to the carbon–carbon triple bond. The industrial chemical 1,1-difluoroethane, for example, is prepared by the reaction of acetylene with hydrogen fluoride.

$$HC \equiv CH \ + \ HF \ \longrightarrow \ CH_2 = CHF \ \overset{HF}{\longrightarrow} \ CH_3CHF_2$$

Acetylene   Hydrogen   Fluoroethylene   1,1-Difluoroethane
fluoride

Hydration

By analogy to the hydration of alkenes, hydration of an alkyne is expected to yield an alcohol. The kind of alcohol, however, would be of a special kind, one in which the hydroxyl group is a substituent on a carbon–carbon double bond. This type of alcohol is called an **enol** (the double bond suffix -*ene* plus the alcohol suffix -*ol*). An important property of enols is their rapid isomerization to aldehydes or ketones under the conditions of their formation.

$$RC \equiv CR' \ + \ H_2O \ \overset{slow}{\longrightarrow} \ RCH = CR' \ \overset{fast}{\longrightarrow} \ RCH_2CR'$$

Alkyne   Water   Enol
(not isolated)

The process by which enols are converted to aldehydes or ketones is called **keto–enol isomerism** or **keto–enol tautomerism.** We will defer discussion of this process until we treat aldehydes and ketones in more detail in Chapter 11.

PROBLEM 5.14 Give the structure of the ketone formed by hydration of 2-butyne. What is the structure of the enol intermediate?

Typically, a combination of sulfuric acid and mercury(II) sulfate is used as the hydration catalyst. Hydration follows Markovnikov's rule, and terminal alkynes yield methyl-substituted ketones.

$$HC \equiv CCH_2CH_2CH_2CH_2CH_3 \ + \ H_2O \ \overset{H_2SO_4}{\underset{HgSO_4}{\longrightarrow}} \ CH_3CCH_2CH_2CH_2CH_2CH_3$$

1-Octyne   2-Octanone
(91%)

[**PROBLEM 5.15**   Write the structure of the enol intermediate in the hydration of 1-octyne.]

Because of the regioselectivity of alkyne hydration, acetylene is the only alkyne structurally capable of yielding an aldehyde under these conditions.

$$HC{\equiv}CH + H_2O \longrightarrow CH_2{=}CHOH \longrightarrow CH_3\overset{\overset{\displaystyle O}{\|}}{C}H$$

Acetylene	Water	Vinyl alcohol	Acetaldehyde
		(not isolated)	

At one time acetaldehyde was prepared on an industrial scale by this method. Modern methods involve direct oxidation of ethylene and are more economical.

LEARNING OBJECTIVES

This chapter introduced the rich chemistry associated with the reactions of carbon–carbon double and triple bonds. Although numerous reactions were included, they can be learned readily if you recognize that most are *addition* reactions, and most of these are *electrophilic* additions. The skills you have learned in this chapter should enable you to:

- Write a chemical equation expressing the addition of each of the following to a representative alkene:

   Hydrogen in the presence of a suitable catalyst
   Hydrogen halide (chloride, bromide, iodide)
   Water containing a trace of acid
   Chlorine or bromine (with and without water present)

- State and give an example of Markovnikov's rule.

- Contrast 1,2 addition and 1,4 addition of a hydrogen halide to a conjugated diene.

- Contrast the acidity of acetylene and terminal alkynes with other hydrocarbons, and describe reaction conditions suitable for converting acetylene and terminal alkynes to their derived carbanions.

- Write a chemical equation for the alkylation of acetylene or a terminal alkyne via the corresponding carbanion.

- Write appropriate chemical equations showing how to convert an alkyne to either a cis alkene or a trans alkene.

- Explain what is meant by the following terms.

Anti addition	Regioselective reaction
Halonium ion	Syn addition
Lindlar palladium	

5.12   SUMMARY

Alkenes, alkadienes, and alkynes are **unsaturated hydrocarbons** and undergo reactions with substances that add to their multiple bonds. Representative addition reactions of alkenes are summarized in Table 5.1. Except for catalytic hydrogenation, these reactions

TABLE 5.1	Reactions of Alkenes

Reaction (Section) and Comments	General Equation and Specific Example
Catalytic hydrogenation (Section 5.1) Alkenes react with hydrogen in the presence of a platinum, palladium, rhodium, or nickel catalyst to form the corresponding alkane.	$R_2C{=}CR_2$ + H_2 $\xrightarrow{\text{Pt, Pd, Rh, or Ni}}$ R_2CHCHR_2 Alkene Hydrogen Alkane 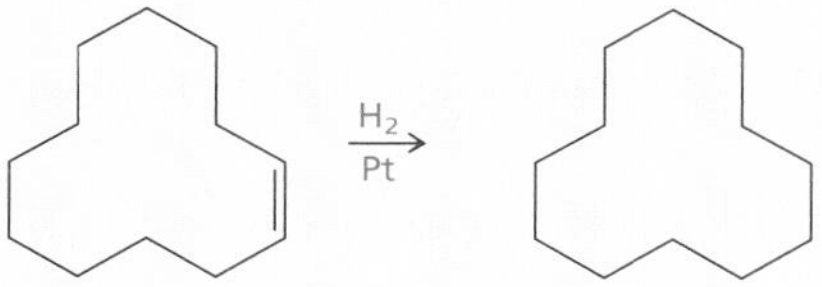 cis-Cyclododecene $\xrightarrow[\text{Pt}]{H_2}$ Cyclododecane (100%)
Addition of hydrogen halides (Sections 5.2–5.4) A proton and a halogen add to the double bond of an alkene to yield an alkyl halide. Addition proceeds in accordance with Markovnikov's rule; hydrogen adds to the carbon that has the greater number of hydrogens, halide to the carbon that has the fewer hydrogens.	$RCH{=}CR'_2$ + HX $\longrightarrow$ $RCH_2{-}\underset{\underset{X}{\vert}}{C}R'_2$ Alkene Hydrogen halide Alkyl halide 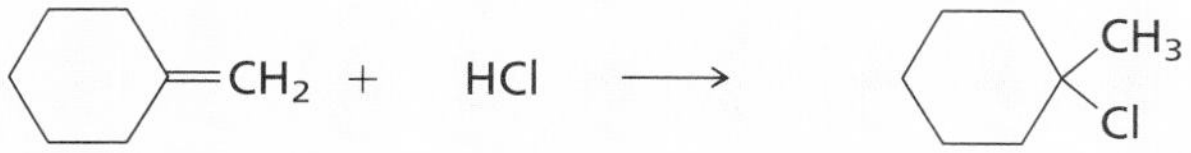 Methylenecyclohexane + HCl (Hydrogen chloride) $\longrightarrow$ 1-Chloro-1-methylcyclohexane (75–80%)
Acid-catalyzed hydration (Section 5.5) Addition of water to the double bond of an alkene takes place in aqueous acid. Addition occurs according to Markovnikov's rule. A carbocation is an intermediate and is captured by a molecule of water acting as a nucleophile.	$RCH{=}CR'_2$ + H_2O $\xrightarrow{H^+}$ $RCH_2\underset{\underset{OH}{\vert}}{C}R'_2$ Alkene Water Alcohol $CH_2{=}C(CH_3)_2$ $\xrightarrow{50\%\ H_2SO_4/H_2O}$ $(CH_3)_3COH$ 2-Methylpropene tert-Butyl alcohol (55–58%)
Addition of halogens (Section 5.6) Bromine and chlorine add to alkenes to form vicinal dihalides. A cyclic halonium ion is an intermediate. Anti addition is observed.	$R_2C{=}CR_2$ + X_2 $\longrightarrow$ $X{-}\underset{\underset{R}{\vert}}{\overset{\overset{R}{\vert}}{C}}{-}\underset{\underset{R}{\vert}}{\overset{\overset{R}{\vert}}{C}}{-}X$ Alkene Halogen Vicinal dihalide $CH_2{=}CHCH_2CH_2CH_2CH_3$ + Br_2 $\longrightarrow$ $BrCH_2\underset{\underset{Br}{\vert}}{C}HCH_2CH_2CH_2CH_3$ 1-Hexene Bromine 1,2-Dibromohexane (100%)

(Continued)

TABLE 5.1	Reactions of Alkenes *(Continued)*

Reaction (Section) and Comments	General Equation and Specific Example
Halohydrin formation (Section 5.6) When treated with bromine or chlorine in aqueous solution, alkenes are converted to vicinal halohydrins. A halonium ion is an intermediate. The halogen adds to the carbon that has the greater number of hydrogens. Addition is anti.	$$RCH{=}CR_2' + X_2 + H_2O \longrightarrow X{-}CH{-}\underset{\underset{R'}{\overset{\|}{R}}}{\overset{\overset{R'}{\|}}{C}}{-}OH + HX$$ Alkene Halogen Water Vicinal Hydrogen halohydrin halide Methylenecyclohexane $\xrightarrow{\underset{H_2O}{Br_2}}$ (1-Bromomethyl)cyclohexanol (89%)

proceed by **electrophilic** attack of the reagent on the π electrons of the double bond. As described in the table, the **regioselectivity** of addition of hydrogen halides to alkenes and the hydration of alkenes can be predicted by applying **Markovnikov's rule.**

Conjugated alkadienes react with many of the same reagents that react with alkenes. What is unusual about conjugated dienes, however, is their capacity to undergo both **1,2 addition** and **1,4 addition** (Section 5.8).

$$CH_2{=}CHCH{=}CH_2 \xrightarrow{HCl} CH_3\underset{\underset{Cl}{\|}}{CH}CH{=}CH_2 + CH_3CH{=}CHCH_2Cl$$

1,3-Butadiene   3-Chloro-1-butene   1-Chloro-2-butene
                                  (78%)   (22%)

via: $CH_3\overset{+}{C}H{-}CH{=}CH_2 \longleftrightarrow CH_3CH{=}CH{-}\overset{+}{C}H_2$

Acetylene and terminal alkynes are more *acidic* than other hydrocarbons (Section 5.9). They have a K_a's for ionization of approximately 10^{-26}, compared with about 10^{-45} for alkenes and about 10^{-60} for alkanes. Sodium amide is a strong enough base to remove a proton from acetylene or a terminal alkyne, but sodium hydroxide is not.

$$CH_3CH_2C{\equiv}CH + NaNH_2 \longrightarrow CH_3CH_2C{\equiv}CNa + NH_3$$

1-Butyne   Sodium amide   Sodium 1-butynide   Ammonia

Table 5.2 summarizes the reactions of alkynes.

ADDITIONAL PROBLEMS

Alkene Reactions

5.16 Write the structure of the major organic product or products formed on treatment of 1-pentene with each of the following:

    (a) Hydrogen chloride   (d) Bromine
    (b) Hydrogen bromide   (e) Bromine in water
    (c) Dilute sulfuric acid

TABLE 5.2	Reactions of Alkynes

Reaction (Section) and Comments	General Equation and Specific Example
Alkylation of acetylene and terminal alkynes (Section 5.10) The acidity of acetylene and terminal alkynes permits them to be converted to their conjugate bases on treatment with sodium amide. These anions are good nucleophiles and react with methyl and primary alkyl halides to form carbon–carbon bonds. Secondary and tertiary alkyl halides cannot be used because they yield only elimination products under these conditions.	$RC{\equiv}CH$ + $NaNH_2$ ⟶ $RC{\equiv}CNa$ + NH_3 Alkyne · Sodium amide · Sodium alkynide · Ammonia $RC{\equiv}CNa$ + $R'CH_2X$ ⟶ $RC{\equiv}CCH_2R'$ + NaX Sodium alkynide · Primary alkyl halide · Alkyne · Sodium halide $(CH_3)_3CC{\equiv}CH$ $\xrightarrow[\text{2. CH}_3\text{I}]{\text{1. NaNH}_2,\ \text{NH}_3}$ $(CH_3)_3CC{\equiv}CCH_3$ 3,3-Dimethyl-1-butyne · 4,4-Dimethyl-2-pentyne (96%)
Hydrogenation of alkynes to alkanes (Section 5.11) Alkynes are completely hydrogenated, yielding alkanes, in the presence of the customary metal hydrogenation catalysts.	$RC{\equiv}CR'$ + $2H_2$ $\xrightarrow{\text{metal catalyst}}$ RCH_2CH_2R' Alkyne · Hydrogen · Alkane 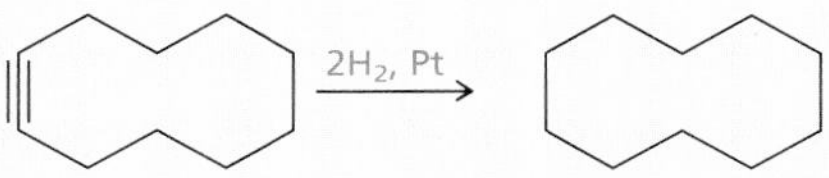 Cyclodecyne $\xrightarrow{2H_2,\ \text{Pt}}$ Cyclodecane (71%)
Hydrogenation of alkynes to alkenes (Section 5.11) Hydrogenation of alkynes may be halted at the alkene stage by using special catalysts. Lindlar palladium is the metal catalyst employed most often. Hydrogenation occurs with syn stereochemistry and yields a cis alkene.	$RC{\equiv}CR'$ + H_2 $\xrightarrow{\text{Lindlar Pd}}$ 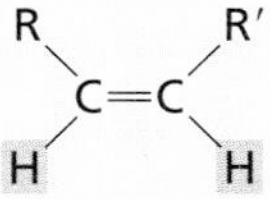 Cis alkene Alkyne · Hydrogen · Cis alkene $CH_3C{\equiv}CCH_2CH_2CH_2CH_3$ $\xrightarrow[\text{Lindlar Pd}]{H_2}$ 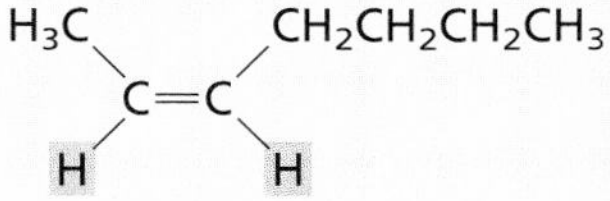 cis-2-Heptene (59%) 2-Heptyne
Metal–ammonia reduction (Section 5.11) Group I metals—sodium is the one usually employed—in liquid ammonia as the solvent convert alkynes to trans alkenes.	$RC{\equiv}CR'$ + $2Na$ + $2NH_3$ 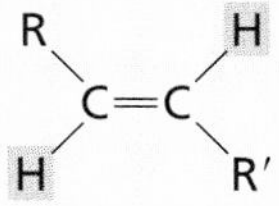 ⟶ Trans alkene + $2NaNH_2$ Alkyne · Sodium · Ammonia · Trans alkene · Sodium amide $CH_3C{\equiv}CCH_2CH_2CH_3$ $\xrightarrow[\text{NH}_3]{\text{Na}}$ 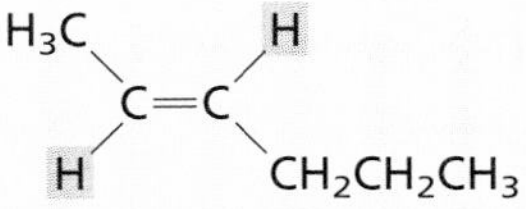 trans-2-Hexene (69%) 2-Hexyne

(Continued)

TABLE 5.2	Reactions of Alkynes *(Continued)*

Reaction (Section) and Comments	General Equation and Specific Example
Acid-catalyzed hydration (Section 5.11) Water adds to the triple bond of alkynes to yield ketones by way of an unstable enol intermediate. The enol arises by Markovnikov hydration of the alkyne. Enol formation is followed by rapid isomerization of the enol to a ketone.	$$RC \equiv CR' + H_2O \xrightarrow[Hg^{2+}]{H_2SO_4} RCH_2\overset{\displaystyle O}{\overset{\|}{C}}R'$$ Alkyne Water Ketone $$HC \equiv CCH_2CH_2CH_2CH_3 + H_2O \xrightarrow[HgSO_4]{H_2SO_4} CH_3\overset{\displaystyle O}{\overset{\|}{C}}CH_2CH_2CH_2CH_3$$ 1-Hexyne Water 2-Hexanone (80%)

5.17   Repeat Problem 5.16 for 2-methyl-2-butene.

5.18   Repeat Problem 5.16 for 1-methylcyclohexene.

5.19   (a) How many alkenes yield 2,2,3,4,4-pentamethylpentane on catalytic hydrogenation?
(b) How many yield 2,3-dimethylbutane?
(c) How many yield methylcyclobutane?

5.20   Two alkenes undergo hydrogenation to yield a mixture of *cis*- and *trans*-1,4-dimethyl-cyclohexane. A third, however, gives only *cis*-1,4-dimethylcyclohexane. What compound is this?

5.21   Write equations describing the preparation of the tertiary alcohol of molecular formula $C_5H_{12}O$ from two different alkenes.

5.22   Write a mechanism for the reaction of hydrogen bromide with 2-methyl-2-butene, using curved arrows to show the flow of electrons.

5.23   Is the electrophilic addition of hydrogen chloride to 2-methylpropene the reverse of the E1 or the E2 elimination reaction of *tert*-butyl chloride?

5.24   The mass 82 isotope of bromine (^{82}Br) is radioactive and is used as a tracer to identify the origin and destination of individual atoms in chemical reactions and biological transformations. A sample of 1,1,2-tribromocyclohexane was prepared by adding ^{82}Br—^{82}Br to ordinary (nonradioactive) 1-bromocyclohexene. How many of the bromine atoms in the 1,1,2-tribromocyclohexane produced are radioactive? Which ones are they?

Conjugated Diene Reactions

5.25   Give the structure, exclusive of stereochemistry, of the principal organic product formed on reaction of 2,3-dimethyl-1,3-butadiene with each of the following:
(a) H_2 (2 mol), platinum catalyst
(b) 1 mol HCl (product of 1,2 addition)
(c) 1 mol HCl (product of 1,4 addition)
(d) 1 mol Br_2 (product of 1,2 addition)
(e) 1 mol Br_2 (product of 1,4 addition)
(f) 2 mol Br_2

5.26   Repeat Problem 5.25 for the reactions of 1,3-cyclohexadiene.

5.27   The products from both parts (b) and (c) of Problem 5.25 are formed from the same carbocation intermediate. Draw its structure and explain the formation of the two products.

Alkyne Reactions

5.28 Write the structure of the major product isolated from the reaction of 1-hexyne with:

(a) Hydrogen (2 mol), platinum
(b) Hydrogen (1 mol), Lindlar palladium
(c) Sodium in liquid ammonia
(d) Sodium amide in liquid ammonia
(e) Product of (d) with 1-bromobutane
(f) Hydrogen chloride (1 mol)
(g) Hydrogen chloride (2 mol)
(h) Aqueous sulfuric acid, mercury(II) sulfate

5.29 Write the structure of the major product isolated from the reaction of 3-hexyne with:

(a) Hydrogen (2 mol), platinum
(b) Hydrogen (1 mol), Lindlar palladium
(c) Sodium in liquid ammonia
(d) Hydrogen chloride (1 mol)
(e) Hydrogen chloride (2 mol)
(f) Aqueous sulfuric acid, mercury(II) sulfate

5.30 When 2-heptyne was treated with aqueous sulfuric acid containing mercury(II) sulfate, two products, each having the molecular formula $C_7H_{14}O$, were obtained in approximately equal amounts. What are these two compounds?

5.31 Write the structures of the enol intermediates that form in the reaction described in Problem 5.30.

Synthesis Using Alkenes and Alkynes

5.32 Suggest a sequence of reactions suitable for preparing each of the following compounds from the indicated starting material. You may use any necessary organic or inorganic reagents.

(a) 2-Methylhexane from 2-methyl-2-hexanol
(b) 2-Propanol from 1-propanol
(c) 1,2-Dibromopropane from 2-bromopropane
(d) 2,2-Dibromopropane from 1,2-dibromopropane
(e) 2-Chloro-2-methylbutane from 2-chloro-3-methylbutane
(f) 3-Bromo-2-methyl-2-pentanol from 2-methyl-2-pentanol

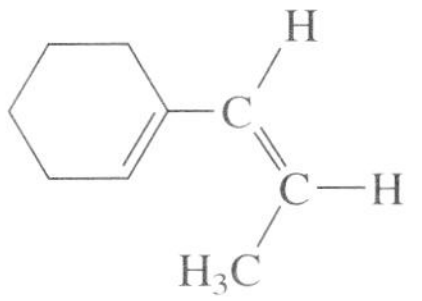

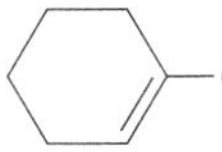

5.33 The ketone 2-heptanone has been identified as contributing to the odor of a number of dairy products, including condensed milk and cheddar cheese. Describe a synthesis of 2-heptanone from acetylene and any necessary organic or inorganic reagents.

$$\underset{\text{2-Heptanone}}{CH_3\overset{\displaystyle O}{\overset{\|}{C}}CH_2CH_2CH_2CH_2CH_3}$$

5.34 (Z)-9-Tricosene [(Z)-$CH_3(CH_2)_7CH{=}CH(CH_2)_{12}CH_3$] is the sex pheromone of the female housefly. Synthetic (Z)-9-tricosene is used as bait to lure male flies to traps that contain insecticide. Using acetylene and alcohols of your choice as starting materials, along with any necessary inorganic reagents, show how you could prepare (Z)-9-tricosene.

5.35   Compound A ($C_7H_{15}Br$) is not a primary alkyl bromide. It yields a single alkene (compound B) on being heated with sodium ethoxide in ethanol. Hydrogenation of compound B yields 2,4-dimethylpentane. Identify compounds A and B.

5.36   Compounds A and B are isomers of molecular formula $C_9H_{19}Br$. Both yield the same alkene C as the exclusive product of elimination on being treated with potassium *tert*-butoxide. Hydrogenation of alkene C gives 2,3,3,4-tetramethylpentane. What are the structures of compounds A and B and alkene C?

5.37   Compound A ($C_{10}H_{18}O$) is a tertiary alcohol that is converted to a mixture of alkenes B and C on being heated in acid. Catalytic hydrogenation of B and C yields the same product. Deduce the structures of alcohol A and alkene C.

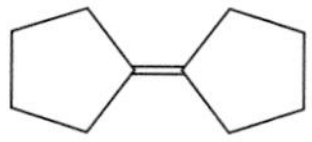

Compound B

CHAPTER 6
AROMATIC COMPOUNDS

This chapter introduces the structure and key reactions of aromatic compounds. A good place to begin is with **aromatic hydrocarbons,** also called **arenes.** Benzene and toluene are the two simplest arenes.

Benzene   Toluene

Although historically the word "aromatic" referred to the origin of many of these compounds in pleasant-smelling plant materials (see the following essay *Aromatic Compounds: History and Some Applications*), today the term has a very different meaning, as you will see when the structure and stability of benzene are discussed.

Aromatic compounds have chemical properties very different from those of unsaturated aliphatic compounds such as alkenes and alkynes. Many of the substances that add to the double bond of alkenes undergo a reaction called **electrophilic aromatic substitution** with aromatic compounds. This reaction is one of the featured topics of the present chapter.

6.1 STRUCTURE AND BONDING OF BENZENE

One of the earliest structural formulas for benzene was proposed by August Kekulé in 1866. Kekulé suggested a structure having alternating single and double bonds in a ring of six carbons, each carbon having one hydrogen attached.

AROMATIC COMPOUNDS: HISTORY AND SOME APPLICATIONS

Benzene was first isolated in 1825 from the gas used to light streetlamps. Nine years later it was prepared from benzoic acid by heating with lime:

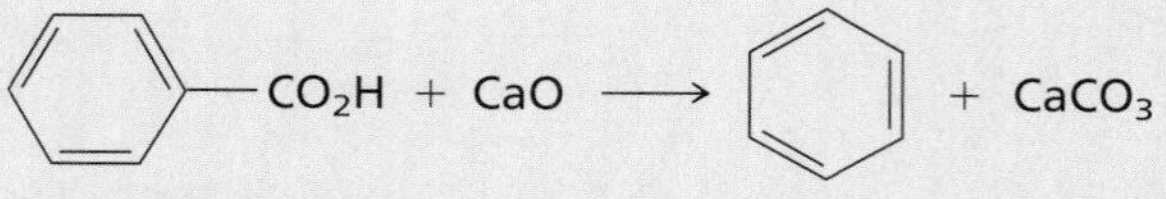

Benzoic acid + CaO ⟶ Benzene (C_6H_6) + CaCO₃

Benzoic acid   Lime   Benzene (C_6H_6)   Calcium carbonate

Benzoic acid had been known for many years and could be isolated from a resinous material found in the bark of certain trees. Although benzene is not a particularly fragrant compound, its origin from a substance found in aromatic plant extracts led to it and related compounds being called **aromatic hydrocarbons.** For the next century (until about 1950) benzene was prepared from coal tar. Since that time methods have been developed for its efficient preparation from petroleum. Production of benzene in the United States in 1999 was about 9 million tons.

The range of substances that are aromatic compounds is exceedingly diverse. The most common analgesics (pain-relievers)—aspirin, acetaminophen, and ibuprofen—all contain benzene rings. Natural products used as flavoring agents include the aromatic compounds vanillin and thymol. Other aromatic compounds include many antibiotics (sulfabenz, for example) and food additives (such as butylated hydroxytoluene, BHT).

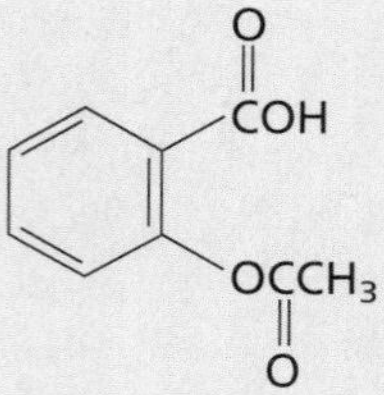

Acetylsalicylic acid
(aspirin)

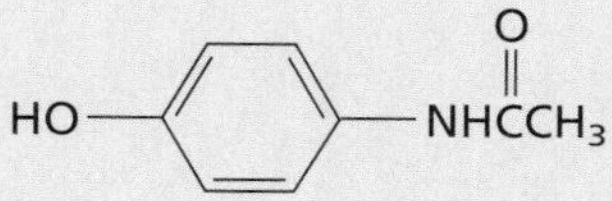

Acetaminophen
(analgesic in Tylenol)

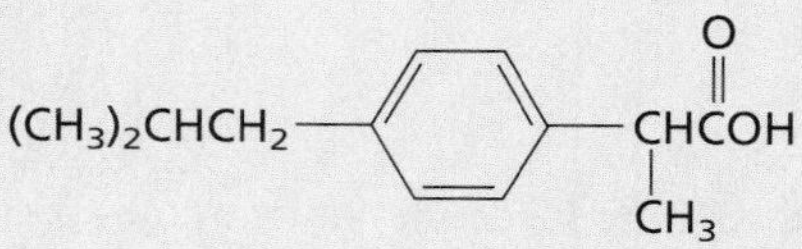

Ibuprofen
(analgesic in Motrin, Advil)

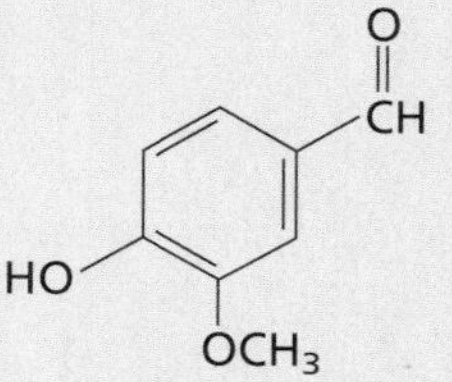

Vanillin
(principal flavor
component of vanilla)

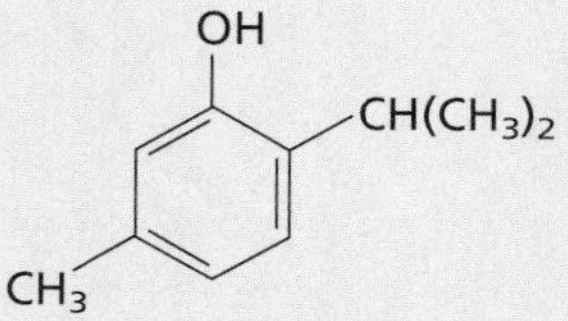

Thymol
(found in the herb thyme)

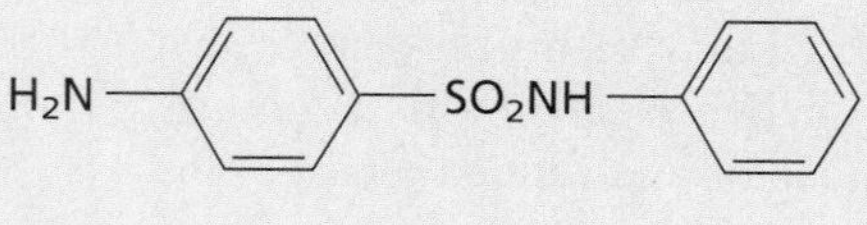

Sulfabenz
(sulfa drug antibiotic)

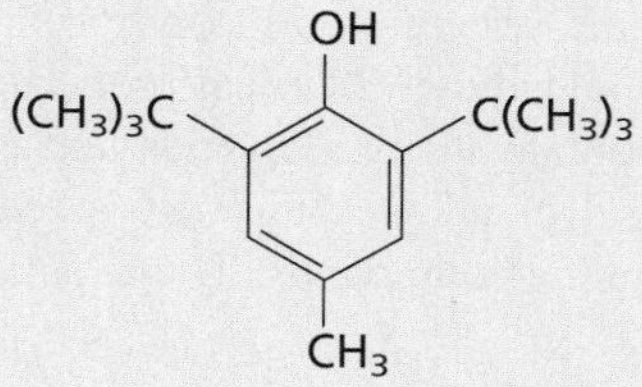

BHT
(antioxidant used as a
food preservative)

$$
\begin{array}{c}
\text{H} \\
\text{H} \overset{|}{\underset{6}{\text{C}}} \overset{1}{\underset{2}{\text{C}}} \text{H} \\
\text{H} \overset{5}{\text{C}} \qquad \overset{3}{\text{C}} \text{H} \\
\overset{|}{\underset{4}{\text{C}}} \\
\text{H}
\end{array}
$$

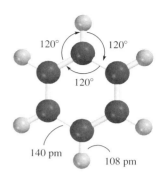

120°   120°
120°

140 pm   108 pm

FIGURE 6.1 Bond distances and bond angles of benzene.

We know today that Kekulé's view does not explain all the observations regarding benzene's structure and reactivity. As shown in Figure 6.1, benzene is a regular hexagon with internal angles of 120°; all six carbon–carbon bonds are the *same* length (140 pm), intermediate between sp^2–sp^2 single bonds (146 pm) and double bonds (134 pm). The fact that all the bond lengths are equal is not explained by the alternating double and single bonds of Kekulé's structure. The structure of benzene is better represented as a hybrid of the two equivalent **resonance forms** shown.

$$
\left[\hspace{2cm} \longleftrightarrow \hspace{2cm} \right] \qquad \text{is equivalent to} \qquad \bigcirc
$$

Inscribing a circle inside the hexagon is a convenient way to represent the two resonance forms of benzene. Remember from Section 1.9 that the double-headed resonance arrow does *not* indicate an equilibrium process. It is a device used to show **electron-delocalized** structures as composites of Lewis structures drawn with localized electrons.

> **PROBLEM 6.1** Write structural formulas for toluene ($C_6H_5CH_3$) and for benzioc acid ($C_6H_5CO_2H$) (a) as resonance hybrids of two Kekulé forms and (b) with the "circle in the ring" symbol.

Benzene is far less reactive than would be expected on the basis of its written structural formulas. If benzene were literally "1,3,5-cyclohexatriene," we would expect it to exhibit reactivity similar to that of an alkene. In fact, benzene is inert or only sluggishly reactive with many substances that react readily with alkenes. For example, hydrogenation of benzene and other arenes is more difficult than hydrogenation of alkenes and alkynes. Low reactivity suggests high stability. Estimates suggest that benzene is 152 kJ/mol (36 kcal/mol) *more* stable than it would be if it were simply a polyene. This increased stability is called **delocalization energy** or **resonance energy.** The "special" stability of benzene and its derivatives is the modern definition of the term **aromaticity.**

In 1861, Johann Josef Loschmidt, who was later to become a professor at the University of Vienna, privately published a book containing a structural formula for benzene similar to that which Kekulé would propose five years later. Loschmidt's book reached few readers, and his ideas were not well known.

6.2 AN ORBITAL HYBRIDIZATION VIEW OF BONDING IN BENZENE

The structural facts that benzene is planar, all of the bond angles are 120°, and each carbon is bonded to three other atoms, suggest sp^2 hybridization for carbon and the framework of σ bonds shown in Figure 6.2a.

In addition to its three sp^2 hybrid orbitals, each carbon has a half-filled $2p$ orbital that can participate in π bonding. Figure 6.2b shows the continuous π system that encompasses all of the carbons resulting from overlap of these $2p$ orbitals. The six π electrons of benzene are delocalized over all six carbons.

As you have seen previously (Sections 1.9 and 4.2), when π electrons are delocalized over several carbon atoms, the molecule is more stable than if these electrons

FIGURE 6.2 (*a*) The framework of bonds shown in the tube model of benzene is one of σ bonds. (*b*) Each carbon is *sp*²-hybridized and has a 2*p* orbital perpendicular to the σ framework. Overlap of the 2*p* orbitals generates a π system encompassing the entire ring.

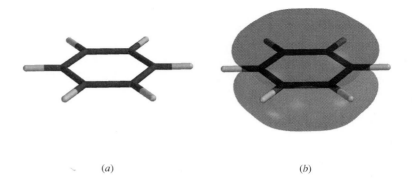

(*a*)   (*b*)

were localized and felt the attractive force of fewer nuclei. The delocalization of the π electrons in a benzene ring results in a cyclic π cloud encompassing all six carbons of the ring. Aromaticity is attributed to this cyclic delocalization of the conjugated system of six π electrons.

6.3   SUBSTITUTED DERIVATIVES OF BENZENE AND THEIR NOMENCLATURE

All compounds that contain a benzene ring are aromatic (have special stability), and substituted derivatives of benzene make up the largest class of aromatic compounds. Many such compounds are named by attaching the name of the substituent as a prefix to *benzene*.

Bromobenzene   *tert*-Butylbenzene   Nitrobenzene

Many simple monosubstituted derivatives of benzene have long-standing common names that have been retained in the IUPAC system. Table 6.1 lists some of the most important ones and some of their commercial uses.

Dimethyl derivatives of benzene are called **xylenes**, and benzene has three xylene isomers: the *ortho* (*o*)-, *meta* (*m*)-, and *para* (*p*)- substituted derivatives.

o-Xylene   *m*-Xylene   *p*-Xylene
(1,2-dimethylbenzene)   (1,3-dimethylbenzene)   (1,4-dimethylbenzene)

The prefix *ortho* signifies a 1,2-disubstituted benzene ring, *meta* signifies 1,3-disubstitution, and *para* signifies 1,4-disubstitution. The abbreviations *o, m,* and *p* can be used when a substance is named as a benzene derivative or when a specific base name (such as acetophenone) is used. For example,

TABLE 6.1	Some Frequently Encountered Derivatives of Benzene	
Structure	**Common Name***	**Commercial Use**
benzaldehyde structure	Benzaldehyde	Used as a chemical intermediate and as a component of synthetic perfumes
benzoic acid structure	Benzoic acid	Its sodium salt is used as a food preservative
styrene structure	Styrene	The basic component of polystyrene and Styrofoam plastics
acetophenone structure	Acetophenone	Used in perfumes to impart an orange-blossom-like odor
phenol structure	Phenol	Used as a germicidal agent and general disinfectant; also a component of epoxy resins
anisole structure	Anisole	Intermediate in the synthesis of perfumes and flavorings
aniline structure	Aniline	Intermediate in preparation of dyes, urethane foams, and photographic chemicals

* These common names are acceptable in IUPAC nomenclature and are the names that will be used in this text.

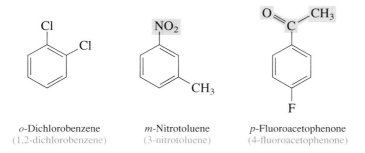

o-Dichlorobenzene
(1,2-dichlorobenzene)

m-Nitrotoluene
(3-nitrotoluene)

p-Fluoroacetophenone
(4-fluoroacetophenone)

PROBLEM 6.2 Write a structural formula for each of the following compounds:
(a) *o*-Ethylanisole   (b) *m*-Chlorostyrene   (c) *p*-Nitroaniline

Sample Solution (a) The parent compound in *o*-ethylanisole is anisole. Anisole, as shown in Table 6.1, has a methoxy (CH_3O-) substituent on the benzene ring.

The ethyl group in *o*-ethylanisole is attached to the carbon adjacent to the one that bears the methoxy substituent.

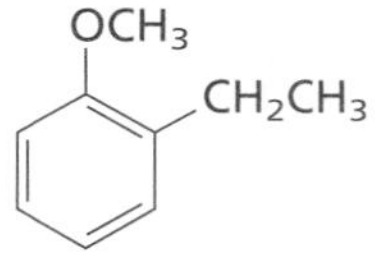

OCH₃
CH₂CH₃

o-Ethylanisole

The prefixes *o*, *m*, and *p* are *not* used when three or more substituents are present on benzene; numerical locants must be used instead.

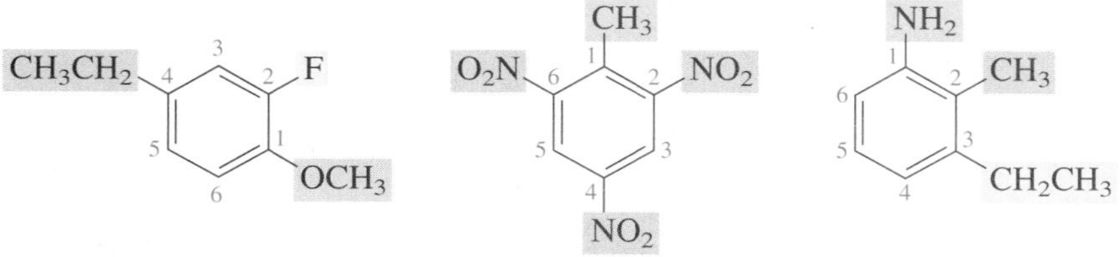

4-Ethyl-2-fluoroanisole   2,4,6-Trinitrotoluene   3-Ethyl-2-methylaniline

In these examples the base name of the benzene derivative determines the carbon at which numbering begins: anisole has its methoxy group at C-1, toluene its methyl group at C-1, and aniline its amino group at C-1. The direction of numbering is chosen to give the next substituted position the lowest number irrespective of what substituent it bears.

The order of appearance of substituents in the name is alphabetical.

When the benzene ring is named as a substituent, the word **phenyl** stands for C₆H₅—. Similarly, an arene named as a substituent is called an **aryl** group. A **benzyl** group is C₆H₅CH₂—.

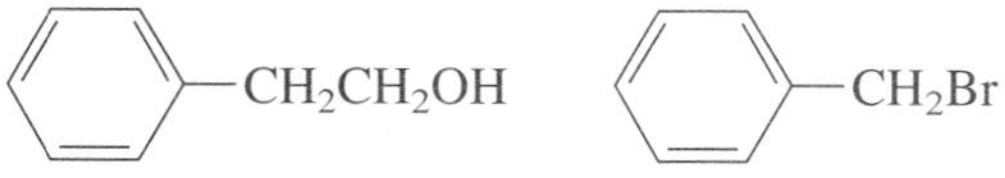

—CH₂CH₂OH   —CH₂Br

2-Phenylethanol   Benzyl bromide

6.4 POLYCYCLIC AROMATIC HYDROCARBONS

Members of a class of arenes called **polycyclic aromatic hydrocarbons** possess substantial resonance energies because each is a collection of benzene rings fused together.

Naphthalene, anthracene, and phenanthrene are the three simplest members of this class. They are all present in **coal tar,** a mixture of organic substances formed when coal is converted to coke by heating at high temperatures (about 1000°C) in the absence of air. Naphthalene is **bicyclic** (has two rings), and its two benzene rings share a common side. Anthracene and phenanthrene are both **tricyclic** aromatic hydrocarbons. Anthracene has three rings fused in a "linear" fashion, and "angular" fusion characterizes phenanthrene. The structural formulas of naphthalene, anthracene, and phenanthrene are shown along with the numbering system used to name their substituted derivatives:

Naphthalene is a white crystalline solid melting at 80°C that sublimes readily. It has a characteristic odor and was formerly used as a moth repellent.

CHEMICAL CARCINOGENS

A carcinogen is a substance or agent capable of causing cancer. A number of chemical carcinogens have received considerable study in recent years, both in an effort to understand the causes of cancer and in an attempt to find ways to prevent its occurrence.

In 1775 Percivall Pott, an English surgeon, wrote of the incidence of scrotal cancer among chimney sweeps, blaming their disease on exposure to soot. Nearly 150 years later experiments showed that mice painted with soot extract developed cancer. These experiments and others in the 1930s led to the identification of numerous polycyclic aromatic hydrocarbons and the isolation of the first chemical identified as a carcinogen, benzo[a]pyrene.

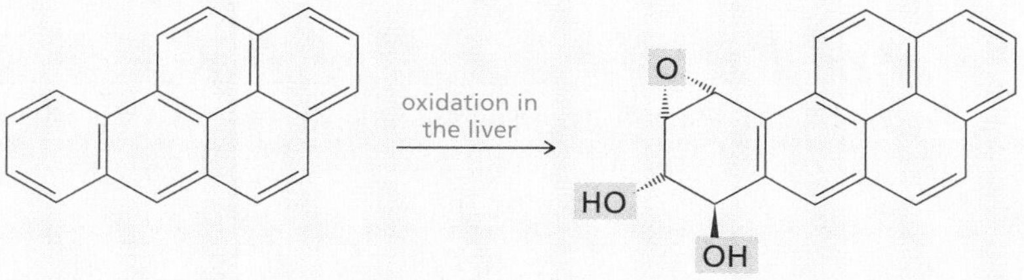

Benzo[a]pyrene   oxidation in the liver   A diol-epoxide

Current theories of carcinogenesis suggest that chemicals such as benzo[a]pyrene are oxidized in the body to form other compounds called **metabolites,** and it is these metabolites that are the actual carcinogens. In the case of benzo[a]-pyrene, enzymes in the liver catalyze its conversion to a diol-epoxide that is able to react with DNA in cells. If these reactions cause a nonlethal change, or **mutation,** in the structure of the cell, uncontrolled cell reproduction may result, giving rise to a tumor.

Benzene has been shown to be a weak carcinogen, and has been implicated as a cause of a rare form of leukemia among industrial workers exposed to large amounts of it. In a manner similar to benzo[a]pyrene, benzene forms an epoxide metabolite. This process is discussed further in the boxed material accompanying Section 10.12.

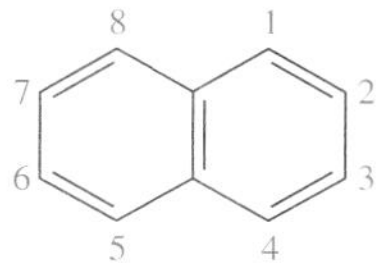

Naphthalene 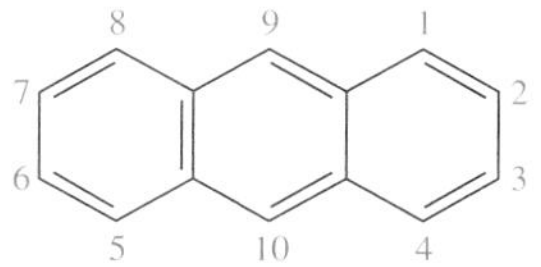   Anthracene 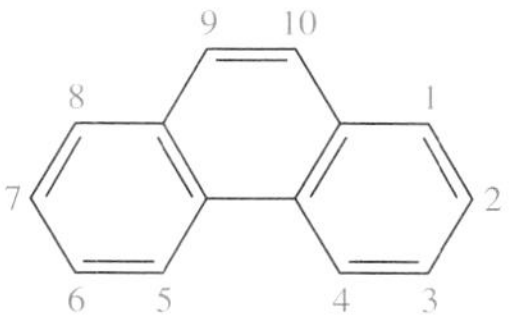   Phenanthrene

A large number of polycyclic aromatic hydrocarbons are known. Many have been synthesized in the laboratory, and several of the others are products of combustion. Benzo[a]pyrene, for example, is present in tobacco smoke, contaminates food cooked on barbecue grills, collects in the soot of chimneys, and is discussed in the boxed essay, *Chemical Carcinogens.*

6.5 AROMATIC SIDE-CHAIN REACTIONS

Before we begin our look at reactions of the benzene ring itself, let us look at how an aromatic ring affects reactions taking place on a side-chain of the benzene ring. In Chapter 5 (Section 5.8) you saw how conjugated dienes undergo addition reactions proceeding through formation of resonance-stabilized allylic carbocation intermediates. A carbocation adjacent to a benzene ring is also resonance-stabilized and is called a **benzylic carbocation.**

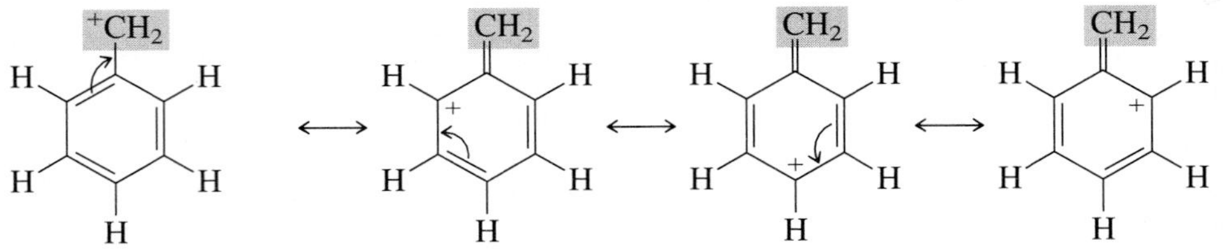

Most stable Lewis structure
of benzyl cation

The regioselectivity of electrophilic addition to a side-chain double bond is governed by the ability of an aromatic ring to stabilize an adjacent carbocation. This is clearly seen in the addition of hydrogen chloride to indene. Only a single chloride is formed.

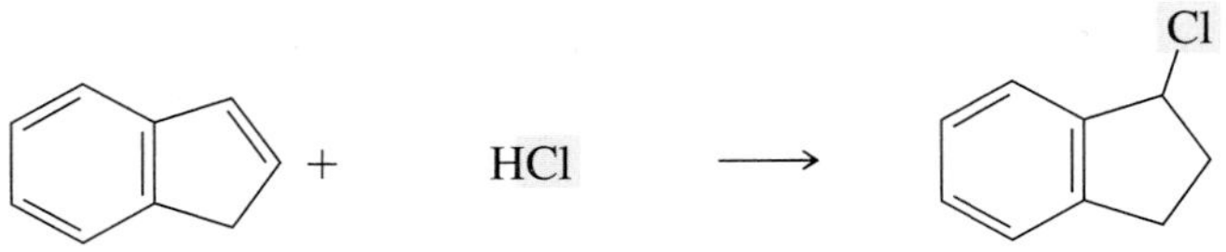

Indene   Hydrogen chloride   1-Chloroindane
(75–84%)

Only the benzylic chloride is formed, because protonation of the double bond occurs in the direction that gives a carbocation that is both secondary and benzylic.

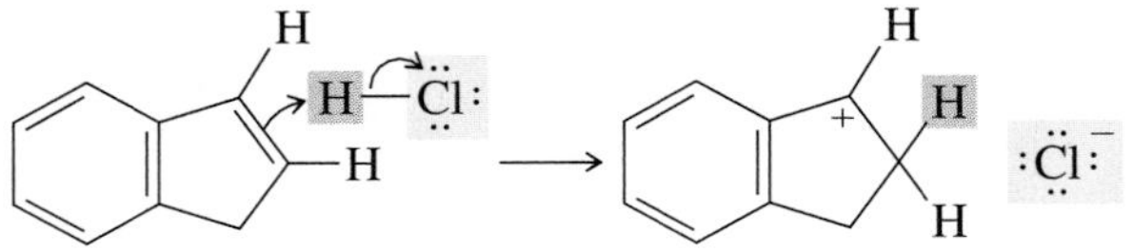

Carbocation that leads to
observed product

Protonation in the opposite direction also gives a secondary carbocation, but it is not benzylic.

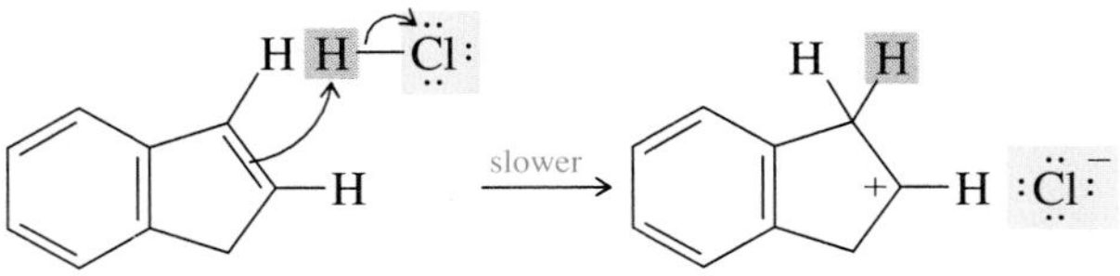

Less stable carbocation

This carbocation does not receive the extra increment of stabilization that its benzylic isomer does and so is formed more slowly. The orientation of addition is controlled by the rate of carbocation formation; the more stable benzylic carbocation is formed faster and is the one that leads to the reaction product.

PROBLEM 6.3 Write the structure of the product of the reaction of 2-phenyl-propene and hydrogen chloride.

A striking example of the activating effect that a benzene ring has on reactions that take place at benzylic positions may be found in the reactions of alkylbenzenes with oxidizing agents. Chromic acid, for example, prepared by adding sulfuric acid to aqueous sodium dichromate, is a strong oxidizing agent but does not react either with benzene or with alkanes.

> An alternative oxidizing agent, similar to chromic acid in its reactions with organic compounds, is potassium permanganate ($KMnO_4$).

$$RCH_2CH_2R' \xrightarrow[\text{H}_2\text{O, H}_2\text{SO}_4,\ \text{heat}]{\text{Na}_2\text{Cr}_2\text{O}_7} \text{no reaction}$$

$$\xrightarrow[\text{H}_2\text{O, H}_2\text{SO}_4,\ \text{heat}]{\text{Na}_2\text{Cr}_2\text{O}_7} \text{no reaction}$$

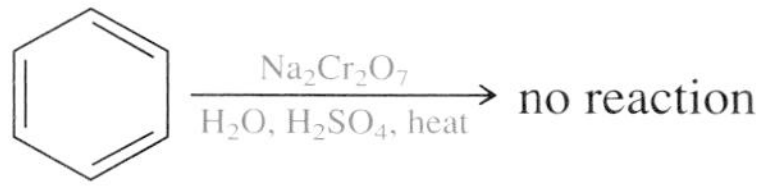

On the other hand, an alkyl side chain on a benzene ring is oxidized on being heated with chromic acid. The product is benzoic acid or a substituted derivative of benzoic acid.

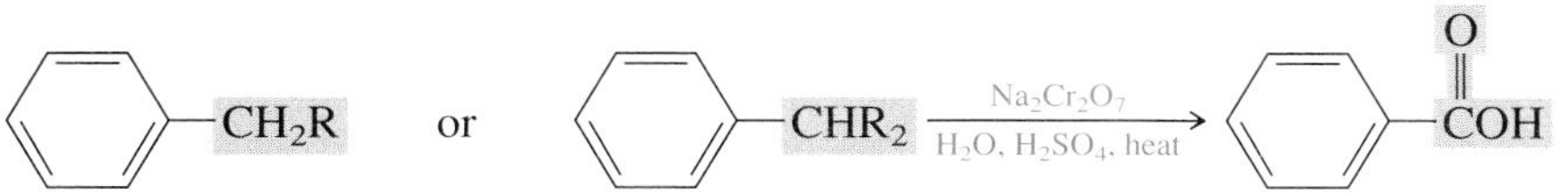

Alkylbenzene   Benzoic acid

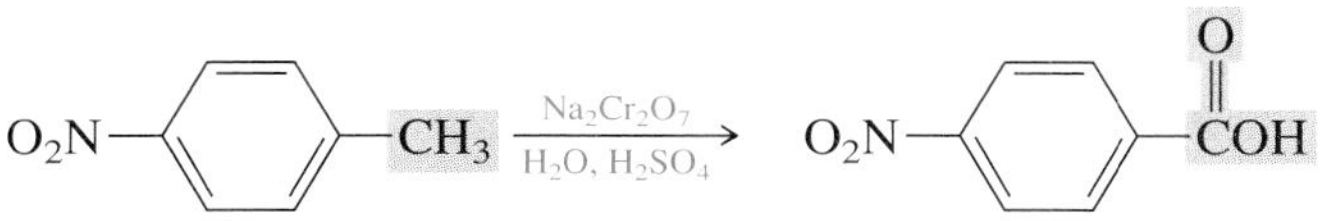

p-Nitrotoluene   p-Nitrobenzoic acid
(82–86%)

When two alkyl groups are present on the ring, both are oxidized.

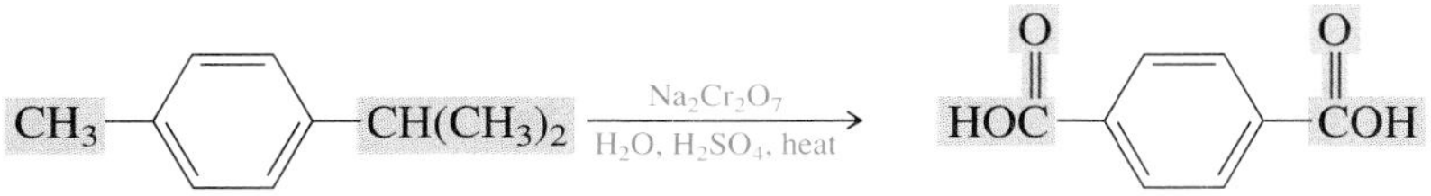

p-Isopropyltoluene   p-Benzenedicarboxylic acid
(45%)

Note that alkyl groups, regardless of their chain length, are converted to carboxyl groups ($—CO_2H$) attached directly to the ring. An exception is a *tert*-alkyl substituent. Because it lacks benzylic hydrogens, a *tert*-alkyl group is not susceptible to oxidation under these conditions.

PROBLEM 6.4 Chromic acid oxidation of 4-*tert*-butyl-1,2-dimethylbenzene yielded a single compound having the molecular formula $C_{12}H_{14}O_4$. What was this compound?

Side-chain oxidation of alkylbenzenes is important in certain metabolic processes. One way in which the body rids itself of foreign substances is by oxidation in the liver to compounds more easily excreted in the urine. Toluene, for example, is oxidized to benzoic acid by this process and is eliminated rather readily.

$$\text{Toluene} \xrightarrow[\substack{\text{cytochrome P-450} \\ \text{(an enzyme in} \\ \text{the liver)}}]{O_2} \text{Benzoic acid}$$

Toluene   Benzoic acid

Benzene, with no alkyl side chain, undergoes a different reaction in the presence of these enzymes, which convert it to a substance capable of inducing mutations in DNA (see boxed essay, *Chemical Carcinogenesis,* in Chapter 10). This difference in chemical behavior seems to be responsible for the fact that benzene is carcinogenic but toluene is not.

Reduction by catalytic hydrogenation of the side-chain double bond of an alkenylbenzene is much easier than hydrogenation of the aromatic ring and can be achieved with high selectivity, leaving the ring unaffected.

$$\text{2-(m-Bromophenyl)-2-butene} + H_2 \xrightarrow{Pt} \text{2-(m-Bromophenyl)butane}$$

2-(*m*-Bromophenyl)-2-butene   Hydrogen   2-(*m*-Bromophenyl)butane
                                                            (92%)

6.6 REACTIONS OF ARENES: ELECTROPHILIC AROMATIC SUBSTITUTION

Many experiments have demonstrated that benzene behaves differently from other unsaturated compounds. For example, conditions under which an alkene undergoes *addition* with a reagent such as bromine give no reaction with benzene. A reaction does take place in the presence of a catalyst, but it is one of *substitution.*

Characteristically, the reagents that react with the aromatic ring of benzene and its derivatives are electron-deficient species called **electrophiles.** We have already discussed (Chapter 5) how electrophiles react with alkenes; electrophilic reagents *add* to alkenes.

$$\underset{\text{Alkene}}{\overset{\diagdown}{\underset{\diagup}{C}}=\overset{\diagup}{\underset{\diagdown}{C}} + \underset{\substack{\text{Electrophilic} \\ \text{reagent}}}{\overset{\delta+ \quad \delta-}{E-Y}} \longrightarrow \underset{\substack{\text{Product of} \\ \text{electrophilic addition}}}{E-\overset{|}{\underset{|}{C}}-\overset{|}{\underset{|}{C}}-Y}$$

Alkene   Electrophilic   Product of
                  reagent   electrophilic addition

A different reaction takes place when electrophiles react with arenes.

Substitution is observed instead of addition.

If we represent an arene by the general formula ArH, where Ar stands for an aryl group, the electrophilic portion of the reagent replaces one of the hydrogens on the ring:

$$\underset{\text{Arene}}{Ar-H} + \underset{\substack{\text{Electrophilic} \\ \text{reagent}}}{\overset{\delta+ \quad \delta-}{E-Y}} \longrightarrow \underset{\substack{\text{Product of} \\ \text{electrophilic aromatic} \\ \text{substitution}}}{Ar-E + H-Y}$$

Arene   Electrophilic   Product of
            reagent   electrophilic aromatic
                                   substitution

TABLE 6.2	Representative Electrophilic Aromatic Substitution Reactions of Benzene

Reaction and Comments	Equation
1. Nitration Warming benzene with a mixture of nitric acid and sulfuric acid gives nitrobenzene. A nitro group (—NO$_2$) replaces one of the ring hydrogens.	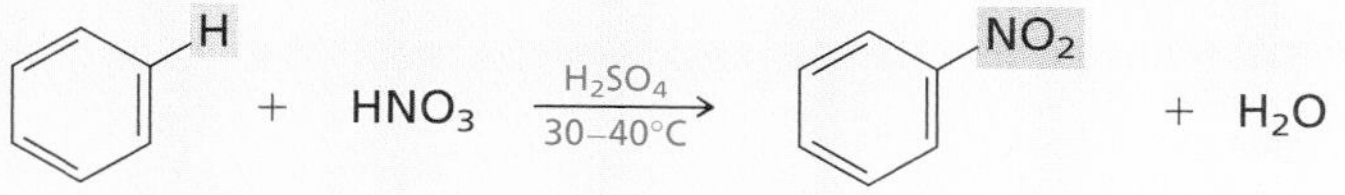 Benzene + HNO$_3$ $\xrightarrow[30-40°C]{H_2SO_4}$ Nitrobenzene (95%) + H$_2$O
2. Sulfonation Treatment of benzene with hot concentrated sulfuric acid gives benzenesulfonic acid. A sulfonic acid group (—SO$_2$OH) replaces one of the ring hydrogens.	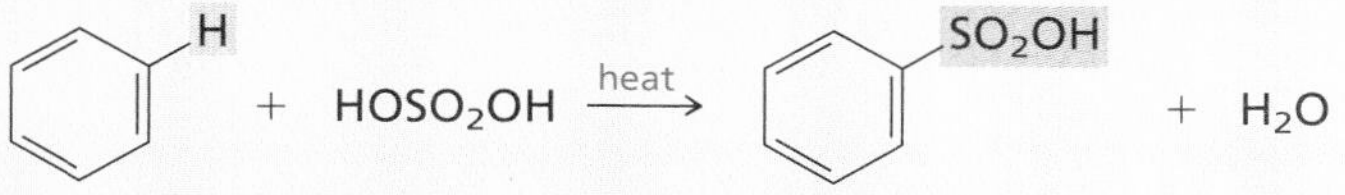 Benzene + HOSO$_2$OH $\xrightarrow{heat}$ Benzenesulfonic acid (100%) + H$_2$O
3. Halogenation Bromine reacts with benzene in the presence of iron(III) bromide as a catalyst to give bromobenzene. Chlorine reacts similarly in the presence of iron(III) chloride to give chlorobenzene.	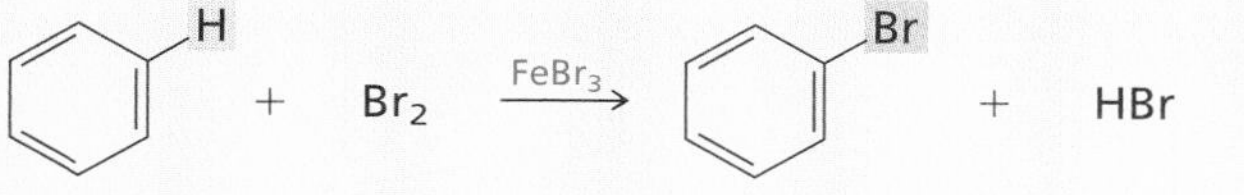 Benzene + Br$_2$ $\xrightarrow{FeBr_3}$ Bromobenzene (65–75%) + HBr (Hydrogen bromide)
4. Friedel–Crafts alkylation Alkyl halides react with benzene in the presence of aluminum chloride to yield alkylbenzenes.	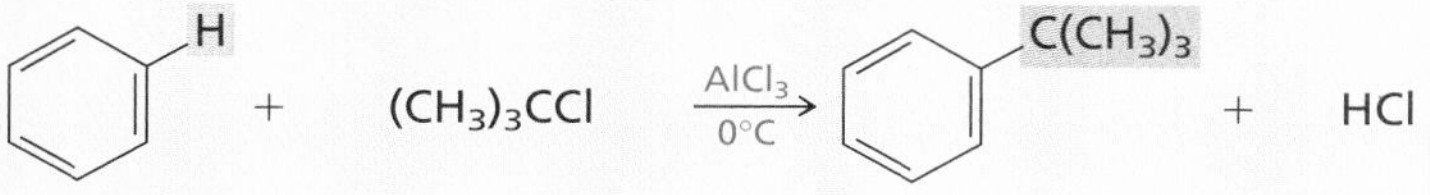 Benzene + (CH$_3$)$_3$CCl $\xrightarrow[0°C]{AlCl_3}$ tert-Butylbenzene (60%) + HCl (Hydrogen chloride)
5. Friedel–Crafts acylation An analogous reaction occurs when acyl halides react with benzene in the presence of aluminum chloride. The products are acylbenzenes.	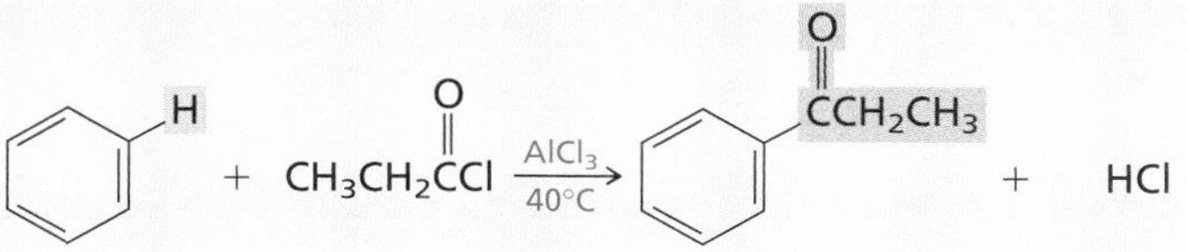 Benzene + CH$_3$CH$_2$CCl $\xrightarrow[40°C]{AlCl_3}$ 1-Phenyl-1-propanone (88%) + HCl (Hydrogen chloride)

This reaction is called **electrophilic aromatic substitution** and is one of the fundamental processes of organic chemistry. The stability of the aromatic ring is such that it tends to be retained in reactions of benzene and its derivatives.

Some of the more frequently encountered electrophilic aromatic substitution reactions are shown in Table 6.2. In each, a group substitutes for one of the hydrogens of benzene. Each of these reactions will be discussed further in this chapter.

6.7 MECHANISM OF ELECTROPHILIC AROMATIC SUBSTITUTION

Electrophilic aromatic substitution is a two-step process. Recall from Chapter 5 that the first step in the electrophilic addition to an alkene is formation of a carbocation by addition of the electrophile to the π bond of the alkene.

The first step in the reaction of electrophilic reagents with benzene is similar. An electrophile accepts an electron pair from the π system of benzene to form a carbocation:

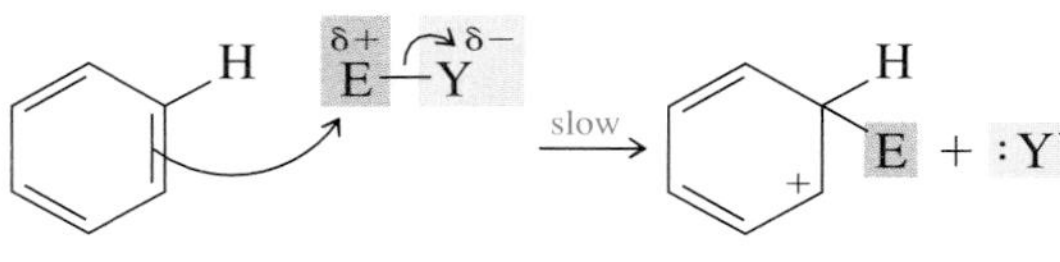

Benzene and electrophile   Carbocation

This particular carbocation is a resonance-stabilized one of the allylic type. It is a **cyclo-hexadienyl cation** (often referred to as an **arenium ion**).

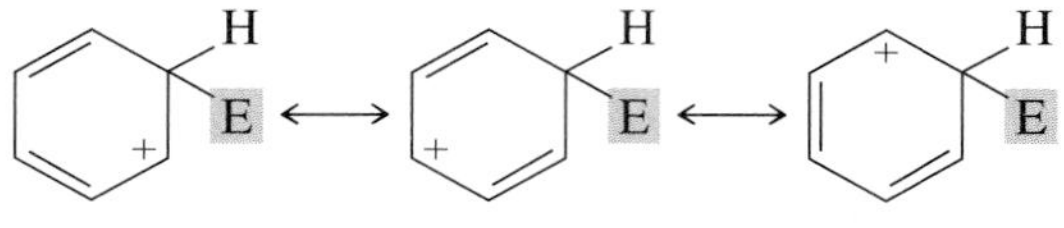

Resonance forms of a cyclohexadienyl cation

Most of the resonance stabilization of benzene is lost when it is converted to the cyclohexadienyl cation intermediate. In spite of being allylic, a cyclohexadienyl cation is *not* aromatic and possesses only a fraction of the resonance stabilization of benzene. Once formed, it rapidly loses a proton, restoring the aromaticity of the ring and giving the product of electrophilic aromatic substitution.

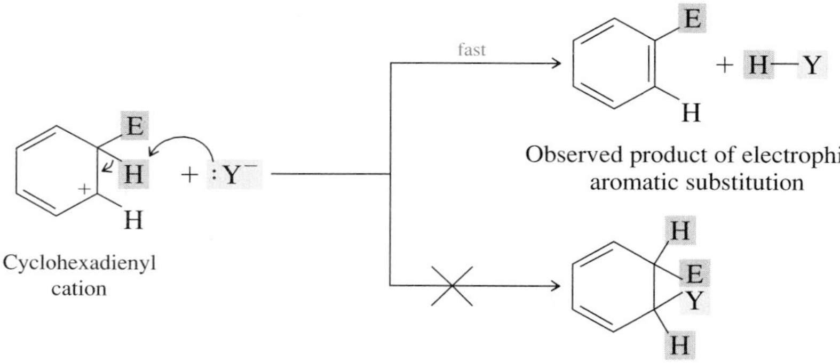

Cyclohexadienyl cation

Observed product of electrophilic aromatic substitution

Not observed—not aromatic

If the Lewis base ($:Y^-$) had acted as a nucleophile and added to carbon, the product would have been a nonaromatic cyclohexadiene derivative. Addition and substitution products arise by alternative reaction paths of a cyclohexadienyl cation. Substitution occurs preferentially because a substantial driving force favors rearomatization.

Figure 6.3 is a potential energy diagram describing the general mechanism of electrophilic aromatic substitution. Notice that the shape is quite different from the potential energy diagram we saw in Chapter 3 (Section 3.6). As described in this section, electrophilic aromatic substitution occurs in two steps. The first step forms the cyclohexadienyl cation **intermediate,** represented by the energy minimum between the two maxima on the diagram. Each reaction step has an energy maximum that is the **transition state** of that elementary step. The first step of electrophilic aromatic substitution is the slow step of the reaction, and its activation energy is higher than that of the second step.

6.8   INTERMEDIATES IN ELECTROPHILIC AROMATIC SUBSTITUTION

We can now describe more fully the electrophilic aromatic substitutions shown in Table 6.2 by looking more closely at the electrophilic species that attacks benzene in each of these reactions.

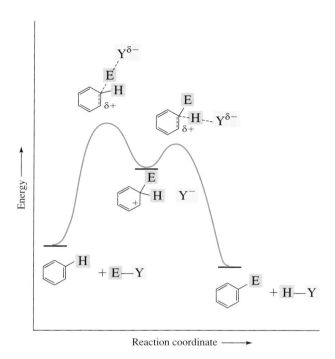

Nitration

The electrophilic species in nitration is NO_2^+, called the **nitronium ion.**

Benzene and nitronium ion   Cyclohexadienyl cation intermediate

Loss of a proton from the cyclohexadienyl cation yields nitrobenzene.

Cyclohexadienyl   Water   Nitrobenzene   Hydronium
cation intermediate   ion

Nitric acid alone does not provide a high enough concentration of nitronium ion for nitration of benzene to occur at an appreciable rate. However, addition of sulfuric acid causes the nitronium ion to form as shown in the following equation:

$$HNO_3 + 2H_2SO_4 \longrightarrow NO_2^+ + H_3O^+ + 2HSO_4^-$$

PROBLEM 6.5 Nitration of 1,4-dimethylbenzene (*p*-xylene) gives a single product having the molecular formula $C_8H_9NO_2$ in high yield. What is this product?

Sulfonation

When sulfuric acid is heated, sulfur trioxide (SO_3) is formed, and it is believed to be the electrophilic species that attacks the aromatic ring.

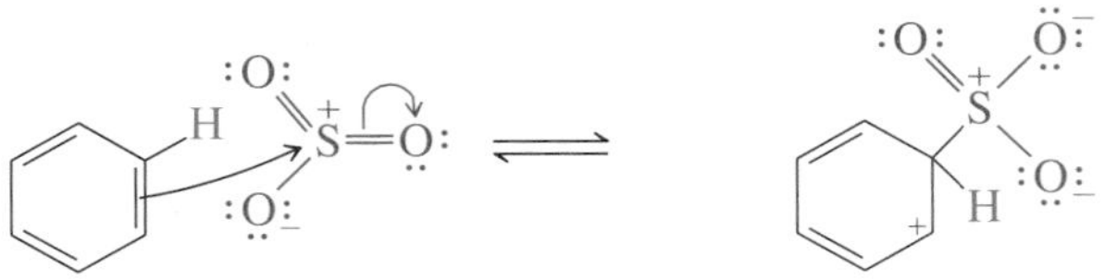

Benzene and sulfur trioxide   Cyclohexadienyl cation intermediate

A proton is lost from the sp^3-hybridized carbon of the intermediate to restore the aromaticity of the ring.

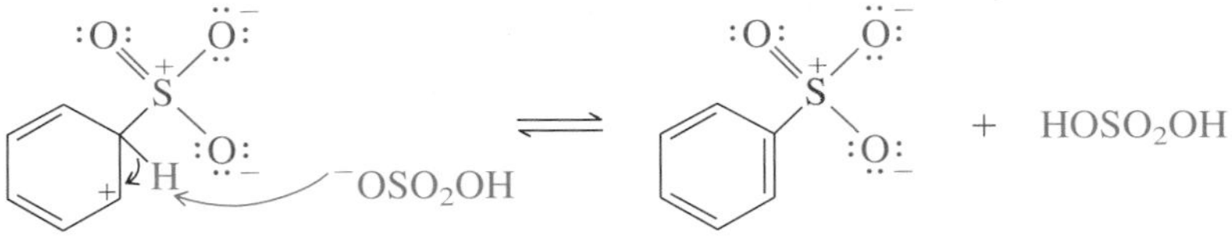

Cyclohexadienyl   Hydrogen   Benzenesulfonate ion   Sulfuric acid
cation intermediate   sulfate ion

A rapid proton transfer from the oxygen of sulfuric acid to the oxygen of benzenesulfonate completes the process.

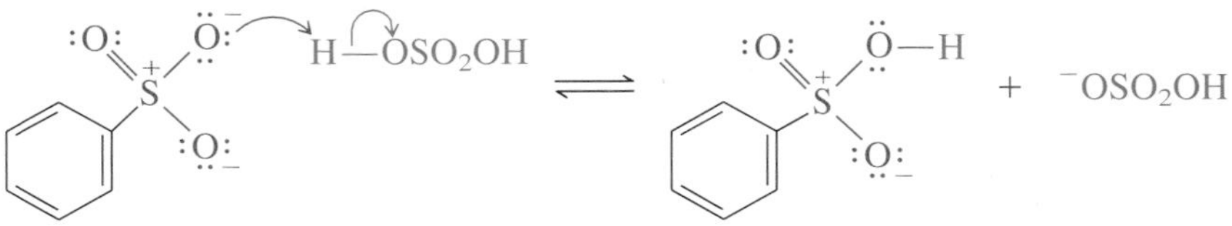

Benzenesulfonate   Sulfuric acid   Benzenesulfonic acid   Hydrogen
ion   sulfate ion

Sulfonations are important industrial reactions used in the manufacture of detergents.

Halogenation

Halogenation is brought about through the reaction of benzene with bromine or chlorine in the presence of a **Lewis acid catalyst.** The catalyst is usually iron(III) bromide ($FeBr_3$) for brominations and iron(III) chloride ($FeCl_3$) for chlorinations.

Iron(III) bromide and chloride are also called **ferric bromide and chloride.**

A catalyst is necessary to increase the electrophilic character of the halogen molecule. This is brought about by formation of a Lewis acid/Lewis base complex, as illustrated for the bromination reaction.

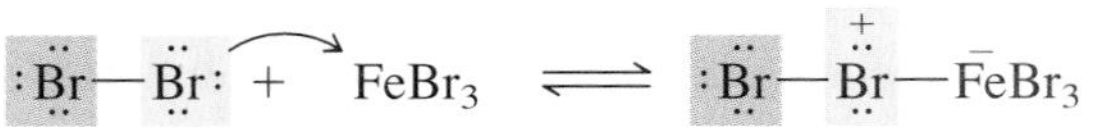

Lewis base   Lewis acid   Lewis acid/Lewis base
complex

The bromine–bromine bond in this complex is polarized, making bromine a better electrophile.

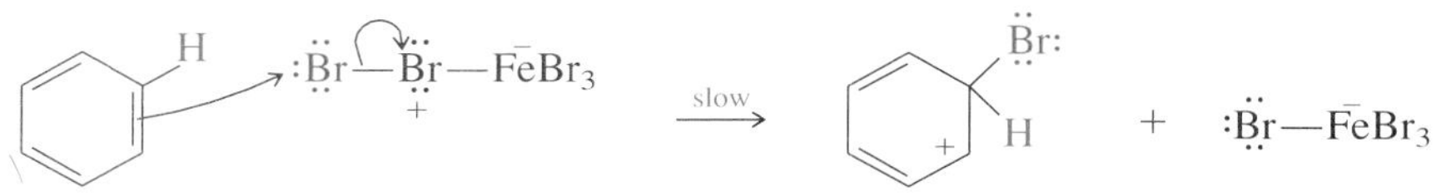

Benzene and bromine–iron(III)   Cyclohexadienyl   Tetrabromoferrate
bromide complex   cation intermediate   ion

Loss of a proton from the cyclohexadienyl cation yields bromobenzene.

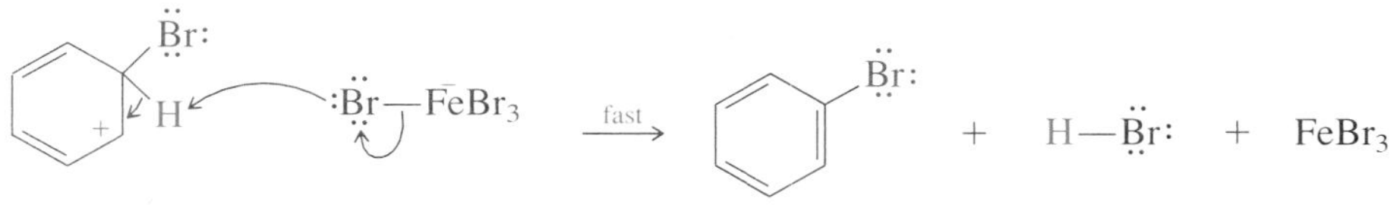

Cyclohexadienyl   Tetrabromoferrate   Bromobenzene   Hydrogen   Iron(III)
cation intermediate   ion   bromide   bromide

Friedel–Crafts Alkylation

This reaction was discovered through the collaborative efforts of Charles Friedel (in France) and James Crafts (an American chemist working in Friedel's laboratory) in 1877. The electrophile is a carbocation formed by reaction of an alkyl halide with aluminum chloride, which acts as a Lewis acid catalyst.

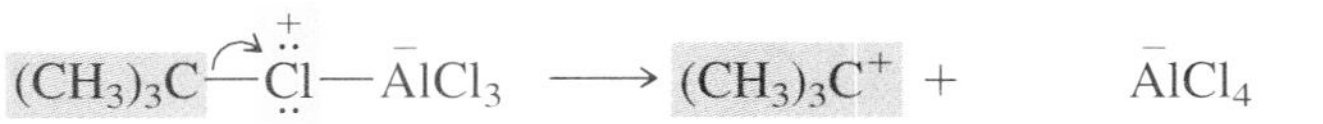

$tert$-Butyl chloride   Aluminum   Lewis acid/Lewis base
                             chloride   complex

$tert$-Butyl chloride/ $tert$-Butyl   Tetrachloroaluminate
aluminum chloride complex   cation   anion

Once formed, the carbocation reacts with benzene in a manner similar to other electrophiles.

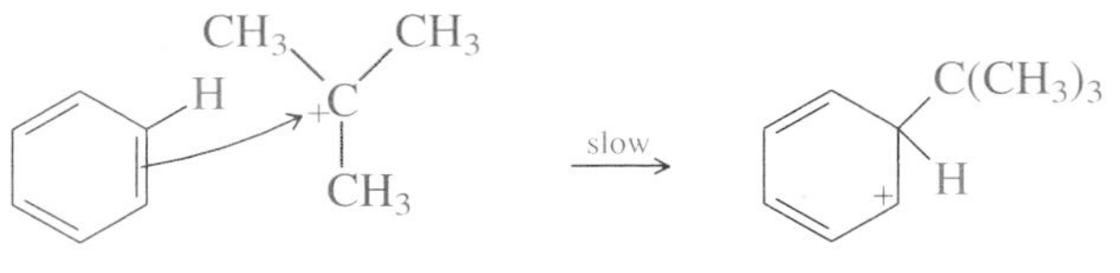

Benzene and $tert$-butyl cation   Cyclohexadienyl
                                       cation intermediate

Loss of a proton from the cyclohexadienyl cation intermediate yields $tert$-butylbenzene.

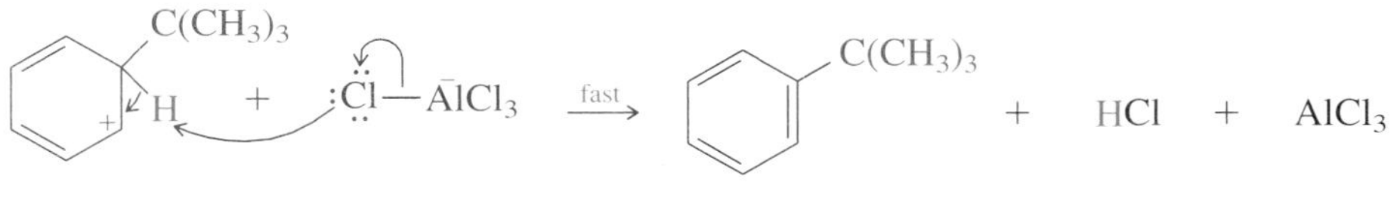

Cyclohexadienyl   Tetrachloroaluminate $tert$-Butylbenzene   Hydrogen   Aluminum
cation intermediate   ion   chloride   chloride

Friedel–Crafts Acylation

Another version of the Friedel–Crafts reaction uses **acyl chlorides** and yields a ketone as the product. The electrophile is an **acyl cation,** also known as an **acylium ion,** which forms from reaction of the acyl halide with aluminum chloride.

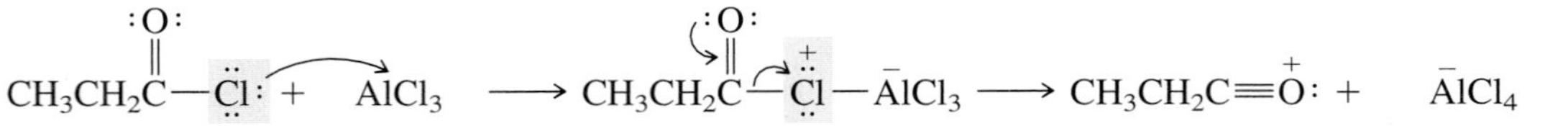

| Propanoyl chloride | Aluminum chloride | Lewis acid/Lewis base complex | Propanoyl cation | Tetrachloro-aluminate ion |

The acyl cation can be represented by the two resonance forms:

$$CH_3CH_2\overset{+}{C}\!=\!\ddot{O}\!: \longleftrightarrow CH_3CH_2C\!\equiv\!\overset{+}{O}\!:$$

Most stable resonance form; oxygen and carbon have octets of electrons

The electrophilic site of an acyl cation is its acyl carbon:

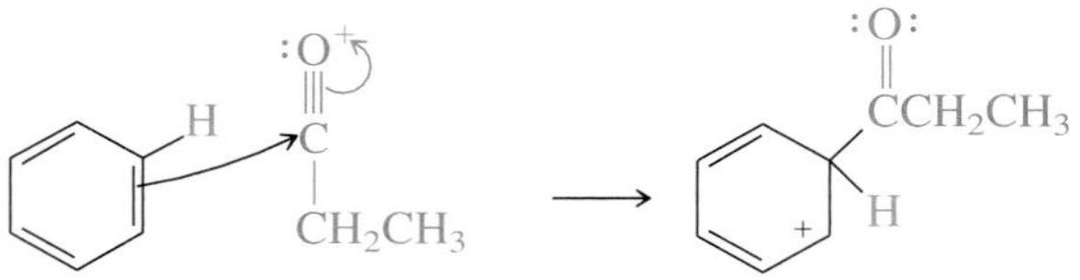

Benzene and propanoyl cation   Cyclohexadienyl cation intermediate

Aromaticity of the ring is restored when it loses a proton to give the acylbenzene.

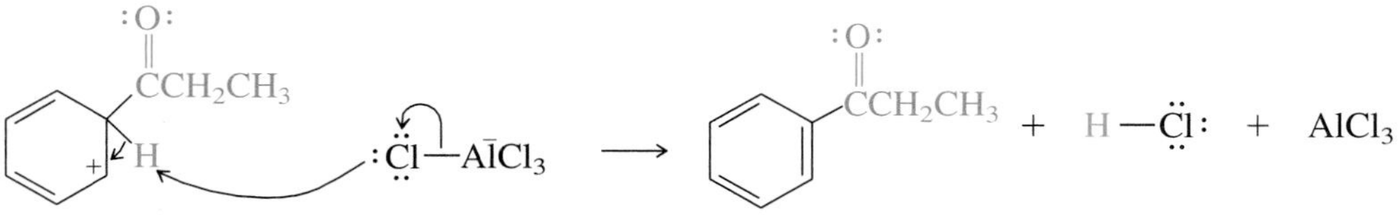

| Cyclohexadienyl cation intermediate | Tetrachloroaluminate ion | 1-Phenyl-1-propanone | Hydrogen chloride | Aluminum chloride |

Carboxylic acid anhydrides, compounds of the type $\overset{\displaystyle O}{\overset{\|}{R}}\overset{\displaystyle O}{\overset{\|}{C}}OCR$, can also serve as sources of acyl cations and, in the presence of aluminum chloride, acylate benzene. One acyl unit of an acid anhydride becomes attached to the benzene ring, and the other becomes part of a carboxylic acid.

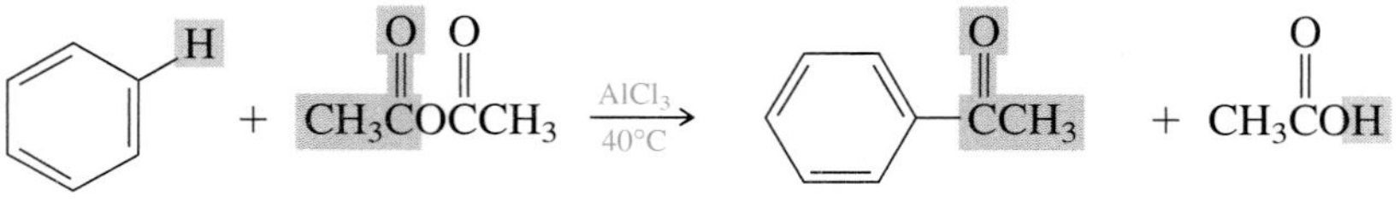

| Benzene | Acetic anhydride | | Acetophenone (76–83%) | Acetic acid |

PROBLEM 6.6 The preceding equation shows one method for preparing acetophenone. Write an equation describing the preparation of acetophenone from benzene using an acyl halide.

6.9 RATE AND REGIOSELECTIVITY IN ELECTROPHILIC AROMATIC SUBSTITUTION

So far we've been concerned only with electrophilic substitution of benzene. Two important questions arise when we turn to analogous substitutions on rings that already bear at least one substituent:

1. What is the effect of a substituent on the *rate* of electrophilic aromatic substitution?

2. What is the effect of a substituent on the *regioselectivity* of electrophilic aromatic substitution?

To illustrate substituent effects on rate, consider the nitration of benzene, toluene, and (trifluoromethyl)benzene.

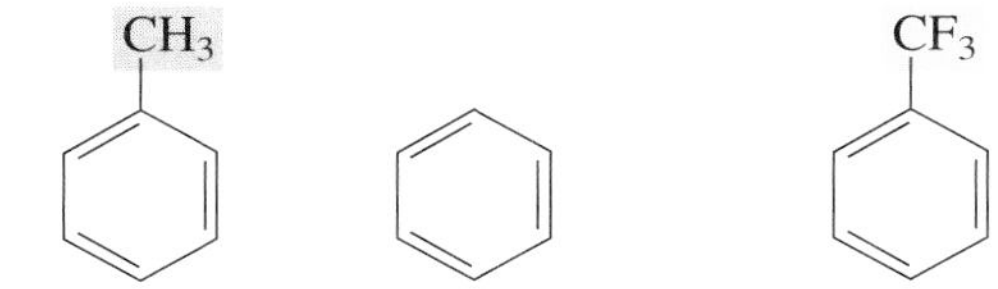

| Toluene | Benzene | (Trifluoromethyl)benzene |
| (most reactive) | | (least reactive) |

Toluene undergoes nitration some 20–25 times faster than benzene. Because toluene is more reactive than benzene, we say that a methyl group *activates* the ring toward electrophilic aromatic substitution. (Trifluoromethyl)benzene, on the other hand, undergoes nitration about 40,000 times more slowly than benzene. We say that a trifluoromethyl group *deactivates* the ring toward electrophilic aromatic substitution.

Just as there is a marked difference in how methyl and trifluoromethyl substituents affect the rate of electrophilic aromatic substitution, so too the way they affect its regioselectivity is markedly different.

Three products are possible from nitration of toluene: *o*-nitrotoluene, *m*-nitrotoluene, and *p*-nitrotoluene. All are formed, but not in equal amounts. Together, the ortho- and para-substituted isomers make up 97% of the product mixture; the meta only 3%.

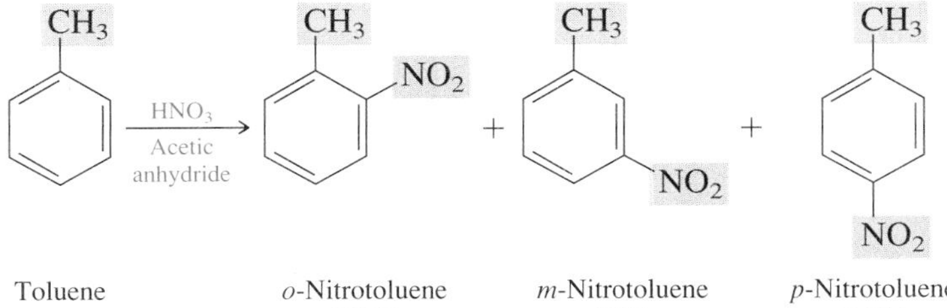

| Toluene | *o*-Nitrotoluene | *m*-Nitrotoluene | *p*-Nitrotoluene |
| | (63%) | (3%) | (34%) |

Because substitution in toluene occurs primarily at positions ortho and para to methyl, we say that

*A methyl substituent is an **ortho, para director**.*

Nitration of (trifluoromethyl)benzene, on the other hand, yields almost exclusively *m*-nitro(trifluoromethyl)benzene (91%). The ortho- and para-substituted isomers are minor components of the reaction mixture.

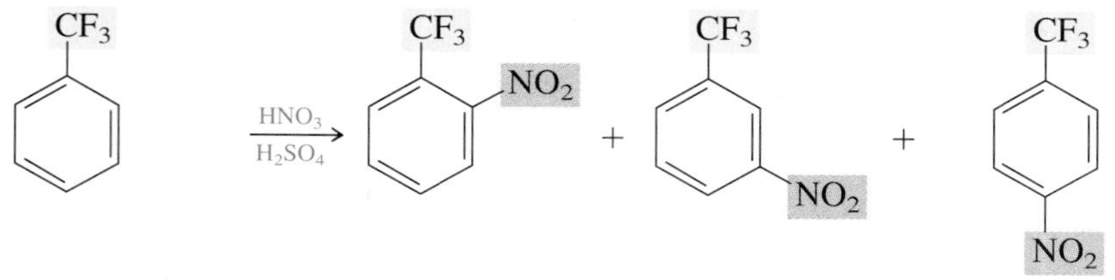

(Trifluoromethyl)benzene   o-Nitro(trifluoro-   m-Nitro(trifluoro-   p-Nitro(trifluoro-
                                   methyl)benzene   methyl)benzene   methyl)benzene
                                       (6%)   (91%)   (3%)

Because substitution in (trifluoromethyl)benzene occurs primarily at positions meta to the substituent, we say that

> A trifluoromethyl group is a **meta director.**

The regioselectivity of substitution, like the rate, is strongly affected by the substituent. In the following several sections we will examine the relationship between the structure of the substituent and its effect on rate and regioselectivity of electrophilic aromatic substitution.

Table 6.3 summarizes orientation and rate effects in electrophilic aromatic substitution reactions for a variety of frequently encountered substituents. It is arranged in order of decreasing activating power: the most strongly activating substituents are at the top, the most strongly deactivating substituents are at the bottom. The main features of the table can be summarized as follows:

1. All activating substituents are ortho, para directors.

2. Halogen substituents are slightly deactivating but are ortho, para-directing.

3. Strongly deactivating substituents are meta directors.

PROBLEM 6.7 Specify whether the major product or products of each of the following reactions is the meta isomer or a mixture of ortho and para isomers and whether the reaction proceeds faster of slower than the same reaction of benzene.

(a) Nitration of anisole ($C_6H_5OCH_3$)   (c) Bromination of chlorobenzene

(b) Nitration of benzoic acid ($C_6H_5\overset{\overset{\textstyle O}{\|}}{C}OH$)

Sample Solution (a) From Table 6.3 you can see that an alkoxy group such as methoxy (—OCH_3) is an activating ortho, para director. The nitration of anisole proceeds faster than the nitration of benzene, and the major products are a mixture of *ortho*-nitroanisole and *para*-nitroanisole.

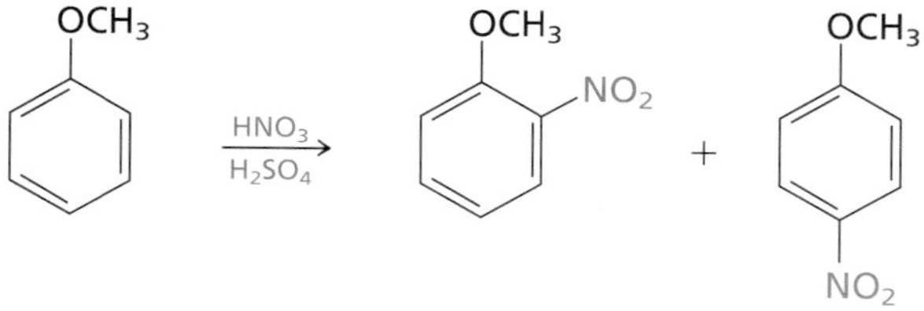

Anisole   o-Nitroanisole   p-Nitroanisole

6.10   SUBSTITUENT EFFECTS: ACTIVATING GROUPS

As you saw in Section 6.7, electrophilic aromatic substitution proceeds through formation of a cyclohexadienyl cation intermediate. The key to understanding the influence of

TABLE 6.3	Classification of Substituents in Electrophilic Aromatic Substitution Reactions	
Effect on Rate	**Substituent**	**Effect on Orientation**
Very strongly activating	—NH$_2$ (amino) —NHR (alkylamino) —NR$_2$ (dialkylamino) —OH (hydroxyl)	Ortho, para-directing
Strongly activating	O ‖ —NHCR (acylamino) —OR (alkoxy) O ‖ —OCR (acyloxy)	Ortho, para-directing
Activating	—R (alkyl) —Ar (aryl) —CH=CR$_2$ (alkenyl)	Ortho, para-directing
Standard of comparison	—H (hydrogen)	
Deactivating	—X (halogen) (X = F, Cl, Br, I) —CH$_2$X (halomethyl)	Ortho, para-directing
Strongly deactivating	O ‖ —CH (formyl) O ‖ —CR (acyl) O ‖ —COH (carboxylic acid) O ‖ —COR (ester) O ‖ —CCl (acyl chloride) —C≡N (cyano) —SO$_3$H (sulfonic acid)	Meta-directing
Very strongly deactivating	—CF$_3$ (trifluoromethyl) —NO$_2$ (nitro)	Meta-directing

a substituent on both rate and regioselectivity of substitution is to further examine the structure and stability of the intermediate. The fundamental principle is that

A more stable carbocation is formed faster than a less stable one.

As noted in the previous section, all activating groups are ortho, para directors. The reason for this is that activating groups are all electron-donating, and thus stabilize the cyclohexadienyl cation intermediate. In addition, the intermediates formed from ortho and para attack are stabilized more than the intermediate formed from meta attack.

Let us look, for example, at the nitration of toluene. Recall from Chapter 3 (Section 3.9) that alkyl groups stabilize carbocations by releasing electron density to the positively charged carbon. Further, by examining the resonance structures that contribute to each of the cyclohexadienyl cation intermediates, we can see that the cations leading to *o*- and *p*-nitrotoluene have tertiary carbocation character. Each has a resonance form in which the positive charge resides on the carbon that bears the methyl group.

Ortho attack

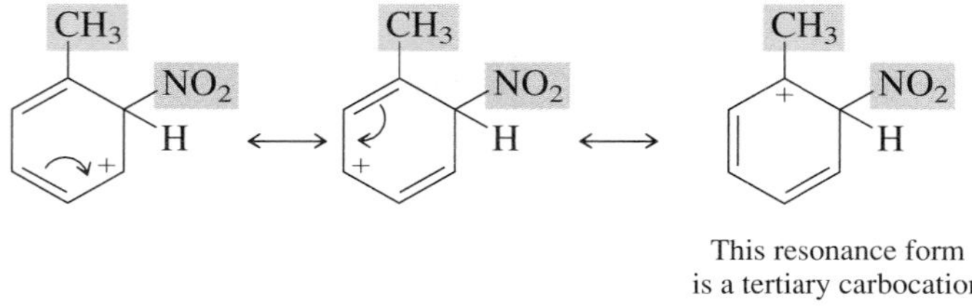

This resonance form
is a tertiary carbocation

Para attack

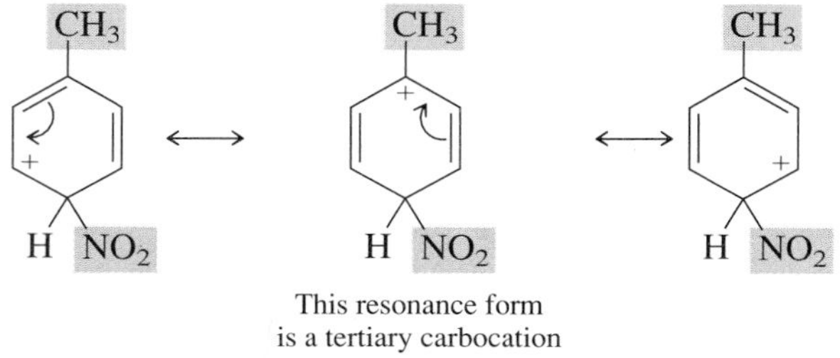

This resonance form
is a tertiary carbocation

The three resonance forms of the intermediate leading to meta substitution are all secondary carbocations.

Meta attack

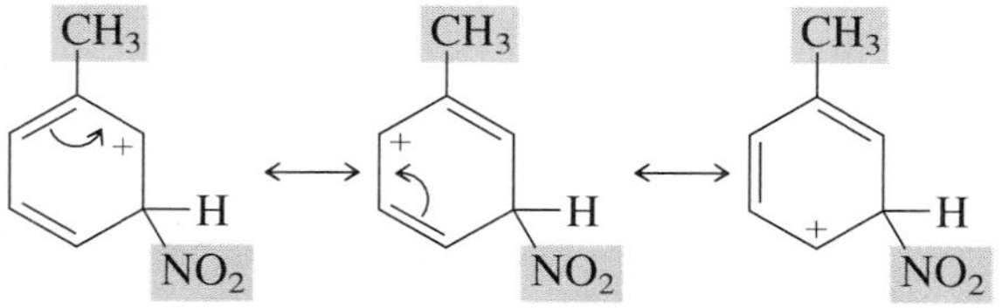

Because of their tertiary carbocation character, the intermediates leading to ortho and to para substitution are more stable and are formed faster than the one leading to meta substitution. They are also more stable than the secondary cyclohexadienyl cation intermediate formed during nitration of benzene. A methyl group is an activating substituent because it stabilizes the carbocation intermediate formed in the rate-determining step more than a hydrogen does. It is ortho, para-directing because it stabilizes the carbocation formed by electrophilic attack at these positions more than it stabilizes the intermediate formed by attack at the meta position.

All alkyl groups, not just methyl, are activating substituents and ortho, para directors. This is because any alkyl group, be it methyl, ethyl, isopropyl, *tert*-butyl, or any other, stabilizes a carbocation site to which it is directly attached. When R = alkyl, where E is any electrophile. All three structures are more stable for R = alkyl than for R = H and are formed more quickly.

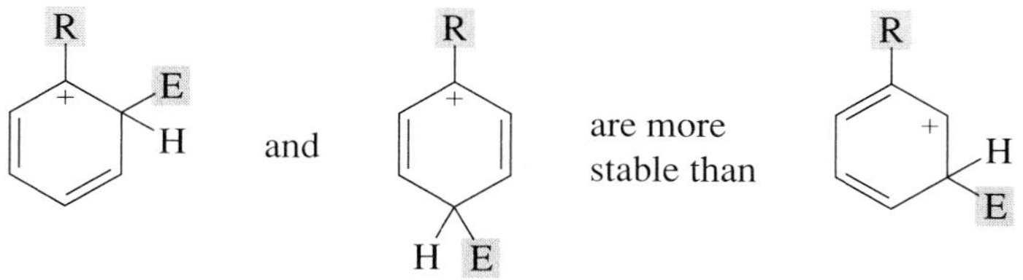

Many of the activating substituents found in Table 6.3 contain an oxygen or nitrogen atom attached to the benzene ring. All can be recognized as electron-donating groups because the atom attached to the ring bears *at least one unshared pair of electrons.* Attack at positions ortho and para to a carbon that bears a substituent —$\ddot{Z}$ (an —$\ddot{O}$— or —$\ddot{N}$—

for example) gives a cation stabilized by delocation of an unshared electron pair. We can illustrate this effect by examining the intermediate formed by ortho attack on a benzene ring bearing an alkoxy group, —$\ddot{O}R$.

Ortho attack

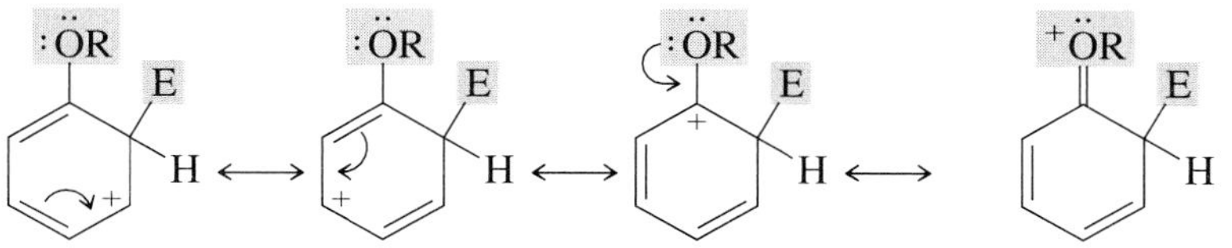

Most stable resonance form; oxygen and all carbons have octets of electrons

Oxygen-stabilized carbocations of this type and that formed from para attack are far more stable than tertiary carbocations. They are best represented by structures in which the positive charge is on oxygen because all the atoms have octets of electrons in such a structure. Their stability permits them to be formed rapidly, resulting in rates of electrophilic aromatic substitution that are much faster than that of benzene. The lone pair on oxygen cannot be directly involved in carbocation stabilization when attack is meta to the substituent.

PROBLEM 6.8 Demonstrate your understanding of the role played by the unshared electron pair in stabilizing the cyclohexadienyl cation intermediate by drawing the resonance structures formed from (a) meta and (b) para attack of an electrophile on an alkoxy-substituted benzene derivative.

Sample Solution (a) Three resonance structures may be drawn that contribute to the intermediate formed by meta attack. The unshared electron pair of the alkoxy group is not directly involved in stabilizing the carbocation.

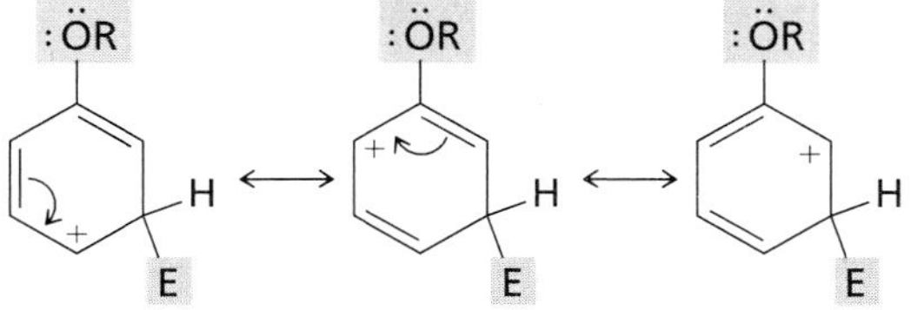

Substituents in which a nitrogen is directly bonded to the ring are even more strongly activating than the corresponding oxygen-containing substituents. The role played by the unshared electron pair on nitrogen in stabilizing the cyclohexadienyl cation intermediate is analogous to that just described for oxygen. Because nitrogen is less electronegative than oxygen, it is a better electron pair donor, and stabilizes the intermediate to a greater extent.

6.11 SUBSTITUENT EFFECTS: STRONGLY DEACTIVATING GROUPS

By looking at the substituent groups on the bottom half of Table 6.3 we can see that

All strongly deactivating groups are meta-directing.

The same principle used to explain activating groups is in place here, but we can state it in reverse:

A less stable carbocation is formed more slowly than a more stable one.

All meta-directing groups are electron-withdrawing. As you will see in the examples that follow, the atom directly bonded to the benzene ring in these substituents has either a partial or a full positive charge ($-\overset{\delta+}{Z}$ or $-\overset{+}{Z}$). The effect of this positive charge is to destabilize the cyclohexadienyl cation intermediate. Thus the intermediate is formed more slowly than in the same reaction of benzene. In addition, the destabilizing effect is greater at positions ortho and para to the substituent, resulting in meta substitution.

Consider, for example, the trifluoromethyl group, CF_3. Because of their high electronegativity the three fluorine atoms polarize the electron distribution in their σ bonds to carbon, so that carbon bears a partial positive charge.

Recall from Section 3.9 that effects that are transmitted by the polarization of σ bonds are called **inductive effects.**

Unlike a methyl group, which is electron-releasing, a trifluoromethyl group is a powerful electron-withdrawing substituent. Consequently, a CF_3 group *destabilizes* a carbocation site to which it is attached.

Methyl group
releases electrons
(stabilizes carbocation)

more stable than

more stable than

Trifluoromethyl
group withdraws electrons
(destabilizes carbocation)

When we examine the cyclohexadienyl cation intermediates involved in the nitration of (trifluoromethyl)benzene, we find that those leading to ortho and para substitution are strongly destabilized.

Ortho attack

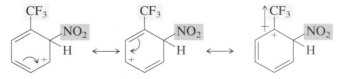

Positive charge on carbon
bearing trifluoromethyl group
(very unstable)

Para attack

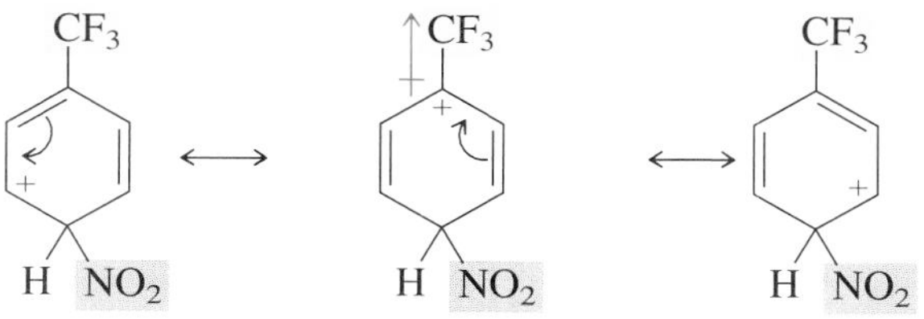

Positive charge on carbon
bearing trifluoromethyl group
(very unstable)

None of the three major resonance forms of the intermediate formed by attack at the meta position has a positive charge on the carbon bearing the trifluoromethyl substituent.

Meta attack

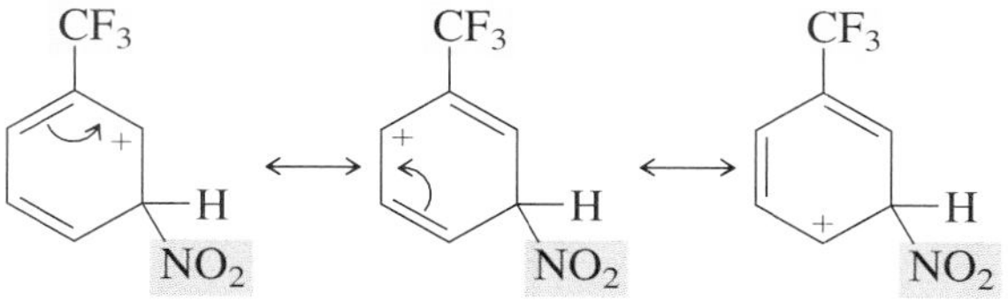

Attack at the meta position leads to a more stable intermediate than attack at either the ortho or the para position, and so meta substitution predominates.

Several other strongly deactivating substituents have a carbonyl group directly attached to the aromatic ring.

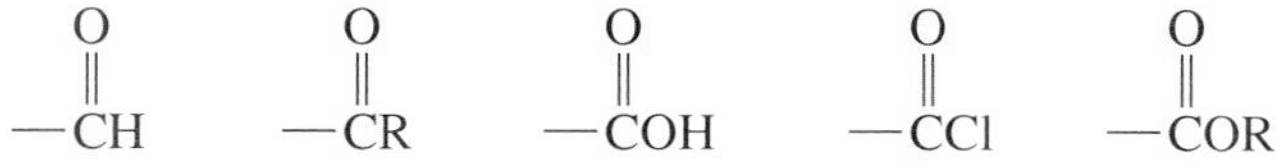

O		$-CH$		

$$\underset{\text{Aldehyde}}{-\overset{\overset{\displaystyle O}{\|}}{C}H} \qquad \underset{\text{Ketone}}{-\overset{\overset{\displaystyle O}{\|}}{C}R} \qquad \underset{\substack{\text{Carboxylic}\\\text{acid}}}{-\overset{\overset{\displaystyle O}{\|}}{C}OH} \qquad \underset{\substack{\text{Acyl}\\\text{chloride}}}{-\overset{\overset{\displaystyle O}{\|}}{C}Cl} \qquad \underset{\text{Ester}}{-\overset{\overset{\displaystyle O}{\|}}{C}OR}$$

The behavior of aromatic aldehydes is typical. Nitration of benzaldehyde takes place several thousand times more slowly than that of benzene and yields *m*-nitrobenzaldehyde as the major product.

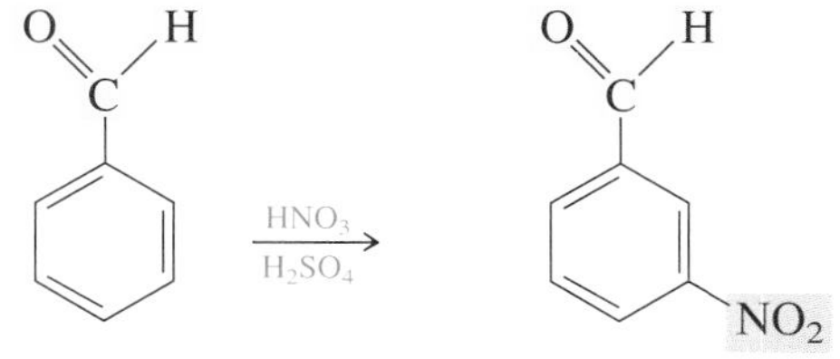

$$\underset{\text{Benzaldehyde}}{} \quad \xrightarrow[\text{H}_2\text{SO}_4]{\text{HNO}_3} \quad \underset{\substack{\textit{m}\text{-Nitrobenzaldehyde}\\(75-84\%)}}{}$$

To understand the effect of a carbonyl group attached directly to the ring, consider its polarization. The electrons in the carbon–oxygen double bond are drawn toward oxygen and away from carbon, leaving the carbon attached to the ring with a partial positive charge. Using benzaldehyde as an example,

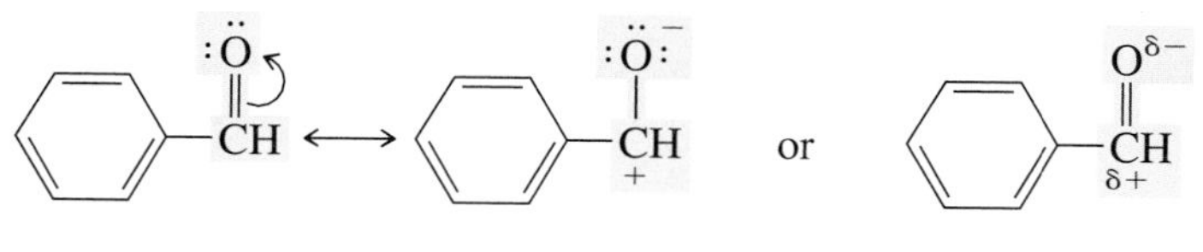

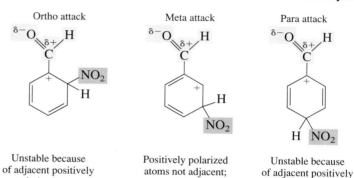

Because the carbon atom attached to the ring is positively polarized, a carbonyl group behaves in much the same way as a trifluoromethyl group and *destabilizes* all the cyclohexadienyl cation intermediates in electrophilic aromatic substitution reactions. The cyclohexadienyl cation intermediates from nitration of benzaldehyde are:

Ortho attack | Meta attack | Para attack

Unstable because of adjacent positively polarized atoms | Positively polarized atoms not adjacent; most stable intermediate | Unstable because of adjacent positively polarized atoms

PROBLEM 6.9 Each of the following reactions has been reported in the chemical literature, and the principal organic product has been isolated in good yield. Write a structural formula for the isolated product of each reaction.

(a) Treatment of benzoyl chloride ($C_6H_5\overset{O}{\underset{\|}{C}}Cl$) with chlorine and iron(III) chloride

(b) Treatment of methyl benzoate ($C_6H_5\overset{O}{\underset{\|}{C}}OCH_3$) with nitric acid and sulfuric acid

(c) Nitration of 1-phenyl-1-propanone ($C_6H_5\overset{O}{\underset{\|}{C}}CH_2CH_3$)

Sample Solution (a) Benzoyl chloride has a carbonyl group attached directly to the ring. A $-\overset{O}{\underset{\|}{C}}Cl$ substituent is meta-directing. The combination of chlorine and iron(III) chloride introduces a chlorine onto the ring. The product is *m*-chlorobenzoyl chloride.

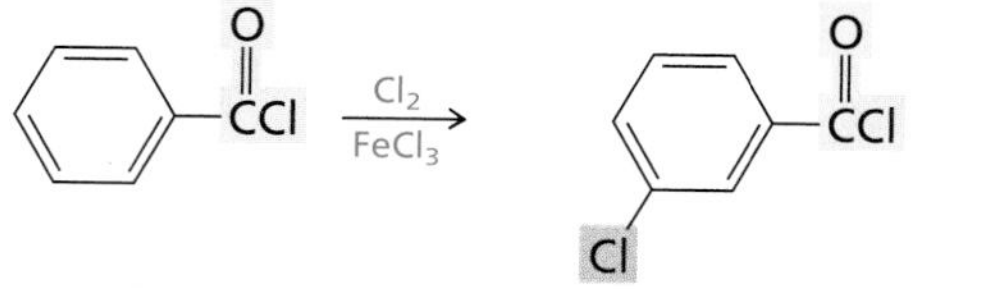

Benzoyl chloride $\xrightarrow[\text{FeCl}_3]{\text{Cl}_2}$ *m*-Chlorobenzoyl chloride
(isolated in 62% yield)

The nitrogen atom of a nitro group bears a full positive charge in its two most stable Lewis structures.

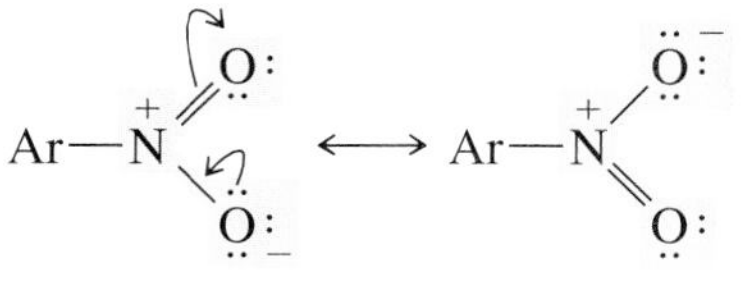

This makes the nitro group a powerful electron-withdrawing deactivating substituent and a meta director.

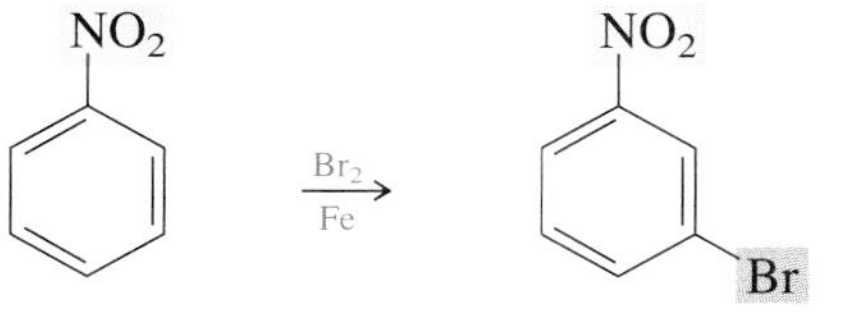

Nitrobenzene

m-Bromonitrobenzene
(60–75%)

> **PROBLEM 6.10**   Would you expect the substituent $-\overset{+}{N}(CH_3)_3$ to more closely resemble $-\ddot{N}(CH_3)_2$ or $-NO_2$ in its effect on rate and regioselectivity in electrophilic aromatic substitution? Why?

6.12   SUBSTITUENT EFFECTS: HALOGENS

The halogens are unique in that they are **deactivating ortho, para directors.** This seeming inconsistency between regioselectivity and rate can be understood by analyzing the two ways that a halogen substituent can affect the stability of a cyclohexadienyl cation. First, halogens are electronegative, and their inductive effect is to draw electrons away from the carbon to which they are bonded in the same way that a trifluoromethyl group does. Thus, all the intermediates formed by electrophilic attack on a halobenzene are less stable than the corresponding cyclohexadienyl cation for benzene, and halobenzenes are less reactive than benzene.

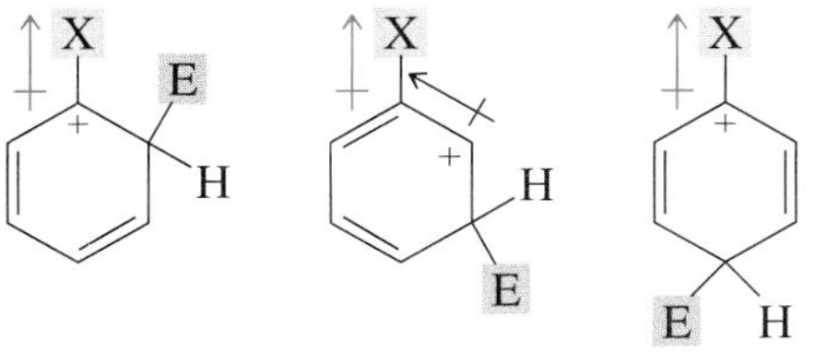

All these ions are less stable when X = F, Cl, Br, or I than when X = H

Like hydroxyl groups and amino groups, however, halogen substituents possess unshared electron pairs that can be donated to a positively charged carbon. This electron donation into the π system stabilizes the intermediates derived from ortho and from para attack.

Ortho attack   Para attack

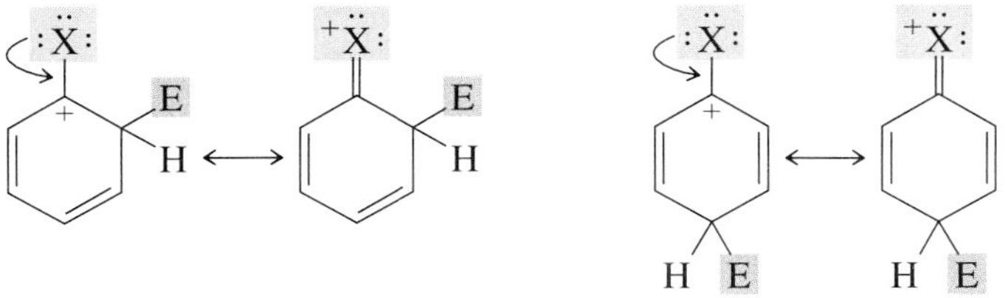

Comparable stabilization of the intermediate leading to meta substitution is not possible.

6.13 REGIOSELECTIVE SYNTHESIS OF DISUBSTITUTED AROMATIC COMPOUNDS

Because the position of electrophilic attack on an aromatic ring is controlled by the directing effects of substituents already present, the preparation of disubstituted aromatic compounds requires that careful thought be given to the order of the introduction of the two groups.

Compare the independent preparations of *m*-bromoacetophenone and *p*-bromoacetophenone from benzene. Both syntheses require a Friedel–Crafts acylation step and a bromination step, but the major product is determined by the *order* in which the two steps are carried out. When the meta-directing acetyl group is introduced first, the final product is *m*-bromoacetophenone.

Aluminum chloride is a stronger Lewis acid than iron(III) bromide and has been used as a catalyst in electrophilic bromination when, as in the example shown, the aromatic ring bears a strongly deactivating substituent.

Benzene

Acetophenone
(76–83%)

m-Bromoacetophenone
(59%)

When the ortho, para-directing bromine is introduced first, the major product is *p*-bromoacetophenone (along with some of its ortho isomer, from which it is separated by distillation).

Benzene

Bromobenzene
(65–75%)

p-Bromoacetophenone
(69–79%)

In planning a synthesis, the *orientation* of the substituents in the desired product determines the *order* of the reactions used to carry out the scheme. If the substituents are meta in the product, then the meta-directing group must be introduced first. Likewise, if the substituents are ortho or para, the first substituent on the ring must be an ortho, para-directing one.

PROBLEM 6.11 Write the chemical equations showing how you could prepare *m*-bromonitrobenzene as the principal organic product, starting with benzene and using any necessary organic or inorganic reagents. How could you prepare *p*-bromonitrobenzene?

6.14 A GENERAL VIEW OF AROMATICITY: HÜCKEL'S RULE

Although most aromatic compounds contain a benzene ring, the presence of such a ring is not a necessary condition for aromaticity. In other words, some aromatic compounds do not contain a benzene ring. In the 1930s the German chemist Erich Hückel proposed that among monocyclic systems, only those containing 2, 6, 10, 14, etc., π electrons could be aromatic. This statement is summarized as **Hückel's rule:**

Among planar, monocyclic, fully conjugated polyenes, only those possessing $(4n + 2)$ *π electrons will be aromatic.*

The most common case of Hückel's rule occurs when $n = 1$, that is, when the molecule has six π electrons. Benzene, of course, fits this example and is aromatic. A number of aromatic compounds and ions that do not contain a benzene ring also satisfy Hückel's rule. As you will see in the next section, Hückel's rule also predicts aromaticity in molecules that are not hydrocarbons.

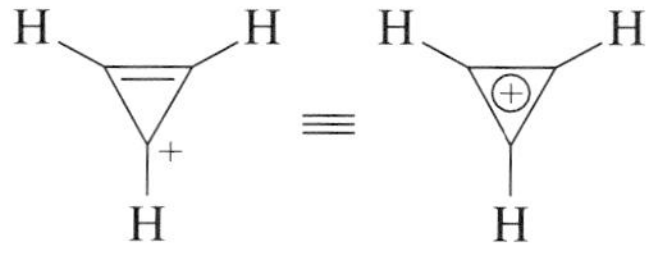

The integer *n* has no physical significance; it is the variable in the mathematical expression that describes the series 2, 6, 10, etc. Hückel's rule only applies to monocyclic systems and should not be applied to polycyclic aromatic hydrocarbons (Section 6.4).

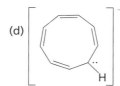

Cyclopropenyl cation
($n = 0$; 2 π electrons)

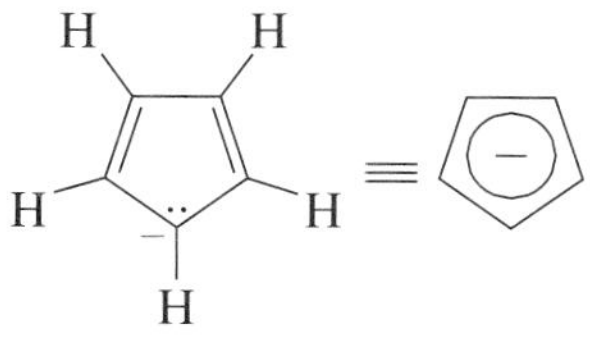

Cyclopentadienide anion
($n = 1$; 6 π electrons)

In counting π electrons, a double bond contributes 2, a positively charged carbon none, and a negatively charged carbon 2.

PROBLEM 6.12 Determine which, if any, of the following molecules or ions is aromatic. Explain your reasoning.

(a)   (b)   (c)   (d)

Sample Solution (a) Although the structure shown has six π electrons from three double bonds, the molecule is not cyclic. A cyclic array of p orbitals is a requirement for aromaticity. 1,3,5-Hexatriene is not aromatic.

6.15 HETEROCYCLIC AROMATIC COMPOUNDS

Cyclic compounds that contain at least one atom other than carbon within their ring are called **heterocyclic compounds.** Nitrogen and oxygen are the most common heteroatoms, although heterocyclic compounds containing other atoms such as sulfur are known. Many heterocyclic compounds are aromatic, as they fit the criteria for aromaticity described in the preceding section. Examples of **heterocyclic aromatic compounds** include pyridine, pyrrole, and furan.

Pyridine   Pyrrole   Furan

Hückel's rule can be extended to heterocyclic aromatic compounds. A single heteroatom can contribute either 0 or 2 of its unshared electrons as needed to the π system so as to satisfy the $(4n + 2)$ π electron requirement. As shown in Figure 6.4a, for example, pyridine is aromatic because it has six π electrons in its ring exclusive of the unshared pair on nitrogen. The nitrogen unshared pair is in an sp^2 hybrid orbital that is perpendicular to the π system, not in a p orbital aligned with it.

FIGURE 6.4 (*a*) Pyridine has six π electrons plus an unshared pair in a nitrogen *sp²* orbital. (*b*) Pyrrole has six π electrons. (*c*) Furan has six π electrons plus an unshared pair in an oxygen *sp²* orbital, which is perpendicular to the π system and does not interact with it.

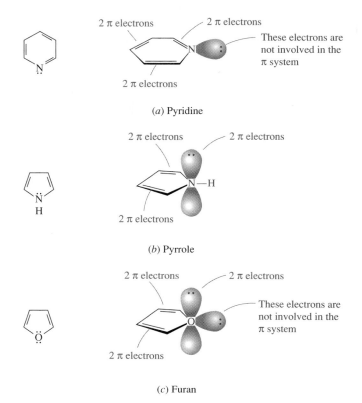

(*a*) Pyridine

(*b*) Pyrrole

(*c*) Furan

Numerous aromatic heterocycles are found in biological systems. A few examples include pyridoxine, serotonin, and adenine.

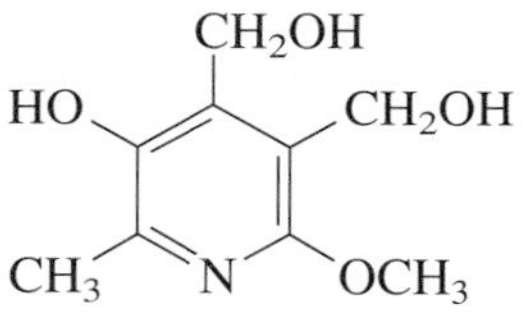

Pyridoxine
(vitamin B$_6$)

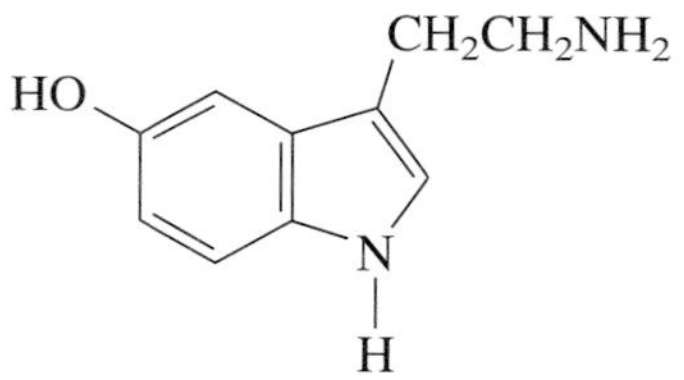

Serotonin
(brain neurotransmitter)

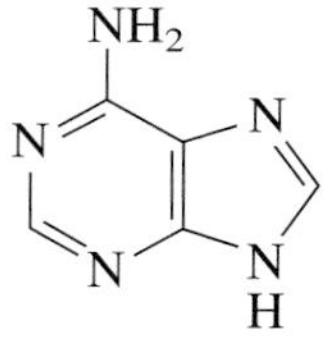

Adenine
(present in DNA and RNA)

LEARNING OBJECTIVES

Aromaticity is one of the fundamental concepts of organic chemistry. You have seen in this chapter how aromaticity leads to the unique chemical reactivity of arenes. The skills you have learned in this chapter should enable you to:

• Describe the structure of benzene using resonance.

• Explain the bonding in benzene using the orbital hybridization model.

• Write a structural formula for an aromatic compound on the basis of its systematic IUPAC name.

Continued

- Write a correct IUPAC name for an aromatic compound on the basis of a given structural formula.
- Write chemical equations for electrophilic addition and oxidation reactions of aromatic side chains.
- Write chemical equations describing the electrophilic aromatic substitution reactions of benzene: halogenation, nitration, sulfonation, Friedel–Crafts alkylation, and Friedel–Crafts acylation.
- Explain the mechanistic basis for the action of activating ortho, para-directing groups.
- Explain the mechanistic basis for the action of deactivating meta-directing groups.
- Explain why the halogens are deactivating ortho, para-directing groups.
- Write chemical equations describing the synthesis of disubstituted aromatic compounds.
- Predict whether a substance is aromatic based on its structure and Hückel's rule.

6.16   SUMMARY

An **aromatic** compound is one that is substantially more stable than expected on the basis of structural formulas that restrict electron pairs to regions between two nuclei. Benzene is an aromatic hydrocarbon. Neither of the two Kekulé formulas for benzene adequately describes its structure or properties. Benzene is regarded as a **resonance hybrid** of these two Kekulé forms (Section 6.1).

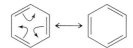

The special stability of benzene is attributed to delocalization of its six π electrons over the six carbons of the ring (Section 6.2).

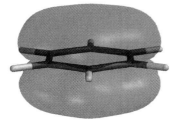

Disubstituted benzene derivatives are named as **ortho, meta,** and **para** according to the relative positions of the substituents (Section 6.3).

Ortho   Meta   Para

Polycyclic aromatic hydrocarbons, of which anthracene is an example, contain two or more benzene rings fused together (Section 6.4).

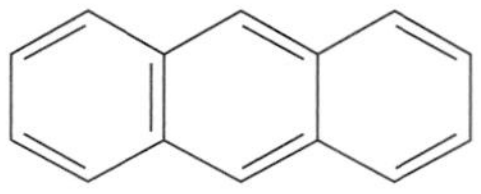

Anthracene

Carbocations adjacent to a benzene ring are resonance-stabilized and are called **benzylic carbocations.** Electrophilic additions to double bonds adjacent to an aromatic ring exhibit regioselectivity resulting from formation of the more stable benzylic carbocations (Section 6.5).

$$C_6H_5CH{=}CHCH_3 + HBr \longrightarrow C_6H_5\overset{\displaystyle Br}{\underset{|}{C}}HCH_2CH_3$$

(formed via $C_6H_5\overset{+}{C}HCH_2CH_3$)

Alkyl side chains on a benzene ring are oxidized on being heated with chromic acid, providing a useful synthesis of substituted benzoic acids.

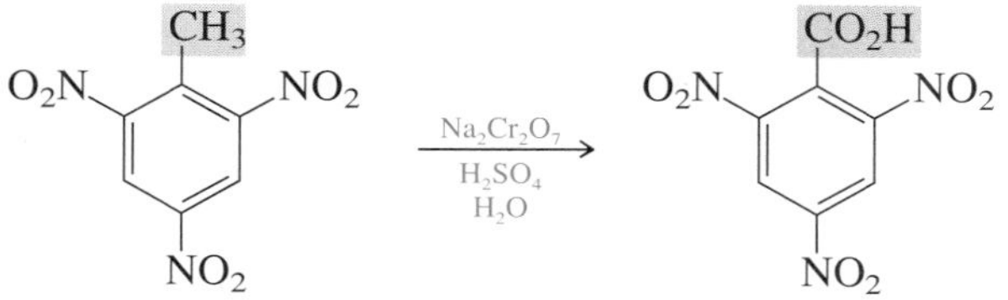

2,4,6-Trinitrotoluene   2,4,6-Trinitrobenzoic acid
(57–69%)

Aromatic compounds undergo substitution rather than addition on reaction with electrophilic reagents (Section 6.6).

Substitution occurs by attack of an electrophile on the π electrons of the aromatic ring in the rate-determining step to form a cyclohexadienyl cation intermediate. Loss of a proton from this intermediate restores the aromaticity of the ring and yields the product of **electrophilic aromatic substitution** (Section 6.7).

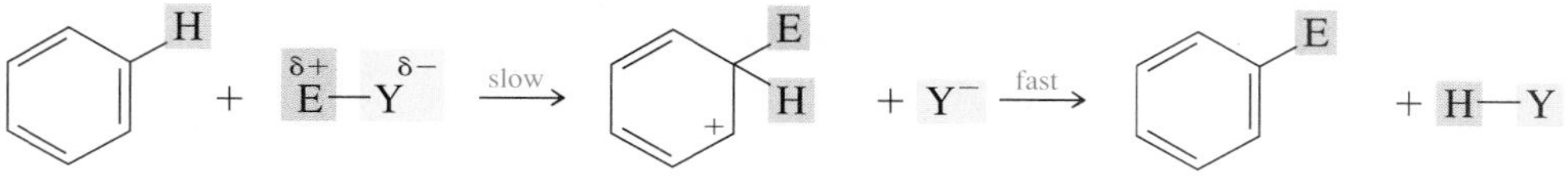

Benzene   Electrophilic   Cyclohexadienyl   Product of
               reagent   cation intermediate   electrophilic aromatic
                                                              substitution

Table 6.2 (Section 6.6) lists some typical examples of electrophilic aromatic substitution reactions.

Substituents on an aromatic ring can influence both the **rate** and **regioselectivity** of further substitution of the ring. Substituents are classified as **activating** or **deactivating** according to whether they cause the ring to react more rapidly or less rapidly than benzene. With respect to regioselectivity, substituents are either **ortho, para-directing,** or **meta-directing** (Section 6.9). Substituents can be arranged into three major categories:

1. **Activating and ortho, para-directing:**   These substituents are **electron-donating** and stabilize the cyclohexadienyl intermediate formed in the reaction (Section 6.10). The effect is greatest at the positions ortho and para to the sub-stituent. Examples include $-\ddot{N}R_2$, $-\ddot{O}R$, $-R$, $-Ar$, and related species. The most strongly activating members of this group are bonded to the ring by a nitro-gen or oxygen atom that bears an unshared pair of electrons.

2. **Deactivating and meta-directing:**   These substituents are **electron-withdrawing** and destabilize the cyclohexadienyl intermediate (Section 6.11). Examples include:

$$-CF_3 \qquad -\overset{\displaystyle O}{\overset{\displaystyle \|}{C}}R \qquad -C\equiv N \qquad -NO_2$$

and related species. All the ring positions are deactivated, but because the meta positions are deactivated less than the ortho and para, meta substitution is favored.

3. **Deactiviating and ortho, para-directing:**   The halogens are the most prominent members of this class (Section 6.12). They withdraw electron density from all the ring positions by an inductive effect, making halobenzenes less reactive than ben-zene. Lone-pair electron donation stabilizes the cyclohexadienyl cations corre-sponding to attack at the ortho and para positions more than those formed by attack at the meta positions, giving rise to the observed regioselectivity.

The order in which substituents are introduced onto a benzene ring needs to be considered to prepare the desired isomer in a multistep synthesis (Section 6.13).

Monocyclic aromatic compounds, including those containing a benzene ring, meet a set of criteria known as **Hückel's rule** (Section 6.14). According to Hückel's rule, pla-nar, monocyclic, completely conjugated polyenes are aromatic if they contain $4n + 2 \pi$ electrons. Aromatic compounds that contain one or more atoms other than carbon in their rings are called **heterocyclic aromatic compounds** (Section 6.15).

ADDITIONAL PROBLEMS

Aromaticity

6.13 Each of the following statements *incorrectly* describes some aspect of benzene's structure. Write a corrected version of each statement.

    (a) Benzene exists as a ring of alternating double and single bonds.
    (b) The most stable conformation of the benzene ring is the chair form.
    (c) All the carbon atoms of benzene are *sp*-hybridized.

6.14 Explain why an aromatic ring cannot contain an sp^3-hybridized carbon atom.

6.15 The most stable resonance structure for a polycyclic aromatic hydrocarbon is the one that has the greatest number of rings that correspond to Kekulé formulations of benzene. Draw the most stable resonance form for:

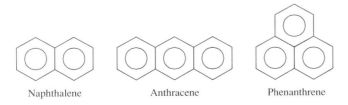

Naphthalene   Anthracene   Phenanthrene

6.16 Which of the following species is aromatic? Assume a planar shape for each.

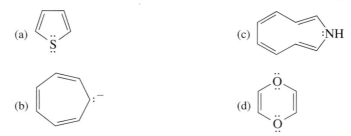

6.17 Write resonance structures for the aromatic cycloheptatrienyl cation sufficient to show the delocalization of the positive charge over all seven carbons.

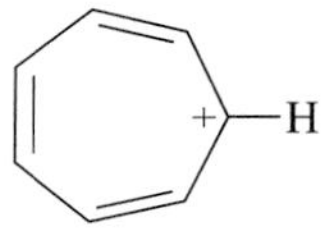

Cycloheptatrienyl cation

Nomenclature

6.18 Write the structures and give a correct systematic name for all the isomeric:

    (a) Nitrotoluenes   (b) Dichlorobenzoic acids   (c) Tribromophenols

6.19 How many monobromo derivatives are possible for (you do not need to name them):

    (a) Anthracene   (b) Naphthalene

6.20 Give a systematic name for each of the following compounds.

    (a) $CH_3CHCH_2CH_2CH_3$
              |
            C_6H_5

    (c) $C_6H_5CH{=}CHCH_3$

    (b) $CH_3CH_2CHCH_2OH$
                |
             C_6H_5

    (d) 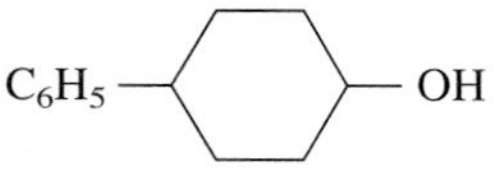 C_6H_5———————OH

6.21 Write structural formulas and give the IUPAC names for all the isomers of $C_6H_5C_4H_9$ that contain a monosubstituted benzene ring.

6.22 Write a structural formula corresponding to each of the following:

    (a) (*E*)-1-Phenyl-1-butene   (e) *p*-Chlorophenol
    (b) (*Z*)-2-Phenyl-2-butene   (f) *p*-Diisopropylbenzene
    (c) (*R*)-1-Phenylethanol   (g) 2,4,6-Tribromoaniline
    (d) *o*-Chlorobenzyl alcohol

6.23 Using numerical locants and the names in Table 6.1 as a guide, give an acceptable IUPAC name for each of the following compounds:

    (a) Diosphenol (used in   (b) *m*-Xylidine (used in
        veterinary medicine   synthesis of lidocaine,
        to control parasites   a local anesthetic)
        in animals)

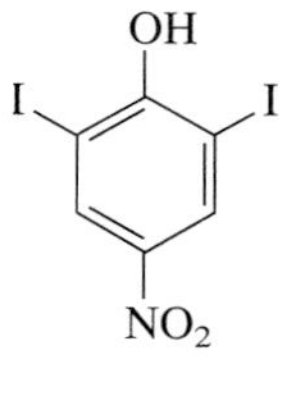

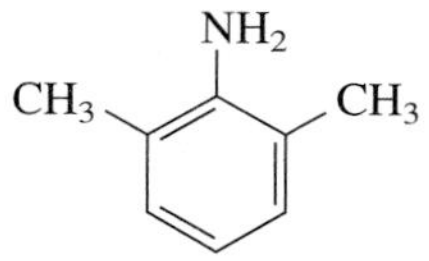

6.24 Give a correct systematic name for each of the following aromatic compounds.

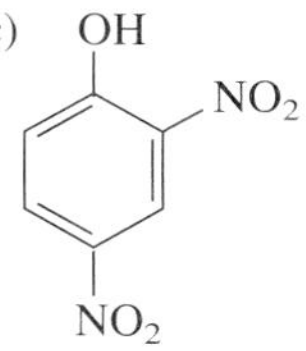

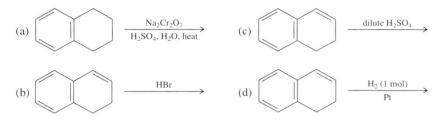

Aromatic Side-Chain Reactions

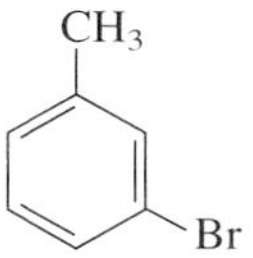

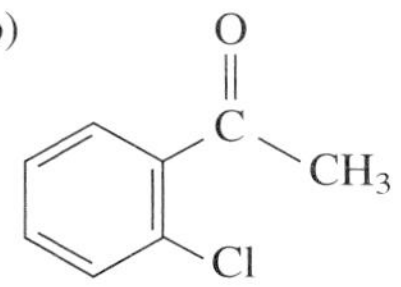

6.25 Give the structure of the major product of each of the following reactions.

(a) $\xrightarrow[\text{H}_2\text{SO}_4, \text{H}_2\text{O, heat}]{\text{Na}_2\text{Cr}_2\text{O}_7}$

(c) $\xrightarrow{\text{dilute H}_2\text{SO}_4}$

(b) $\xrightarrow{\text{HBr}}$

(d) $\xrightarrow[\text{Pt}]{\text{H}_2 \text{ (1 mol)}}$

6.26 Write chemical equations for each of the following industrial oxidation reactions, carried out using air in the presence of a catalyst.

(a) Oxidation of *o*-xylene to give *o*-phthalic acid, used in the synthesis of plasticizers
(b) Oxidation of *p*-xylene to give terephthalic acid, used in the preparation of polyester fibers and films

Substitution Reactions

6.27 Give reagents for effecting each of the following reactions and write the principal products. If an ortho, para mixture is expected, show both. If the meta isomer is the expected major product, write only that isomer.

(a) Nitration of benzene
(b) Nitration of the product of (a)
(c) Bromination of toluene
(d) Bromination of (trifluoromethyl)benzene
(e) Sulfonation of anisole
(f) Sulfonation of acetanilide ($C_6H_5NH\overset{\overset{\displaystyle O}{\|}}{C}CH_3$)
(g) Chlorination of bromobenzene

6.28 Give the major product or products of the following Friedel–Crafts reactions.

(a) Alkylation of benzene with benzyl chloride ($C_6H_5CH_2Cl$)
(b) Alkylation of anisole with benzyl chloride
(c) Acylation of benzene with benzoyl chloride ($C_6H_5\overset{\overset{\displaystyle O}{\|}}{C}Cl$)
(d) Acylation of the product of (a) with benzoyl chloride

6.29 Write the structure of the product from monobromination of the following compound with Br_2 and $FeBr_3$.

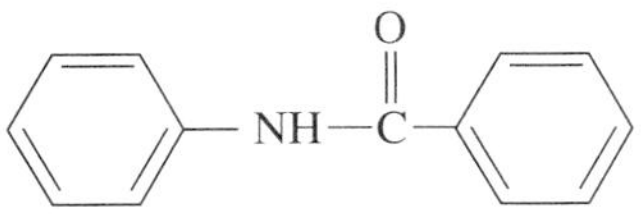

6.30   What combination of acyl chloride or acid anyhdride and arene would you choose to pre-pare the following compound by a Friedel–Crafts acylation reaction?

Substitution Mechanisms

6.31   Write the structures of the electrophilic species present in each of the types of electrophilic aromatic substitution reactions: halogenation, nitration, sulfonation, and Friedel–Crafts alkylation and acylation.

6.32   In each of the following pairs of compounds choose which one will react faster with the indicated reagent, and write a chemical equation for the faster reaction:

(a) Toluene or chlorobenzene with a mixture of nitric acid and sulfuric acid

(b) Fluorobenzene or (trifluoromethyl)benzene with benzyl chloride and aluminum chloride

(c) Methyl benzoate ($C_6H_5\overset{\overset{\displaystyle O}{\|}}{C}OCH_3$) or phenyl acetate ($C_6H_5O\overset{\overset{\displaystyle O}{\|}}{C}CH_3$) with bromine in acetic acid

(d) Acetanilide ($C_6H_5NH\overset{\overset{\displaystyle O}{\|}}{C}CH_3$) or nitrobenzene with sulfur trioxide in sulfuric acid

6.33   Write a step-by-step mechanism describing the reaction of benzene with chlorine in the presence of iron(III) chloride ($FeCl_3$). Be sure to write all the resonance structures contributing to the intermediate in the reaction.

6.34   Write structural formulas for the cyclohexadienyl cations formed from aniline ($C_6H_5NH_2$) during:

(a) Ortho bromination (four resonance structures)

(b) Meta bromination (three resonance structures)

(c) Para bromination (four resonance structures)

6.35   Write a structural formula for the most stable cyclohexadienyl cation intermediate formed in both the following reactions. Is the cyclohexadienyl cation more or less stable than the corresponding intermediate formed by electrophilic attack on benzene?

(a) Nitration of isopropylbenzene

(b) Bromination of nitrobenzene

6.36   Identify the following aromatic substituents as being activating and ortho, para-directing or deactivating and meta-directing:

(a) $-\overset{..}{\underset{..}{S}}H$ (thiol)

(b) $-\overset{+}{S}(CH_3)_2$ (dimethylsulfonium)

Synthesis of Aromatic Compounds

6.37   Write equations showing how you could prepare each of the following from benzene and any necessary organic or inorganic reagents. If an ortho, para mixture is formed in any step of your synthesis, assume that you can separate the two isomers.

(a) Isopropylbenzene

(b) *p*-Isopropylbenzenesulfonic acid

(c) *m*-Chloroacetophenone

(d) *p*-Chloroacetophenone

6.38   Two students wish to prepare *m*-chloronitrobenzene beginning with benzene. One student carries out the chlorination step first; the other begins with the nitration step. Write equations for both these reaction schemes. Which one gives the correct product?

6.39 Beginning with toluene and any other necessary organic or inorganic reagents, outline reaction schemes that will prepare

    (a) *m*-Bromobenzoic acid   (b) *p*-Bromobenzoic acid

6.40 In an attempt to prepare the compound shown, three synthetic reaction schemes were attempted. Only one of these worked; which one was it? What are the products from the other two pathways?

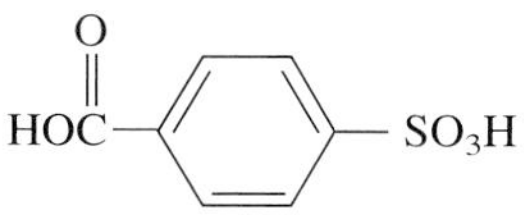

I: C_6H_6 $\xrightarrow[\text{SO}_3]{\text{H}_2\text{SO}_4}$ $\xrightarrow[\text{AlCl}_3]{\text{CH}_3\text{CH}_2\text{Br}}$ $\xrightarrow[\text{H}_2\text{SO}_4, \text{H}_2\text{O, heat}]{\text{Na}_2\text{Cr}_2\text{O}_7}$

II: C_6H_6 $\xrightarrow[\text{AlCl}_3]{\text{CH}_3\text{CH}_2\text{Br}}$ $\xrightarrow[\text{SO}_3]{\text{H}_2\text{SO}_4}$ $\xrightarrow[\text{H}_2\text{SO}_4, \text{H}_2\text{O, heat}]{\text{Na}_2\text{Cr}_2\text{O}_7}$

III: C_6H_6 $\xrightarrow[\text{AlCl}_3]{\text{CH}_3\text{CH}_2\text{Br}}$ $\xrightarrow[\text{H}_2\text{SO}_4, \text{H}_2\text{O, heat}]{\text{Na}_2\text{Cr}_2\text{O}_7}$ $\xrightarrow[\text{SO}_3]{\text{H}_2\text{SO}_4}$

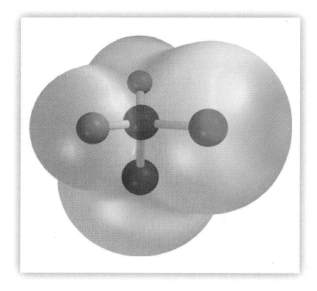

CHAPTER 7
STEREOCHEMISTRY

We live in a three-dimensional world; in fact, a dictionary definition of three-dimensional is "life-like." Molecules are three-dimensional, as we saw in Chapter 1 when we discussed bonding and molecular shapes. The chemical changes that take place all around us, especially those in living organisms, are influenced by the three-dimensional relationships among molecules. The study of the spatial arrangement of atoms and molecules is known as **stereochemistry.** Isomers that have the same constitution but differ in the spatial arrangement of their atoms are called **stereoisomers.** We have already had considerable experience with certain types of stereoisomers—those involving cis and trans substitution patterns in alkenes and in cycloalkanes.

Our major objectives in this chapter are to develop a feeling for molecules as three-dimensional objects and to become familiar with stereochemical principles, terms, and notation. A full understanding of organic and biological chemistry requires an awareness of the spatial requirements for interactions between molecules; this chapter provides the basis for that understanding.

7.1 MOLECULAR CHIRALITY: ENANTIOMERS

Everything has a mirror image, but not all things are superimposable on their mirror images. Mirror-image superimposability characterizes many objects we use every day. Cups and saucers, forks and spoons, chairs and beds are all identical with their mirror images. Many other objects though—and this is the more interesting case—are not. Your left hand and your right hand, for example, are mirror images of each other but can't be made to coincide point for point in three dimensions. In 1894, William Thomson coined a word for this property. He defined an object as **chiral** if it is not superimposable on its mirror image. Applying Thomson's term to chemistry, we say that

William Thomson is better known as Lord Kelvin and is remembered for his contributions to the field of thermodynamics.

A molecule is chiral if its two mirror-image forms are not superimposable in three dimensions.

The word *chiral* is derived from the Greek word *cheir,* meaning "hand," and it is entirely appropriate to speak of the "handedness" of molecules. The opposite of chiral is **achiral.** A molecule that *is* superimposable on its mirror image is achiral.

In organic chemistry, chirality most often occurs in molecules that contain a carbon that is attached to four different groups. An example is bromochlorofluoromethane (BrClFCH).

Bromochlorofluoromethane

As shown in Figure 7.1, the two mirror images of bromochlorofluoromethane cannot be superimposed on each other.

Because the two mirror images of bromochlorofluoromethane are not super-imposable, BrClFCH is chiral.

The two mirror images of bromochlorofluoromethane have the same constitution, but they differ in the arrangement of their atoms in space; they are **stereoisomers.** Stereoisomers that are related as an object and its nonsuperimposable mirror image are classified as **enantiomers.** Just as an object has one, and only one, mirror image, a chiral molecule can have one, and only one, enantiomer.

Consider next a molecule such as chlorodifluoromethane (ClF_2CH), in which two of the atoms attached to carbon are the same. Figure 7.2 (page 185) shows two molecular models of ClF_2CH drawn so as to be mirror images. As is evident from these drawings, it is a simple matter to merge the two models so that all the atoms match.

Because mirror-image representations of chlorodifluoromethane are super-imposable on each other, ClF_2CH is achiral.

The two mirror-image representations of ClF_2CH are identical and are *not* enantiomers.

The surest test for chirality is a careful examination of mirror-image forms for superimposability. Working with models provides the best practice in dealing with molecules as three-dimensional objects and is strongly recommended.

7.2 THE STEREOGENIC CENTER

As we've just seen, molecules of the general type

$$w \longrightarrow \overset{\textstyle x}{\underset{\textstyle z}{\mathrm{C}}} \longrightarrow y$$

are chiral when *w, x, y,* and *z* are different substituents. A tetrahedral carbon atom that bears four different substituents is called a **stereogenic center.**

Noting the presence of a single stereogenic center in a molecule is a simple, rapid way to determine that it is chiral. For example, C-2 is a stereogenic center in 2-butanol;

Older terms for stereogenic center include **chiral carbon atom** and **asymmetric carbon atom.** A newer, but not yet widely used, term is **chirality center.**

FIGURE 7.1 A molecule with four different groups attached to a single carbon is chiral. Its two mirror-image forms are not superimposable.

(*a*) Structures A and B are mirror-image representations of bromochlorofluoromethane (BrClFCH).

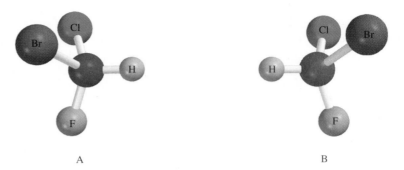

A   B

(*b*) To test for superimposability, reorient B by turning it 180°.

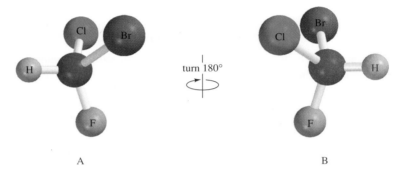

turn 180°

A   B

(*c*) Compare A and B. The two do not match. A and B cannot be superimposed on each other. Bromochlorofluoromethane is therefore a chiral molecule. The two mirror-image forms are enantiomers of each other.

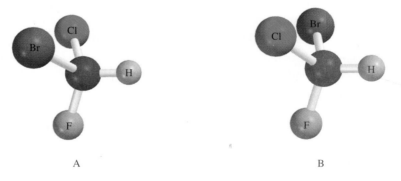

A   B

it bears a hydrogen atom and methyl, ethyl, and hydroxyl groups as its four different substituents. By way of contrast, none of the carbon atoms bears four different groups in the achiral alcohol 2-propanol.

$$CH_3-\underset{\underset{OH}{|}}{\overset{\overset{H}{|}}{C}}-CH_2CH_3 \qquad\qquad CH_3-\underset{\underset{OH}{|}}{\overset{\overset{H}{|}}{C}}-CH_3$$

2-Butanol   2-Propanol
chiral; four different substituents at C-2   achiral; two of the substituents at C-2 are the same

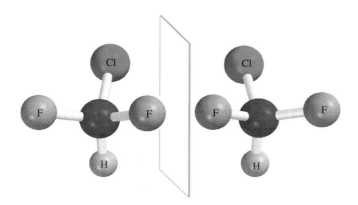

FIGURE 7.2 Mirror-image forms of chlorodifluoromethane are superimposable on each other. Chlorodifluoromethane is achiral.

PROBLEM 7.1   Examine the following for stereogenic centers:

(a) 2-Bromopentane   (c) 1-Bromo-2-methylbutane

(b) 3-Bromopentane   (d) 2-Bromo-2-methylbutane

Sample Solution   A stereogenic carbon has four different substituents. (a) In 2-bromopentane, C-2 satisfies this requirement. (b) None of the carbons in 3-bromopentane has four different substituents, and so none of its atoms is a stereogenic center.

$$CH_3-\underset{\underset{Br}{|}}{\overset{\overset{H}{|}}{C}}-CH_2CH_2CH_3 \qquad CH_3CH_2-\underset{\underset{Br}{|}}{\overset{\overset{H}{|}}{C}}-CH_2CH_3$$

2-Bromopentane   3-Bromopentane

Molecules containing a single stereogenic center are very common and include many naturally occurring substances. Linalool and limonene, shown in the following formulas, are but two examples; several more are shown in Table 7.1. Carbons that are part of a double bond or a triple bond can't be stereogenic centers; however, a carbon atom in a ring can be if it bears two different substituents and the path traced around the ring from that carbon in one direction is different from that traced in the other. C-3 is the stereogenic center in linalool; it is C-4 in limonene.

$$(CH_3)_2C=CHCH_2CH_2-\underset{\underset{OH}{|}}{\overset{\overset{CH_3}{|}}{\underset{3}{C}}}-\overset{2}{\underset{}{C}}H=\overset{1}{\underset{}{C}}H_2$$

Linalool
(a pleasant-smelling oil obtained from orange flowers)

Limonene
(a constituent of lemon oil)

7.3   SYMMETRY IN ACHIRAL STRUCTURES

Recognizing the presence of a plane of symmetry can enable us to determine, by inspection, if a molecule is chiral or achiral.

TABLE 7.1	Some Naturally Occuring Chiral Molecules Having One Stereogenic Center	
Compound Name	**Structural Formula***	**Where Found**
Lactic acid	$\underset{\underset{\displaystyle OH}{\overset{\overset{\displaystyle O}{\parallel}}{CH_3\overset{*}{C}HCOH}}}{}$	In muscles; can also be obtained from milk
Malic acid	$\underset{OH}{HO_2C\overset{*}{C}HCH_2CO_2H}$	Apples and other fruits
Glyceraldehyde	$\underset{\underset{\displaystyle OH}{\overset{\overset{\displaystyle O}{\parallel}}{HOCH_2\overset{*}{C}HCH}}}{}$	Carbohydrate formed during energy-producing breakdown of sugars in the body
α-Phellandrene		Component of eucalyptus oil; used in perfumes

* An asterisk (*) indicates location of the stereogenic center.

A **plane of symmetry** bisects a molecule so that one half of the molecule is the mirror image of the other half. The achiral molecule chlorodifluoromethane (Figure 7.2), for example, has the plane of symmetry shown in Figure 7.3. Any molecule having a plane of symmetry is achiral. A chiral molecule cannot have a plane of symmetry.

Why worry about planes of symmetry when stereogenic centers are easy to find in a molecule? As you will see in Sections 7.10 and 7.11, a molecule with two or more stereogenic centers may be chiral *or* achiral, and locating a plane of symmetry is often a convenient way to distinguish between these two possibilities.

7.4   PROPERTIES OF CHIRAL MOLECULES: OPTICAL ACTIVITY

The foundations of organic stereochemistry were independently established by Jacobus Van't Hoff and Joseph Le Bel in 1874. The experimental facts that led van't Hoff and Le Bel to propose that molecules having the same constitution could differ in the arrangement of their atoms in space concerned the physical property of optical activity. **Optical activity** is the ability of a chiral substance to rotate the plane of **plane-polarized light** and is measured using an instrument called a **polarimeter.** (Figure 7.4).

The phenomenon of optical activity was discovered by the French physicist Jean-Baptiste Biot in 1815.

FIGURE 7.3 A plane of symmetry defined by the atoms H—C—Cl divides chlorodifluoromethane into two mirror-image halves. The carbon, chlorine, and hydrogen atoms lie within the plane.

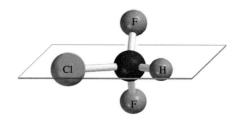

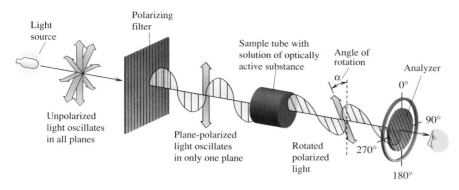

FIGURE 7.4   A schematic diagram of a polarimeter.

(Adapted, with permission, from M. Silberberg, *Chemistry*, 2nd ed., McGraw-Hill, New York, 2000, p. 616.)

The light used to measure optical activity has two properties: it consists of a single wavelength and it is plane-polarized. The wavelength most often used corresponds to the yellow light produced by a sodium vapor lamp (called the **D line**). As shown in Figure 7.4, the beam of unpolarized light from the lamp is transformed to plane-polarized light by passing through a polarizing filter. This plane-polarized light now passes through the sample, which is either a pure liquid or a solid dissolved in a suitable solvent such as water, ethanol, or chloroform. The sample is "optically active" if it rotates the plane of polarized light. The direction and magnitude of rotation are measured using a second polarizing filter (the "analyzer") and cited as α, the observed rotation.

> *To be optically active, the sample must contain a chiral substance and one enantiomer must be present in excess of the other.*

A substance that does not rotate the plane of polarized light is said to be optically inactive.

> *All achiral substances are optically inactive.*

Optical activity is a physical property, just as melting point, boiling point, and density are. Rotation of the plane of polarized light to the right (clockwise) is taken as positive (+), and rotation to the left (counterclockwise) is taken as negative (−). The magnitude of the observed rotation α depends on how many molecules the light beam encounters. To account for the effects of path length and concentration, chemists have defined the term **specific rotation**, given the symbol $[\alpha]$. Specific rotation is calculated from the observed rotation according to the expression

> An older designation of positive and negative optical rotation was **dextrorotatory(d)** and **levorotatory(l)**, respectively, from the Latin prefixes *dextro-* ("to the right") and *levo-* ("to the left").

$$[\alpha] = \frac{100\alpha}{cl}$$

where c is the concentration of the sample in grams per 100 mL of solution, and l is the length of the polarimeter tube in decimeters. (One decimeter is 10 cm.) The temperature in degrees Celsius and the wavelength of light (usually the sodium D line) are indicated as a superscript and subscript, respectively. Thus specific rotation is cited in the form $[\alpha]_D^{25} = +15°$.

PROBLEM 7.2   Cholesterol, when isolated from natural sources, is obtained as a single enantiomer. The observed rotation α of a 0.3-g sample of cholesterol in 15 mL of chloroform solution contained in a 10-cm polarimeter tube is −0.78°. Calculate the specific rotation of cholesterol.

Our discussion has focused on the optical rotation of one enantiomer of a chiral substance. What if we determine the optical rotation of the *other* enantiomer, that is, the mirror image of the substance? The angle of rotation will be the *same,* but it will be in the *opposite* direction. For example, if one enantiomer of a chiral substance has a specific rotation $[\alpha]$ of $+37°$, the other enantiomer will have $[\alpha] = -37°$.

A mixture of *equal amounts* of the two enantiomers of a chiral substance has *no* optical rotation ($\alpha = 0$), because the positive and negative rotations of the individual enantiomers cancel. A mixture containing equal quantities of the enantiomers of a chiral substance is called **racemic mixture** and is *optically inactive*.

7.5   ABSOLUTE AND RELATIVE CONFIGURATION

The spatial arrangement of substituents at a stereogenic center is its **absolute configuration**. Neither the sign nor the magnitude of optical rotation by itself can tell us the absolute configuration of a substance. Thus, one of the following structures is $(+)$-2-butanol and the other is $(-)$-2-butanol, but without additional information we can't tell which is which.

> In several places throughout the chapter we will use red and blue frames to call attention to structures that are enantiomeric.

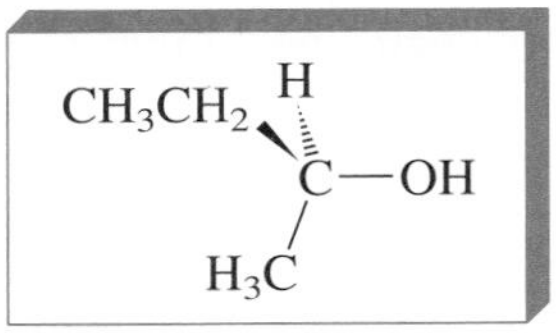

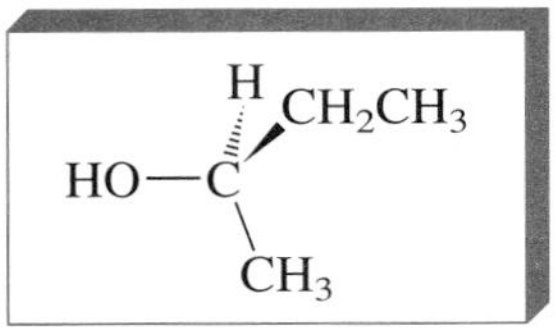

$$CH_3CH_2 \overset{H}{\underset{H_3C}{\overset{|}{C}}} {-}OH \qquad HO{-}\overset{H}{\underset{CH_3}{\overset{|}{C}}} CH_2CH_3$$

Although no absolute configuration was known for any substance before 1951, organic chemists had experimentally determined the configurations of thousands of compounds relative to one another (their **relative configurations**) through chemical interconversion. When, in 1951, the absolute configuration of a salt of $(+)$-tartaric acid was determined, the absolute configurations of all the compounds whose configurations had been related to $(+)$-tartaric acid stood revealed as well. Thus, returning to the pair of 2-butanol enantiomers that introduced this section, their absolute configurations are now known to be as shown.

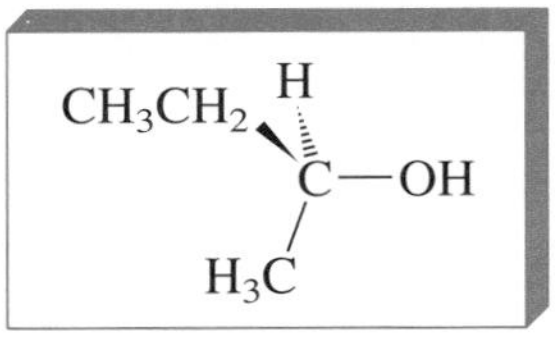

$$CH_3CH_2 \overset{H}{\underset{H_3C}{\overset{|}{C}}} {-}OH \qquad HO{-}\overset{H}{\underset{CH_3}{\overset{|}{C}}} CH_2CH_3$$

(+)-2-Butanol 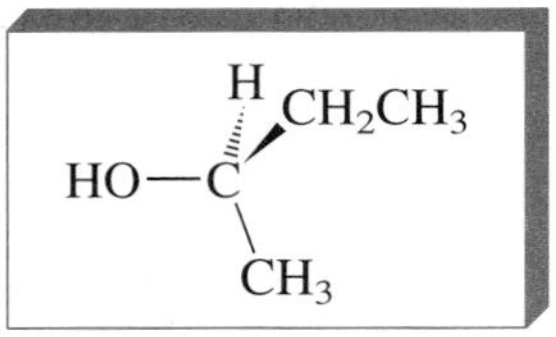   (−)-2-Butanol

PROBLEM 7.3 Does the molecular model shown represent $(+)$-2-butanol or $(-)$-2-butanol?

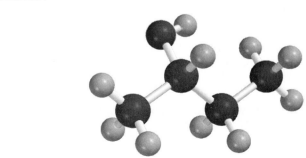

7.6   THE CAHN–INGOLD–PRELOG *R–S* NOTATIONAL SYSTEM

Just as it makes sense to have a nomenclature system by which we can specify the constitution of a molecule in words rather than pictures, so too is it helpful to have one that lets us describe stereochemistry. We have already had some experience with this idea when we distinguished between *E* and *Z* stereoisomers of alkenes.

In the *E–Z* system, substituents are ranked by atomic number according to a set of rules devised by R. S. Cahn, Sir Christopher Ingold, and Vladimir Prelog (Section 4.4). Actually, Cahn, Ingold, and Prelog first developed their ranking system to deal with the problem of the absolute configuration at a stereogenic center, and this is the system's major application. Table 7.2 shows how the Cahn–Ingold–Prelog system, called the **sequence rules,** is used to specify the absolute configuration at the stereogenic center in (+)-2-butanol.

The enantiomers of 2-butanol were shown in the previous section. As outlined in Table 7.1, (+)-2-butanol has the *S* configuration. Its enantiomer, (−)-2-butanol, has the *R* configuration. A general statement can be made:

TABLE 7.2	Absolute Configuration According to the Cahn–Ingold–Prelog Notational System

Step Number	Example
	Given that the absolute configuration of (+)-2-butanol is CH_3CH_2 \ / H C—OH H_3C (+)-2-Butanol
1. Identify the substituents at the stereogenic center, and rank them in order of decreasing precedence according to the system described in Section 4.4. Precedence is determined by atomic number, working outward from the point of attachment at the stereogenic center.	In order of decreasing precedence, the four substituents attached to the stereogenic center of 2-butanol are HO— > CH_3CH_2— > CH_3— > H— Highest Lowest
2. Orient the molecule so that the lowest ranked substituent points away from you.	As represented in the wedge-and-dash drawing at the top of this table, the molecule is already appropriately oriented. Hydrogen is the lowest ranked substituent attached to the stereogenic center and points away from us.
3. Draw the three highest ranked substituents as they appear to you when the molecule is oriented so that the lowest ranked group points away from you.	CH_3CH_2 \ / OH \ / CH_3
4. If the order of decreasing precedence of the three highest ranked substituents appears in a clockwise sense, the absolute configuration is *R* (Latin *rectus,* "right," "correct"). If the order of decreasing precedence is counterclockwise, the absolute configuration is *S* (Latin *sinister,* "left").	The order of decreasing precedence is *counterclockwise*. The configuration at the stereogenic center is *S*. Second highest CH_3CH_2 \ / OH Highest \ / CH_3 Third highest

One mirror image of a stereogenic center is R; *the other is* S. A stereogenic center can *only* be *R* or *S*.

PROBLEM 7.4   Assign absolute configurations as *R* or *S* to each of the following compounds:

(a)

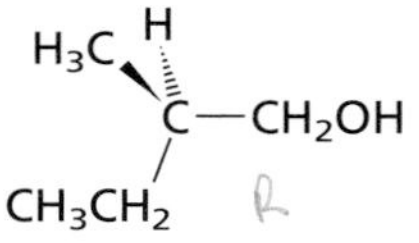

$$H_3C \quad H$$
$$C—CH_2OH$$
$$CH_3CH_2$$

(+)-2-Methyl-1-butanol

(c)

$$H \quad CH_3$$
$$C—CH_2Br$$
$$CH_3CH_2$$

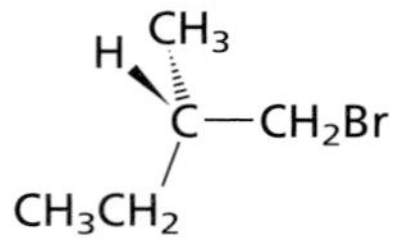

(+)-1-Bromo-2-methylbutane

(b)

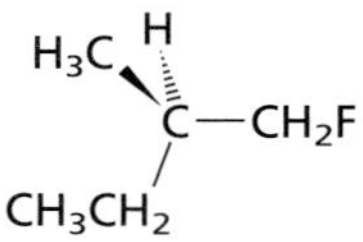

$$H_3C \quad H$$
$$C—CH_2F$$
$$CH_3CH_2$$

(+)-1-Fluoro-2-methylbutane

(d)

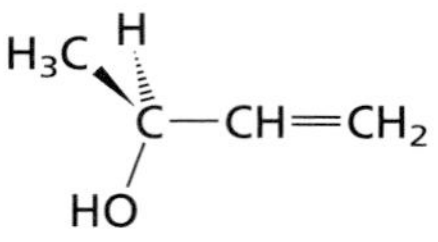

$$H_3C \quad H$$
$$C—CH=CH_2$$
$$HO$$

(+)-3-Buten-2-ol

Sample Solution   (a) The highest ranking substituent at the stereogenic center of 2-methyl-1-butanol is CH_2OH; the lowest is H. Of the remaining two, ethyl outranks methyl.

Order of precedence: $CH_2OH > CH_3CH_2 > CH_3 > H$

The lowest ranking substituent (hydrogen) points away from us in the drawing. The three highest ranking groups trace a clockwise path from $CH_2OH \longrightarrow CH_3CH_2 \longrightarrow CH_3$.

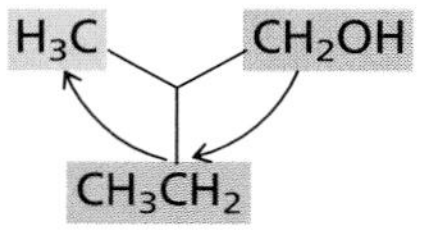

$$H_3C \qquad CH_2OH$$
$$CH_3CH_2$$

This compound therefore has the *R* configuration. It is (*R*)-(+)-2-methyl-1-butanol.

7.7   FISCHER PROJECTIONS

Stereochemistry deals with the three-dimensional arrangement of a molecule's atoms, and we have attempted to show stereochemistry with wedge-and-dash drawings and computer-generated models. It is possible, however, to convey stereochemical information in an abbreviated form using a method devised by the German chemist Emil Fischer.

Let's return to bromochlorofluoromethane as a simple example of a chiral molecule. The two enantiomers of BrClFCH are shown as ball-and-stick models, as wedge-and-dash drawings, and as **Fischer projections** in Figure 7.5. Fischer projections are always generated the same way: the molecule is oriented so that the vertical bonds at the stereogenic center are directed away from you and the horizontal bonds point toward you. A projection of the bonds onto the page is a cross. The stereogenic carbon lies at the center of the cross but is not explicitly shown.

It is customary to orient the molecule so that the carbon chain is vertical with the lowest numbered carbon at the top as shown for the Fischer projection of (*R*)-2-butanol.

Fischer was the foremost organic chemist of the late nineteenth century. He won the 1902 Nobel Prize in chemistry for his pioneering work in carbohydrate and protein chemistry.

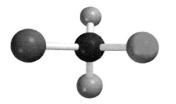
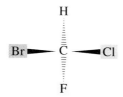
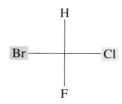

FIGURE 7.5   Ball-and-stick models *(left)*, wedge-and-dash drawings *(center)*, and Fischer projections *(right)* of the *R* and *S* enantiomers of bromochlorofluoromethane.

(*R*)-Bromochlorofluoromethane

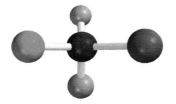
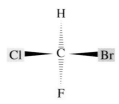
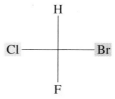

(*S*)-Bromochlorofluoromethane

The Fischer projection   HO——$|$——H   corresponds to   HO—C—H

(*R*)-2-Butanol

When specifying a configuration as *R* or *S*, the safest procedure is to convert a Fischer projection to a three-dimensional representation, remembering that the horizontal bonds always point toward you.

PROBLEM 7.5   Write Fischer projections for each of the compounds of Problem 7.4.

Sample Solution   (a) The structure of (*R*)-(+)-2-methyl-1-butanol is shown in the structure that follows at the left. View the structural formula from a position chosen so that the $HOCH_2$—C—CH_2CH_3 segment is aligned vertically, with the vertical bonds pointing away from you. Replace the wedge-and-dash bonds by lines to give the Fischer projection shown at the right.

CH_3 H
C—CH_2OH is the CH_2OH which becomes the CH_2OH
CH_3CH_2 same as   H—C—CH_3 Fischer projection   H——CH_3
                           CH_2CH_3 CH_2CH_3

7.8   PHYSICAL PROPERTIES OF ENANTIOMERS

The usual physical properties such as density, melting point, and boiling point are identical within experimental error for both enantiomers of a chiral compound.

Enantiomers can have striking differences, however, in properties that depend on the arrangement of atoms in space. Take, for example, the enantiomeric forms of carvone. (*R*)-(−)-Carvone is the principal component of spearmint oil. Its enantiomer, (*S*)-(+)-carvone, is the principal component of caraway seed oil. The odors of the two enantiomers are distinctly different.

CHIRAL DRUGS

A recent estimate places the number of prescription and over-the-counter drugs marketed throughout the world at about 2000. Approximately one third of these are either naturally occurring substances themselves or are prepared by chemical modification of natural products. Most of the drugs derived from natural sources are chiral and are almost always obtained as a single enantiomer rather than as a racemic mixture. Not so with the over 500 chiral substances represented among the more than 1300 drugs that are the products of synthetic organic chemistry. Until recently, such substances were, with few exceptions, prepared, sold, and administered as racemic mixtures even though the desired therapeutic activity resided in only one of the enantiomers. Spurred by a number of factors ranging from safety and efficacy to synthetic methodology and economics, this practice is undergoing rapid change as more and more chiral synthetic drugs become available in enantiomerically pure form.

Because of the high degree of chiral recognition inherent in most biological processes (Section 7.8), it is unlikely that both enantiomers of a chiral drug will exhibit the same level, or even the same kind, of effect. At one extreme, one enantiomer has the desired effect, and the other exhibits no biological activity at all. In this case, which is relatively rare, the racemic form is simply a drug that is 50% pure and contains 50% "inert ingredients." Real cases are more complicated. For example, the *S* enantiomer is responsible for the pain-relieving properties of ibuprofen, normally sold as a racemic mixture. The 50% of racemic ibuprofen that is the *R* enantiomer is not completely wasted, however, because enzyme-catalyzed reactions in our body convert much of it to active (*S*)-ibuprofen.

A much more serious drawback to using chiral drugs as racemic mixtures is illustrated by thalidomide, briefly employed as a sedative and antinausea drug in Europe and Great Britain during the period 1959–1962. The desired properties are those of (*R*)-thalidomide. (*S*)-Thalidomide, however, has a very different spectrum of biological activity and was shown to be responsible for over 2000 cases of serious birth defects in children born to women who took it while pregnant.

Thalidomide

Basic research directed toward understanding the factors that control the stereochemistry of chemical reactions has led to new synthetic methods that make it practical to prepare chiral molecules in enantiomerically pure form. Recognizing this, most major pharmaceutical companies are examining their existing drugs to see which ones are the best candidates for synthesis as single enantiomers and, when preparing a new drug, design its synthesis so as to provide only the desired enantiomer. In 1992, the United States Food and Drug Administration (FDA) issued guidelines that encouraged such an approach, but left open the door for approval of new drugs as racemic mixtures when special circumstances warrant. One incentive to developing enantiomerically pure versions of existing drugs is that the novel production methods they require may make them eligible for patent protection separate from that of the original drugs. Thus the temporary monopoly position that patent law views as essential to fostering innovation can be extended by transforming a successful chiral, but racemic, drug into an enantiomerically pure version.

Ibuprofen

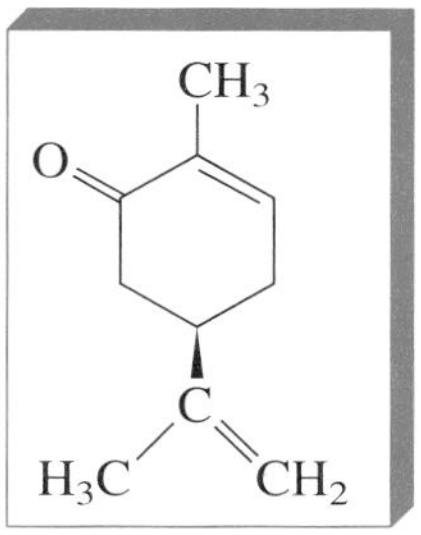

(R)-(−)-Carvone
(from spearmint oil)

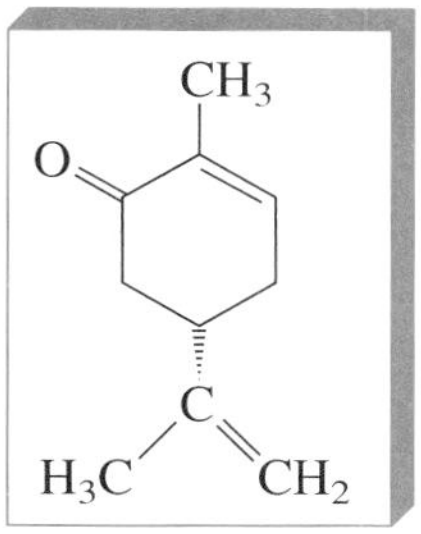

(S)-(+)-Carvone
(from caraway seed oil)

The difference in odor between (R)- and (S)-carvone results from their different behavior toward receptor sites in the nose. It is believed that volatile molecules occupy only those odor receptors that have the proper shape to accommodate them. Because the receptor sites are themselves chiral, one enantiomer may fit one kind of receptor, whereas the other enantiomer fits a different kind. An analogy that can be drawn is to hands and gloves. Your left hand and your right hand are enantiomers. You can place your left hand into a left glove but not into a right one. The receptor (the glove) can accommodate one enantiomer of a chiral object (your hand) but not the other.

The term **chiral recognition** refers to the process whereby some chiral receptor or reagent interacts selectively with one of the enantiomers of a chiral molecule. Very high levels of chiral recognition are common in biological processes. (−)-Nicotine, for example, is much more toxic than (+)-nicotine, and (+)-adrenaline is more active in the constriction of blood vessels than (−)-adrenaline. (−)-Thyroxine is an amino acid of the thyroid gland, which speeds up metabolism and causes nervousness and loss of weight. It is one of the most widely used of all prescription drugs—about 10 million people in the United States take (−)-thyroxine on a daily basis. Its enantiomer (+)-thyroxine has none of the metabolism-regulating effects, but is sometimes given to heart patients to lower their cholesterol levels.

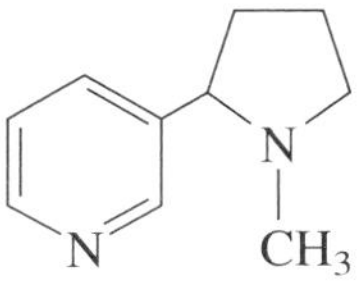

Nicotine

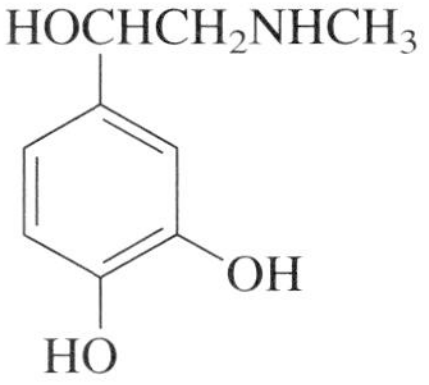

Adrenaline

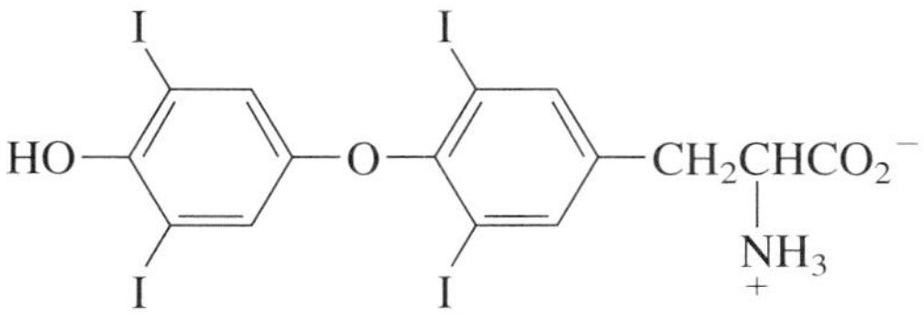

Thyroxine

(Can you find the stereogenic center in each of these?)

7.9   REACTIONS THAT CREATE A STEREOGENIC CENTER

Many of the reactions we have already encountered in earlier chapters can produce a chiral product from an achiral starting material. A large number of the reactions of alkenes fall into this category. For example, the reaction of 2-butene with hydrogen bromide converts the achiral alkene into a product that contains a stereogenic center.

(E)- or (Z)-2-butene
(achiral)

2-Bromobutane
(chiral)

Carbocation
(achiral)

Figure 7.6 shows why equal amounts of (R)- and (S)-2-bromobutane are formed in this reaction, giving an optically inactive racemic mixture. Electrophilic addition to an alkene (Section 5.2) proceeds by way of carbocation. The bonds to the positively charged carbon are coplanar and define a plane of symmetry in the carbocation, which is achiral. The rates at which bromide ion attacks the carbocation at its two mirror-image faces are equal, and the product, 2-bromobutane, although chiral, is optically inactive because it is formed as a racemic mixture.

It is a general principle that

> *Optically active products cannot be formed when optically inactive substrates react with optically inactive reagents.*

This principle holds regardless of whether the starting materials are achiral or a racemic mixture of a chiral substance.

Optically inactive starting materials can give optically active products *only* if they are treated with an optically active reagent or if the reaction is catalyzed by an optically active substance. The best examples are found in living systems. Most biochemical reactions are catalyzed by enzymes. **Enzymes** are chiral and exist as a single enantiomer; they provide an asymmetric environment in which chemical reactions can take place. Thus, in most enzyme-catalyzed reactions only one enantiomer of the product is formed. The enzyme fumarase, for example, catalyzes the hydration of fumaric acid, which is achiral, to malic acid in apples and other fruits. Only the S enantiomer of malic acid is formed in this reaction.

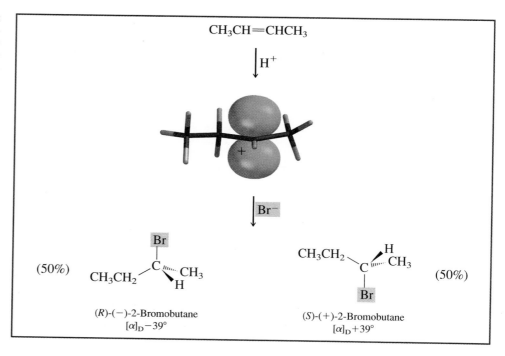

FIGURE 7.6 Electrophilic addition of hydrogen bromide to (E) and (Z)-2-butene proceeds by way of an achiral carbocation, which leads to equal quantities of (R)- and (S)-2-bromobutane.

PROBLEM 7.6 Biological reduction of pyruvic acid, catalyzed by the enzyme lactate dehydrogenase, gives (+)-lactic acid, represented by the Fischer projection shown. What is the configuration of (+)-lactic acid according to the Cahn–Ingold–Prelog *R–S* notational system? Making a molecular model of the Fischer projection will help.

$$CH_3\overset{\displaystyle O}{\overset{\displaystyle \|}{C}}CO_2H \xrightarrow{\text{biological reduction}} \quad HO\!-\!\!\!\overset{\displaystyle CO_2H}{\underset{\displaystyle CH_3}{|}}\!\!\!-H$$

Pyruvic acid   (+)-Lactic acid

The presence of optical activity was viewed by Louis Pasteur as perhaps *the* unique criterion of life. He remarked in 1860 that the difference between the chemistry of living and dead matter is "the molecular asymmetry of organic natural products."

7.10   CHIRAL MOLECULES WITH TWO STEREOGENIC CENTERS

When a molecule contains two stereogenic centers, as does 2,3-dihydroxybutanoic acid, how many stereoisomers are possible?

$$\underset{4}{CH_3}\underset{3}{CH}\underset{2}{CH}\underset{1}{C}\overset{\displaystyle O}{\underset{\displaystyle OH}{\diagup}}$$
$$\underset{\displaystyle HO}{\,}\underset{\displaystyle OH}{\,}$$

2,3-Dihydroxybutanoic acid

We can use straightforward reasoning to come up with the answer. The absolute configuration at C-2 may be *R* or *S*. Likewise, C-3 may have either the *R* or the *S* configuration. The four possible combinations of these two stereogenic centers are

(2*R*,3*R*)	(stereoisomer I)	(2*S*,3*S*)	(stereoisomer II)
(2*R*,3*S*)	(stereoisomer III)	(2*S*,3*R*)	(stereoisomer IV)

Stereoisomers I and II are enantiomers of each other; the enantiomer of (*R*,*R*) is (*S*,*S*). Likewise stereoisomers III and IV are enantiomers of each other, the enantiomer of (*R*,*S*) being (*S*,*R*).

Stereoisomer I is not a mirror image of III or IV, so is not an enantiomer of either one. Stereoisomers that are not related as an object and its mirror image are called **diastereomers;**

Diastereomers are stereoisomers that are not enantiomers.

Thus, stereoisomer I is a diastereomer of III and a diastereomer of IV. Similarly, II is a diastereomer of III and IV.

To convert a molecule with two stereogenic centers to its enantiomer, the configuration at both centers must be changed. Reversing the configuration at only one stereogenic center converts it to a diastereomeric structure.

We can represent the stereoisomers of a molecule with more than one stereogenic center using Fischer projections. The molecule is arranged in an *eclipsed* conformation for projection onto the page, as shown in Figure 7.7. As in all Fischer projections, horizontal lines represent bonds coming toward you; vertical bonds point away.

FIGURE 7.7 Representations of (2R,3R)-dihydroxybutanoic acid. (a) The staggered conformation is the most stable but is not properly arranged to show stereochemistry according to the Fischer projection method. (b) Rotation about the C-2—C-3 bond gives the eclipsed conformation, and projection of the eclipsed conformation onto the page gives (c) a correct Fischer projection.

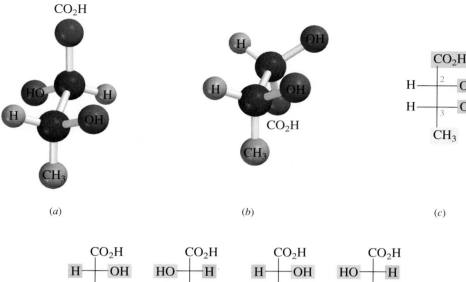

(a)   (b)   (c)

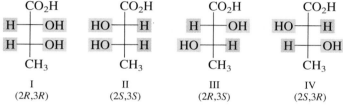

I   II   III   IV
(2R,3R)   (2S,3S)   (2R,3S)   (2S,3R)

Because diastereomers are not mirror images of each other, they can have quite different physical and chemical properties. For example, the (2R,3R) stereoisomer of 3-amino-2-butanol is a liquid, but the (2R,3S) diastereomer is a crystalline solid.

3-Amino-2-butanol

> **PROBLEM 7.7** Draw Fischer projections or make molecular models of the four stereoisomeric 3-amino-2-butanols, and label each stereogenic center as R or S.

> **PROBLEM 7.8** One other stereoisomer of 3-amino-2-butanol is a crystalline solid. Which one?

7.11 ACHIRAL MOLECULES WITH TWO STEREOGENIC CENTERS

Now think about a molecule, such as 2,3-butanediol, which has two stereogenic centers that are equivalently substituted.

$$CH_3CHCHCH_3$$
$$\quad\ |\ \ |$$
$$\quad HO\ \ OH$$

2,3-Butanediol

Only *three*, not four, stereoisomeric 2,3-butanediols are possible. These three are shown in Figure 7.8. The (2R,3R) and (2S,3S) forms are enantiomers of each other and have equal and opposite optical rotations. A third combination of stereogenic centers, (2R,3S),

Alkene stereoisomers are another example of diastereomers. (E)- and (Z)-2-butene are stereoisomers, but they are not mirror images; they are diastereomers.

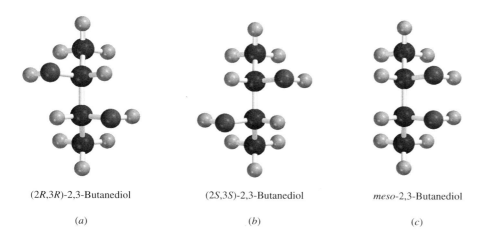

(2R,3R)-2,3-Butanediol

(a)

(2S,3S)-2,3-Butanediol

(b)

meso-2,3-Butanediol

(c)

FIGURE 7.8 Stereoisomeric 2,3-butanediols shown in their eclipsed conformations for convenience. Stereoisomers (a) and (b) are enantiomers of each other. Structure (c) is a diastereomer of (a) and (b), and is achiral. It is called *meso*-2,3-butanediol.

however, gives an *achiral* structure that is superimposable on its (2S,3R) mirror image. Because it is achiral, this third stereoisomer is *optically inactive*. We call achiral molecules that have stereogenic centers **meso forms.** The meso form in Figure 7.8 is known as *meso*-2,3-butanediol. The meso form has a **plane of symmetry** and is achiral.

Fischer projection formulas can help us identify meso forms. Of the three stereoisomeric 2,3-butanediols, notice that only in the meso stereoisomer does a dashed line through the center of the Fischer projection divide the molecule into two mirror-image halves.

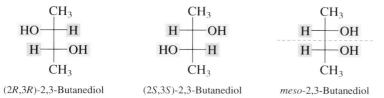

(2R,3R)-2,3-Butanediol

(2S,3S)-2,3-Butanediol

meso-2,3-Butanediol

When using Fischer projections for this purpose, however, be sure to remember what three-dimensional objects they stand for. Do not, for example, test for superposition of the two chiral stereoisomers by a procedure that involves moving any part of a Fischer projection out of the plane of the paper in any step.

7.12 MOLECULES WITH MULTIPLE STEREOGENIC CENTERS

Many naturally occurring compounds contain several stereogenic centers. By an analysis similar to that described for the case of two stereogenic centers, it can be shown that the maximum number of stereoisomers for a particular constitution is 2^n, where n is equal to the number of stereogenic centers.

> **PROBLEM 7.9** Using *R* and *S* descriptors, write all the possible combinations for a molecule with three stereogenic centers.

When two or more of a molecule's stereogenic centers are equivalently substituted, meso forms are possible, and the number of stereoisomers is then less than 2^n. Thus, 2^n represents the *maximum* number of stereoisomers for a molecule containing n stereogenic centers.

The best examples of substances with multiple stereogenic centers are the carbohydrates (Chapter 15). One class of carbohydrates, called hexoses, has the constitution

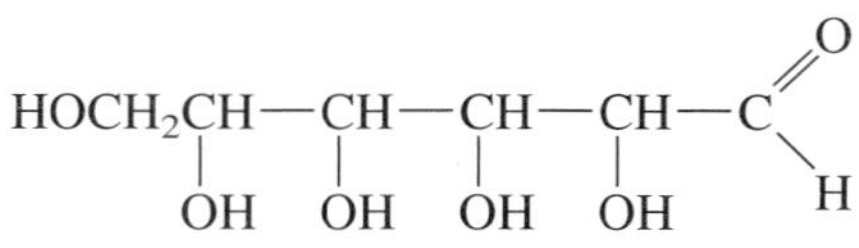

$$HOCH_2CH-CH-CH-CH-C \overset{O}{\underset{H}{\diagup}}$$

$$\underset{OH \quad OH \quad OH \quad OH}{|\quad\quad|\quad\quad|\quad\quad|}$$

A hexose

Because there are four stereogenic centers and no possibility of meso forms, the number of stereoisomeric hexoses is 2^4, or 16. All 16 are known, having been isolated either as natural products or as the products of chemical synthesis.

7.13   RESOLUTION OF ENANTIOMERS

The separation of a racemic mixture into its enantiomeric components is termed **resolution.** The first resolution, that of tartaric acid, was carried out by Louis Pasteur in 1848. Tartaric acid is a byproduct of wine making and is almost always found as its dextrorotatory *2R,3R* stereoisomer, shown here in a perspective drawing and in a Fischer projection.

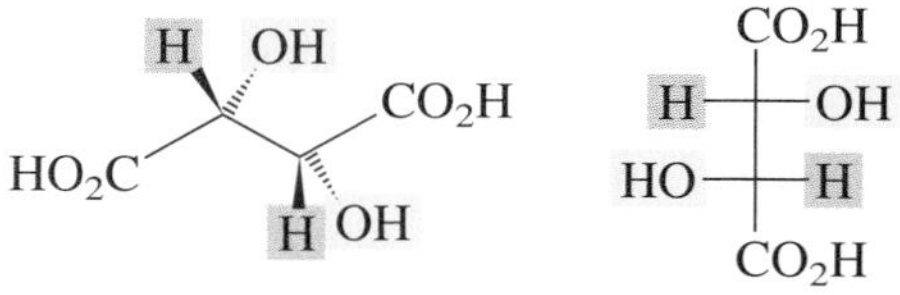

(2R,3R)-Tartaric acid
(mp 170°C, $[\alpha]_D$ +12°)

> **PROBLEM 7.10** There are two other stereoisomeric tartaric acids. Write their Fischer projections, and specify the configuration at their stereogenic centers.

Occasionally, an optically inactive sample of tartaric acid was obtained. Pasteur noticed that the sodium ammonium salt of optically inactive tartaric acid was a mixture of two mirror-image crystal forms. With microscope and tweezers, Pasteur carefully separated the two. He found that one kind of crystal (in aqueous solution) was dextrorotatory, whereas the mirror-image crystals rotated the plane of polarized light an equal amount but were levorotatory.

Although Pasteur was unable to provide a structural explanation—that had to wait for van't Hoff and Le Bel a quarter of a century later—he correctly deduced that the enantiomeric quality of the crystals was the result of enantiomeric molecules. The rare form of tartaric acid was optically inactive because it contained equal amounts of (+)-tartaric acid and (−)-tartaric acid. It had earlier been called racemic acid (from Latin *racemus,* "a bunch of grapes"), a name that subsequently gave rise to our present term *racemic mixture* for an equal mixture of enantiomers.

> **PROBLEM 7.11** Could the unusual, optically inactive form of tartaric acid studied by Pasteur have been *meso*-tartaric acid?

Pasteur's technique of separating enantiomers not only is laborious but requires that the crystal forms of enantiomers be distinguishable. This happens very rarely. Consequently, alternative and more general approaches for resolving enantiomers have been developed. Most are based on a strategy of temporarily converting the enantiomers of a racemic mixture to diastereomeric derivatives, separating these diastereomers, then regenerating the enantiomeric starting materials.

Whenever possible, the chemical reactions involved in the formation of diastereomers and their conversion to separate enantiomers are simple acid–base reactions. For example, naturally occurring (S)-(−)-malic acid is often used to resolve amines. One such amine that has been resolved in this way is 1-phenylethylamine. Amines are bases, and malic acid is an acid. Proton transfer from (S)-(−)-malic acid to a racemic mixture of (R)- and (S)-1-phenylethylamine gives a mixture of diastereomeric salts.

$$C_6H_5\underset{\underset{CH_3}{|}}{C}HNH_2 \quad + \quad HO_2CCH_2\underset{\underset{OH}{|}}{C}HCO_2H \quad \longrightarrow \quad C_6H_5\underset{\underset{CH_3}{|}}{C}H\overset{+}{N}H_3 \quad {}^-O_2CCH_2\underset{\underset{OH}{|}}{C}HCO_2H$$

| 1-Phenylethylamine | (S)-(−)-Malic acid | 1-Phenylethylammonium (S)-malate |
| (racemic mixture) | (resolving agent) | (mixture of diastereomeric salts) |

The diastereomeric salts are separated and the individual enantiomers of the amine liberated by treatment with a base:

$$C_6H_5\underset{\underset{CH_3}{|}}{C}H\overset{+}{N}H_3 \quad {}^-O_2CCH_2\underset{\underset{OH}{|}}{C}HCO_2H \quad + \quad 2OH^- \quad \longrightarrow$$

1-Phenylethylammonium (S)-malate   Hydroxide
(a single diastereomer)

$$C_6H_5\underset{\underset{CH_3}{|}}{C}HNH_2 \quad + \quad {}^-O_2CCH_2\underset{\underset{OH}{|}}{C}HCO_2^- \quad + \quad 2H_2O$$

1-Phenylethylamine   (S)-(−)-Malic acid   Water
(a single enantiomer)   (recovered resolving agent)

This method is widely used for the resolution of chiral amines and carboxylic acids. Analogous methods based on the formation and separation of diastereomers have been developed for other functional groups; the precise approach depends on the kind of chemical reactivity associated with the functional groups present in the molecule.

The rapidly increasing demand for enantiomerically pure starting materials and intermediates in the pharmaceutical industry (see the boxed essay entitled *Chiral Drugs* in this chapter) has increased interest in developing methods for resolving racemic mixtures.

LEARNING OBJECTIVES

This chapter focused on the spatial relationships of atoms and molecules. The skills you have learned in this chapter should enable you to:

- Identify a chiral molecule by locating a stereogenic center.
- Explain what is meant by the terms *enantiomer* and *diastereomer.*
- Draw perspective (wedge-and-dash) representations of chiral molecules.
- Describe how a plane of symmetry relates to whether or not a molecule is chiral.
- Explain optical activity as a property of chiral molecules.
- Specify the absolute configuration of a molecule using the *R–S* notational system.
- Draw Fischer projections of chiral molecules.

Continued

- Explain why addition reactions to achiral alkenes give racemic mixtures of products.
- Explain the meaning of the term *meso form.*
- Give the number of stereoisomers possible for a molecule having more than one stereogenic center.
- Describe how a racemic mixture may be resolved into individual enantiomers.

7.14 SUMMARY

Chemistry in three dimensions is known as **stereochemistry.** At its most fundamental level, stereochemistry deals with molecular structure; at another level, it is concerned with chemical reactivity. Table 7.3 summarizes some basic definitions relating to molecular structure and stereochemistry.

TABLE 7.3	Classification of Isomers*	

Definition	Example
1. Constitutional isomers differ in the order in which their atoms are connected.	Three constitutionally isomeric compounds have the molecular formula C_3H_8O:

$$CH_3CH_2CH_2OH \qquad \underset{\displaystyle OH}{CH_3CHCH_3} \qquad CH_3CH_2OCH_3$$

	1-Propanol 2-Propanol Ethyl methyl ether
2. Stereoisomers have the same constitution but differ in the arrangement of their atoms in space.	
(a) Enantiomers are stereoisomers that are related as an object and its nonsuperimposable mirror image.	The two enantiomeric forms of 2-chlorobutane are

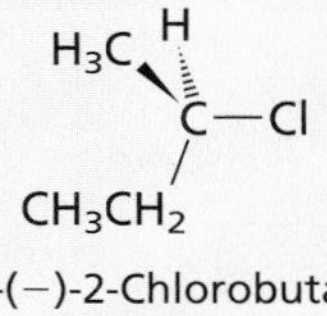

(R)-(−)-2-Chlorobutane   (S)-(+)-2-Chlorobutane

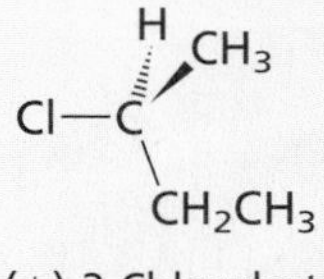

| **(b) Diastereomers** are stereoisomers that are not enantiomers. | The cis and trans isomers of 4-methylcyclohexanol are stereoisomers, but they are not related as an object and its mirror image; they are diastereomers. |

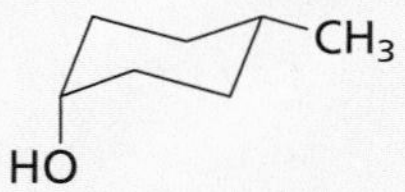

cis-4-Methylcyclohexanol   trans-4-Methylcyclohexanol

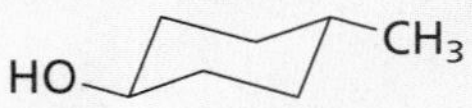

* Isomers are different compounds that have the same molecular formula. They may be either constitutional isomers or stereoisomers.

A molecule is **chiral** if it cannot be superimposed on its mirror image. Nonsuperimposable mirror images are **enantiomers** of one another. Molecules in which mirror images are superimposable are achiral (Section 7.1).

$$CH_3CHCH_2CH_3 \qquad CH_3CHCH_3$$
$$\vert \qquad\qquad\qquad \vert$$
$$Cl \qquad\qquad\qquad Cl$$

2-Chlorobutane   2-Chloropropane
  (chiral)   (achiral)

The most common kind of chiral molecule contains a carbon atom that bears four different atoms of groups, Such an atom is called a **stereogenic center** (Section 7.2). Table 7.3 shows the enantiomers of 2-chlorobutane. C-2 is a stereogenic center in 2-chlorobutane. A molecule that has a plane of symmetry cannot be chiral (Section 7.3).

Optical activity, or the degree to which a substance rotates the plane of polarized light, is a physical property used to characterize chiral substances (Section 7.4). Enantiomers have equal and opposite **optical rotations.** To be optically active a substance must be chiral, and one enantiomer must be present in excess of the other. A **racemic mixture** is optically inactive and contains equal quantities of enantiomers.

Absolute configuration is an exact description of the arrangement of atoms in space (Section 7.5). Absolute configuration in chiral molecules is best specified using the prefixes R and S of the Cahn–Ingold–Prelog notational system (Section 7.6). Substituents at a stereogenic center are ranked in order of decreasing precedence. If the three highest ranked substituents trace a clockwise path (highest→second highest→third highest) when the lowest ranked substituent is pointed away from you, the configuration is R. If the path is counterclockwise, the configuration is S. Table 7.3 shows the R and S enantiomers of 2-chlorobutane.

A **Fischer projection** shows how a molecule would look if its bonds were projected onto a flat surface (Section 7.7). Horizontal lines represent bonds coming toward you; vertical bonds point away from you. The projection is normally drawn so that the carbon chain is vertical, with the lowest numbered carbon at the top.

(R)-2-Chlorobutane   (S)-2-Chlorobutane

Both enantiomers of the same substance are identical in most of their physical properties (Section 7.8). The most prominent differences are biological ones, such as taste and odor, in which the substance interacts with a chiral receptor site in a living system. Enantiomers also have important consequences in medicine, in which the two enantiomeric forms of a drug can have much different effects on a patient.

A chemical reaction can convert an achiral substance to a chiral one. If the product contains a single stereogenic center, it is formed as a racemic mixture. Optically active products can be formed from optically inactive starting materials only if some optically active agent is present (Section 7.9).

When a molecule has two stereogenic centers that are not equivalently substituted, four stereoisomers are possible (Section 7.10). Stereoisomers that are not enantiomers are classified as **diastereomers.** Achiral molecules that contain stereogenic centers are called **meso forms** (Section 7.11). For a particular constitution, the maximum number of stereoisomers is 2^n, where n is the number of structural units capable of stereochemical variation—usually this is the number of stereogenic centers (Section 7.12).

Resolution is the separation of a racemic mixture into its enantiomers (Section 7.13). It is normally carried out by converting the mixture of enantiomers to a mixture of diastereomers, separating the diastereomers, then regenerating the enantiomers.

ADDITIONAL PROBLEMS

Stereogenic Centers and Stereoisomers

7.12 Which of the isomeric halides having the molecular formula $C_5H_{11}Br$ are chiral? Which are achiral?

7.13 Draw both enantiomers of each chiral stereoisomer identified in Problem 7.12 using perspective (wedge-and-dash) drawings.

7.14 Repeat Problem 7.13 using Fischer projections.

7.15 In each of the following pairs of compounds one is chiral and the other is achiral. Identify each compound as chiral or achiral, as appropriate.

(a) $ClCH_2CHCH_2OH$ and $HOCH_2CHCH_2OH$
         | |
         OH Cl

(b) $CH_3CH{=}CHCH_2Br$ and $CH_3CHCH{=}CH_2$
                                                         |
                                                        Br

(c)
```
         CH3                              CH3
H2N ——————— H                   H ——————— NH2
                     and
  H ——————— NH2                   H ——————— NH2
         CH3                              CH3
```

7.16 Identify the relationship in each of the following pairs. Do the drawings represent constitutional isomers or stereoisomers, or are they just different ways of drawing the same compound? If they are stereoisomers, are they enantiomers or diastereomers? (Molecular models may prove useful in this problem.)

(a)
```
   H  CH3                         H3C  H
      \                               \
       C—CH2Br        and             C—CH2OH
      /                               /
   HO                             Br
```

(b)
```
     H  CH3                         H  Br
        \                              \
         C—Br          and             C—CH3
        /                              /
 CH3CH2                         CH3CH2
```

(c)
```
     H  CH3                        H3C  H
        \                              \
         C—Br          and             C—CH2CH3
        /                              /
 CH3CH2                            Br
```

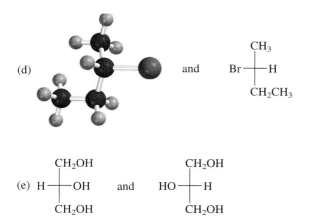

(d) and

$$CH_3$$
$$Br\text{—}\!\!|\!\!\text{—}H$$
$$CH_2CH_3$$

(e) $H\text{—}\!\!|\!\!\text{—}OH$ and $HO\text{—}\!\!|\!\!\text{—}H$

$$CH_2OH \qquad\qquad CH_2OH$$
$$CH_2OH \qquad\qquad CH_2OH$$

7.17 Redraw each of the molecules in Problem 7.16 (a) to (c) using Fischer projections with the methyl group attached to the stereogenic center at the top and a vertical arrangement of the carbon chain.

Molecules Having Two or More Stereogenic Centers

7.18 Identify any planes of symmetry in each of the following compounds. Which of the compounds is chiral? Which is achiral?

(a) *cis*-1,2-Dichlorocyclopropane   (c) *cis*-2-Chlorocyclopropane
(b) *trans*-1,2-Dichlorocyclopropane   (d) *trans*-2-Chlorocyclopropane

7.19 Write the structures of all the isomers, including stereoisomers, of trichlorocyclopropane. Which of these is (are) chiral?

7.20 Using Fischer projections, draw all the stereoisomers of 2,3-dibromopentane. Specify the relationship between pairs of compounds as being enantiomeric or diastereomeric.

7.21 2,4-Dibromopentane has fewer stereoisomers than 2,3-dibromopentane. Explain.

7.22 One of the stereoisomers of 1,3-dimethylcyclohexane is a meso isomer. Identify which one (cis or trans?).

7.23 Cholesterol is an important molecule in the formation of steroid hormones in our bodies, yet an excess in the blood has been implicated as a contributing factor to certain types of heart disease. Cholesterol has eight stereogenic centers. How many stereoisomers are possible having the same constitution as cholesterol, assuming there are no meso isomers?

7.24 Identify the stereogenic centers in each of the following naturally occurring substances.

(a)   (b)

Limonene
(a constituent of lemon oil)

Biotin
(a nutrient essential for normal growth)

(c)

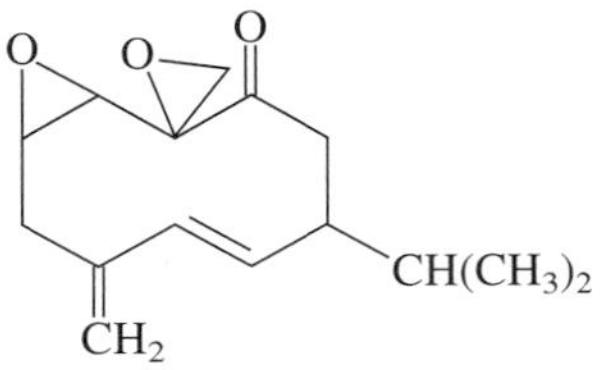

Periplanone B
(sex attractant of the American cockroach)

(d)

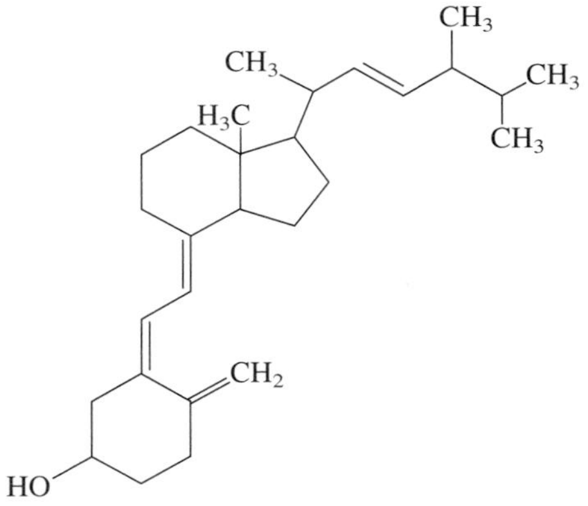

Calciferol
(a hormone also called vitamin D_2,
which is involved in calcium deposition in bones)

7.25 Carbohydrates having four carbons and an aldehyde group are called aldotetroses and have the general formula

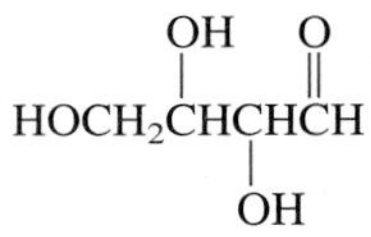

$$\underset{\underset{OH}{|}}{HOCH_2CH}\overset{\overset{OH}{|}}{C}H\overset{\overset{O}{||}}{C}H$$

Draw all the aldotetrose stereoisomers using Fischer projections. Are there any meso isomers?

Absolute Configuration of Chiral Molecules

7.26 Specify the absolute configurations of the enantiomers in Problem 7.13 as R or S.

7.27 Specify the absolute configuration of each stereogenic center in Problem 7.16 (a) to (c) using the $R–S$ notational system.

7.28 Specify the absolute configuration of each stereogenic center in Problem 7.20 using the $R–S$ notational system.

7.29 Draw each of the following molecules using both a perspective view and a Fischer projection.

(a) (S)-Pentanol
(b) (R)-3-Chloro-2-methylpentane

(c) ($2S,3S$)-2-Bromo-3-pentanol
(d) (R)-3-Bromo-1-pentene

7.30 Draw the R and S enantiomers of a chiral alkene having the molecular formula C_6H_{12}.

7.31 In 1996, it was determined that the absolute configuration of (−)-bromochlorofluoromethane is *R*. Which of the following is (are) (−)-BrClFCH?

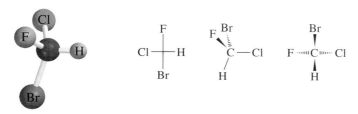

7.32 Specify the configuration at *R* or *S* in each of the following.

(a) (−)-2-Octanol

(b) Monosodium L-glutamate (only this stereoisomer is of any value as a flavor-enhancing agent)

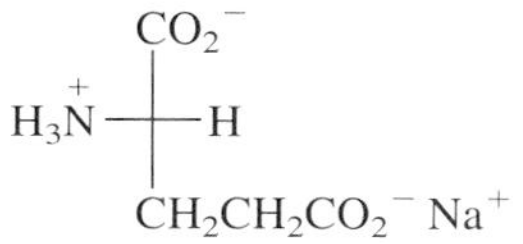

$$\begin{array}{c} CO_2^- \\ | \\ H_3\overset{+}{N}\!\!-\!\!\!\!|\!\!-\!\!H \\ | \\ CH_2CH_2CO_2^-\ Na^+ \end{array}$$

Miscellaneous Problems

7.33 An aqueous solution containing 10 g of fructose (a sugar used to sweeten candy and other foods) was diluted to 500 mL with water and placed in a polarimeter tube 20 cm long. The measured optical rotation was −5.20°. Calculate the specific rotation of fructose.

7.34 A certain natural product having $[\alpha]_D$ + 40.3° was isolated. Two structures have been independently proposed for this compound. Which one do you think is more likely to be correct? Why?

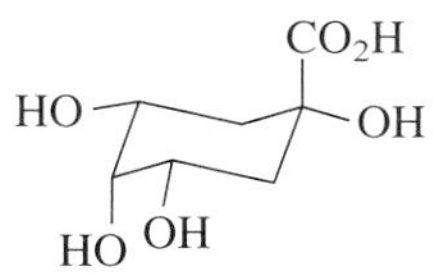

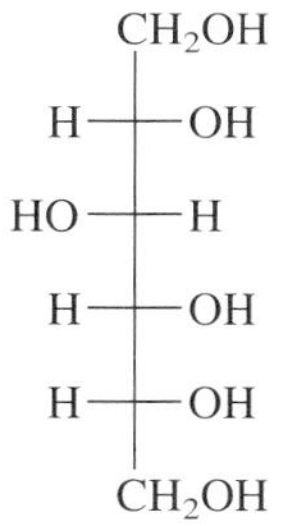

$$\begin{array}{c} CH_2OH \\ H\!-\!\!|\!\!-\!OH \\ HO\!-\!\!|\!\!-\!H \\ H\!-\!\!|\!\!-\!OH \\ H\!-\!\!|\!\!-\!OH \\ CH_2OH \end{array}$$

7.35 Write a mechanism for the acid-catalyzed hydration of 1-butene that explains why the product is obtained as a racemic mixture.

7.36 How many stereoisomeric products are obtained when (*S*)-3-chloro-1-butene reacts with HCl by Markovnikov addition? Are these products enantiomers or diastereomers of each other?

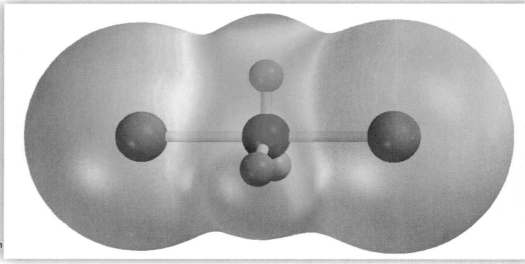

CHAPTER 8
NUCLEOPHILIC SUBSTITUTION

When we discussed elimination reactions in Chapter 4, we learned that a Lewis base can react with an alkyl halide to form an alkene. In the present chapter, you will find that the same kinds of reactants can also undergo a different reaction, one in which the Lewis base acts as a **nucleophile** to substitute for the halide substituent on carbon.

$$R\!-\!\ddot{X}\!: \;+\;\; Y\!:^- \;\;\longrightarrow\;\; R\!-\!Y \;\;+\;\; :\ddot{X}:^-$$

Alkyl   Lewis base   Product of   Halide
halide   nucleophilic   anion
                             substitution

We first encountered nucleophilic substitution in Chapter 3, in the reaction of alcohols with hydrogen halides to form alkyl halides. Now we'll see how alkyl halides can themselves be converted to other classes of organic compounds by nucleophilic substitution.

8.1 FUNCTIONAL GROUP TRANSFORMATION BY NUCLEOPHILIC SUBSTITUTION

Nucleophilic substitution reactions of alkyl halides are related to elimination reactions in that the halogen acts as a **leaving group** on carbon and is lost as an anion. Table 8.1 illustrates several examples of functional group transformations. Each of the nucleophiles is typically used as a sodium or potassium salt.

$$M^+ \; {}^-Y\!: \;+\; R\!-\!\ddot{X}\!: \;\longrightarrow\; R\!-\!Y \;+\; M^+ \; :\ddot{X}:^-$$

Nucleophilic   Alkyl   Product of   Metal halide
reagent   halide   nucleophilic
                         substitution

TABLE 8.1	Representative Functional Group Transformations by Nucleophilic Substitution Reactions of Alkyl Halides

Nucleophile and Comments	General Equation and Specific Example
Alkoxide ion (RO:⁻) The oxygen atom of a metal alkoxide acts as a nucleophile to replace the halogen of an alkyl halide. The product is an *ether*.	$R'O:^-$ + $R{-}X:$ ⟶ $R'OR$ + $:X:^-$ Alkoxide ion Alkyl halide Ether Halide ion $(CH_3)_2CHCH_2ONa + CH_3CH_2Br \longrightarrow (CH_3)_2CHCH_2OCH_2CH_3 +$ NaBr Sodium isobutoxide Ethyl bromide Ethyl isobutyl ether (66%) Sodium bromide
Hydrogen sulfide ion (HS:⁻) Use of hydrogen sulfide as a nucleophile permits the conversion of alkyl halides to compounds of the type RSH. These compounds are the sulfur analogs of alcohols and are known as *thiols*.	$HS:^-$ + $R{-}X:$ ⟶ RSH + $:X:^-$ Hydrogen sulfide ion Alkyl halide Thiol Halide ion KSH + $CH_3CH(CH_2)_6CH_3$ (Br) ⟶ $CH_3CH(CH_2)_6CH_3$ (SH) + KBr Potassium hydrogen sulfide 2-Bromononane 2-Nonanethiol (74%) Potassium bromide
Cyanide ion (:C≡N:) The negatively charged carbon atom of cyanide ion is usually the site of its nucleophilic character. Use of cyanide ion as a nucleophile permits the extension of a carbon chain by carbon–carbon bond formation. The product is an *alkyl cyanide*, or *nitrile*.	$:N{\equiv}C:^-$ + $R{-}X:$ ⟶ $RC{\equiv}N:$ + $:X:^-$ Cyanide ion Alkyl halide Alkyl cyanide Halide ion NaCN + ⬠–Cl ⟶ ⬠–CN + NaCl Sodium cyanide Cyclopentyl chloride Cyclopentyl cyanide (70%) Sodium chloride
Azide ion (:N=N=N:) Sodium azide is a reagent used for carbon–nitrogen bond formation. The product is an *alkyl azide*.	$:N{=}N{=}N:^-$ + $R{-}X:$ ⟶ $RN{=}N{=}N:$ + $:X:^-$ Azide ion Alkyl halide Alkyl azide Halide ion NaN₃ + $CH_3(CH_2)_4I$ ⟶ $CH_3(CH_2)_4N_3$ + NaI Sodium azide Pentyl iodide Pentyl azide (52%) Sodium iodide
Iodide ion (:I:⁻) Alkyl chlorides and bromides are converted to *alkyl iodides* by treatment with sodium iodide in acetone. NaI is soluble in acetone, but NaCl and NaBr are insoluble and crystallize from the reaction mixture, driving the reaction to completion.	$:I:^-$ + $R{-}X:$ $\xrightarrow{acetone}$ $R{-}I:$ + $:X:^-$ Iodide ion Alkyl chloride or bromide Alkyl iodide Chloride or bromide ion CH_3CHCH_3 (Br) + NaI $\xrightarrow{acetone}$ CH_3CHCH_3 (I) + NaBr (solid) 2-Bromopropane Sodium iodide 2-Iodopropane (63%) Sodium bromide

Notice that all the examples in Table 8.1 involve **alkyl halides,** that is, compounds in which the halogen is attached to an sp^3-hybridized carbon. **Alkenyl halides** and **aryl halides,** compounds in which the halogen is attached to sp^2-hybridized carbons, are essentially unreactive under these conditions, and the principles to be developed in this chapter do not apply to them.

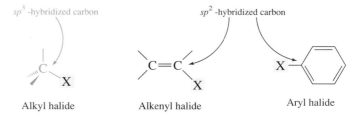

Alkyl halide   Alkenyl halide   Aryl halide

PROBLEM 8.1 Write a structural formula for the principal organic product formed in the reaction of methyl bromide with each of the following compounds:

(a) NaOH (sodium hydroxide)

(b) KOCH$_2$CH$_3$ (potassium ethoxide)

(c) LiN$_3$ (lithium azide)

(d) KCN (potassium cyanide)

(e) NaSH (sodium hydrogen sulfide)

(f) NaI (sodium iodide)

Sample Solution (a) The nucleophile in sodium hydroxide is the negatively charged hydroxide ion. The reaction that occurs is nucleophilic substitution of bromide by hydroxide. The product is methyl alcohol.

$$ \text{HO}{:}^- \quad + \quad CH_3{-}\ddot{\text{B}}\text{r}{:} \quad \longrightarrow \quad CH_3{-}\ddot{\text{O}}\text{H} \quad + \quad {:}\ddot{\text{B}}\text{r}{:}^- $$

Hydroxide ion   Methyl bromide   Methyl alcohol   Bromide ion
(nucleophile)   (substrate)   (product)   (leaving group)

8.2   THE S$_N$2 MECHANISM OF NUCLEOPHILIC SUBSTITUTION

The reaction of methyl bromide with sodium hydroxide to form methyl alcohol is an example of a nucleophilic substitution.

$$ CH_3Br \quad + \quad HO^- \quad \longrightarrow \quad CH_3OH \quad + \quad Br^- $$

Methyl bromide   Hydroxide ion   Methyl alcohol   Bromide ion

The rate of this reaction is observed to be directly proportional to the concentration of both methyl bromide and sodium hydroxide. It is first order in each reactant, or **second order** overall.

$$ \text{Rate} = k[CH_3Br][HO^-] $$

The second-order kinetic behavior is interpreted to mean that the rate-determining step is **bimolecular,** that is, both hydroxide ion and methyl bromide are involved at the transition state. The symbol given to this mechanism is **S$_N$2,** standing for **substitution nucleophilic bimolecular.**

The S$_N$2 mechanism is a **concerted** (single-step) process in which cleavage of the bond between carbon and the leaving group (the halogen) is assisted by formation of a bond between carbon and the nucleophile.

The S$_N$2 mechanism for the hydrolysis of methyl bromide may be represented by a single elementary step:

$$\ddot{H\ddot{O}}:^{-} \; + \; CH_3\ddot{Br}: \longrightarrow \overset{\delta-}{H\ddot{O}} \text{---} CH_3 \text{---} \overset{\delta-}{\ddot{Br}}: \longrightarrow H\ddot{O}CH_3 \; + \; :\ddot{Br}:^{-}$$

| Hydroxide ion | Methyl bromide | Transition state | Methyl alcohol | Bromide ion |

Carbon is partially bonded to both the incoming nucleophile and the departing halide at the transition state. Progress is made toward the transition state as the nucleophile begins to share a pair of its electrons with carbon and the halide ion leaves, taking with it the pair of electrons in its bond to carbon.

> **PROBLEM 8.2**   Is the two-step sequence depicted in the following equations consistent with the second-order kinetic behavior observed for the hydrolysis of methyl bromide?
>
> $$CH_3Br \xrightarrow{\text{slow}} CH_3^+ + Br^-$$
>
> $$CH_3^+ + HO^- \xrightarrow{\text{fast}} CH_3OH$$

8.3   HOW S$_N$2 REACTIONS OCCUR: STEREOCHEMISTRY

A variety of stereochemical studies reveal that S$_N$2 reactions proceed by **inversion of configuration.** The nucleophile attacks carbon from the side opposite the bond to the leaving group.

NUCLEOPHILIC SUBSTITUTION AND CANCER

Agents that can cause cancer in either laboratory animals or humans are called **carcinogens** (see boxed essay, *Chemical Carcinogens,* in Section 6.4). Many different types of chemicals may act as carcinogens in addition to the aromatic hydrocarbons described in Chapter 6. Among these are certain **alkylating agents,** compounds that can act as substrates in nucleophilic substitutions. For example, chloromethyl methyl ether and bis(2-chloroethyl)-amine are both active carcinogens.

$$ClCH_2OCH_3 \qquad\qquad (ClCH_2CH_2)_2NH$$

| Chloromethyl methyl ether | Bis(2-chloroethyl)amine |

One theory of **carcinogenesis,** or how these compounds interact with cells, is that cellular nucleophiles displace the leaving group (chloride in these compounds) to form a **bound adduct.**

$$RCH_2Cl \quad + \quad Nu: \longrightarrow \quad Nu\text{---}CH_2R$$

| Carcinogen | Nucleophile | Bound adduct |

Cellular nucleophiles may include functional groups such as —$\ddot{N}H_2$, —$\ddot{S}H$, and —$\ddot{O}H$, all of which are abundant in the molecules that make up living systems, such as proteins and nucleic acids. The bound adduct changes the shape of the biomolecule by changing its molecular composition. A change in shape will often lead to a change in the function of the molecule, and this can, in turn, lead to a cancer-causing mutation of the cell.

An interesting point can be made about the carcinogenic amine ($ClCH_2CH_2)_2NH$, known as nitrogen mustard. This compound is also used to treat certain forms of cancer as a **chemotherapy** agent. The same mechanism by which a cell becomes cancerous—disruption of the normal functioning of the cell through a mutation—is used to kill cancer cells. Because cancer cells are rapidly dividing, they are more susceptible to mutation-causing chemicals such as nitrogen mustard than normal body cells.

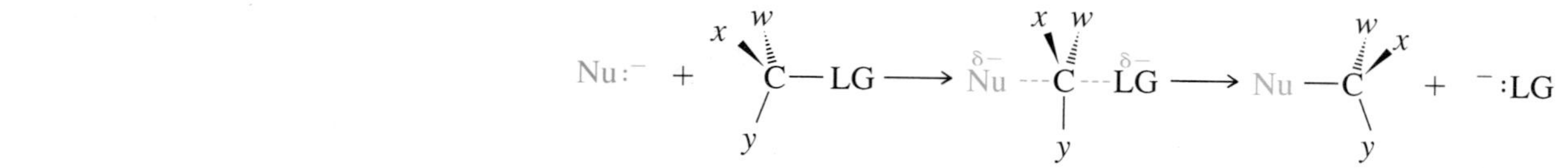

Notice that the groups w, x, and y have flipped from the left to the right, much as an umbrella would flip inside out in a windstorm. Organic chemists often refer to this as a **Walden inversion**, after the German chemist Paul Walden, who described the earliest experiments in this area in the 1890s. We can illustrate inversion of configuration in the hydrolysis of methyl bromide as follows:

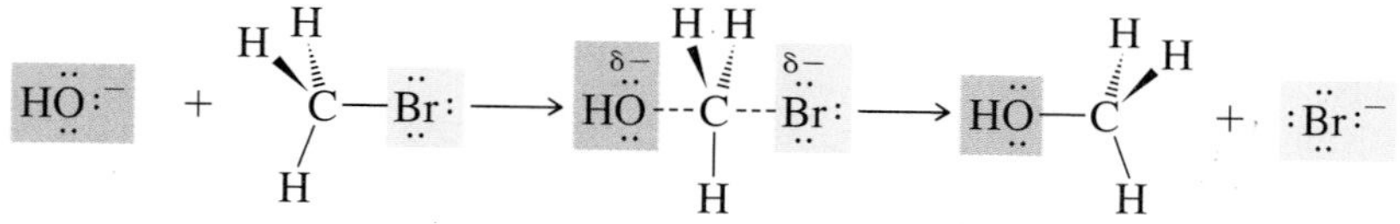

| Hydroxide ion | Methyl bromide | Transition state | Methyl alcohol | Bromide ion |

This picture of the S_N2 mechanism is based on many experiments using optically active alkyl halides as substrates. By comparing the relative configurations of the reactants and products in S_N2 reactions, it was determined that the product configuration is always opposite that of the reactant. In other words, inversion of configuration has taken place. For example, reaction of (S)-(+)-2-bromooctane with hydroxide ion gives the 2-octanol enantiomer having a configuration *opposite* to that of the starting alkyl halide.

Although the alkyl halide and alcohol given in this example have opposite signs of rotation when they have opposite configurations, it cannot be assumed that this will be true for all alkyl halide/alcohol pairs. (See Section 7.5)

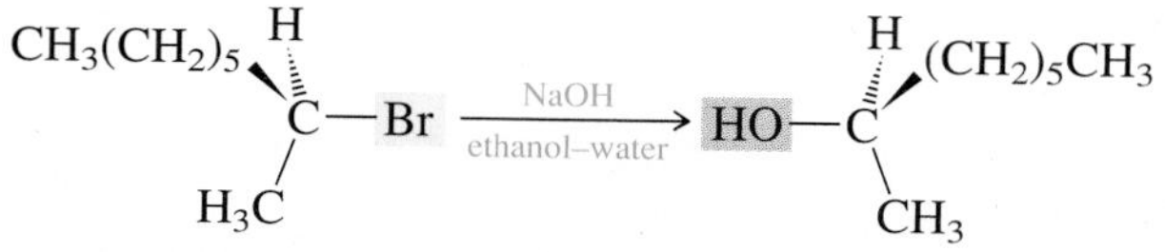

(S)-(+)-2-Bromooctane   (R)-(−)-2-Octanol

Nucleophilic substitution has occurred with inversion of configuration, consistent with the following transition state:

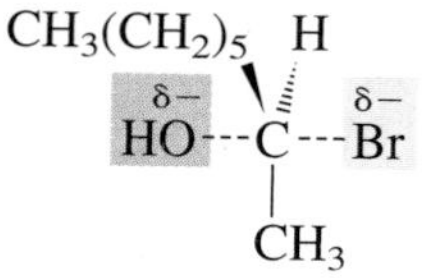

PROBLEM 8.3   The Fischer projection formula for (+)-2-bromooctane is shown. Write the Fischer projection of the (−)-2-octanol formed from it by nucleophilic substitution with inversion of configuration.

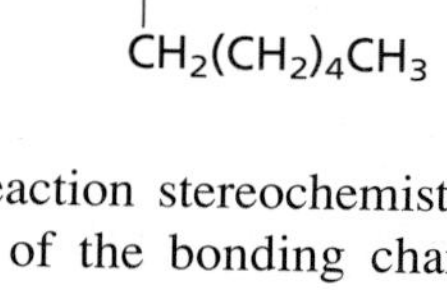

When we consider the overall reaction stereochemistry along with the kinetic data, a fairly complete picture emerges of the bonding changes that take place during S_N2 reactions. The potential energy diagram of Figure 8.1 for the hydrolysis of (S)-(+)-2-bromooctane is one that is consistent with the experimental observations.

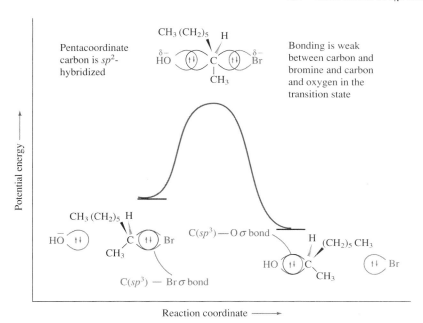

FIGURE 8.1 Hybrid orbital description of the bonding changes that take place at carbon during nucleophilic substitution by the S$_N$2 mechanism.

Hydroxide ion acts as a nucleophile, using an unshared electron pair to attack carbon from the side opposite the bond to the leaving group. The hybridization of the carbon at which substitution occurs changes from sp^3 in the alkyl halide to sp^2 in the transition state. Both the nucleophile (hydroxide) and the leaving group (bromide) are partially bonded to this carbon in the transition state. We say that the S$_N$2 transition state is **pentacoordinate;** carbon is fully bonded to three substituents and partially bonded to both the leaving group and the incoming nucleophile. The bonds to the nucleophile and the leaving group are relatively long and weak at the transition state.

Once past the transition state, the leaving group is expelled and carbon becomes tetracoordinate, its hybridization returning to sp^3.

8.4   STERIC EFFECTS IN S$_N$2 REACTIONS

Very large differences occur in the rates at which the various kinds of alkyl halides—methyl, primary, secondary, or tertiary—undergo nucleophilic substitution. For example, in the reaction

$$\underset{\text{Alkyl bromide}}{\text{RBr}} \quad + \quad \underset{\text{Lithium iodide}}{\text{LiI}} \quad \xrightarrow{\text{acetone}} \quad \underset{\text{Alkyl iodide}}{\text{RI}} \quad + \quad \underset{\text{Lithium bromide}}{\text{LiBr}}$$

the rates of nucleophilic substitution of a series of alkyl bromides differ by a factor of over 10^6 when comparing the most reactive member of the group (methyl bromide) and the least reactive member (*tert*-butyl bromide).

The large rate difference between methyl, ethyl, isopropyl, and *tert*-butyl bromides reflects the **steric hindrance** each offers to nucleophilic attack. The nucleophile must approach the alkyl halide from the side opposite the bond to the leaving group, and, as illustrated in Figure 8.2, this approach is hindered by alkyl substituents on the carbon that is being attacked. The three hydrogens of methyl bromide offer little resistance to approach of the nucleophile, and a rapid reaction occurs. Replacing one

FIGURE 8.2 Ball-and-spoke and space-filling models of alkyl bromides, showing how substituents shield the carbon atom that bears the leaving group from attack by a nucleophile. The nucleophile must attack from the side opposite the bond to the leaving group.

Least crowded–
most reactive

Most crowded–
least reactive

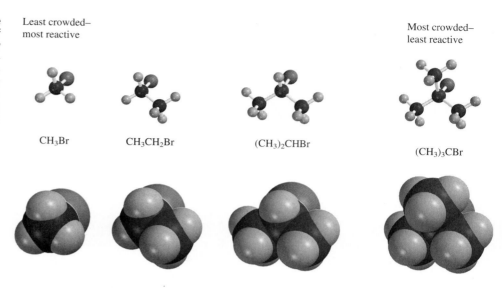

CH₃Br   CH₃CH₂Br   (CH₃)₂CHBr   (CH₃)₃CBr

of the hydrogens by a methyl group somewhat shields the carbon from attack by the nucleophile and causes ethyl bromide to be less reactive than methyl bromide. Replacing all three hydrogen substituents by methyl groups almost completely blocks backside approach to the tertiary carbon of (CH₃)₃CBr and shuts down bimolecular nucleophilic substitution.

In general, S$_N$2 reactions exhibit the following dependence of rate on substrate structure:

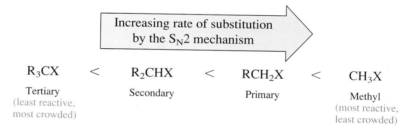

Increasing rate of substitution
by the S$_N$2 mechanism

R₃CX   <   R₂CHX   <   RCH₂X   <   CH₃X

Tertiary   Secondary   Primary   Methyl
(least reactive,   (most reactive,
most crowded)   least crowded)

PROBLEM 8.4 Identify the compound in each of the following pairs that reacts with sodium iodide in acetone at the faster rate:

(a) 1-Chlorohexane or cyclohexyl chloride

(b) 1-Bromopentane or 3-bromopentane

(c) 2-Bromo-2-methylhexane or 2-bromo-5-methylhexane

(d) 2-Bromopropane or 1-bromodecane

Sample Solution (a) Compare the structures of the two chlorides. 1-Chlorohexane is a primary alkyl chloride; cyclohexyl chloride is secondary. Primary alkyl halides are less crowded at the site of substitution than secondary ones and react faster in substitution by the S$_N$2 mechanism. 1-Chlorohexane is more reactive.

CH₃CH₂CH₂CH₂CH₂CH₂Cl

1-Chlorohexane
(primary, more reactive)

Cyclohexyl chloride
(secondary, less reactive)

8.5   THE S$_N$1 MECHANISM OF NUCLEOPHILIC SUBSTITUTION

Having just learned that tertiary alkyl halides are practically inert to substitution by the S$_N$2 mechanism because of steric hindrance, we might wonder whether they undergo nucleophilic substitution at all. We'll see in this section that they do, but by a mechanism different from S$_N$2.

The hydrolysis of *tert*-butyl bromide in water occurs readily and is characterized by a first-order rate law:

$$(CH_3)_3CBr \quad + \quad H_2O \quad \longrightarrow \quad (CH_3)_3COH \quad + \quad HBr$$

tert-Butyl bromide   Water   *tert*-Butyl alcohol   Hydrogen bromide

$$Rate = k[(CH_3)_3CBr]$$

The rate of hydrolysis depends only on the concentration of the substrate, *tert*-butyl bromide. The first-order rate law is interpreted as evidence for a **unimolecular** rate-determining step—a step that involves only the alkyl halide. The proposed mechanism is outlined in Figure 8.3, and is called **S$_N$1**, standing for **substitution nucleophilic unimolecular**. The first step, unimolecular dissociation of the alkyl halide to form a carbocation intermediate, is rate-determining.

FIGURE 8.3   The S$_N$1 mechanism for hydrolysis of *tert*-butyl bromide.

The Overall Reaction:

$$(CH_3)_3CBr \quad + \quad 2H_2O \quad \longrightarrow \quad (CH_3)_3COH \quad + \quad H_3O^+ \quad + \quad Br^-$$

tert-Butyl bromide   Water   *tert*-Butyl alcohol   Hydronium ion   Bromide ion

Step 1: The alkyl halide dissociates to a carbocation and a halide ion.

$$(CH_3)_3C\!-\!\ddot{B}r: \quad \xrightarrow{slow} \quad (CH_3)_3C^+ \quad + \quad :\ddot{B}r:^-$$

tert-Butyl bromide   *tert*-Butyl cation   Bromide ion

Step 2: The carbocation formed in step 1 reacts rapidly with a water molecule. Water is a nucleophile. This step completes the nucleophilic substitution stage of the mechanism and yields an alkyloxonium ion.

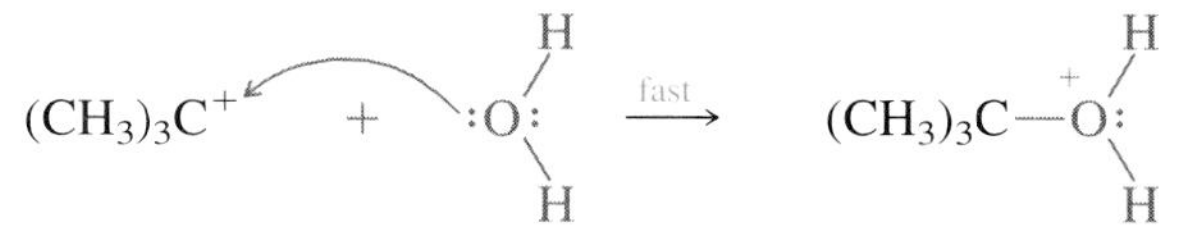

$$(CH_3)_3C^+ \quad + \quad :\ddot{O}\!\!\begin{smallmatrix}H\\\\H\end{smallmatrix} \quad \xrightarrow{fast} \quad (CH_3)_3C\!-\!\overset{+}{\ddot{O}}\!\!\begin{smallmatrix}H\\\\H\end{smallmatrix}$$

tert-Butyl cation   Water   *tert*-Butyloxonium ion

Step 3: This step is a fast acid–base reaction that follows the nucleophilic substitution. Water acts as a base to remove a proton from the alkyloxonium ion to give the observed product of the reaction, *tert*-butyl alcohol.

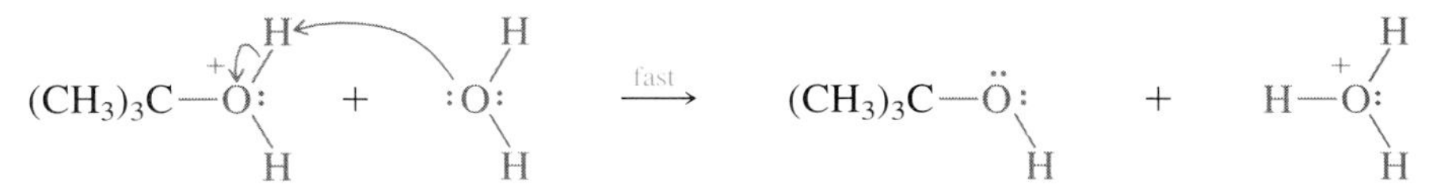

$$(CH_3)_3C\!-\!\overset{+}{\ddot{O}}\!\!\begin{smallmatrix}H\\\\H\end{smallmatrix} \quad + \quad :\ddot{O}\!\!\begin{smallmatrix}H\\\\H\end{smallmatrix} \quad \xrightarrow{fast} \quad (CH_3)_3C\!-\!\ddot{O}\!\!\begin{smallmatrix}\\\\H\end{smallmatrix} \quad + \quad H\!-\!\overset{+}{\ddot{O}}\!\!\begin{smallmatrix}H\\\\H\end{smallmatrix}$$

tert-Butyloxonium ion   Water   *tert*-Butyl alcohol   Hydronium ion

Figure 8.4 shows a potential energy diagram for the reaction of *tert*-butyl bromide and water. Notice how the shape of the diagram differs substantially from that of an S_N2 process, as you saw in Figure 8.1. Each elementary step of the S_N1 reaction has its own transition state. The rate-determining step, formation of the carbocation, has the largest activation energy of any step in the reaction. The second and third steps of the reaction, capture of the carbocation by water and proton transfer from the alkyloxonium ion, are both fast reactions and have small activation barriers.

The nucleophile in S_N1 reactions is usually the solvent. Substitutions in which the nucleophile is the solvent are called **solvolysis reactions.** Solvolysis occurs most readily in protic solvents such as water and alcohols. The nucleophile in Figure 8.3 is water, and the reaction is called **hydrolysis.**

> **PROBLEM 8.5** Suggest a structure for the product of nucleophilic substitution obtained on solvolysis of *tert*-butyl bromide in methanol, and outline a reasonable mechanism for its formation.

8.6   CARBOCATION STABILITY AND S_N1 REACTION RATES

The relative rate order in S_N1 reactions is exactly the opposite of that seen in S_N2 reactions:

S_N1 reactivity: methyl < primary < secondary < tertiary
S_N2 reactivity: tertiary < secondary < primary < methyl

Clearly, the steric crowding that influences reaction rates in S_N2 processes plays no role in S_N1 reactions.

The order of alkyl halide reactivity in S_N1 reactions is the same as the order of carbocation stability: *The more stable the carbocation, the more reactive the alkyl halide.*

FIGURE 8.4 Energy diagram illustrating the S_N1 mechanism for hydrolysis of *tert*-butyl bromide.

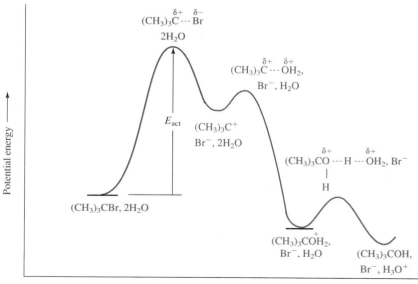

We have seen this situation before in the reaction of alcohols with hydrogen halides (Section 3.9), in the acid-catalyzed dehydration of alcohols (Section 4.8), and in the conversion of alkyl halides to alkenes by the E1 mechanism (Section 4.11). As in these other reactions, an electronic effect, specifically, the stabilization of the carbocation intermediate by alkyl substituents, is the decisive factor.

PROBLEM 8.6 Identify the compound in each of the following pairs that reacts at the faster rate in an S_N1 reaction:

(a) Isopropyl bromide or isobutyl bromide

(b) Cyclopentyl iodide or 1-methylcyclopentyl iodide

(c) Cyclopentyl bromide or 1-bromo-2,2-dimethylpropane

Sample Solution (a) Isopropyl bromide, $(CH_3)_2CHBr$, is a secondary alkyl halide, whereas isobutyl bromide, $(CH_3)_2CHCH_2Br$, is primary. Because the rate-determining step in an S_N1 reaction is carbocation formation and because secondary carbocations are more stable than primary carbocations, isopropyl bromide is more reactive than isobutyl bromide in nucleophilic substitution by the S_N1 mechanism.

Compounds capable of forming an allylic carbocation (Section 5.8) or a benzylic carbocation (Section 6.5), both of which are resonance-stabilized, readily undergo substitution by the S_N1 mechanism.

$$CH_2{=}CH{-}\overset{+}{C}H_2 \qquad \text{(benzyl cation)}{-}\overset{+}{C}H_2$$

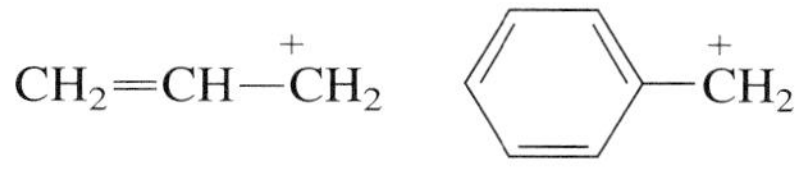

Allyl cation   Benzyl cation

Hydrolysis of 3-chloro-3-methyl-1-butene, for example, proceeds rapidly and gives a mixture of two allylic alcohols.

$$(CH_3)_2\underset{Cl}{C}CH{=}CH_2 \xrightarrow[Na_2CO_3]{H_2O} (CH_3)_2\underset{OH}{C}CH{=}CH_2 + (CH_3)_2C{=}CHCH_2OH$$

3-Chloro-3-methyl-   2-Methyl-3-buten-2-ol   3-Methyl-2-buten-1-ol
1-butene   (85%)   (15%)

Both alcohols are formed from the same carbocation. Water may react with the carbocation to give either a primary alcohol or a tertiary alcohol.

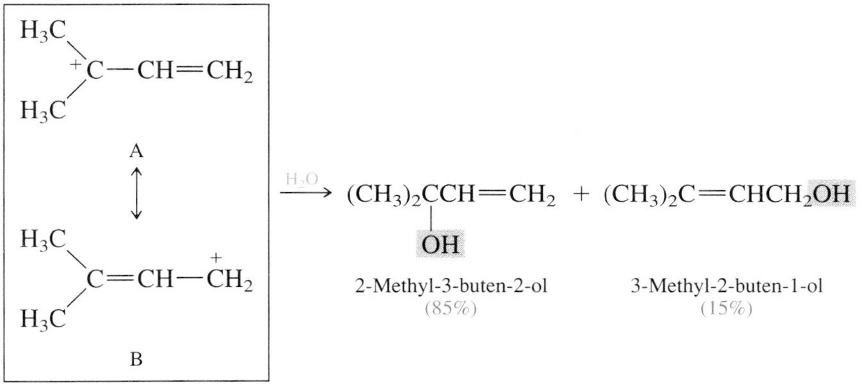

$$\xrightarrow{H_2O} (CH_3)_2\underset{OH}{C}CH{=}CH_2 + (CH_3)_2C{=}CHCH_2OH$$

2-Methyl-3-buten-2-ol   3-Methyl-2-buten-1-ol
(85%)   (15%)

> **PROBLEM 8.7** A second compound of molecular formula C_5H_9Cl undergoes hydrolysis to give the same mixture of alcohols as the preceding equation. Write the structure of this compound.

8.7   STEREOCHEMISTRY OF S_N1 REACTIONS

As we saw earlier (Section 8.3), S_N2 reactions proceed with 100% inversion of configuration. The situation is quite different for S_N1 reactions. When the leaving group is attached to the stereogenic center of an optically active alkyl halide, ionization gives a carbocation intermediate that is achiral. It is achiral because the three bonds to the positively charged carbon lie in the same plane—a plane of symmetry (Section 7.3) for the carbocation.

Figure 8.5 shows how such a carbocation can react with a nucleophile at either its top face, giving one enantiomer, or at its bottom face, giving the other enantiomer. You might have expected that these pathways would be equally likely, but that is rarely the case. As you can see in Figure 8.5, the carbocation is not completely "free" when it is attacked by the nucleophile. Ionization of the alkyl halide gives a carbocation–halide ion pair, and the halide ion shields one face of the carbocation. Thus more product results from inversion of configuration than from retention. The products of S_N1 reactions are partially racemic, and we can say that **partial racemization** has taken place. For example, hydrolysis of optically active 2-bromooctane gives both (*S*)-2-octanol and (*R*)-2-octanol. The product is *not* a racemic mixture of 50% *S* and 50% *R*. Rather, it is 83% *S* and 17% *R*.

CH₃ H   H CH₃   CH₃ H

C—Br $\xrightarrow[\text{ethanol}]{H_2O}$ HO—C + C—OH

CH₃(CH₂)₅   (CH₂)₅CH₃   CH₃(CH₂)₅

(*R*)-(−)-2-Bromooctane   (*S*)-(+)-2-Octanol   (*R*)-(−)-2-Octanol

FIGURE 8.5 Inversion of configuration predominates in S_N1 reactions because one face of the carbocation is shielded by the leaving group (red).

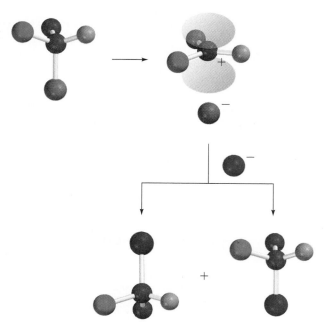

More than 50%   Less than 50%

8.8 SUBSTITUTION AND ELIMINATION AS COMPETING REACTIONS

We have seen in this chapter and in Chapter 4 that an alkyl halide and a Lewis base can react together in either a substitution or an elimination reaction.

Substitution can take place by the S_N1 or the S_N2 mechanism, elimination by E1 or E2.

How can we predict whether substitution or elimination will be the principal reaction observed with a particular combination of reactants?

The two most important factors are the *structure of the alkyl halide* and the *basicity of the anion.*

Secondary alkyl halides illustrate the competition between substitution and elimination. A secondary alkyl halide such as isopropyl bromide reacts with a strong base like sodium ethoxide to yield mainly the product of elimination.

$$CH_3CHCH_3 \xrightarrow[CH_3CH_2OH,\ 55°C]{NaOCH_2CH_3} CH_3CH{=}CH_2 \ + \ CH_3CHCH_3$$
$$\underset{Br}{|} \qquad\qquad\qquad\qquad\qquad\qquad \underset{OCH_2CH_3}{|}$$

| Isopropyl bromide | Propene (87%) | Ethyl isopropyl ether (13%) |

For this combination of alkyl halide and Lewis base, elimination by the E2 mechanism is faster than substitution by S_N2. Figure 8.6 illustrates the relationship between the S_N2 and E2 pathways.

Although elimination is favored for the reaction of secondary alkyl halides with strong bases such as alkoxides and hydroxide, *less basic* anions will tend to favor substitution. Thus the cyanide ion, which is a much weaker base than hydroxide, gives mainly the product of substitution with secondary alkyl halides.

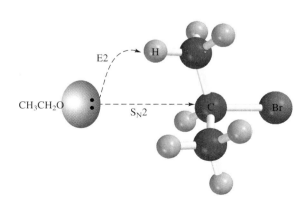

FIGURE 8.6 When a Lewis base reacts with an alkyl halide, either substitution or elimination can occur. Substitution (S_N2) occurs when the nucleophile attacks carbon to displace bromide. Elimination occurs when the Lewis base abstracts a proton. The alkyl halide shown is isopropyl bromide. The carbon atom that bears the leaving group is somewhat sterically hindered, and elimination (E2) predominates over substitution with alkoxide bases.

$$CH_3CH(CH_2)_5CH_3 \xrightarrow[\text{DMSO}]{\text{KCN}} CH_3CH(CH_2)_5CH_3$$

Cl	CN
2-Chlorooctane	2-Cyanooctane (70%)

Primary alkyl halides are less hindered than secondary ones, and the predominant reaction is substitution by the S_N2 mechanism, regardless of the basicity of the nucleophile. Thus substitution predominates even with strong bases such as hydroxide and alkoxides. For example,

$$CH_3CH_2CH_2Br \xrightarrow[\text{CH}_3\text{CH}_2\text{OH, 55°C}]{\text{NaOCH}_2\text{CH}_3} CH_3CH=CH_2 + CH_3CH_2CH_2OCH_2CH_3$$

Propyl bromide	Propene (9%)	Ethyl propyl ether (91%)

Tertiary alkyl halides are so sterically hindered to nucleophilic attack that the presence of any anionic Lewis base favors elimination. Usually substitution predominates over elimination in tertiary alkyl halides only when anionic Lewis bases are absent. In the solvolysis of the tertiary bromide 2-bromo-2-methylbutane, for example, the ratio of substitution to elimination is 64:36 in pure ethanol but falls to 1:99 in the presence of 2 M sodium ethoxide.

$$\underset{\substack{| \\ Br}}{\overset{\substack{CH_3 \\ |}}{CH_3CCH_2CH_3}} \xrightarrow[\text{25°C}]{\text{ethanol}} \underset{\substack{| \\ OCH_2CH_3}}{\overset{\substack{CH_3 \\ |}}{CH_3CCH_2CH_3}} + (CH_3)_2C=CHCH_3 + \underset{}{\overset{\substack{CH_3 \\ |}}{CH_2=CCH_2CH_3}}$$

2-Bromo-2-methyl-butane	2-Ethoxy-2-methylbutane (major product in absence of sodium ethoxide)	2-Methyl-2-butene	2-Methyl-1-butene
		(alkene mixture is major product in presence of sodium ethoxide)	

PROBLEM 8.8 Predict the major organic product of each of the following reactions:

(a) Cyclohexyl bromide and $KOCH_2CH_3$

(b) Ethyl bromide and ⬡—OK

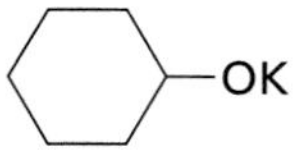

(c) *sec*-Butyl bromide solvolysis in methanol

(d) *sec*-Butyl bromide solvolysis in methanol containing 2 M $NaOCH_3$

Sample Solution (a) Cyclohexyl bromide is a secondary halide and reacts with alkoxide bases by elimination rather than substitution. The major organic products are cyclohexene and ethanol.

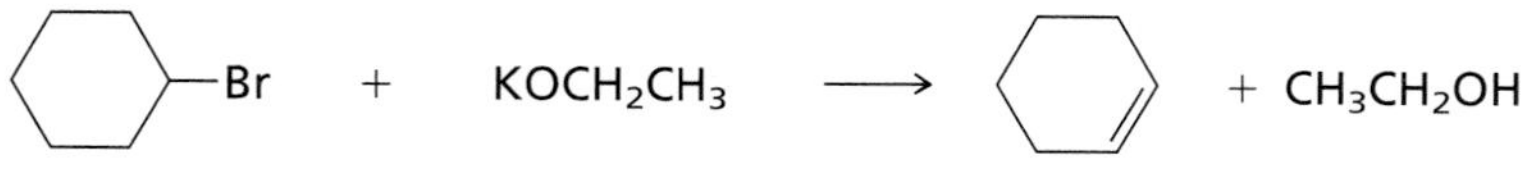

$$\text{⬡—Br} + KOCH_2CH_3 \longrightarrow \text{⬡} + CH_3CH_2OH$$

Cyclohexyl bromide	Potassium ethoxide	Cyclohexene	Ethanol

LEARNING OBJECTIVES

This chapter emphasized reaction mechanisms and focused on nucleophilic substitutions. The skills you have learned in this chapter should enable you to:

- Identify a reaction as being a nucleophilic substitution and predict the products.
- Explain what is meant by the term *bimolecular nucleophilic substitution.*
- Predict the stereochemical result of an S_N2 reaction.
- Predict the effect of steric crowding on an S_N2 reaction.
- Explain what is meant by the term *unimolecular nucleophilic substitution.*
- Explain how carbocation stability affects the rate of an S_N1 reaction.
- Predict the stereochemical result of an S_N1 reaction.
- Predict whether elimination or substitution will predominate in the reaction of a specific alkyl halide with a Lewis base.

8.9 SUMMARY

Nucleophilic substitution plays an important role in functional group transformations (Section 8.1). Examples of synthetically useful nucleophilic substitutions were given in Table 8.1. Nucleophilic substitution can be classified by two mechanisms: **S_N2 (substitution nucleophilic bimolecular)** and **S_N1 (substitution nucleophilic unimolecular)**. The details of these mechanisms are covered in Sections 8.2 through 8.7, and are summarized in Table 8.2.

When an alkyl halide is capable of undergoing elimination, the elimination process can compete with nucleophilic substitution (Section 8.8). Substitution predominates with primary alkyl halides. Secondary alkyl halides undergo substitution effectively with bases weaker than hydroxide. The normal reaction of a secondary alkyl halide with a base as strong as or stronger than hydroxide is elimination. Elimination predominates when tertiary halides react with any anion. Solvolysis in the absence of an anionic base causes nucleophilic substitution by the S_N1 mechanism to be the major reaction of tertiary alkyl halides.

ADDITIONAL PROBLEMS

Predicting Products of Substitution and Elimination Reactions

8.9 Write the structure of the major organic product to be expected from the reaction of 1-bromopropane with each of the following:

(a) Sodium iodide in acetone
(b) Sodium ethoxide (CH_3CH_2ONa)
(c) Sodium cyanide (NaCN)
(d) Sodium azide (NaN_3)
(e) Sodium hydrogen sulfide (NaSH)
(f) Sodium methanethiolate ($NaSCH_3$)

	TABLE 8.2	Comparison of S_N2 and S_N1 Mechanisms of Nucleophilic Substitution in Alkyl Halides

	S_N2	S_N1
Characteristics of mechanism	Single step:	Two elementary steps:
	$^-Nu:\ \ R\!-\!X: \longrightarrow Nu\!-\!R + :X:^-$	Step 1: $R\!-\!X: \rightleftharpoons R^+ + :X:^-$
	Nucleophile displaces leaving group; bonding to the incoming nucleophile accompanies cleavage of the bond to the leaving group. (Sections 8.2 and 8.3)	Step 2: $R^+ + :Nu^- \longrightarrow R\!-\!Nu$
		Ionization of alkyl halide (step 1) is rate-determining. (Section 8.5)
Rate-determining transition state	$^{\delta-}Nu\text{-}\text{-}\text{-}R\text{-}\text{-}\text{-}\ddot{X}:^{\delta-}$	$^{\delta+}R\text{-}\text{-}\text{-}\ddot{X}:^{\delta-}$
	(Sections 8.2 and 8.3)	(Section 8.5)
Molecularity	Bimolecular (Section 8.2)	Unimolecular (Section 8.5)
Kinetics and rate law	Second order: Rate = k[alkyl halide][nucleophile] (Section 8.2)	First order: Rate = k[alkyl halide] (Section 8.5)
Effect of structure on rate	$CH_3X > RCH_2X > R_2CHX > R_3CX$	$R_3CX > R_2CHX > RCH_2X > CH_3X$
	Rate is governed by steric effects (crowding in transition state). Methyl and primary alkyl halides can react only by the S_N2 mechanism; they never react by the S_N1 mechanism. (Section 8.4)	Rate is governed by stability of carbocation that is formed in ionization step. Tertiary alkyl halides can react only by the S_N1 mechanism; they never react by the S_N2 mechanism. (Section 8.6)
Stereochemistry	Inversion of configuration at reaction site. Nucleophile attacks carbon from side opposite bond to leaving group (Section 8.3).	Partial racemization when leaving group is located at a stereogenic center (Section 8.7).

8.10 All of the reactions of 1-bromopropane in the preceding problem give the product of nucleophilic substitution in high yield. High yields of substitution products are also obtained in all but one of the analogous reactions using 2-bromopropane as the substrate. In one case, however, 2-bromopropane is converted to propene, especially when the reaction is carried out at elevated temperature ($\approx 55°C$). Which reactant is most effective in converting 2-bromopropane to propene?

8.11 Each of the reagents in Problem 8.9 converts 2-bromo-2-methylpropane to the same major product. What is this product?

8.12 Write the structure of the major product, including stereochemistry, to be expected from reaction of (R)-2-bromopentane with each of the following reagents:
  (a) Sodium iodide in acetone   (c) Solvolysis in ethanol
  (b) Sodium cyanide   (d) Sodium ethoxide ($NaOCH_2CH_3$)

8.13 Write an equation, clearly showing the stereochemistry of the starting material and the product, for the reaction of (S)-1-bromo-2-methylbutane with sodium iodide in acetone. What is the configuration (R or S) of the product?

8.14 Identify the product in each of the following reactions:

(a) $ClCH_2CH_2CHCH_2CH_3$ $\xrightarrow[\text{acetone}]{\text{NaI (1.0 equiv)}}$ $C_5H_{10}ClI$
    (with Cl below the CH carbon)

(b) $BrCH_2CH_2Br$ + $NaSCH_2CH_2SNa$ $\longrightarrow$ $C_4H_8S_2$

(c) $ClCH_2CH_2CH_2CH_2Cl$ + Na_2S $\longrightarrow$ C_4H_8S

8.15 Each of the reactions shown involves nucleophilic substitution. The product of reaction (a) is an isomer of the product of reaction (b). What kind of isomer? By what mechanism does nucleophilic substitution occur? Write the structural formula of the product of each reaction.

(a) 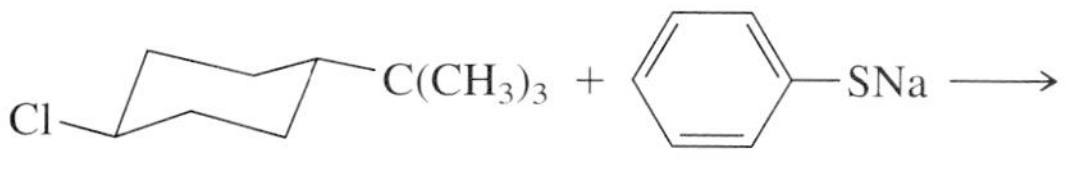 Cl—(cyclohexane ring)—$C(CH_3)_3$ + (benzene ring)—SNa $\longrightarrow$

(b) 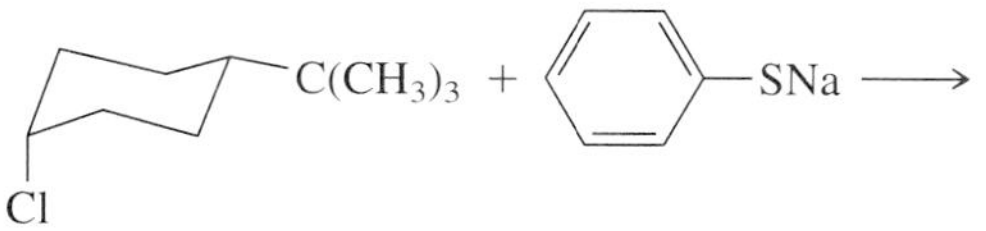 (cyclohexane ring with Cl)—$C(CH_3)_3$ + (benzene ring)—SNa $\longrightarrow$

8.16 What two stereoisomeric substitution products would you expect from the hydrolysis (S_N1) of *cis*-1,4-dimethylcyclohexyl bromide? From *trans*-1,4-dimethylcyclohexyl bromide?

8.17 Each of the following nucleophilic substitution reactions has been reported in chemical research journals. Many of them involve reactants that are somewhat more complex than those we have dealt with to this point. Nevertheless, you should be able to predict the product by analogy to what you know about nucleophilic substitution in simple systems.

(a) $BrCH_2\overset{\displaystyle O}{\overset{\displaystyle \|}{C}}OCH_2CH_3$ $\xrightarrow[\text{acetone}]{\text{NaI}}$

(b) 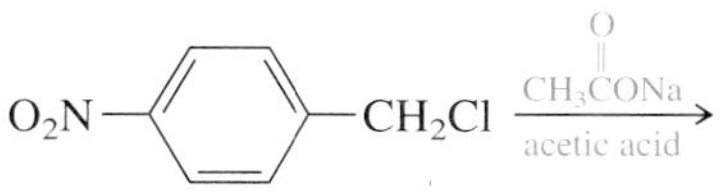 O_2N—(benzene ring)—CH_2Cl $\xrightarrow[\text{acetic acid}]{CH_3\overset{\displaystyle O}{\overset{\displaystyle \|}{C}}ONa}$

(c) $CH_3CH_2OCH_2CH_2Br$ $\xrightarrow[\text{ethanol–water}]{\text{NaCN}}$

(d) 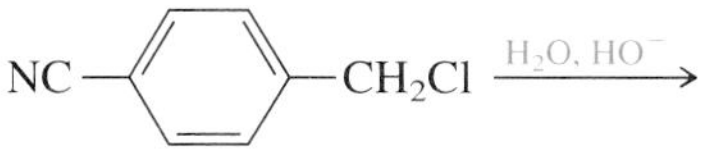 NC—(benzene ring)—CH_2Cl $\xrightarrow{H_2O, \ HO^-}$

(e) $ClCH_2\overset{\displaystyle O}{\overset{\displaystyle \|}{C}}OC(CH_3)_3$ $\xrightarrow[\text{acetone–water}]{\text{NaN}_3}$

S_N2 and S_N1 Mechanisms

8.18 Arrange the isomers of molecular formula C_4H_9Cl in order of decreasing rate of reaction with sodium iodide in acetone by the S_N2 mechanism.

8.19 In each of the following indicate which S_N2 reaction will occur faster. Explain your reasoning.
   (a) Hexyl chloride or cyclohexyl chloride with sodium azide (NaN_3) in aqueous ethanol.
   (b) 1-Chloro-2,2-dimethylbutane or 1-chlorohexane with sodium iodide in acetone.

8.20 There is an overall 29-fold difference in reactivity of 1-chlorohexane, 2-chlorohexane, and 3-chlorohexane toward potassium iodide in acetone.

(a) Which one is the most reactive? Why?

(b) Two of the isomers differ by only a factor of 2 in reactivity. Which two? Which one is the more reactive? Why?

8.21 Indicate which compound from each of the following pairs will undergo S_N1 solvolysis in ethanol at the faster rate. Explain your reasoning.

(a) 1-Bromo-2-methylpropane or 2-bromobutane

(b) 1-Chlorocyclohexene or 3-chlorocyclohexene

(c) Chlorocyclohexane or 3-chlorocyclohexene

(d) CH_3—⬡—Br   or   ⬡—CH_2Br

(e)   ⬡—CH_2CHCH_3 or   ⬡—$CHCH_2CH_3$
                    | |
                    I   I

8.22 Give the mechanistic symbol (S_N1, S_N2, E1, or E2) most consistent with each of the following statements:

(a) Methyl halides react with sodium ethoxide in ethanol only by this mechanism.

(b) Unhindered primary halides react with sodium ethoxide in ethanol mainly by this mechanism.

(c) When cyclohexyl bromide is treated with sodium ethoxide in ethanol, the major product is formed by this mechanism.

(d) The major substitution product obtained by solvolysis of *tert*-butyl bromide in ethanol arises by this mechanism.

(e) These reaction mechanisms proceed in a single step.

(f) This substitution mechanism involves a carbocation intermediate.

8.23 The ratio of elimination to substitution is exactly the same (26% elimination) for 2-bromo-2-methylbutane and 2-iodo-2-methylbutane in 80% ethanol/20% water at 25°C.

(a) By what mechanism does substitution most likely occur in these compounds under these conditions?

(b) By what mechanism does elimination most likely occur in these compounds under these conditions?

(c) Which substrate undergoes substitution faster?

(d) Which substrate undergoes elimination faster?

(e) What two substitution products are formed from each substrate?

(f) What two elimination products are formed from each substrate?

(g) Why do you suppose the ratio of elimination to substitution is the same for the two substrates?

8.24 The compound KSCN is a source of thiocyanate ion.

(a) Write the two most stable Lewis structures for thiocyanate ion, and identify the atom in each that bears a formal charge of −1.

(b) Two constitutionally isomeric products of molecular formula C_5H_9NS were isolated in a combined yield of 87% in the reaction shown. Suggest reasonable structures for these two compounds.

$$CH_3CH_2CH_2CH_2Br \xrightarrow{\text{KSCN}}$$

Synthesis Using Nucleophilic Substitution Reactions

8.25 Select the combination of alkyl bromide and potassium alkoxide that would be the most effective in the syntheses of the following ethers:

    (a) $CH_3OC(CH_3)_3$

    (b)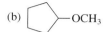

    (c) $(CH_3)_3CCH_2OCH_2CH_3$

8.26 Outline an efficient synthesis of each of the following compounds from the indicated starting material and any necessary organic or inorganic reagents. In most cases more than one step is required.

    (a) Ethanethiol (CH_3CH_2SH) from ethyl alcohol
    (b) Cyclopentyl cyanide from cyclopentene
    (c) Cyclopentyl cyanide from cyclopentanol
    (d) $NCCH_2CH_2CN$ from ethyl alcohol
    (e) Isobutyl iodide from isobutyl chloride

8.27 Describe the preparation of each of the following alkynes from acetylene using an alkylation reaction (Section 5.10).

    (a) 1-Hexyne
    (b) 2-Hexyne
    (c) 3-Hexyne
    (d) 4-Methyl-1-pentyne

Molecular Modeling

8.28 Illustrate the stereochemistry associated with unimolecular nucleophilic substitution by constructing molecular models of *cis*-4-*tert*-butylcyclohexyl bromide, its derived carbocation, and the alcohols formed from it by hydrolysis under S_N1 conditions.

8.29 Given the molecular formula $C_6H_{11}Br$, construct a molecular model of the isomer that is a primary alkyl bromide yet relatively unreactive toward bimolecular nucleophilic substitution.

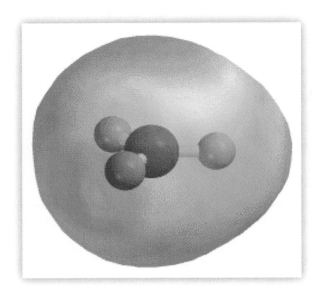

CHAPTER 9
FREE RADICALS

Several of the reactions we have examined proceed by multistep mechanisms that involve carbocation intermediates. These reactions include

Preparation of alkyl halides from alcohols (Chapter 3)

Dehydration of alcohols (Chapter 4)

Electrophilic addition to alkenes (Chapter 5)

Electrophilic aromatic substitution (Chapter 6)

Unimolecular (S_N1) nucleophilic substitution (Chapter 8)

In this chapter we will explore a different type of reactive intermediate—a neutral species known as a free radical.

9.1 STRUCTURE AND STABILITY OF FREE RADICALS

Free radicals are species that contain unpaired electrons. The octet rule notwithstanding, not all compounds have all of their electrons paired. Oxygen (O_2) is the most familiar example of a compound with unpaired electrons; it has two of them. Compounds that have an odd number of electrons, such as nitrogen dioxide (NO_2), must have at least one unpaired electron.

$$:\ddot{O}-\ddot{O}: \qquad :\ddot{O}=\ddot{N}-\ddot{O}: \qquad \cdot\ddot{N}=\ddot{O}:$$

Oxygen   Nitrogen dioxide   Nitrogen monoxide

Nitrogen monoxide ("nitric oxide") is another stable free radical. Although known for hundreds of years, NO has only recently been discovered to be an extremely important biochemical messenger and moderator of so many biological processes that it might be better to ask "Which ones is it not involved in?"

The free radicals that we usually see in carbon chemistry are much less stable than these. Simple alkyl radicals, for example, require special procedures for their isolation and study. We will encounter them here only as reactive intermediates, formed in one step of a reaction mechanism and consumed in the next. Alkyl radicals are classified as primary, secondary, or tertiary according to the number of carbon atoms directly attached to the carbon that bears the unpaired electron.

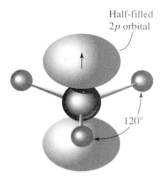

FIGURE 9.1 Orbital hybridization model of bonding in methyl radical. Carbon is sp^2-hybridized with an unpaired electron in a 2p orbital.

An alkyl radical is neutral and has one more electron than the corresponding carbocation. Thus, bonding in methyl radical may be approximated by simply adding an electron to the vacant 2p orbital of sp^2-hybridized carbon in methyl cation (Figure 9.1).

> **PROBLEM 9.1** Verify that methyl radical is neutral by calculating the formal charge at carbon and at each hydrogen.

Free radicals, like carbocations, have an unfilled 2p orbital and are stabilized by substituents, such as alkyl groups, that release electrons. Consequently, the order of free-radical stability parallels that of carbocations.

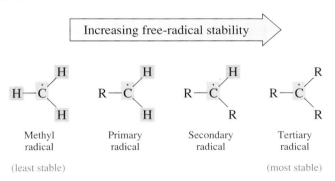

> **PROBLEM 9.2** Write a structural formula for the most stable of the free radicals that have the formula C_5H_{11}.

9.2 BOND DISSOCIATION ENERGIES

Some of the evidence indicating that alkyl substituents stabilize free radicals comes from bond energies. The strength of a bond is measured by the energy required to break it. A covalent bond can be broken in two ways. In a **homolytic cleavage** a bond between two atoms is broken so that each of them retains one of the electrons in the bond.

A curved arrow shown as a single-barbed fishhook ⌒ signifies the movement of *one* electron. "Normal" curved arrows ⌒ track the movement of a *pair* of electrons.

$$X \colon Y \longrightarrow X \cdot + \cdot Y$$

Homolytic bond cleavage

In contrast, in a **heterolytic cleavage** one fragment retains both electrons.

$$X \!:\! Y \longrightarrow X^+ + :Y^-$$

Heterolytic bond cleavage

FREE RADICALS AND BIOLOGY

Free-radical intermediates are more than laboratory curiosities. They are involved in a number of biological processes as well. Cell damage from free radicals, for example, is thought to involve formation of organic hydroperoxides. Many lipids (Chapter 16), including some found in cell membranes, contain unsaturated hydrocarbon chains that are susceptible to free-radical oxidation.

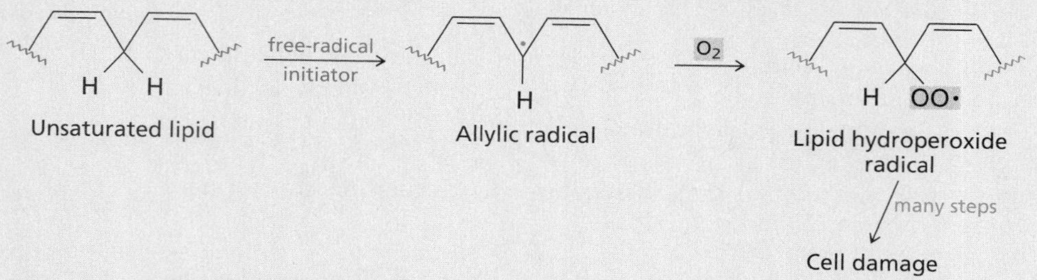

Oxygen plays a role in this destructive sequence, and may in fact serve as the free-radical initiator that begins the process. Molecular oxygen (O_2) is the most abundant free radical in nature. Oxygen in its most stable state has two unpaired electrons and is a stable **diradical.**

Other potential sources of free radicals include the enzymatic steps that reduce molecular oxygen to water in cellular respiration. Intermediates that form include the superoxide radical ($O_2^{-\cdot}$), hydroperoxide radical ($HO_2\cdot$), and hydroxyl radical ($HO\cdot$); these species can induce cell damage by serving as free-radical initiators in the lipid oxidation sequence described earlier. In the liver, metabolism of toxic substances such as carbon tetrachloride (CCl_4) and chloroform ($CHCl_3$) also produce free radicals.

Potential free-radical initiators include compounds found in photochemical smog. The nitrogen oxides NO and NO_2, for example, have an odd number of electrons and exist as free radicals. Interestingly, NO produced by enzymes is involved in several biological processes in the body.

What defenses do we have against free-radical damage? The body has natural repair mechanisms involving compounds known as **antioxidants** to limit the extent of oxidative damage to cells. One natural antioxidant is vitamin E.

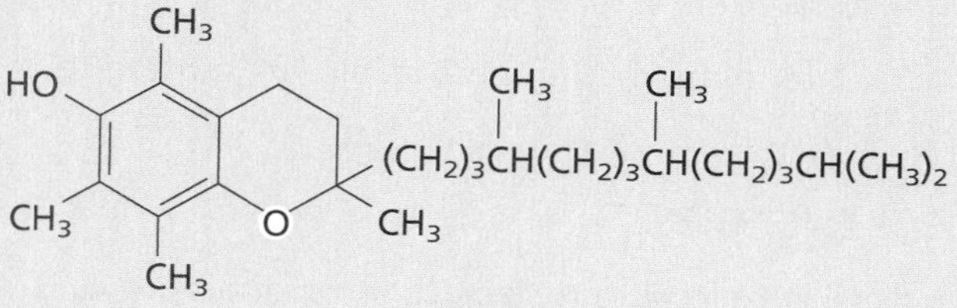

Vitamin E

Vitamin E, also known as α-tocopherol, interrupts the free-radical chain mechanism that leads to cell damage.

Oxidative damage through lipid peroxidation in cells appears to play an important role in the aging process. One theory of aging is that while young the body's ability to repair cell damage is most efficient. As we age, we are exposed cumulatively to more "challenge" by free-radical sources, while the repair mechanisms in our body become less efficient, leading to greater amounts of cell damage.

We assess the relative stability of alkyl radicals by measuring the enthalpy change ($\Delta H°$) for the homolytic cleavage of a C—H bond in an alkane:

$$R\overset{\frown}{}\overset{\frown}{}H \longrightarrow R\cdot + \cdot H$$

The more stable the radical, the lower the energy required to generate it by C—H bond homolysis.

The energy required for homolytic bond cleavage is called the **bond dissociation energy (BDE)**. A list of some bond dissociation energies is given in Table 9.1.

As the table indicates, C—H bond dissociation energies in alkanes are approximately 375 to 435 kJ/mol (90–105 kcal/mol). Homolysis of the H—CH_3 bond in methane gives methyl radical and requires 435 kJ/mol (104 kcal/mol). The dissociation energy of the H—CH_2CH_3 bond in ethane, which gives a primary radical, is somewhat less (410 kJ/mol, or 98 kcal/mol) and is consistent with the notion that ethyl radical (primary) is more stable than methyl.

TABLE 9.1	Bond Dissociation Energies of Some Representative Compounds*				
	Bond Dissociation Energy			**Bond Dissociation Energy**	
Bond	**kJ/mol**	**(kcal/mol)**	**Bond**	**kJ/mol**	**(kcal/mol)**
Diatomic molecules					
H—H	435	(104)			
F—F	159	(38)	H—F	568	(136)
Cl—Cl	242	(58)	H—Cl	431	(103)
Br—Br	192	(46)	H—Br	366	(87.5)
I—I	150	(36)	H—I	297	(71)
Alkanes					
CH_3—H	435	(104)	CH_3—CH_3	368	(88)
CH_3CH_2—H	410	(98)	CH_3CH_2—CH_3	355	(85)
$CH_3CH_2CH_2$—H	410	(98)			
$(CH_3)_2CH$—H	397	(95)			
$(CH_3)_2CHCH_2$—H	410	(98)	$(CH_3)_2CH$—CH_3	351	(84)
$(CH_3)_3C$—H	380	(91)	$(CH_3)_3C$—CH_3	334	(80)
Alkyl halides					
CH_3—F	451	(108)	$(CH_3)_2CH$—F	439	(105)
CH_3—Cl	349	(83.5)	$(CH_3)_2CH$—Cl	339	(81)
CH_3—Br	293	(70)	$(CH_3)_2CH$—Br	284	(68)
CH_3—I	234	(56)	$(CH_3)_3C$—Cl	330	(79)
CH_3CH_2—Cl	338	(81)	$(CH_3)_3C$—Br	263	(63)
$CH_3CH_2CH_2$—Cl	343	(82)			
Water and alcohols					
HO—H	497	(119)	CH_3CH_2—OH	380	(91)
CH_3O—H	426	(102)	$(CH_3)_2CH$—OH	385	(92)
CH_3—OH	380	(91)	$(CH_3)_3C$—OH	380	(91)

* Bond dissociation energies refer to bond indicated in structural formula for each substance.

The dissociation energy of the terminal C—H bond in propane is exactly the same as that of ethane. The resulting free radical is primary ($R\dot{C}H_2$) in both cases.

$$CH_3CH_2CH_2\text{—}H \longrightarrow CH_3CH_2\dot{C}H_2 + \quad H\cdot \qquad \Delta H° = +410 \text{ kJ}$$

			(98 kcal)
Propane	n-Propyl radical (primary)	Hydrogen atom	

Note, however, that Table 9.1 includes two entries for propane. The second entry corresponds to the cleavage of a bond to one of the hydrogens of the methylene (CH_2) group. It requires slightly less energy to break a C—H bond in the methylene group than in the methyl group.

$$CH_3\underset{\underset{H}{|}}{C}HCH_3 \longrightarrow CH_3\dot{C}HCH_3 + \quad H\cdot \qquad \Delta H° = +397 \text{ kJ}$$

			(95 kcal)
Propane	Isopropyl radical (secondary)	Hydrogen atom	

Because the starting material (propane) and one of the products (H·) are the same in both processes, the difference in bond dissociation energies is equal to the energy difference between an n-propyl radical (primary) and an isopropyl radical (secondary). As depicted in Figure 9.2, the secondary radical is 13 kJ/mol (3 kcal/mol) more stable than the primary radical.

Similarly, by comparing the bond dissociation energies of the two different types of C—H bonds in 2-methylpropane, we see that a tertiary radical is 30 kJ/mol (7 kcal/mol) more stable than a primary radical.

$$CH_3\underset{\underset{CH_3}{|}}{C}HCH_2\text{—}H \longrightarrow CH_3\underset{\underset{CH_3}{|}}{C}H\dot{C}H_2 + \quad H\cdot \qquad \Delta H° = +410 \text{ kJ}$$

			(98 kcal)
2-Methylpropane	Isobutyl radical (primary)	Hydrogen atom	

FIGURE 9.2 Diagram showing how bond dissociation energies of methylene and methyl C—H bonds in propane reveal a difference in stabilities between two isomeric free radicals. The secondary radical is more stable than the primary.

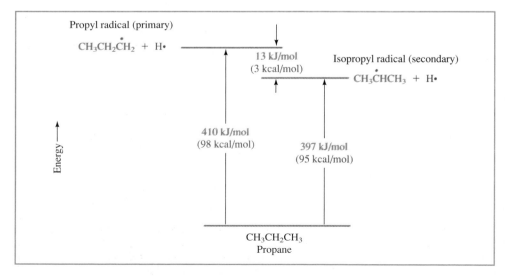

$$
\underset{\substack{\text{2-Methylpropane}}}{\text{CH}_3\overset{\displaystyle H}{\underset{\displaystyle \text{CH}_3}{\text{C}}}\text{CH}_3} \longrightarrow \underset{\substack{tert\text{-Butyl}\\ \text{radical}\\ \text{(tertiary)}}}{\text{CH}_3\overset{\displaystyle \cdot}{\underset{\displaystyle \text{CH}_3}{\text{C}}}\text{CH}_3} + \underset{\substack{\text{Hydrogen}\\ \text{atom}}}{\text{H}\cdot} \qquad \Delta H^\circ = +380\ \text{kJ}\\ (91\ \text{kcal})
$$

PROBLEM 9.3 Carbon–carbon bond dissociation energies have been measured for alkanes. Without referring to Table 9.1, identify the alkane in each of the following pairs that has the lower carbon–carbon bond dissociation energy, and explain the reason for your choice.

(a) Ethane or propane

(b) Propane or 2-methylpropane

(c) 2-Methylpropane or 2,2-dimethylpropane

Sample Solution (a) First write the equations that describe homolytic carbon–carbon bond cleavage in each alkane.

$$
\underset{\text{Ethane}}{\text{CH}_3\text{—CH}_3} \longrightarrow \underset{\text{Two methyl radicals}}{\cdot\text{CH}_3 + \cdot\text{CH}_3}
$$

$$
\underset{\text{Propane}}{\text{CH}_3\text{CH}_2\text{—CH}_3} \longrightarrow \underset{\text{Ethyl radical}}{\text{CH}_3\dot{\text{C}}\text{H}_2} + \underset{\text{Methyl radical}}{\cdot\text{CH}_3}
$$

Cleavage of the carbon–carbon bond in ethane yields two methyl radicals, whereas propane yields an ethyl radical and one methyl radical. Ethyl radical is more stable than methyl, and so less energy is required to break the carbon–carbon bond in propane than in ethane. The measured carbon–carbon bond dissociation energy in ethane is 368 kJ/mol (88 kcal/mol), and that in propane is 355 kJ/mol (85 kcal/mol).

9.3 CHLORINATION OF METHANE

Most alkyl halides are prepared in the laboratory from alcohols (Section 3.7). However, a number of industrially important alkyl chlorides are prepared by direct chlorination of alkanes. In the chlorination of alkanes, chlorine atoms replace one or more hydrogens of the alkane.

$$
\underset{\text{Alkane}}{\text{R—H}} + \underset{\text{Halogen}}{\text{X}_2} \longrightarrow \underset{\text{Alkyl halide}}{\text{R—X}} + \underset{\text{Hydrogen halide}}{\text{H—X}}
$$

For example, methane reacts with chlorine at high temperature in a gas-phase reaction.

$$
\underset{\text{Methane}}{\text{CH}_4} + \underset{\text{Chlorine}}{\text{Cl}_2} \xrightarrow{400–440^\circ\text{C}} \underset{\substack{\text{Chloromethane}\\ \text{(Methyl chloride)}}}{\text{CH}_3\text{Cl}} + \underset{\substack{\text{Hydrogen}\\ \text{chloride}}}{\text{HCl}}
$$

Chlorination of methane provides approximately one third of the annual U.S. production of chloromethane. The reaction of methanol with hydrogen chloride is the major synthetic method for the preparation of chloromethane.

PROBLEM 9.4 Each of the hydrogens of methane may be replaced by chlorine in a series of sequential substitutions. The first reaction, formation of chloromethane, was just described. Write balanced equations describing substitution of the remaining three hydrogens of methane by chlorine.

Sample Solution   Chloromethane reacts with chlorine to give dichloromethane and hydrogen chloride.

$$CH_3Cl \quad + \quad Cl_2 \quad \xrightarrow{400-440°C} \quad CH_2Cl_2 \quad + \quad HCl$$

Chloromethane   Chlorine   Dichloromethane   Hydrogen
                                      (Methylene chloride)   chloride

9.4   MECHANISM OF METHANE CHLORINATION

The generally accepted mechanism for the chlorination of methane is presented in Figure 9.3. As we just saw (Section 9.3), the reaction is normally carried out in the gas phase at high temperature. The reaction itself is strongly exothermic, but energy must be put into the system to get it going. This energy goes into breaking the weakest bond in the system, which, as we see from the bond dissociation energy data in Table 9.1, is the Cl—Cl bond with a bond dissociation energy of 242 kJ/mol (58 kcal/mol). The step in which Cl—Cl bond homolysis occurs is called the **initiation step.**

Each chlorine atom formed in the initiation step has seven valence electrons and is very reactive. Once formed, a chlorine atom abstracts a hydrogen atom from methane as shown in step 2 in Figure 9.3. Hydrogen chloride, one of the isolated products from

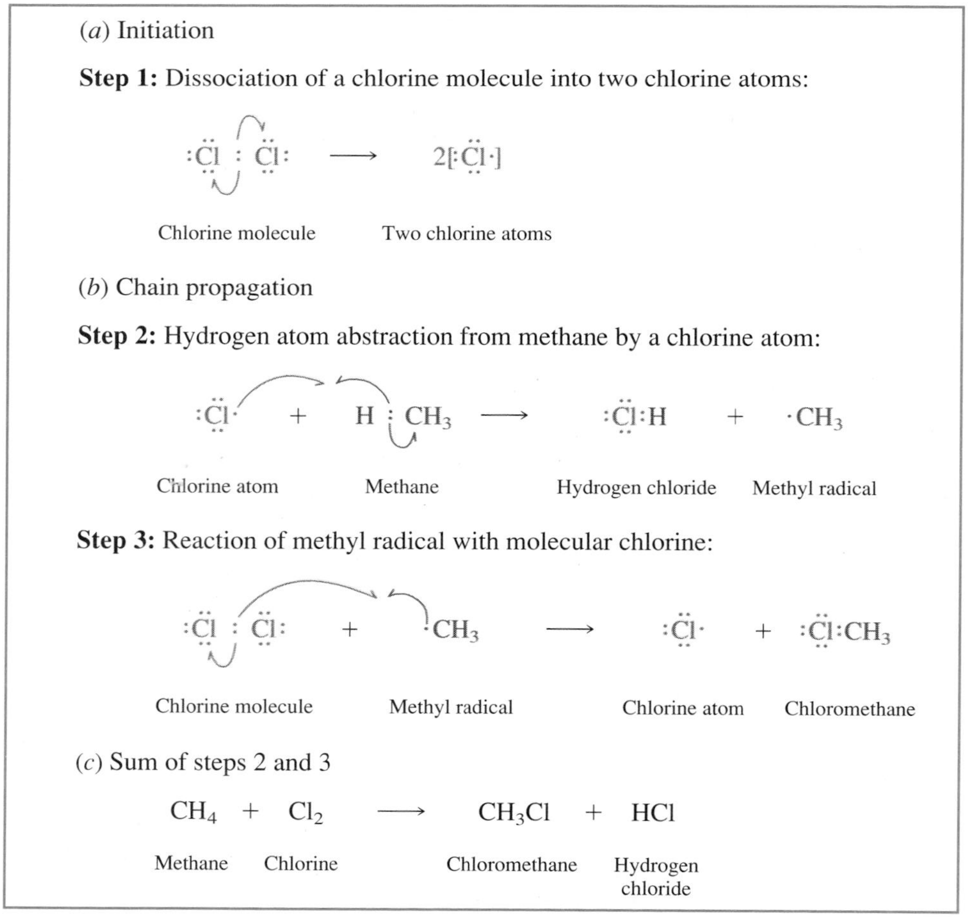

The bond dissociation energy of the other reactant, methane, is much higher. It is 435 kJ/mol (104 kcal/mol).

FIGURE 9.3   The initiation and propagation steps in the free-radical mechanism for the chlorination of methane. Together the two propagation steps give the overall equation for the reaction.

(*a*) Initiation

Step 1: Dissociation of a chlorine molecule into two chlorine atoms:

$$:\!\ddot{C}l\!:\!\ddot{C}l\!: \quad \longrightarrow \quad 2[:\!\ddot{C}l\!\cdot]$$

Chlorine molecule   Two chlorine atoms

(*b*) Chain propagation

Step 2: Hydrogen atom abstraction from methane by a chlorine atom:

$$:\!\ddot{C}l\!\cdot \quad + \quad H\!:\!CH_3 \quad \longrightarrow \quad :\!\ddot{C}l\!:\!H \quad + \quad \cdot CH_3$$

Chlorine atom   Methane   Hydrogen chloride   Methyl radical

Step 3: Reaction of methyl radical with molecular chlorine:

$$:\!\ddot{C}l\!:\!\ddot{C}l\!: \quad + \quad \cdot CH_3 \quad \longrightarrow \quad :\!\ddot{C}l\!\cdot \quad + \quad :\!\ddot{C}l\!:\!CH_3$$

Chlorine molecule   Methyl radical   Chlorine atom   Chloromethane

(*c*) Sum of steps 2 and 3

$$CH_4 \quad + \quad Cl_2 \quad \longrightarrow \quad CH_3Cl \quad + \quad HCl$$

Methane   Chlorine   Chloromethane   Hydrogen
                                              chloride

the overall reaction, is formed in this step. A methyl radical is also formed, which then attacks a molecule of Cl_2 in step 3. Attack of methyl radical on Cl_2 gives chloromethane, the other product of the overall reaction, along with a chlorine atom which then cycles back to step 2, repeating the process. Steps 2 and 3 are called the **propagation steps** of the reaction and, when added together, give the overall equation for the reaction. Because one initiation step can result in a great many propagation cycles, the overall process is called a **free-radical chain reaction.**

> Chlorination of methane is an example of a **substitution** reaction, but the mechanism is very different from the substitutions we saw in Chapter 8.

PROBLEM 9.5   Write equations for the initiation and propagation steps for the formation of dichloromethane by free-radical chlorination of chloromethane.

In practice, side reactions intervene to reduce the efficiency of the propagation steps. The chain sequence is interrupted whenever two odd-electron species combine to give an even-electron product. Reactions of this type are called **chain-terminating steps.** Some commonly observed chain-terminating steps in the chlorination of methane are shown in the following equations.

Combination of a methyl radical with a chlorine atom:

$$\overset{\cdot}{C}H_3 \quad \cdot \overset{\cdot\cdot}{\underset{\cdot\cdot}{C}l}: \quad \longrightarrow \quad CH_3\overset{\cdot\cdot}{\underset{\cdot\cdot}{C}l}:$$

Methyl radical   Chlorine atom   Chloromethane

HALOGENATED HYDROCARBONS AND THE ENVIRONMENT

Until scientists started looking specifically for them, it was widely believed that naturally occurring organohalogen compounds were rare. We now know that more than 2000 such compounds occur naturally, with the oceans being a particularly rich source. Over 50 organohalogen compounds, including $CHBr_3$, $CHBrClI$, $BrCH_2CH_2I$, CH_2I_2, $Br_2CHCH{=}O$, I_2CHCO_2H, and $(Cl_3C)_2C{=}O$, have been found in a single species of Hawaiian red seaweed, for example. It is not surprising that organisms living in the oceans have adapted to their halide-rich environment by incorporating chlorine, bromine, and iodine into their metabolic processes. Chloromethane (CH_3Cl), bromomethane (CH_3Br), and iodomethane (CH_3I) are all produced by marine algae and kelp, but land-based plants and fungi also contribute their share to the more than 5 million tons of the methyl halides formed each year by living systems. The ice plant, which grows in arid regions throughout the world and is cultivated as a ground cover along coastal highways in California, biosynthesizes CH_3Cl, for example.

Although for the most part they have been replaced by safer alternatives, at one time many pesticides were halogenated hydrocarbons. Some, like DDT, were used on a massive scale.

Dichlorodiphenyltrichloroethane
(DDT)

DDT was introduced as a pesticide to control the mosquitoes that transmit malaria and was exceedingly successful in this role for about 30 years. Eventually, however, resistance to DDT emerged in the pests it was supposed to control. In the 1960s it became apparent that species of birds such as the peregrine falcon and the bald eagle were threatened because DDT had accumulated in their fatty tissue causing them to lay eggs with thin shells that broke prematurely. The use of DDT then became subject to strict regulation.

The effect on the environment of the class of halogenated hydrocarbons known as chlorofluorocarbons is discussed in Section 9.6.

Combination of two methyl radicals:

$$\overset{\cdot}{C}H_3 \quad \overset{\cdot}{C}H_3 \longrightarrow CH_3CH_3$$

Two methyl radicals   Ethane

Combination of two chlorine atoms:

$$:\overset{\cdot\cdot}{\underset{\cdot\cdot}{Cl}}\cdot \quad \cdot\overset{\cdot\cdot}{\underset{\cdot\cdot}{Cl}}: \longrightarrow \quad Cl_2$$

Two chlorine atoms   Chlorine molecule

Termination steps are, in general, less likely to occur than the propagation steps. Each of the termination steps requires two free radicals to encounter each other in a medium that contains far greater quantities of other materials (methane and chlorine molecules) with which they can react. Although some chloromethane undoubtedly arises via direct combination of methyl radicals with chlorine atoms, most of it is formed by the propagation sequence shown in Figure 9.3.

9.5   HALOGENATION OF HIGHER ALKANES

Like the chlorination of methane, chlorination of ethane is carried out on an industrial scale as a high-temperature gas-phase reaction.

$$CH_3CH_3 + \quad Cl_2 \quad \xrightarrow{420^{\circ}C} \quad CH_3CH_2Cl \quad + \quad HCl$$

Ethane   Chlorine   Chloromethane   Hydrogen chloride
                                        (78%)

As in the chlorination of methane, it is often difficult to limit the reaction to monochlorination, and derivatives having more than one chlorine atom are also formed.

> **PROBLEM 9.6** Chlorination of ethane yields, in addition to ethyl chloride, a mixture of two isomeric dichlorides. What are the structures of these two dichlorides?

In the laboratory it is more convenient to use light, either visible or ultraviolet, as the source of energy to initiate the reaction. Reactions that occur when light energy is absorbed by a molecule are called **photochemical reactions.** Photochemical techniques permit the reaction of alkanes with chlorine to be performed at room temperature.

Photochemical energy is indicated by writing "light" or "*hv*" above the arrow. The symbol *hv* is equal to the energy of a light photon.

$$\diamondsuit \quad + \quad Cl_2 \quad \xrightarrow{hv} \quad \diamondsuit\!-Cl \quad + \quad HCl$$

Cyclobutane   Chlorine   Chlorocyclobutane   Hydrogen
                                        (73%)   chloride

Methane, ethane, and cyclobutane share the common feature that each one can give only a *single* monochloro derivative. All the hydrogens of cyclobutane, for example, are equivalent, and substitution of any one gives the same product as substitution of any other. Chlorination of alkanes in which all the hydrogens are not equivalent is more complicated in that a mixture of every possible monochloro derivative is formed, as the chlorination of butane illustrates:

$$CH_3CH_2CH_2CH_3 \xrightarrow[hv,\ 35^\circ C]{Cl_2} CH_3CH_2CH_2CH_2Cl \ + \ CH_3\underset{\underset{Cl}{|}}{C}HCH_2CH_3$$

| Butane | 1-Chlorobutane (28%) | 2-Chlorobutane (72%) |

The percentages cited in this equation reflect the composition of the monochloride fraction of the product mixture rather than the isolated yield of each component.

Bromine reacts with alkanes by a free-radical chain mechanism analogous to that of chlorine. Unlike chlorination, however, bromination is a highly selective process. When an alkane contains primary, secondary, and tertiary hydrogens, it is usually only the tertiary hydrogen that is replaced by bromine.

$$CH_3\underset{\underset{CH_3}{|}}{\overset{\overset{H}{|}}{C}}CH_2CH_2CH_3 \ + \ Br_2 \xrightarrow{\underset{60^\circ C}{hv}} CH_3\underset{\underset{CH_3}{|}}{\overset{\overset{Br}{|}}{C}}CH_2CH_2CH_3 \ + \ HBr$$

| 2-Methylpentane | Bromine | 2-Bromo-2-methylpentane (76% isolated yield) | Hydrogen bromide |

The yield cited in this reaction is the isolated yield of purified product. Isomeric bromides constitute only a tiny fraction of the crude product.

PROBLEM 9.7 Give the structure of the principal organic product formed by free-radical bromination of each of the following:

(a) Methylcyclopentane   (c) 2,2,4-Trimethylpentane

(b) 1-Isopropyl-1-methylcyclopentane

Sample Solution (a) Write the structure of the starting hydrocarbon, and identify any tertiary hydrogens that are present. The only tertiary hydrogen in methylcyclopentane is the one attached to C-1. This is the one replaced by bromine.

| Methylcyclopentane | 1-Bromo-1-methylcyclopentane |

This difference in selectivity between chlorination and bromination of alkanes needs to be kept in mind when one wishes to prepare an alkyl halide from an alkane:

1. Because chlorination of an alkane yields every possible monochloride, it is used only when all the hydrogens in an alkane are equivalent.

2. Bromination is normally used only to prepare tertiary alkyl bromides from alkanes.

9.6   CHLOROFLUOROCARBONS AND THE ENVIRONMENT

For many years halogenated derivatives of methane and ethane have been used as aerosol propellants and refrigerants. Typical examples of these **chlorofluorocarbons,** or **CFCs,** include:

CFCs are widely known by the DuPont trade name Freon.

$$CCl_2F_2 \qquad\qquad CClF_2CClF_2$$

| Dichlorodifluoromethane (CFC-12) | 1,2-Dichloro-1,1,2,2-tetrafluoroethane (CFC-114) |

These compounds have the advantage of being chemically inert and nontoxic. Their lack of chemical reactivity allows them to persist in the environment, however, and as gases they are able to diffuse into the stratosphere. There they play a role in depletion of the ozone layer. Under the influence of sunlight, the carbon–chlorine bond cleaves to give a chlorine atom.

$$R—Cl \xrightarrow{hv} R\cdot \ + \ Cl\cdot$$

This chlorine atom triggers the decomposition of ozone by a free-radical chain reaction. By international agreement, the manufacture of CFCs was halted in January 1996. CFCs are being replaced by environmentally safer alternatives, hydrochlorofluorocarbons, HCFCs. Examples include:

$$CHCl_2CF_3 \qquad\qquad CHClFCF_3$$

2,2-Dichloro-1,1,1-trifluoroethane   2-Chloro-1,1,1,2-tetrafluoroethane
(HCFC-123); a replacement for CFC-11   (HCFC-124); a replacement for CFC-114

Notice that HCFCs possess a hydrogen that CFCs lack. This causes HCFCs to be destroyed by air oxidation before they reach the ozone layer. The ozone depletion potential of a typical HCFC is estimated at less than 2% of that of a CFC. Even HCFCs are scheduled to be phased out by 2030, however.

9.7   FREE-RADICAL ADDITION OF HYDROGEN BROMIDE TO ALKENES

In Chapter 5 (Sections 5.2 through 5.4) you learned that hydrogen halides (HCl, HBr, HI) undergo electrophilic addition to alkenes. Carbocations are intermediates, and the reactions proceed by an ionic mechanism. The more stable carbocation is formed, causing hydrogen halide addition to be regioselective and favoring formation of the more substituted alkyl halide (Markovnikov's rule).

Alkenes may also undergo free-radical addition under certain conditions. An example is the peroxide-induced addition of hydrogen bromide. **Peroxides** are compounds of the type ROOR that can initiate a free-radical chain reaction. In the presence of peroxides, hydrogen bromide (but none of the other hydrogen halides) adds to alkenes with a regioselectivity *opposite* to that of Markovnikov's rule.

Addition of HBr in the absence of peroxides:

$$CH_2\!\!=\!\!CHCH_2CH_3 + \qquad HBr \quad \xrightarrow[\text{peroxides}]{\text{no}} \quad CH_3CHCH_2CH_3$$
$$\overset{|}{Br}$$

1-Butene   Hydrogen bromide   2-Bromobutane
                                              (only product; 90% yield)

Addition of HBr in the presence of peroxides:

$$CH_2\!\!=\!\!CHCH_2CH_3 + \qquad HBr \quad \xrightarrow{\text{peroxides}} \quad BrCH_2CH_2CH_2CH_3$$

1-Butene   Hydrogen bromide   1-Bromobutane
                                              (only product; 95% yield)

> **PROBLEM 9.8** Write an equation for the addition of hydrogen bromide to 2-methyl-2-butene in the presence of peroxides.

The mechanism of the free-radical addition of hydrogen bromide to 1-butene is presented in Figure 9.4. The regioselectivity of addition is determined in step 3 of the mechanism in which a bromine atom adds to the alkene. The bromine adds to the double bond in the direction that leads to the more stable of two possible free-radical intermediates. In the case of 1-butene, the bromine atom adds to C-1 because this generates a secondary radical. Had bromine added to C-2, the free radical produced would have been a less stable primary radical. Once the carbon–bromine bond is formed in step 3, hydrogen abstraction from HBr in step 4 completes the process.

The overall reaction:

$$CH_3CH_2CH=CH_2 \quad + \quad HBr \quad \xrightarrow[\text{light or heat}]{ROOR} \quad CH_3CH_2CH_2CH_2Br$$

1-Butene   Hydrogen bromide   1-Bromobutane

The mechanism:

(*a*) Initiation

Step 1: Dissociation of a peroxide into two alkoxy radicals:

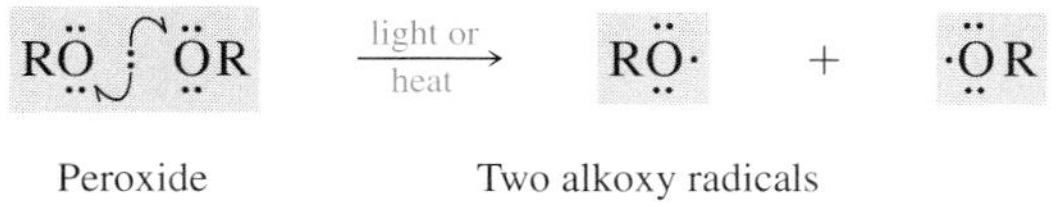

$$R\ddot{\underset{\cdot\cdot}{O}} \overset{\frown}{\underset{\cdot\cdot}{O}}R \quad \xrightarrow[\text{heat}]{\text{light or}} \quad R\ddot{\underset{\cdot\cdot}{O}}\cdot \quad + \quad \cdot\ddot{\underset{\cdot\cdot}{O}}R$$

Peroxide   Two alkoxy radicals

Step 2: Hydrogen atom abstraction from hydrogen bromide by an alkoxy radical:

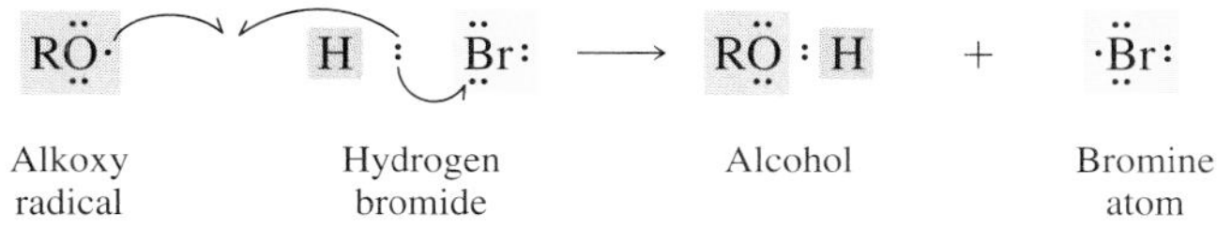

$$R\ddot{\underset{\cdot\cdot}{O}}\cdot \quad \overset{\frown}{\quad} \quad H \overset{\cdot\cdot}{:} \ddot{\underset{\cdot\cdot}{Br}}: \quad \longrightarrow \quad R\ddot{\underset{\cdot\cdot}{O}}:H \quad + \quad \cdot\ddot{\underset{\cdot\cdot}{Br}}:$$

Alkoxy   Hydrogen   Alcohol   Bromine
radical   bromide   atom

(*b*) Chain propagation

Step 3: Addition of a bromine atom to the alkene:

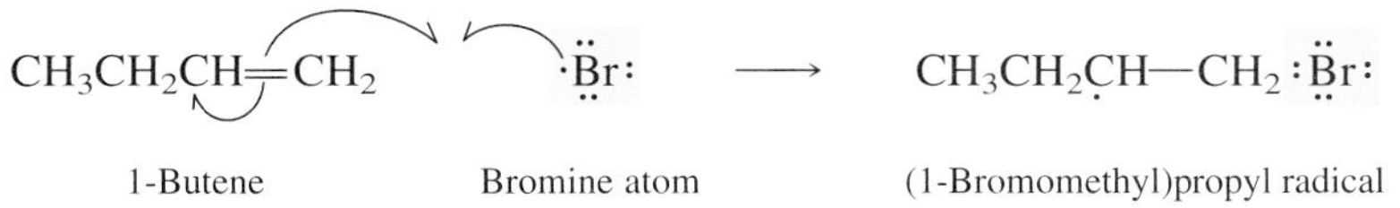

$$CH_3CH_2CH{=}CH_2 \quad \overset{\frown}{\quad} \quad \cdot\ddot{\underset{\cdot\cdot}{Br}}: \quad \longrightarrow \quad CH_3CH_2\dot{C}H{-}CH_2:\ddot{\underset{\cdot\cdot}{Br}}:$$

1-Butene   Bromine atom   (1-Bromomethyl)propyl radical

Step 4: Abstraction of a hydrogen atom from hydrogen bromide by the free radical formed in step 3:

$$CH_3CH_2\dot{C}H{-}CH_2Br \quad \overset{\frown}{\quad} \quad H \overset{\cdot\cdot}{:} \ddot{\underset{\cdot\cdot}{Br}}: \quad \longrightarrow \quad CH_3CH_2CH_2CH_2Br \quad + \quad \cdot\ddot{\underset{\cdot\cdot}{Br}}:$$

(1-Bromomethyl)propyl   Hydrogen   1-Bromobutane   Bromine
radical   bromide   atom

FIGURE 9.4   Initiation and propagation steps in the free-radical addition of hydrogen bromide to 1-butene.

The regioselectivity of addition of hydrogen bromide to alkenes under normal (ionic addition) conditions is controlled by the tendency of a **proton** to add to the double bond so as to produce the more stable **carbocation.** Under free-radical conditions the regioselectivity is governed by addition of a **bromine atom** to give the more stable **alkyl radical.**

PROBLEM 9.9 Write the propagation steps for the free-radical addition of hydrogen bromide to 2-methyl-2-butene.

9.8   POLYMERIZATION OF ALKENES

Peroxides can initiate other free-radical addition reactions of alkenes. One that is exceptionally important is free-radical **polymerization** [from the Greek *poly* ("many") and *meros* ("parts")], a reaction in which alkene molecules add to one another repeatedly to

One of the most frequently recycled plastics is high-density polyethylene, HDPE, used in milk jugs, trash bags, detergent bottles, etc. A container made of HDPE can be identified by the number "2" found on the bottom.

form long chains. Most of the enormous volume of ethylene produced by the petrochemical industry (see boxed material in Sections 4.1 and 5.6) is used to prepare **polyethylene,** a high-molecular-weight polymer formed by heating ethylene at high pressure in the presence of oxygen or a peroxide.

$$n CH_2{=}CH_2 \xrightarrow[\substack{O_2 \text{ or} \\ peroxides}]{\substack{200°C \\ 2000 \text{ atm}}} {-}CH_2{-}CH_2{-}(CH_2{-}CH_2)_{n-2}{-}CH_2{-}CH_2{-}$$

Ethylene   Polyethylene

In this reaction n can have a value of thousands.

The mechanism of free-radical polymerization of ethylene is outlined in Figure 9.5. Dissociation of a peroxide initiates the process in step 1. The resulting peroxy radical adds to the carbon–carbon double bond in step 2, giving a new radical, which then adds to a second molecule of ethylene in step 3. The carbon–carbon bond-forming process in step 3 can be repeated thousands of times to give long carbon chains.

In spite of the *-ene* ending to its name, polyethylene is much more closely related to alk*anes* than to alk*enes*. It is simply a long chain of CH_2 groups bearing at its ends an alkoxy group (from the initiator) or a carbon–carbon double bond.

A large number of compounds with carbon–carbon double bonds have been polymerized to yield materials having useful properties. Some of the more important or familiar of these are listed in Table 9.2. Not all these monomers are effectively polymerized under free-radical conditions, and much research has been carried out to develop alternative polymerization techniques. One of these, **coordination polymerization,** employs a mixture of titanium tetrachloride, $TiCl_4$, and triethylaluminum, $(CH_3CH_2)_3Al$, as a catalyst. Polyethylene produced by coordination polymerization has a higher density than that

FIGURE 9.5 Mechanism of peroxide-initiated free-radical polymerization of ethylene.

Step 1: Homolytic dissociation of a peroxide produces alkoxy radicals that serve as free-radical initiators:

RÖ : ÖR ⟶ RÖ· + ·ÖR

Peroxide   Two alkoxy radicals

Step 2: An alkoxy radical adds to the carbon–carbon double bond:

RÖ· + $CH_2{=}CH_2$ ⟶ RÖ—CH_2—$\dot{C}H_2$

Alkoxy   Ethylene   2-Alkoxyethyl
radical   radical

Step 3: The radical produced in step 2 adds to a second molecule of ethylene:

RÖ—CH_2—$\dot{C}H_2$ + $CH_2{=}CH_2$ ⟶ RÖ—CH_2—CH_2—CH_2—$\dot{C}H_2$

2-Alkoxyethyl   Ethylene   4-Alkoxybutyl radical
radical

The radical formed in step 3 then adds to a third molecule of ethylene, and the process continues, forming a long chain of methylene groups.

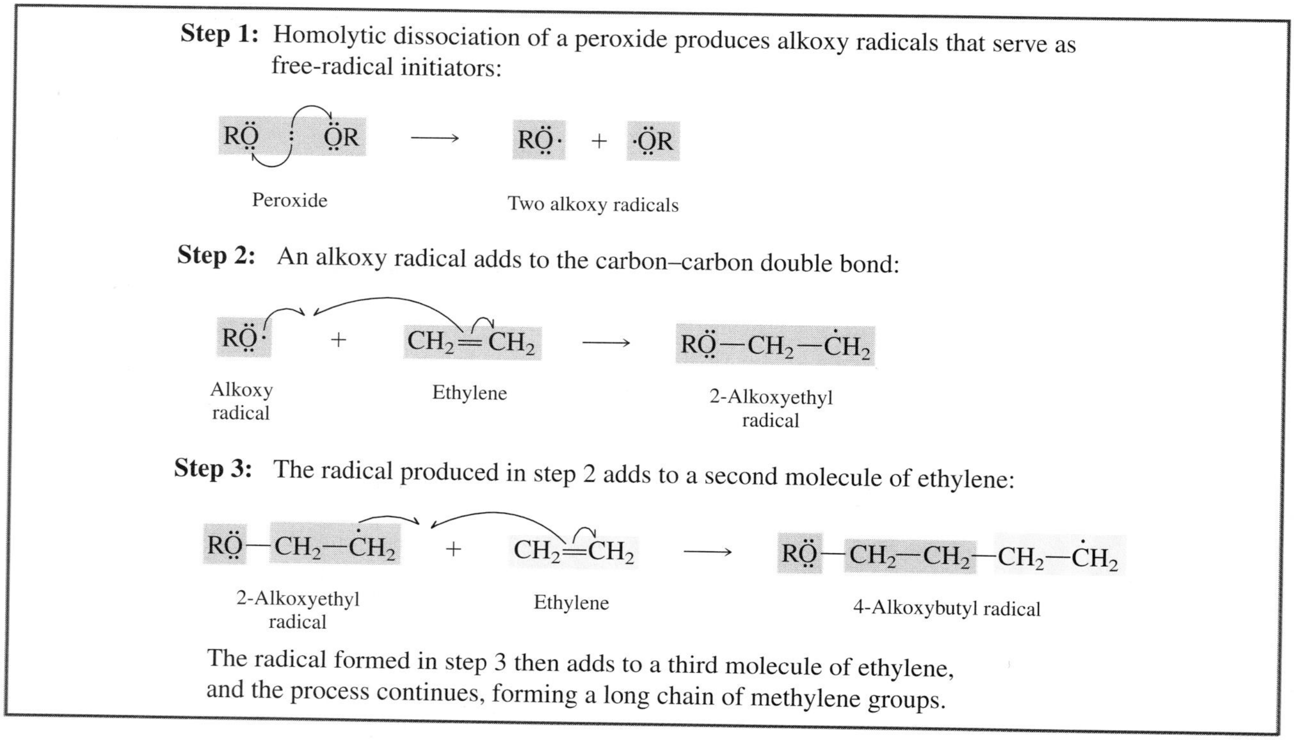

TABLE 9.2 Some Compounds with Carbon–Carbon Double Bonds Used to Prepare Polymers

A. Alkenes of the type $CH_2{=}CH{-}X$ used to form polymers of the type $(-CH_2{-}\underset{X}{CH}-)_n$

Compound	Structure	—X in Polymer	Application
Ethylene	$CH_2{=}CH_2$	—H	Polyethylene films as packaging material; "plastic" squeeze bottles are molded from high-density polyethylene
Propene	$CH_2{=}CH{-}CH_3$	—CH$_3$	Polypropylene fibers for use in carpets and automobile tires; consumer items (luggage, appliances, etc.); packaging material
Styrene	$CH_2{=}CH{-}$⬡ 	 ⬡	Polystyrene packaging, housewares, luggage, radio and television cabinets
Vinyl chloride	$CH_2{=}CH{-}Cl$	—Cl	Poly(vinyl chloride) (PVC) has replaced leather in many of its applications; PVC tubes and pipes are often used in place of copper
Acrylonitrile	$CH_2{=}CH{-}C{\equiv}N$	$-C{\equiv}N$	Wool substitute in sweaters, blankets, etc.

B. Alkenes of the type $CH_2{=}CX_2$ used to form polymers of the type $(-CH_2{-}CX_2-)_n$

Compound	Structure	X in Polymer	Application
1,1-Dichloroethene (vinylidene chloride)	$CH_2{=}CCl_2$	Cl	Saran used as air- and water-tight packaging film
2-Methylpropene	$CH_2{=}C(CH_3)_2$	CH$_3$	Polyisobutene is component of "butyl rubber," one of earliest synthetic rubber substitutes

C. Others

Compound	Structure	Polymer	Application
Tetrafluoroethene	$CF_2{=}CF_2$	$(-CF_2{-}CF_2-)_n$ (Teflon)	Nonstick coating for cooking utensils; bearings, gaskets, and fittings
Methyl methacrylate	$CH_2{=}\underset{CH_3}{\overset{}{C}}CO_2CH_3$	$(-CH_2{-}\underset{CH_3}{\overset{CO_2CH_3}{C}}-)_n$	When cast in sheets, is transparent; used as glass substitute (Lucite, Plexiglas)
2-Methyl-1,3-butadiene	$CH_2{=}\underset{CH_3}{\overset{}{C}}CH{=}CH_2$	$(-CH_2C{=}\underset{CH_3}{\overset{}{CH}}{-}CH_2-)_n$ (Polyisoprene)	Synthetic rubber

DIENE POLYMERS

Some 500 years ago during Columbus's second voyage to what are now the Americas, he and his crew saw children playing with balls made from the latex of trees that grew there. Later, Joseph Priestley called this material "rubber" to describe its ability to erase pencil marks by rubbing, and in 1823 Charles Macintosh demonstrated how rubber could be used to make waterproof coats and shoes. Shortly thereafter Michael Faraday determined an empirical formula of C_5H_8 for rubber. It was eventually determined that rubber is a polymer of 2-methyl-1,3-butadiene.

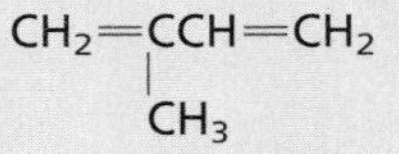

$$CH_2=CCH=CH_2$$
$$|$$
$$CH_3$$

2-Methyl-1,3-butadiene
(common name: **isoprene**)

The structure of rubber corresponds to 1,4 addition of several thousand isoprene units to one another:

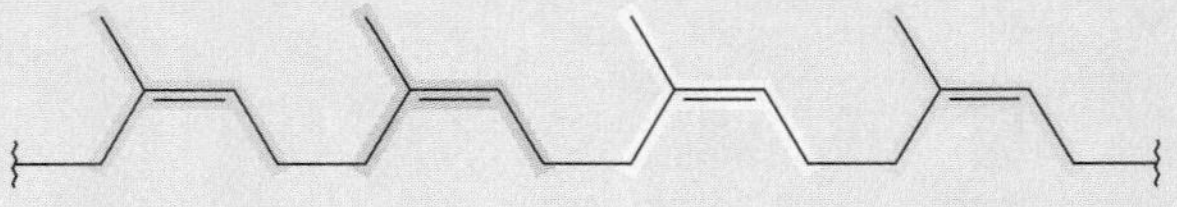

All the double bonds in rubber have the Z (or cis) configuration. A different polymer of isoprene, called *gutta-percha*, has shorter polymer chains and E (or trans) double bonds. Gutta-percha is a tough, horn-like substance once used as a material for golf ball covers.

In natural rubber the attractive forces between neighboring polymer chains are relatively weak, and little overall structural order exists. The chains slide easily past one another when stretched and return, in time, to their disordered state when the distorting force is removed. The ability of a substance to recover its original shape after distortion is its **elasticity.** The elasticity of natural rubber is satisfactory only within a limited temperature range; it is too rigid when cold and too sticky when warm to be very useful. Rubber's elasticity is improved by **vulcanization,** a process discovered by Charles Goodyear in 1839. When natural rubber is heated with sulfur, a chemical reaction occurs in which neighboring polyisoprene chains become connected through covalent bonds to sulfur. Although these sulfur "bridges" permit only limited

movement of one chain with respect to another, their presence ensures that the rubber will snap back to its original shape once the distorting force is removed.

As the demand for rubber increased, so did the chemical industry's efforts to prepare a synthetic substitute. One of the first **elastomers** (a synthetic polymer that possesses elasticity) to find a commercial niche was **neoprene,** discovered by chemists at Du Pont in 1931. Neoprene is produced by free-radical polymerization of 2-chloro-1,3-butadiene and has the greatest variety of applications of any elastomer. Some uses include electrical insulation, conveyer belts, hoses, and weather balloons.

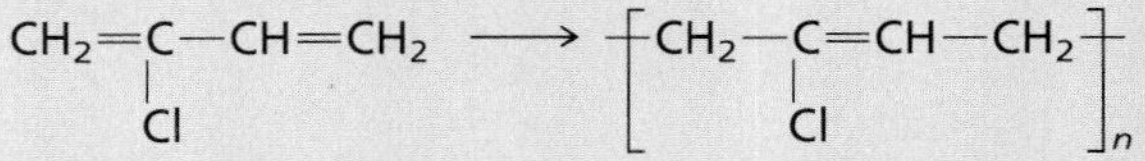

2-Chloro-1,3-butadiene   Neoprene

The elastomer produced in greatest amount is **styrene-butadiene rubber (SBR).** Annually, just under 10^9 lb of SBR is produced in the United States, and almost all of it is used in automobile tires. As its name suggests, SBR is prepared from styrene and 1,3-butadiene. It is an example of a **copolymer,** a polymer assembled from two or more different monomers. Free-radical polymerization of a mixture of styrene and 1,3-butadiene gives SBR.

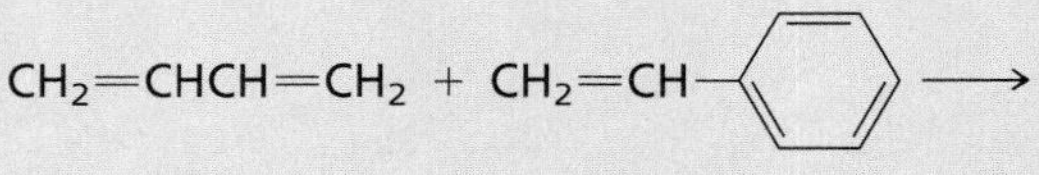

1,3-Butadiene   Styrene

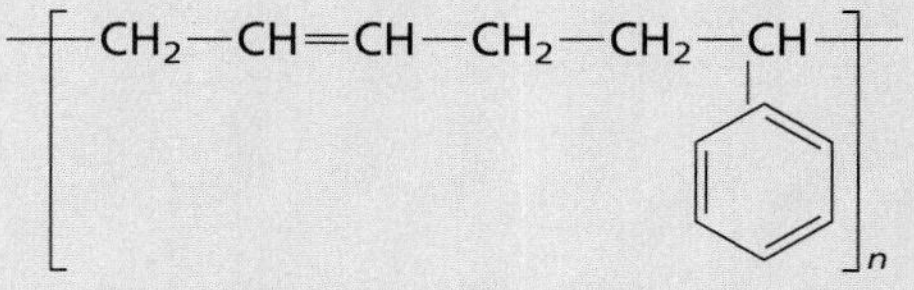

Styrene-butadiene rubber

Coordination polymerization of isoprene using Ziegler–Natta catalyst systems (Section 9.8) gives a material similar in properties to natural rubber, as does polymerization of 1,3-butadiene. Poly(1,3-butadiene) is produced in about two thirds the quantity of SBR each year. It, too, finds its principal use in tires.

produced by free-radical polymerization and somewhat different—in many applications, more desirable—properties. The catalyst system used in coordination polymerization was developed independently by Karl Ziegler in Germany and Giulio Natta in Italy in the early 1950s. They shared the Nobel Prize in chemistry in 1963 for this work. The Ziegler–Natta catalyst system gives a form of **polypropylene** suitable for plastics and fibers. When propene is polymerized under free-radical conditions, the polypropylene has physical properties (such as a low melting point) that make it useless for most applications.

LEARNING OBJECTIVES

This chapter has focused on intermediates known as free radicals. The skills you have learned in this chapter should enable you to:

- Describe the bonding model for a free radical.
- Explain what is meant by the term *bond dissociation energy,* and discuss how it relates to free-radical stability.
- Give the stability trend for a series of alkyl free radicals.
- Write chemical equations describing the mechanism of the reaction of an alkane with chlorine.
- Explain what is meant by the terms *initiation, propagation,* and *termination.*
- Give the major product from bromination of an alkane.
- Write a chemical equation describing the addition of hydrogen bromide to an alkene in the presence of a peroxide.

9.9 SUMMARY

Free radicals are neutral intermediates characterized by an unpaired electron (Section 9.1). Carbon free radicals are sp^2-hybridized as depicted in the figure.

Like carbocations, free radicals are stabilized by alkyl substituents. The order of free-radical stability parallels that of carbocation stability.

The energy required for homolytic bond cleavage is the **bond dissociation energy (BDE)** (Section 9.2). The elementary steps (1) through (3) describe a **free-radical chain mechanism** for the reaction of an alkane with a halogen (Section 9.4).

1. Initiation step $\qquad X_2 \quad \longrightarrow \quad 2X\cdot$

               Halogen molecule   Two halogen atoms

2. Propagation step $\qquad RH \ + \ X\cdot \ \longrightarrow \ R\cdot \ + \ HX$

                   Alkane   Halogen   Alkyl   Hydrogen

                            atom   radical   halide

3. Propagation step   R· + X$_2$ ⟶ RX + ·X

Alkyl   Halogen   Alkyl   Halogen
radical   molecule   halide   atom

The combination of any two radical species is a **termination step** that interrupts the chain process.

Termination step   P· + Q· ⟶ P—Q

Chlorination of alkanes occurs with a low degree of selectivity (Section 9.5). A mixture of every possible monochloride is usually formed from alkanes with nonequivalent hydrogens. Thus chlorination is most often used when all the hydrogens of the alkane are equivalent.

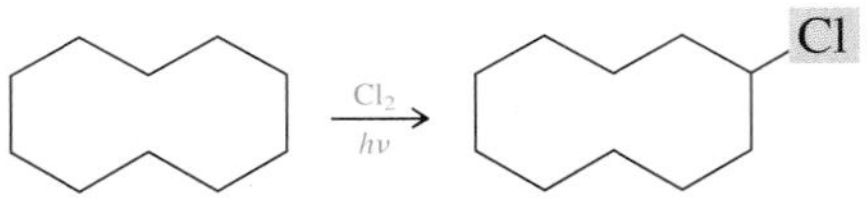

Cyclodecane   Cyclodecyl chloride
(64%)

Bromination, however, is highly selective for replacing tertiary hydrogens.

$$(CH_3)_2CHC(CH_3)_3 \xrightarrow[h\nu]{Br_2} (CH_3)_2CC(CH_3)_3$$
$$\underset{Br}{|}$$

2,2,3-Trimethylbutane   2-Bromo-2,3,3-
trimethylbutane
(80%)

Hydrogen bromide reacts with alkenes in the presence of peroxides by a free-radical mechanism (Section 9.7). The regioselectivity of the reaction is *opposite* to that predicted by Markovnikov's rule. That is, the less-substituted alkyl bromide forms as the major product.

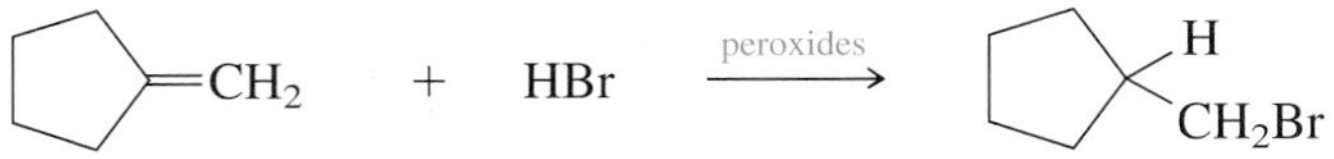

Methylenecyclopentane   Hydrogen   (Bromomethyl)cyclopentane
bromide   (60%)

ADDITIONAL PROBLEMS

Free-Radical Halogenation

9.10 Write structural formulas for all the isomeric alkyl free radicals of formula C_4H_9. Which one is the most stable?

9.11 Write the structures of all the possible monochlorides formed during the photochemical chlorination of $(CH_3)_3CCH_2CH_3$.

9.12 Among the isomeric alkanes of molecular formula C_5H_{12}, identify the one that on reaction with chlorine in the presence of light yields only:

    (a) A single monochloride
    (b) Three isomeric monochlorides
    (c) Four isomeric monochlorides
    (d) Two isomeric dichlorides

9.13 Write out the steps describing the initiation and propagation steps of the free-radical bromination of ethane.

9.14 Cyclopropyl chloride has been prepared by the free-radical chlorination of cyclopropane. Write a stepwise mechanism for this reaction.

9.15 Photochemical chlorination of 2,2,4-trimethylpentane gives four isomeric monochlorides, whereas bromination yields only a single monobromide. Write structural formulas for these products and explain the difference in reactivity between chlorination and bromination.

9.16 In a search for fluorocarbons having anesthetic properties, 1,2-dichloro-1,1-difluoropropane was subjected to photochemical chlorination. Two isomeric products were obtained, one of which was identified as 1,2,3-trichloro-1,1-difluoropropane. What is the structure of the second compound?

9.17 Two isomeric compounds A and B have the molecular formula C_3H_7Cl. Chlorination of A gave a mixture of two dichlorides of formula $C_3H_6Cl_2$. Chlorination of B gave three different compounds of formula $C_3H_6Cl_2$ (although these may not all be different from the dichlorides from A). What are the structural formulas of A and B and the dichlorides obtained from each?

Free-Radical Additions

9.18 Write the structure of the major product formed in the reaction of each of the following alkenes with hydrogen bromide in the presence of peroxides.
  (a) 1-Pentene
  (b) 1-Methylcyclohexene

9.19 Outline the mechanism for the reaction in part (a) of the previous problem. Explain how the reaction differs from addition of hydrogen bromide in the absence of peroxides.

9.20 Suggest a sequence of reactions suitable for preparing each of the following compounds from the indicated starting material. You may use any necessary organic or inorganic reagents.
  (a) 1-Bromopropane from 2-bromopropane
  (b) 2-Bromo-3-ethylpentane from 3-ethylpentane

  (c)  CN from cyclopentane

9.21 Write the steps that describe the mechanism for the first two carbon–carbon bond-forming processes in the free-radical polymerization of styrene, $C_6H_5CH{=}CH_2$. Assume the initiator is RO·.

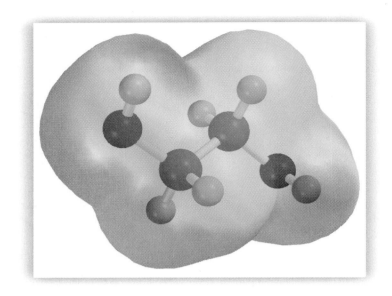

CHAPTER 10
ALCOHOLS, ETHERS, AND PHENOLS

The next several chapters deal with the chemistry of various oxygen-containing functional groups. The interplay of these important classes of compounds—alcohols, ethers, aldehydes, ketones, carboxylic acids, and derivatives of carboxylic acids—is fundamental to organic chemistry and biochemistry.

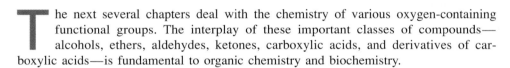

We'll start by discussing a class of compound familiar to you from Chapter 3, alcohols. With this chapter we will extend our knowledge of alcohols, particularly with respect to their relationship to carbonyl-containing compounds. In addition we will discuss two related classes of compounds: ethers and phenols.

10.1 SOURCES OF ALCOHOLS

At one time the major source of **methanol** was as a byproduct in the production of charcoal from wood—hence its common name **wood alcohol.** Today most of the 10 billion pounds of methanol used annually in the United States is synthetic in origin. Methanol is highly poisonous; drinking as little as 30 mL can be fatal. Ingestion of smaller amounts can lead to blindness.

When vegetable matter ferments, its carbohydrates are converted to **ethanol** and carbon dioxide by enzymes present in yeast. Fermentation of barley produces beer; grapes give wine. The maximum ethanol content is about 15%, because higher concentrations inactivate the enzymes, halting fermentation. Because ethanol boils at 78°C and water at 100°C, distillation of the fermentation broth can be used to give "distilled spirits" of increased ethanol content. Whiskey is the aged distillate of fermented grain and contains slightly less than 50% ethanol. Brandy and cognac are made by aging the distilled spirits from fermented grapes and other fruits. The characteristic flavors, odors, and colors of the various alcoholic beverages depend on both their origin and the way they are aged.

Synthetic ethanol is derived from petroleum by hydration of ethylene. In the United States, some 700 million pounds of synthetic ethanol is produced annually. It is relatively inexpensive and useful for industrial applications. To make it unfit for drinking, it is *denatured* by adding any of a number of noxious materials, a process that exempts it from the high taxes most governments impose on ethanol used in beverages.

Our bodies are reasonably well equipped to metabolize ethanol, making it less dangerous than methanol. Alcohol abuse and alcoholism, however, have been and remain persistent problems.

Isopropyl alcohol is prepared from petroleum by hydration of propene. With a boiling point of 82°C, isopropyl alcohol evaporates quickly from the skin, producing a cooling effect. Often containing dissolved oils and fragrances, it is the major component of rubbing alcohol. Isopropyl alcohol possesses weak antibacterial properties and is used to maintain medical instruments in a sterile condition and to clean the skin before minor surgery.

Methanol, ethanol, and isopropyl alcohol are readily available starting materials used in organic synthesis. Many other alcohols are commercially available at low cost. Some occur naturally; others are synthetic. Figure 10.1 presents the structures of a few naturally occurring alcohols.

> Some of the substances used to denature ethanol include methanol, benzene, pyridine, castor oil, and gasoline.

10.2 REACTIONS THAT YIELD ALCOHOLS: A REVIEW AND A PREVIEW

Some of the reactions used for the laboratory synthesis of alcohols have already been discussed and are reviewed in Table 10.1. Additional methods will be introduced in the sections and chapters that follow.

Several methods for preparing alcohols involve **reduction** of carbonyl groups.

> Table 10.1 illustrates a theme that you will see throughout organic chemistry: A reaction that is characteristic of *one* functional group often serves as a synthetic method for preparing *another*.

Before discussing specific reduction methods, let us review the oxidation states of carbon in organic compounds. Among one-carbon compounds, as summarized in Table 10.2, methane and carbon dioxide represent extremes in the oxidation state of carbon. Methane is the most reduced form of carbon, and carbon dioxide contains carbon in its most oxidized state. A useful generalization from Table 10.2 is the following:

Oxidation of carbon corresponds to an increase in the number of bonds between carbon and oxygen or to a decrease in the number of carbon–hydrogen bonds. Conversely, *reduction corresponds to an increase in the number of carbon–hydrogen bonds or to a decrease in the number of carbon–oxygen bonds.*

From Table 10.2 it can be seen that each successive increase in oxidation state increases the number of bonds between carbon and oxygen and decreases the number of carbon–hydrogen bonds. Methane has four C—H bonds and no C—O bonds; carbon dioxide has four C—O bonds and no C—H bonds.

FIGURE 10.1 Some naturally occurring alcohols.

Menthol (obtained from oil of
peppermint and used to flavor
tobacco and food)

Glucose (a carbohydrate)

Cholesterol (principal constituent of
gallstones and biosynthetic precursor
of the steroid hormones)

Citronellol (found in rose and
geranium oil and used in perfumery)

Retinol (vitamin A, an important
substance in vision)

The task of choosing the correct reagents for a particular functional group transformation is simplified if you keep in mind the progression of oxidation states of common organic substances. A change from a lower to a higher oxidation state is an **oxidation.** A reagent that brings about the oxidation of a substance is called an **oxidizing agent.** Conversely, a change from a higher to a lower oxidation state is a **reduction,** brought about with a reagent that is a **reducing agent.**

As an example using the one-carbon compounds in Table 10.2, a reduction would convert formaldehyde (an aldehyde) to methanol (an alcohol). Summarizing the relationship between functional groups:

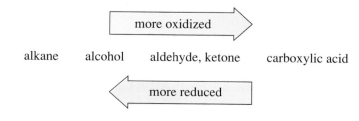

more oxidized

alkane   alcohol   aldehyde, ketone   carboxylic acid

more reduced

TABLE 10.1	Summary of Reactions Discussed in Earlier Chapters That Yield Alcohols

Reaction (section) and Comments	General Equation and Specific Example
Acid-catalyzed hydration of alkenes (Section 5.5) The elements of water add to the double bond in accordance with Markovnikov's rule.	$R_2C{=}CR_2 \ + \ H_2O \ \xrightarrow{H^+} \ R_2CHCR_2$ $\qquad\qquad\qquad\qquad\qquad$ OH Alkene $\qquad$ Water $\qquad$ Alcohol $(CH_3)_2C{=}CHCH_3 \ \xrightarrow[H_2SO_4]{H_2O} \ CH_3\overset{\displaystyle CH_3}{\underset{\displaystyle OH}{C}}CH_2CH_3$ 2-Methyl-2-butene $\qquad\qquad$ 2-Methyl-2-butanol $\qquad\qquad\qquad\qquad\qquad\qquad\qquad$ (90%)
Hydrolysis of alkyl halides (Section 8.1) A reaction useful only with substrates that do not undergo E2 elimination readily.	$RX \ + \ HO^- \ \longrightarrow \ ROH \ + \ X^-$ Alkyl $\quad$ Hydroxide $\qquad$ Alcohol $\quad$ Halide halide $\qquad$ ion $\qquad\qquad\qquad\qquad$ ion 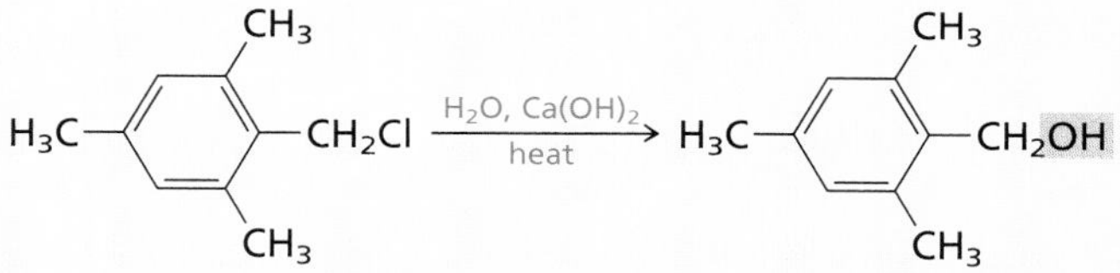 2,4,6-Trimethylbenzyl chloride $\xrightarrow[\text{heat}]{H_2O,\ Ca(OH)_2}$ 2,4,6-Trimethylbenzyl alcohol (78%)

TABLE 10.2	Oxidation State of Carbon in One-Carbon Compounds

			Number of Carbon–Oxygen Bonds	Number of Carbon–Hydrogen Bonds
Highest oxidation state	Carbon dioxide	$O{=}C{=}O$	4	0
	Formic acid	$\overset{\displaystyle O}{\overset{\displaystyle \|}{HCOH}}$	3	1
	Formaldehyde	$\overset{\displaystyle O}{\overset{\displaystyle \|}{HCH}}$	2	2
	Methanol	CH_3OH	1	3
Lowest oxidation state	Methane	CH_4	0	4

Increasing oxidation state →

PROBLEM 10.1   One type of breath analyzer used to detect drunk drivers causes the ethanol in the person's breath to undergo the chemical change shown by reaction with a chromium reagent. A color change in the chromium compound from orange to green is used to determine the amount of alcohol present.

$$CH_3CH_2OH \xrightarrow{\text{chromium reagent}} CH_3\overset{\overset{\displaystyle O}{\|}}{C}OH$$

Ethanol   Acetic acid

Classify the chemical change of ethanol as an oxidation or a reduction. Is the chromium reagent an oxidizing agent or a reducing agent?

10.3   PREPARATION OF ALCOHOLS BY REDUCTION OF ALDEHYDES AND KETONES

As described in the preceding section, transforming an aldehyde or ketone into an alcohol requires a reduction because the number of hydrogens bonded to carbon is greater in the alcohol than in the carbonyl-containing aldehyde or ketone. Reduction of an aldehyde gives a primary alcohol:

$$R\overset{\overset{\displaystyle O}{\|}}{C}H \xrightarrow{\text{reducing agent}} RCH_2OH$$

Aldehyde   Primary alcohol

Ketones yield secondary alcohols:

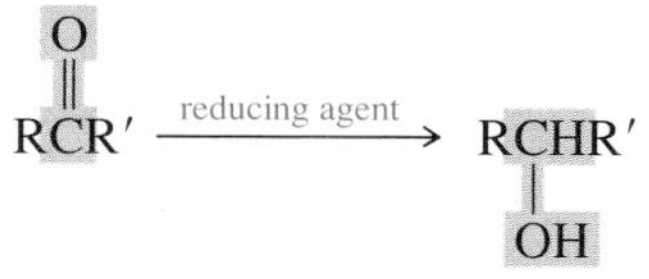

$$R\overset{\overset{\displaystyle O}{\|}}{C}R' \xrightarrow{\text{reducing agent}} R\underset{\underset{\displaystyle OH}{|}}{C}HR'$$

Ketone   Secondary alcohol

In the same way that catalytic hydrogenation reduces carbon–carbon double bonds (Section 5.1), it also reduces carbon–oxygen double bonds. Finely divided metals such as platinum, palladium, nickel, and ruthenium are effective for the hydrogenation of aldehydes and ketones. For example,

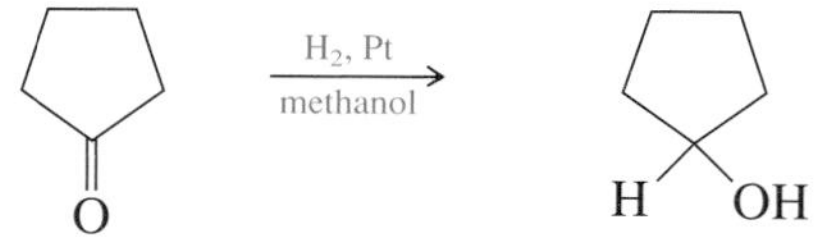

Cyclopentanone   Cyclopentanol
                                     (93–95%)

PROBLEM 10.2   Which of the isomeric $C_4H_{10}O$ alcohols can be prepared by hydrogenation of aldehydes? Which can be prepared by hydrogenation of ketones? Which cannot be prepared by hydrogenation of a carbonyl compound?

For most laboratory-scale reductions of aldehydes and ketones, catalytic hydrogenation has been replaced by methods based on **metal hydride** reducing agents. The two most common reagents are sodium borohydride and lithium aluminum hydride.

$$Na^+ \begin{bmatrix} H \\ | \\ H{-}B{=}H \\ | \\ H \end{bmatrix} \qquad Li^+ \begin{bmatrix} H \\ | \\ H{-}Al{=}H \\ | \\ H \end{bmatrix}$$

Sodium borohydride   Lithium aluminum hydride
($NaBH_4$)   ($LiAlH_4$)

Sodium borohydride is especially easy to use, needing only to be added to an aqueous or alcoholic solution of an aldehyde or a ketone:

$$\underset{\substack{\text{4,4-Dimethyl-2-pentanone}}}{CH_3\overset{\overset{\displaystyle O}{\|}}{C}CH_2C(CH_3)_3} \xrightarrow[\text{ethanol}]{NaBH_4} \underset{\substack{\text{4,4-Dimethyl-2-pentanol}\\(85\%)}}{CH_3\overset{\overset{\displaystyle OH}{|}}{C}HCH_2C(CH_3)_3}$$

Lithium aluminum hydride must be used in a different manner, however. $LiAlH_4$ reacts violently with water and alcohols and so must be used in **anhydrous** solvents such as diethyl ether. Following reduction, a separate hydrolysis step is required to liberate the alcohol product.

Anhydrous means "without water."

$$\underset{\text{Heptanal}}{CH_3(CH_2)_5\overset{\overset{\displaystyle O}{\|}}{C}H} \xrightarrow[\text{2. } H_2O]{\text{1. } LiAlH_4,\ \text{diethyl ether}} \underset{\substack{\text{1-Heptanol}\\(86\%)}}{CH_3(CH_2)_5CH_2OH}$$

Neither sodium borohydride nor lithium aluminum hydride reduces isolated carbon–carbon double bonds. This makes possible the selective reduction of a carbonyl group in a molecule that contains both carbon–carbon and carbon–oxygen double bonds.

$$\underset{\text{6-Methyl-5-hepten-2-one}}{(CH_3)_2C{=}CHCH_2CH_2\overset{\overset{\displaystyle O}{\|}}{C}CH_3} \xrightarrow[\text{2. } H_2O]{\text{1. } LiAlH_4,\ \text{diethyl ether}} \underset{\substack{\text{6-Methyl-5-hepten-2-ol}\\(90\%)}}{(CH_3)_2C{=}CHCH_2CH_2\overset{\overset{\displaystyle OH}{|}}{C}HCH_3}$$

Catalytic hydrogenation would not be suitable for this transformation, because H_2 adds to carbon–carbon double bonds faster than it reduces carbonyl groups.

PROBLEM 10.3   What is the product of each of the following reductions?

(a) $(C_6H_5)_2CH\overset{\overset{\displaystyle O}{\|}}{C}CH_3 \xrightarrow[\text{2. } H_2O]{\text{1. } LiAlH_4,\ \text{diethyl ether}}$

(b) $(CH_3)_2CHCH_2\overset{\overset{\displaystyle O}{\|}}{C}H \xrightarrow[\text{ethanol}]{H_2,\ Pt}$

(c) $\overset{\overset{\displaystyle O}{\|}}{C}H \xrightarrow[\text{CH}_3\text{OH}]{NaBH_4}$

(d) $\overset{\overset{\displaystyle O}{\|}}{C}H \xrightarrow[\text{ethanol}]{H_2\ (2\ mol),\ Pt}$

Sample Solution   Lithium aluminum hydride converts the ketone to the corresponding secondary alcohol. In this case the product is 1,1-diphenyl-2-propanol.

$$\underset{\substack{\text{1,1-Diphenyl-2-}\\\text{propanone}}}{(C_6H_5)_2CH\overset{\overset{\displaystyle O}{\|}}{C}CH_3} \xrightarrow[\text{2. } H_2O]{\text{1. } LiAlH_4,\ \text{diethyl ether}} \underset{\substack{\text{1,1-Diphenyl-2-}\\\text{propanol}\\(84\%)}}{(C_6H_5)_2CH\overset{\overset{\displaystyle OH}{|}}{C}HCH_3}$$

TABLE 10.3	Summary of Reactions of Alcohols Discussed in Earlier Chapters

Reaction (section) and Comments	General Equation and Specific Example
Reaction with hydrogen halides (Section 3.7) The order of alcohol reactivity parallels the order of carbocation stability: $R_3C^+ > R_2CH^+ > RCH_2^+$ $> CH_3^+$. Benzylic alcohols react readily.	$ROH \ + \ HX \ \longrightarrow \ RX \ + \ H_2O$ Alcohol Hydrogen halide Alkyl halide Water 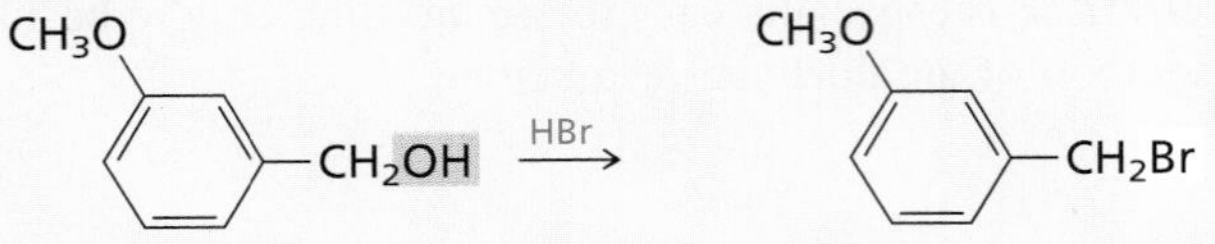 _m_-Methoxybenzyl alcohol $\xrightarrow{HBr}$ _m_-Methoxybenzyl bromide (98%)

Acid-catalyzed dehydration (Section 4.7) This is a frequently used procedure for the preparation of alkenes. The order of alcohol reactivity parallels the order of carbocation stability: $R_3C^+ > R_2CH^+ > RCH_2^+$. Benzylic alcohols react readily.	$R_2CCHR_2 \xrightarrow[\text{heat}]{H^+} R_2C{=}CR_2 \ + \ H_2O$
	           $\mid$
	           OH
	Alcohol Alkene Water 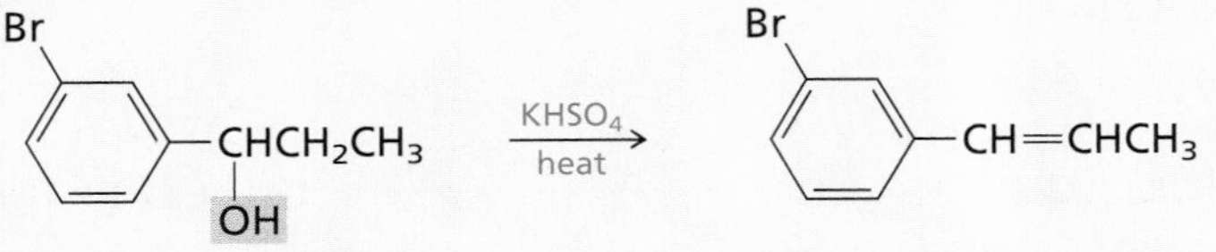
	
	1-(_m_-Bromophenyl)-1-propanol $\xrightarrow[\text{heat}]{KHSO_4}$ 1-(_m_-Bromophenyl)propene (71%)

10.4   REACTIONS OF ALCOHOLS: A REVIEW AND A PREVIEW

Alcohols are versatile starting materials for the preparation of a variety of organic compounds. We have already discussed, in previous chapters, two reactions of alcohols, which are summarized in Table 10.3.

Alcohols undergo reactions involving various combinations of the carbon–oxygen bond and the O—H bond of the hydroxyl group, illustrated as follows

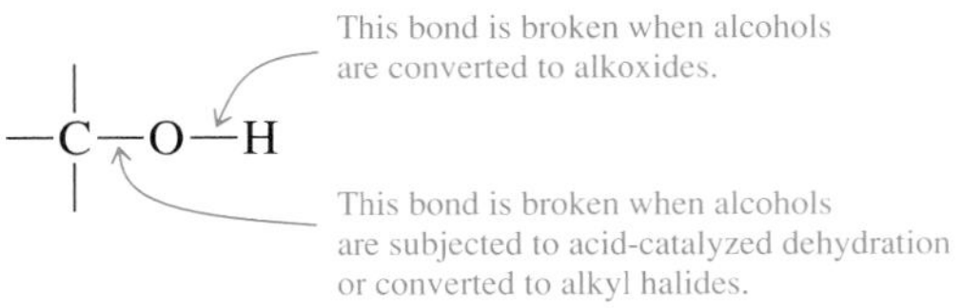

This bond is broken when alcohols are converted to alkoxides.

$$-\overset{\mid}{\underset{\mid}{C}}-O-H$$

This bond is broken when alcohols are subjected to acid-catalyzed dehydration or converted to alkyl halides.

As you will see in the next section, primary and secondary alcohols can exhibit a third reaction type, oxidation, in which a carbon–oxygen double bond is formed by cleavage of both an O—H bond and a C—H bond.

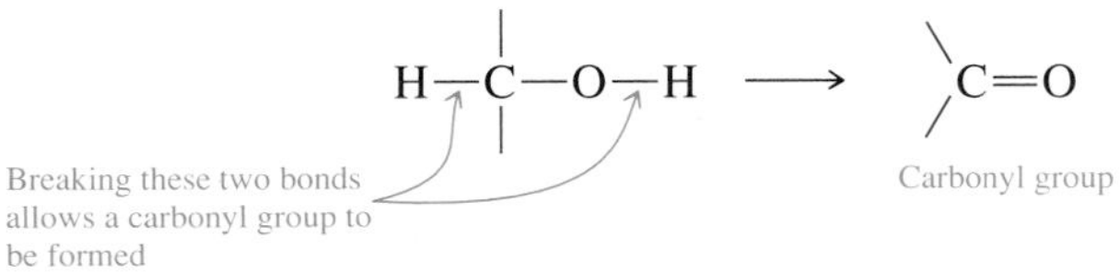

$$H-\overset{\mid}{\underset{\mid}{C}}-O-H \ \longrightarrow \ \overset{\diagdown}{\underset{\diagup}{C}}{=}O$$

Breaking these two bonds allows a carbonyl group to be formed

Carbonyl group

Other reactions of alcohols will be encountered in the chapters on aldehydes and ketones (Chapter 11) and carboxylic acid derivatives (Chapter 13).

10.5   OXIDATION OF ALCOHOLS

Oxidation of an alcohol yields a carbonyl compound (Section 10.2). Whether the resulting carbonyl compound is an aldehyde, a ketone, or a carboxylic acid depends on the alcohol and on the oxidizing agent.

Primary alcohols may be oxidized either to an aldehyde or to a carboxylic acid:

$$RCH_2OH \xrightarrow{\text{oxidize}} \underset{\text{Aldehyde}}{\overset{\overset{\displaystyle O}{\|}}{RCH}} \xrightarrow{\text{oxidize}} \underset{\text{Carboxylic acid}}{\overset{\overset{\displaystyle O}{\|}}{RCOH}}$$

Primary alcohol

Vigorous oxidation leads to the formation of a carboxylic acid, but a number of methods permit us to stop the oxidation at the intermediate aldehyde stage. The reagents most commonly used for oxidizing alcohols are based on high-oxidation-state transition metals, particularly chromium(VI).

Chromic acid (H_2CrO_4) is a good oxidizing agent and is formed when solutions containing chromate (CrO_4^{2-}) or dichromate $(Cr_2O_7^{2-})$ are acidified. Sometimes it is possible to obtain aldehydes in satisfactory yield before they are further oxidized, but in most cases carboxylic acids are the major products isolated on treatment of primary alcohols with chromic acid.

$$FCH_2CH_2CH_2OH \xrightarrow[\text{H}_2\text{SO}_4,\ \text{H}_2\text{O}]{\text{K}_2\text{Cr}_2\text{O}_7} \underset{\substack{\text{3-Fluoropropanoic acid}\\(74\%)}}{\overset{\overset{\displaystyle O}{\|}}{FCH_2CH_2COH}}$$

3-Fluoro-1-propanol

> Potassium permanganate $(KMnO_4)$ will also oxidize primary alcohols to carboxylic acids. What is the oxidation state of manganese in $KMnO_4$?

Conditions that do permit the easy isolation of aldehydes in good yield by oxidation of primary alcohols employ various Cr(VI) species as the oxidant in anhydrous media. Two such reagents are **pyridinium chlorochromate (PCC),** $C_5H_5NH^+\ ClCrO_3^-$, and **pyridinium dichromate (PDC),** $(C_5H_5NH)_2^{2+}\ Cr_2O_7^{2-}$; both are used in dichloromethane.

$$CH_3(CH_2)_5CH_2OH \xrightarrow[\text{CH}_2\text{Cl}_2]{\text{PCC}} \underset{\substack{\text{Heptanal}\\(78\%)}}{\overset{\overset{\displaystyle O}{\|}}{CH_3(CH_2)_5CH}}$$

1-Heptanol

$$(CH_3)_3C-\!\!\!\left\langle\!\!\!\bigcirc\!\!\!\right\rangle\!\!\!-CH_2OH \xrightarrow[\text{CH}_2\text{Cl}_2]{\text{PDC}} (CH_3)_3C-\!\!\!\left\langle\!\!\!\bigcirc\!\!\!\right\rangle\!\!\!-\overset{\overset{\displaystyle O}{\|}}{CH}$$

p-tert-Butylbenzyl alcohol   *p-tert*-Butylbenzaldehyde
                                              (94%)

Secondary alcohols are oxidized to ketones by the same reagents that oxidize primary alcohols:

$$\underset{\underset{\text{Secondary alcohol}}{RCHR'}}{\overset{OH}{|}} \xrightarrow{\text{oxidize}} \underset{\underset{\text{Ketone}}{RCR'}}{\overset{O}{||}}$$

Cyclohexanol $\xrightarrow[\text{H}_2\text{SO}_4,\ \text{H}_2\text{O}]{\text{Na}_2\text{Cr}_2\text{O}_7}$ Cyclohexanone (85%)

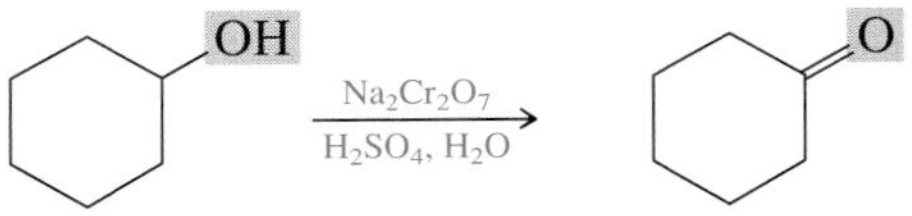

$$\underset{\text{1-Octen-3-ol}}{CH_2{=}CH\overset{\overset{OH}{|}}{C}HCH_2CH_2CH_2CH_2CH_3} \xrightarrow[\text{CH}_2\text{Cl}_2]{\text{PDC}} \underset{\underset{\text{(80\%)}}{\text{1-Octen-3-one}}}{CH_2{=}CH\overset{\overset{O}{||}}{C}CH_2CH_2CH_2CH_2CH_3}$$

Tertiary alcohols have no hydrogen on their hydroxyl-bearing carbon and do not undergo oxidation readily:

$$\underset{\underset{R''}{|}}{\overset{\overset{R'}{|}}{R{-}C{-}OH}} \xrightarrow{\text{oxidize}} \text{no reaction except under forcing conditions}$$

In the presence of strong oxidizing agents at elevated temperatures, oxidation of tertiary alcohols leads to cleavage of the various carbon–carbon bonds at the hydroxyl-bearing carbon atom, and a complex mixture of products results.

PROBLEM 10.4   Predict the principal organic product of each of the following reactions:

(a) $ClCH_2CH_2CH_2CH_2OH \xrightarrow[\text{H}_2\text{SO}_4,\ \text{H}_2\text{O}]{\text{K}_2\text{Cr}_2\text{O}_7}$

(b) $CH_3\underset{\underset{OH}{|}}{C}HCH_2CH_2CH_2CH_2CH_2CH_3 \xrightarrow[\text{H}_2\text{SO}_4,\ \text{H}_2\text{O}]{\text{Na}_2\text{Cr}_2\text{O}_7}$

(c) $CH_3CH_2CH_2CH_2CH_2CH_2CH_2OH \xrightarrow[\text{CH}_2\text{Cl}_2]{\text{PCC}}$

Sample Solution   (a) The reactant is a primary alcohol and so can be oxidized either to an aldehyde or to a carboxylic acid. Aldehydes are the major products only when the oxidation is carried out in anhydrous media. Carboxylic acids are formed when water is present. The reaction shown produced 4-chlorobutanoic acid in 56% yield.

$$\underset{\text{4-Chloro-1-butanol}}{ClCH_2CH_2CH_2CH_2OH} \xrightarrow[\text{H}_2\text{SO}_4,\ \text{H}_2\text{O}]{\text{K}_2\text{Cr}_2\text{O}_7} \underset{\text{4-Chlorobutanoic acid}}{ClCH_2CH_2CH_2\overset{\overset{O}{||}}{C}OH}$$

BIOLOGICAL OXIDATION OF ALCOHOLS

Many of the transformations that we describe as laboratory reactions have counterparts that occur in living systems. For example, oxidation of alcohols to carbonyl compounds or the reverse, reduction of carbonyl compounds to alcohols, occur in many biological processes. Ethanol, for example, is oxidized in the liver to acetaldehyde. The enzyme that catalyzes the oxidation of ethanol is called **alcohol dehydrogenase.**

$$CH_3CH_2OH \xrightarrow{\text{alcohol dehydrogenase}} CH_3\overset{\displaystyle O}{\overset{\|}{C}}H$$

Ethanol   Acetaldehyde

The rate of oxidation of ethanol in a particular person is constant, regardless of the concentration of alcohol present. Thus ingestion of ethanol from alcoholic beverages at a rate greater than the rate of oxidation results in a buildup of ethanol in the bloodstream, leading to intoxication.

Acetaldehyde is further oxidized to acetate in a reaction catalyzed by the enzyme **aldehyde dehydrogenase.**

$$CH_3\overset{\displaystyle O}{\overset{\|}{C}}H \xrightarrow{\text{aldehyde dehydrogenase}} CH_3\overset{\displaystyle O}{\overset{\|}{C}}O^-$$

Acetaldehyde   Acetate

Acetate is used by the body in the synthesis of fatty acids and cholesterol (see Chapter 16).

As noted in Section 10.1, methanol (CH_3OH) is a poisonous substance that can cause blindness or death from the ingestion of even small amounts. Methanol is oxidized to formaldehyde in the body by the same enzyme system that oxidizes ethanol, alcohol dehydrogenase.

$$CH_3OH \xrightarrow{\text{alcohol dehydrogenase}} \overset{\displaystyle O}{\overset{\|}{H}}CH$$

Methanol   Formaldehyde

Formaldehyde is toxic to the retina of the eye and is the cause of blindness from methanol ingestion. Formaldehyde is further oxidized in the body to formate.

$$\overset{\displaystyle O}{\overset{\|}{H}}CH \xrightarrow{\text{aldehyde dehydrogenase}} \overset{\displaystyle O}{\overset{\|}{H}}CO^-$$

Formaldehyde   Formate

Formate is not used rapidly by the body and collects in the blood as formic acid. This lowers the pH of the blood below the normal physiological range, resulting in a fatal condition called **acidosis.**

One treatment for methanol poisoning makes use of the fact that both methanol and ethanol are oxidized in the body by the same enzyme systems. A solution containing ethanol is given to a victim of methanol poisoning. Ethanol is considerably less toxic than methanol, and the ethanol oxidation "ties up" the enzymes, preventing the more harmful methanol oxidation from occurring.

10.6   THIOL NOMENCLATURE

Sulfur lies just below oxygen in the periodic table, and many oxygen-containing organic compounds have sulfur analogs. The sulfur analogs of alcohols (ROH) are **thiols (RSH).** Thiols are given substitutive IUPAC names by appending the suffix *-thiol* to the name of the corresponding alkane, numbering the chain in the direction that gives the lower locant to the carbon that bears the —SH group and retaining the final *-e* of the alkane name. When the —SH group is named as a substituent, it is called a **mercapto** group. It is also referred to as a **sulfhydryl** group, but this is a generic term, not used in systematic nomenclature.

$$(CH_3)_2CHCH_2CH_2SH \qquad HSCH_2CH_2OH \qquad HSCH_2\overset{\displaystyle SH}{\overset{|}{C}}HCH_2OH$$

3-Methyl-1-butanethiol   2-Mercaptoethanol   2,3-Dimercapto-1-propanol

Thiols have a marked tendency to bond to mercury, and the word *mercaptan* comes from the Latin *mercurium captans,* which means "seizing mercury." The drug **dimercaprol** is used to treat mercury and lead poisoning; it is 2,3-dimercapto-1-propanol.

At one time thiols were named **mercaptans.** Thus, CH_3CH_2SH was called "ethyl mercaptan" according to this system, but this nomenclature has been abandoned although it may be encountered in older books or journals.

10.7 PROPERTIES OF THIOLS

When one encounters a thiol for the first time, especially a low-molecular-weight thiol, its most obvious property is its foul odor. Ethanethiol is added to natural gas so that leaks can be detected without special equipment—your nose is so sensitive that it can detect less than one part of ethanethiol in 10,000,000,000 parts of air! The odor of thiols weakens with the number of carbons, because both the volatility and the sulfur content decrease. 1-Dodecanethiol, for example, has only a faint odor.

> **PROBLEM 10.5** The main components of a skunk's scent fluid are 3-methyl-1-butanethiol and *cis-* and *trans-*2-butene-1-thiol. Write structural formulas for each of these compounds.

Compare the boiling points of H_2S ($-60°C$) and H_2O ($100°C$).

The S—H bond is less polar than the O—H bond, and hydrogen bonding in thiols is much weaker than that of alcohols. Thus, methanethiol (CH_3SH) is a gas at room temperature (bp 6°C), and methanol (CH_3OH) is a liquid (bp 65°C).

Thiols are weak acids, but are far more acidic than alcohols. We have seen (Section 3.5) that most alcohols have K_a values in the range 10^{-16} to 10^{-18} ($pK_a = 16$ to 18). The corresponding values for thiols are about $K_a = 10^{-10}$ ($pK_a = 10$). The significance of this difference is that a thiol can be quantitatively converted to its conjugate base (RS^-), called an **alkanethiolate** anion, by hydroxide:

$$RS{-}H \; + \; :OH^- \longrightarrow RS:^- \; + \; H{-}OH$$

Alkanethiol	Hydroxide ion	Alkanethiolate ion	Water
(stronger acid)	(stronger base)	(weaker base)	(weaker acid)
($pK_a = 10$)			($pK_a = 15.7$)

Thiols, therefore, dissolve in aqueous media when the pH is greater than 10.

Another difference between thiols and alcohols concerns their oxidation. We have seen earlier in this chapter that oxidation of alcohols gives compounds having carbonyl groups. Analogous oxidation of thiols to compounds with C=S functions does *not* occur. Only sulfur is oxidized, not carbon, and compounds containing sulfur in various oxidation states are possible. These include a series of acids classified as **sulfenic, sulfinic,** and **sulfonic** according to the number of oxygens attached to sulfur.

$$RS{-}H \longrightarrow RS{-}OH \longrightarrow RS^{+}{-}OH \longrightarrow RS^{2+}{-}OH$$

Thiol	Sulfenic acid	Sulfinic acid	Sulfonic acid

Of these the most important are the sulfonic acids. In general, however, sulfonic acids are not prepared by oxidation of thiols. Arenesulfonic acids ($ArSO_3H$), for example, are prepared by sulfonation of arenes (Section 6.6).

One of the most important oxidative processes, especially from a biochemical perspective, is the oxidation of thiols to **disulfides.**

$$2RSH \underset{reduce}{\overset{oxidize}{\rightleftharpoons}} RSSR$$

Thiol   Disulfide

Although a variety of oxidizing agents are available for this transformation, it occurs so readily that thiols are slowly converted to disulfides by the oxygen in the air. Dithiols give cyclic disulfides by intramolecular sulfur–sulfur bond formation. An example of a cyclic disulfide is the coenzyme α-lipoic acid. The last step in the laboratory synthesis of α-lipoic acid is an iron(III)-catalyzed oxidation of the dithiol shown here:

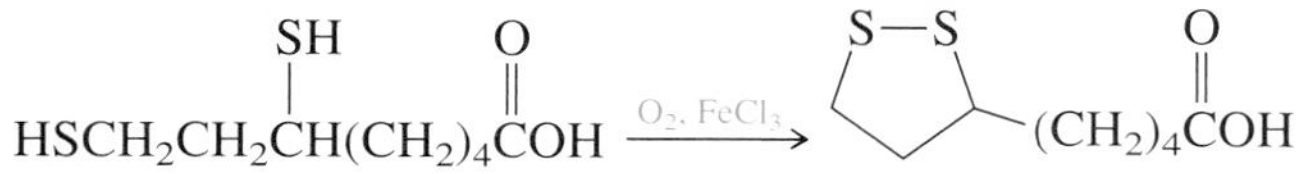

6,8-Dimercaptooctanoic acid   α-Lipoic acid (78%)

Rapid and reversible making and breaking of the sulfur–sulfur bond is essential to the biological function of α-lipoic acid.

10.8   INTRODUCTION TO ETHERS

Ethers contain a C—O—C unit. In contrast to alcohols, ethers undergo relatively few chemical reactions. Lacking the O—H group, ethers are unable to undergo the deprotonation and oxidation reactions typical of alcohols. This lack of reactivity makes ethers valuable as solvents in a number of synthetically useful chemical reactions.

Epoxides are an important exception to the chemical inertness of ethers. **Epoxides,** ethers in which the C—O—C unit forms a three-membered ring, are very reactive substances. Epoxides are quite common in nature. The female gypsy moth, for example, attracts the male by emitting an epoxide known as **disparlure.** On detecting the presence of this pheromone, the male follows the scent to its origin and mates with the female.

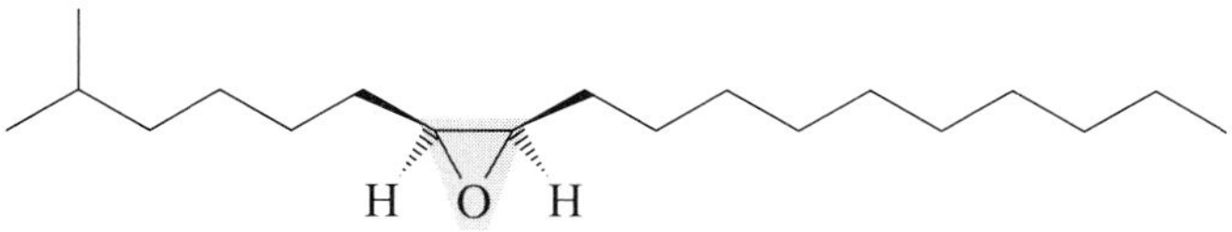

Disparlure

In one strategy designed to control the spread of the gypsy moth, infested areas are sprayed with synthetic disparlure. With the sex attractant everywhere, male gypsy moths become hopelessly confused as to the actual location of individual females. Many otherwise fertile female gypsy moths then live out their lives without producing hungry gypsy moth caterpillars.

10.9   ETHER NOMENCLATURE

Ethers are named as **alkoxy** derivatives of alkanes. Alternatively, ethers may be named by specifying the two alkyl groups in alphabetical order as separate words, then adding the word *ether* at the end. When both alkyl groups are the same, the prefix *di-* precedes the name of the alkyl group.

$$CH_3CH_2OCH_2CH_3 \qquad CH_3CH_2OCH_3$$

Ethoxyethane   Methoxyethane
Diethyl ether   Ethyl methyl ether

The role played by disulfides in the structure of peptides and proteins will be examined in Chapter 17.

Often diethyl ether is called just **ether**, or **anesthesia ether**. The latter name derives from its former use as a hospital anesthetic. Diethyl ether is extremely flammable, and its use in hospitals has largely been replaced by less hazardous alternatives.

Ethers are described as **symmetrical** or **unsymmetrical** depending on whether the two groups bonded to oxygen are the same or different. Diethyl ether is a symmetrical ether; ethyl methyl ether is unsymmetrical.

Cyclic ethers have their oxygen as part of a ring—they are **heterocyclic compounds.** Tetrahydrofuran and tetrahydropyran are cyclic ethers.

Tetrahydrofuran   Tetrahydropyran

In general, the properties of cyclic ethers are very much like those of their noncyclic counterparts. Epoxides are an exception, however. At one time epoxides were named as oxides of alkenes. Ethylene oxide and propylene oxide, for example, are the common names of two industrially important epoxides. The industrial use of ethylene oxide is discussed in Section 10.12.

$$H_2C\underset{O}{\overset{\frown}{\longrightarrow}}CH_2 \qquad H_2C\underset{O}{\overset{\frown}{\longrightarrow}}CHCH_3$$

Ethylene oxide   Propylene oxide

Epoxides may be named as epoxy derivatives of alkanes. According to this system, ethylene oxide becomes epoxyethane and propylene oxide becomes 1,2-epoxypropane. The prefix *epoxy-* always immediately precedes the alkane ending; it is not listed in alphabetical order as are other substituents.

$$H_3C\backslash \qquad CH_3$$
$$H_3C / \underset{O}{} \backslash H$$

1,2-Epoxycyclohexane   2-Methyl-2,3-epoxybutane

> **PROBLEM 10.6** The following ethers are suspected of being **mutagens,** which means that they can induce genetic mutations in cells. Write the structures of each of these ethers.
>
> (a) Chloromethyl methyl ether   (b) 3,4-Epoxy-1-butene
>
> **Sample Solution** (a) Chloromethyl methyl ether has a chloromethyl group ($ClCH_2$—) and a methyl group (CH_3—) attached to oxygen. Its structure is $ClCH_2OCH_3$.

10.10 PREPARATION OF ETHERS

A long-standing method for the preparation of ethers is the **Williamson ether synthesis.** Nucleophilic substitution of an alkyl halide by an alkoxide gives the carbon–oxygen bond of an ether:

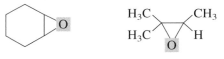

Alkoxide   Alkyl   Ether   Halide ion
ion   halide

The reaction is named for Alexander Williamson, a British chemist who used it to prepare diethyl ether in 1850.

POLYETHER ANTIBIOTICS

One way in which pharmaceutical companies search for new drugs is by growing colonies of microorganisms in nutrient broths and assaying the substances produced for their biological activity. This method has yielded thousands of antibiotic substances, of which hundreds have been developed into effective drugs. Antibiotics are, by definition, toxic (*anti* = "against"; *bios* = "life"), and the goal is to find substances that are more toxic to infectious organisms than to their human hosts.

A number of **polyether antibiotics** have been discovered using fermentation technology. They are characterized by the presence of several cyclic ether structural units, as illustrated for the case of monensin in Figure 10.2. Monensin and other naturally occurring polyethers are able to form stable complexes with metal ions. The structure of the monensin–sodium bromide complex is depicted in Figure 10.2b, which shows that four ether oxygens and two hydroxyl groups surround a sodium ion. The alkyl groups are oriented toward the outside of the complex, and the polar oxygens and the metal ion are on the inside. The hydrocarbon-like surface of the complex permits it to carry its sodium ion through the hydrocarbon-like interior of a cell membrane. This disrupts the normal balance of sodium ions within the cell and interferes with important processes of cellular respiration. Small amounts of monensin are added to poultry feed to kill parasites that live in the intestines of chickens. Compounds such as monensin and the crown ethers that affect metal ion transport are referred to as **ionophores** ("ion carriers").

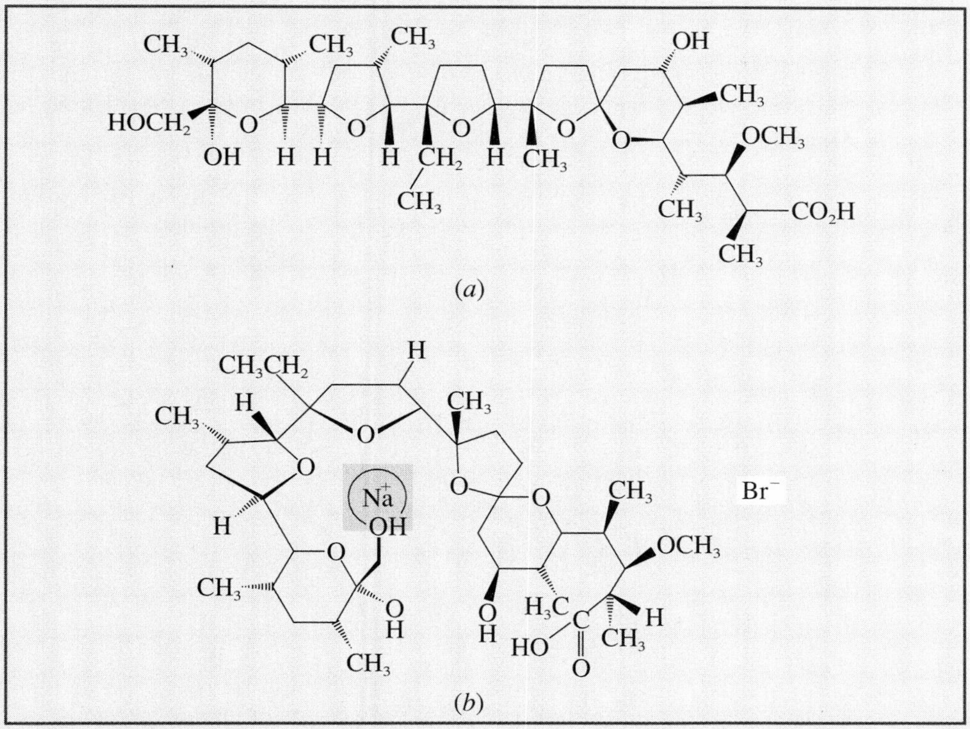

FIGURE 10.2 (a) The structure of monensin; (b) the structure of the monensin–sodium bromide complex showing coordination of sodium ion by oxygen atoms of monensin.

Preparation of ethers by the Williamson ether synthesis is most successful when the alkyl halide is one that is reactive toward S_N2 substitution. Methyl halides and primary alkyl halides are the best substrates.

$$CH_3CH_2CH_2CH_2ONa + CH_3CH_2I \longrightarrow CH_3CH_2CH_2CH_2OCH_2CH_3 + NaI$$

| Sodium butoxide | Iodoethane | Butyl ethyl ether (71%) | Sodium iodide |

> **PROBLEM 10.7** Write equations describing two different ways in which benzyl ethyl ether ($C_6H_5CH_2OCH_2CH_3$) could be prepared by a Williamson ether synthesis.

Secondary and tertiary alkyl halides are not suitable because they tend to react with alkoxide bases by E2 elimination rather than by S_N2 substitution. Whether the alkoxide base is primary, secondary, or tertiary is much less important than the nature of the alkyl halide. Thus benzyl isopropyl ether is prepared in high yield from benzyl chloride, a primary chloride that is incapable of undergoing elimination, and sodium isopropoxide:

$$(CH_3)_2CHONa + \text{〈benzene〉}-CH_2Cl \longrightarrow (CH_3)_2CHOCH_2-\text{〈benzene〉} + NaCl$$

| Sodium isopropoxide | Benzyl chloride | Benzyl isopropyl ether (84%) | Sodium chloride |

The alternative synthetic route using the sodium salt of benzyl alcohol and an isopropyl halide would be much less effective because of increased competition from elimination as the alkyl halide becomes more sterically hindered.

> **PROBLEM 10.8** Only one combination of alkyl halide and alkoxide is appropriate for the preparation of each of the following ethers by the Williamson ether synthesis. What is the correct combination in each case?
>
> (a) $CH_3CH_2O-\text{〈cyclopentyl〉}$ (c) $(CH_3)_3COCH_2C_6H_5$
>
> (b) $CH_2{=}CHCH_2OCH(CH_3)_2$
>
> **Sample Solution** (a) The ether linkage of cyclopentyl ethyl ether involves a primary carbon and a secondary one. Choose the alkyl halide corresponding to the primary alkyl group, leaving the secondary alkyl group to arise from the alkoxide nucleophile.
>
> $$\text{〈cyclopentyl〉}-ONa + CH_3CH_2Br \xrightarrow{S_N2} \text{〈cyclopentyl〉}-OCH_2CH_3$$
>
> | Sodium cyclopentanolate | Ethyl bromide | Cyclopentyl ethyl ether |
>
> The alternative combination, cyclopentyl bromide and sodium ethoxide, is not appropriate because elimination will be the major reaction:
>
> $$CH_3CH_2ONa + \text{〈cyclopentyl〉}-Br \xrightarrow{E2} CH_3CH_2OH + \text{〈cyclopentene〉}$$
>
> | Sodium ethoxide | Bromocyclopentane | Ethanol | Cyclopentene |
> | | | (major products) | |

Both reactants in the Williamson ether synthesis usually originate in alcohol precursors. Sodium and potassium alkoxides are prepared by reaction of an alcohol with the appropriate metal, and alkyl halides are most commonly made from alcohols by reaction with a hydrogen halide (Section 3.7).

10.11 PREPARATION OF EPOXIDES

Epoxides are very easy to prepare via the reaction of an alkene with a peroxy acid. This process is known as **epoxidation.**

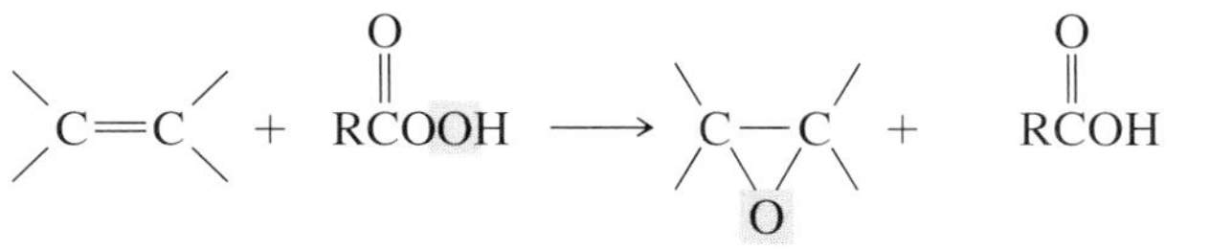

$$\overset{\diagdown}{\underset{\diagup}{C}}=\overset{\diagup}{\underset{\diagdown}{C}} \ + \ \overset{O}{\overset{\|}{RCOOH}} \ \longrightarrow \ \overset{\diagdown}{\underset{\diagup}{C}}\underset{\diagdown O \diagup}{-}\overset{\diagup}{\underset{\diagdown}{C}} \ + \ \overset{O}{\overset{\|}{RCOH}}$$

Alkene   Peroxy acid   Epoxide   Carboxylic acid

A commonly used peroxy acid is peroxyacetic acid (CH_3CO_2OH). Peroxyacetic acid is normally used in acetic acid as the solvent, but epoxidation reactions tolerate a variety of solvents and are often carried out in dichloromethane or chloroform.

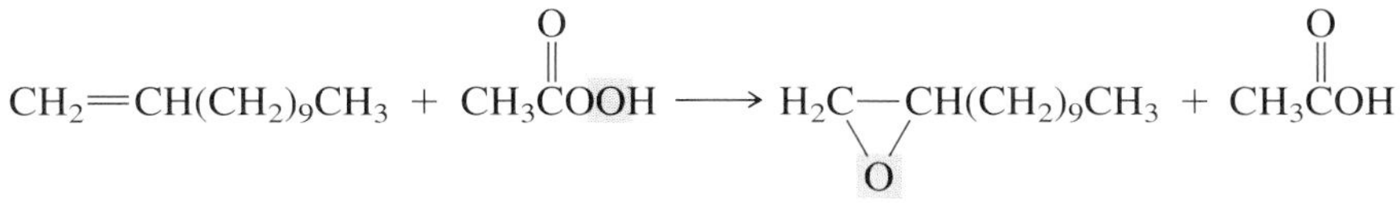

$$CH_2{=}CH(CH_2)_9CH_3 \ + \ CH_3\overset{O}{\overset{\|}{C}}OOH \ \longrightarrow \ H_2C{-}CH(CH_2)_9CH_3 \ + \ CH_3\overset{O}{\overset{\|}{C}}OH$$
$$\underset{O}{\diagdown \diagup}$$

1-Dodecene   Peroxyacetic   1,2-Epoxydodecane   Acetic
                    acid   (52%)   acid

Epoxidation of alkenes with peroxy acids is a syn addition to the double bond. Substituents that are cis to each other in the alkene remain cis in the epoxide; substituents that are trans in the alkene remain trans in the epoxide.

The mechanism of alkene epoxidation is believed to be a concerted process involving a single bimolecular elementary step, as shown in Figure 10.3.

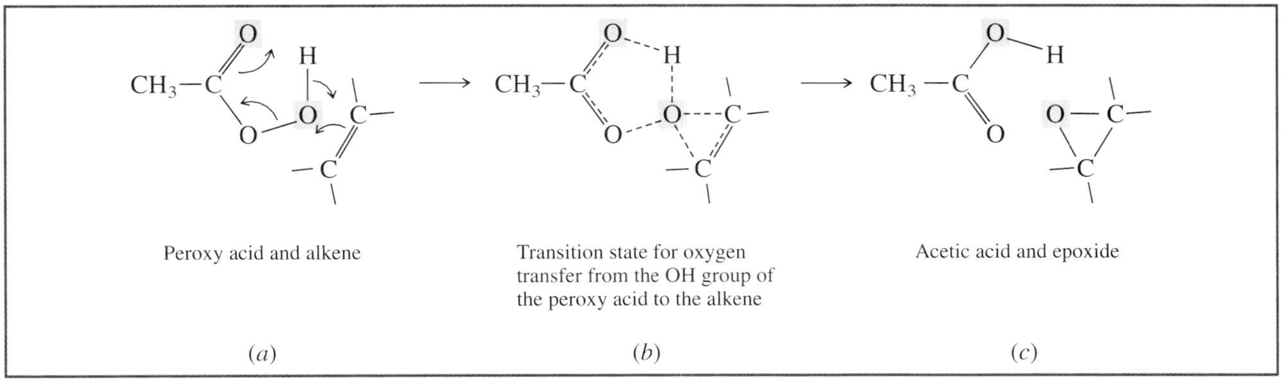

Peroxy acid and alkene   Transition state for oxygen   Acetic acid and epoxide
                                      transfer from the OH group of
                                      the peroxy acid to the alkene

(*a*)   (*b*)   (*c*)

FIGURE 10.3   A one-step mechanism for epoxidation of alkenes by peroxyacetic acid. In part (*a*) the starting peroxy acid is shown in a conformation in which the proton of the OH group is hydrogen bonded to the oxygen of the C=O group. (*b*) The weak O—O bond of the peroxy acid breaks, and both C—O bonds of the epoxide form in the same transition state leading to products (*c*).

$$
\underset{\substack{\text{(E)-1,2-Diphenylethene}}}{\underset{\text{H}}{\overset{C_6H_5}{\diagdown}}C=C\overset{H}{\underset{C_6H_5}{\diagup}}} \quad + \quad \underset{\substack{\text{Peroxyacetic}\\\text{acid}}}{CH_3\overset{O}{\overset{\|}{C}}OOH} \quad \longrightarrow \quad \underset{\substack{\text{trans-1,2-Diphenylepoxyethane}\\(78\text{–}83\%)}}{C_6H_5\diagup\overset{O}{\diagup\diagdown}\diagdown \overset{H}{\underset{C_6H_5}{}}} \quad + \quad \underset{\substack{\text{Acetic}\\\text{acid}}}{CH_3\overset{O}{\overset{\|}{C}}OH}
$$

> **PROBLEM 10.9** The structure of disparlure, the sex attractant of the female gypsy moth, was given in Section 10.8. Draw the structure of the alkene that could be used to prepare synthetic disparlure by an epoxidation reaction. Is this alkene the *E* or *Z* isomer?

10.12 REACTIONS OF EPOXIDES

The chemical property that most distinguishes epoxides from other ethers is their greater reactivity toward nucleophilic reagents. Reactions that lead to opening of the three-membered ring relieve the angle strain present in epoxides.

Angle strain is the main source of strain in epoxides, but torsional strain that results from the eclipsing of bonds on adjacent carbons is also present. Both kinds of strain are relieved when a ring-opening reaction occurs.

$$
\underset{\text{Nucleophile}}{HNu:} \quad + \quad \underset{\text{Epoxide}}{R_2C\overset{O}{\overline{\quad\quad}}CR_2} \quad \longrightarrow \quad \underset{\substack{\text{Product}}}{R_2C\overset{OH}{\underset{\underset{Nu}{|}}{-}}CR_2}
$$

Ethylene oxide is a typical epoxide, widely used in the chemical industry (almost 9 billion pounds in 1999). It is a synthetic intermediate for a wide variety of familiar end products. For instance, ethylene glycol, used in automotive antifreeze and to make polyester fabrics, is prepared by the hydrolysis of ethylene oxide in dilute sulfuric acid.

$$
\underset{\text{Ethylene oxide}}{H_2C\underset{O}{\overline{\quad}}CH_2} \quad + \quad \underset{\text{Water}}{H_2O} \quad \xrightarrow{H_2SO_4} \quad \underset{\substack{\text{2-Ethanediol}\\\text{(ethylene glycol)}}}{HOCH_2CH_2OH}
$$

Water is the nucleophile in this reaction, which is catalyzed by the sulfuric acid.

Other nucleophiles react with ethylene oxide in an analogous manner to yield 2-substituted derivatives of ethanol. 2-Ethoxyethanol is used as a lacquer thinner and varnish remover as well as an anti-icing additive in aviation fuels. The nucleophile in this reaction is ethanol.

$$
\underset{\text{Ethylene oxide}}{H_2C\underset{O}{\overline{\quad}}CH_2} \quad \xrightarrow[H_2SO_4,\ 25°C]{CH_3CH_2OH} \quad \underset{\substack{\text{2-Ethoxyethanol}\\(85\%)}}{CH_3CH_2OCH_2CH_2OH}
$$

Reaction of ethylene oxide with aqueous ammonia (ammonia is the nucleophile) gives 2-aminoethanol, used commercially as a corrosion inhibitor and as an emulsifying agent in some paints.

$$
\underset{\text{Ethylene oxide}}{H_2C\underset{O}{\overline{\quad}}CH_2} \quad \xrightarrow{NH_3,\ H_2O} \quad \underset{\text{2-Aminoethanol}}{H_2NCH_2CH_2OH}
$$

EPOXIDES AND CHEMICAL CARCINOGENESIS

Epoxides appear to play a role in the interactions of certain chemical carcinogens with cells. A key step in chemical carcinogenesis is the reaction of a cellular nucleophile with a molecule that can undergo nucleophilic attack. A number of carcinogenic molecules form epoxides in the body; these epoxides are attacked by the cellular nucleophiles. This process of transforming an unreactive molecule into a reactive one is called **bioactivation.**

One example of a carcinogenic molecule that undergoes bioactivation to form an epoxide is benzo[a]pyrene, which was described in the boxed essay accompanying Section 6.4. Another example is benzene.

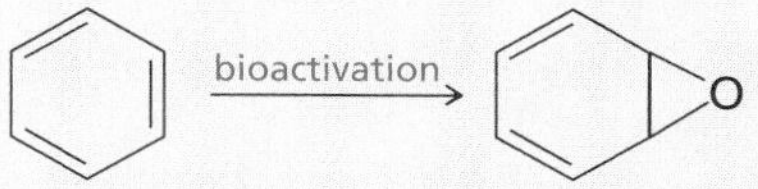

The study of carcinogenesis is difficult because metabolic reactions similar to the one shown for benzene are normal parts of cellular metabolism. The biosynthesis of naturally occurring phenols such as tyrosine involves reactions that proceed through epoxide intermediates. And yet these reactions don't cause cancer!

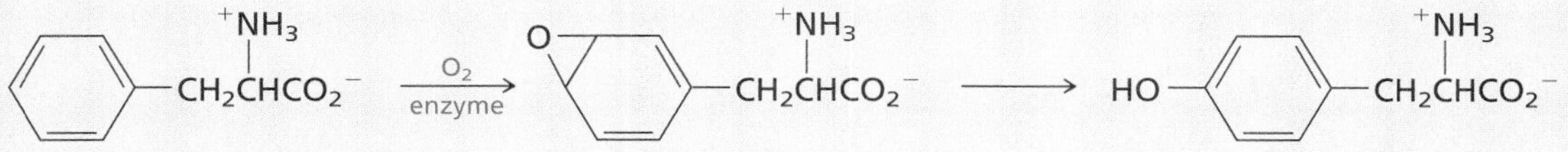

| Phenylalanine | Epoxide intermediate | | Tyrosine |

PROBLEM 10.10   A commercial material known as butyl cellosolve, used as a lacquer solvent and in varnish remover, is made by reaction of ethylene oxide with 1-butanol. Draw the structure of this product.

10.13 INTRODUCTION TO PHENOLS: NOMENCLATURE

Phenols are compounds that have a hydroxyl group bonded directly to a benzene ring. The parent compound is called simply **phenol.** Several phenol derivatives have common names acceptable in the IUPAC system. For example, the o-, m-, and p-methylphenols are called **cresols.** More highly substituted compounds are named as derivatives of phenol. Numbering of the ring begins at the hydroxyl-substituted carbon and proceeds in the direction that gives the lower number to the next substituted carbon. Substituents are cited in alphabetical order.

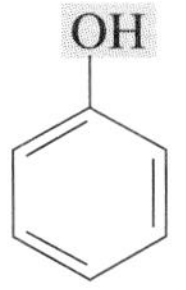

| Phenol | m-Cresol | 5-Chloro-2-methylphenol |

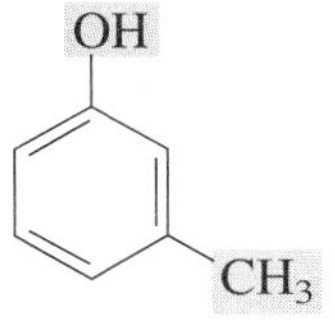

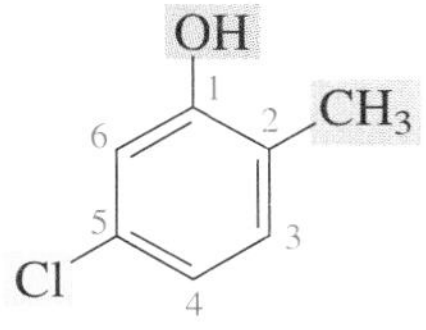

The three dihydroxy derivatives of benzene may be named as 1,2-, 1,3-, and 1,4-benzenediol, respectively, but each is more familiarly known by its common name.

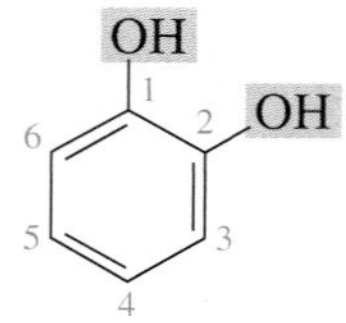

1,2-Benzenediol
(pyrocatechol)

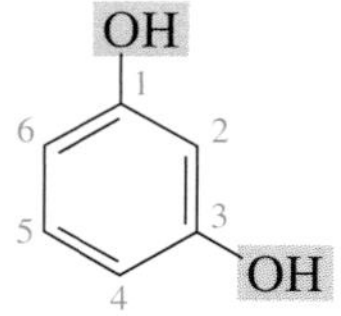

1,3-Benzenediol
(resorcinol)

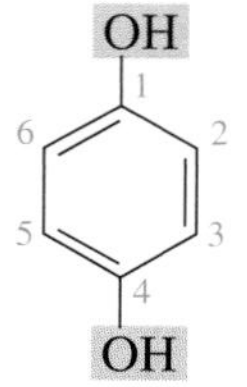

1,4-Benzenediol
(hydroquinone)

Pyrocatechol is often called catechol.

PROBLEM 10.11 Write a structural formula for each of the following compounds:

(a) 1,2,3-Benzenetriol (pyrogallol, a photographic developer)

(b) 2,4,6-Trinitrophenol (picric acid, an explosive)

(c) 2,4,5-Trichlorophenol (material used in the synthesis of a pesticide)

Sample Solution   (a) Pyrogallol is the common name for 1,2,3-benzenetriol. The three hydroxyl groups occupy adjacent positions on a benzene ring.

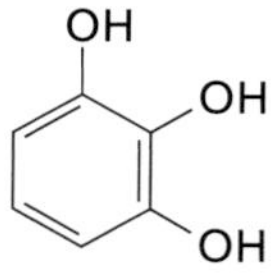

1,2,3-Benzenetriol
(pyrogallol)

10.14 SYNTHETIC AND NATURALLY OCCURRING PHENOL DERIVATIVES

Phenol and the cresols have antiseptic properties and are used in dilute aqueous solution as household disinfectants. Lysol is one brand-name example. Phenol is also used as the starting material for products as varied as the analgesic aspirin and the pesticide 2,4-D. Pentachlorophenol is used as a wood preservative.

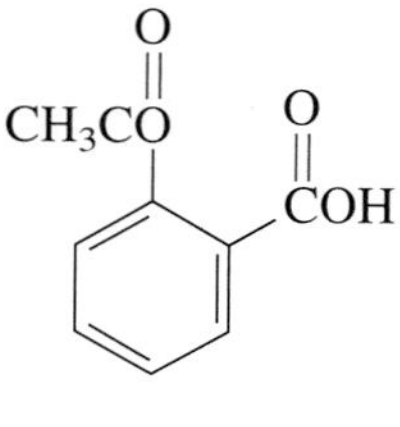

Aspirin
(Acetylsalicylic acid)

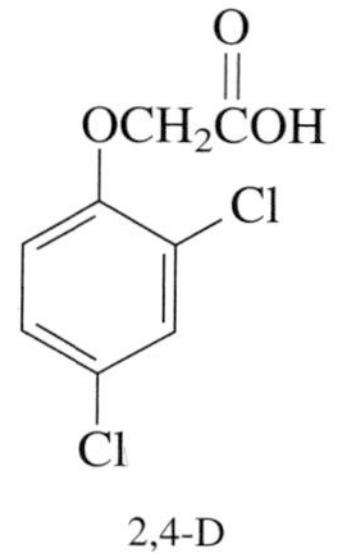

2,4-D
(2,4-Dichloro-phenoxyacetic acid)

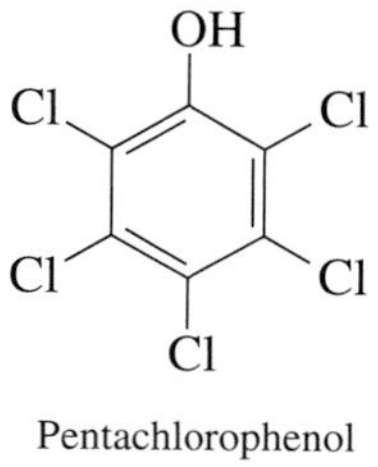

Pentachlorophenol

Phenolic compounds are quite common natural products and are found in both plants and animals. As the examples shown in Figure 10.4 illustrate, the phenolic group may be part of molecules that are chemically quite different.

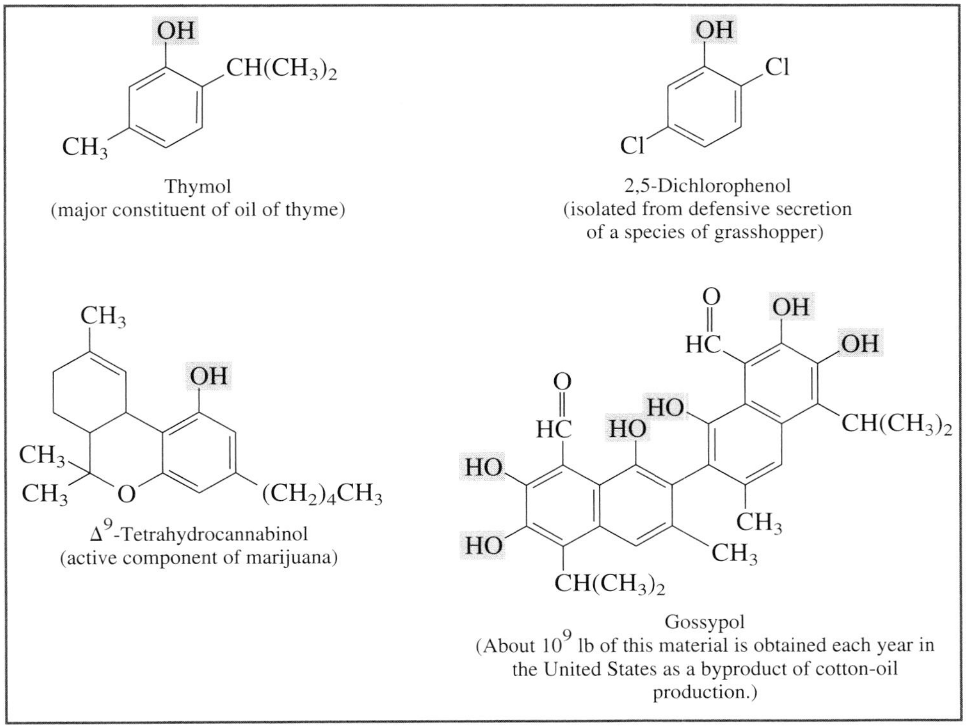

FIGURE 10.4 Some naturally occurring phenols.

10.15 ACIDITY OF PHENOLS

The most characteristic property of phenols is their acidity. Phenols are 10^6–10^8 times more acidic than alcohols. Recall that alcohols have ionization constants K_a in the 10^{-16}–10^{-18} range (pK_a 16–18), whereas the K_a for most phenols is about 10^{-10} (pK_a 10).

To help us understand why phenols are more acidic than alcohols, let's compare the ionization equilibria for phenol and ethanol. In particular, consider the differences in charge delocalization in ethoxide ion and in phenoxide ion. The negative charge in ethoxide ion is localized on oxygen.

$$CH_3CH_2\ddot{O}-H \rightleftharpoons H^+ + CH_3CH_2\ddot{O}:^- \qquad K_a = 10^{-16}\ (pK_a = 16)$$

Ethanol   Proton   Ethoxide ion

The negative charge in phenoxide ion is stabilized by electron delocalization into the ring.

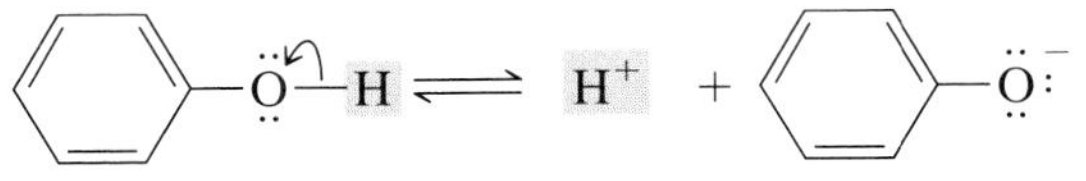

$$K_a = 10^{-10}\ (pK_a = 10)$$

Phenol   Proton   Phenoxide ion

Because of its acidity, phenol was known as **carbolic acid** when Joseph Lister introduced it as an antiseptic in 1865 to prevent postoperative bacterial infections that were then a life-threatening hazard in even minor surgical procedures.

TABLE 10.4	Acidities of Some Phenols	

Compound Name	Ionization Constant K_a	pK_a
Monosubstituted phenols		
Phenol	1.0×10^{-10}	10.0
o-Cresol	4.7×10^{-11}	10.3
m-Cresol	8.0×10^{-11}	10.1
p-Cresol	5.2×10^{-11}	10.3
o-Methoxyphenol	1.0×10^{-10}	10.0
m-Methoxyphenol	2.2×10^{-10}	9.6
p-Methoxyphenol	6.3×10^{-11}	10.2
o-Nitrophenol	5.9×10^{-8}	7.2
m-Nitrophenol	4.4×10^{-9}	8.4
p-Nitrophenol	6.9×10^{-8}	7.2
Di- and trinitrophenols		
2,4-Dinitrophenol	1.1×10^{-4}	4.0
3,5-Dinitrophenol	2.0×10^{-7}	6.7
2,4,6-Trinitrophenol	4.2×10^{-1}	0.4

Recall from Section 10.13 that cresols are methyl-substituted derivatives of phenol.

Electron delocalization in phenoxide is represented by resonance among the structures:

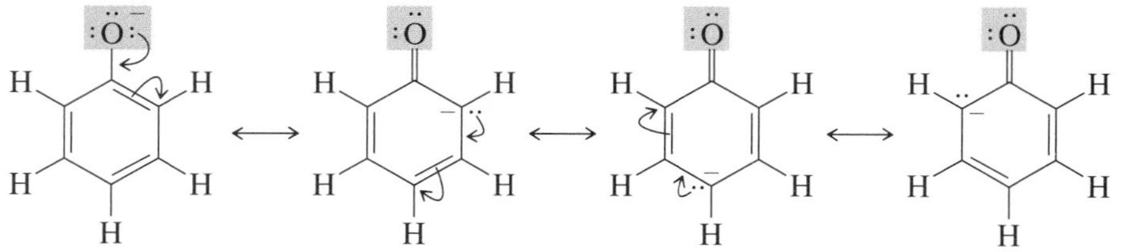

The negative charge in phenoxide ion is shared by the oxygen and the carbons that are ortho and para to it. Delocalization of its negative charge strongly stabilizes phenoxide ion.

As Table 10.4 shows, most phenols have ionization constants similar to that of phenol itself. Alkyl substitution produces negligible changes in acidities, as do weakly electronegative groups attached to the ring. Only when the substituent is strongly electron-withdrawing, as in a nitro group, is a substantial change in acidity noted. The ionization constants of o- and p-nitrophenol are several hundred times greater than that of phenol. An o- or p-nitro group greatly stabilizes the phenoxide ion by permitting a portion of the negative charge to be borne by its own oxygens.

Electron delocalization in o-*nitrophenoxide ion:*

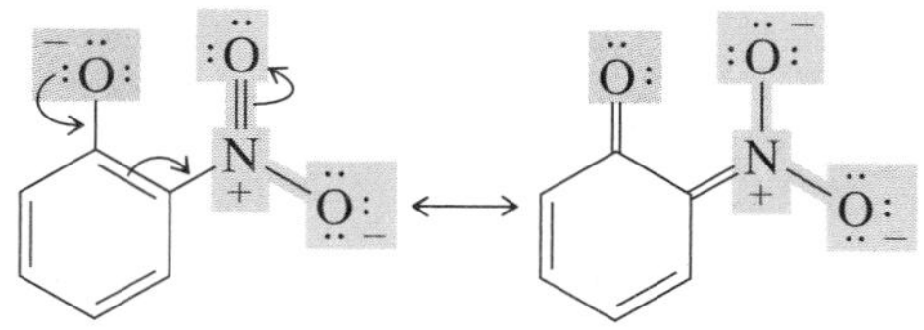

4-Methylpyrocatechol
(4-methyl-1,2-benzenediol)

4-Methyl-1,2-benzoquinone
(68%)

Silver oxide is a weak oxidizing agent.

Quinones are colored; *p*-benzoquinone, for example, is yellow. Many occur naturally and have been used as dyes. **Alizarin** is a red pigment extracted from the roots of the madder plant. Its preparation from anthracene, a coal tar derivative, in 1868 was a significant step in the development of the synthetic dyestuff industry.

Alizarin

Quinones based on the anthracene ring system are called **anthraquinones.** Alizarin is one example of an **anthraquinone dye.**

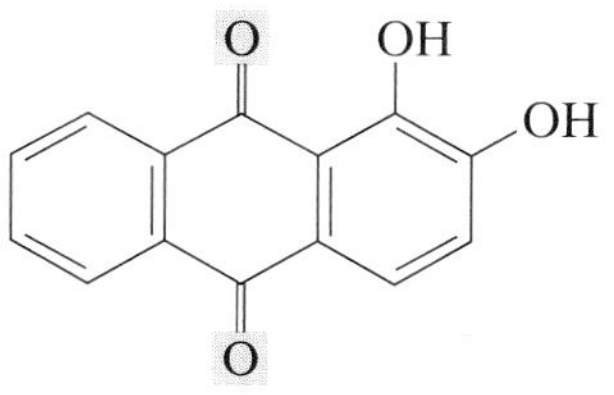

The oxidation–reduction process that connects hydroquinone and benzoquinone involves two 1-electron transfers and is rapidly reversible. The ready reversibility of this reaction is essential to the role that quinones play in cellular respiration, the process by which an organism uses molecular oxygen to convert its food to carbon dioxide, water, and energy. Electrons are not transferred directly from the substrate molecule to oxygen but instead are transferred by way of an **electron-transport chain** involving a succession of oxidation–reduction reactions. A key component of this electron-transport chain is the substance known as **ubiquinone,** or coenzyme Q:

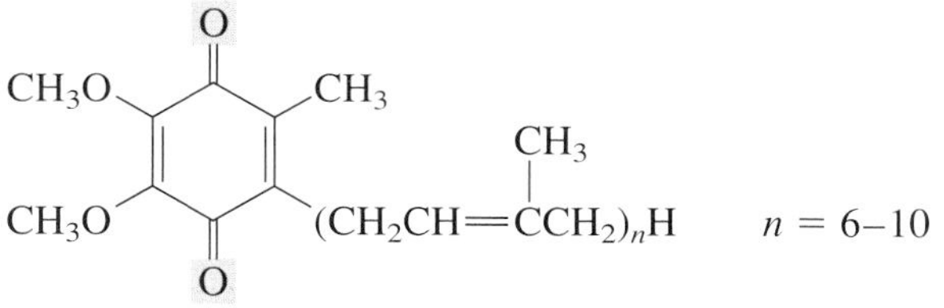

$n = 6\text{--}10$

Ubiquinone
(coenzyme Q)

The name *ubiquinone* is a shortened form of *ubiquitous quinone,* a term coined to describe the observation that this substance can be found in all cells. The length of its side chain varies among different organisms; the most common form in vertebrates has $n = 10$, and ubiquinones in which $n = 6$ to 9 are found in yeasts and plants.

Another physiologically important quinone is vitamin K. Here "K" stands for *koagulation* (Danish) because this substance was first identified as essential for the normal clotting of blood.

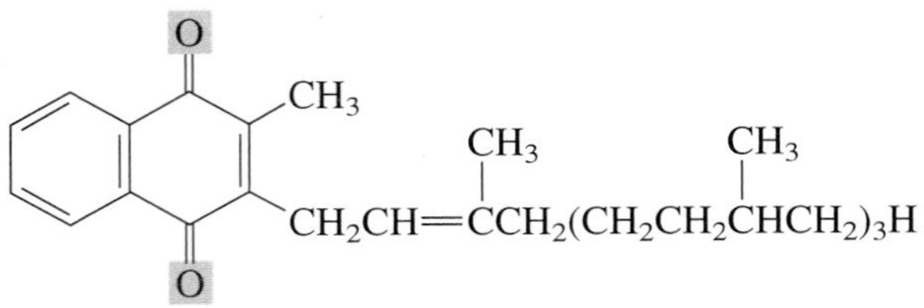

Vitamin K

Intestinal flora is a general term for the bacteria, yeast, and fungi that live in the large intestine.

Some vitamin K is provided in the normal diet, but a large proportion of that required by humans is produced by their intestinal flora.

LEARNING OBJECTIVES

This chapter focused on the chemical and physical properties of four classes of organic compounds: alcohols, thiols, ethers, and phenols. The skills you have learned in this chapter should enable you to:

- Explain how oxidation and reduction relate to organic compounds.
- Write a chemical equation for the oxidation of a primary alcohol to either an aldehyde or a carboxylic acid.
- Write a chemical equation for the oxidation of a secondary alcohol to a ketone.
- Draw the structure and give a systematic name for a thiol.
- Give a systematic name for an ether or an epoxide.
- Write a chemical equation for the preparation of an ether by the Williamson method.
- Write a chemical equation for the preparation of an epoxide from an alkene.
- Predict the product of the reaction of an epoxide with a nucleophile.
- Give a systematic name for a phenol derivative.
- Explain the acidity of phenol derivatives.
- Write a chemical equation for the formation of a quinone by the oxidation of a 1,4-benzenediol.

10.18 SUMMARY

In this chapter we have examined four important classes of organic molecules: alcohols, thiols, ethers, and phenols.

Alcohols are most often prepared by reduction using either hydrogen in the presence of a catalyst or a metal hydride reagent such as sodium borohydride or lithium aluminum hydride. Primary alcohols are prepared from aldehydes; secondary alcohols from the corresponding ketone (Section 10.3)

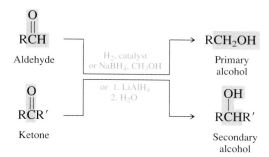

Primary and secondary alcohols may be used as the starting materials for the preparation of carbonyl compounds (Section 10.5). Whether the product is an aldehyde, a ketone, or a carboxylic acid depends on the alcohols undergoing oxidation and the conditions used. Table 10.5 summarizes these reactions.

Thiols are compounds of the type RSH (Section 10.6). Thiols are more acidic than alcohols and are readily deprotonated by reaction with aqueous base (Section 10.7). Thiols can be oxidized to disulfides (RSSR), sulfenic acids (RSOH), sulfinic acids (RSO_2H), and sulfonic acids (RSO_3H).

Ethers are compounds that contain a C—O—C linkage (Section 10.8). They may be named in the IUPAC system as **alkoxy** derivatives of alkanes or by naming each alkyl group as a separate word (in alphabetical order) followed by the word *ether* (Section 10.9).

$$CH_3OCH_2CH_2CH_2CH_2CH_2CH_3$$

1-Methoxyhexane
(hexyl methyl ether)

TABLE 10.5	Oxidation of Alcohols	
Class of Alcohol	**Desired Product**	**Suitable Oxidizing Agent(s)**
RCH_2OH Primary	O ‖ RCH Aldehyde	PCC* PDC
RCH_2OH Primary	O ‖ RCOH Carboxylic acid	$Na_2Cr_2O_7$, H_2SO_4, H_2O H_2CrO_4
RCHR′ | OH Secondary	O ‖ RCR′ Ketone	PCC PDC $Na_2Cr_2O_7$, H_2SO_4, H_2O, H_2CrO_4

* PCC is pyridinium chlorochromate; PDC is pyridinium dichromate. Both are used in dichloromethane.

Epoxides are normally named as **epoxy** derivatives of alkanes.

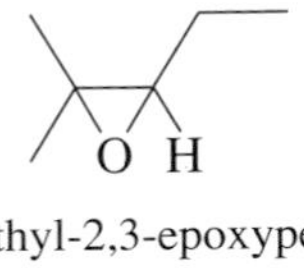

2-Methyl-2,3-epoxypentane

Ethers may be prepared by the Williamson ether synthesis (Section 10.10). An alkoxide ion displaces a halide leaving group in an S_N2 reaction.

$$RO^- \ + \ R'CH_2X \ \longrightarrow \ ROCH_2R' \ + \ X^-$$

Alkoxide ion	Primary alkyl halide	Ether	Halide ion

$$(CH_3)_2CHCH_2ONa \ + \ CH_3CH_2Br \ \longrightarrow \ (CH_3)_2CHCH_2OCH_2CH_3 \ + \ NaBr$$

Sodium isobutoxide	Ethyl bromide	Ethyl isobutyl ether (66%)	Sodium bromide

The reaction proceeds best with methyl or primary alkyl halides. Elimination, not substitution, predominates when the alkyl halide is secondary or tertiary.

Epoxides may be prepared from alkenes by epoxidation with a peroxy acid (Section 10.11).

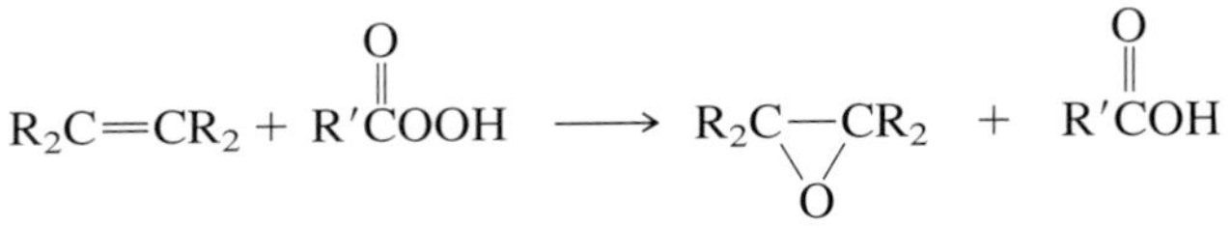

$$R_2C{=}CR_2 + R'\overset{O}{\overset{\|}{C}}OOH \ \longrightarrow \ R_2C\overset{}{-}CR_2 \ + \ R'\overset{O}{\overset{\|}{C}}OH$$

Alkene	Peroxy acid	Epoxide	Carboxylic acid

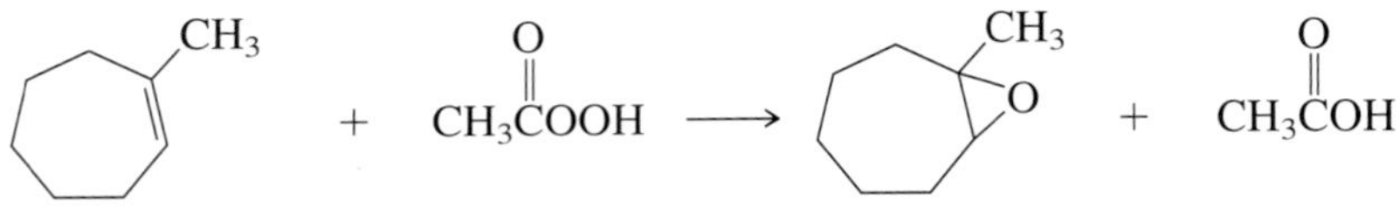

1-Methylcycloheptene	Peroxyacetic acid	1-Methyl-1,2-epoxycycloheptane (65%)	Acetic acid

Epoxides react with nucleophiles to give products in which the three-membered ring has opened (Section 10.12).

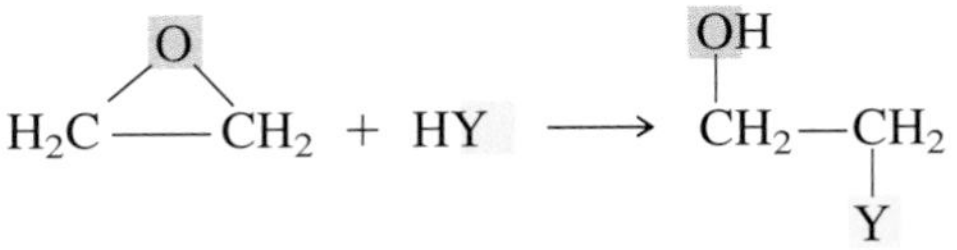

$$H_2C\overset{O}{\overline{\quad\quad}}CH_2 + HY \ \longrightarrow \ \overset{OH}{\underset{}{CH_2}}{-}\underset{Y}{CH_2}$$

Phenols are more acidic than alcohols (Section 10.15) because the phenoxide ion is stabilized by delocalization of the negative charge into the aromatic ring. A typical K_a is about 10^{-10} (pK_a = 10).

$$ArOH \ \rightleftharpoons \ H^+ \ + \ ArO^-$$

Phenol		Phenoxide anion

Phenoxide ions, prepared from phenols in base, react with alkyl halides to give alkyl aryl ethers (Section 10.16).

$$ArO^- \ + \ RX \ \longrightarrow \ ArOR \ + \ X^-$$

| Phenoxide anion | Alkyl halide | Alkyl aryl ether | Halide anion |

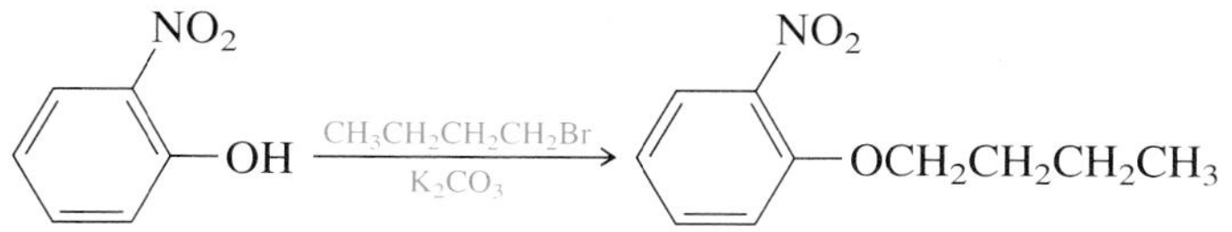

o-Nitrophenol

Butyl o-nitrophenyl
ether
(75–80%)

Oxidation of 1,4-benzenediols gives colored compounds known as quinones (Section 10.17).

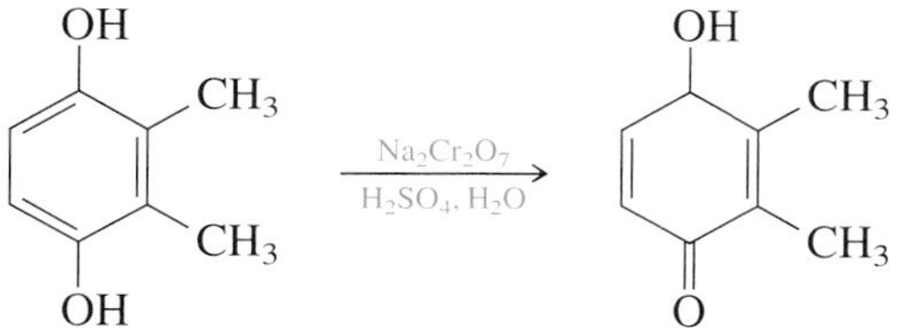

ADDITIONAL PROBLEMS

Alcohols: Preparation and Reactions

10.13 Write chemical equations showing all necessary reagents for the preparation of 1-butanol by each of the following methods:

(a) Reduction with sodium borohydride
(b) Reduction with lithium aluminum hydride
(c) Catalytic hydrogenation
(d) Nucleophilic substitution of an alkyl halide

10.14 Repeat parts (a) to (c) of Problem 10.13 for the preparation of cyclohexanol.

10.15 Give the structure of the major organic product from the reaction of 1-butanol with each of the following reagents.

(a) PCC in CH_2Cl_2 (c) Sodium amide ($NaNH_2$)
(b) $K_2Cr_2O_7$, H_2SO_4, H_2O

10.16 Repeat Problem 10.15 with 2-butanol as the reactant.

10.17 By writing the appropriate chemical equations, show how 1-hexanol could be converted into:

(a) Hexanal $CH_3CH_2CH_2CH_2CH_2\overset{\overset{\displaystyle O}{\|}}{C}H$

(b) Hexanoic acid $CH_3CH_2CH_2CH_2CH_2\overset{\overset{\displaystyle O}{\|}}{C}OH$

10.18 Which of the isomeric $C_5H_{12}O$ alcohols can be prepared by sodium borohydride reduction of a carbonyl compound?

10.19 Evaluate the feasibility of the route

$$RH \xrightarrow[\substack{\text{light} \\ \text{or heat}}]{Br_2} RBr \xrightarrow{KOH} ROH$$

as a method for preparing 2-methyl-2-propanol from 2-methylpropane.

10.20 Write equations showing how 1-phenylethanol $C_6H_5\underset{\underset{OH}{|}}{C}HCH_3$ could be prepared from each of the following starting materials:

(a) Acetophenone $C_6H_5\overset{\overset{O}{\|}}{C}CH_3$ (b) Benzene

Ethers and Epoxides

10.21 Write the structures of all the constitutionally isomeric ethers of molecular formula $C_5H_{12}O$, and give an acceptable name for each.

10.22 Many ethers, including diethyl ether, are effective as general anesthetics. Because simple ethers are quite flammable, their place in medical practice has been taken by highly halogenated nonflammable ethers. Two such general anesthetic agents are isoflurane and enflurane. These compounds are isomeric: isoflurane is 1-chloro-2,2,2-trifluoroethyl difluoromethyl ether, and enflurane is 2-chloro-1,1,2-trifluoroethyl difluoromethyl ether. Write the structural formulas of isoflurane and enflurane.

10.23 Give a systematic name for disparlure, the gypsy moth pheromone whose structure was shown in Section 10.8.

10.24 The octane rating of unleaded gasoline may be boosted by adding a small amount of *tert*-butyl methyl ether (known commercially as MTBE). MTBE is the subject of some controversy due to groundwater contamination. Write the structure of this substance, and outline a method for its preparation using the Williamson method starting with an alcohol and an alkyl halide.

10.25 Show how each of the following ethers could be prepared by the Williamson ether synthesis:

(a) $CH_3CH_2OCH_2CH_3$ (b) $CH_3OCH_2CH_2CH_3$ (two ways)

10.26 Write the structure of the major organic products A to D formed from the following sequence of steps:

(a) 2-Methyl-1-butene $+ H_2O \xrightarrow{H^+}$ A
(b) A $+ NaNH_2 \longrightarrow$ B (B is the conjugate base of A)
(c) $C_6H_5CH_2OH + HBr \longrightarrow$ C
(d) B $+$ C $\longrightarrow$ D

10.27 Predict the major organic product of each of the following reactions. Specify stereochemistry where appropriate.

(a) $CH_3CH{=}CHCH_2Cl + (CH_3)_3CO^-K^+ \longrightarrow$

(b) $CH_3CH_2I +$
$$\underset{H}{\overset{CH_3CH_2}{\diagdown}}\underset{}{\overset{\vdots CH_3}{\underset{|}{C}}}{-}ONa \longrightarrow$$

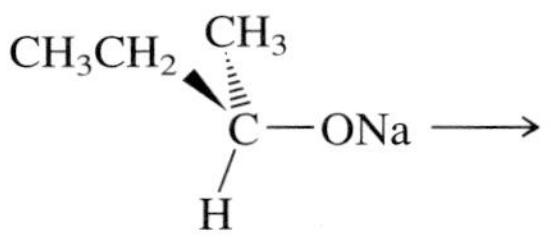

(c)
$$\underset{H}{\overset{C_6H_5}{\diagdown}}C{=}C\underset{H}{\overset{CH_3}{\diagup}} + \underset{}{\overset{O}{\diagdown}}C_6H_5{-}\overset{\overset{O}{\|}}{C}OOH \longrightarrow$$

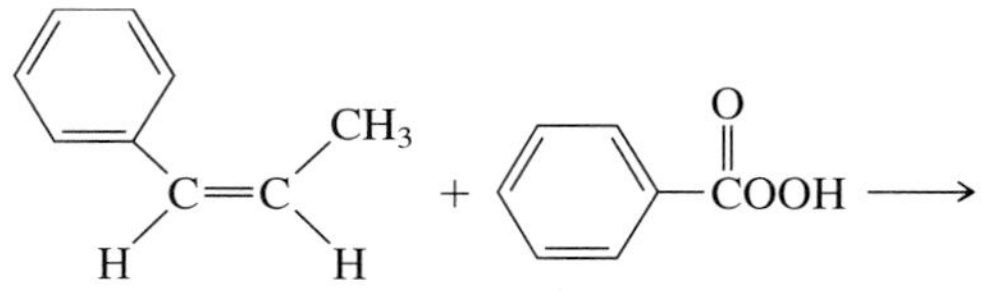

(d)

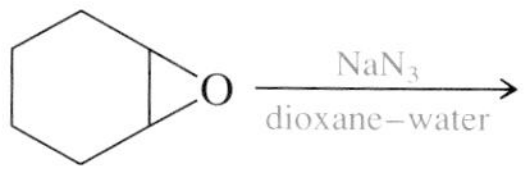

$$\xrightarrow[\text{dioxane-water}]{\text{NaN}_3}$$

Phenols and Aryl Ethers

10.28 The IUPAC rules permit the use of common names for a number of familiar phenols and aryl ethers. These common names are listed here along with their systematic names. Write the structure of each compound.

(a) Vanillin (4-hydroxy-3-methoxybenzaldehyde): a component of vanilla bean oil, which contributes to its characteristic flavor.

(b) Thymol (2-isopropyl-5-methylphenol): obtained from oil of thyme.

(c) Carvacrol (5-isopropyl-2-methylphenol): present in oil of thyme and marjoram.

(d) Salicyl alcohol (*o*-hydroxybenzyl alcohol): obtained from bark of poplar and willow trees. (*Hint:* Benzyl alcohol has the formula $C_6H_5CH_2OH$.)

10.29 Name each of the following compounds:

(a)

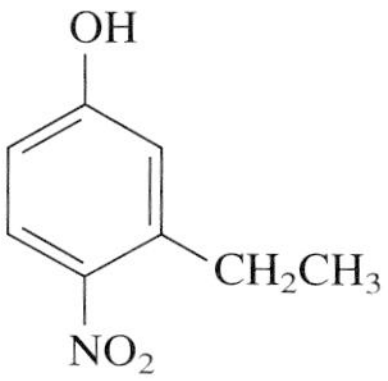

(c)

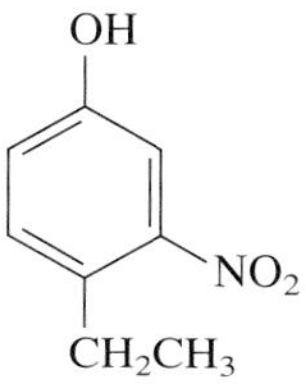

(b)

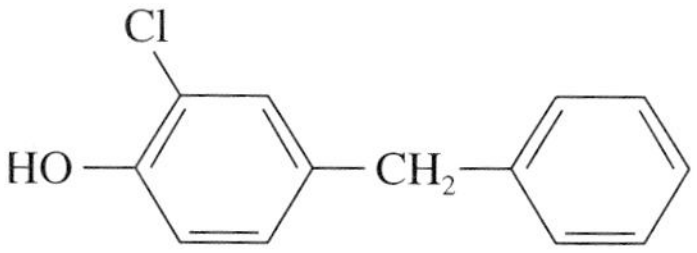

10.30 Fats and oils in food turn rancid by reaction with the oxygen in air. Small amounts of **antioxidants,** compounds that inhibit reaction with oxygen, are added to most prepackaged foods as preservatives. Two phenols, BHA and BHT, are widely used for this purpose.

Give acceptable IUPAC names for each of these compounds.

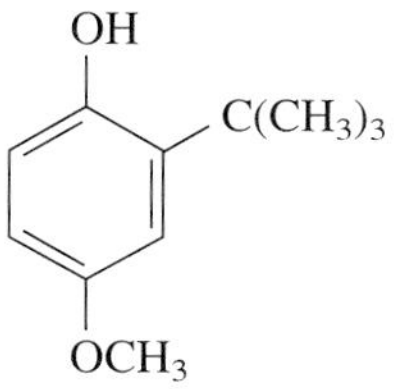

BHA
(*butylated
hydroxy anisole*)

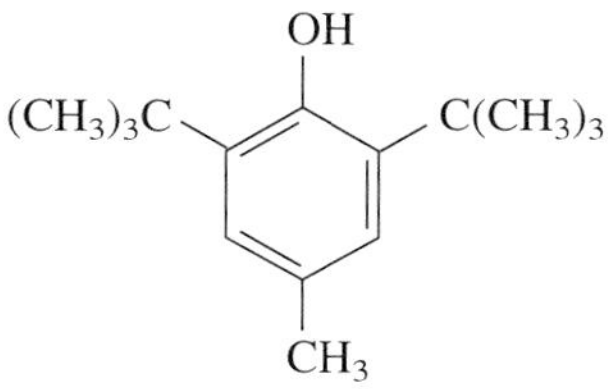

BHT
(*butylated
hydroxy toluene*)

10.31 Write a balanced chemical equation for each of the following reactions:

(a) Phenol + sodium hydroxide $\longrightarrow$
(b) Product of (a) + ethyl bromide $\longrightarrow$
(c) *m*-Cresol + ethylene oxide $\longrightarrow$

Miscellaneous Problems

10.32 The cis isomer of 3-hexen-1-ol ($CH_3CH_2CH{=\!=}CHCH_2CH_2OH$) has the characteristic odor of green leaves and grass. Suggest a synthesis for this compound from acetylene and any necessary organic or inorganic reagents. (*Hint:* You will find reviewing Sections 5.9 and 5.10 helpful for this problem.)

10.33 Each of the following reactions has been carried out by research chemists and described in a chemical journal. Although the molecules may be somewhat more complex than those we are used to seeing, the reaction types are ones we have encountered. Identify the major organic product in each case.

(a) 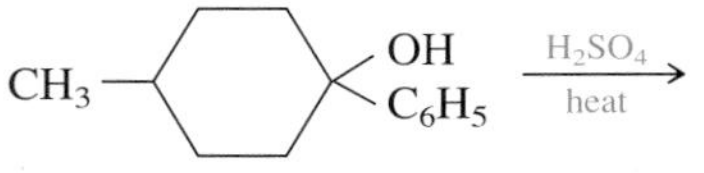 CH$_3$——⟨cyclohexane ring with OH and C$_6$H$_5$⟩ $\xrightarrow[\text{heat}]{\text{H}_2\text{SO}_4}$

(b) CH$_3$CHC≡C(CH$_2$)$_3$CH$_3$ $\xrightarrow[\text{H}_2\text{SO}_4,\ \text{H}_2\text{O}]{\text{H}_2\text{CrO}_4}$
       |
      OH

(c) CH$_3$CCH$_2$CH=CHCH$_2$CCH$_3$ $\xrightarrow[\text{2. H}_2\text{O}]{\text{1. LiAlH}_4,\ \text{diethyl ether}}$
      ‖   ‖
      O   O

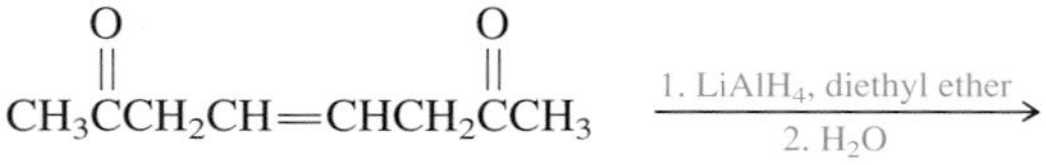

(d) 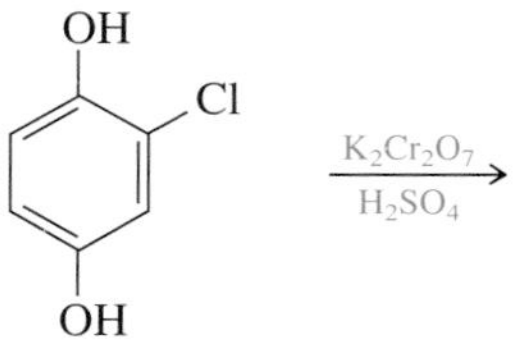 ⟨benzene ring with OH (top), Cl, OH (bottom)⟩ $\xrightarrow[\text{H}_2\text{SO}_4]{\text{K}_2\text{Cr}_2\text{O}_7}$

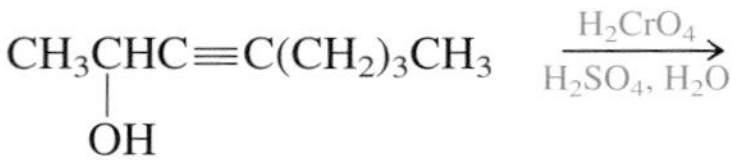

(e) 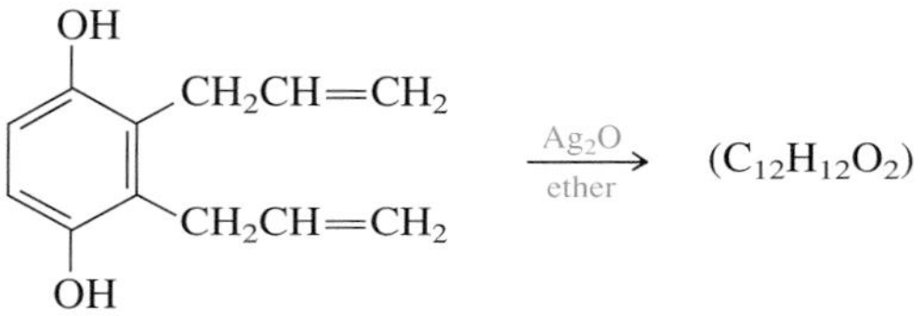 ⟨benzene ring with OH (top), CH$_2$CH=CH$_2$, CH$_2$CH=CH$_2$, OH (bottom)⟩ $\xrightarrow[\text{ether}]{\text{Ag}_2\text{O}}$ (C$_{12}$H$_{12}$O$_2$)

10.34 Sorbitol is a sweetener often substituted for cane sugar because it is better tolerated by diabetics. It is also an intermediate in the commercial synthesis of vitamin C. Sorbitol is prepared by high-pressure hydrogenation of glucose over a nickel catalyst. What is the structure (including stereochemistry) of sorbitol?

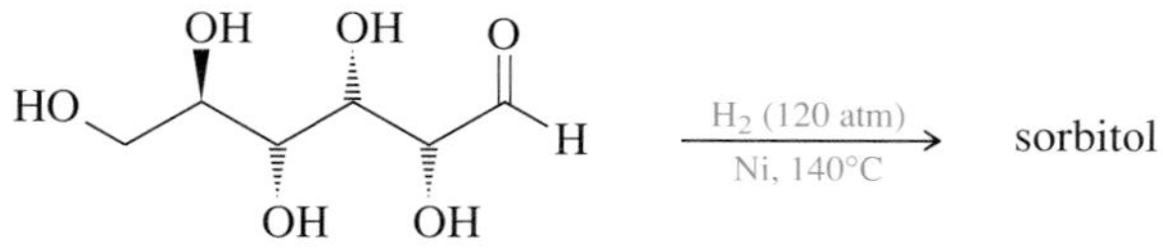

HO⟨chain with OH, OH, OH, OH groups and CHO⟩H $\xrightarrow[\text{Ni, 140°C}]{\text{H}_2\ (120\ \text{atm})}$ sorbitol

Glucose

10.35 R. B. Woodward was one of the leading organic chemists of the twentieth century. Known primarily for his achievements in the systhesis of complex natural products, he was awarded the Nobel Prize in chemistry in 1965. He entered Massachusetts Institute of Technology as a 16-year-old freshman in 1933 and four years later was awarded the Ph.D. While a student there he carried out a synthesis of **estrone,** a female sex hormone. The early stages of Woodward's estrone synthesis required the conversion of *m*-methoxybenzaldehyde to *m*-methoxybenzyl cyanide, which was accomplished in three steps.

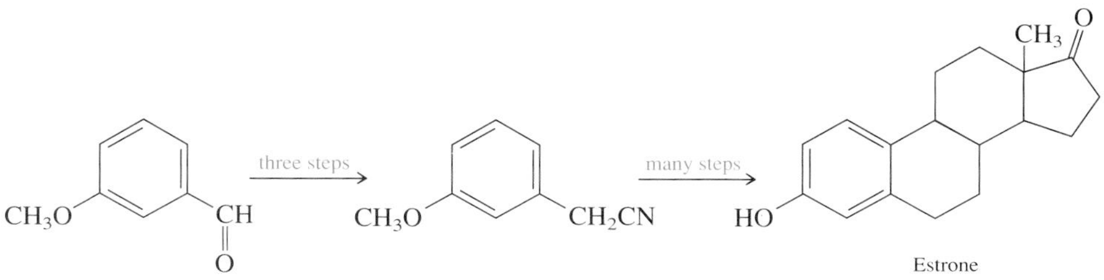

Estrone

Suggest a reasonable three-step sequence, showing all necessary reagents, for the preparation of *m*-methoxybenzyl cyanide from *m*-methoxybenzaldehyde.

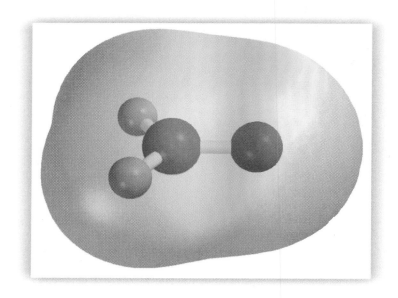

CHAPTER 11
ALDEHYDES AND KETONES

I n this chapter we continue our study of oxygen-containing organic compounds by examining aldehydes and ketones. Aldehydes and ketones occupy a key position among compounds that contain what is probably the most important functional group in organic and biological chemistry, the **carbonyl group.**

Carbonyl group   Aldehyde   Ketone

Aldehydes differ from ketones in having at least one hydrogen attached to the carbonyl group. **Ketones** have two alkyl or aryl (aromatic) groups (or one of each) attached to the carbon of the carbonyl group.

As you will see throughout this and the remaining chapters, the carbonyl group is found in numerous substances of biological interest, from flavorings in fruits to human sex hormones. Also, you will find that aldehydes and ketones are widely used as starting materials for the preparation of other classes of organic compounds.

11.1 NOMENCLATURE

The longest continuous chain that contains the $-\overset{\displaystyle O}{\overset{\displaystyle \|}{C}}H$ group provides the base name for aldehydes. The *-e* ending of the corresponding alkane name is replaced by *-al*, and substituents are specified in the usual way. It is not necessary to specify the location of the

$-\overset{\displaystyle O}{\overset{\displaystyle \|}{C}}H$ group in the name because the chain must be numbered by starting with this group as C-1.

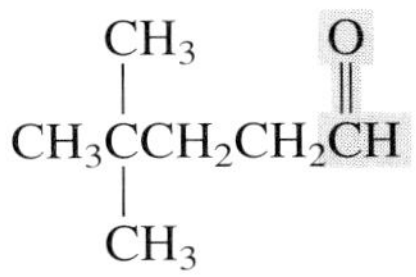

$$\underset{\underset{\displaystyle CH_3}{|}}{\overset{\overset{\displaystyle CH_3}{|}}{CH_3CCH_2CH_2}}\overset{\overset{\displaystyle O}{\|}}{CH}$$

4,4-Dimethylpentanal

$$CH_2{=}CHCH_2CH_2CH_2\overset{\overset{\displaystyle O}{\|}}{CH}$$

5-Hexenal

Certain common names of familiar aldehydes are acceptable as IUPAC names. A few examples include

$$\overset{\overset{\displaystyle O}{\|}}{HCH}$$

Formaldehyde
(methanal)

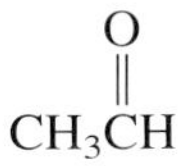

$$CH_3\overset{\overset{\displaystyle O}{\|}}{CH}$$

Acetaldehyde
(ethanal)

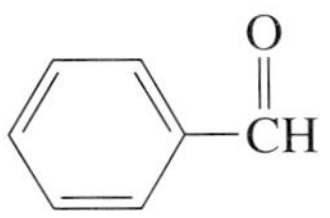

Benzaldehyde
(benzenecarbaldehyde)

PROBLEM 11.1 The common names and structural formulas of a few aldehydes follow. Provide an IUPAC name.

(a) 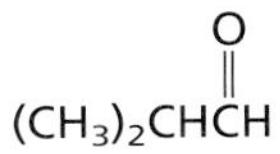 $(CH_3)_2CH\overset{\overset{\displaystyle O}{\|}}{CH}$

(isobutyraldehyde)

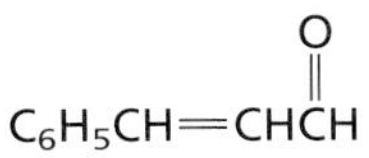

(b) 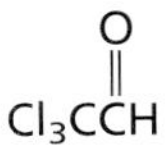 $Cl_3C\overset{\overset{\displaystyle O}{\|}}{CH}$

(chloral)

(c) $C_6H_5CH{=}CH\overset{\overset{\displaystyle O}{\|}}{CH}$

(cinnamaldehyde)

(d) 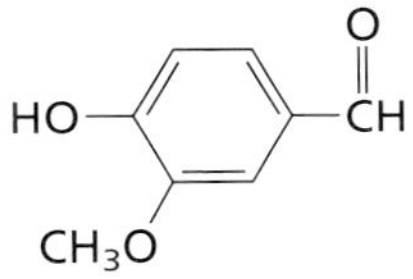   HO— ... —$\overset{\overset{\displaystyle O}{\|}}{CH}$
        CH_3O

(vanillin)

Sample Solution   (a) Don't be fooled by the fact that the common name is isobutyraldehyde. The longest continuous chain has three carbons, and so the base name is **propanal.** A methyl group occurs at C-2; thus the compound is 2-methylpropanal.

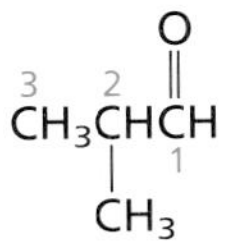

$$\underset{\underset{\displaystyle CH_3}{|}}{\overset{3}{CH_3}\overset{2}{CH}\overset{\overset{\displaystyle O}{\|}}{\underset{1}{CH}}}$$

2-Methylpropanal
(isobutyraldehyde)

With ketones, the -*e* ending of an alkane is replaced by -*one* in the longest continuous chain containing the carbonyl group. The chain is numbered in the direction that provides the lower number for this group.

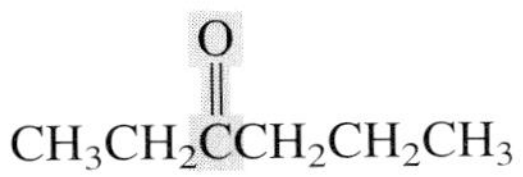

$$CH_3CH_2\overset{\overset{\displaystyle O}{\|}}{C}CH_2CH_2CH_3$$

3-Hexanone

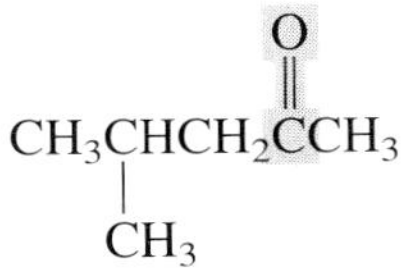

$$\underset{\underset{\displaystyle CH_3}{|}}{CH_3CHCH_2}\overset{\overset{\displaystyle O}{\|}}{C}CH_3$$

4-Methyl-2-pentanone

$$CH_3{-}\bigcirc{=}O$$

4-Methylcyclohexanone

The IUPAC rules also permit ketones to be named by identifying the groups attached to the carbonyl group as separate words followed by the word *ketone*. The groups are listed alphabetically.

$$CH_3CH_2\overset{\overset{\displaystyle O}{\|}}{C}CH_2CH_2CH_3$$

Ethyl propyl
ketone

Benzyl ethyl ketone

PROBLEM 11.2   As with aldehydes, a number of ketones are known by their common names. For each of the following give a systematic name.

(a) $CH_3\overset{\overset{\displaystyle O}{\|}}{C}CH_3$

(b)

(c) $(CH_3)_3\overset{\overset{\displaystyle O}{\|}}{C}CH_3$

Acetone

Acetophenone

Pinacolone

Sample Solution   (a) Acetone is a common solvent used as nail polish remover and in organic chemistry laboratories to rinse glassware. The chain is three carbons long with the carbonyl at C-2; thus the systematic name for acetone is 2-propanone. Acetone has two methyl groups attached to the carbonyl and may also be called dimethyl ketone.

$$CH_3\overset{\overset{\displaystyle O}{\|}}{C}CH_3$$

2-Propanone
(dimethyl ketone)

Common names sometimes tell us that a carbonyl group is present. Two examples in human biochemistry are retin*al,* an aldehyde important in vision (see Figure 11.6), and progester*one,* a ketone that is a female sex hormone (Section 16.9)

11.2   STRUCTURE AND BONDING: THE CARBONYL GROUP

Two notable aspects of the carbonyl group are its geometry and its polarity. The carbonyl group and the atoms directly attached to it lie in the same plane. Formaldehyde, for example, is planar. The bond angles involving the carbonyl group of aldehydes and ketones are close to 120°.

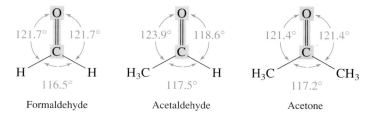

Formaldehyde   Acetaldehyde   Acetone

You have previously seen the *sp²*-hybridization bonding model in Sections 1.14, 3.9, and 4.2.

Using formaldehyde as an example, we can describe the bonding of a carbonyl group according to an sp^2-hybridization model analogous to that of ethylene, as shown in Figure 11.1.

Because oxygen is more electronegative than carbon (Section 1.5), the electron density in both the σ and π components of the carbon–oxygen double bond is displaced toward oxygen. The carbonyl group is polarized so that carbon is partially positive and oxygen is partially negative.

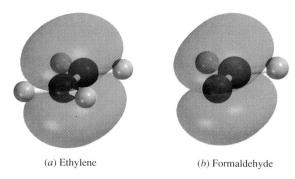

(*a*) Ethylene   (*b*) Formaldehyde

FIGURE 11.1   Similarities between the orbital hybridization models of bonding in (*a*) ethylene and (*b*) formaldehyde. Both molecules have the same number of electrons, and carbon is sp^2-hybridized in both. In formaldehyde, one of the carbons is replaced by an sp^2-hybridized oxygen (*shown in red*). Oxygen has two unshared electron pairs; each pair occupies an sp^2-hybridized orbital. Like the carbon–carbon double bond of ethylene, the carbon–oxygen double bond of formaldehyde is composed of a two-electron σ component and a two-electron π component.

$$\overset{\delta+}{\underset{}{C}}{=}\overset{\delta-}{O} \qquad \text{or} \qquad C{=}O$$

In resonance terms, electron delocalization in the carbonyl group is represented by contributions from two principal resonance structures:

$$C{=}\ddot{O}: \longleftrightarrow \ {}^+C{-}\ddot{\underset{..}{O}}:{}^-$$

A   B

Of these two, A, having one more covalent bond and avoiding the separation of positive and negative charges that characterizes B, better describes the bonding in a carbonyl group.

> The chemistry of the carbonyl group is considerably simplified if you remember that carbon is partially positive (has carbocation character) and oxygen is partially negative (weakly basic).

11.3   PHYSICAL PROPERTIES

In general, aldehydes and ketones have higher boiling points than alkenes because they are more polar and the dipole–dipole attractive forces between molecules are stronger. But they have lower boiling points than alcohols because, unlike alcohols, two carbonyl groups can't form hydrogen bonds to each other.

	$CH_3CH_2CH{=}CH_2$	$CH_3CH_2CH{=}O$	$CH_3CH_2CH_2OH$
	1-Butene	Propanal	1-Propanol
bp (1 atm)	$-6°C$	$49°C$	$97°C$
Solubility in water (g/100 mL)	Negligible	20	Miscible in all proportions

Aldehydes and ketones can form hydrogen bonds with the protons of OH groups. This makes them more soluble in water than alkenes, but less soluble than alcohols.

11.4   SOURCES OF ALDEHYDES AND KETONES

As we'll see later in this chapter and the next, aldehydes and ketones are involved in many of the most-used reactions in synthetic organic chemistry. Where do aldehydes and ketones come from?

FIGURE 11.2 Some naturally occurring aldehydes and ketones.

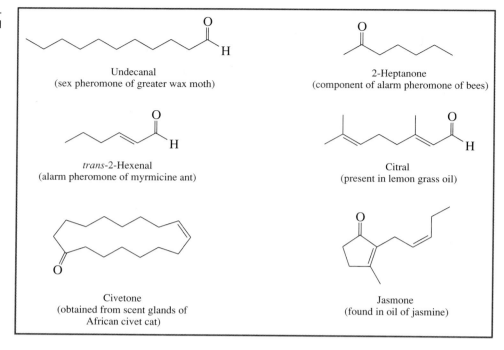

Undecanal
(sex pheromone of greater wax moth)

2-Heptanone
(component of alarm pheromone of bees)

trans-2-Hexenal
(alarm pheromone of myrmicine ant)

Citral
(present in lemon grass oil)

Civetone
(obtained from scent glands of
African civet cat)

Jasmone
(found in oil of jasmine)

Many occur naturally. In terms of both variety and quantity, aldehydes and ketones rank among the most common and familiar natural products. Several are shown in Figure 11.2.

Many are made in the laboratory from alkenes, alkynes, arenes, and alcohols by reactions that you already know about and are summarized in Table 11.1.

Many low-molecular-weight aldehydes and ketones are important industrial chemicals. Formaldehyde, a starting material for a number of plastics, is prepared by oxidation of methanol over a silver or iron oxide/molybdenum oxide catalyst at elevated temperature.

> The name *aldehyde* was invented to stand for *al*cohol *dehyd*rogenatum, indicating that aldehydes are related to alcohols by loss of hydrogen.

$$CH_3OH \ + \ \tfrac{1}{2}O_2 \ \xrightarrow[\text{500°C}]{\text{catalyst}} \ \overset{\displaystyle O}{\overset{\displaystyle \|}{HCH}} \ + \ H_2O$$

Methanol   Oxygen   Formaldehyde   Water

Similar processes are used to convert ethanol to acetaldehyde and isopropyl alcohol to acetone.

11.5 REACTIONS OF ALDEHYDES AND KETONES: A REVIEW AND A PREVIEW

As you have seen in Chapter 10 (Section 10.3), one characteristic reaction of aldehydes and ketones is their reduction to alcohols.

Aldehyde or
ketone

Primary or
secondary alcohol

TABLE 11.1 Summary of Reactions Discussed in Earlier Chapters That Yield Aldehydes and Ketones

Reaction (section) and Comments	General Equation and Specific Example

Hydration of alkynes (Section 5.11) Reaction occurs by way of an enol intermediate formed by Markovnikov addition of water to the triple bond.

$$RC{\equiv}CR' + H_2O \xrightarrow[\text{HgSO}_4]{\text{H}_2\text{SO}_4} \overset{\displaystyle O}{\overset{\|}{RCCH_2R'}}$$

Alkyne   Ketone

$$HC{\equiv}C(CH_2)_5CH_3 + H_2O \xrightarrow[\text{HgSO}_4]{\text{H}_2\text{SO}_4} \overset{\displaystyle O}{\overset{\|}{CH_3C(CH_2)_5CH_3}}$$

1-Octyne   2-Octanone (91%)

Friedel–Crafts acylation of aromatic compounds (Section 6.6) Acyl chlorides and carboxylic acid anhydrides acylate aromatic rings in the presence of aluminum chloride. The reaction is electrophilic aromatic substitution in which acylium ions are generated and attack the ring.

$$ArH + \overset{\displaystyle O}{\overset{\|}{RCCl}} \xrightarrow{\text{AlCl}_3} \overset{\displaystyle O}{\overset{\|}{ArCR}} + HCl \quad \text{or}$$

$$ArH + \overset{\displaystyle O\ \ O}{\overset{\|\ \ \|}{RCOCR}} \xrightarrow{\text{AlCl}_3} \overset{\displaystyle O}{\overset{\|}{ArCR}} + RCO_2H$$

CH₃O—〈 〉 + CH₃COCCH₃ (with two C=O) $\xrightarrow{\text{AlCl}_3}$ CH₃O—〈 〉—CCH₃ (with C=O)

Anisole   Acetic anhydride   p-Methoxyacetophenone (90–94%)

Oxidation of primary alcohols to aldehydes (Section 10.5) Pyridinium dichromate (PDC) or pyridinium chlorochromate (PCC) in anhydrous media such as dichloromethane oxidizes primary alcohols to aldehydes while avoiding overoxidation to carboxylic acids.

$$RCH_2OH \xrightarrow[\text{CH}_2\text{Cl}_2]{\text{PDC or PCC}} \overset{\displaystyle O}{\overset{\|}{RCH}}$$

Primary alcohol   Aldehyde

$$CH_3(CH_2)_8CH_2OH \xrightarrow[\text{CH}_2\text{Cl}_2]{\text{PDC}} \overset{\displaystyle O}{\overset{\|}{CH_3(CH_2)_8CH}}$$

1-Decanol   Decanal (98%)

Oxidation of secondary alcohols to ketones (Section 10.5) Many oxidizing agents are available for converting secondary alcohols to ketones. PDC or PCC may be used, as well as other Cr(VI)-based agents such as chromic acid or potassium dichromate and sulfuric acid.

$$\underset{\underset{\displaystyle OH}{|}}{RCHR'} \xrightarrow{\text{Cr(VI)}} \overset{\displaystyle O}{\overset{\|}{RCR'}}$$

Secondary alcohol   Ketone

$$\underset{\underset{\displaystyle OH}{|}}{C_6H_5CHCH_2CH_2CH_2CH_3} \xrightarrow[\substack{\text{acetic acid/} \\ \text{water}}]{\text{CrO}_3} \overset{\displaystyle O}{\overset{\|}{C_6H_5CCH_2CH_2CH_2CH_3}}$$

1-Phenyl-1-pentanol   1-Phenyl-1-pentanone (93%)

Appropriate reducing agents include hydrogen in the presence of a metal catalyst, sodium borohydride, or lithium aluminum hydride.

The polar nature of the carbonyl group is an important factor in almost all aldehyde and ketone reactions. Nucleophiles attack the electron-deficient carbon of the carbonyl group; electrophiles attack the electron-rich oxygen of the carbonyl.

| Nucleophiles attack carbon | $\overset{\delta+}{C} = \overset{\delta-}{O}$ | Electrophiles, such as H^+, bond to oxygen |

Many of the reactions presented in the remaining sections of this chapter involve **nucleophilic addition** to the carbonyl group and may be represented by the equation:

$$\overset{\delta-}{\underset{\delta+}{C}} = \overset{\delta-}{O} \;+\; \overset{\delta+}{X} - \overset{\delta-}{Y} \;\longrightarrow\; C \overset{O-X}{\underset{Y}{\Big\backslash}}$$

Aldehyde   Product of
or ketone   nucleophilic addition

The next section explores the mechanism of nucleophilic addition to aldehydes and ketones. There we'll discuss their **hydration,** a reaction in which water adds to the $C = O$ group. After we use this reaction to develop some general principles, we'll then survey a number of related reactions of synthetic, mechanistic, or biological interest.

11.6 PRINCIPLES OF NUCLEOPHILIC ADDITION: HYDRATION OF ALDEHYDES AND KETONES

Effects of Structure on Equilibrium

Aldehydes and ketones react with water in a rapid equilibrium:

$$\underset{\substack{\text{Aldehyde} \\ \text{or ketone}}}{\overset{\overset{\displaystyle O}{\|}}{RCR'}} \;+\; \underset{\text{Water}}{H_2O} \;\overset{fast}{\rightleftharpoons}\; \underset{\substack{\text{Geminal diol} \\ \text{(hydrate)}}}{\overset{\displaystyle OH}{\underset{\displaystyle OH}{RCR'}}} \qquad K_{hydr} = \frac{[\text{hydrate}]}{[\text{carbonyl compound}][\text{water}]}$$

Overall, the reaction is classified as an **addition.** The elements of water add to the carbonyl group. Hydrogen becomes bonded to the negatively polarized carbonyl oxygen, hydroxyl to the positively polarized carbon.

The amount of hydrate present at equilibrium varies greatly and is normally much greater for aldehydes than for ketones. Formaldehyde ($H_2C = O$) exists in aqueous solution almost completely in the hydrate form (99+%), whereas only about 0.1% of acetone, $(CH_3)_2C = O$, is present as the hydrate. The position of equilibrium depends strongly on the nature of the carbonyl group and is influenced by a combination of **electronic** and **steric** effects.

Consider first the inductive effect of substituents on the stabilization of the carbonyl group. Substitutents, such as alkyl groups, that stabilize the carbonyl group decrease the extent of hydration.

A striking example of an electronic effect on carbonyl group stability and its relation to the equilibrium constant for hydration is seen in the case of hexafluoroacetone.

In contrast to the almost negligible hydration of acetone, hexafluoroacetone is completely hydrated.

$$CF_3\overset{\overset{\displaystyle O}{\|}}{C}CF_3 \ + \ H_2O \ \rightleftharpoons \ CF_3\overset{\overset{\displaystyle OH}{|}}{\underset{\underset{\displaystyle OH}{|}}{C}}CF_3 \qquad K_{hydr} = 22,000$$

Hexafluoroacetone   Water   1,1,1,3,3,3-Hexafluoro-
                                    2,2-propanediol

Instead of stabilizing the carbonyl group by electron donation as alkyl substituents do, trifluoromethyl groups destabilize it by withdrawing electrons. A less stabilized carbonyl group is associated with a greater equilibrium constant for addition.

> **PROBLEM 11.3** Chloral is a common name for trichloroethanal (see Problem 11.1b). A solution of chloral in water is called chloral hydrate; this material has featured prominently in countless detective stories as the notorious "Mickey Finn" knockout drops. Write a structural formula for chloral hydrate.

Now let's turn our attention to steric effects by looking at how the size of the groups that were attached to C=O affect K_{hydr}. The bond angles at carbon shrink from $\approx 120°$ to $\approx 109.5°$ as the hybridization changes from sp^2 in the reactant (aldehyde or ketone) to sp^3 in the product (hydrate). The increased crowding this produces in the hydrate is better tolerated, and K_{hydr} is greater when the groups are small (hydrogen) than when they are large (alkyl).

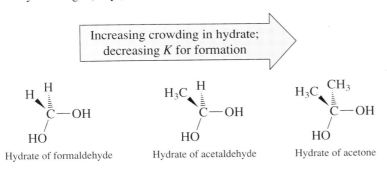

Increasing crowding in hydrate;
decreasing K for formation

Hydrate of formaldehyde   Hydrate of acetaldehyde   Hydrate of acetone

Electronic and steric effects operate in the same direction. Both cause the equilibrium constants for hydration of aldehydes to be greater than those of ketones.

Mechanism of Hydration

Hydration of aldehydes and ketones is a rapid reaction, quickly reaching equilibrium, but fac... in acid or base than in neutral solution. Thus instead of a single mechanism for hydration, we'll look at two mechanisms, one for basic and the other for acidic solution.

The base-catalyzed mechanism (Figure 11.3) is a two-step process in which the first step is rate-determining. In it, the nucleophile, a hydroxide ion, attacks the carbon of the carbonyl group and bonds to it. The product of this step is an alkoxide ion, which abstracts a proton from water in the second step, yielding the geminal diol. The second step, like all the other proton transfers between oxygens that we have seen, is fast.

The role of the basic catalyst (HO⁻) is to increase the rate of the nucleophilic addition step. Hydroxide ion, the nucleophile in the base-catalyzed reaction, is much more reactive than a water molecule, the nucleophile in neutral media.

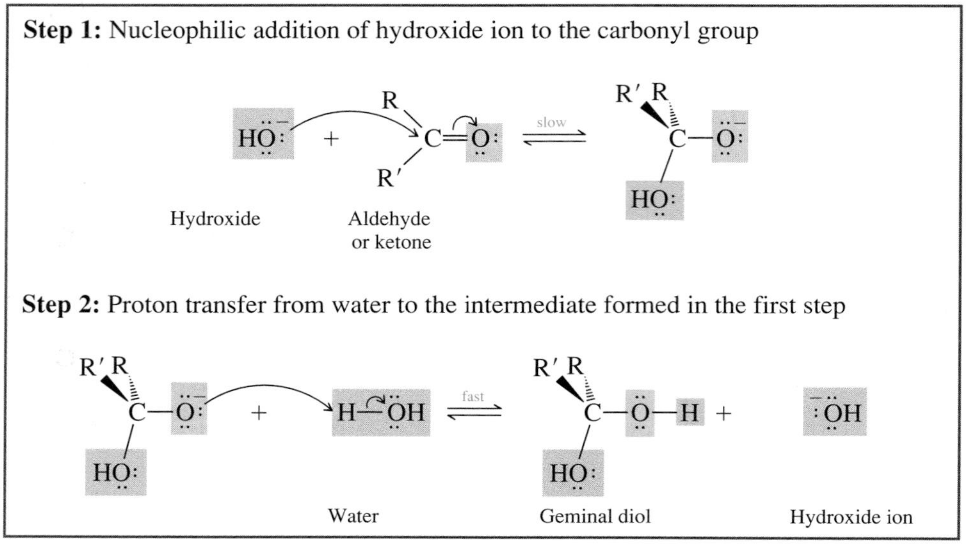

FIGURE 11.3 The mechanism of hydration of an aldehyde or ketone in basic solution. Hydroxide ion is a catalyst; it is consumed in the first step and regenerated in the second.

Step 1: Nucleophilic addition of hydroxide ion to the carbonyl group

Hydroxide   Aldehyde
                   or ketone

Step 2: Proton transfer from water to the intermediate formed in the first step

Water   Geminal diol   Hydroxide ion

Three steps are involved in the acid-catalyzed hydration reaction, as shown in Figure 11.4. The first and last are rapid proton-transfer processes. The second is the nucleophilic addition step. The acid catalyst activates the carbonyl group toward attack by a weakly nucleophilic water molecule. Protonation of oxygen makes the carbonyl carbon of an aldehyde or a ketone much more electrophilic. Expressed in resonance terms, the protonated carbonyl has a greater degree of carbocation character than an unprotonated carbonyl.

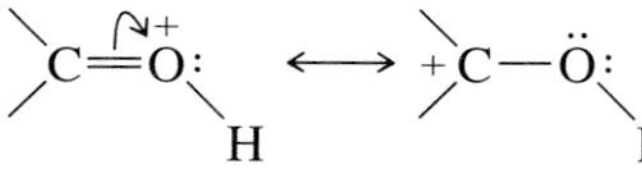

> **PROBLEM 11.4** Compared with the reaction in pure water, does the percentage hydrate in equilibrium with an aldehyde or ketone increase, decrease, or remain unchanged under conditions of acid catalysis? Basic catalysis? Explain.

The hydration of aldehydes and ketones illustrates some important features of nucleophilic additions.

Under basic conditions, the first step is *addition of the nucleophile to the positively polarized carbon of the carbonyl group.* Under acidic conditions, however, the first step is *protonation of the negatively polarized oxygen of the carbonyl group.*

The protonated species is better able to undergo nucleophilic addition by a weak nucleophile such as water.

With this as background, let us now examine how the principles of nucleophilic addition apply to the characteristic reactions of aldehydes and ketones. We'll begin with the addition of hydrogen cyanide.

11.7   CYANOHYDRIN FORMATION

The product of addition of hydrogen cyanide to an aldehyde or a ketone contains both a hydroxyl group and a cyano group bonded to the same carbon. Compounds of this type are called **cyanohydrins.**

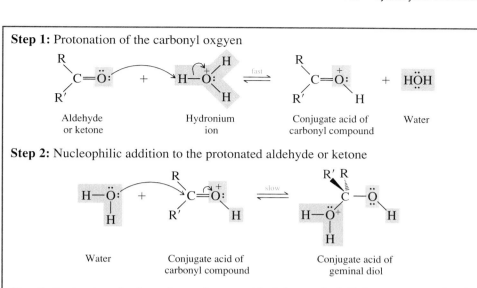

FIGURE 11.4 The mechanism of hydration of an aldehyde or ketone in acidic solution. Hydronium ion is a catalyst; it is consumed in the first step and regenerated in the third.

The mechanism of this reaction is outlined in Figure 11.5. It is analogous to the mechanism of base-catalyzed hydration in that the nucleophile (cyanide ion) attacks the carbonyl carbon in the first step of the reaction, followed by proton transfer to the carbonyl oxygen in the second step.

The addition of hydrogen cyanide is catalyzed by cyanide ion, but HCN is too weak an acid to provide enough :C≡N: for the reaction to proceed at a reasonable rate. Cyanohydrins are therefore normally prepared by adding an acid to a solution containing the carbonyl compound and sodium or potassium cyanide. This procedure ensures that free cyanide ion is always present in amounts sufficient to increase the rate of the reaction.

FIGURE 11.5 The mechanism of cyanohydrin formation from an aldehyde or a ketone. Cyanide ion is a catalyst; it is consumed in the first step, and regenerated in the second.

The overall reaction:

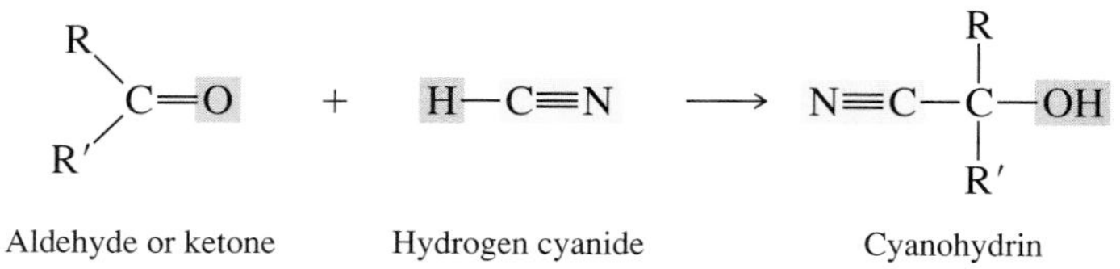

Aldehyde or ketone   Hydrogen cyanide   Cyanohydrin

Step 1: Nucleophilic attack by the negatively charged carbon of cyanide ion at the carbonyl carbon of the aldehyde or ketone. Hydrogen cyanide itself is not very nucleophilic and does not ionize to form cyanide ion to a significant extent. Thus, a source of cyanide ion such as NaCN or KCN is used.

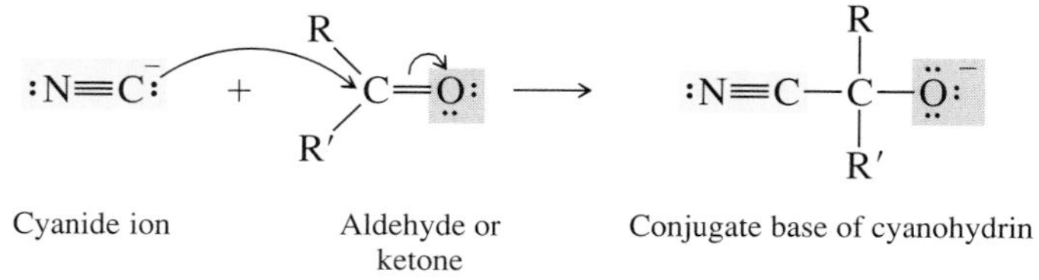

Cyanide ion   Aldehyde or   Conjugate base of cyanohydrin
                     ketone

Step 2: The alkoxide ion formed in the first step abstracts a proton from hydrogen cyanide. This step yields the cyanohydrin product and regenerates cyanide ion.

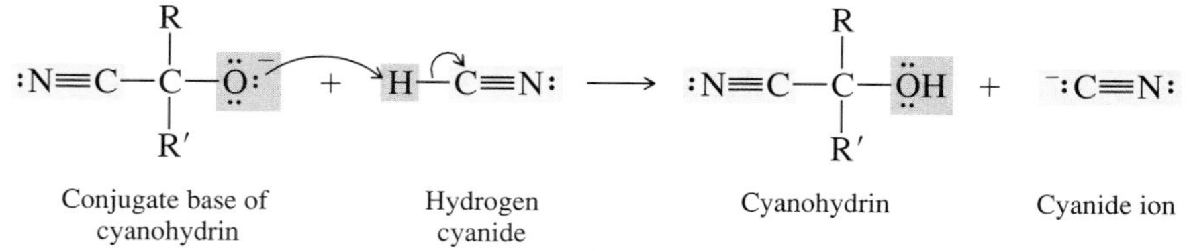

Conjugate base of   Hydrogen   Cyanohydrin   Cyanide ion
cyanohydrin   cyanide

Converting aldehydes and ketones to cyanohydrins is of synthetic value for two reasons: (1) a new carbon–carbon bond is formed, and (2) the cyano group in the product can be converted to a carboxylic acid function (CO_2H) by hydrolysis (to be discussed in Section 12.10) or to an amine of the type CH_2NH_2 by reduction (to be discussed in Section 14.6).

> **PROBLEM 11.5** The hydroxyl group of a cyanohydrin is also a potentially reactive site. Methacrylonitrile is an industrial chemical used in the production of plastics and fibers. One method for its preparation is the acid-catalyzed dehydration of acetone cyanohydrin. Deduce the structure of methacrylonitrile.

A few cyanohydrins and ethers of cyanohydrins occur naturally. One species of millipede stores benzaldehyde cyanohydrin, along with an enzyme that catalyzes its cleavage to benzaldehyde and hydrogen cyanide, in separate compartments above its legs. When attacked, the insect ejects a mixture of the cyanohydrin and the enzyme, repelling the invader by spraying it with hydrogen cyanide.

> **PROBLEM 11.6** Draw the structure of benzaldehyde cyanohydrin.

11.8   ACETAL FORMATION

Many of the most interesting and useful reactions of aldehydes and ketones involve transformation of the initial product of nucleophilic addition to some other substance under
the reaction conditions. An example is the reaction of aldehydes with alcohols under conditions of acid catalysis. The expected product of nucleophilic addition of the alcohol
to the carbonyl group is called a **hemiacetal.** The product actually isolated, however,
corresponds to reaction of one mole of the aldehyde with *two* moles of alcohol to give
geminal diethers known as **acetals:**

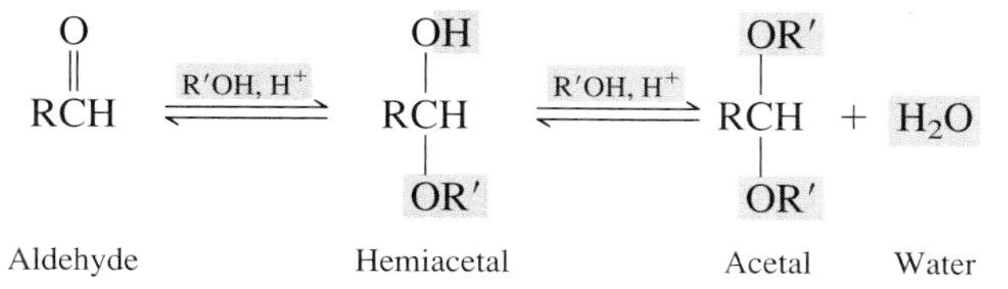

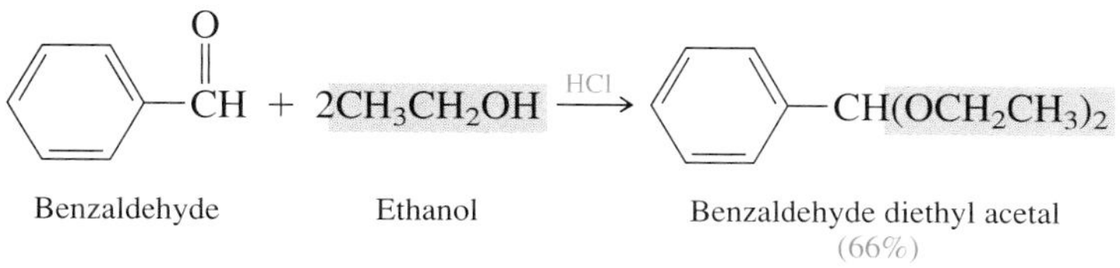

$$
\underset{\text{Aldehyde}}{\overset{\overset{\textstyle O}{\|}}{RCH}} \underset{}{\overset{\text{R'OH, H}^+}{\rightleftharpoons}} \underset{\text{Hemiacetal}}{\overset{\overset{\textstyle OH}{|}}{\underset{\underset{\textstyle OR'}{|}}{RCH}}} \underset{}{\overset{\text{R'OH, H}^+}{\rightleftharpoons}} \underset{\text{Acetal}}{\overset{\overset{\textstyle OR'}{|}}{\underset{\underset{\textstyle OR'}{|}}{RCH}}} + \underset{\text{Water}}{H_2O}
$$

Benzaldehyde + Ethanol → Benzaldehyde diethyl acetal (66%)

(reaction: benzaldehyde + 2CH₃CH₂OH → with HCl → C₆H₅CH(OCH₂CH₃)₂)

The overall reaction proceeds in two stages. The hemiacetal is formed in the first
stage by nucleophilic addition of the alcohol to the carbonyl group. The mechanism of
hemiacetal formation is exactly analogous to that of acid-catalyzed hydration of aldehydes and ketones (Section 11.6):

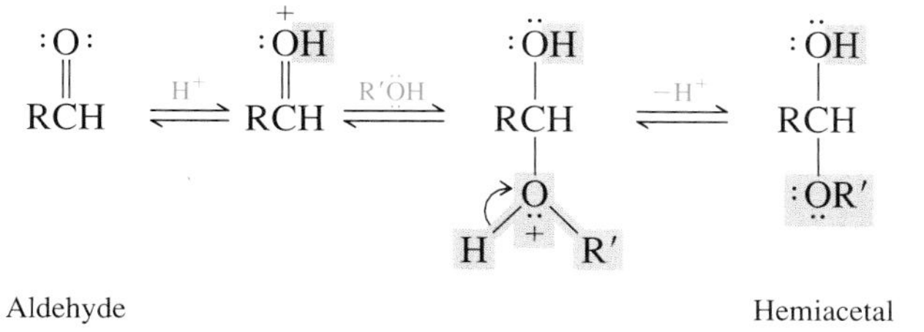

Aldehyde   Hemiacetal

Under the acidic conditions of its formation, the hemiacetal is converted to an acetal by
way of a carbocation intermediate:

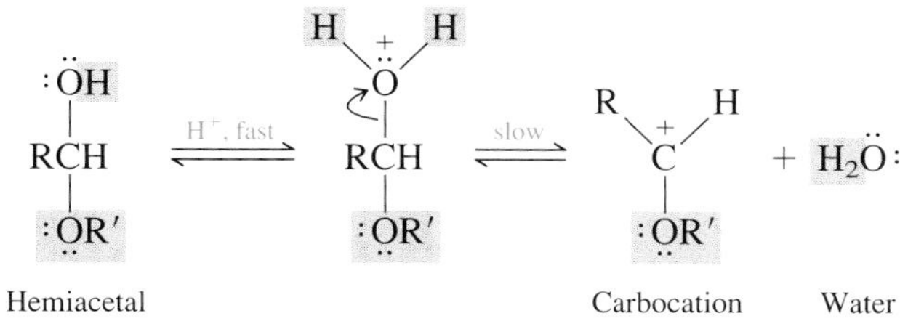

Hemiacetal   Carbocation   Water

This carbocation is stabilized by electron release from its oxygen substituent:

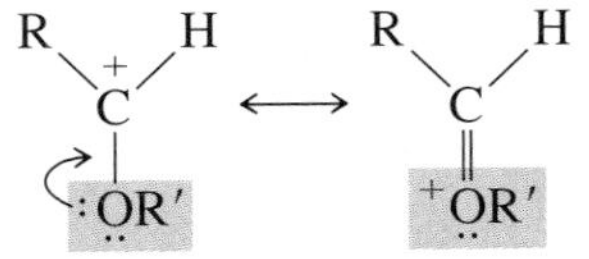

A particularly stable resonance form; both carbon and oxygen have octets of electrons.

Nucleophilic capture of the carbocation intermediate by an alcohol molecule leads to an acetal:

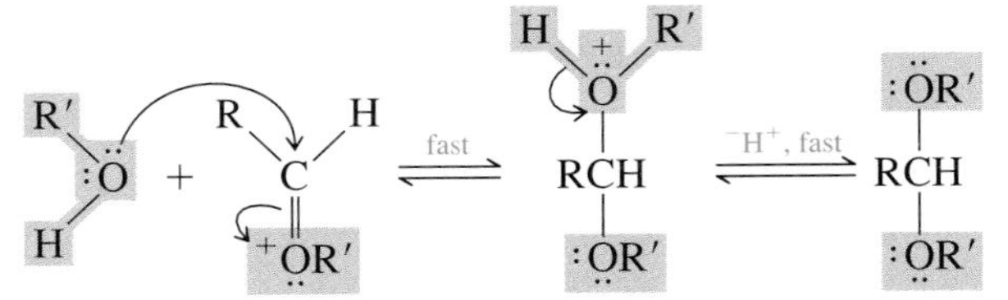

Alcohol   Acetal

> **PROBLEM 11.7**   Write a stepwise mechanism for the formation of benzaldehyde diethyl acetal from benzaldehyde and ethanol under conditions of acid catalysis.

Acetal formation is reversible in acid. An equilibrium is established between the reactants, that is, the carbonyl compound and the alcohol, and the acetal product. The position of equilibrium is favorable for acetal formation from most aldehydes, especially when excess alcohol is present as the reaction solvent. For most ketones the position of equilibrium is unfavorable, and other methods must be used for the preparation of acetals from ketones.

You will encounter hemiacetals and acetals again in Chapter 15 because they are central to understanding the structure and properties of carbohydrates.

> At one time it was customary to designate the products of addition of alcohols to ketones as **ketals.** This term has been dropped from the IUPAC system of nomenclature, and the term **acetal** is now applied to the adducts of both aldehydes and ketones.

11.9   REACTION WITH PRIMARY AMINES: IMINES

A second two-stage reaction that begins with nucleophilic addition to aldehydes and ketones is their reaction with primary amines, compounds of the type RNH_2 or $ArNH_2$. In the first stage of the reaction the amine adds to the carbonyl group to give a species known as a **carbinolamine.** Once formed, the carbinolamine undergoes dehydration to yield the product of the reaction, an N-alkyl- or N-aryl-substituted **imine:**

> N-Substituted imines are sometimes called **Schiff's bases,** after Hugo Schiff, a German chemist who described their formation in 1864.

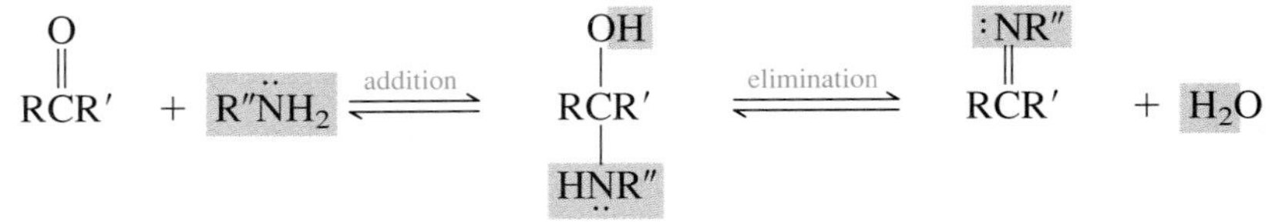

Aldehyde   Primary   Carbinolamine N-Substituted   Water
or ketone   amine   imine

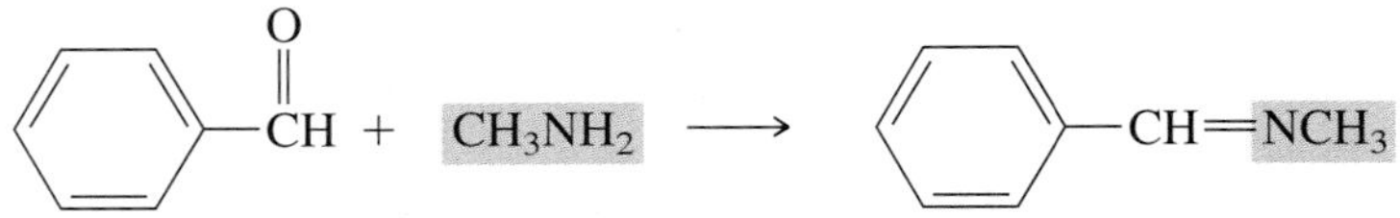

Benzaldehyde   Methylamine N-Benzylidenemethylamine
                                       (70%)

PROBLEM 11.8 Write the structure of the carbinolamine intermediate and the imine product formed in the reaction of each of the following:
(a) Acetaldehyde and benzylamine, $C_6H_5CH_2NH_2$
(b) Benzaldehyde and butylamine, $CH_3CH_2CH_2CH_2NH_2$
(c) Cyclohexanone and *tert*-butylamine, $(CH_3)_3CNH_2$

Sample Solution   The carbinolamine is formed by nucleophilic addition of the amine to the carbonyl group. Its dehydration gives the imine product.

$$CH_3\overset{\displaystyle O}{\overset{\|}{C}}H \;+\; C_6H_5CH_2NH_2 \longrightarrow CH_3\underset{\underset{\displaystyle H}{|}}{\overset{\overset{\displaystyle OH}{|}}{C}}H{-}NCH_2C_6H_5 \xrightarrow{-H_2O} CH_3CH{=}NCH_2C_6H_5$$

| Acetaldehyde | Benzylamine | Carbinolamine intermediate | Imine product (*N*-ethylidenebenzylamine) |

Imine formation is a reversible reaction. Both imine formation and hydrolysis have been studied extensively because of their relevance to biochemical processes. Many biological reactions involve initial binding of a carbonyl compound to an enzyme or coenzyme by way of imine formation (see the essay titled *Imines in Biological Chemistry* following this section).

We will conclude our discussion of nucleophilic additions with one of the most synthetically useful reactions of aldehydes and ketones, their reaction with Grignard reagents to form alcohols. Let us first take a brief look at Grignard reagents and why they are so valuable in organic chemistry.

11.10 REACTIONS THAT INTRODUCE NEW CARBON–CARBON BONDS: ORGANOMETALLIC COMPOUNDS

One of the challenges facing an organic chemist is how to construct the carbon skeleton of a desired compound. For reasons of availability and cost, one or more reactions that will change the number of carbons are often part of a synthesis.

As you will see in the next several sections, aldehydes and ketones react with substances called Grignard reagents to give alcohols having a greater number of carbons than the starting material. Before discussing this reaction, let us first examine some general aspects of carbon–carbon bond formation in a chemical reaction.

In what types of reactions are new carbon–carbon bonds formed? Friedel–Crafts alkylation and acylation (Section 6.6) are quite useful for this purpose, but these reactions are limited to the synthesis of aromatic compounds. A more general approach is a carbon nucleophile attacking another carbon as a substrate.

$$R{:}^- \;+\; R'{-}X \longrightarrow R{-}R' \;+\; X^-$$

Nucleophilic substitution by cyanide ion ($^-{:}C{\equiv}N{:}$) on an alkyl halide (Section 8.1) offers one example of this approach. Another example is the synthesis of substituted alkynes by alkylation of an acetylide anion, for example, sodium acetylide ($NaC{\equiv}CH$), with an alkyl halide (Section 5.10).

For a carbon atom to act as a nucleophile, it must have a pair of electrons that can be used to form a bond to a carbon atom of the substrate. In the case of cyanide or acetylide ion, the nucleophilic carbon atom bears a full negative charge. A nucleophilic

IMINES IN BIOLOGICAL CHEMISTRY

Many biological processes involve an "association" between two species in a step prior to some subsequent transformation. This association can take many forms. It can be a weak association of the attractive van der Waals type or a stronger interaction such as a hydrogen bond. It can be an electrostatic attraction between a positively charged atom of one molecule and a negatively charged atom of another. Covalent bond formation between two species of complementary chemical reactivity represents an extreme kind of association. It often occurs in biological processes in which aldehydes or ketones react with amines via imine intermediates.

An example of a biologically important aldehyde is **pyridoxal phosphate.** Pyridoxal phosphate is the active form of **vitamin B₆** and is a coenzyme for many of the reactions of α-amino acids. In these reactions the amino acid binds to the coenzyme by reacting with it to form an imine of the kind shown in the equation. Reactions then take place at the amino acid portion of the imine, modifying the amino acid. In the last step, enzyme-catalyzed hydrolysis cleaves the imine to pyridoxal and the modified amino acid.

A key step in the chemistry of vision is binding of an aldehyde to an enzyme via an imine. An outline of the steps involved is presented in Figure 11.6. It starts with **β-carotene,** a pigment that occurs naturally in several fruits and vegetables, including carrots. β-Carotene undergoes oxidative cleavage in the liver to give an alcohol known as **retinol** or **vitamin A.** Oxidation of vitamin A, followed by isomerization of one of its double bonds, gives the aldehyde 11-*cis*-retinal. In the eye, the aldehyde function of 11-*cis*-retinal combines with an amino group of the protein **opsin** to form an imine called **rhodopsin.** When rhodopsin absorbs a photon of visible light, the cis double bond of the retinal unit undergoes a photochemical cis-to-trans isomerization, which is attended by a dramatic change in its shape and a change in the conformation of rhodopsin. This conformational change is translated into a nerve impulse perceived by the brain as a visual image. Enzyme-promoted hydrolysis of the photochemically isomerized rhodopsin regenerates opsin and a molecule of all-*trans*-retinal. Once all-*trans*-retinal has been enzymatically converted to its 11-cis isomer, it and opsin reenter the cycle.

Pyridoxal phosphate   α-Amino acid   Imine

carbon atom may also have a *partial* negative charge, as is the case in compounds in which carbon is bonded to certain metals.

When carbon is covalently bonded to an atom *more* electronegative than itself, such as oxygen, nitrogen, or the halogens, the electron distribution in the bond is polarized so that carbon is slightly positive and the more electronegative atom is slightly negative. Conversely, when carbon is bonded to a less electronegative element, such as a metal, the electrons in the bond are more strongly attracted toward carbon.

X is more electronegative   M is less electronegative
than carbon   than carbon

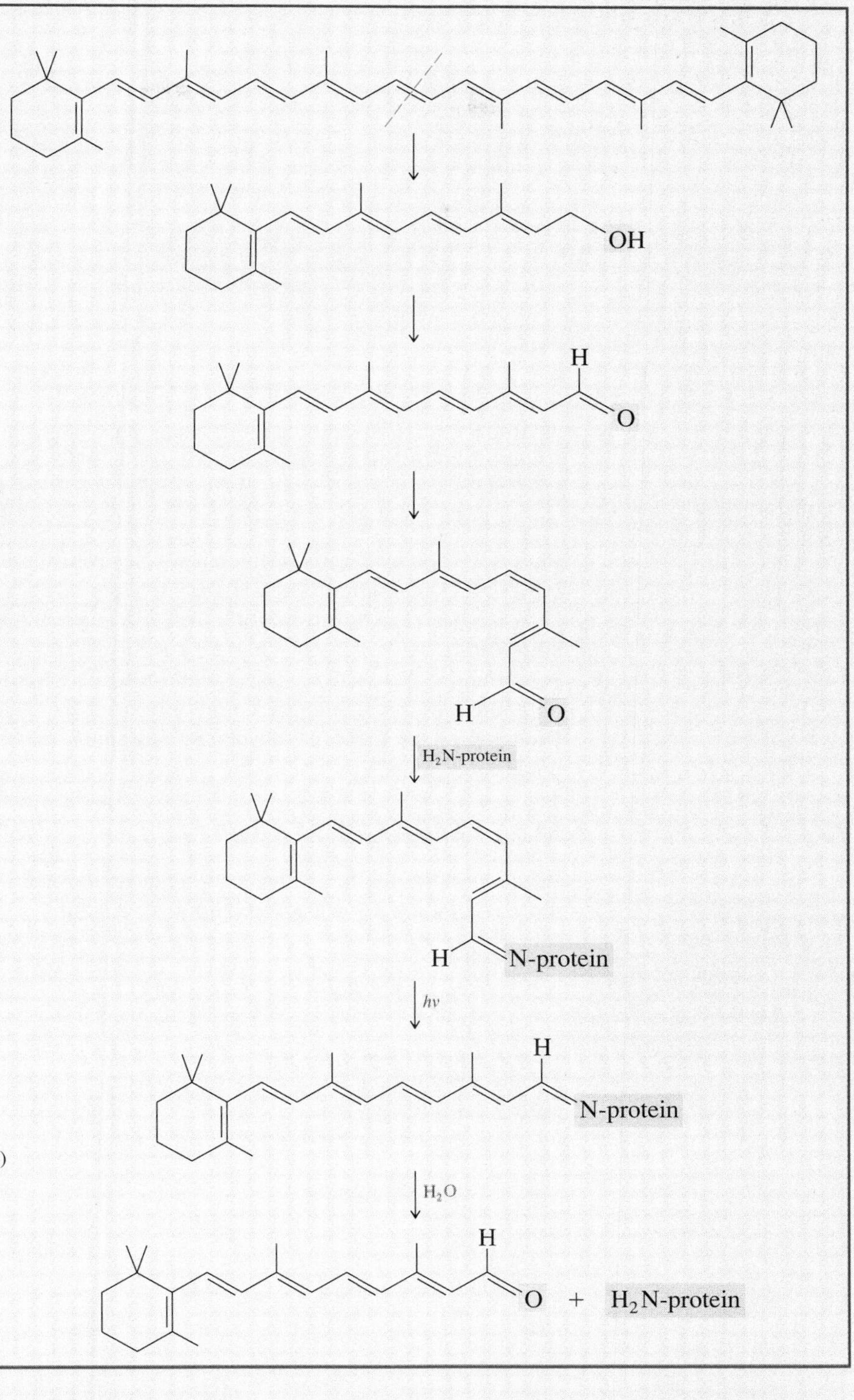
β-Carotene obtained from the diet is cleaved at its central carbon–carbon bond to give vitamin A (retinol)
OH
Oxidation of retinol converts it to the corresponding aldehyde, retinal.
H
O
The double bond at C-11 is isomerized from the trans to the cis configuration
H
O
11-*cis*-Retinal is the biologically active stereoisomer and reacts with the protein opsin to form an imine. The covalently bound complex between 11-*cis*-retinal and ospin is called **rhodopsin.**
H_2N-protein
H
N-protein
Rhodopsin absorbs a photon of light, causing the cis double-bond at C-11 to undergo a photochemical transformation to trans, which triggers a nerve impulse detected by the brain as a visual image.
$h\nu$
H
N-protein
Hydrolysis of the isomerized (inactive) form of rhodopsin liberates opsin and the all-trans isomer of retinal.
H_2O
H
O + H_2N-protein

FIGURE 11.6 Imine formation between the aldehyde function of 11-*cis*-retinal and an amino group of a protein (opsin) is involved in the chemistry of vision. The numbering scheme used in retinal is based on one specifically developed for carotenes and compounds derived from them.

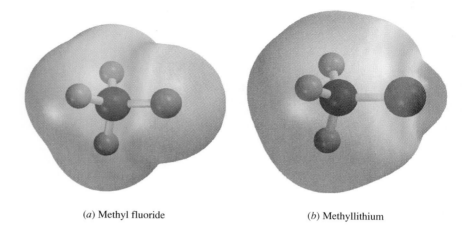

(*a*) Methyl fluoride   (*b*) Methyllithium

FIGURE 11.7   Electrostatic potential maps of (*a*) methyl fluoride and (*b*) methyllithium. The electron distribution is reversed in the two compounds. Carbon is electron-poor (*blue*) in methyl fluoride, but electron-rich (*red*) in methyllithium.

Figure 11.7 shows how different the electron distribution is between methyl fluoride (CH_3F) and methyllithium (CH_3Li).

A species in which carbon is directly bonded to a metal is called an **organometallic compound.** A negatively charged carbon is called a **carbanion,** and organometallic compounds are said to have **carbanionic character.** Although several types of organometallic reagents are commonly used by organic chemists, the most frequently encountered of these is the class of organomagnesium compounds called Grignard reagents.

11.11 GRIGNARD REAGENTS

Reaction of an organic halide with magnesium metal gives an organomagnesium halide, more commonly known as a **Grignard reagent.**

> Grignard reagents are named after the French chemist Victor Grignard. For his studies of organomagnesium compounds Grignard shared the 1912 Nobel Prize in chemistry.

$$RX \quad + \quad Mg \quad \xrightarrow{\text{diethyl ether}} \quad RMgX$$

Organic halide   Magnesium   Grignard reagent; an
                                               organomagnesium halide

The R group may be methyl or primary, secondary, or tertiary; it may also be a cycloalkyl, alkenyl, or aryl group. The halide may be an iodide, chloride, or bromide; fluorides are too unreactive toward magnesium to be useful synthetically.

Cyclohexyl chloride   Magnesium   Cyclohexylmagnesium chloride
                                                    (96%)

Bromobenzene   Magnesium   Phenylmagnesium bromide
                                             (95%)

Anhydrous diethyl ether is the customary solvent used when preparing organo-magnesium compounds. Sometimes the reaction does not begin readily, but once started, it is exothermic and maintains the temperature of the reaction mixture at the boiling point of diethyl ether (35°C).

PROBLEM 11.9   Write the structure of the Grignard reagent formed from each of the following compounds on reaction with magnesium in diethyl ether:

(a) *p*-Bromofluorobenzene   (c)   Iodocyclobutane

(b) Allyl chloride ($H_2C=CHCH_2Cl$)   (d) 1-Bromocyclohexene

Sample Solution   (a) Of the two halogen substituents on the aromatic ring, bromine reacts much faster than fluorine with magnesium. Therefore, fluorine is left intact on the ring, whereas the carbon–bromine bond is converted to a carbon–magnesium bond.

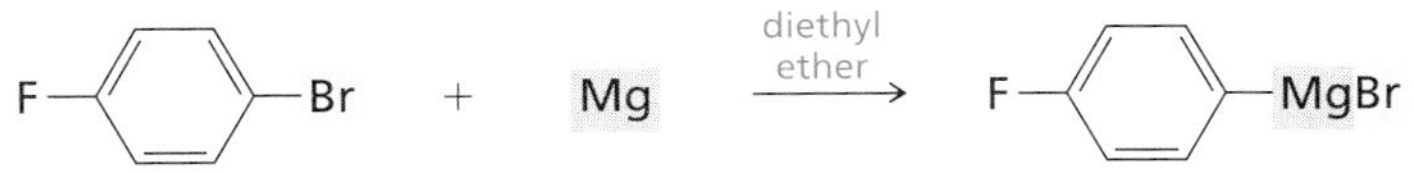

p-Bromofluorobenzene	Magnesium	*p*-Fluorophenylmagnesium bromide

---

Grignard reagents are stable species when prepared in suitable solvents such as anhydrous diethyl ether. They are strongly basic, however, and react instantly with proton donors even as weakly acidic as water and alcohols. A proton is transferred from the hydroxyl group to the negatively polarized carbon of the organometallic compound to form a hydrocarbon.

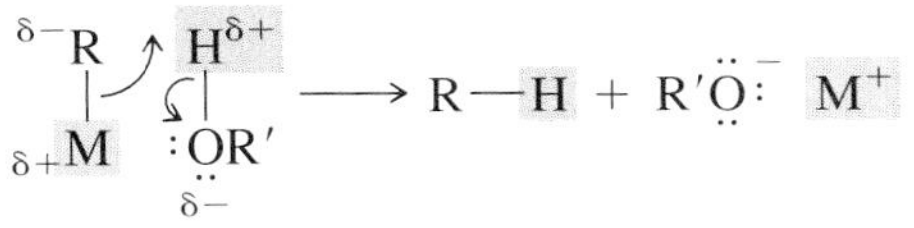

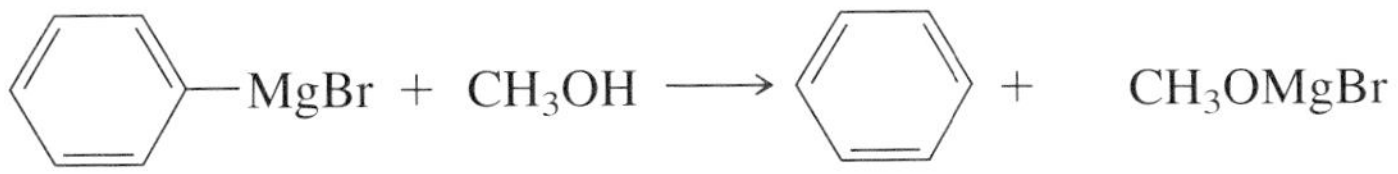

Phenylmagnesium bromide	Methanol	Benzene (100%)	Methoxymagnesium bromide

11.12 SYNTHESIS OF ALCOHOLS USING GRIGNARD REAGENTS

The main synthetic application of Grignard reagents is their reaction with certain carbonyl-containing compounds to produce alcohols. Carbon–carbon bond formation is rapid when a Grignard reagent reacts with an aldehyde or ketone.

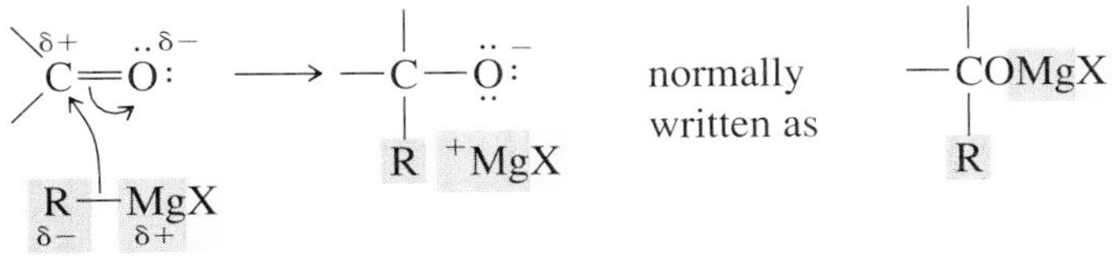

A carbonyl group is quite polar, and its carbon atom is electrophilic. Grignard reagents are nucleophilic and add to carbonyl groups, forming a new carbon–carbon bond. This **nucleophilic addition** step leads to an alkoxymagnesium halide, which in the second stage of the synthesis is converted to an alcohol by adding aqueous acid.

$$R-\overset{|}{\underset{|}{C}}-OMgX \; + \; H_3O^+ \; \longrightarrow \; R-\overset{|}{\underset{|}{C}}-OH \; + \; Mg^{2+} \; + \; X^- \; + \; H_2O$$

Alkoxymagnesium   Hydronium   Alcohol   Magnesium   Halide   Water
halide   ion   ion   ion

The type of alcohol produced depends on the carbonyl compound. Substituents present on the carbonyl group of an aldehyde or ketone stay there—they become substituents on the carbon that bears the hydroxyl group in the product. Thus as shown in Table 11.2, formaldehyde reacts with Grignard reagents to yield primary alcohols, aldehydes yield secondary alcohols, and ketones yield tertiary alcohols.

PROBLEM 11.10 Write the structure of the product of the reaction of propylmagnesium bromide with each of the following. Assume that the reactions are worked up by the addition of dilute aqueous acid.

(a) Formaldehyde $\overset{O}{\overset{\|}{HCH}}$

(b) Benzaldehyde $\overset{O}{\overset{\|}{C_6H_5CH}}$

(c) Cyclohexanone $\langle\bigcirc\rangle\!=\!O$

(d) 2-Butanone $CH_3\overset{O}{\overset{\|}{C}}CH_2CH_3$

Sample Solution (a) Grignard reagents react with formaldehyde to give primary alcohols having one more carbon atom than the alkyl halide from which the Grignard reagent was prepared. The product is 1-butanol.

$$CH_3CH_2CH_2-MgBr \;\; \xrightarrow{\text{diethyl ether}} \;\; CH_3CH_2CH_2-\underset{H}{\overset{H}{\underset{|}{\overset{|}{C}}}}-OMgBr \;\; \xrightarrow{H_3O^+} \;\; CH_3CH_2CH_2CH_2OH$$

$$\overset{H}{\underset{H}{\diagdown}}C\!=\!O$$

Propylmagnesium bromide   1-Butanol
+ formaldehyde

11.13 GRIGNARD REAGENTS IN SYNTHESIS

As you saw in the preceding section, Grignard reagents provide a straightforward method for preparing alcohols. What if you are given the structure of a target alcohol and asked to choose the starting materials? Which Grignard reagent would you choose? For that matter, could a Grignard reaction be used at all? Remember, alcohols can also be prepared by reduction of an aldehyde or ketone (Section 10.3). The key to answering these questions is to work the problem *backward*. A phrase used to describe the process of reasoning backward in a synthesis is **retrosynthetic analysis.** The symbol used to indicate a retrosynthetic step is an open arrow written from product to suitable precursors.

Target molecule $\Longrightarrow$ precursors

Consider for example the preparation of 1-pentanol. One method uses pentanal as the starting material.

| TABLE 11.2 | Reactions of Grignard Reagents with Aldehydes and Ketones |

Reaction	General Equation and Specific Example
Reaction with formaldehyde Grignard reagents react with formaldehyde ($H_2C=O$) to give **primary alcohols** having one more carbon than the Grignard reagent.	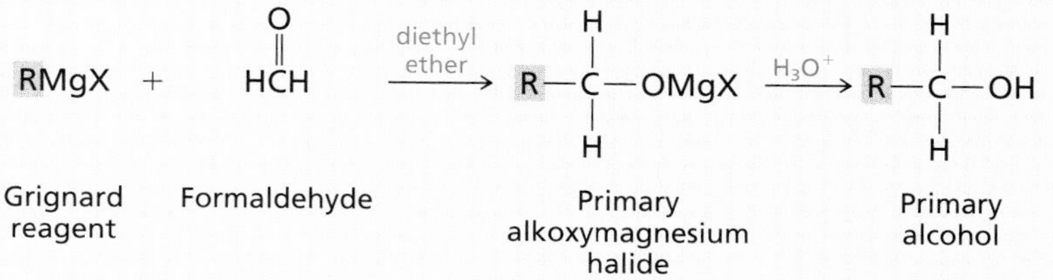 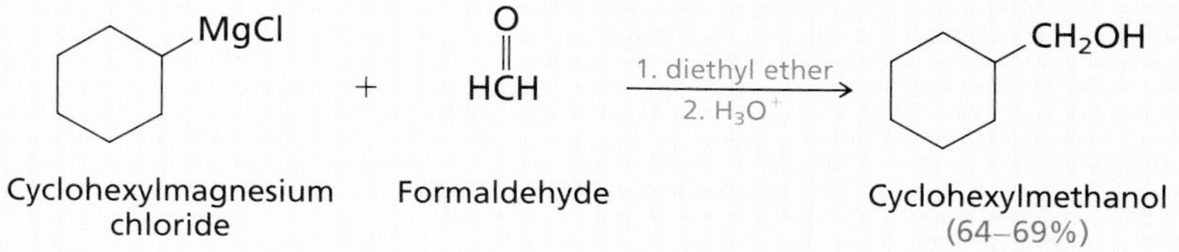 $$RMgX \ + \ HCH \xrightarrow{\text{diethyl ether}} R-\overset{H}{\underset{H}{C}}-OMgX \xrightarrow{H_3O^+} R-\overset{H}{\underset{H}{C}}-OH$$ Grignard reagent — Formaldehyde — Primary alkoxymagnesium halide — Primary alcohol Cyclohexylmagnesium chloride — Formaldehyde — Cyclohexylmethanol (64–69%)
Reaction with aldehydes Grignard reagents react with aldehydes ($RCH=O$) to give **secondary alcohols**.	$$RMgX \ + \ R'CH \xrightarrow{\text{diethyl ether}} R-\overset{H}{\underset{R'}{C}}-OMgX \xrightarrow{H_3O^+} R-\overset{H}{\underset{R'}{C}}-OH$$ Grignard reagent — Aldehyde — Secondary alkoxymagnesium halide — Secondary alcohol $$CH_3(CH_2)_4CH_2MgBr \ + \ CH_3CH \xrightarrow[\text{2. } H_3O^+]{\text{1. diethyl ether}} CH_3(CH_2)_4CH_2\underset{\underset{OH}{|}}{C}HCH_3$$ Hexylmagnesium bromide — Ethanal (acetaldehyde) — 2-Octanol (84%)
Reaction with ketones Grignard reagents react with ketones (RCR') to give **tertiary alcohols**.	$$RMgX \ + \ R'CR'' \xrightarrow{\text{diethyl ether}} R-\overset{R''}{\underset{R'}{C}}-OMgX \xrightarrow{H_3O^+} R-\overset{R''}{\underset{R'}{C}}-OH$$ Grignard reagent — Ketone — Tertiary alkoxymagnesium halide — Tertiary alcohol $$CH_3MgCl \ + \ \text{Cyclopentanone} \xrightarrow[\text{2. } H_3O^+]{\text{1. diethyl ether}} \text{1-Methylcyclopentanol}$$ Methylmagnesium chloride — Cyclopentanone — 1-Methylcyclopentanol (62%)

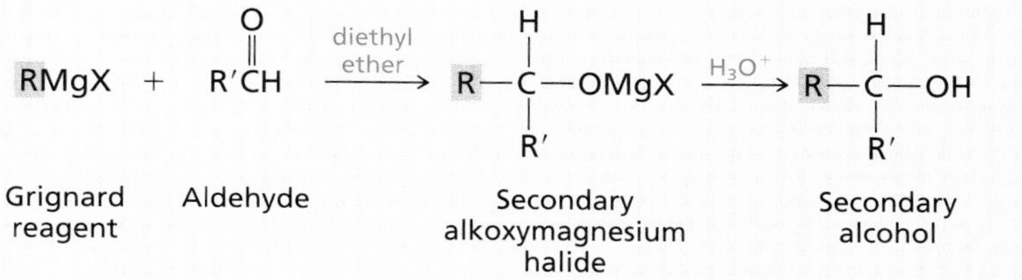

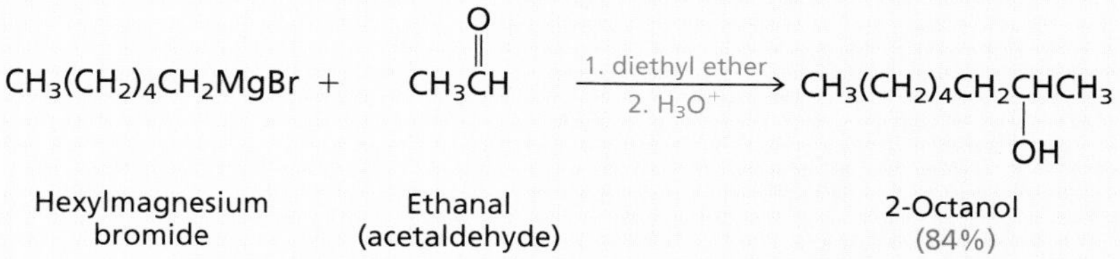

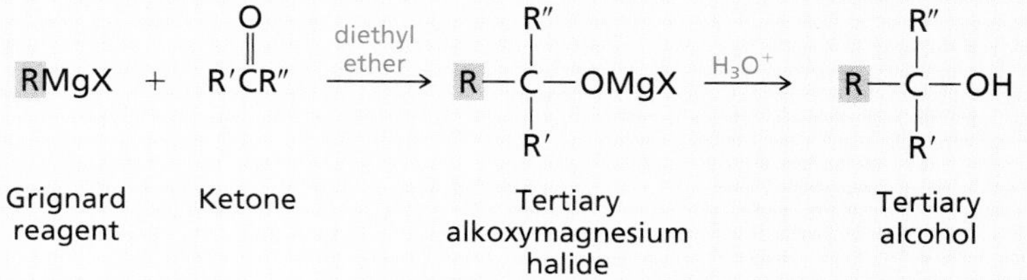

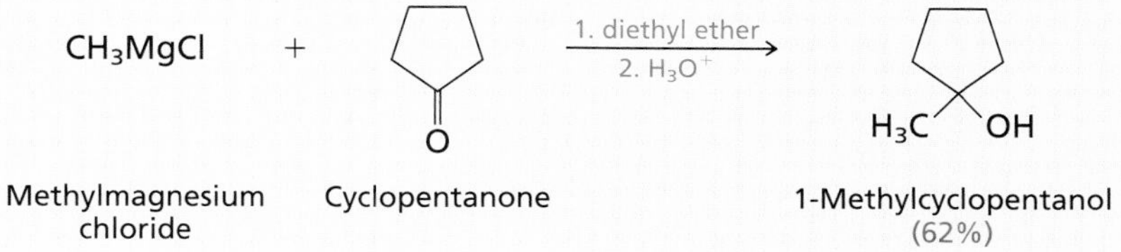

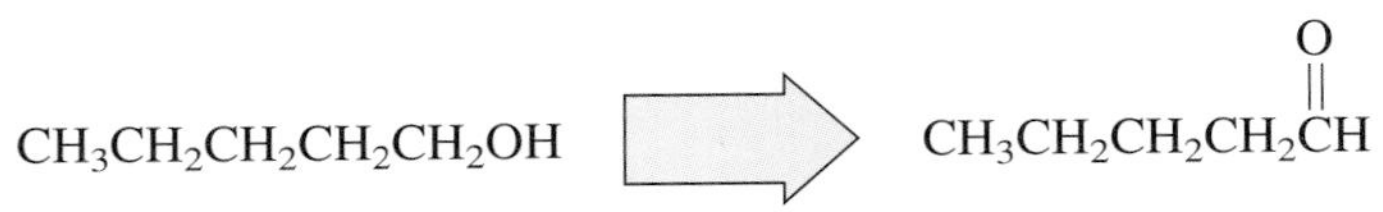

$$CH_3CH_2CH_2CH_2CH_2OH \quad \Longrightarrow \quad CH_3CH_2CH_2CH_2\overset{\overset{\displaystyle O}{\|}}{C}H$$

Reduction of pentanal with either sodium borohydride or lithium aluminum hydride will accomplish the desired transformation.

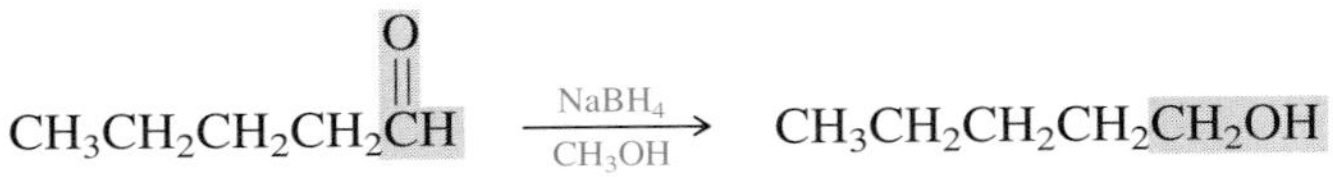

$$CH_3CH_2CH_2CH_2\overset{\overset{\displaystyle O}{\|}}{C}H \quad \xrightarrow[CH_3OH]{NaBH_4} \quad CH_3CH_2CH_2CH_2CH_2OH$$

Notice that the starting material and product have the *same number* of carbon atoms. Reduction of an aldehyde (or ketone) does not change its carbon skeleton.

Now consider how we could use the Grignard reaction to prepare 1-pentanol. Focus your attention on the hydroxyl-bearing carbon of the target molecule because this carbon was part of the carbonyl group in the starting material. Mentally disconnect the alkyl group attached to the carbon bearing the hydroxyl group—in essence reversing the carbon–carbon bond formation of the Grignard step.

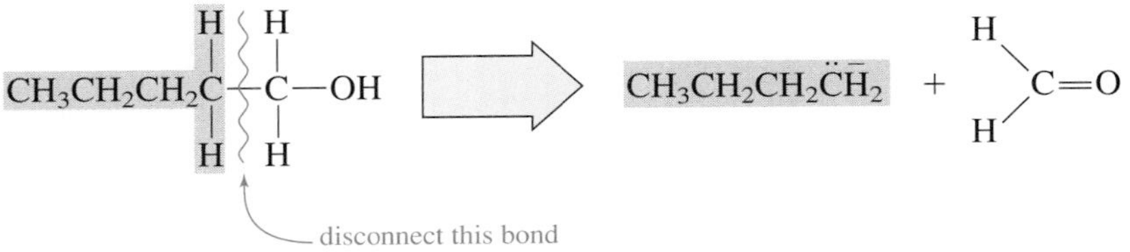

disconnect this bond

This disconnection reveals the structures of the Grignard reagent and the carbonyl compound that can serve as precursors to the target molecule. Notice that we have written the Grignard reagent (RMgX) as a carbanion (R:$^-$) to emphasize its chemical nature.

Thus retrosynthetic analysis tells us that 1-pentanol may be prepared by reaction of formaldehyde and a butylmagnesium halide such as butylmagnesium iodide.

$$CH_3CH_2CH_2CH_2MgI \; + \; H_2C{=}O \quad \xrightarrow[2.\; H_3O^+]{1.\; diethyl\; ether} \quad CH_3CH_2CH_2CH_2CH_2OH$$

Two possible combinations of Grignard reagent and aldehyde will give a secondary alcohol. These can be revealed by disconnecting each of the two alkyl groups of the target alcohol.

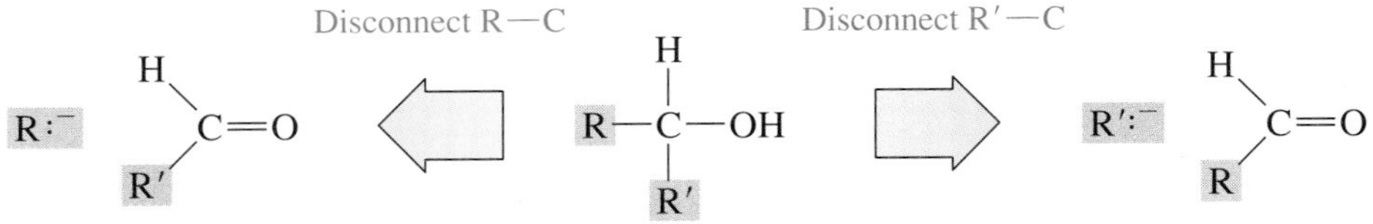

Disconnect R—C   Disconnect R′—C

PROBLEM 11.11 Suggest how each of the following alcohols might be prepared using a Grignard reagent.

(a) 2-Methyl-1-propanol $(CH_3)_2CHCH_2OH$

(b) 2-Hexanol $CH_3CHCH_2CH_2CH_2CH_3$ (2 methods)
$\quad\quad\quad\quad\quad\quad\quad\; |$
$\quad\quad\quad\quad\quad\quad\quad OH$

(c) 1-Phenyl-1-propanol $C_6H_5CHCH_2CH_3$ (2 methods)
                                   |
                                   OH

Sample Solution Mentally disconnect the bond between the alkyl group and the hydroxyl-bearing carbon.

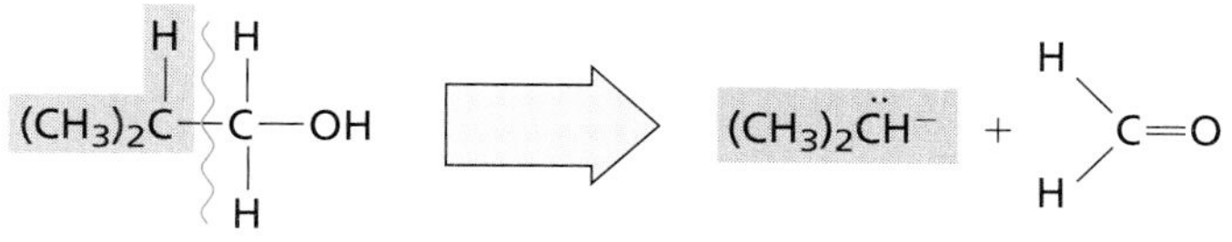

$$(CH_3)_2C \overset{H}{\underset{H}{-}} C-OH \implies (CH_3)_2\ddot{C}H^- + \overset{H}{\underset{H}{\diagdown}}C{=}O$$

This reveals the appropriate precursors to be formaldehyde and the Grignard reagent derived from 2-bromopropane.

$$(CH_3)_2CHMgBr \ + \ H_2C{=}O \ \xrightarrow[\text{2. } H_3O^+]{\text{1. diethyl ether}} \ (CH_3)_2CHCH_2OH$$

Three combinations of Grignard reagent and ketone give rise to tertiary alcohols:

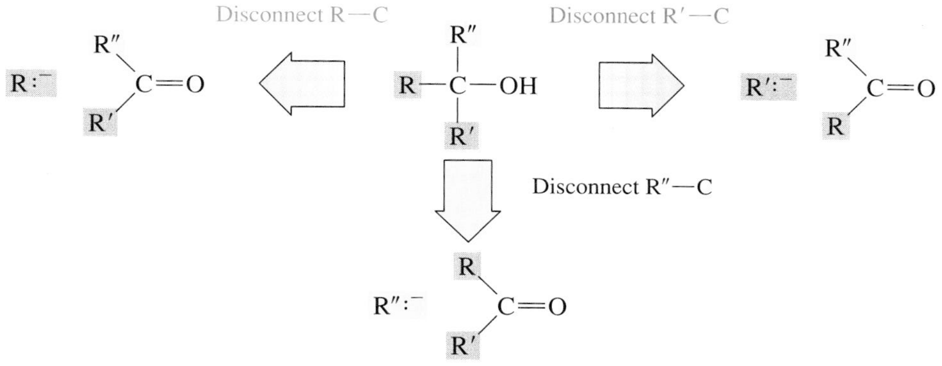

Disconnect R—C   Disconnect R′—C

$$R:^- \quad \overset{R''}{\underset{R'}{\diagdown}}C{=}O \quad \Longleftarrow \quad R-\overset{R''}{\underset{R'}{\overset{|}{\underset{|}{C}}}}-OH \quad \Longrightarrow \quad R':^- \quad \overset{R''}{\underset{R}{\diagdown}}C{=}O$$

Disconnect R″—C

$$R'':^- \quad \overset{R}{\underset{R'}{\diagdown}}C{=}O$$

PROBLEM 11.12 Suggest three ways in which 2-phenyl-2-butanol,

$$\overset{\overset{\textstyle OH}{\textstyle |}}{C_6H_5\underset{\underset{\textstyle CH_3}{\textstyle |}}{C}CH_2CH_3}$$

can be prepared using Grignard reagents.

11.14 OXIDATION OF ALDEHYDES

Aldehydes are readily oxidized to carboxylic acids by a number of reagents, including those based on Cr(VI) in aqueous media.

$$\overset{O}{\overset{\|}{RCH}} \xrightarrow{\text{oxidize}} \overset{O}{\overset{\|}{RCOH}}$$

Aldehyde   Carboxylic acid

$$\text{Furfural} \xrightarrow[\text{H}_2\text{SO}_4, \ \text{H}_2\text{O}]{\text{K}_2\text{Cr}_2\text{O}_7} \text{Furoic acid}$$

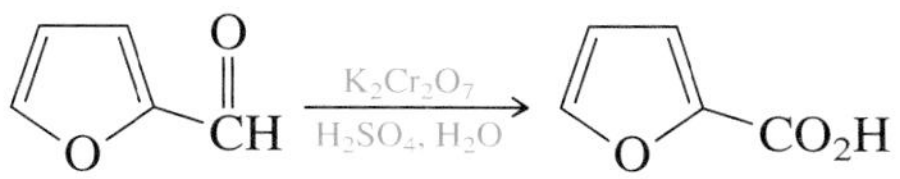

Furfural   Furoic acid
                    (75%)

11.15 THE α-CARBON ATOM AND ITS HYDROGENS

It is convenient to use the Greek letters α, β, γ, and so forth, to locate the carbons in a molecule in relation to the carbonyl group. The carbon atom adjacent to the carbonyl is the α-carbon atom, the next one down the chain is the β carbon, and so on. Butanal, for example, has an α carbon, a β carbon, and a γ carbon.

$$
\underset{\gamma\quad\beta\quad\alpha}{CH_3CH_2CH_2}\overset{\displaystyle O \atop \displaystyle \|}{CH}
\qquad
\begin{array}{l}\text{Carbonyl group is reference point;}\\ \text{no Greek letter assigned to it.}\end{array}
$$

Hydrogens take the same Greek letter as the carbon atom to which they are attached. A hydrogen connected to the α-carbon atom is an α hydrogen. Butanal has two α protons, two β protons, and three γ protons. No Greek letter is assigned to the hydrogen attached directly to the carbonyl group of an aldehyde.

PROBLEM 11.13 How many α hydrogens are in each of the following?

(a) 3,3-Dimethyl-2-butanone (c) Benzyl methyl ketone
(b) 2,2-Dimethylpropanal   (d) Cyclohexanone

Sample Solution (a) This ketone has two different α carbons, but only one of them has hydrogens attached. There are three equivalent α hydrogens. The other nine hydrogens are attached to β-carbon atoms.

$$
\underset{\alpha}{CH_3}-\overset{\displaystyle O \atop \displaystyle \|}{C}-\underset{\underset{\beta}{CH_3}}{\overset{\overset{\beta}{CH_3}}{\underset{\alpha}{C}}}-\underset{\beta}{CH_3}
$$

3,3-Dimethyl-2-butanone

Other than nucleophilic addition to the carbonyl group, the most important reactions of aldehydes and ketones involve substitution of an α hydrogen. Although we will discuss only simple examples, reactions that involve the α carbon and its hydrogens are frequently encountered in biochemistry.

11.16 ENOLS AND ENOLIZATION

Aldehydes and ketones that have at least one α hydrogen are in equilibrium with an isomer called an **enol**.

$$
\underset{\substack{\text{Aldehyde or}\\\text{ketone}}}{\overset{\displaystyle O \atop \displaystyle \|}{R_2CHCR'}}
\;\rightleftharpoons\;
\underset{\text{Enol}}{\overset{\displaystyle OH \atop \displaystyle |}{R_2C{=}CR'}}
$$

> The keto and enol forms are constitutional isomers. Using older terminology they are referred to as **tautomers** of each other.

Enols are related to an aldehyde or a ketone by a proton-transfer equilibrium known as **keto–enol tautomerism**. (*Tautomerism* refers to an interconversion between two structures that differ by the placement of an atom or a group.)

The amount of enol present at equilibrium, the enol content, is quite small for simple aldehydes and ketones. The equilibrium constants for enolization, as shown by the following examples, are much less than 1.

$$\underset{\substack{\text{Acetaldehyde}\\\text{(keto form)}}}{CH_3\overset{\displaystyle O}{\overset{\|}{C}}H} \;\rightleftharpoons\; \underset{\substack{\text{Vinyl alcohol}\\\text{(enol form)}}}{CH_2{=}CHOH} \qquad K \approx 3 \times 10^{-7}$$

$$\underset{\substack{\text{Acetone}\\\text{(keto form)}}}{CH_3\overset{\displaystyle O}{\overset{\|}{C}}CH_3} \;\rightleftharpoons\; \underset{\substack{\text{Propen-2-ol}\\\text{(enol form)}}}{CH_2{=}\overset{\displaystyle OH}{\overset{|}{C}}CH_3} \qquad K \approx 6 \times 10^{-9}$$

With unsymmetrical ketones, enolization may occur in either of two directions:

$$\underset{\substack{\text{1-Buten-2-ol}\\\text{(enol form)}}}{CH_2{=}\overset{\displaystyle OH}{\overset{|}{C}}CH_2CH_3} \;\rightleftharpoons\; \underset{\substack{\text{2-Butanone}\\\text{(keto form)}}}{CH_3\overset{\displaystyle O}{\overset{\|}{C}}CH_2CH_3} \;\rightleftharpoons\; \underset{\substack{\text{2-Buten-2-ol}\\\text{(enol form)}}}{CH_3\overset{\displaystyle OH}{\overset{|}{C}}{=}CHCH_3}$$

The ketone is by far the most abundant species present at equilibrium. Both enols are also present, but in very small concentrations.

PROBLEM 11.14 Write structural formulas corresponding to

(a) The enol form of 2,4-dimethyl-3-pentanone

(b) The enol form of acetophenone

(c) The two enol forms of 2-methylcyclohexanone

Sample Solution (a) Remember that enolization involves the α-carbon atom. The ketone 2,4-dimethyl-3-pentanone gives a single enol because the two α carbons are equivalent.

$$\underset{\substack{\text{2,4-Dimethyl-3-pentanone}\\\text{(keto form)}}}{(CH_3)_2CH\overset{\displaystyle O}{\overset{\|}{C}}CH(CH_3)_2} \;\rightleftharpoons\; \underset{\substack{\text{2,4-Dimethyl-2-penten-3-ol}\\\text{(enol form)}}}{(CH_3)_2C{=}\overset{\displaystyle OH}{\overset{|}{C}}CH(CH_3)_2}$$

It is important to recognize that an enol is a real substance, capable of independent existence. An enol is *not* a resonance form of a carbonyl compound; the two are constitutional isomers of each other.

11.17 BASE-CATALYZED ENOLIZATION: ENOLATE IONS

Enolization can be catalyzed by a base such as hydroxide ion, as shown in Figure 11.8. The base abstracts a proton from the α-carbon atom to form the conjugate base of the carbonyl compound. This anion is called an **enolate ion.** Hydroxide ion is a sufficiently strong base to remove an α hydrogen because these protons are far more acidic than would be expected for a simple hydrocarbon.

$$\underset{\substack{|\\ \boxed{H}}}{R_2\overset{\displaystyle O}{\overset{\|}{C}}CR'}$$

This proton is far more acidic than a hydrogen in an alkane.

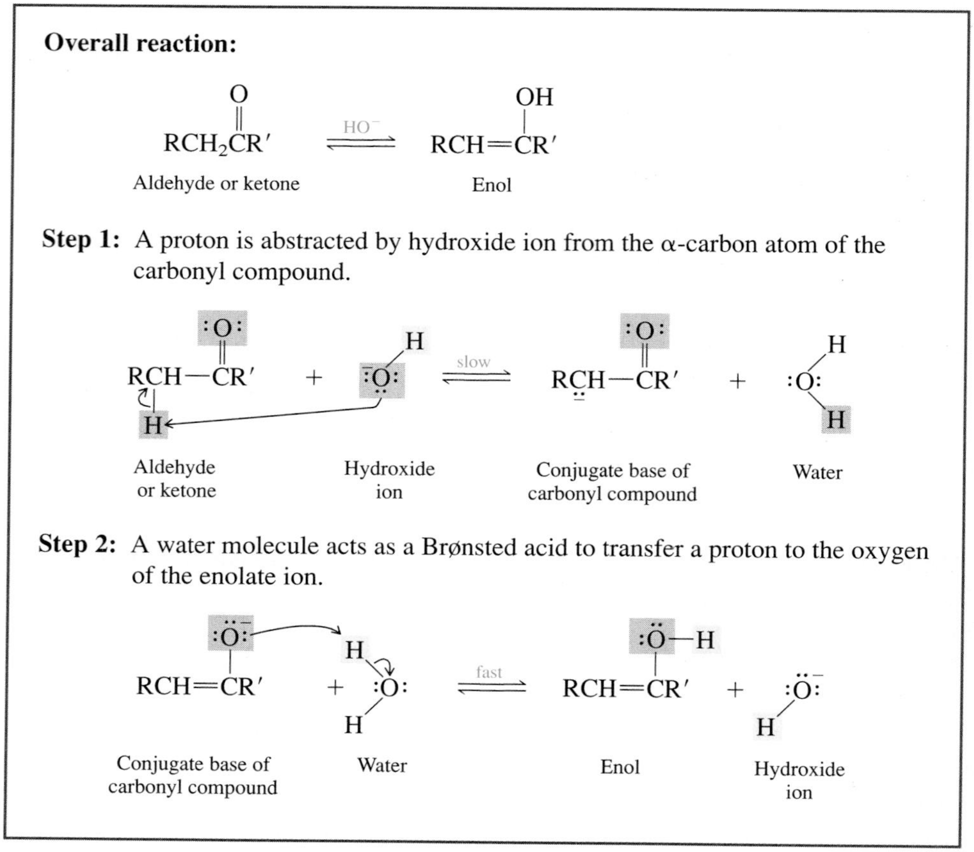

FIGURE 11.8 Mechanism of the base-catalyzed enolization of an aldehyde or ketone in aqueous solution.

Overall reaction:

Step 1: A proton is abstracted by hydroxide ion from the α-carbon atom of the carbonyl compound.

Step 2: A water molecule acts as a Brønsted acid to transfer a proton to the oxygen of the enolate ion.

The enolate ion is a resonance-stabilized species. Its negative charge is shared by the α-carbon atom and the carbonyl oxygen.

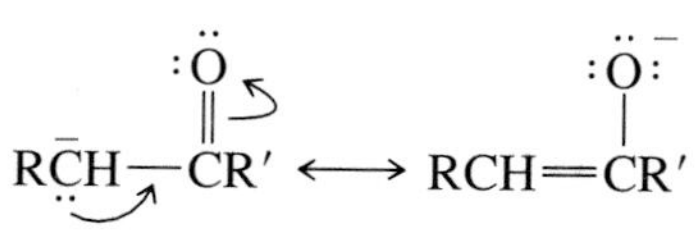

Electron delocalization
in conjugate base of ketone

Delocalization of the negative charge onto the electronegative oxygen, as represented by the preceding resonance structures, is responsible for the enhanced acidity of aldehydes and ketones. With K_a's in the 10^{-16}–10^{-20} range, aldehydes and ketones are about as acidic as water and alcohols. Thus, hydroxide ion and alkoxide ions are sufficiently strong bases to produce solutions containing significant concentrations of enolate ions at equilibrium.

11.18 THE ALDOL CONDENSATION

As you have just seen, an aldehyde is partially converted to its enolate anion by bases such as hydroxide ion and alkoxide ions.

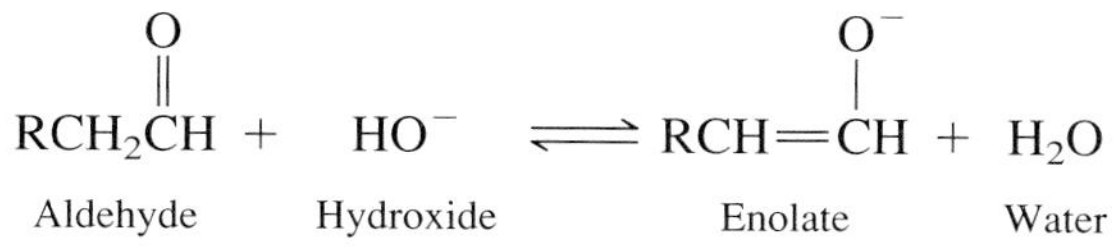

$$RCH_2\overset{O}{\overset{\|}{C}}H + HO^- \rightleftharpoons RCH=\overset{O^-}{\overset{|}{C}}H + H_2O$$

Aldehyde   Hydroxide   Enolate   Water

In a solution that contains both an aldehyde and its enolate ion, the enolate undergoes nucleophilic addition to the carbonyl group. This addition is analogous to the addition reactions of other nucleophilic reagents to aldehydes and ketones described in the chapter.

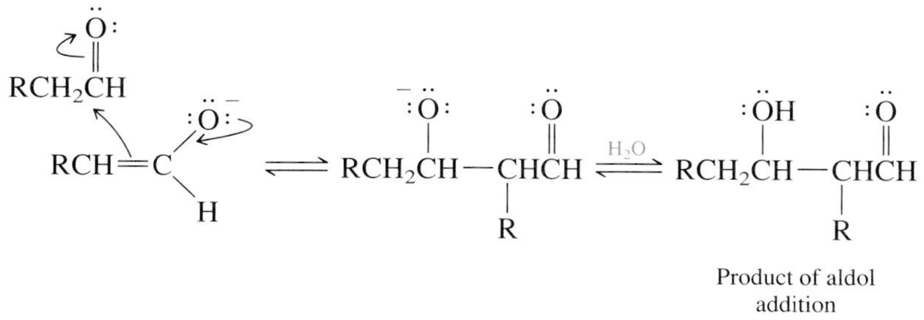

The alkoxide formed in the nucleophilic addition step then abstracts a proton from the solvent (usually water or ethanol) to yield the product of **aldol addition.** This product is known as an aldol because it contains both an aldehyde function and a hydroxyl group (*ald + ol = aldol*).

An important feature of aldol addition is that

Carbon–carbon bond formation occurs between the α-carbon atom of one aldehyde and the carbonyl group of another.

This is because carbanion (enolate) generation can involve proton abstraction *only* from the α-carbon atom. The overall transformation can be represented schematically, as shown in Figure 11.9.

Aldol addition occurs readily with aldehydes:

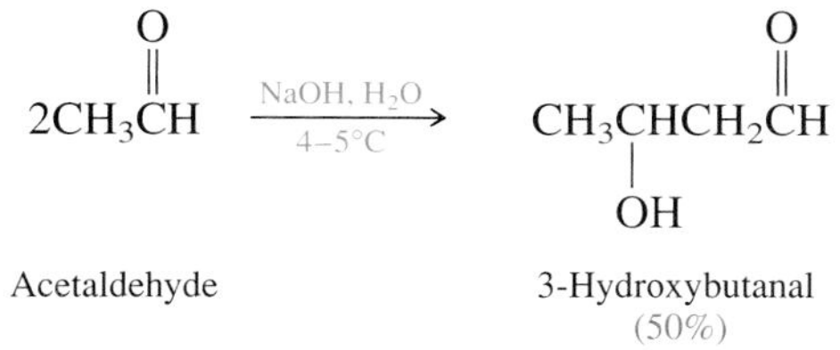

$$2CH_3\overset{O}{\overset{\|}{C}}H \xrightarrow[\text{4–5°C}]{\text{NaOH, H}_2\text{O}} CH_3\overset{}{\underset{OH}{\overset{}{C}H}}CH_2\overset{O}{\overset{\|}{C}}H$$

Acetaldehyde   3-Hydroxybutanal
(50%)

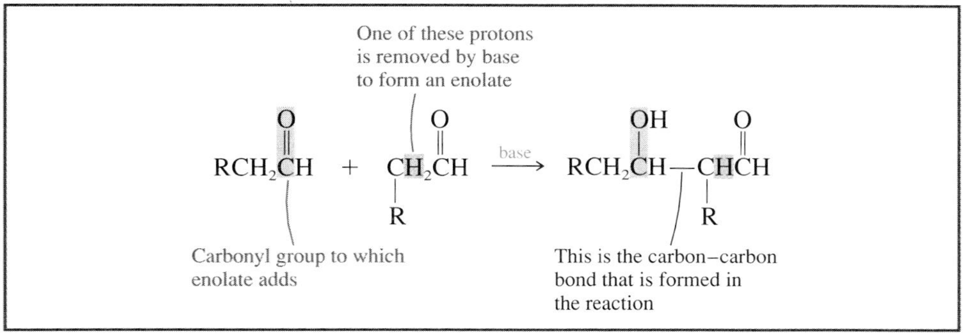

Some of the earliest studies of the aldol reaction were carried out by Aleksander Borodin. Though a physician by training and a chemist by profession, Borodin is remembered as the composer of some familiar works in Russian music, for example, *The Polovtsian Dances.*

FIGURE 11.9 The reactive sites in aldol addition are the carbonyl group of one aldehyde molecule and the α-carbon atom of another.

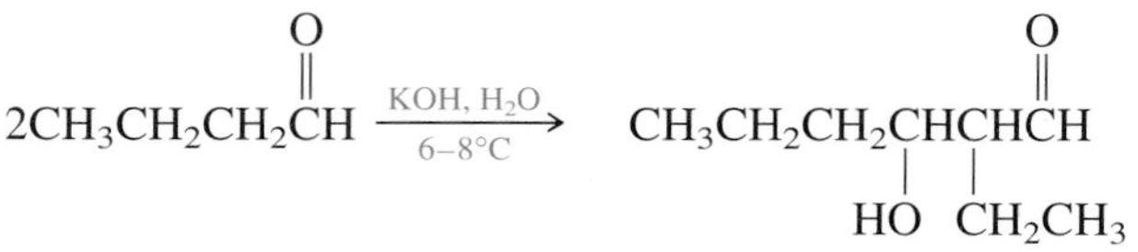

$$2CH_3CH_2CH_2\overset{\displaystyle O}{\overset{\|}{C}}H \xrightarrow[6-8°C]{KOH, H_2O} CH_3CH_2CH_2\overset{\displaystyle O}{\overset{\|}{C}H}CHCH$$

Butanal

2-Ethyl-3-hydroxyhexanal
(75%)

PROBLEM 11.15 Write the structure of the aldol addition product of

(a) Pentanal   (c) 3-Methylbutanal

(b) 2-Methylbutanal

Sample Solution (a) A good way to correctly identify the aldol addition product of any aldehyde is to work through the process mechanistically. Remember that the first step is enolate formation and that this *must* involve proton abstraction from the α carbon.

$$CH_3CH_2CH_2\overset{\alpha}{C}H_2\overset{\displaystyle O}{\overset{\|}{C}}H + HO^- \rightleftharpoons CH_3CH_2CH_2\overset{\displaystyle O}{\overset{\|}{C}}\ddot{C}HCH \longleftrightarrow CH_3CH_2CH_2CH=CH$$

Pentanal   Hydroxide   Enolate of pentanal

Now use the negatively charged α carbon of the enolate to form a new carbon–carbon bond to the carbonyl group. Proton transfer from the solvent completes the process.

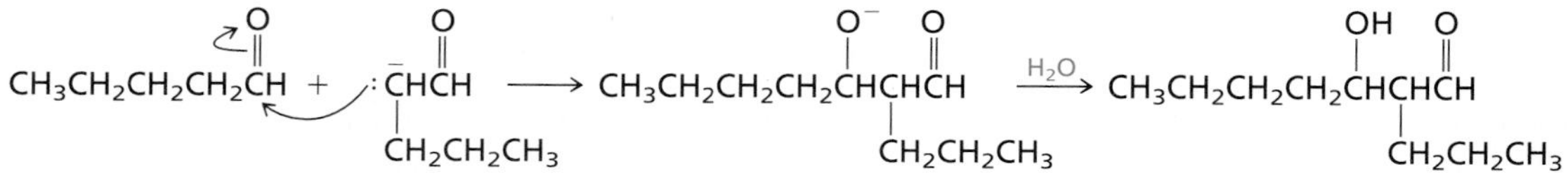

3-Hydroxy-2-propylheptanal
(aldol addition product of pentanal)

Pentanal   Enolate of pentanal

The β-hydroxy aldehyde products of aldol addition undergo dehydration on heating, to yield **α,β-unsaturated aldehydes:**

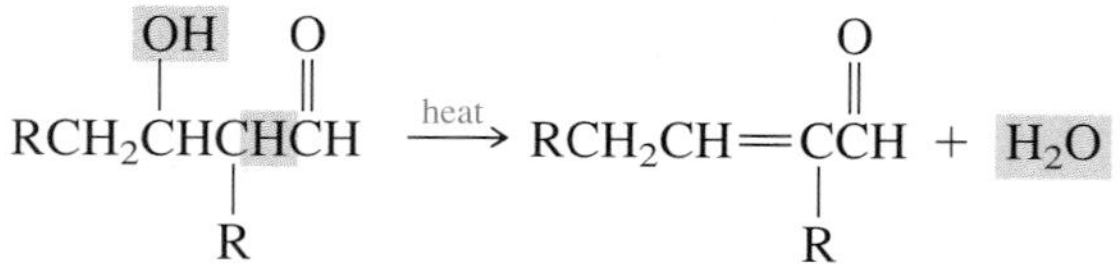

$$RCH_2\overset{OH}{\overset{|}{C}}HCH\overset{\displaystyle O}{\overset{\|}{C}}H \xrightarrow{heat} RCH_2CH=C\overset{\displaystyle O}{\overset{\|}{C}}H + H_2O$$

β-Hydroxy aldehyde   α,β-Unsaturated aldehyde   Water

It might be helpful to review the concept of conjugation presented in Section 4.2.

Conjugation of the newly formed double bond with the carbonyl group stabilizes the α,β-unsaturated aldehyde, provides the driving force for the dehydration, and controls its regioselectivity. If the α,β-unsaturated aldehyde is the desired product, the base-catalyzed aldol addition reaction is carried out at elevated temperature. Under these conditions the aldol addition product rapidly loses water to form the α,β-unsaturated aldehyde.

$$2CH_3CH_2CH_2\overset{\displaystyle O}{\overset{\|}{C}}H \xrightarrow[80-100°C]{NaOH, H_2O} CH_3CH_2CH_2CH=C\overset{\displaystyle O}{\overset{\|}{C}}H \quad via \quad CH_3CH_2CH_2\overset{OH}{\overset{|}{C}}HCH\overset{\displaystyle O}{\overset{\|}{C}}H$$

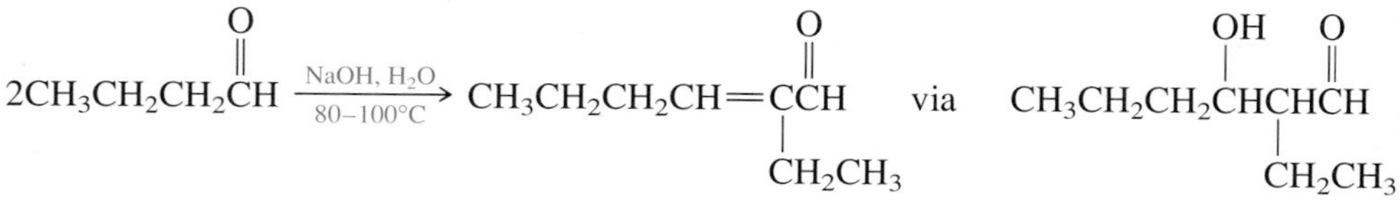

Butanal   2-Ethyl-2-hexenal
(86%)

2-Ethyl-3-hydroxyhexanal
(not isolated; dehydrates under reaction conditions)

Reactions in which two molecules of an aldehyde combine to form an α,β-unsaturated aldehyde and a molecule of water are called **aldol condensations.**

PROBLEM 11.16 Write the structure of the aldol condensation product of each of the aldehydes in Problem 11.15. One of these aldehydes can undergo aldol addition, but not aldol condensation. Which one? Why?

Sample Solution (a) Dehydration of the product of aldol addition of pentanal introduces the double bond between C-2 and C-3 to give an α,β-unsaturated aldehyde.

$$\underset{\substack{\displaystyle CH_2CH_2CH_3}}{CH_3CH_2CH_2CH_2\overset{\displaystyle OH}{\underset{|}{CH}}\overset{\displaystyle O}{\overset{\|}{CHCH}}} \xrightarrow{-H_2O} \underset{\substack{\displaystyle CH_2CH_2CH_3}}{CH_3CH_2CH_2CH_2CH=\overset{\displaystyle O}{\overset{\|}{CCH}}}$$

Product of aldol addition of   Product of aldol condensation
pentanal   of pentanal
(3-hydroxy-2-propylheptanal)   (2-propyl-2-heptenal)

Aldol condensations and the reverse, called retro-aldol reactions, are very important in the biochemistry of cells. By reactions such as these living organisms break down sugars to smaller molecules and produce the energy that allows the organism to function. The process, known as **glycoysis,** has been studied extensively.

LEARNING OBJECTIVES

This chapter introduced the chemical and physical properties of two related classes of compounds: aldehydes and ketones. The central themes of this chapter were the properties and reactions of the carbonyl group. The skills you have learned in this chapter should enable you to

• Provide an acceptable IUPAC name for an aldehyde or a ketone.
• Explain how the polarity of the carbonyl group affects physical properties of aldehydes and ketones.
• Describe how the polar nature of the carbonyl group affects nucleophilic additions to aldehydes and ketones.
• Write an equation for the hydration of an aldehyde or a ketone, and write mechanisms for the process under acid- and base-catalyzed conditions.
• Write a chemical equation for the formation of a cyanohydrin.
• Write a chemical equation for the formation of an acetal.
• Write a chemical equation for the reaction of an aldehyde or a ketone with a primary amine.
• Show how an organometallic compound can be used to form new carbon–carbon bonds.
• Use chemical equations to describe the reaction of a Grignard reagent with an aldehyde or a ketone.
• Plan the synthesis of a specific alcohol using a Grignard reagent by working the problem backward.

Continued

- Write a chemical equation for the oxidation of an aldehyde.
- Explain the process of enolization, and draw the enol form of a specific aldehyde of ketone.
- Write a chemical equation for the aldol condensation of an aldehyde.

11.19 SUMMARY

Systematic names for both aldehydes and ketones are based on the alkane having the same number of carbons as the longest chain containing the carbonyl group (Section 11.1). The *-e* ending is replaced with *-al* for aldehydes and *-one* for ketones. The carbonyl group is always C-1 in aldehydes. The chain is numbered in the direction that gives the lowest locant to the carbon of the carbonyl group of ketones.

3-Methylbutanal   3-Methyl-2-butanone

Ketones may also be named by citing the two groups attached to the carbonyl in alphabetical order followed by the word *ketone*. Thus, 3-methyl-2-butanone becomes isopropyl methyl ketone.

The carbonyl carbon is sp^2-hybridized, and it and the atoms attached to it are coplanar (Section 11.2).

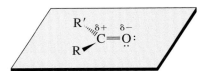

Aldehydes and ketones are polar molecules (Section 11.3). Nucleophiles attack C=O at carbon (positively polarized), and electrophiles, especially protons, are attacked by oxygen (negatively polarized) (Section 11.5).

The characteristic reaction of aldehydes and ketones is **nucleophilic addition** to the carbon of the carbonyl group (Section 11.5). A summary of the nucleophilic addition reactions introduced in this chapter is presented in Table 11.3.

Aldehyde   Product of nucleophilic
or ketone   addition to carbonyl group

The last entry is an example of the use of an **organometallic compound** (Section 11.10) called a **Grignard reagent** for the formation of new carbon–carbon single bonds. Aldehydes are easily oxidized to carboxylic acids in aqueous solution (Section 11.14).

TABLE 11.3	Nucleophilic Addition to Aldehydes and Ketones

Reaction (section) and Comments	General Equation and Typical Example
Hydration (Section 11.6) Can be either acid- or base-catalyzed. Equilibrium constant is normally unfavorable for hydration of ketones unless R, R′, or both are strongly electron-withdrawing.	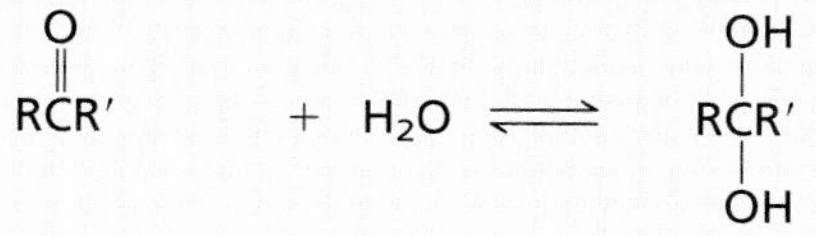 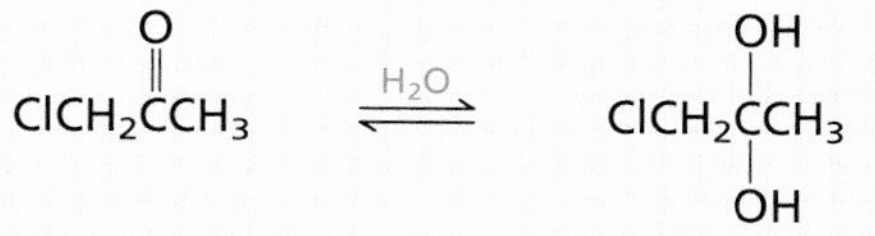 $$\underset{\text{Aldehyde or ketone}}{RCR'} + \underset{\text{Water}}{H_2O} \rightleftharpoons \underset{\text{Geminal diol}}{RCR'}$$ (O double bond on RCR′; product has two OH groups) $$\underset{\substack{\text{Chloroacetone}\\(90\%\text{ at equilibrium})}}{ClCH_2CCH_3} \xrightarrow{H_2O} \underset{\substack{\text{Chloroacetone hydrate}\\(10\%\text{ at equilibrium})}}{ClCH_2CCH_3}$$
Cyanohydrin formation (Section 11.7) Reaction is catalyzed by cyanide ion. Cyanohydrins are useful synthetic intermediates; cyano group can be hydrolyzed to $-CO_2H$ or reduced to $-CH_2NH_2$.	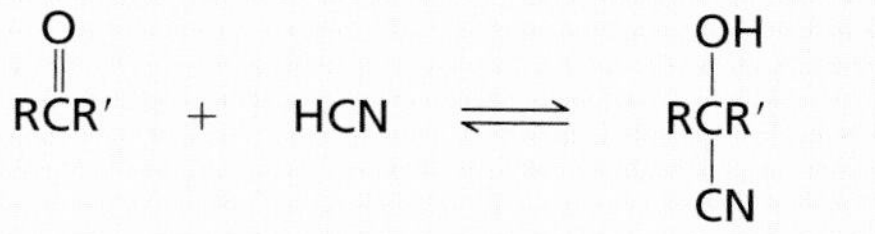 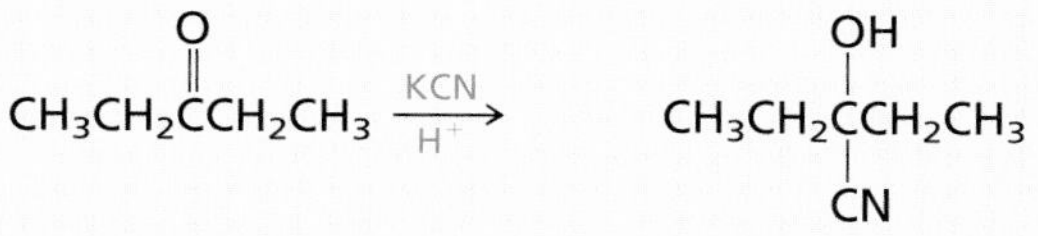 $$\underset{\substack{\text{Aldehyde}\\\text{or ketone}}}{RCR'} + \underset{\substack{\text{Hydrogen}\\\text{cyanide}}}{HCN} \rightleftharpoons \underset{\text{Cyanohydrin}}{RCR'}$$ (product has OH and CN groups) $$\underset{\text{3-Pentanone}}{CH_3CH_2CCH_2CH_3} \xrightarrow[H^+]{KCN} \underset{\substack{\text{3-Pentanone cyanohydrin}\\(75\%)}}{CH_3CH_2CCH_2CH_3}$$ (product has OH and CN groups)
Acetal formation (Section 11.8) Reaction is acid-catalyzed. Equilibrium constant normally favorable for aldehydes, unfavorable for ketones.	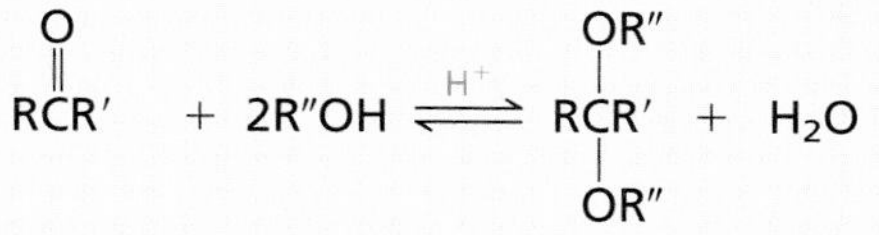 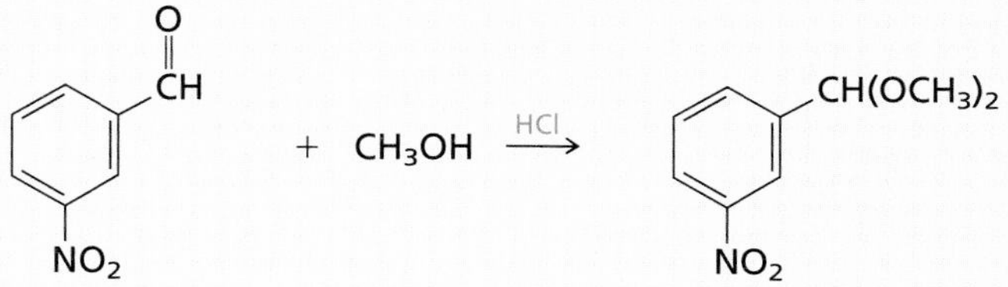 $$\underset{\substack{\text{Aldehyde}\\\text{or ketone}}}{RCR'} + \underset{\text{Alcohol}}{2R''OH} \xrightarrow{H^+} \underset{\text{Acetal}}{RCR'} + \underset{\text{Water}}{H_2O}$$ (product has two OR″ groups) $$\underset{m\text{-Nitrobenzaldehyde}}{} + \underset{\text{Methanol}}{CH_3OH} \xrightarrow{HCl} \underset{\substack{m\text{-Nitrobenzaldehyde}\\\text{dimethyl acetal}\\(76\text{–}85\%)}}{CH(OCH_3)_2}$$

(Continued)

TABLE 11.3	Nucleophilic Addition to Aldehydes and Ketones *(Continued)*

Reaction (section) and Comments	General Equation and Typical Example
Reaction with primary amines (Section 11.9)　Isolated product is an imine (Schiff's base). A carbinolamine intermediate is formed, which undergoes dehydration to an imine.	$$\underset{\text{Aldehyde or ketone}}{\overset{\overset{\displaystyle O}{\|}}{RCR'}} + \underset{\text{Primary amine}}{R''NH_2} \rightleftharpoons \underset{\text{Imine}}{\overset{\overset{\displaystyle NR''}{\|}}{RCR'}} + \underset{\text{Water}}{H_2O}$$ $$\underset{\text{2-Methylpropanal}}{(CH_3)_2CH\overset{\overset{\displaystyle O}{\|}}{C}H} + \underset{\textit{tert}\text{-Butylamine}}{(CH_3)_3CNH_2} \longrightarrow \underset{\substack{\textit{N}\text{-(2-Methyl-1-propylidene)-}\\ \textit{tert}\text{-butylamine}\\ (50\%)}}{(CH_3)_2CHCH{=}NC(CH_3)_3}$$
Alcohol synthesis via the reaction of Grignard reagents with carbonyl compounds (Section 11.12)　This is one of the most useful reactions in synthetic organic chemistry. Grignard reagents react with formaldehyde to yield primary alcohols, with aldehydes to give secondary alcohols, and with ketones to form tertiary alcohols.	$$\underset{\substack{\text{Grignard}\\\text{reagent}}}{RMgX} + \underset{\substack{\text{Aldehyde}\\\text{or ketone}}}{\overset{\overset{\displaystyle O}{\|}}{R'CR''}} \xrightarrow[\text{2. H}_3\text{O}^+]{\text{1. diethyl ether}} \underset{\text{Alcohol}}{\overset{\displaystyle R'}{\underset{\displaystyle R''}{RCOH}}}$$ $$\underset{\substack{\text{Methylmagnesium}\\\text{iodide}}}{CH_3MgI} + \underset{\text{Butanal}}{CH_3CH_2CH_2\overset{\overset{\displaystyle O}{\|}}{C}H} \xrightarrow[\text{2. H}_3\text{O}^+]{\text{1. diethyl ether}} \underset{\substack{\text{2-Pentanol}\\(82\%)}}{CH_3CH_2CH_2\underset{\underset{\displaystyle OH}{\|}}{C}HCH_3}$$

$$\underset{\text{Aldehyde}}{\overset{\overset{\displaystyle O}{\|}}{RCH}} \xrightarrow[\text{H}_2\text{O}]{\text{Cr(VI)}} \underset{\text{Carboxylic acid}}{\overset{\overset{\displaystyle O}{\|}}{RCOH}}$$

Aldehydes and ketones exist in equilibrium with their corresponding enol forms (Section 11.16)

$$\underset{\substack{\text{Aldehyde}\\\text{or ketone}}}{R_2CH{-}\overset{\overset{\displaystyle O}{\|}}{C}R'} \rightleftharpoons \underset{\text{Enol}}{R_2C{=}\overset{\overset{\displaystyle OH}{\|}}{C}R'}$$

$$\underset{\text{Cyclopentanone}}{\text{(structure)}{=}O} \underset{}{\overset{K}{\rightleftharpoons}} \underset{\text{Cyclopenten-1-ol}}{\text{(structure)}{-}OH} \qquad K = 1 \times 10^{-8}$$

The rate at which equilibrium is achieved is increased by acidic or basic catalysts. The enol content of simple aldehydes and ketones is quite small.

A proton on the α carbon of an aldehyde or ketone is more acidic than most other protons bound to carbon. A base such as hydroxide ion may be used to abstract an α proton to give an **enolate ion** (Section 11.17). Enolate ions are stabilized by resonance involving the carbonyl group.

$$\underset{\substack{\text{Aldehyde} \\ \text{or ketone}}}{R_2CHCR'} + \underset{\substack{\text{Hydroxide} \\ \text{ion}}}{HO^-} \rightleftharpoons \underset{\substack{\text{Enolate} \\ \text{anion}}}{R_2C=CR'} + \underset{\text{Water}}{H_2O}$$

$$\underset{\text{3-Pentanone}}{CH_3CH_2CCH_2CH_3} + \underset{\text{Hydroxide ion}}{HO^-} \rightleftharpoons \underset{\substack{\text{Enolate anion} \\ \text{of 3-pentanone}}}{CH_3CH=CCH_2CH_3} + \underset{\text{Water}}{H_2O}$$

The **aldol condensation** is a useful synthetic reaction for carbon–carbon bond formation (Section 11.18). Nucleophilic addition of an enolate ion to a carbonyl group, followed by dehydration, yields an α,β-unsaturated carbonyl compound.

$$\underset{\text{Aldehyde}}{2RCH_2CR'} \xrightarrow{HO^-} \underset{\substack{\alpha,\beta\text{-Unsaturated} \\ \text{aldehyde}}}{RCH_2C=CCR'} + \underset{\text{Water}}{H_2O}$$

$$\underset{\text{Octanal}}{CH_3(CH_2)_6CH} \xrightarrow[CH_3CH_2OH]{NaOCH_2CH_3} \underset{\substack{\text{2-Hexyl-2-decenal} \\ (79\%)}}{CH_3(CH_2)_6CH=C(CH_2)_5CH_3}$$

ADDITIONAL PROBLEMS

Nomenclature and Physical Properties

11.17 Each of the following aldehydes or ketones is known by a nonsystematic name. Its IUPAC name is provided in parentheses. Write a structural formula for each one.

(a) Pivaldehyde (2,2-dimethylpropanal)
(b) Acrolein (2-propenal)
(c) Deoxybenzoin (benzyl phenyl ketone)
(d) Diacetone alcohol (4-hydroxy-4-methyl-2-pentanone)
(e) Mesityl oxide (4-methyl-3-penten-2-one)
(f) Citral [(E)-3,7-dimethyl-2,6-octadienal]

11.18 Give an acceptable IUPAC name for each of the following aldehydes or ketones.

(a) $(CH_3)_3CCHCH_2CCH_2CH_3$ (with Cl and O substituents)

(c) $(CH_3)_2C=CHCH_2CH$ (with O)

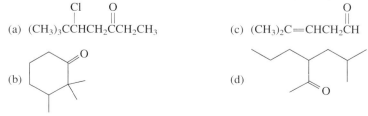
(b)

(d)

11.19 Although the following are all permissible IUPAC names for ketones, each can be named in a different way based on the longest carbon chain. Convert each name into one based on the latter system.

 (a) Dibenzyl ketone
 (b) Benzyl *tert*-butyl ketone
 (c) Ethyl isopropyl ketone
 (d) Isobutyl phenyl ketone
 (e) Allyl methyl ketone

11.20 What is wrong with each of the following chemical names? Write the correct systematic name or explain why the compound cannot exist.

 (a) Cyclohexanal
 (b) 2,2-Dimethyl-4-hexanone
 (c) 1-Chloro-3-propanone
 (d) Trimethylacetaldehyde

11.21 Correctly associate each of the following compounds of similar molecular weight with its boiling point:

Compounds: pentanal; 1-pentanol; hexane
Boiling points (°C): 68.8; 103.4; 138

11.22 Briefly explain why benzyl alcohol is more soluble in water than benzaldehyde.

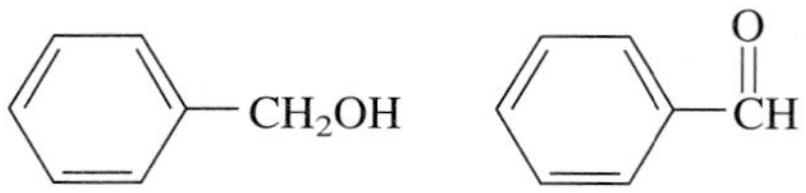

Benzyl alcohol   Benzaldehyde

Aldehyde and Ketone Reactions

11.23 Predict the product of the reaction of propanal with each of the following:

 (a) Lithium aluminum hydride, followed by treatment with water
 (b) Sodium borohydride in methanol
 (c) Hydrogen (nickel catalyst)
 (d) Methylmagnesium iodide (CH_3MgI), followed by dilute acid
 (e) Methanol containing a trace of acid
 (f) Aniline ($C_6H_5NH_2$)
 (g) Sodium cyanide with addition of sulfuric acid
 (h) Chromic acid

11.24 Repeat Problem 11.23 (a) through (g) for cyclopentanone.

11.25 (a) Write structural formulas and provide IUPAC names for all the isomeric aldehydes and ketones that have the molecular formula $C_5H_{10}O$. Include stereoisomers.

 (b) Which of the isomers in part (a) yield chiral alcohols on reaction with sodium borohydride?

 (c) Which of the isomers in part (a) yield chiral alcohols on reaction with methylmagnesium iodide?

11.26 Write the structure of the major organic product of each of the following reactions.

 (a) 2-Iodopropane with magnesium in diethyl ether
 (b) Product of part (a) with formaldehyde in diethyl ether, followed by dilute acid
 (c) Product of part (a) with benzaldehyde ($C_6H_5\overset{\underset{\textstyle O}{\textstyle \|}}{C}H$) in diethyl ether, followed by dilute acid
 (d) Product of part (a) with cyclopentanone in diethyl ether, followed by dilute acid

11.27 Equilibrium constants for the dissociation (K_{diss}) of cyanohydrins according to the equation

$$
\underset{\substack{| \\ \text{CN}}}{\overset{\substack{\text{OH} \\ |}}{R\overset{|}{C}R'}} \quad \underset{\Longleftarrow}{\overset{K_{\text{diss}}}{}} \quad \overset{\overset{\text{O}}{\|}}{R\overset{}{C}R'} \quad + \quad \text{HCN}
$$

Cyanohydrin   Aldehyde   Hydrogen
                   or ketone   cyanide

have been measured for a number of cyanohydrins. Which cyanohydrin in each of the following pairs has the greater dissociation constant?

(a) $\underset{\substack{| \\ \text{OH}}}{CH_3CH_2\overset{\overset{\text{OH}}{|}}{C}HCN}$ or $(CH_3)_2\overset{\overset{\text{OH}}{|}}{C}CN$

(b) $C_6H_5\overset{\overset{\text{OH}}{|}}{C}HCN$ or $C_6H_5\underset{\substack{| \\ CH_3}}{\overset{\overset{\text{OH}}{|}}{C}}CN$

11.28 Write the structure of the carbinolamine intermediate and the imine product formed in the reaction of each of the following:

(a) Cyclohexanone and *tert*-butylamine $(CH_3)_3CNH_2$

(b) Acetophenone and cyclohexylamine   ⬡—NH_2

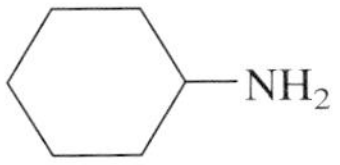

11.29 Analyze the following structures and determine all the combinations of Grignard reagent and carbonyl compound that will give rise to each.

(a) $CH_3CH_2\underset{\substack{| \\ OH}}{C}HCH_2CH(CH_3)_2$ (c)

(b) $(CH_3)_3CCH_2OH$

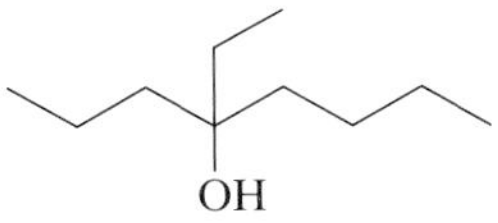

11.30 Which of the compounds in Problem 11.29 could be prepared by reduction of an aldehyde or ketone? Write the reactions that would take place, using lithium aluminum hydride as the reducing agent.

11.31 Supply the structure of the reactant missing from each of the following reactions.

(a) $? + 2CH_3CH_2OH \xrightarrow{\text{H}^+} CH_3CH(OCH_2CH_3)_2$

(b) 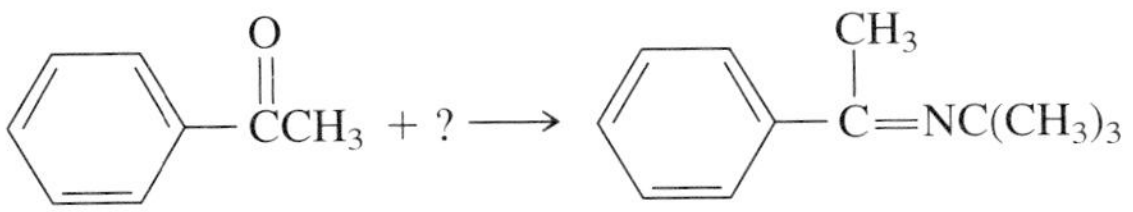 ⬡—$\overset{\overset{\text{O}}{\|}}{C}CH_3 + ? \longrightarrow$ ⬡—$\underset{\substack{| \\ }}{\overset{\overset{CH_3}{|}}{C}}=NC(CH_3)_3$

(c) $C_6H_5MgBr + ? \xrightarrow[\text{2. H}_3\text{O}^+]{\text{1. diethyl ether}} C_6H_5\underset{\substack{| \\ CH_3}}{\overset{\overset{\text{OH}}{|}}{C}}CH_2C_6H_5$

11.32 Cyclohexanone oxime ($C_6H_{11}NO$) is an industrial chemical used in the preparation of nylon. It is prepared by the reaction of cyclohexanone with hydroxylamine (H_2NOH). Suggest a reasonable structure for cyclohexanone oxime.

Enols and Enolates

11.33 In each of the following pairs of compounds, choose the one that is able to enolize, and write the structure of its enol form.

$$\text{(a)} \quad (CH_3)_3C\overset{\displaystyle O}{\overset{\|}{C}}H \quad \text{or} \quad (CH_3)_2CH\overset{\displaystyle O}{\overset{\|}{C}}H$$

$$\text{(b)} \quad C_6H_5\overset{\displaystyle O}{\overset{\|}{C}}C_6H_5 \quad \text{or} \quad C_6H_5CH_2\overset{\displaystyle O}{\overset{\|}{C}}CH_2C_6H_5$$

11.34 Give the structure of the expected organic product in the reaction of propanal with each of the following.

(a) Sodium hydroxide in cold ethanol
(b) Sodium hydroxide in hot ethanol
(c) Product of part (b) with sodium borohydride in ethanol

11.35 Butanal is the starting material for the active ingredient in a popular bug repellent. The product of aldol addition of butanal is reduced by catalytic hydrogenation to give the final product, which has the formula $C_8H_{18}O_2$. Write a series of equations to outline this procedure.

Synthesis

11.36 Using 1-bromobutane and any necessary organic or inorganic reagents, suggest efficient methods for the preparation of each of the following alcohols. More than one step may be necessary. (*Hint:* Work backward from the target alcohol.)

(a) 1-Pentanol   (c) 1-Phenyl-1-pentanol
(b) 2-Hexanol

11.37 Show how 2-butanone could be prepared by a procedure in which all of the carbons originate in ethanol.

11.38 The fungus responsible for Dutch elm disease is spread by European bark beetles when they burrow into the tree. Other beetles congregate at the site, attracted by the scent of a mixture of chemicals, some emitted by other beetles and some coming from the tree. One of the compounds given off by female bark beetles is 4-methyl-3-heptanol. Suggest a synthesis of this pheromone from alcohols of five or fewer carbon atoms.

$$\overset{\displaystyle OH}{\overset{\displaystyle |}{CH_3CH_2CH\underset{\displaystyle \underset{\displaystyle CH_3}{|}}{C}HCH_2CH_2CH_3}}$$

4-Methyl-3-heptanol

Miscellaneous Problems

11.39 Addition of phenylmagnesium bromide to 4-*tert*-butylcyclohexanone gives two isomeric tertiary alcohols as products. Both alcohols yield the same alkene when subjected to acid-catalyzed dehydration. Suggest reasonable structures for these two alcohols.

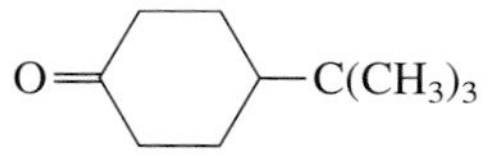

4-*tert*-Butylcyclohexanone

11.40 Compounds that contain both carbonyl and alcohol functional groups are often more stable as cyclic hemiacetals or cyclic acetals than as open-chain compounds. Examples of several of these are shown here. Deduce the structure of the open-chain form of each.

(a)

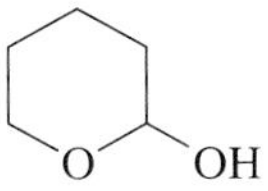

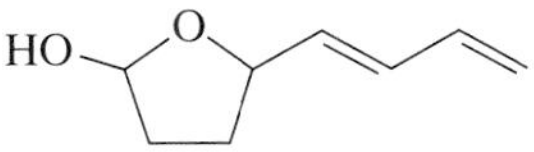

(c)

H₃C ... CH₃

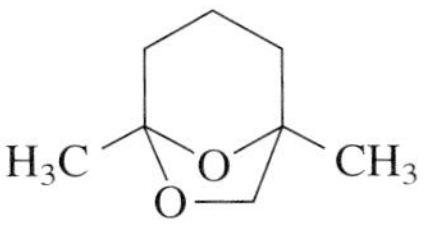

Frontalin (aggregating pheromone of the Southern pine beetle)

(b) HO

11.41 An unknown ketone, compound A, reacted with sodium cyanide to which acid had been added to give compound B, $C_6H_{11}NO$. Reaction of A with sodium borohydride in methanol gave compound C, an achiral substance that underwent dehydration with concentrated sulfuric acid to give 2-pentene as the *only* product. Identify compounds A to C, and write an equation for each chemical reaction described.

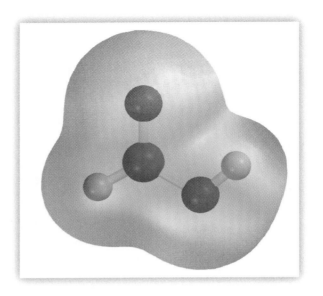

CHAPTER 12
CARBOXYLIC ACIDS

$$\begin{array}{c} O \\ \parallel \end{array}$$

Carboxylic acids, compounds of the type $RCOH$, constitute one of the most frequently encountered classes of organic compounds. Countless natural products are carboxylic acids or are derived from them. Some carboxylic acids, such as acetic acid, have been known for centuries. Others, such as the prostaglandins, which are powerful regulators of numerous biological processes, remained unknown until relatively recently. Still others, aspirin for example, are the products of chemical synthesis. The therapeutic effects of aspirin, welcomed long before the discovery of prostaglandins, are now understood to result from aspirin's ability to inhibit the biosynthesis of prostaglandins (see Boxed Essay *Prostaglandins,* following Section 12.3).

Acetic acid
(present in
vinegar)

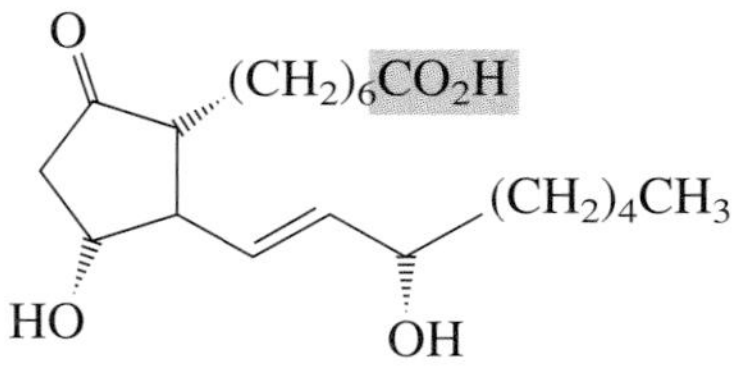

PGE_1
(a prostaglandin; a small amount of PGE_1
lowers blood pressure significantly)

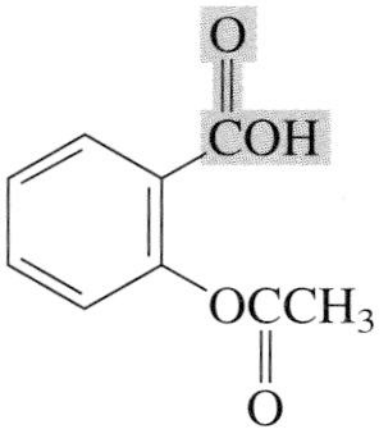

Aspirin

The chemistry of carboxylic acids is the central theme of this chapter. The importance of carboxylic acids is magnified when we realize that they are the parent compounds of a large group of derivatives that includes acyl chlorides, acid anhydrides, esters, and amides. Those classes of compounds will be discussed in Chapter 13.

Together, this chapter and the next tell the story of some of the most fundamental structural types and functional group transformations in organic and biological chemistry.

12.1 CARBOXYLIC ACID NOMENCLATURE

Nowhere in organic chemistry are common names used more often than with the carboxylic acids. Many carboxylic acids are better known by common names than by their systematic names, and the IUPAC nomenclature rules accept these common names as permissible alternatives to the systematic ones. Table 12.1 lists both the common and the systematic names of a number of important carboxylic acids.

Systematic names for carboxylic acids are derived by counting the number of carbons in the longest continuous chain that includes the carboxyl group and replacing the -e ending of the corresponding alkane by -oic acid. The first three acids in the table, methanoic (1 carbon), ethanoic (2 carbons), and octadecanoic acid (18 carbons), illustrate this point. When substituents are present, their locations are identified by number; numbering of the carbon chain always begins at the carboxyl group. This is illustrated in entries 4 and 5 in the table.

Notice that compounds 4 and 5 are named as hydroxy derivatives of carboxylic acids, rather than as carboxyl derivatives of alcohols. We have seen earlier that hydroxyl groups take precedence over double bonds, and double bonds take precedence over halogens and alkyl groups, in naming compounds. Carboxylic acids outrank all the common groups we have encountered to this point.

TABLE 12.1	Systematic and Common Names of Some Carboxylic Acids		
	Structural Formula	**Systematic Name**	**Common Name**
1.	HCO_2H	Methanoic acid	Formic acid
2.	CH_3CO_2H	Ethanoic acid	Acetic acid
3.	$CH_3(CH_2)_{16}CO_2H$	Octadecanoic acid	Stearic acid
4.	CH_3CHCO_2H 　　$\|$ 　　OH	2-Hydroxypropanoic acid	Lactic acid
5.	⬡—$CHCO_2H$ 　　$\|$ 　　OH	2-Hydroxy-2-phenylethanoic acid	Mandelic acid
6.	$CH_2{=}CHCO_2H$	Propenoic acid	Acrylic acid
7.	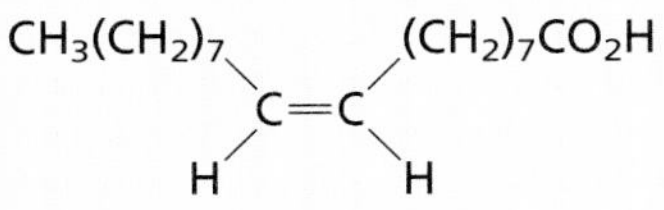 $CH_3(CH_2)_7$　$(CH_2)_7CO_2H$ 　　　＼C＝C／ 　　　H　　H	(Z)-9-Octadecenoic acid	Oleic acid
8.	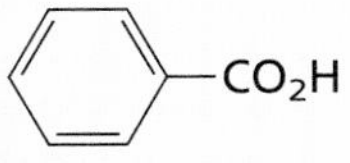 ⬡—CO_2H	Benzenecarboxylic acid	Benzoic acid
9.	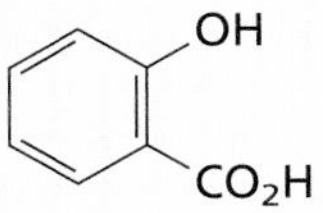 ⬡ with OH and CO_2H	o-Hydroxybenzenecarboxylic acid	Salicylic acid

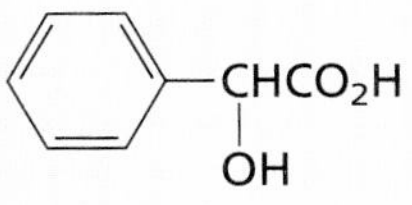

Double bonds in the main chain are signaled by the ending *-enoic acid*, and their position is designated by a numerical prefix. Entries 6 and 7 are representative carboxylic acids that contain double bonds. Double-bond stereochemistry is specified by using either the cis–trans or the *E–Z* notation.

PROBLEM 12.1 The list of carboxylic acids in Table 12.1 is by no means exhaustive insofar as common names are concerned. Many others are known by their common names, a few of which follow. Give a systematic IUPAC name for each.

(a) $CH_2\!\!=\!\!CCO_2H$
        |
       CH_3

   Methacrylic acid

(b) H_3C H
        \   /
         $C\!\!=\!\!C$
        /   \
       H CO_2H

   Crotonic acid

(c) CH_3———CO_2H

   p-Toluic acid

Sample Solution (a) Methacrylic acid is an industrial chemical used in the preparation of transparent plastics such as Lucite and Plexiglas. The carbon chain that includes both the carboxylic acid and the double bond is three carbon atoms in length. The compound is named as a derivative of propenoic acid. It is not necessary to locate the position of the double bond by number, as in "2-propenoic acid," because no other positions are structurally possible for it. The methyl group is at C-2, and so the correct systematic name for methacrylic acid is 2-methylpropenoic acid.

12.2 STRUCTURE AND BONDING

The structural features of the carboxyl group are most apparent in formic acid. Formic acid is planar, with one of its carbon–oxygen bonds shorter than the other, and with bond angles at carbon close to 120°.

120 pm

$$H\!-\!C\begin{smallmatrix}O\\ \\O\end{smallmatrix}\!\!-\!\!H$$

134 pm

Formic acid

This suggests sp^2 hybridization at carbon, and a $\sigma + \pi$ carbon–oxygen double bond analogous to that of aldehydes and ketones.

Additionally, sp^2 hybridization of the hydroxyl oxygen allows one of its unshared electron pairs to be delocalized by orbital overlap with the π system of the carbonyl group (Figure 12.1). In resonance terms, this electron delocalization is represented as:

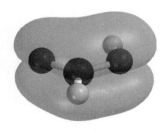

FIGURE 12.1 Carbon and both oxygens are sp^2-hybridized in formic acid. The π component of the C=O group and the *p* orbital of the OH oxygen overlap to form an extended π system that includes carbon and the two oxygens.

Lone-pair donation from the hydroxyl oxygen makes the carbonyl group less electrophilic than that of an aldehyde or ketone. The graphic that opened this chapter is an electrostatic potential map of formic acid that shows the most electron-rich site to be the oxygen of the carbonyl group and the most electron-poor one to be, as expected, the OH proton.

12.3 PHYSICAL PROPERTIES

The melting points and boiling points of carboxylic acids are higher than those of hydrocarbons and oxygen-containing organic compounds of comparable size and shape and indicate strong intermolecular attractive forces.

	2-Methyl-1-butene	2-Butanone	2-Butanol	Propanoic acid
bp (1 atm):	31°C	80°C	99°C	141°C

A unique hydrogen-bonding arrangement, shown here and in Figure 12.2, contributes to these attractive forces.

$$CH_3-C\underset{O-H\cdots\cdots O}{\overset{O\cdots\cdots H-O}{}}C-CH_3$$

The hydroxyl group of one carboxylic acid molecule acts as a proton donor toward the carbonyl oxygen of a second. Likewise, the hydroxyl proton of the second carboxyl function interacts with the carbonyl oxygen of the first. The result is that the two carboxylic acid molecules are held together by *two* hydrogen bonds. So efficient is this hydrogen bonding that some carboxylic acids exist as hydrogen-bonded dimers even in the gas phase. In the pure liquid a mixture of hydrogen-bonded dimers and higher aggregates is present.

In aqueous solution intermolecular association between carboxylic acid molecules is replaced by hydrogen bonding to water. The solubility properties of carboxylic acids are similar to those of alcohols. Carboxylic acids of four carbon atoms or fewer are miscible with water in all proportions.

> **PROBLEM 12.2** Vinegar is a 5% solution of acetic acid in water. Draw a diagram showing the hydrogen bonding forces present in vinegar.

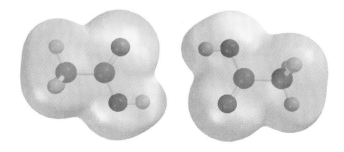

FIGURE 12.2 Attractions between regions of positive (*blue*) and negative (*red*) electrostatic potential are responsible for intermolecular hydrogen bonding between two molecules of acetic acid.

PROSTAGLANDINS

A prominent example of naturally occurring carboxylic acids is the class of compounds called prostaglandins. **Prostaglandins** were first isolated in minute amounts from sheep prostate glands in the 1930s. However, since the 1960s scientists have realized that prostaglandins exist in almost all cells and participate in a variety of regulatory functions in the human body.

More than a dozen prostaglandins have been identified as regulators of biological processes. All are 20-carbon carboxylic acids that contain a cyclopentane ring. Two representative prostaglandins, known as PGE$_1$ and PGF$_{1\alpha}$, are shown.

The range of physiological effects of prostaglandins is remarkable. Some prostaglandins relax bronchial tissue, others contract it. PGE$_1$ dilates

Prostaglandin E$_1$
(PGE$_1$)

Prostaglandin F$_{1\alpha}$
(PGF$_{1\alpha}$)

blood vessels, lowering blood pressure, and offers promise as a drug to reduce the formation of blood clots.

One result of prostaglandin research has been an answer to the question "How does aspirin work?" Although aspirin has been used as a pain reliever since 1899, little was known about its mechanism of action. Research now shows that the body makes prostaglandins in response to tissue damage. The consequence of this prostaglandin production is pain and inflammation. Aspirin interferes with prosta-

glandin biosynthesis, thus lowering prostaglandin levels and reducing pain and inflammation.

Prostaglandins continue to be an important focus of health care research. It is hoped that new drugs, and a greater understanding of human physiology, may develop from this work. In 1982 three European scientists, Sune Bergström and Bengt Samuelsson of Sweden and John Vane of Great Britain, shared the Nobel Prize in physiology or medicine for their pioneering research in prostaglandins.

12.4   ACIDITY OF CARBOXYLIC ACIDS

Carboxylic acids are the most acidic class of compounds that contain only carbon, hydrogen, and oxygen.

With ionization constants K_a on the order of 10^{-5} ($pK_a \approx 5$), they are much stronger acids than water and alcohols. The case should not be overstated, however. Carboxylic acids are weak acids; a 0.1 M solution of acetic acid in water, for example, is only 1.3% ionized.

PROBLEM 12.3 Acetylsalicylic acid (aspirin) has an ionization constant $K_a = 3.3 \times 10^{-4}$. Calculate the pK_a of acetylsalicylic acid. Is acetylsalicylic acid a stronger or weaker acid than benzoic acid ($pK_a = 4.2$)?

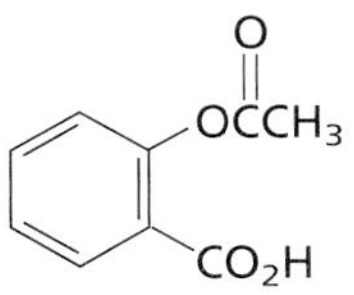

Acetylsalicylic acid
(aspirin)

To understand the greater acidity of carboxylic acids compared with water and alcohols, compare the structural changes that accompany the ionization of a representative alcohol (ethanol) and a representative carboxylic acid (acetic acid). The equilibria that define K_a are

Ionization of ethanol:

$$CH_3CH_2OH \rightleftharpoons H^+ + CH_3CH_2O^- \qquad K_a = \frac{[H^+][CH_3CH_2O^-]}{[CH_3CH_2OH]} = 10^{-16}$$

Ethanol   Ethoxide ion

Ionization of acetic acid:

$$CH_3COH \rightleftharpoons H^+ + CH_3CO^- \qquad K_a = \frac{[H^+][CH_3CO_2^-]}{[CH_3CO_2H]} = 1.8 \times 10^{-5}$$

Acetic acid   Acetate ion

Recall from Chapter 3 (Section 3.5) that the stronger the acid, the weaker the conjugate base. Acetate ion is stabilized by electron delocalization not available to ethoxide ion. This delocalization, expressed by resonance between the following Lewis structures, causes the negative charge in acetate to be shared equally by both oxygens.

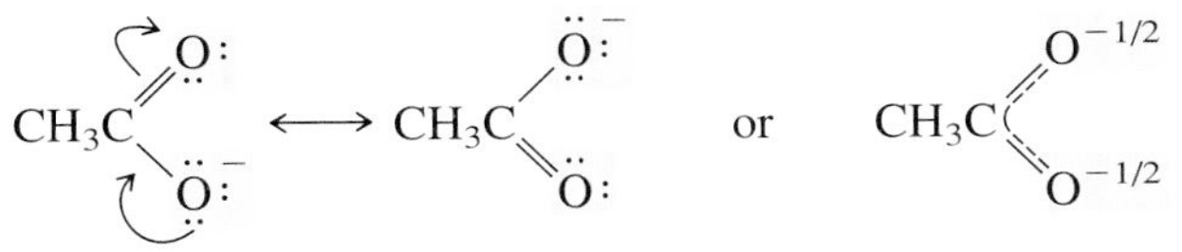

Another effect that plays a role in stabilizing acetate ion is the inductive effect of the carbonyl group. The carbonyl group is electron-withdrawing, and by attracting electrons away from the negatively charged oxygen, acetate ion is stabilized.

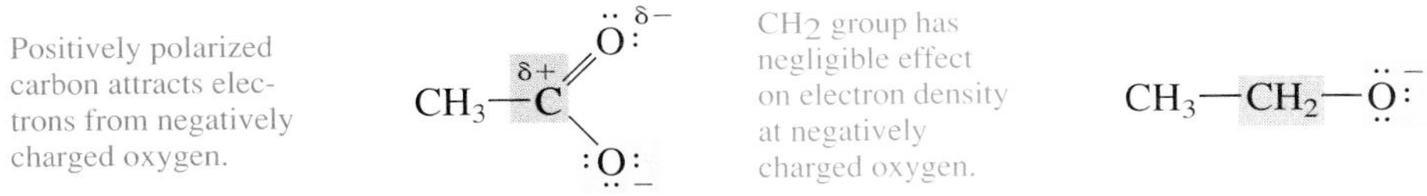

12.5   SUBSTITUENTS AND ACID STRENGTH

Alkyl groups have little effect on the acidity of a carboxylic acid. The ionization constants of all acids that have the general formula $C_nH_{2n+1}CO_2H$ are very similar to one another and equal approximately 10^{-5} (pK_a 5). Table 12.2 gives a few examples.

An electronegative substituent, particularly if it is attached to the α carbon, increases the acidity of a carboxylic acid. As the data in Table 12.2 show, all the monohaloacetic acids are about 100 times more acidic than acetic acid. Multiple halogen substitution increases the acidity even more; trichloroacetic acid is 7000 times more acidic than acetic acid!

The acid-strengthening effect of electronegative atoms or groups is easily seen as an inductive effect of the substituent transmitted through the σ bonds of the molecule. According to this model, the σ electrons in the carbon–chlorine bond of chloroacetate ion are drawn toward chlorine, leaving the α-carbon atom with a slight positive charge. The α carbon, because of this positive character, attracts electrons from the negatively charged carboxylate, thus dispersing the charge and stabilizing the anion. The more stable the anion, the greater the equilibrium constant for its formation.

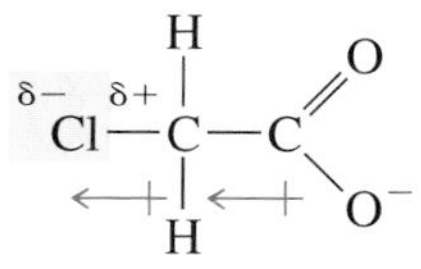

Chloroacetate anion is stabilized by electron-withdrawing effect of chlorine.

TABLE 12.2   Effect of Substituents on Acidity of Carboxylic Acids

Name of Acid	Structure	Ionization Constant K_a*	pK_a
Standard of comparison			
Acetic acid	CH_3CO_2H	1.8×10^{-5}	4.7
Alkyl substituents have a negligible effect on acidity.			
Propanoic acid	$CH_3CH_2CO_2H$	1.3×10^{-5}	4.9
2-Methylpropanoic acid	$(CH_3)_2CHCO_2H$	1.6×10^{-5}	4.8
2,2-Dimethylpropanoic acid	$(CH_3)_3CCO_2H$	0.9×10^{-5}	5.1
Heptanoic acid	$CH_3(CH_2)_5CO_2H$	1.3×10^{-5}	4.9
α-Halogen substituents increase acidity.			
Fluoroacetic acid	FCH_2CO_2H	2.5×10^{-3}	2.6
Chloroacetic acid	$ClCH_2CO_2H$	1.4×10^{-3}	2.9
Bromoacetic acid	$BrCH_2CO_2H$	1.4×10^{-3}	2.9
Dichloroacetic acid	Cl_2CHCO_2H	5.0×10^{-2}	1.3
Trichloroacetic acid	Cl_3CCO_2H	1.3×10^{-1}	0.9
Electron-attracting groups increase acidity.			
Methoxyacetic acid	$CH_3OCH_2CO_2H$	2.7×10^{-4}	3.6
Cyanoacetic acid	$N\equiv CCH_2CO_2H$	3.4×10^{-3}	2.5
Nitroacetic acid	$O_2NCH_2CO_2H$	2.1×10^{-2}	1.7

* In water at 25°C.

Inductive effects fall off rapidly as the number of σ bonds between the carboxyl group and the substituent increases. Consequently, the acid-strengthening effect of a halogen decreases as it becomes more remote from the carboxyl group:

ClCH$_2$CO$_2$H	ClCH$_2$CH$_2$CO$_2$H	ClCH$_2$CH$_2$CH$_2$CO$_2$H
Chloroacetic acid	3-Chloropropanoic acid	4-Chlorobutanoic acid
$K_a = 1.4 \times 10^{-3}$	$K_a = 1.0 \times 10^{-4}$	$K_a = 3.0 \times 10^{-5}$
p$K_a = 2.9$	p$K_a = 4.0$	p$K_a = 4.5$

PROBLEM 12.4 Which is the stronger acid in each of the following pairs?

(a) (CH$_3$)$_3$CCH$_2$CO$_2$H   or   (CH$_3$)$_3\overset{+}{N}$CH$_2$CO$_2$H

(b) CH$_3$CH$_2$CO$_2$H   or   CH$_3$CHCO$_2$H
                                                          |
                                                         OH

(c) CH$_3\overset{\text{O}}{\overset{\|}{\text{C}}}CO_2$H   or   (CH$_3$)$_2$CHCO$_2$H

Sample Solution (a) Think of the two compounds as substituted derivatives of acetic acid. A *tert*-butyl group is slightly electron-releasing and has only a modest effect on acidity. The compound (CH$_3$)$_3$CCH$_2$CO$_2$H is expected to have an acid strength similar to that of acetic acid. A trimethylammonium substituent, on the other hand, is positively charged and is a powerful electron-withdrawing substituent. The compound (CH$_3$)$_3\overset{+}{N}$CH$_2$CO$_2$H is expected to be a much stronger acid than (CH$_3$)$_3$CCH$_2$CO$_2$H. The measured ionization constants, shown as follows, confirm this prediction.

(CH$_3$)$_3$CCH$_2$CO$_2$H	(CH$_3$)$_3\overset{+}{N}$CH$_2$CO$_2$H
Weaker acid	Stronger acid
$K_a = 5 \times 10^{-6}$	$K_a = 1.5 \times 10^{-2}$
(p$K_a = 5.3$)	(p$K_a = 1.8$)

12.6 IONIZATION OF SUBSTITUTED BENZOIC ACIDS

A considerable body of data is available on the acidity of substituted benzoic acids. Benzoic acid itself is a somewhat stronger acid than acetic acid. Its carboxyl group is attached to an sp^2-hybridized carbon and ionizes to a greater extent than one that is attached to an sp^3-hybridized carbon. Remember, carbon becomes more electron-withdrawing as its s character increases.

CH$_3$CO$_2$H	CH$_2$=CHCO$_2$H	⬡—CO$_2$H
Acetic acid	Acrylic acid	Benzoic acid
$K_a = 1.8 \times 10^{-5}$	$K_a = 5.5 \times 10^{-5}$	$K_a = 6.3 \times 10^{-5}$
(pK_a 4.8)	(pK_a 4.3)	(pK_a 4.2)

Table 12.3 lists the ionization constants of some substituted benzoic acids. The largest effects are observed when strongly electron-withdrawing substituents are ortho to the carboxyl group. An *o*-nitro substituent, for example, increases the acidity of benzoic acid 100-fold. Substituent effects are small at positions meta and para to the carboxyl group. In those cases the pK_a values are clustered in the range 3.5–4.5.

TABLE 12.3	Acidity of Some Substituted Benzoic Acids		
Substituent in	K_a (pK_a)* for different positions of substituent X		
XC$_6$H$_4$CO$_2$H	**Ortho**	**Meta**	**Para**
1. H	6.3×10^{-5} (4.2)	6.3×10^{-5} (4.2)	6.3×10^{-5} (4.2)
2. CH$_3$	1.2×10^{-4} (3.9)	5.3×10^{-5} (4.3)	4.2×10^{-5} (4.4)
3. F	5.4×10^{-4} (3.3)	1.4×10^{-4} (3.9)	7.2×10^{-5} (4.1)
4. Cl	1.2×10^{-3} (2.9)	1.5×10^{-4} (3.8)	1.0×10^{-4} (4.0)
5. Br	1.4×10^{-3} (2.8)	1.5×10^{-4} (3.8)	1.1×10^{-4} (4.0)
6. I	1.4×10^{-3} (2.9)	1.4×10^{-4} (3.9)	9.2×10^{-5} (4.0)
7. CH$_3$O	8.1×10^{-5} (4.1)	8.2×10^{-5} (4.1)	3.4×10^{-5} (4.5)
8. O$_2$N	6.7×10^{-3} (2.2)	3.2×10^{-4} (3.5)	3.8×10^{-4} (3.4)

* In water at 25°C.

12.7   SALTS OF CARBOXYLIC ACIDS

In the presence of strong bases such as sodium hydroxide, carboxylic acids are neutralized rapidly and quantitatively:

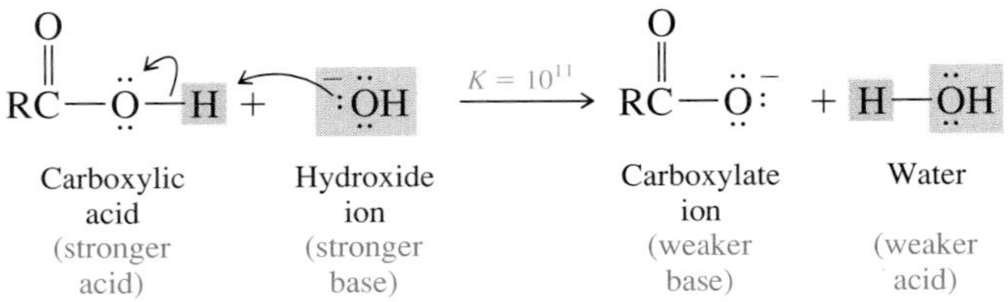

Recall from Chapter 3 (Section 3.5) that in any acid–base reaction the position of equilibrium favors formation of the weaker acid and the weaker base.

PROBLEM 12.5   Write an ionic equation for the reaction of acetic acid with each of the following, and specify whether the equilibrium favors starting materials or products (K_a values can be found in Table 3.2):

(a) Sodium ethoxide

(b) Potassium *tert*-butoxide

(c) Sodium bromide

(d) Sodium acetylide

(e) Potassium nitrate

(f) Lithium amide

Sample Solution   (a) This is an acid–base reaction; ethoxide ion is the base.

$$CH_3CO_2H \quad + \quad CH_3CH_2O^- \quad \longrightarrow \quad CH_3CO_2^- \quad + \quad CH_3CH_2OH$$

Acetic acid	Ethoxide ion	Acetate ion	Ethanol
(stronger acid)	(stronger base)	(weaker base)	(weaker acid)

The position of equilibrium lies well to the right. Ethanol, with a K_a of 10^{-16} (pK_a16), is a much weaker acid than acetic acid.

The metal carboxylate salts formed on neutralization of carboxylic acids are named by first specifying the metal ion and then adding the name of the acid modified by replacing -*ic acid* by -*ate*.

$$CH_3COLi$$

Lithium
acetate

$$Cl—\!\!\!\!\!\!\langle\ \rangle\!\!\!\!\!\!—CONa$$

Sodium *p*-chlorobenzoate

Metal carboxylates are ionic, and when the molecular weight isn't too high, the sodium and potassium salts of carboxylic acids are soluble in water. Carboxylic acids therefore may be extracted from ether solutions into aqueous sodium or potassium hydroxide.

The solubility behavior of salts of carboxylic acids having 12–18 carbons is unusual and can be illustrated by considering sodium stearate:

hydrophobic

$$O^- \ Na^+$$

Sodium stearate
(sodium octadecanoate)

hydrophilic

Sodium stearate has a polar carboxylate group at one end of a long hydrocarbon chain. The carboxylate group is **hydrophilic** ("water-loving") and tends to confer water solubility on the molecule. The hydrocarbon chain is **hydrophobic** (literally "water-hating") or **lipophilic** ("fat-loving") and tends to associate with other hydrocarbon chains. The compromise achieved by sodium stearate when it is placed in water is to form a colloidal dispersion of spherical aggregates called **micelles.** Each micelle is composed of 50–100 individual molecules. Micelles form spontaneously when the carboxylate concentration exceeds a certain minimum value called the **critical micelle concentration.** A representation of a micelle is shown in Figure 12.3.

Polar carboxylate groups dot the surface of the micelle. There they bind to water molecules and to sodium ions. The nonpolar hydrocarbon chains are directed toward the interior of the micelle, where individually weak but cumulatively significant induced-dipole/induced-dipole forces bind them together. Micelles are approximately spherical because a sphere encloses the maximum volume of material for a given surface area and disrupts the water structure least. Because their surfaces are negatively charged, two micelles repel each other rather than clustering to form higher aggregates.

It is the formation and properties of micelles that are responsible for the cleansing action of **soap.** Water that contains sodium stearate removes grease by enclosing it in the

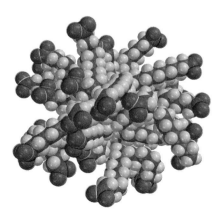

FIGURE 12.3 A space-filling model of a micelle formed by association of carboxylate ions derived from a fatty acid. In general, the hydrophobic carbon chains are inside and the carboxylate ions on the surface, but the micelle is irregular, and contains voids, channels, and tangled carbon chains. Each carboxylate is associated with a metal ion such as Na$^+$ (not shown).

hydrocarbon-like interior of the micelles. The grease is washed away with the water, not because it dissolves in the water but because it dissolves in the micelles that are dispersed in the water. Sodium stearate is an example of a soap; sodium and potassium salts of other C_{12}–C_{18} unbranched carboxylic acids possess similar properties.

Detergents are substances, including soaps, that cleanse by micellar action. A large number of synthetic detergents are known. One example is sodium lauryl sulfate. Sodium lauryl sulfate has a long hydrocarbon chain terminating in a polar sulfate ion and forms soap-like micelles in water.

Sodium lauryl sulfate
(sodium dodecyl sulfate)

Detergents are designed to be effective in hard water, meaning water containing calcium or magnesium ions that form insoluble carboxylate salts with soaps. These precipitates rob the soap of its cleansing power and form an unpleasant scum. The calcium and magnesium salts of synthetic detergents such as sodium lauryl sulfate, however, are soluble and retain their micelle-forming ability in water.

12.8   SOURCES OF CARBOXYLIC ACIDS

Many carboxylic acids were first isolated from natural sources and were given names based on their origin. Formic acid (Latin *formica*, "ant") was obtained by distilling ants. Since ancient times acetic acid (Latin *acetum*, "vinegar") has been known to be present in wine that has turned sour. Butyric acid (Latin *butyrum*, "butter") contributes to the odor of both rancid butter and ginkgo berries, and lactic acid (Latin *lac*, "milk") has been isolated from sour milk.

Although these humble origins make interesting historical notes, in most cases the large-scale preparation of carboxylic acids relies on chemical synthesis. Virtually none of the 3×10^9 lb of acetic acid produced in the United States each year is obtained from vinegar. Instead, most industrial acetic acid comes from the reaction of methanol with carbon monoxide.

$$CH_3OH \; + \; CO \;\; \xrightarrow[\text{heat, pressure}]{\substack{\text{cobalt or} \\ \text{rhodium catalyst}}} \; CH_3CO_2H$$

Methanol   Carbon   Acetic acid
              monoxide

The principal end use of acetic acid is in the production of vinyl acetate for paints and adhesives.

The carboxylic acid produced in the greatest amounts is 1,4-benzenedicarboxylic acid (terephthalic acid). About 5×10^9 lb/year is produced in the United States as a starting material for the preparation of polyester fibers. One important process converts *p*-xylene to terephthalic acid by oxidation with nitric acid:

p-Xylene   1,4-Benzenedicarboxylic acid
                                  (terephthalic acid)

You will recognize the side-chain oxidation of *p*-xylene to terephthalic acid as a reaction type discussed previously (Section 6.5). Examples of other reactions encountered earlier that can be applied to the synthesis of carboxylic acids are collected in Table 12.4.

The examples in the table give carboxylic acids that have the same number of carbon atoms as the starting material. The reactions to be described in the next two sections permit carboxylic acids to be prepared by extending a chain by one carbon atom and are of great value in laboratory syntheses of carboxylic acids.

TABLE 12.4	Summary of Reactions Discussed in Earlier Chapters That Yield Carboxylic Acids

Reaction (section) and Comments	General Equation and Specific Example
Side-chain oxidation of alkylbenzenes (Section 6.5) A primary or secondary alkyl side chain on an aromatic ring is converted to a carboxyl group by reaction with a strong oxidizing agent such as potassium permanganate or chromic acid.	$ArCHR_2 \xrightarrow[K_2Cr_2O_7,\ H_2SO_4]{KMnO_4\ or} ArCO_2H$ Alkylbenzene Arenecarboxylic acid 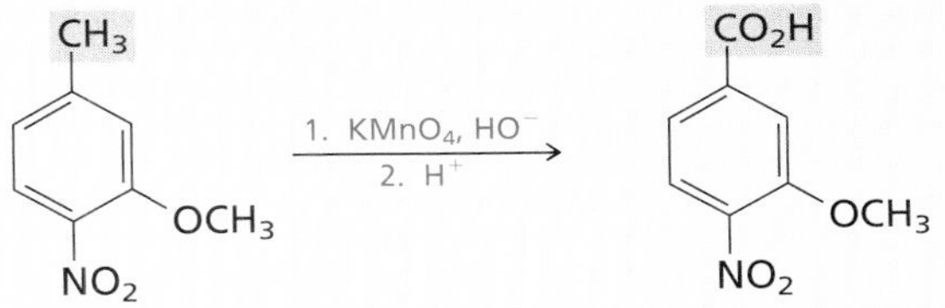 3-Methoxy-4-nitrotoluene → 3-Methoxy-4-nitrobenzoic acid (100%)
Oxidation of primary alcohols (Section 10.5) Potassium permanganate and chromic acid convert primary alcohols to carboxylic acids by way of the corresponding aldehyde.	$RCH_2OH \xrightarrow[K_2Cr_2O_7,\ H_2SO_4]{KMnO_4\ or} RCO_2H$ Primary alcohol Carboxylic acid $(CH_3)_3CCHC(CH_3)_3 \xrightarrow[H_2O,\ H_2SO_4]{H_2CrO_4} (CH_3)_3CCHC(CH_3)_3$ with CH$_2$OH → CO$_2$H 2-*tert*-Butyl-3,3-dimethyl-1-butanol → 2-*tert*-Butyl-3,3-dimethylbutanoic acid (82%)
Oxidation of aldehydes (Section 11.14) Aldehydes are particularly sensitive to oxidation and are converted to carboxylic acids by a number of oxidizing agents, including potassium permanganate and chromic acid.	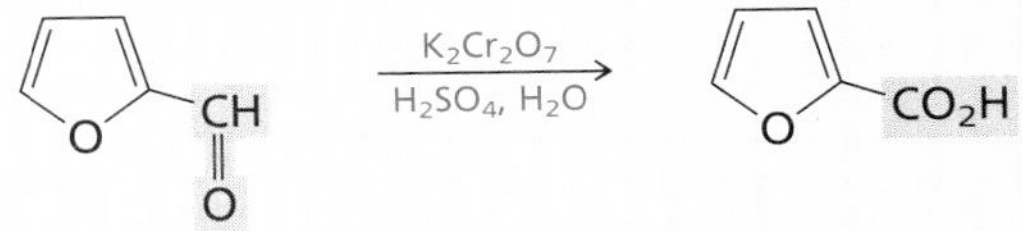 $\underset{\text{Aldehyde}}{RCH{=}O} \xrightarrow{\text{oxidizing agent}} \underset{\text{Carboxylic acid}}{RCO_2H}$ Furan-2-carbaldehyde (furfural) → Furan-2-carboxylic acid (furoic acid) (75%)

12.9 SYNTHESIS OF CARBOXYLIC ACIDS BY THE CARBOXYLATION OF GRIGNARD REAGENTS

We've seen how Grignard reagents add to the carbonyl group of aldehydes and ketones. Grignard reagents react in much the same way with carbon dioxide to yield magnesium salts of carboxylic acids in a reaction called **carboxylation.** Acidification converts these magnesium salts to the desired carboxylic acids.

$$\underset{\substack{\text{Grignard reagent}\\\text{acts as a nucleophile}\\\text{toward carbon dioxide}}}{\overset{\delta-}{R}\overset{\ddot{:}O\!:}{\underset{\delta+}{\underset{MgX\ :\ddot{O}\!:}{C}}}} \longrightarrow \underset{\substack{\text{Halomagnesium}\\\text{carboxylate}}}{R\overset{:O:}{C}OMgX} \xrightarrow[\text{H}_2\text{O}]{\text{H}^+} \underset{\substack{\text{Carboxylic}\\\text{acid}}}{R\overset{:O:}{C}\ddot{O}H}$$

Overall, the carboxylation of Grignard reagents transforms an alkyl or aryl halide to a carboxylic acid in which the carbon skeleton has been extended by one carbon atom.

$$\underset{\text{2-Chlorobutane}}{\underset{\underset{\text{Cl}}{|}}{CH_3CHCH_2CH_3}} \xrightarrow[\substack{\text{2. CO}_2\\\text{3. H}_3\text{O}^+}]{\text{1. Mg, diethyl ether}} \underset{\substack{\text{2-Methylbutanoic acid}\\(76-86\%)}}{\underset{\underset{\text{CO}_2\text{H}}{|}}{CH_3CHCH_2CH_3}}$$

> **PROBLEM 12.6** Outline procedures for the following conversions, using carboxylation of a Grignard reagent.
> (a) 1-Chlorobutane $\longrightarrow$ pentanoic acid
> (b) 2-Bromopropane $\longrightarrow$ 2-methylpropanoic acid
> (c) Bromobenzene $\longrightarrow$ benzoic acid
>
> **Sample Solution** (a) Reaction of the four-carbon alkyl halide 1-chlorobutane with magnesium metal gives the corresponding Grignard reagent. Subsequent reaction with carbon dioxide, followed by acid hydrolysis, yields the desired five-carbon carboxylic acid, pentanoic acid.
>
> $$\underset{\text{1-Chlorobutane}}{CH_3CH_2CH_2CH_2Cl} \xrightarrow[\substack{\text{2. CO}_2\\\text{3. H}_3\text{O}^+}]{\text{1. Mg, diethyl ether}} \underset{\text{Pentanoic acid}}{CH_3CH_2CH_2CH_2CO_2H}$$

The major limitation to carboxylation is that the alkyl or aryl halide must not bear substituents that are incompatible with Grignard reagents, such as OH, NH, SH, or C=O.

12.10 SYNTHESIS OF CARBOXYLIC ACIDS BY THE PREPARATION AND HYDROLYSIS OF NITRILES

Primary and secondary alkyl halides may be converted to the next higher carboxylic acid by a two-step synthetic sequence involving the preparation and hydrolysis of nitriles. **Nitriles,** also known as **alkyl cyanides,** are prepared by nucleophilic substitution (Section 8.1).

$$:\overset{..}{\underset{..}{X}}—R \quad + \quad :\overline{C}≡N: \quad \longrightarrow \quad RC≡N \quad + \quad :\overset{..}{\underset{..}{X}}:^{-}$$

Primary or secondary alkyl halide   Cyanide ion   Nitrile (alkyl cyanide)   Halide ion

The reaction is of the S_N2 type and works best with primary and secondary alkyl halides. Elimination is the only reaction observed with tertiary alkyl halides. Aryl and vinyl halides do not react.

Once the cyano group has been introduced, the nitrile is subjected to hydrolysis. Usually this is carried out in aqueous acid at reflux.

$$RC≡N \quad + \quad 2H_2O \quad + \quad H^+ \quad \xrightarrow{heat} \quad \overset{O}{\overset{\|}{RCOH}} \quad + \quad NH_4^+$$

Nitrile   Water   Carboxylic acid   Ammonium ion

Benzyl chloride   Benzyl cyanide (92%)   Phenylacetic acid (77%)

> **PROBLEM 12.7** Repeat Problem 12.6 using hydrolysis of a nitrile as a key step. Which of the conversions of Problem 12.6 cannot be carried out by this method?
>
> **Sample Solution** (a) Reaction of 1-chlorobutane with sodium cyanide gives the five-carbon nitrile, pentanenitrile. Hydrolysis of pentanenitrile yields the desired carboxylic acid, pentanoic acid.
>
> $$CH_3CH_2CH_2CH_2Cl \xrightarrow{NaCN} CH_3CH_2CH_2CH_2CN \xrightarrow[\text{heat}]{H_2O, H_2SO_4} CH_3CH_2CH_2CH_2CO_2H$$
>
> 1-Chlorobutane   Pentanenitrile   Pentanoic acid

Nitrile groups in cyanohydrins are hydrolyzed under conditions similar to those of alkyl cyanides. Cyanohydrin formation followed by hydrolysis provides a route to the preparation of α-hydroxy carboxylic acids.

Recall the preparation of cyanohydrins in Section 11.7.

$$\overset{O}{\overset{\|}{CH_3CCH_2CH_2CH_3}} \xrightarrow[\text{2. } H^+]{\text{1. NaCN}} \underset{CN}{\overset{OH}{CH_3\overset{|}{\underset{|}{C}}CH_2CH_2CH_3}} \xrightarrow[\text{heat}]{H_2O, HCl} \underset{CO_2H}{\overset{OH}{CH_3\overset{|}{\underset{|}{C}}CH_2CH_2CH_3}}$$

2-Pentanone   2-Pentanone cyanohydrin   2-Hydroxy-2-methyl-pentanoic acid (60% from 2-pentanone)

12.11 REACTIONS OF CARBOXYLIC ACIDS

The most apparent chemical property of carboxylic acids, their ability to act as acids, has already been examined in earlier sections of this chapter. Carboxylic acids can be reduced, and this provides a means of preparing primary alcohols. Reduction of a carboxylic acid is

Recall that primary alcohols may also be prepared by the reduction of aldehydes (Section 10.3).

difficult, and only the powerful reducing agent lithium aluminum hydride (Section 10.3) will accomplish the task.

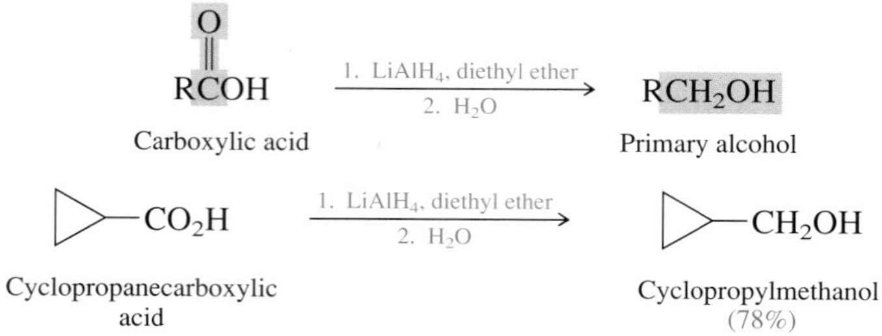

$$\underset{\substack{\text{Carboxylic acid}}}{\text{RCOH}} \xrightarrow[\text{2. } H_2O]{\text{1. LiAlH}_4\text{, diethyl ether}} \underset{\substack{\text{Primary alcohol}}}{\text{RCH}_2\text{OH}}$$

$$\underset{\substack{\text{Cyclopropanecarboxylic}\\\text{acid}}}{\triangleright\!\!-\!\text{CO}_2\text{H}} \xrightarrow[\text{2. } H_2O]{\text{1. LiAlH}_4\text{, diethyl ether}} \underset{\substack{\text{Cyclopropylmethanol}\\(78\%)}}{\triangleright\!\!-\!\text{CH}_2\text{OH}}$$

Sodium borohydride is not nearly as potent a hydride donor as lithium aluminum hydride and does not reduce carboxylic acids.

Other reactions of carboxylic acids involve their conversion to acyl derivatives. Two of these conversions will be introduced here; they and others will be explored in more detail in Chapter 13.

Carboxylic acids react with thionyl chloride ($SOCl_2$) to yield acyl chlorides.

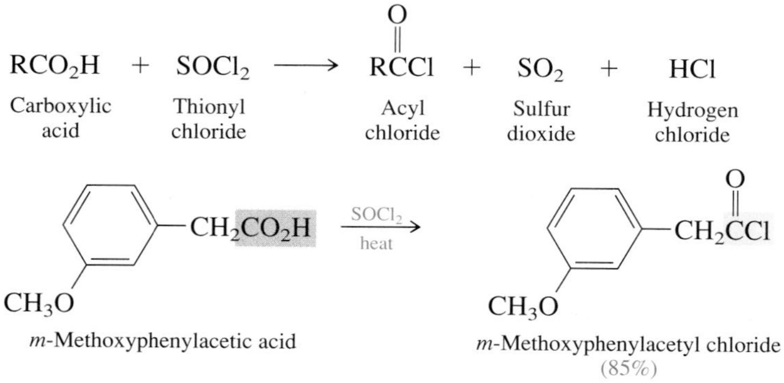

Acyl chlorides are reagents in Friedel–Crafts acylation (Sections 6.6 and 6.8).

$$\underset{\substack{\text{Carboxylic}\\\text{acid}}}{\text{RCO}_2\text{H}} + \underset{\substack{\text{Thionyl}\\\text{chloride}}}{\text{SOCl}_2} \longrightarrow \underset{\substack{\text{Acyl}\\\text{chloride}}}{\text{RCCl}} + \underset{\substack{\text{Sulfur}\\\text{dioxide}}}{\text{SO}_2} + \underset{\substack{\text{Hydrogen}\\\text{chloride}}}{\text{HCl}}$$

m-Methoxyphenylacetic acid $\xrightarrow[\text{heat}]{\text{SOCl}_2}$ *m*-Methoxyphenylacetyl chloride (85%)

Carboxylic acids react with alcohols in the presence of an acid catalyst to form esters.

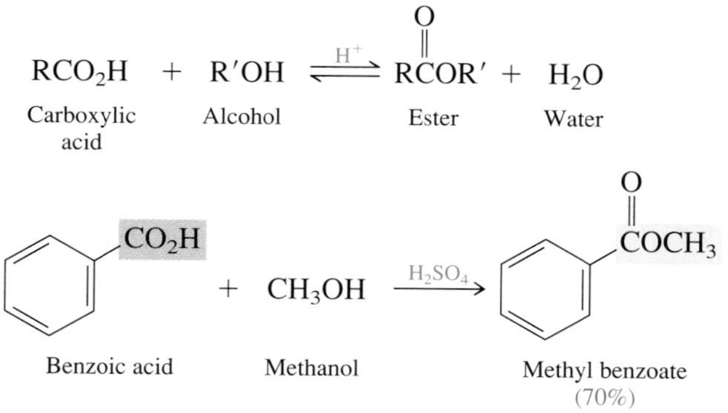

$$\underset{\substack{\text{Carboxylic}\\\text{acid}}}{\text{RCO}_2\text{H}} + \underset{\substack{\text{Alcohol}}}{\text{R'OH}} \underset{}{\overset{\text{H}^+}{\rightleftharpoons}} \underset{\substack{\text{Ester}}}{\text{RCOR'}} + \underset{\substack{\text{Water}}}{\text{H}_2\text{O}}$$

Benzoic acid   + CH_3OH $\xrightarrow{H_2SO_4}$ Methyl benzoate (70%)

Acid-catalyzed esterification is one of the fundamental reactions of organic chemistry. Its mechanism will be explored in some detail in Chapter 13.

PROBLEM 12.8 Predict the product from the reaction of phenylacetic acid, $C_6H_5CH_2CO_2H$, with each of the following reagents.

(a) KOH (c) $LiAlH_4$, then H_2O
(b) $SOCl_2$ (d) CH_3CH_2OH, H^+

Sample Solution (a) Carboxylic acids react with strong bases to yield metal carboxylate salts.

$$\underset{\substack{\text{Phenylacetic}\\\text{acid}}}{C_6H_5CH_2\overset{\displaystyle O}{\overset{\|}{C}}OH} \; + \; \underset{\substack{\text{Potassium}\\\text{hydroxide}}}{KOH} \; \longrightarrow \; \underset{\substack{\text{Potassium}\\\text{phenylacetate}}}{C_6H_5CH_2\overset{\displaystyle O}{\overset{\|}{C}}O^-K^+} \; + \; \underset{\text{Water}}{H_2O}$$

LEARNING OBJECTIVES

This chapter has focused on the properties of the most acidic class of organic compounds: carboxylic acids. The skills you have learned in this chapter should enable you to

- Provide an acceptable systematic IUPAC name for a carboxylic acid.
- Use chemical formulas to describe the hydrogen-bonding forces between carboxylic acid molecules in the liquid state and in aqueous solution.
- Write a chemical equation describing the acidity of a carboxylic acid.
- Explain the greater acidity of carboxylic acids compared with alcohols by using resonance structures.
- Explain how substituents affect the acidity of a carboxylic acid.
- Write an equation describing the formation of a carboxylate salt from the corresponding carboxylic acid.
- Explain the structural features that enable soaps and detergents to act as cleansing agents.
- Write a chemical equation describing the preparation of a carboxylic acid by carboxylation of a Grignard reagent.
- Write a chemical equation describing the preparation of a carboxylic acid by hydrolysis of a nitrile.
- Write a chemical equation describing the reduction of a carboxylic acid to give a primary alcohol.
- Write a chemical equation describing the formation of an acyl chloride from a carboxylic acid.
- Write a chemical equation describing the formation of an ester by reaction of a carboxylic acid with an alcohol in the presence of an acid catalyst.

12.12 SUMMARY

Carboxylic acids take their names from the alkane that contains the same number of carbons as the longest continuous chain that contains the —CO_2H group. The -*e* ending is replaced by -*oic acid*. Numbering begins at the carbon of the —CO_2H group (Section 12.1).

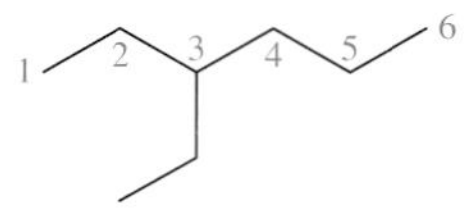

3-Ethylhexane 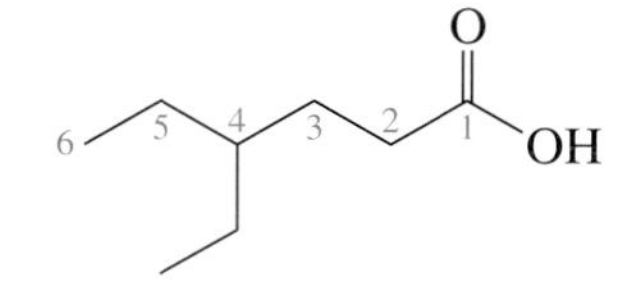   4-Ethylhexanoic acid

Like the carbonyl group of aldehydes and ketones, the carbon of a C=O unit in a carboxylic acid is sp^2-hybridized. Compared with the carbonyl group of an aldehyde or ketone, the C=O unit of a carboxylic acid receives an extra degree of stabilization from its attached OH group (Section 12.2).

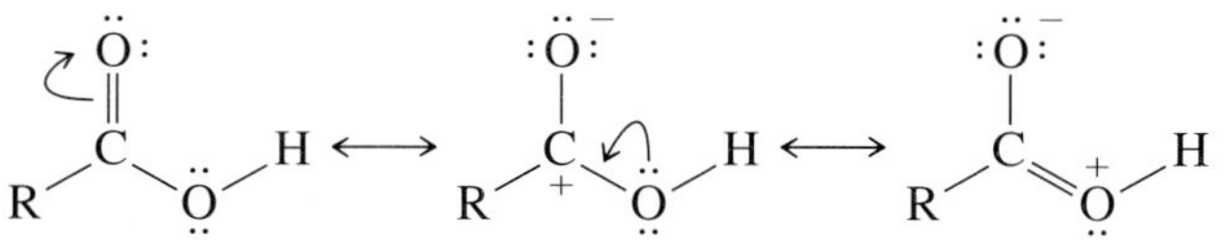

Hydrogen bonding in carboxylic acids raises their melting points and boiling points above those of comparably constituted alkanes, alcohols, aldehydes, and ketones (Section 12.3).

Carboxylic acids are weak acids and, in the absence of electron-attracting substituents, have dissociation constants K_a of approximately 10^{-5} ($pK_a = 5$) (Section 12.4). Carboxylic acids are much stronger acids than alcohols because of the electron-withdrawing power of the carbonyl group (inductive effect) and its ability to delocalize negative charge in the carboxylate anion (resonance effect).

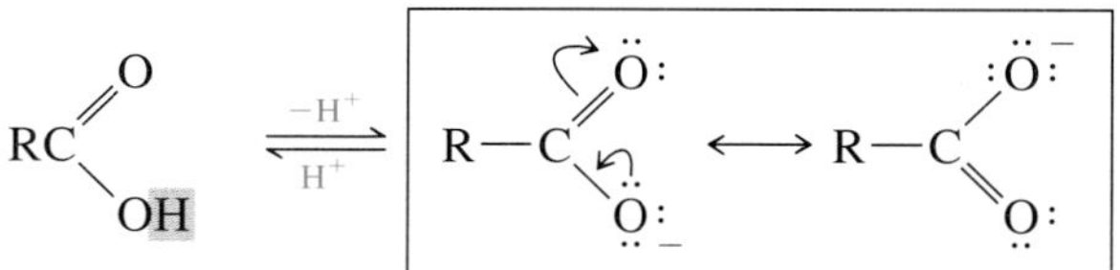

Carboxylic acid

Resonance description of electron delocalization in carboxylate anion

Electronegative substituents, especially those within a few bonds of the carboxyl group, increase the acidity of carboxylic acids (Sections 12.5, 12.6).

CF_3CO_2H

Trifluoroacetic acid
$K_a = 5.9 \times 10^{-1}$
($pK_a = 0.2$)

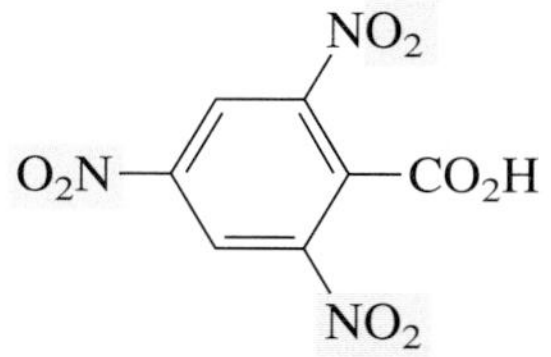

2,4,6-Trinitrobenzoic acid
$K_a = 2.2 \times 10^{-1}$
($pK_a = 0.6$)

Although carboxylic acids dissociate to only a small extent in water, they are deprotonated almost completely in basic solution.

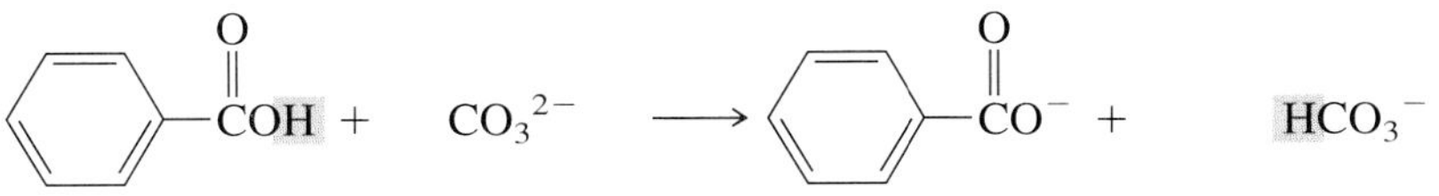

Benzoic acid   Carbonate ion   Benzoate ion   Hydrogen carbonate ion
$K_a = 6.3 \times 10^{-5}$ $K_a = 5 \times 10^{-11}$
(stronger acid)   (weaker acid)

Several reactions introduced in earlier chapters for the preparation of carboxylic acids are summarized in Table 12.4. Two new methods are introduced: carboxylation of Grignard reagents and hydrolysis of nitriles. Both methods *add one carbon* to the skeleton of the starting material.

Carboxylic acids can be prepared by the reaction of Grignard reagents with carbon dioxide (Section 12.9).

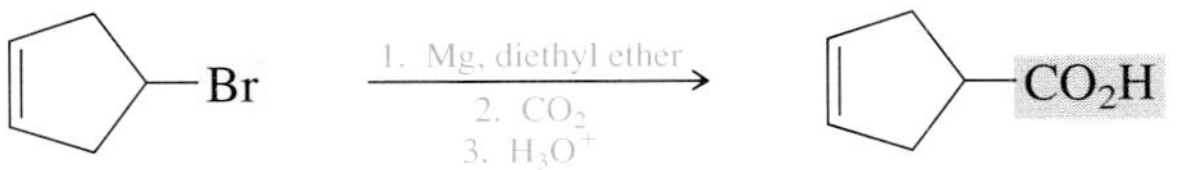

4-Bromocyclopentene

1. Mg, diethyl ether
2. CO_2
3. H_3O^+

Cyclopentene-
4-carboxylic acid
(66%)

Nitriles, which can be prepared from primary and secondary alkyl halides by nucleophilic substitution with cyanide ion, can be converted to carboxylic acids by hydrolysis (Section 12.10).

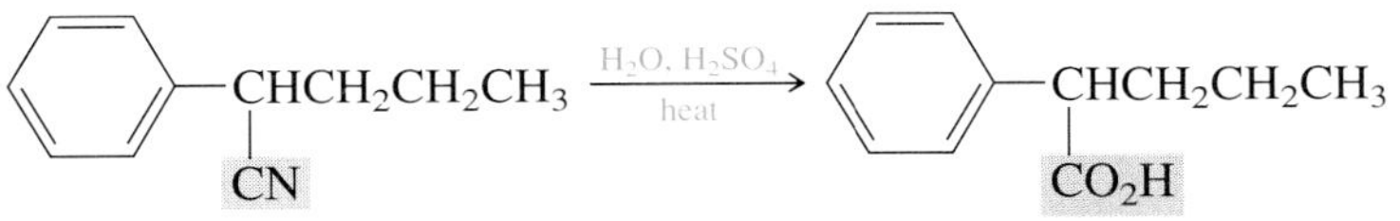

2-Phenylpentanenitrile

H_2O, H_2SO_4
heat

2-Phenylpentanoic acid
(52%)

Likewise, the cyano group of a cyanohydrin can be hydrolyzed to —CO_2H.

Reactions of carboxylic acids include their reduction to primary alcohols with lithium aluminum hydride, their conversion to acyl chlorides by reaction with thionyl chloride, and their conversion to esters by reaction with an alcohol in the presence of an acid catalyst (Section 12.11).

$$RCO_2H \xrightarrow[\text{2. } H_2O]{\text{1. LiAlH}_4} RCH_2OH$$

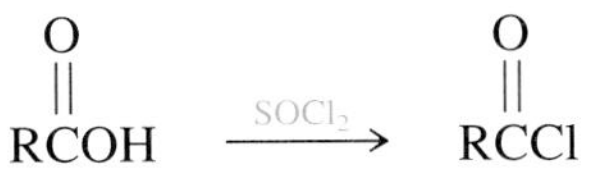

Carboxylic
acid

SOCl₂

Acyl
chloride

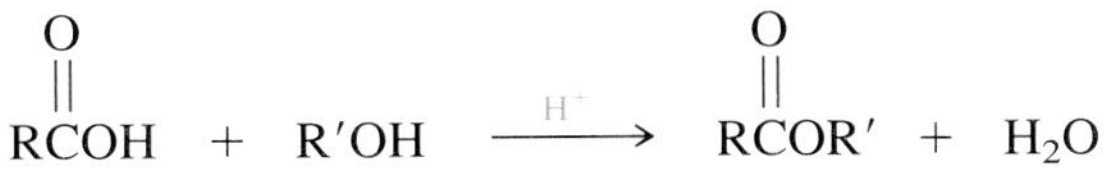

Carboxylic   Alcohol   Ester   Water
acid

ADDITIONAL PROBLEMS

Nomenclature, Physical Properties, and Acidity

12.9　Many carboxylic acids are much better known by their common names than by their systematic names. Some of these follow. Provide a structural formula for each one on the basis of its systematic name.

　　(a) 2-Hydroxypropanoic acid (better known as lactic acid, it is found in sour milk and is formed in the muscles during exercise)

　　(b) 2-Hydroxy-2-phenylethanoic acid (also known as mandelic acid, it is obtained from plums, peaches, and other fruits)

　　(c) Tetradecanoic acid (also known as myristic acid, it can be obtained from a variety of fats)

　　(d) 10-Undecenoic acid (also called undecylenic acid, it is used, in combination with its zinc salt, to treat fungal infections such as athlete's foot)

　　(e) 3,5-Dihydroxy-3-methylpentanoic acid (also called mevalonic acid, it is an important intermediate in the biosynthesis of terpenes and steroids)

　　(f) (*E*)-2-Methyl-2-butenoic acid (also known as tiglic acid, it is a constituent of various natural oils)

12.10　Give an acceptable IUPAC name for each of the following:

　　(a) $CH_3(CH_2)_6CO_2H$

　　(b) $CH_3(CH_2)_6CO_2K$

　　(c) $CH_2{=}CH(CH_2)_5CO_2H$

　　(d)

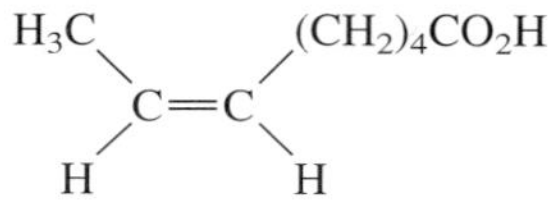

　　(e)

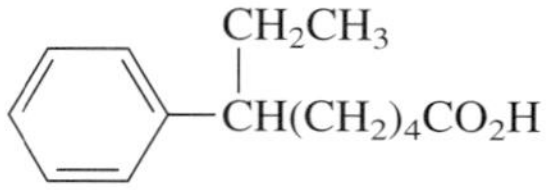

12.11　The following compounds are used as food preservatives. They are called **antioxidants** and retard oxidation that leads to spoilage of fats and oils. Write the chemical structure of each.

　　(a) Sodium benzoate

　　(b) Calcium propionate (IUPAC name, calcium propanoate)

　　(c) Potassium sorbate (sorbic acid is $CH_3CH{=}CHCH{=}CHCO_2H$)

12.12　Rank the compounds in each of the following groups in order of decreasing acidity:

　　(a) Acetic acid, ethane, ethanol

　　(b) Benzene, benzoic acid, benzyl alcohol

　　(c) Acetic acid, ethanol, trifluoroacetic acid, 2,2,2-trifluoroethanol

12.13　Identify the more acidic compound in each of the following pairs:

　　(a) $CF_3CH_2CO_2H$　　or　　$CF_3CH_2CH_2CO_2H$

　　(b) $CH_3CH_2CH_2CO_2H$　　or　　$CH_3C{\equiv}CCO_2H$

　　(c) 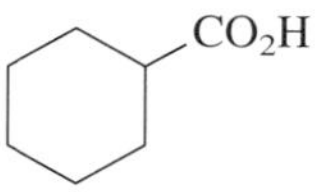
　　or

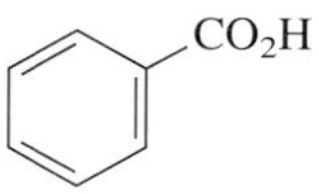

　　(d) 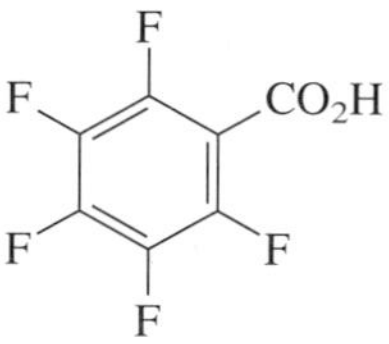
　　or

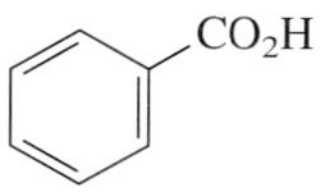

Preparation of Carboxylic Acids

12.14 Write a chemical reaction for the preparation of each of the compounds in Problem 12.11 from the appropriate carboxylic acid.

12.15 Propose methods for preparing butanoic acid from each of the following:
(a) 1-Butanol
(b) Butanal
(c) 1-Chloropropane (two methods)
(d) 2-Propanol
(e) Acetaldehyde

12.16 Describe two methods for the preparation of 5-methylhexanoic acid from 4-methyl-1-pentanol.

Reactions of Carboxylic Acids

12.17 Give the product of the reaction of pentanoic acid with each of the following reagents:
(a) Sodium hydroxide
(b) Sodium bicarbonate
(c) Thionyl chloride
(d) Lithium aluminum hydride, then hydrolysis with water
(e) 2-Propanol, acid (catalyst)

12.18 Show how butanoic acid may be converted to each of the following compounds using any other necessary organic or inorganic reagents.
(a) 1-Butanol
(b) 1-Bromobutane
(c) Pentanoic acid

(d) Butanoyl chloride ($CH_3CH_2CH_2\overset{\displaystyle O}{\overset{\|}{C}}Cl$)
(e) Butanal

(f) 1-Phenyl-1-butanone ($C_6H_5\overset{\displaystyle O}{\overset{\|}{C}}CH_2CH_2CH_3$)

(g) Butyl butanoate ($CH_3CH_2CH_2\overset{\displaystyle O}{\overset{\|}{C}}OCH_2CH_2CH_2CH_3$)

12.19 Outline a sequence of reactions that will allow pentanal to be converted into
(a) Pentanoic acid
(b) Hexanoic acid
(c) 2-Hydroxyhexanoic acid

12.20 Unbranched carboxylic acids obtained from natural sources, called fatty acids, usually contain an even number of carbon atoms. To prepare an acid having an odd number of carbons, the necessary conversion is

$$RCO_2H \longrightarrow RCH_2CO_2H$$

Show how this might be done by describing the preparation of nonanoic acid from octanoic acid.

Miscellaneous Problems

12.21 In the presence of the enzyme aconitase, the double bond of aconitic acid undergoes hydration. The reaction is reversible, and the following equilibrium is established:

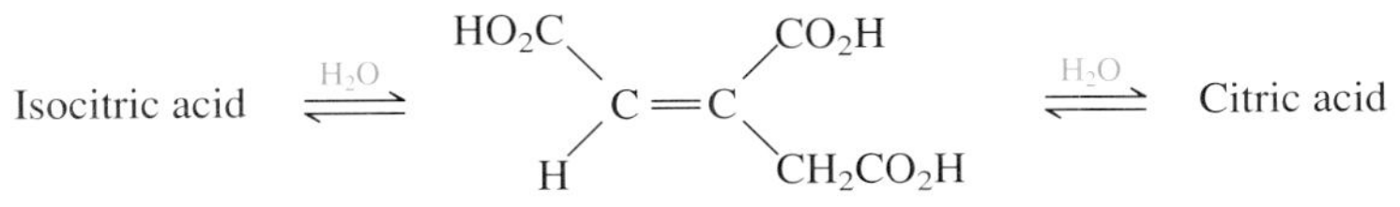

Isocitric acid $\underset{H_2O}{\rightleftharpoons}$
$$\underset{H}{\overset{HO_2C}{\diagdown}}C=C\underset{CH_2CO_2H}{\overset{CO_2H}{\diagup}}$$
$\underset{H_2O}{\rightleftharpoons}$ Citric acid

Isocitric acid	Aconitic acid	Citric acid
($C_6H_8O_7$)	($C_6H_8O_7$)	($C_6H_8O_7$)
(6% at equilibrium)	(4% at equilibrium)	(90% at equilibrium)

(a) The major tricarboxylic acid present is citric acid, the substance responsible for the tart taste of citrus fruits. Citric acid is achiral. What is its structure?

(b) What is the constitution of isocitric acid? How many stereoisomers of isocitric acid are possible?

12.22 Suggest reasonable explanations for each of the following observations:

(a) Both hydrogens are anti to each other in the most stable conformation of formic acid.

(b) The dissociation constant of *o*-hydroxybenzoic acid is greater (by a factor of 12) than that of *o*-methoxybenzoic acid.

(c) Ascorbic acid (vitamin C), although not a carboxylic acid, is sufficiently acidic to cause carbon dioxide liberation on being dissolved in aqueous sodium bicarbonate.

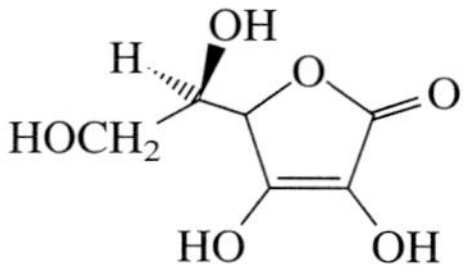

Ascorbic acid

12.23 It is sometimes necessary to prepare isotopically labeled samples of organic substances for probing biological transformations and reaction mechanisms. Various sources of the radioactive carbon-14 isotope are available. Describe synthetic procedures by which benzoic acid, labeled with ^{14}C at its carbonyl carbon, could be prepared from benzene and the following ^{14}C-labeled precursors. You may use any necessary organic or inorganic reagents. (In the formulas shown, an asterisk indicates ^{14}C.)

(a) $\overset{*}{C}H_3Cl$

(b)
$$\overset{\displaystyle O}{\underset{\displaystyle *}{\overset{\displaystyle \|}{H\overset{}{C}H}}}$$

(c) $\overset{*}{C}O_2$

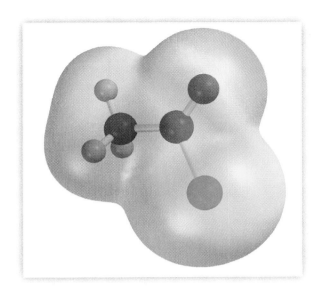

CHAPTER 13
CARBOXYLIC ACID DERIVATIVES

T he previous two chapters described three classes of organic compounds characterized by the presence of a carbonyl $\left(\,\diagup C{=}O\right)$ group: aldehydes, ketones, and carboxylic acids. We will now expand our discussion to include the principal classes of carboxylic acid derivatives. Two of these, **esters** and **amides,** are especially important in both organic and biochemistry and will receive particular emphasis.

13.1 NOMENCLATURE OF CARBOXYLIC ACID DERIVATIVES

Each of the **carboxylic acid derivatives** we will encounter possesses an **acyl group,**
$$\underset{\substack{\|\\O}}{RC}{-} \quad \text{or} \quad \underset{\substack{\|\\O}}{ArC}{-}$$
attached to a halogen, oxygen, or nitrogen atom. The four classes of carboxylic acid derivatives are

1. Acyl chlorides, $\underset{\substack{\|\\O}}{RC}Cl$

2. Anhydrides, $\underset{\substack{\|\\O}}{RC}O\underset{\substack{\|\\O}}{C}R$

3. Esters, $\underset{\substack{\|\\O}}{RC}OR'$

4. Amides, $\underset{\substack{\|\\O}}{RC}NH_2,\ \underset{\substack{\|\\O}}{RC}NHR',\ \text{and}\ \underset{\substack{\|\\O}}{RC}NR'_2$

The systematic names of each class are based on the name of the corresponding carboxylic acid. **Acyl chlorides** are named by replacing the *-ic acid* ending of the carboxylic acid name with *-yl chloride*.

$$CH_3\overset{O}{\overset{\|}{C}}Cl$$

$$\langle \text{benzene ring} \rangle -\overset{O}{\overset{\|}{C}}Cl$$

Acetyl chloride
(from acetic acid)

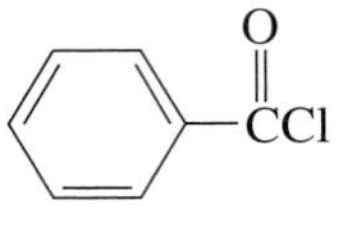

Benzoyl chloride
(from benzoic acid)

Anhydrides are named in a similar manner. The word *acid* is replaced with the word *anhydride*.

$$CH_3\overset{O}{\overset{\|}{C}}O\overset{O}{\overset{\|}{C}}CH_3$$

$$C_6H_5\overset{O}{\overset{\|}{C}}O\overset{O}{\overset{\|}{C}}C_6H_5$$

Acetic anhydride

Benzoic anhydride

The alkyl group and the acyl group of an ester are specified independently. **Esters** are named as alkyl alkanoates. The alkyl group R′ of $R\overset{O}{\overset{\|}{C}}OR'$ is cited first, followed by the acyl portion. The acyl portion is named by substituting the suffix *-ate* for the *-ic* ending of the corresponding acid.

$$CH_3\overset{O}{\overset{\|}{C}}OCH_2CH_3$$

$$CH_3CH_2\overset{O}{\overset{\|}{C}}OCH_3$$

$$\langle \text{benzene ring} \rangle -\overset{O}{\overset{\|}{C}}OCH_2CH_2Cl$$

Ethyl acetate

Methyl propanoate

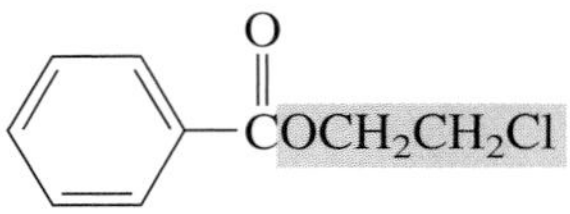

2-Chloroethyl benzoate

Aryl esters, that is, compounds of the type $R\overset{O}{\overset{\|}{C}}OAr$, are named in an analogous way.

The names of **amides** of the type $R\overset{O}{\overset{\|}{C}}NH_2$ are derived from carboxylic acids by replacing the suffix *-oic acid* or *-ic acid* by *-amide*.

$$CH_3\overset{O}{\overset{\|}{C}}NH_2$$

$$C_6H_5\overset{O}{\overset{\|}{C}}NH_2$$

$$(CH_3)_2CHCH_2\overset{O}{\overset{\|}{C}}NH_2$$

Acetamide

Benzamide

3-Methylbutanamide

We name compounds of the type $R\overset{O}{\overset{\|}{C}}NHR'$ and $R\overset{O}{\overset{\|}{C}}HNR_2$ as *N*-alkyl- and *N,N*-dialkyl-substituted derivatives of a parent amide.

$$CH_3\overset{O}{\overset{\|}{C}}NHCH_3$$

$$C_6H_5\overset{O}{\overset{\|}{C}}N(CH_2CH_3)_2$$

$$CH_3CH_2CH_2\overset{O}{\overset{\|}{C}}NCH(CH_3)_2$$
$$|$$
$$CH_3$$

N-Methylacetamide

N,N-Diethylbenzamide

N-Isopropyl-*N*-methylbutanamide

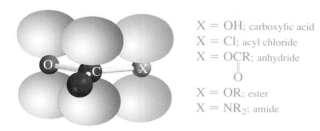

X = OH; carboxylic acid
X = Cl; acyl chloride
X = OCR; anhydride
         ‖
         O
X = OR; ester
X = NR$_2$; amide

FIGURE 13.1 The three σ bonds originating at the carbonyl carbon are coplanar. The *p* orbital of the carbonyl carbon, its oxygen, and the atom by which group X is attached to the acyl group overlap to form an extended π system through which the π electrons are delocalized.

PROBLEM 13.1 Write a structural formula for each of the following compounds:

(a) 2-Phenylbutanoyl chloride
(b) 2-Phenylbutanoic anhydride
(c) Butyl 2-phenylbutanoate
(d) 2-Phenylbutyl butanoate
(e) 2-Phenylbutanamide
(f) *N*-Ethyl-2-phenylbutanamide

Sample Solution (a) A 2-phenylbutanoyl group is a four-carbon acyl unit that bears a phenyl substituent at C-2. When the name of an acyl group is followed by the name of a halide, it designates an **acyl halide.**

$$CH_3CH_2\underset{\underset{\displaystyle C_6H_5}{|}}{C}H\overset{\overset{\displaystyle O}{\|}}{C}Cl$$

2-Phenylbutanoyl chloride

13.2 STRUCTURE OF CARBOXYLIC ACID DERIVATIVES

Like other carbonyl-containing compounds we have studied—aldehydes, ketones, and carboxylic acids—derivatives of carboxylic acids have a planar arrangement of bonds to the carbonyl group. An important structural feature of acyl chlorides, anhydrides, esters, and amides is that the atom, $\ddot{X}$, attached to the acyl group bears an unshared pair of electrons that can interact with the carbonyl π system, as shown in Figure 13.1.

This electron delocalization can be represented in resonance terms by contributions from the following resonance structures:

$$R-\overset{\displaystyle \ddot{O}:}{\underset{\displaystyle \ddot{X}}{C}} \longleftrightarrow R-\overset{\displaystyle :\ddot{O}:^-}{\underset{\displaystyle \overset{+}{X}}{C}} \longleftrightarrow R-\overset{\displaystyle :\ddot{O}:^-}{\underset{\displaystyle \overset{+}{X}}{C}}$$

Electron release from the substituent stabilizes the carbonyl group and decreases its electrophilic character. The extent of this electron delocalization depends on the electron-donating properties of the substituent X. Generally, the less electronegative X is, the better it donates electrons to the carbonyl group and the greater its stabilizing effect.

Resonance stabilization in acyl chlorides is not nearly as pronounced as in other derivatives of carboxylic acids:

$$R-\overset{\displaystyle \ddot{O}:}{\underset{\displaystyle :\ddot{Cl}:}{C}} \longleftrightarrow R-\overset{\displaystyle :\ddot{O}:^-}{\underset{\displaystyle :\overset{+}{Cl}:}{C}}$$

Weak resonance
stabilization

The carbon–chlorine bond is long and overlap between the $3p$ orbitals of chlorine and the π orbital of the carbonyl group is poor. Consequently, little delocalization occurs in the electron pairs of chlorine into the π system. The carbonyl group of an acyl chloride feels the normal electron-withdrawing inductive effect of a chlorine substituent without a significant compensating electron-releasing effect due to lone-pair donation by chlorine. This makes the carbonyl carbon of an acyl chloride more susceptible to attack by nucleophiles than that of other carboxylic acid derivatives.

Anhydrides are better stabilized by electron delocalization than are acyl chlorides. The lone-pair electrons of oxygen are delocalized more effectively into the carbonyl group. Resonance involves both carbonyl groups of an anhydride.

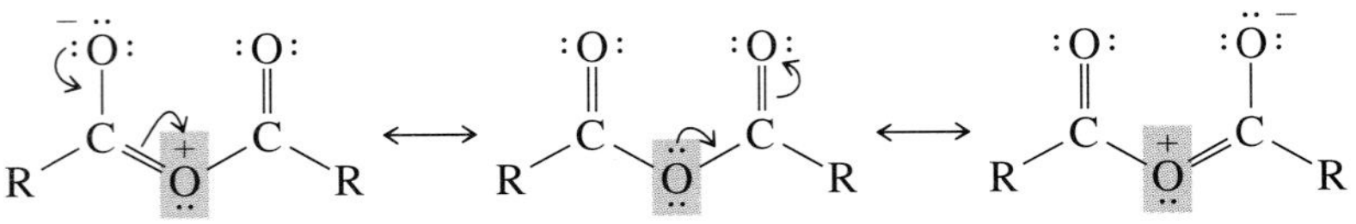

The carbonyl group of an ester is stabilized more than is that of an anhydride. Because both acyl groups of an anhydride compete for the oxygen lone pair, each carbonyl is stabilized less than the single carbonyl group of an ester.

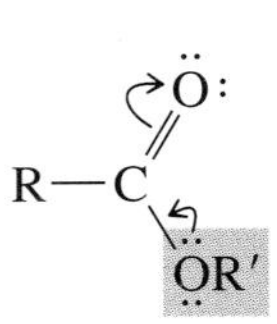

Ester   is more effective than   Anhydride

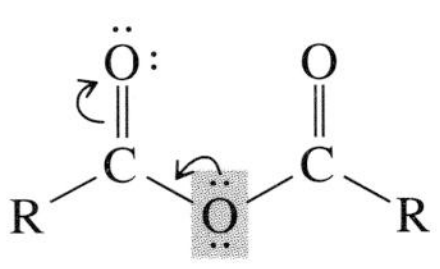

Esters are stabilized by resonance to about the same extent as carboxylic acids but not as much as amides. Nitrogen is less electronegative than oxygen and is a better electron-pair donor.

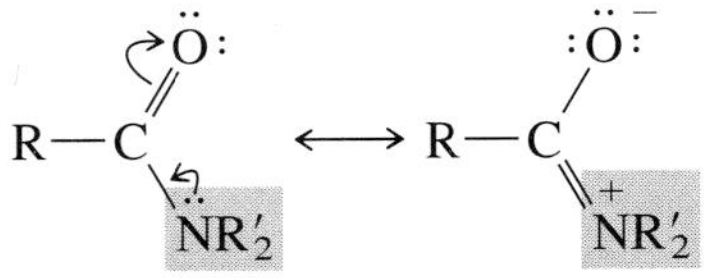

Very effective
resonance stabilization

Electron release from nitrogen stabilizes the carbonyl group of amides and decreases the rate at which nucleophiles attack the carbonyl carbon. Nucleophilic reagents attack electrophilic sites in a molecule; if electrons are donated to an electrophilic site in a molecule by a substituent, then the tendency of that molecule to react with external nucleophiles is moderated.

13.3   NUCLEOPHILIC ACYL SUBSTITUTION: HYDROLYSIS

The most important reaction of carboxylic acid derivatives is **nucleophilic acyl substitution** as represented by the general equation:

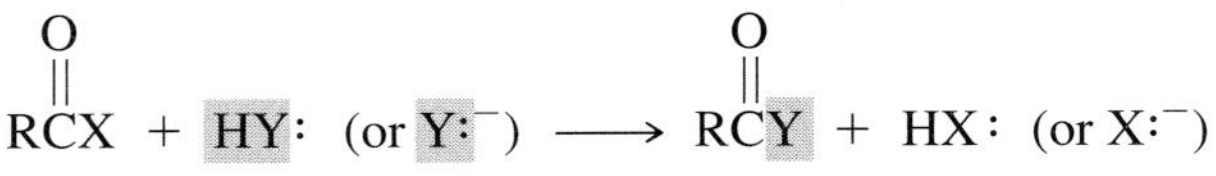

$$RCX + HY: \text{ (or } Y:^-) \longrightarrow RCY + HX: \text{ (or } X:^-)$$

Nucleophilic acyl substitution is the main process by which the various carboxylic acid derivatives are interconverted, both in the laboratory and in biological systems. To illustrate the mechanism of the reaction, let us examine the hydrolysis of an acyl derivative. **Hydrolysis** is the reaction with water, and it converts the various carboxylic acid derivatives to the corresponding carboxylic acid.

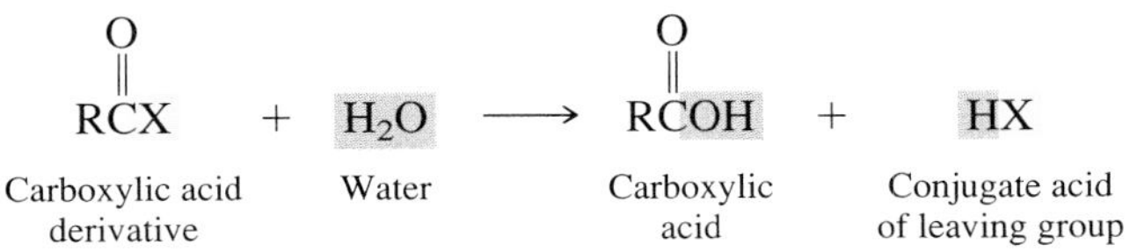

$$
\underset{\substack{\text{Carboxylic acid} \\ \text{derivative}}}{\overset{\overset{\textstyle O}{\|}}{RCX}} \; + \; \underset{\text{Water}}{H_2O} \; \longrightarrow \; \underset{\substack{\text{Carboxylic} \\ \text{acid}}}{\overset{\overset{\textstyle O}{\|}}{RCOH}} \; + \; \underset{\substack{\text{Conjugate acid} \\ \text{of leaving group}}}{HX}
$$

Acyl chlorides (X = Cl) react rapidly with water. The mechanism for hydrolysis of an acyl chloride is outlined in Figure 13.2.

In the first stage of the mechanism, water undergoes nucleophilic addition to the carbonyl group to form a **tetrahedral intermediate.** This stage of the process is analogous to the hydration of aldehydes and ketones discussed in Section 11.6.

The tetrahedral intermediate has three potential leaving groups: two hydroxyl groups and a chlorine. In the second stage of the reaction, the tetrahedral intermediate dissociates, eliminating chloride ion. Loss of chloride ion from the tetrahedral intermediate is faster than loss of hydroxide ion; chloride ion is less basic than hydroxide ion and is a better leaving group. The tetrahedral intermediate dissociates because this dissociation restores the resonance-stabilized carbonyl group.

Nucleophilic substitution in acyl chlorides occurs much more readily than nucleophilic substitution in alkyl chlorides. Benzoyl chloride ($C_6H_5\overset{\overset{\textstyle O}{\|}}{C}Cl$), for example, reacts with water almost 1000 times faster than benzyl chloride ($C_6H_5CH_2Cl$) does. The sp^2-hybridized carbon of an acyl chloride is less sterically hindered than the sp^3-hybridized

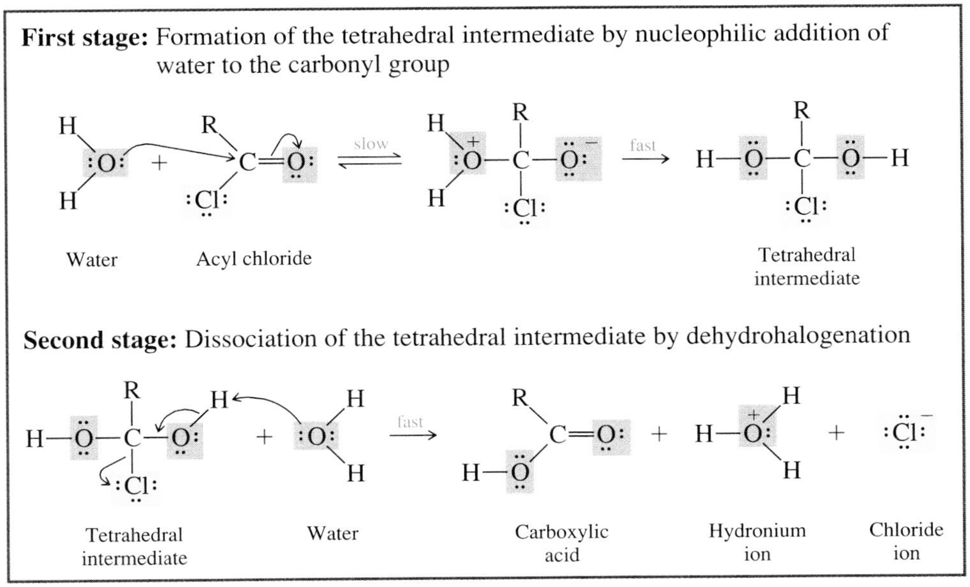

First stage: Formation of the tetrahedral intermediate by nucleophilic addition of water to the carbonyl group

| Water | Acyl chloride | | Tetrahedral intermediate |

Second stage: Dissociation of the tetrahedral intermediate by dehydrohalogenation

| Tetrahedral intermediate | Water | Carboxylic acid | Hydronium ion | Chloride ion |

FIGURE 13.2 Hydrolysis of acyl chloride proceeds by way of a tetrahedral intermediate. Formation of the tetrahedral intermediate is rate-determining.

carbon of an alkyl chloride, making an acyl chloride more open to nucleophilic attack. Also, unlike the S_N2 transition state or a carbocation intermediate in an S_N1 reaction, the tetrahedral intermediate in nucleophilic acyl substitution has a stable arrangement of bonds and can be formed via a lower energy transition state.

> **PROBLEM 13.2**   Acyl chlorides are called **lachrymators** because of the strong irritating effect they have on the eyes. This effect comes about from the reaction with moisture naturally present in our eyes. Write a balanced equation showing the reaction of benzoyl chloride with water, and outline a mechanism for this reaction.

The relative ease of hydrolysis of the various acid derivatives can be used as a measure of the chemical reactivity of these compounds. Acyl chlorides undergo hydrolysis 100 *billion* times faster than esters. In fact, the reaction of an acyl chloride with water can occur with explosive force, producing large quantities of heat.

Anhydrides are less reactive than acyl chlorides but still undergo hydrolysis about 10 million times faster than esters. Amides, on the other hand, are *less* reactive than esters. Their rate of hydrolysis is less than one one-hundredth that of an ester.

The trend in reactivities can be summarized:

$$\underset{\substack{\text{Acyl chloride}\\ \text{(most reactive)}}}{\underset{\text{RCCl}}{\overset{\overset{\text{O}}{\|}}{}}} \quad > \quad \underset{\text{Anhydride}}{\underset{\text{RCOCR}}{\overset{\overset{\text{O}\quad\text{O}}{\|\quad\|}}{}}} \quad > \quad \underset{\text{Ester}}{\underset{\text{RCOR}'}{\overset{\overset{\text{O}}{\|}}{}}} \quad > \quad \underset{\substack{\text{Amide}\\ \text{(least reactive)}}}{\underset{\text{RCNR}'_2}{\overset{\overset{\text{O}}{\|}}{}}}$$

The reactivity series of acid derivatives can be explained by considering the resonance stabilization of the acyl groups presented in Section 13.2. Acyl chlorides are stabilized the *least* and are the *most* reactive acid derivatives. Conversely, amides, the *most* stabilized derivative, are the *least* reactive.

One consequence of the reactivity trend is that a *more* reactive derivative can be used to prepare a *less* reactive one, but not the reverse. For example, anhydrides may be used to prepare esters and amides, but not acyl chlorides.

13.4   NATURAL SOURCES OF ESTERS

Many esters occur naturally. Those of low molecular weight are fairly volatile, and many have pleasing odors. Esters often form a significant fraction of the fragrant oil of fruits and flowers. The aroma of oranges, for example, contains 30 different esters along with 10 carboxylic acids, 34 alcohols, 34 aldehydes and ketones, and 36 hydrocarbons.

$$\underset{\substack{\text{3-Methylbutyl acetate}\\ \text{(contributes to characteristic}\\ \text{odor of bananas)}}}{\text{CH}_3\overset{\overset{\text{O}}{\|}}{\text{C}}\text{OCH}_2\text{CH}_2\text{CH(CH}_3)_2} \qquad \underset{\substack{\text{Methyl salicylate}\\ \text{(principal component of oil}\\ \text{of wintergreen)}}}{}$$

3-Methylbutyl acetate
(contributes to characteristic
odor of bananas)

Methyl salicylate
(principal component of oil
of wintergreen)

3-Methylbutyl acetate is more commonly known as isoamyl acetate.

Among the chemicals used by insects to communicate with one another, esters occur frequently.

BIOLOGICAL ACYL TRANSFER

Biochemists often refer to nucleophilic acyl substitution as **acyl transfer** because an acyl group is transferred from one atom to another.

$$\underset{X}{\overset{O}{\underset{\|}{R C}}} X + HY \longrightarrow \underset{}{\overset{O}{\underset{\|}{R C}}} Y + HX$$

What kinds of species of the type $\overset{O}{\underset{\|}{R C}}X$ characterize acyl transfers in living systems? In the laboratory, an organic chemist uses acyl chlorides and anhydrides to prepare other acyl compounds because of the greater reactivity of these materials. Nucleophilic acyl substitutions are also common in biochemistry, but acyl chlorides are not found in living systems, and the only known example of a naturally occurring carboxylic acid anhydride is cantharidin, isolated from a beetle and once believed to be an aphrodisiac.

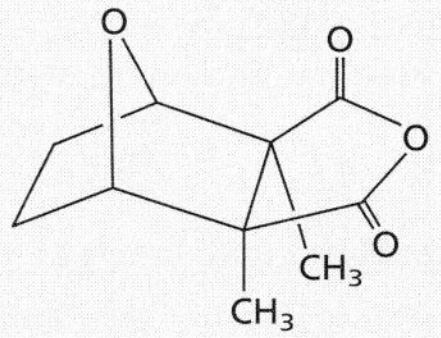

Cantharidin

Many biological transfers use acyl **thioesters,** compounds of the type $\overset{O}{\underset{\|}{R C}}SR'$. Thioesters are more reactive than simple esters, $\overset{O}{\underset{\|}{R C}}OR'$. One thioester, acetyl coenzyme A, has the relatively complicated structure shown in Figure 16.4. Acetyl coenzyme A is abbreviated $CH_3\overset{O}{\underset{\|}{C}}SCoA$. One of the reactions of acetyl coenzyme A in living systems is the formation of acetate esters by the acyl transfer reaction:

$$CH_3\overset{O}{\underset{\|}{C}}SCoA + ROH \longrightarrow CH_3\overset{O}{\underset{\|}{C}}OR + CoASH$$

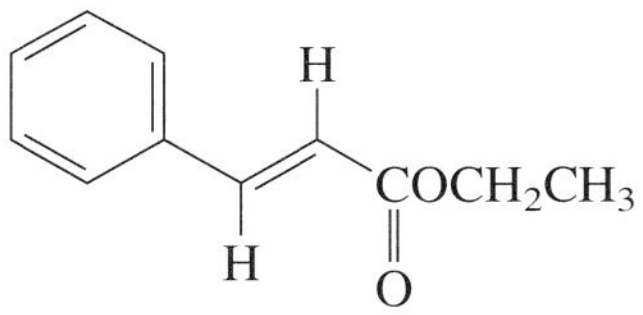

Ethyl cinnamate
(one of the constituents of
the sex pheromone of the
male oriental fruit moth)

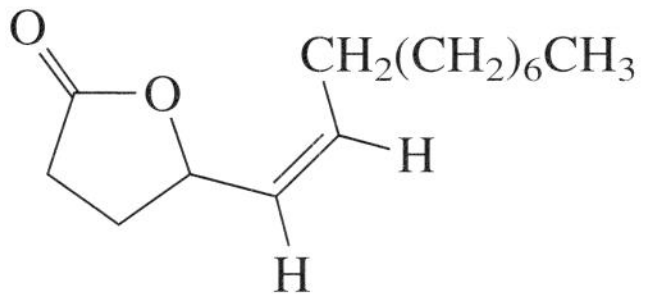

(Z)-5-Tetradecen-4-olide
(sex pheromone of female
Japanese beetle)

Esters of glycerol, called **glycerol triesters, triacylglycerols,** or **triglycerides,** are abundant natural products. The most important group of glycerol triesters includes those in which each acyl group is unbranched and has 14 or more carbon atoms.

$$CH_3(CH_2)_{16}\overset{\overset{\displaystyle O}{\|}}{C}O\diagup\diagdown O\overset{\overset{\displaystyle O}{\|}}{C}(CH_2)_{16}CH_3$$

$$O\overset{\displaystyle \|}{C}(CH_2)_{16}CH_3$$

Tristearin
(a trioctadecanoyl ester of glycerol found
in many animal and vegetable fats)

A molecular model of tri-stearin is shown in Figure 16.1.

Fats and **oils** are naturally occurring mixtures of glycerol triesters. Fats are mixtures that are solids at room temperature; oils are liquids. The long-chain carboxylic acids obtained from fats and oils by hydrolysis are known as **fatty acids.**

13.5   PREPARATION OF ESTERS: FISCHER ESTERIFICATION

In Chapter 12 (Section 12.11) you learned that esters can be prepared by the reaction of a carboxylic acid with an alcohol in the presence of an acid catalyst. This reaction is known as the **Fischer esterification.**

$$ROH \ + \ R'\overset{\overset{\displaystyle O}{\|}}{C}OH \ \underset{}{\overset{H^+}{\rightleftharpoons}} \ R'\overset{\overset{\displaystyle O}{\|}}{C}OR \ + \ H_2O$$

Alcohol   Carboxylic acid   Ester   Water

$$CH_3CH_2\overset{\overset{\displaystyle O}{\|}}{C}OH \ + \ CH_3CH_2CH_2CH_2OH \ \xrightarrow{H_2SO_4} \ CH_3CH_2\overset{\overset{\displaystyle O}{\|}}{C}OCH_2CH_2CH_2CH_3 \ + \ H_2O$$

Propanoic   1-Butanol   Butyl propanoate   Water
acid   (85%)

Why an equilibrium shifts in response to an excess of a reagent can be understood by recalling Le Châtelier's Principle (Section 5.5).

Fischer esterification is reversible, and the position of equilibrium lies slightly to the side of products when the reactants are simple alcohols and carboxylic acids. When the Fischer esterification is used for preparative purposes, the position of equilibrium can be made more favorable by using either the alcohol or the carboxylic acid in excess. Alternatively, the equilibrium can be shifted to favor formation of the ester by removing water from the reaction mixture. This can be accomplished by adding benzene as a cosolvent and distilling the azeotropic mixture of benzene and water.

An **azeotropic mixture** contains two or more substances that distill together at a constant boiling point. The benzene–water azeotrope contains 9% water and boils at 69°C.

The mechanism of an acid-catalyzed esterification is worth examining because it illustrates the principles of nucleophilic acyl substitution outlined in Section 13.3, especially the role played by the tetrahedral intermediate.

An important aspect of the mechanism is that the ester oxygen (—OR) is the oxygen of the starting alcohol, not the carboxylic acid. The mechanism must account for this observation.

$$R'\overset{\overset{\displaystyle O}{\|}}{C}{-}OH \ + \ RO{-}H \ \longrightarrow \ R'\overset{\overset{\displaystyle O}{\|}}{C}{-}OR \ + \ H_2O$$

This bond is broken in conversion of an alcohol to an ester.

This oxygen is the same one that was part of the starting alcohol.

This oxygen was part of the starting carboxylic acid.

For example, consider the formation of methyl benzoate from the reaction of benzoic acid and methanol.

$$\underset{\text{Benzoic acid}}{C_6H_5\overset{O}{\overset{\|}{C}}OH} + \underset{\text{Methanol}}{CH_3OH} \xrightarrow{H^+} \underset{\text{Methyl benzoate}}{C_6H_5\overset{O}{\overset{\|}{C}}OCH_3} + \underset{\text{Water}}{H_2O}$$

The mechanism is best viewed as a combination of two stages: formation of a tetrahedral intermediate, and dissociation of this intermediate as outlined in Figure 13.3.

PROBLEM 13.3 Give the structure of the ester formed in the following reaction, and show the mechanism of its formation. Be sure to show the structure of the tetrahedral intermediate.

$$CH_3CH_2CH_2CH_2OH + CH_3CH_2\overset{O}{\overset{\|}{C}}OH \xrightarrow[\text{heat}]{H_2SO_4}$$

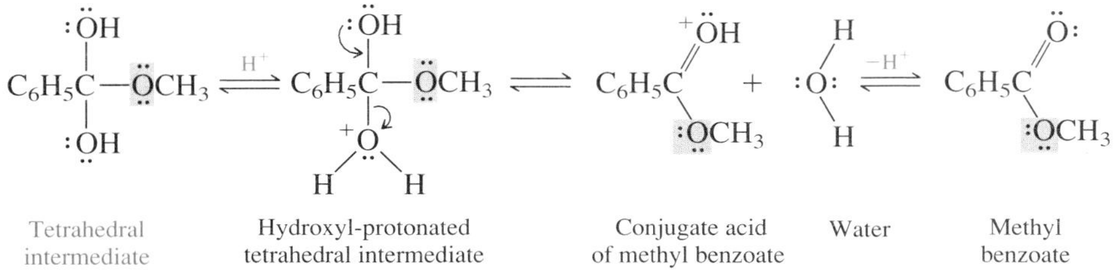

FIGURE 13.3 The mechanism of acid-catalyzed esterification of benzoic acid with methanol.

Initially, the carboxylic acid is protonated on its carbonyl oxygen.

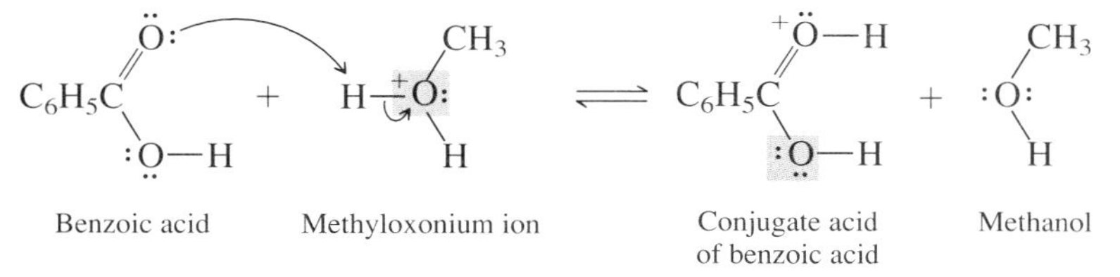

Protonation of the carboxylic acid increases the positive character of its carbonyl group. A molecule of the alcohol acts as a nucleophile and attacks the carbonyl carbon. Loss of a proton gives the tetrahedral intermediate.

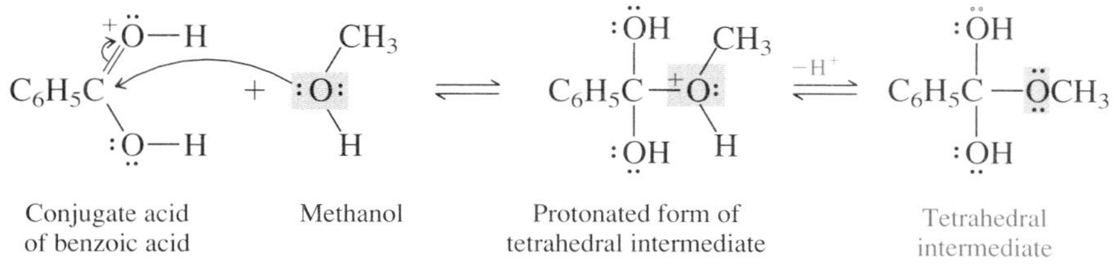

The second stage begins with protonation of the tetrahedral intermediate on one of its hydroxyl oxygens. This intermediate loses water to give the conjugate acid of the ester product. Deprotonation gives the neutral form of the ester product.

13.6 PREPARATION OF ESTERS: ADDITIONAL METHODS

You have seen acyl chlorides before in the Friedel–Crafts acylation of aromatic rings (Section 6.6).

In addition to Fischer esterification, esters may also be prepared from acyl chlorides and anhydrides. Acyl chlorides are prepared by reaction of a carboxylic acid with thionyl chloride (Section 12.11). Reaction of the acyl chloride with an alcohol produces the desired ester.

$$
\underset{\text{Alcohol}}{\text{ROH}} + \underset{\text{Acyl chloride}}{\overset{\overset{\displaystyle O}{\|}}{R'CCl}} \longrightarrow \underset{\text{Ester}}{\overset{\overset{\displaystyle O}{\|}}{R'COR}} + \underset{\substack{\text{Hydrogen} \\ \text{chloride}}}{\text{HCl}}
$$

This reaction is normally carried out in the presence of a weak base such as pyridine, which reacts with the hydrogen chloride that is formed.

$$
(CH_3)_2CHCH_2OH + \text{3,5-Dinitrobenzoyl chloride} \xrightarrow{\text{pyridine}} \text{Isobutyl 3,5-dinitrobenzoate} \ (86\%)
$$

Isobutyl alcohol   3,5-Dinitrobenzoyl chloride   Isobutyl 3,5-dinitrobenzoate (86%)

PROBLEM 13.4 Write chemical equations that show how each of the following esters could be prepared by using an acyl chloride in the presence of pyridine.

(a) Ethyl benzoate

(b) Benzyl acetate

(c) Isopropyl 2-methylpropanoate

Sample Solution (a) From the name of the ester, we can determine that the acyl portion is derived from benzoic acid, and the alkyl group from ethanol. Thus, first prepare the acyl chloride, benzoyl chloride. This is then allowed to react with ethanol in the presence of pyridine.

$$
\underset{\text{Benzoic acid}}{\overset{\overset{\displaystyle O}{\|}}{C_6H_5CH_2COH}} \xrightarrow{SOCl_2} \underset{\substack{\text{Benzoyl} \\ \text{chloride}}}{\overset{\overset{\displaystyle O}{\|}}{C_6H_5CCl}} \xrightarrow[\text{pyridine}]{CH_3CH_2OH} \underset{\text{Ethyl benzoate}}{\overset{\overset{\displaystyle O}{\|}}{C_6H_5COCH_2CH_3}}
$$

Esters can also be prepared by reaction of an anhydride with an alcohol.

$$
\underset{\text{Anhydride}}{\overset{\overset{\displaystyle O \ \ O}{\| \ \ \|}}{RCOCR'}} + \underset{\text{Alcohol}}{R'OH} \longrightarrow \underset{\text{Ester}}{\overset{\overset{\displaystyle O}{\|}}{RCOR'}} + \underset{\substack{\text{Carboxylic} \\ \text{acid}}}{\overset{\overset{\displaystyle O}{\|}}{RCOH}}
$$

The most common use of this reaction is in the preparation of acetate esters from the reaction of acetic anhydride with an alcohol.

$$
\underset{\substack{\text{Acetic anhydride}}}{\overset{\substack{O \quad O \\ \parallel \quad \parallel}}{CH_3COCCH_3}} + \underset{\substack{\textit{sec}\text{-Butyl alcohol} \\ \quad \\ CH_3}}{HOCHCH_2CH_3} \longrightarrow \underset{\substack{\textit{sec}\text{-Butyl acetate} \\ (60\%) \\ CH_3}}{\overset{\substack{O \\ \parallel}}{CH_3COCHCH_2CH_3}} + \underset{\substack{\text{Acetic acid}}}{\overset{\substack{O \\ \parallel}}{CH_3COH}}
$$

> **PROBLEM 13.5** Acetylsalicylic acid is the chemical name for aspirin. Acetylsalicylic acid is the acetate ester of salicylic acid and is made by treating salicylic acid with acetic anhydride. Write a chemical equation for this reaction. 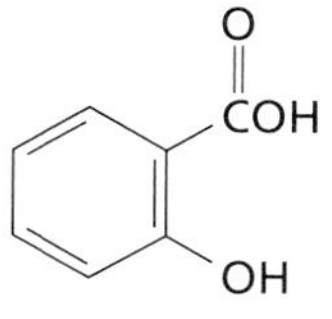
>
> Salicylic acid

Esterifications using acyl chlorides or anhydrides proceed by mechanisms analogous to that described for hydrolysis of acid derivatives in Section 13.3. That is, addition to the carbonyl carbon by an alcohol molecule to give a tetrahedral intermediate is followed by elimination of a leaving group to give the ester product. The net result is that the acyl group of the acyl chloride or anhydride is transferred to the oxygen of the alcohol. This fact is clearly evident in the esterification of chiral alcohols, where, because none of the bonds to the stereogenic center is broken in the process, *retention of configuration is observed.*

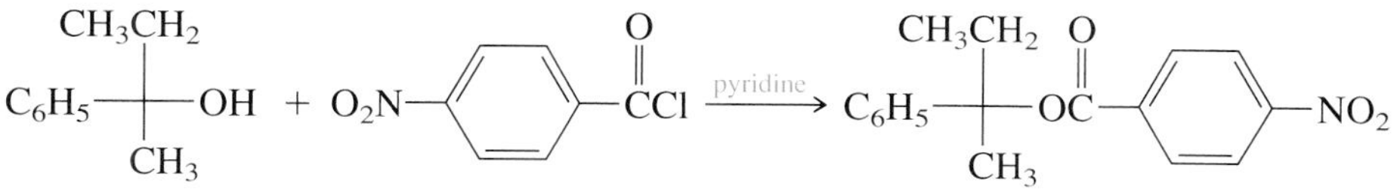

(R)-(+)-2-Phenyl-2-butanol + p-Nitrobenzoyl chloride →(pyridine)→ (R)-(−)-1-Methyl-1-phenylpropyl p-nitrobenzoate (63% yield)

> **PROBLEM 13.6** A similar conclusion may be drawn by considering the reactions of the cis and trans isomers of 4-*tert*-butylcyclohexanol with acetic anhydride. On the basis of the information just presented, predict the product formed from each stereoisomer.

13.7 REACTION OF ESTERS: HYDROLYSIS

Esters are fairly stable in neutral aqueous solution, but undergo hydrolysis (Section 13.3) when heated with water in the presence of strong acids or bases. The mechanism of hydrolysis of esters in dilute aqueous acid is the reverse of the Fischer esterification (Section 13.5):

$$
\underset{\substack{\text{Ester}}}{\overset{\substack{O \\ \parallel}}{RCOR'}} + \underset{\substack{\text{Water}}}{H_2O} \underset{}{\overset{H^+}{\rightleftharpoons}} \underset{\substack{\text{Carboxylic} \\ \text{acid}}}{\overset{\substack{O \\ \parallel}}{RCOH}} + \underset{\substack{\text{Alcohol}}}{R'OH}
$$

AN ESTER OF AN INORGANIC ACID: NITROGLYCERIN

n 1847 an Italian scientist, Ascanio Sobrero, discovered that glycerol, obtained as a by-product from the manufacture of soap, could be converted into a nitrate triester called nitroglycerin.

$$\underset{\underset{\displaystyle OH}{|}}{HOCH_2CHCH_2OH} + 3HONO_2 \longrightarrow \underset{\underset{\displaystyle ONO_2}{|}}{O_2NOCH_2CHCH_2ONO_2} + 3H_2O$$

Glycerol   Nitric acid   Glycerol trinitrate
                                                   (nitroglycerin)

Nitroglycerin is an explosive and extremely unstable liquid, and for years this property made it dangerous to manufacture and use. Alfred Nobel, a Swedish inventor, discovered in 1867 that the explosive power of nitroglycerin could be controlled by absorbing the liquid on an inert powder—producing **dynamite.**

When Nobel died, he left his entire fortune to a foundation that he established in his will. The foundation was to give an award each year, recognizing outstanding accomplishments toward world peace and in the fields of chemistry, physics, literature, and physiology or medicine. These awards are what we know as the Nobel Prizes. the Nobel Peace Prize is presented each year in Oslo, Norway; the others are presented in Stockholm, Sweden. Since 1969 the Central Bank of Sweden has also awarded a Nobel Memorial Prize in Economic Science. The income from the trust fund established by Alfred Nobel is the source of the monetary award that accompanies each Nobel Prize except the economics award.

Nitroglycerin has another, very different, use. Nitroglycerin is a blood vessel dilator and is used to treat a type of heart pain called angina. Nitroglycerin medication is absorbed rapidly, and the pill, which contains less than a milligram of nitroglycerin, is placed under the tongue for almost immediate relief of the angina symptoms.

When esterification is the objective, water is removed from the reaction mixture to encourage ester formation. When ester hydrolysis is the objective, the reaction is carried out in the presence of a generous excess of water.

Ethyl   Water   2-Chloro-2-phenylacetic   Ethyl
2-chloro-2-phenylacetate   acid   alcohol
                                               (80–82%)

PROBLEM 13.7 One component of beeswax is the ester triacontyl hexadecanoate. Write a balanced equation for the acid-catalyzed hydrolysis of this compound.

$$\underset{\displaystyle \text{Triacontyl hexadecanoate}}{CH_3(CH_2)_{14}\overset{\displaystyle O}{\overset{\displaystyle \|}{C}}OCH_2(CH_2)_{28}CH_3}$$

Unlike its acid-catalyzed counterpart, ester hydrolysis in aqueous base is *irreversible.*

$$
\underset{\text{Ester}}{\text{RCOR}'} + \underset{\text{Hydroxide ion}}{\text{HO}^-} \longrightarrow \underset{\substack{\text{Carboxylate} \\ \text{ion}}}{\text{RCO}^-} + \underset{\text{Alcohol}}{\text{R}'\text{OH}}
$$

Because it is consumed, hydroxide ion is a reactant, not a catalyst.

This is because carboxylic acids are converted to their corresponding carboxylate anions under these conditions, and these anions are incapable of acyl transfer to alcohols. To isolate the carboxylic acid, a separate acidification step following hydrolysis is necessary. Acidification converts the carboxylate salt to the free acid.

$$
\underset{\substack{\text{Methyl 2-methylpropenoate} \\ \text{(methyl methacrylate)}}}{\text{CH}_2{=}\overset{\displaystyle\text{O}}{\overset{\|}{\text{C}}}\text{COCH}_3 \atop \underset{\text{CH}_3}{|}} \xrightarrow[\text{2. H}_2\text{SO}_4]{\text{1. NaOH, H}_2\text{O, heat}} \underset{\substack{\text{2-Methylpropenoic} \\ \text{acid} \\ (87\%) \text{ (methacrylic acid)}}}{\text{CH}_2{=}\overset{\displaystyle\text{O}}{\overset{\|}{\text{C}}}\text{COH} \atop \underset{\text{CH}_3}{|}} + \underset{\text{Methyl alcohol}}{\text{CH}_3\text{OH}}
$$

Ester hydrolysis in base is called **saponification,** which means "soap making." Over 2000 years ago, the Phoenicians made soap by heating animal fat with wood ashes. Animal fat is rich in glycerol triesters, and wood ashes are a source of potassium carbonate. Basic cleavage of the fats produced a mixture of long-chain carboxylic acids as their potassium salts.

$$
\text{CH}_3(\text{CH}_2)_x\overset{\displaystyle\text{O}}{\overset{\|}{\text{C}}}\text{O} \diagdown \underset{\underset{\text{O}}{\|}}{\underset{\text{OC}(\text{CH}_2)_y\text{CH}_3}{|}} \diagup \overset{\displaystyle\text{O}}{\overset{\|}{\text{OC}}}(\text{CH}_2)_z\text{CH}_3 \xrightarrow[\text{heat}]{\text{K}_2\text{CO}_3, \text{H}_2\text{O}}
$$

$$
\underset{\text{Glycerol}}{\text{HOCH}_2\text{CHCH}_2\text{OH} \atop \underset{\text{OH}}{|}} + \underset{\text{Potassium carboxylate salts}}{\text{KOC}(\text{CH}_2)_x\text{CH}_3 + \text{KOC}(\text{CH}_2)_y\text{CH}_3 + \text{KOC}(\text{CH}_2)_z\text{CH}_3}
$$

Potassium and sodium salts of long-chain carboxylic acids form micelles that dissolve grease (Section 12.7) and have cleansing properties. The carboxylic acids obtained by saponification of fats are called **fatty acids.**

> **PROBLEM 13.8** Trimyristin is obtained from coconut oil and has the molecular formula $C_{45}H_{86}O_6$. On being heated with aqueous sodium hydroxide followed by acidification, trimyristin was converted to glycerol and tetradecanoic acid as the only products. What is the structure of trimyristin?

The mechanism of basic ester hydrolysis is another example of nucleophilic acyl substitution by way of a tetrahedral intermediate and is outlined in Figure 13.4. Unlike reactions in acidic solution, nucleophilic addition is *not* preceded by a protonation step. All the steps are reversible except the last one. The equilibrium constant for proton abstraction from the carboxylic acid by hydroxide ion is so large that step 4 is essentially irreversible.

> **PROBLEM 13.9**   On the basis of the general mechanism for basic ester hydrolysis shown in Figure 13.4, write an analogous sequence of steps for the saponification of ethyl benzoate.

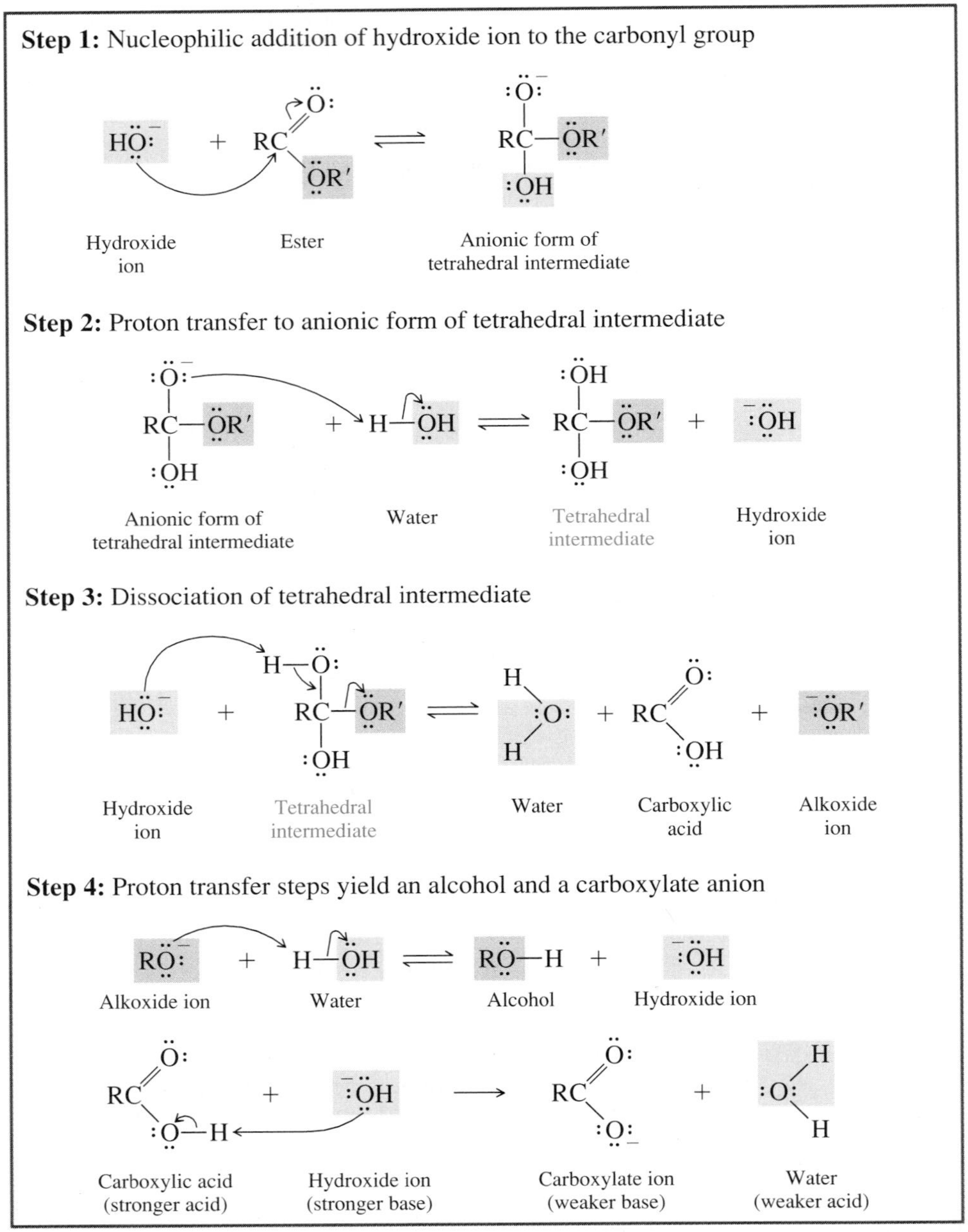

Step 1: Nucleophilic addition of hydroxide ion to the carbonyl group

Hydroxide   Ester   Anionic form of
ion   tetrahedral intermediate

Step 2: Proton transfer to anionic form of tetrahedral intermediate

Anionic form of   Water   Tetrahedral   Hydroxide
tetrahedral intermediate   intermediate   ion

Step 3: Dissociation of tetrahedral intermediate

Hydroxide   Tetrahedral   Water   Carboxylic   Alkoxide
ion   intermediate   acid   ion

Step 4: Proton transfer steps yield an alcohol and a carboxylate anion

Alkoxide ion   Water   Alcohol   Hydroxide ion

Carboxylic acid   Hydroxide ion   Carboxylate ion   Water
(stronger acid)   (stronger base)   (weaker base)   (weaker acid)

FIGURE 13.4   The mechanism of ester hydrolysis in basic solution.

13.8 PREPARATION OF TERTIARY ALCOHOLS FROM ESTERS AND GRIGNARD REAGENTS

Tertiary alcohols can be prepared by a variation of the Grignard synthesis that employs an ester as the carbonyl component. Methyl and ethyl esters are readily available and are the types most often used. Two moles of a Grignard reagent are required per mole of ester; the first mole reacts with the ester, converting it to a ketone.

In Chapter 11 (Sections 11.12 and 11.13) you saw how Grignard reagents can be used to prepare alcohols by reaction with aldehydes and ketones.

$$
\underset{\substack{\text{Grignard}\\\text{reagent}}}{\text{RMgX}} + \underset{\substack{\text{Methyl}\\\text{ester}}}{\text{R}'\overset{\overset{\text{O}}{\|}}{\text{C}}\text{OCH}_3} \xrightarrow{\text{diethyl ether}} \text{R}'\overset{\overset{\text{O—MgX}}{|}}{\underset{\underset{\text{R}}{|}}{\text{C}}}\text{—OCH}_3 \longrightarrow \underset{\text{Ketone}}{\text{R}'\overset{\overset{\text{O}}{\|}}{\text{C}}\text{R}} + \underset{\substack{\text{Methoxymagnesium}\\\text{halide}}}{\text{CH}_3\text{OMgX}}
$$

The ketone is not isolated, but reacts rapidly with the Grignard reagent to give, after adding aqueous acid, a tertiary alcohol. Ketones are more reactive than esters toward Grignard reagents, and so it is not normally possible to interrupt the reaction at the ketone stage even if only one equivalent of the Grignard reagent is used.

$$
\underset{\text{Ketone}}{\text{R}'\overset{\overset{\text{O}}{\|}}{\text{C}}\text{R}} + \underset{\substack{\text{Grignard}\\\text{reagent}}}{\text{RMgX}} \xrightarrow[\text{2. H}_3\text{O}^+]{\text{1. diethyl ether}} \underset{\substack{\text{Tertiary}\\\text{alcohol}}}{\text{R}'\overset{\overset{\text{OH}}{|}}{\underset{\underset{\text{R}}{|}}{\text{C}}}\text{R}}
$$

Two of the groups bonded to the hydroxyl-bearing carbon of the alcohol are the same because they are derived from the Grignard reagent. For example,

$$
\underset{\substack{\text{Methylmagnesium}\\\text{bromide}}}{2\text{CH}_3\text{MgBr}} + \underset{\substack{\text{Methyl}\\\text{2-methylpropanoate}}}{(\text{CH}_3)_2\text{CH}\overset{\overset{\text{O}}{\|}}{\text{C}}\text{OCH}_3} \xrightarrow[\text{2. H}_3\text{O}^+]{\text{1. diethyl ether}} \underset{\substack{\text{2,3-Dimethyl-}\\\text{2-butanol}\\(73\%)}}{(\text{CH}_3)_2\text{CH}\overset{\overset{\text{OH}}{|}}{\underset{\underset{\text{CH}_3}{|}}{\text{C}}}\text{CH}_3} + \underset{\text{Methanol}}{\text{CH}_3\text{OH}}
$$

PROBLEM 13.10 What combination of ester and Grignard reagent could you use to prepare each of the following tertiary alcohols?

(a) $\text{C}_6\text{H}_5\underset{\underset{\text{OH}}{|}}{\overset{}{\text{C}}}(\text{CH}_2\text{CH}_3)_2$

(b) $(\text{C}_6\text{H}_5)_2\underset{\underset{\text{OH}}{|}}{\text{C}}{\triangleleft}$

Sample Solution (a) To apply the principles of retrosynthetic analysis to this case, we disconnect both ethyl groups from the tertiary carbon and identify them as arising from the Grignard reagent. The phenyl group originates in an ester of the type $\text{C}_6\text{H}_5\text{CO}_2\text{R}$ (a benzoate ester).

$$
\text{C}_6\text{H}_5\underset{\underset{\text{OH}}{|}}{\text{C}}(\text{CH}_2\text{CH}_3)_2 \implies \text{C}_6\text{H}_5\overset{\overset{\text{O}}{\|}}{\text{C}}\text{OR} + 2\text{CH}_3\text{CH}_2\text{MgX}
$$

An appropriate synthesis would be

$$2CH_3CH_2MgBr \; + \; C_6H_5\overset{\displaystyle O}{\overset{\displaystyle \|}{C}}OCH_3 \; \xrightarrow[\text{2. } H_3O^+]{\text{1. diethyl ether}} \; C_6H_5\underset{\displaystyle OH}{C}(CH_2CH_3)_2$$

Ethylmagnesium bromide	Methyl benzoate	3-Phenyl-3-pentanol

13.9 REDUCTION OF ESTERS

Esters are readily reduced using lithium aluminum hydride (Section 10.3) as the reducing agent. Two alcohols are formed from each ester molecule. The acyl group of the ester is cleaved, giving a primary alcohol.

$$\overset{\displaystyle O}{\overset{\displaystyle \|}{R C}}OR' \longrightarrow RCH_2OH \; + \; R'OH$$

Ester	Primary alcohol	Alcohol

For example,

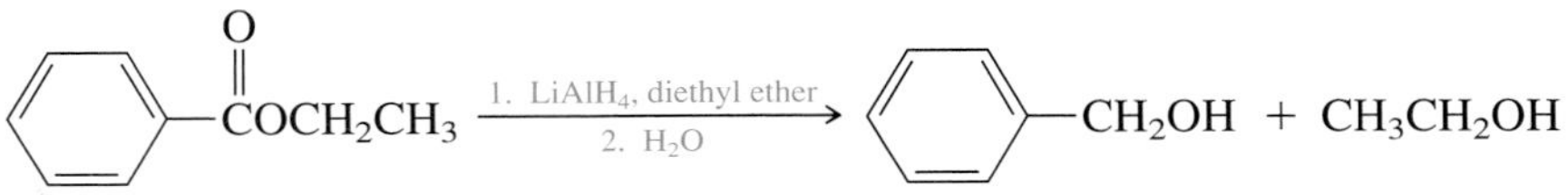

$$\text{C}_6\text{H}_5{-}\overset{\displaystyle O}{\overset{\displaystyle \|}{C}}OCH_2CH_3 \; \xrightarrow[\text{2. } H_2O]{\text{1. LiAlH}_4, \text{ diethyl ether}} \; \text{C}_6\text{H}_5{-}CH_2OH \; + \; CH_3CH_2OH$$

Ethyl benzoate	Benzyl alcohol (90%)	Ethanol

> **PROBLEM 13.11** Give the structure of an ester that will yield a mixture containing equimolar amounts of 1-propanol and 2-propanol on reduction with lithium aluminum hydride.

13.10 NATURALLY OCCURRING AMIDES

As with esters, a large number of amides occur naturally. The penicillin and cephalosporin **antibiotics,** which are so useful in treating bacterial infections, are among the best known products of the pharmaceutical industry. Penicillin G and Cephalexin are two examples.

Penicillin G and Cephalexin both contain two amide functions. Cyclic amides are called **lactams,** and these are both examples of **β-lactam antibiotics.**

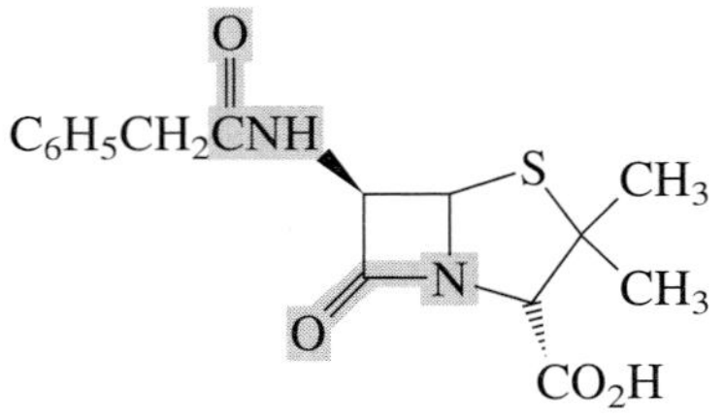

Penicillin G	Cephalexin

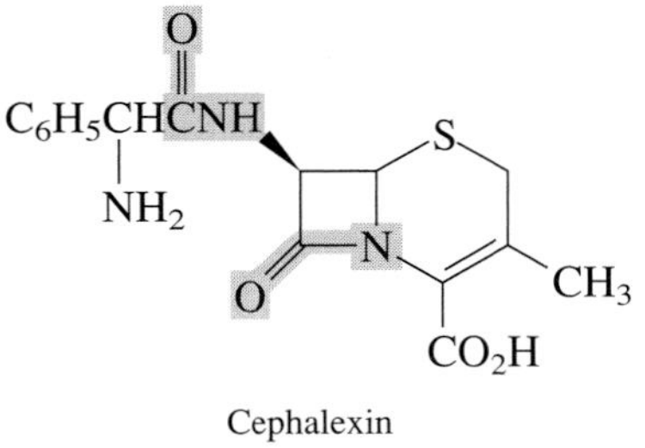

Other examples of natural amides include nicotinamide, a component of the coenzyme nicotinamide adenine dinucleotide (NAD) and *N*-acetyl-D-glucosamine, an amino sugar that is the main component of **chitin,** the substance that makes up the tough outer skeleton of arthropods and insects.

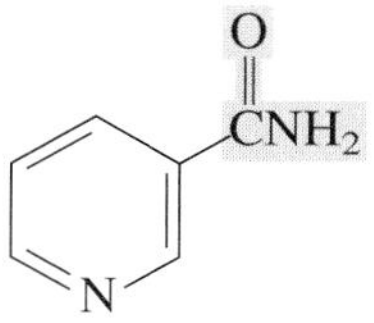

Nicotinamide 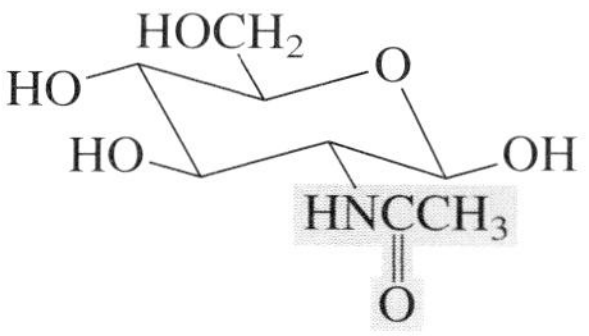   N-Acetyl-D-glucosamine

By far the most prevalent examples of natural amides are the polymers of amides that make up peptides and proteins in biological systems. These will be explored in Chapter 17.

13.11 PREPARATION OF AMIDES

Amides are readily prepared by acylation of ammonia and amines with acyl chlorides, anhydrides, or esters.

Acylation of ammonia (NH_3) yields an amide ($R'\overset{\displaystyle O}{\overset{\|}{C}}NH_2$).

Primary amines (RNH_2) yield *N*-substituted amides ($R'\overset{\displaystyle O}{\overset{\|}{C}}NHR$).

Secondary amines (R_2NH) yield *N,N*-disubstituted amides ($R'\overset{\displaystyle O}{\overset{\|}{C}}NR_2$).

Two molar equivalents of amine are required in the reaction with acyl chlorides and acid anhydrides; one molecule of amine acts as a nucleophile, the second as a Brønsted base.

$2R_2NH$ +	$R'\overset{O}{\overset{\|}{C}}Cl$	⟶	$R'\overset{O}{\overset{\|}{C}}NR_2$ +	$R_2\overset{+}{N}H_2\ Cl^-$
Amine	Acyl chloride		Amide	Hydrochloride salt of amine

$2R_2NH$ +	$R'\overset{O}{\overset{\|}{C}}O\overset{O}{\overset{\|}{C}}R'$	⟶	$R'\overset{O}{\overset{\|}{C}}NR_2$ +	$R_2\overset{+}{N}H_2\ ^-O\overset{O}{\overset{\|}{C}}R'$
Amine	Acid anhydride		Amide	Carboxylate salt of amine

For example,

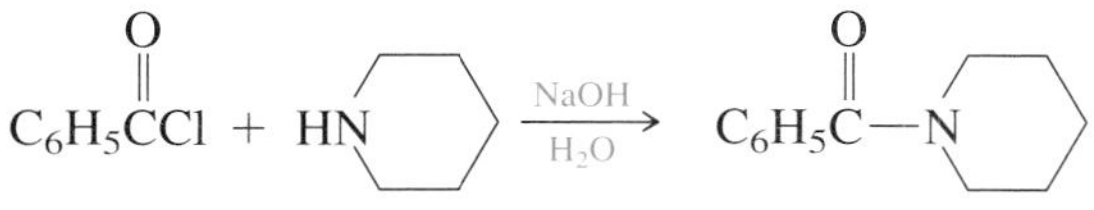

Benzoyl   Piperidine   *N*-Benzoylpiperidine
chloride   (87–91%)

$$CH_3\overset{O}{\underset{\parallel}{C}}O\overset{O}{\underset{\parallel}{C}}CH_3 + H_2N-\!\!\!\!\bigcirc\!\!\!\!-CH(CH_3)_2 \longrightarrow CH_3\overset{O}{\underset{\parallel}{C}}NH-\!\!\!\!\bigcirc\!\!\!\!-CH(CH_3)_2$$

| Acetic | p-Isopropylaniline | p-Isopropylacetanilide |
| anhydride | | (98%) |

It is possible to use only one molar equivalent of amine in these reactions if some other base, such as sodium hydroxide, is present in the reaction mixture to react with the hydrogen chloride or carboxylic acid that is formed. This is a useful procedure in those cases in which the amine is a valuable one or is available only in small quantities.

Esters and amines react in a 1:1 molar ratio to give amides. No acidic product is formed from the ester, and so no additional base is required.

$$R_2NH + R'\overset{O}{\underset{\parallel}{C}}OCH_3 \longrightarrow R'\overset{O}{\underset{\parallel}{C}}NR_2 + CH_3OH$$

| Amine | Methyl ester | Amide | Methanol |

$$FCH_2\overset{O}{\underset{\parallel}{C}}OCH_2CH_3 + NH_3 \xrightarrow{H_2O} FCH_2\overset{O}{\underset{\parallel}{C}}NH_2 + CH_3CH_2OH$$

| Ethyl | Ammonia | Fluoroacetamide | Ethanol |
| fluoroacetate | | (90%) | |

PROBLEM 13.12 Write an equation showing the preparation of the following amides from the indicated carboxylic acid derivative:

(a) $(CH_3)_2CH\overset{O}{\underset{\parallel}{C}}NH_2$ from an acyl chloride

(b) $CH_3\overset{O}{\underset{\parallel}{C}}NHCH_3$ from an anhydride

(c) $H\overset{O}{\underset{\parallel}{C}}N(CH_3)_2$ from a methyl ester

Sample Solution (a) Amides of the type $R\overset{O}{\underset{\parallel}{C}}NH_2$ are derived by acylation of ammonia.

$$(CH_3)_2CH\overset{O}{\underset{\parallel}{C}}Cl + 2NH_3 \longrightarrow (CH_3)_2CH\overset{O}{\underset{\parallel}{C}}NH_2 + NH_4Cl$$

| 2-Methylpropanoyl | Ammonia | 2-Methylpropanamide | Ammonium chloride |
| chloride | | | |

Two molecules of ammonia are needed because its acylation produces, in addition to the desired amide, a molecule of hydrogen chloride. Hydrogen chloride (an acid) reacts with ammonia (a base) to give ammonium chloride.

Amides are sometimes prepared directly from carboxylic acids and amines by a two-step process. The first step is an acid–base reaction in which the acid and the amine combine to form an ammonium carboxylate salt. On heating, the ammonium carboxylate salt loses water to form an amide.

$$\underset{\substack{\text{Carboxylic}\\\text{acid}}}{\text{RCOH}} \ + \ \underset{\text{Amine}}{\boxed{\text{R}_2'\text{NH}}} \ \longrightarrow \ \underset{\substack{\text{Ammonium}\\\text{carboxylate salt}}}{\text{RCO}^- \ \overset{+}{\boxed{\text{R}_2'\text{NH}_2}}} \ \overset{\text{heat}}{\longrightarrow} \ \underset{\text{Amide}}{\text{RCNR}_2'} \ + \ \underset{\text{Water}}{\text{H}_2\text{O}}$$

In practice, both steps may be combined in a single operation by simply heating a carboxylic acid and an amine together:

$$\underset{\text{Benzoic acid}}{\text{C}_6\text{H}_5\text{COH}} \ + \ \underset{\text{Aniline}}{\text{C}_6\text{H}_5\text{NH}_2} \ \xrightarrow{225^\circ\text{C}} \ \underset{\substack{N\text{-Phenylbenzamide}\\(80-84\%)}}{\text{C}_6\text{H}_5\text{CNHC}_6\text{H}_5} \ + \ \underset{\text{Water}}{\text{H}_2\text{O}}$$

13.12 HYDROLYSIS OF AMIDES

Recall from Section 13.3 that amides are the least reactive of the carboxylic acid derivatives. As a result, hydrolysis is the only nucleophilic acyl substitution that amides undergo. Amides are fairly stable in water, but the amide bond is cleaved on heating in the presence of strong acids or bases.

In acid solution, the products of hydrolysis are the carboxylic acid and the protonated amine, an ammonium ion.

$$\underset{\text{Amide}}{\boxed{\text{RCNR}_2'}} \ + \ \underset{\text{Hydronium ion}}{\text{H}_3\text{O}^+} \ \longrightarrow \ \underset{\substack{\text{Carboxylic}\\\text{acid}}}{\text{RCOH}} \ + \ \underset{\text{Ammonium ion}}{\boxed{\text{R}'\!-\!\overset{+}{\underset{\text{H}}{\overset{\text{H}}{\text{N}}}}\!-\!\text{R}'}}$$

In base the carboxylic acid is deprotonated, giving a carboxylate ion:

$$\underset{\text{Amide}}{\boxed{\text{RCNR}_2'}} \ + \ \underset{\text{Hydroxide ion}}{\text{HO}^-} \ \longrightarrow \ \underset{\text{Carboxylate ion}}{\text{RCO}^-} \ + \ \underset{\text{Amine}}{\boxed{\text{R}'\!-\!\overset{\text{R}'}{\underset{\text{H}}{\ddot{\text{N}}}}}}$$

The acid–base reactions that occur after the amide bond is broken make the overall hydrolysis irreversible in both cases. The amine product is protonated in acid; the carboxylic acid is deprotonated in base.

$$\underset{\text{2-Phenylbutanamide}}{\text{CH}_3\text{CH}_2\text{CHCNH}_2} \ \xrightarrow[\text{heat}]{\text{H}_2\text{O, H}_2\text{SO}_4} \ \underset{\substack{\text{2-Phenylbutanoic}\\\text{acid}\\(88-90\%)}}{\text{CH}_3\text{CH}_2\text{CHCOH}} \ + \ \underset{\substack{\text{Ammonium hydrogen}\\\text{sulfate}}}{\overset{+}{\text{NH}_4} \ \text{HSO}_4^-}$$

CONDENSATION POLYMERS: POLYAMIDES AND POLYESTERS

All fibers are polymers of one kind or another. Cotton, for example, is cellulose, and cellulose is a naturally occurring polymer of glucose. Silk and wool are naturally occurring polymers of amino acids. An early goal of inventors and entrepreneurs was to produce fibers from other naturally occurring polymers. Their earliest efforts consisted of chemically modifying the short cellulose fibers obtained from wood so that they could be processed into longer fibers more like cotton and silk. These efforts were successful, and the resulting fibers of modified cellulose, known generically as **rayon**, have been produced by a variety of techniques since the late nineteenth century.

A second approach involved direct chemical synthesis of polymers by connecting appropriately chosen small molecules together into a long chain. In 1938, E. I. Du Pont de Nemours and Company announced the development of **nylon,** the first synthetic polymer fiber.

The leader of Du Pont's effort was Wallace H. Carothers, who reasoned that he could reproduce the properties of silk by constructing a polymer chain held together, as is silk, by amide bonds. The necessary amide bonds were formed by heating a dicarboxylic acid with a diamine. Hexanedioic acid (adipic acid) and 1,6-hexanediamine (hexamethylenediamine) react to give a salt that, when heated, gives a **polyamide** called **nylon 6,6.** The amide bonds form by a condensation reaction, and nylon 6,6 is an example of a **condensation polymer.**

$$\underset{\text{Adipic acid}}{\text{HOC(CH}_2)_4\text{COH}} + \underset{\text{Hexamethylenediamine}}{\text{H}_2\text{N(CH}_2)_6\text{NH}_2} \longrightarrow {}^-\text{OC(CH}_2)_4\text{CO}^- \ \ {}^+\text{H}_3\text{N(CH}_2)_6\text{NH}_3^+$$

Nylon 66

The first "6" in nylon 6,6 stands for the number of carbons in the diamine, the second for the number of carbons in the dicarboxylic acid. Nylon 6,6 was an immediate success and fostered the development of a large number of related polyamides, many of which have also found their niche in the marketplace.

A slightly different class of polyamides is the **aramids (aromatic polyamides).** Like the nylons, the aramids are prepared from a dicarboxylic acid and a diamine, but the functional groups are anchored to benzene rings. An example of an aramid is **Kevlar,** which is a polyamide derived from 1,4-benzenedicarboxylic acid (terephthalic acid) and 1,4-benzenediamine (*p*-phenylenediamine):

Kevlar
(a polyamide of the aramid class)

Kevlar fibers are very strong, which makes Kevlar a popular choice in applications where the ratio of strength to weight is important. For example, a cable made from Kevlar weighs only one fifth as much as a steel one but is just as strong. Kevlar is also used to make lightweight bulletproof vests.

Nomex is another aramid fiber. Kevlar and Nomex differ only in that the substitution pattern in the aromatic rings is para in Kevlar but meta in Nomex. Nomex is best known for its fire-resistant properties and is used in protective clothing for firefighters, astronauts, and race-car drivers.

Polyesters are a second class of condensation polymers, and the principles behind their synthesis parallel those of polyamides. Ester formation between the functional groups of a dicarboxylic acid and a diol serve to connect small molecules together into a long polyester. The most familiar example of a polyester is **Dacron,** which is prepared from 1,4-benzenedicarboxylic acid and 1,2-ethanediol (ethylene glycol):

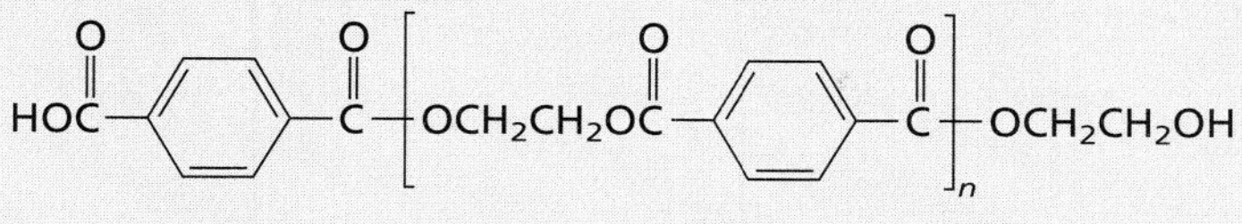

$$\text{HOC} \overset{\text{O}}{\underset{\|}{}} -\bigcirc- \overset{\text{O}}{\underset{\|}{\text{C}}} \left[\text{OCH}_2\text{CH}_2\text{OC} -\bigcirc- \overset{\text{O}}{\underset{\|}{\text{C}}} \right]_n \text{OCH}_2\text{CH}_2\text{OH}$$

Dacron
(a polyester)

The production of polyester fibers leads that of all other types. Annual United States production of polyester fibers is 1.6 million tons versus 1.4 million tons for cotton and 1.0 million tons for nylon. Wool and silk trail far behind at 0.04 and 0.01 million tons, respectively.

Not all synthetic polymers are used as fibers. **Mylar,** for example, is chemically the same as Dacron, but is prepared in the form of a thin film instead of a fiber. **Lexan** is a polyester that, because of its impact resistance, is used as a shatterproof substitute for glass. It is a **polycarbonate** having the structure shown:

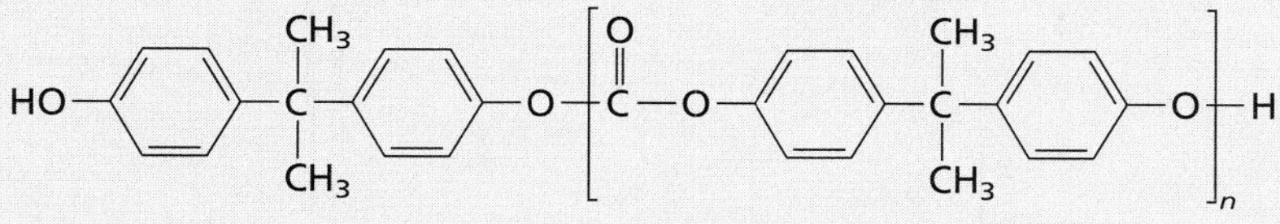

$$\text{HO}-\bigcirc- \overset{\text{CH}_3}{\underset{\text{CH}_3}{\text{C}}} -\bigcirc- \text{O} \left[\overset{\text{O}}{\underset{\|}{\text{C}}} -\text{O}-\bigcirc- \overset{\text{CH}_3}{\underset{\text{CH}_3}{\text{C}}} -\bigcirc- \text{O} \right]_n \text{H}$$

Lexan
(a polycarbonate)

In terms of the number of scientists and engineers involved, research and development in polymer chemistry is the principal activity of the chemical industry. The initial goal of making synthetic materials that are the equal of natural fibers has been more than met; it has been far exceeded. What is also important is that all of this did not begin with a chance discovery. It began with management decisions to do basic research in a specific area and to support it in the absence of any guarantee that success would be quickly achieved.

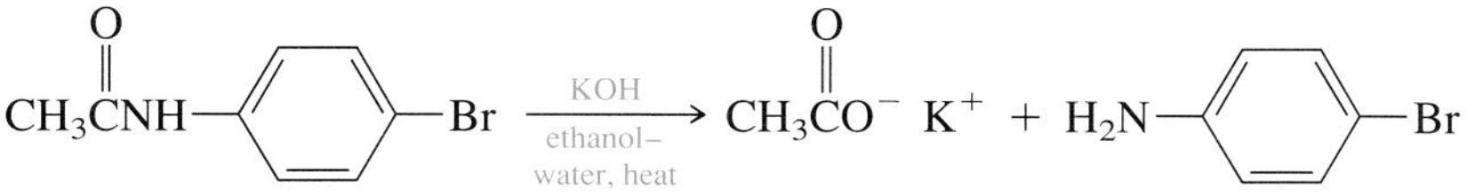

$$\text{CH}_3\overset{\text{O}}{\underset{\|}{\text{C}}}\text{NH}-\bigcirc-\text{Br} \xrightarrow[\substack{\text{ethanol-}\\\text{water, heat}}]{\text{KOH}} \text{CH}_3\overset{\text{O}}{\underset{\|}{\text{C}}}\text{O}^-\ \text{K}^+ + \text{H}_2\text{N}-\bigcirc-\text{Br}$$

N-(4-Bromophenyl)acetamide   Potassium   p-Bromoaniline
(p-bromoacetanilide)   acetate   (95%)

PROBLEM 13.13 Write a chemical equation for the hydrolysis of each of the following amides, using the conditions indicated.

(a) $(\text{CH}_3)_2\text{CH}\overset{\text{O}}{\underset{\|}{\text{C}}}\text{NH}_2$ (heat in aqueous hydrochloric acid)

(b) $\text{CH}_3\overset{\text{O}}{\underset{\|}{\text{C}}}\text{NHCH}_3$ (heat in aqueous sodium hydroxide)

Sample Solution (a) Hydrolysis under acidic conditions yields the carboxylic acid and the salt of the amine. Thus, hydrolysis of 2-methylpropanamide in aqueous hydrochloric acid yields propanoic acid and ammonium chloride.

$$\begin{array}{ccccccc} & \overset{\displaystyle O}{\underset{\displaystyle \|}{}} & & & & \overset{\displaystyle O}{\underset{\displaystyle \|}{}} & \\ (CH_3)_2CHCNH_2 & + & H_2O & \xrightarrow[\text{heat}]{\text{HCl}} & (CH_3)_2CHCOH & + & NH_4Cl \\ \text{2-Methylpropanamide} & & \text{Water} & & \text{2-Methylpropanoic} & & \text{Ammonium} \\ & & & & \text{acid} & & \text{chloride} \end{array}$$

The mechanism for amide hydrolysis in basic solution is shown in Figure 13.5.

PROBLEM 13.14 On the basis of the general mechanism for basic hydrolysis of an amide shown in Figure 13.5, write an analogous sequence for the hydrolysis of N,N-dimethylformamide $H\overset{\displaystyle O}{\overset{\displaystyle \|}{C}}N(CH_3)_2$.

One other important reaction of amides, reduction to give an amine, will be presented in the next chapter.

LEARNING OBJECTIVES

This chapter has focused on the most common classes of carboxylic acid derivatives: acyl chlorides, anhydrides, esters, and amides. The skills you have learned in this chapter should enable you to

- Give an acceptable systematic IUPAC name for a carboxylic acid derivative.
- Write the structure of a carboxylic acid derivative given its sytematic name.
- Explain by using chemical equations the general mechanism for a nucleophilic acyl substitution, and describe the role of the tetrahedral intermediate.
- Write a chemical equation and give the mechanism for the preparation of an ester by the Fischer esterification method.
- Write chemical equations for the preparation of esters using acyl chlorides or acetic anhydride.
- Write chemical equations for the acid-catalyzed and basic hydrolysis of an ester.
- Show how the reaction of an ester with a Grignard reagent can be used to prepare a tertiary alcohol.
- Write a chemical equation for the reduction of an ester.
- Use chemical equations to describe the preparation of an amide by the reaction of an amine with a carboxylic acid, an acyl chloride, acetic anhydride, or an ester.
- Write chemical equations for the hydrolysis of an amide under both acidic and basic conditions.

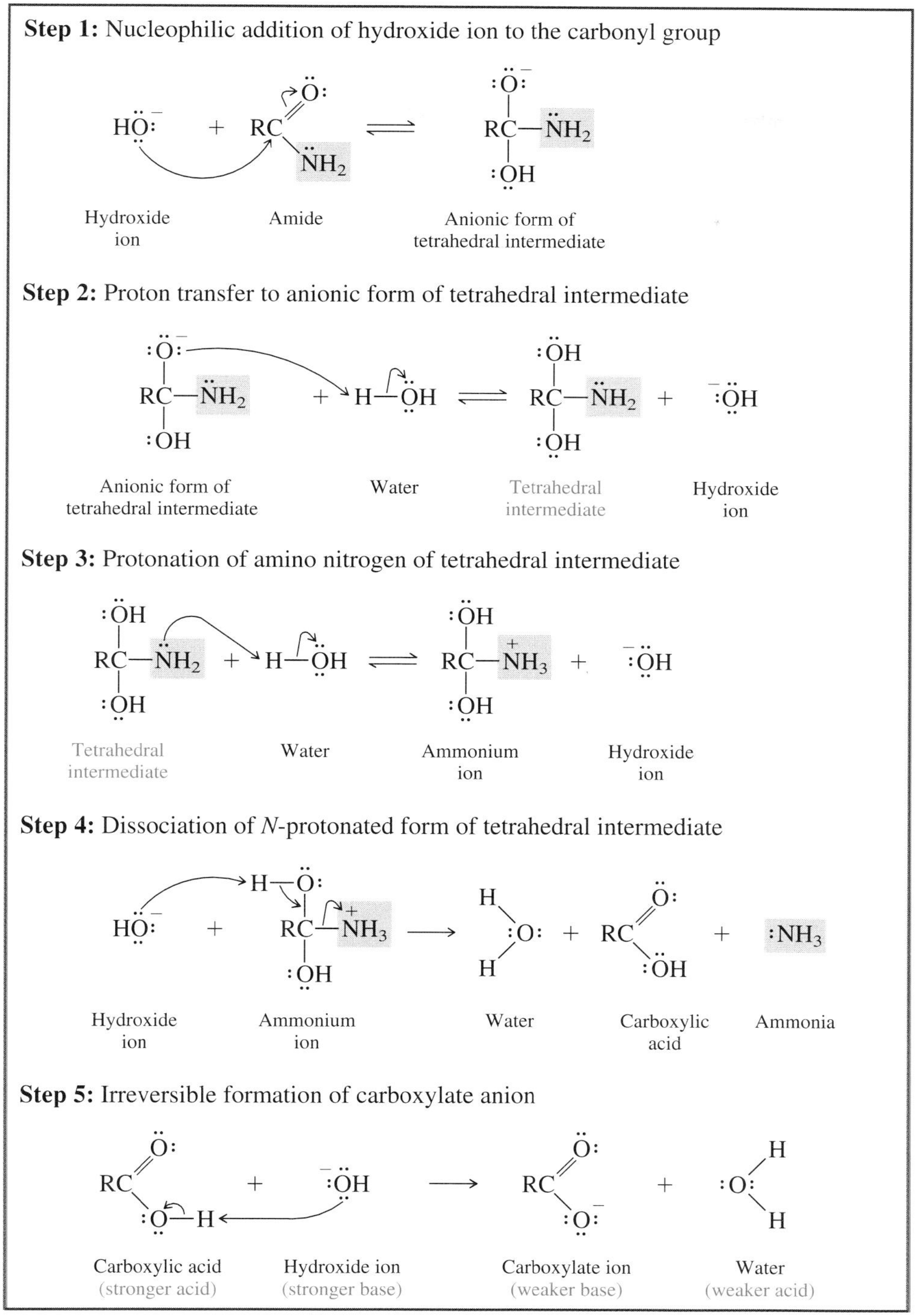

Step 1: Nucleophilic addition of hydroxide ion to the carbonyl group

| Hydroxide ion | Amide | Anionic form of tetrahedral intermediate |

Step 2: Proton transfer to anionic form of tetrahedral intermediate

| Anionic form of tetrahedral intermediate | Water | Tetrahedral intermediate | Hydroxide ion |

Step 3: Protonation of amino nitrogen of tetrahedral intermediate

| Tetrahedral intermediate | Water | Ammonium ion | Hydroxide ion |

Step 4: Dissociation of *N*-protonated form of tetrahedral intermediate

| Hydroxide ion | Ammonium ion | Water | Carboxylic acid | Ammonia |

Step 5: Irreversible formation of carboxylate anion

| Carboxylic acid (stronger acid) | Hydroxide ion (stronger base) | Carboxylate ion (weaker base) | Water (weaker acid) |

FIGURE 13.5 The mechanism of amide hydrolysis in basic solution.

13.13 SUMMARY

This chapter concerns the preparation and reactions of **acyl chlorides, anhydrides, esters,** and **amides.** These compounds are generally classified as carboxylic acid derivatives, and their nomenclature is based on that of carboxylic acids (Section 13.1).

$$\underset{\substack{\text{Acyl}\\\text{chloride}}}{\overset{\displaystyle O}{\overset{\|}{RCCl}}} \qquad \underset{\text{Anhydride}}{\overset{\displaystyle O\ \ \ O}{\overset{\|\ \ \ \ \|}{RCOCR}}} \qquad \underset{\text{Ester}}{\overset{\displaystyle O}{\overset{\|}{RCOR'}}} \qquad \underset{\text{Amide}}{\overset{\displaystyle O}{\overset{\|}{RCNR'_2}}}$$

The structure and reactivity of carboxylic acid derivatives depend on how well the atom bonded to the carbonyl group donates electrons to it (Section 13.2).

Electron-pair donation stabilizes the carbonyl group and makes it less reactive toward nucleophilic acyl substitution.

$$\underset{\substack{\text{Least stabilized}\\\text{carbonyl group}}}{\overset{\text{Most reactive}}{\overset{\displaystyle O}{\overset{\|}{RCCl}}}} > \overset{\displaystyle O\ \ \ O}{\overset{\|\ \ \ \ \|}{RCOCR}} > \overset{\displaystyle O}{\overset{\|}{RCOR'}} > \underset{\substack{\text{Most stabilized}\\\text{carbonyl group}}}{\overset{\text{Least reactive}}{\overset{\displaystyle O}{\overset{\|}{RCNR'_2}}}}$$

Nitrogen is a better electron-pair donor than oxygen, and amides have a more stabilized carbonyl than esters and anhydrides. Chlorine is the poorest electron-pair donor, and acyl chlorides have the least stabilized carbonyl group and are the most reactive.

The characteristic reaction of acyl chlorides, anhydrides, esters, and amides is **nucleophilic acyl substitution** (Section 13.3). Addition of a nucleophilic reagent HY: to the carbonyl group leads to a tetrahedral intermediate that dissociates to give the product of substitution:

$$\underset{\substack{\text{Carboxylic}\\\text{acid derivative}}}{\overset{\displaystyle O}{\overset{\|}{RC-X}}} + \underset{\text{Nucleophile}}{HY:} \ \rightleftharpoons\ \underset{\substack{\text{Tetrahedral}\\\text{intermediate}}}{\overset{\displaystyle OH}{\underset{\displaystyle Y}{RC-X}}} \ \rightleftharpoons\ \underset{\substack{\text{Product of}\\\text{nucleophilic}\\\text{acyl substitution}}}{\overset{\displaystyle O}{\overset{\|}{RC-Y}}} + \underset{\substack{\text{Conjugate acid}\\\text{of leaving}\\\text{group}}}{HX:}$$

Each of the derivatives—acyl chlorides, anhydrides, esters, and amides—may be converted to a carboxylic acid by hydrolysis.

$$\underset{\substack{\text{Carboxylic}\\\text{acid derivative}}}{\overset{\displaystyle O}{\overset{\|}{RCX}}} + \underset{\text{Water}}{H_2O} \longrightarrow \underset{\substack{\text{Carboxylic}\\\text{acid}}}{\overset{\displaystyle O}{\overset{\|}{RCOH}}} + \underset{\substack{\text{Conjugate acid}\\\text{of leaving group}}}{HX}$$

Acyl chlorides and anhydrides are the most reactive acyl derivatives; amides are the least reactive.

Esters may be prepared from the reaction of a carboxylic acid with an alcohol in the presence of an acid catalyst. The reaction is known as **Fischer esterification** (Section 13.5).

$$\underset{\substack{\text{Carboxylic} \\ \text{acid}}}{\text{RCOH}} \quad + \quad \underset{\text{Alcohol}}{\text{R'OH}} \quad \overset{\text{H}^+}{\rightleftharpoons} \quad \underset{\text{Ester}}{\text{RCOR'}} \quad + \quad \underset{\text{Water}}{\text{H}_2\text{O}}$$

The mechanism of Fischer esterification illustrates three important aspects of nucleophilic acyl substitution:

1. Activation of the carbonyl group by protonation of the carbonyl oxygen
2. Nucleophilic addition to the protonated carbonyl, forming a tetrahedral intermediate
3. Elimination from the tetrahedral intermediate to restore the carbonyl group

Esters may also be prepared by the reaction of an acyl chloride or anhydride with an alcohol in the presence of a base such as pyridine (Section 13.6).

$$\underset{\text{Acyl chloride}}{\text{RCCl}} \quad + \quad \underset{\text{Alcohol}}{\text{R'OH}} \quad \longrightarrow \quad \underset{\text{Ester}}{\text{RCOR'}} \quad + \quad \underset{\substack{\text{Hydrogen} \\ \text{chloride}}}{\text{HCl}}$$

$$\underset{\text{Anhydride}}{\text{RCOCR}} \quad + \quad \underset{\text{Alcohol}}{\text{R'OH}} \quad \longrightarrow \quad \underset{\text{Ester}}{\text{RCOR'}} \quad + \quad \underset{\substack{\text{Carboxylic} \\ \text{acid}}}{\text{RCOH}}$$

A principal reaction of esters is hydrolysis (Section 13.7). Under acidic conditions, hydrolysis proceeds by a mechanism that is the reverse of Fischer esterification. Basic hydrolysis, also known as **saponification,** is not reversible due to carboxylate salt formation.

Acid hydrolysis:

$$\underset{}{\text{RCOR'}} \quad + \quad \text{H}_2\text{O} \quad \overset{\text{H}^+}{\rightleftharpoons} \quad \underset{}{\text{RCOH}} \quad + \quad \text{R'OH}$$

Basic hydrolysis:

$$\underset{}{\text{RCOR'}} \quad + \quad \text{HO}^- \quad \overset{\text{H}_2\text{O}}{\rightleftharpoons} \quad \underset{}{\text{RCO}^-} \quad + \quad \text{R'OH}$$

Tertiary alcohols in which two of the substituents are the same may be prepared by reaction of an ester with two moles of Grignard reagent (Section 13.8).

$$2C_6H_5MgBr \ + \ CH_3\overset{O}{\overset{\|}{C}}OCH_2CH_3 \ \xrightarrow[\text{2. } H_3O^+]{\text{1. diethyl ether}} \ (C_6H_5)_2\overset{OH}{\overset{|}{C}}CH_3$$

Phenylmagnesium bromide   Ethyl acetate   1,1-Diphenyl-ethanol

Reduction of esters may be accomplished using the reducing agent lithium aluminum hydride (Section 13.9). Two alcohols are formed from each ester molecule.

$$RCO_2R' \ \xrightarrow[\text{2. } H_2O]{\text{1. LiAlH}_4} \ RCH_2OH \ + \ R'OH$$

Amides may be prepared (Section 13.11) by heating an amine with a carboxylic acid or by reaction of an amine with an acyl chloride. Acetamides are frequently prepared by reaction of an amine with acetic anhydride.

$$R_2NH \ + \ R'\overset{O}{\overset{\|}{C}}OH \ \xrightarrow{\text{heat}} \ R_2N\overset{O}{\overset{\|}{C}}R' \ + \ H_2O$$

Amine   Carboxylic acid   Amide   Water

$$2R_2NH \ + \ R'\overset{O}{\overset{\|}{C}}Cl \ \longrightarrow \ R_2N\overset{O}{\overset{\|}{C}}R' \ + \ R_2\overset{+}{N}H_2 \ Cl^-$$

Amine   Acyl chloride   Amide   Ammonium salt

$$CH_3\overset{O}{\overset{\|}{C}}O\overset{O}{\overset{\|}{C}}CH_3 \ + \ 2RNH_2 \ \longrightarrow \ CH_3\overset{O}{\overset{\|}{C}}NHR \ + \ CH_3\overset{O}{\overset{\|}{C}}O^- \ H_3\overset{+}{N}R$$

Acetic anhydride   Amine   An acetamide   An ammonium acetate salt

Like ester hydrolysis, amide hydrolysis can be achieved in either aqueous acid or aqueous base. The process is irreversible in both media. In base, the carboxylic acid is converted to the carboxylate anion; in acid, the amine is protonated to an ammonium ion:

$$RC\overset{O}{\overset{\|}{N}}R'_2 \ + \ H_2O \quad
\begin{cases}
\xrightarrow{H_3O^+} \ R\overset{O}{\overset{\|}{C}}OH \ + \ R'_2\overset{+}{N}H_2 \\
\\
\xrightarrow{HO^-} \ R\overset{O}{\overset{\|}{C}}O^- \ + \ R'_2NH
\end{cases}$$

Amide   Water   Carboxylic acid   Ammonium ion   Carboxylate ion   Amine

ADDITIONAL PROBLEMS

Nomenclature

13.15 Write a structural formula for each of the following compounds:

(a) *m*-Chlorobenzoyl bromide
(b) 4-Methylpentanoyl chloride
(c) Trifluoroacetic anhydride
(d) 1-Phenylethyl acetate
(e) Butyl 2-methylbutanoate
(f) *N*-Ethylbenzamide
(g) *N*,*N*-Diphenylacetamide

13.16 Give an acceptable systematic (IUPAC) name for each of the following compounds:

(a) CH_3CHCH_2CCl (with =O on the C, and Cl below the first CH)

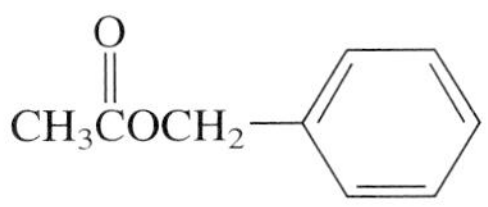

(b) CH_3COCH_2— (benzyl, with =O)

(c) CH_3OCCH_2— (benzyl, with =O)

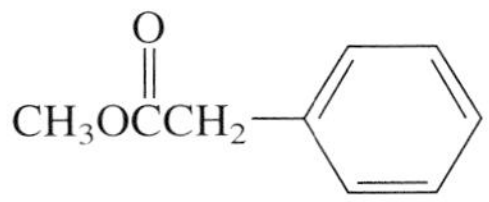

(d) $(CH_3)_2CHCH_2CH_2CNH_2$ (with =O)

Preparation and Reaction of Acyl Derivatives

13.17 Write a structural formula for the principal organic product or products of each of the following reactions:

(a) Phenylacetic acid and thionyl chloride
(b) Product of part (a) and water
(c) Product of part (a) and benzyl alcohol ($C_6H_5CH_2OH$)
(d) Product of part (a) and benzylamine ($C_6H_5CH_2NH_2$)
(e) Product of part (c) and aqueous hydrochloric acid, heat
(f) Product of part (c) and aqueous sodium hydroxide, heat
(g) Product of part (d) and aqueous hydrochloric acid, heat
(h) Product of part (d) and aqueous sodium hydroxide, heat
(i) Acetic anhydride and cyclohexanol
(j) Acetic anhydride and dimethylamine [$(CH_3)_2NH$]
(k) Ethyl formate and ethylamine
(l) Methyl phenylacetate and lithium aluminum hydride, then water

13.18 Write a balanced chemical equation describing the reaction of benzoyl chloride with

(a) 2-Propanol
(b) Methylamine, CH_3NH_2 (2 mol)
(c) Dimethylamine, $(CH_3)_2NH$ (2 mol)
(d) Water

13.19 Supply the formula of the reactant missing from each of the following equations:

(a) ? + benzoic acid $\xrightarrow{heat}$ benzamide + water

(b) ? + 1-butanol $\longrightarrow$ butyl acetate + acetic acid

(c) ? + H_3O^+ $\xrightarrow{heat}$ $HCO_2H + NH_4^+$

(d) ? + propanoic acid $\xrightarrow{H^+}$ H_2O + butyl propanoate

(e) ? $\xrightarrow[\text{2. } H_2O]{\text{1. LiAlH}_4}$ $CH_3CH_2CH_2OH$ + $(CH_3)_3COH$

Nucleophilic Acyl Substitution Mechanisms

13.20 For each of the following nucleophilic acyl substitutions write the structure of the tetrahedral intermediate in the process.

(a) Benzoyl chloride and ethanol   (c) Saponification of methyl benzoate
(b) Acetic anhydride and ethylamine   (d) Methyl benzoate and ammonia

13.21 Write a mechanism for the preparation of ethyl acetate by Fischer esterification.

13.22 Write a mechanism for the hydrolysis in acid solution of *N*-phenylbenzamide.

13.23 One method used by chemists to probe a reaction mechanism is to employ a reagent in which an atom is enriched in an isotope present naturally only in trace amounts. The fate of this isotope can be determined by using a technique known as mass spectrometry. Water enriched in oxygen-18 (^{18}O) has been used to study the hydrolysis of esters. If the basic hydrolysis of ethyl acetate is carried out in ^{18}O-labeled HO^- in ^{18}O-labeled water, will the ^{18}O isotope be found in the carboxylate ion product or in the alcohol?

Synthesis

13.24 Using ethanol as the source of all the carbon atoms, along with any necessary inorganic reagents, show how you could prepare each of the following:
     (a) Acetyl chloride
     (b) Ethyl acetate
     (c) Acetamide

13.25 Using toluene as the ultimate source of all the carbon atoms, along with any necessary inorganic reagents, show how you could prepare each of the following:
     (a) Benzoic acid   (e) Phenylacetic acid
     (b) Benzoyl chloride   (f) *p*-Nitrobenzoyl chloride
     (c) Benzyl benzoate   (g) *m*-Nitrobenzoyl chloride
     (d) Benzamide

13.26 Using bromobenzene and any other necessary organic or inorganic reagents, suggest an efficient synthesis of:
     (a) 1,1-Diphenyl-1-propanol
     (b) 3-Phenyl-3-pentanol
     (c) Benzamide

Miscellaneous Problems

13.27 Cyclic carboxylic acid derivatives usually undergo the same chemical reactions as their counterparts that lack a ring. Thus, cyclic esters and amides undergo reactions typical of acyclic esters and amides. With this in mind, give the structure of the organic product of each of the following reactions:

(a) 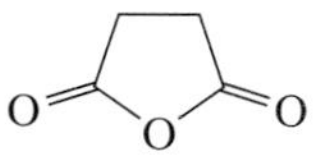   and water

(b) 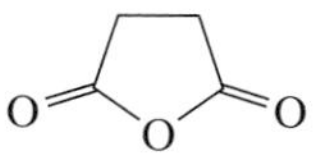   and aqueous sodium hydroxide

(c) 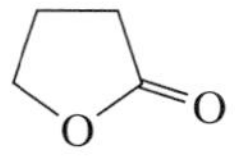   and aqueous sodium hydroxide

(d) 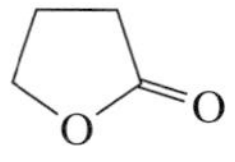   and aqueous ammonia

(e) 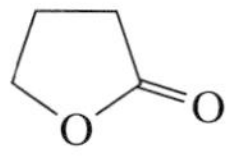   and lithium aluminum hydride, then H_2O

(f) 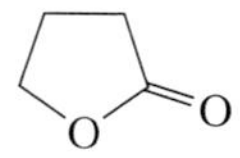 [structure: γ-butyrolactone]   and excess methylmagnesium bromide, then H_3O^+

(g) 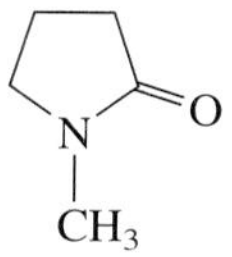 [structure: N-methyl pyrrolidinone]   and aqueous sodium hydroxide

(h) 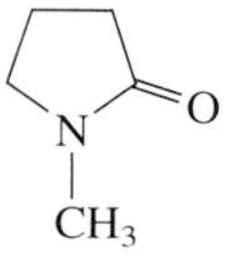 [structure: N-methyl pyrrolidinone]   and aqueous hydrochloric acid, heat

13.28 The cyclic ester (lactone) 4-butanolide was allowed to stand in acidic solution in which the water had been labeled with ^{18}O. When the lactone was extracted from the solution after 4 days, it was found to contain ^{18}O. Which oxygen of the lactone do you think became isotopically labeled?

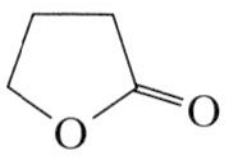

[structure: 4-Butanolide]

4-Butanolide

13.29 The compound having the structure shown was heated with dilute sulfuric acid to give a product having the molecular formula $C_5H_{12}O_3$ in 63–71% yield. Propose a reasonable structure for this product. What other organic compound is formed in this reaction?

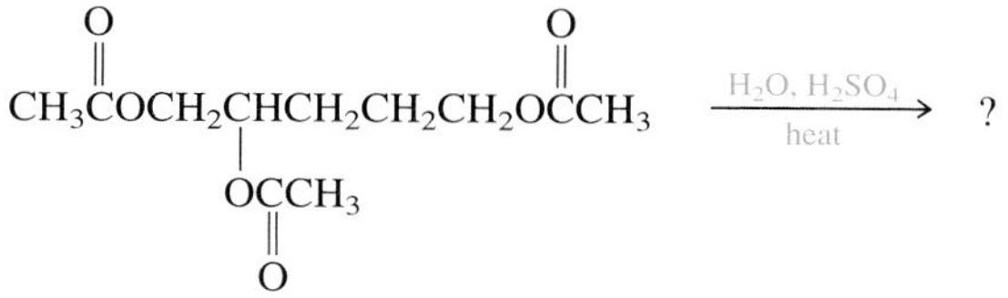

$$CH_3COCH_2CHCH_2CH_2CH_2OCCH_3 \xrightarrow[\text{heat}]{H_2O,\ H_2SO_4} ?$$
$$\underset{\underset{O}{\overset{|}{\underset{\|}{OCCH_3}}}}{}$$

13.30 Procaine (Novocain) is a local anesthetic often used by dentists. Supply the reagents that are missing from the first two steps of the synthesis shown.

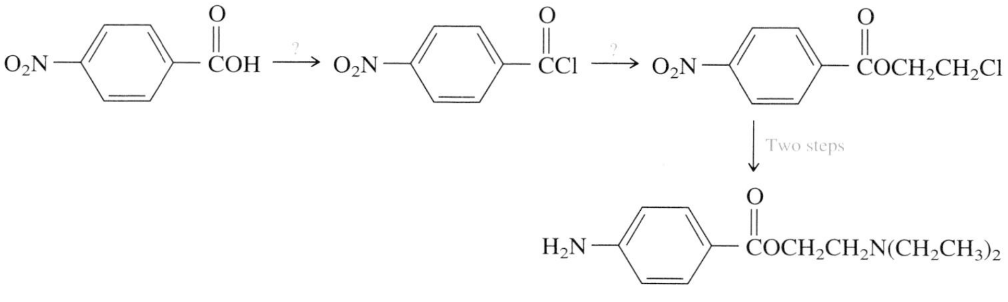

$$O_2N-\langle\text{ring}\rangle-COH \xrightarrow{?} O_2N-\langle\text{ring}\rangle-CCl \xrightarrow{?} O_2N-\langle\text{ring}\rangle-COCH_2CH_2Cl$$

↓ Two steps

$$H_2N-\langle\text{ring}\rangle-COCH_2CH_2N(CH_2CH_3)_2$$

Novocain

13.31 Outline reasonable mechanisms for each of the following reactions:

(a) 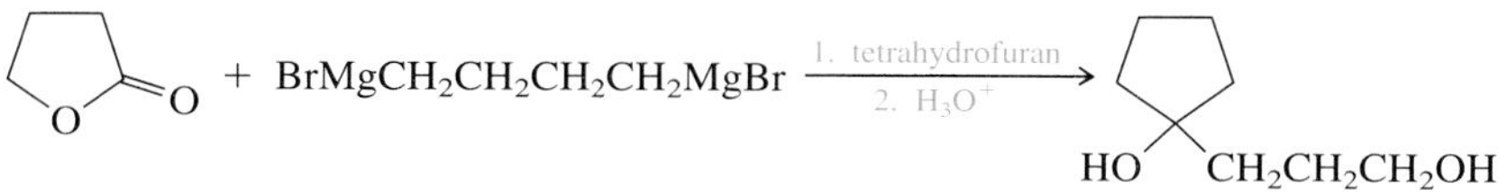 [structure: γ-butyrolactone] $+ BrMgCH_2CH_2CH_2CH_2MgBr \xrightarrow[\text{2. } H_3O^+]{\text{1. tetrahydrofuran}}$ [structure: cyclopentane ring]

$$HO \quad CH_2CH_2CH_2OH$$

(b)

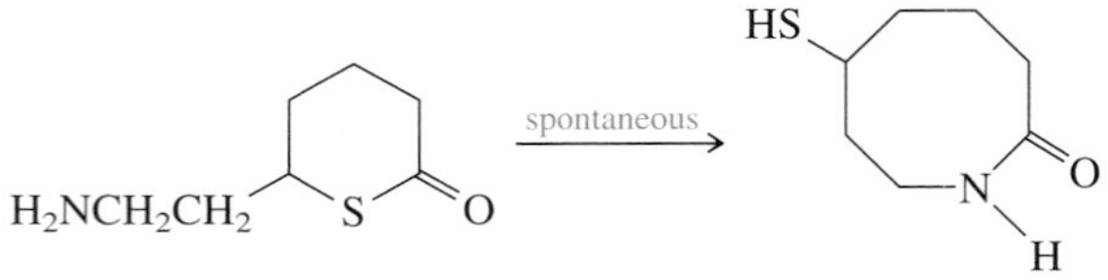

13.32 When compounds of the type represented by A are allowed to stand in pentane, they are converted to a constitutional isomer.

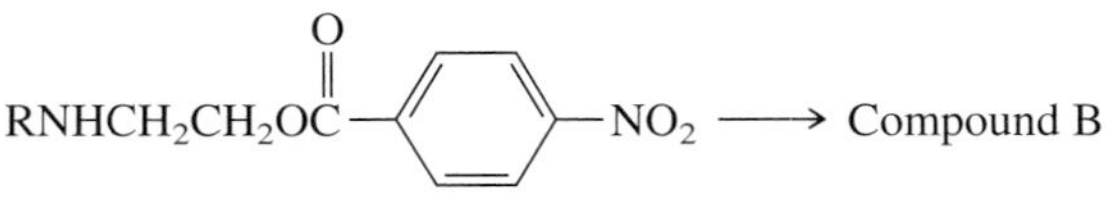

Compound A

Hydrolysis of either A or B yields $RNHCH_2CH_2OH$ and *p*-nitrobenzoic acid. Suggest a reasonable structure for compound B, and demonstrate your understanding of the mechanism of this reaction by writing the structure of the key intermediate in the conversion of compound A to compound B.

13.33 (a) In the presence of dilute hydrochloric acid, compound A is converted to a constitutional isomer, compound B.

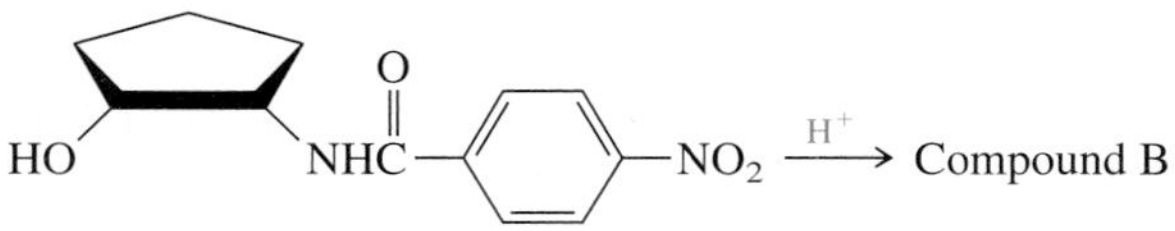

Compound A

Suggest a reasonable structure for compound B.
(b) The trans stereoisomer of compound A is stable under the reaction conditions. Why does it not rearrange?

13.34 Poly(vinyl alcohol) is a useful water-soluble polymer. It cannot be prepared directly from vinyl alcohol, because of the rapidity with which vinyl alcohol (CH_2=CHOH) isomerizes to acetaldehyde. Vinyl acetate, however, does not rearrange and can be polymerized to poly(vinyl acetate). How could you make use of this fact to prepare poly(vinyl alcohol)?

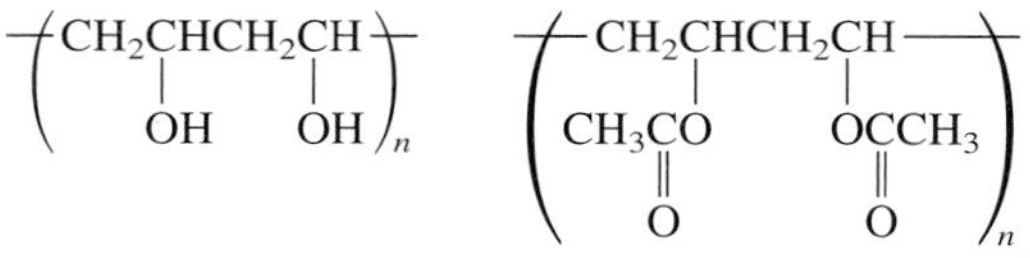

Poly(vinyl alcohol)   Poly(vinyl acetate)

13.35 Lucite is a polymer of methyl methacrylate.
(a) Assuming the first step in the polymerization of methyl methacrylate is as shown,

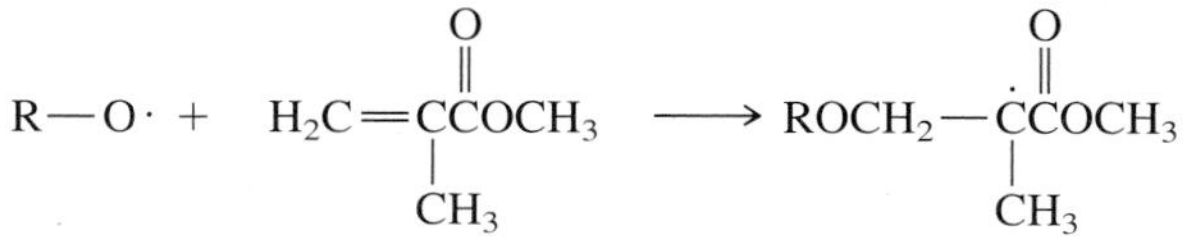

Methyl methacrylate

write a structural formula for the free radical produced after the next two propagation steps.
(b) Outline a synthesis of methyl methacrylate from acetone, sodium cyanide, and any necessary organic or inorganic reagents.

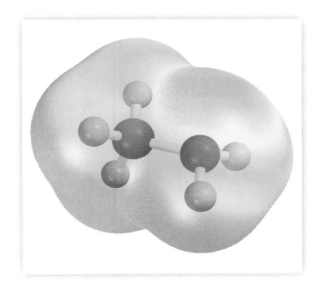

CHAPTER 14
AMINES

itrogen-containing compounds are essential to life. Their ultimate source is atmospheric nitrogen which, by a process known as **nitrogen fixation,** is reduced to ammonia, then converted to organic nitrogen compounds. This chapter describes the chemistry of **amines,** organic derivatives of ammonia. **Alkylamines** have their nitrogen attached to sp^3-hybridized carbon; **arylamines** have their nitrogen attached to an sp^2-hybridized carbon of a benzene or benzene-like ring.

$$R — \ddot{N}\diagdown \qquad Ar — \ddot{N}\diagdown$$

R = alkyl group:
alkylamine

Ar = aryl group:
arylamine

Amines, like ammonia, are weak bases. They are, however, the strongest uncharged bases found in significant quantities under physiological conditions. Amines are usually the bases involved in biological acid–base reactions; they are often the nucleophiles in biological nucleophilic substitutions.

Our word "vitamin" was coined in 1912 in the belief that the substances present in the diet that prevented scurvy, pellagra, beriberi, rickets, and other diseases were "vital amines." In many cases, that belief was confirmed; certain vitamins did prove to be amines. In many other cases, however, vitamins were not amines. Nevertheless, the name **vitamin** entered our language and stands as a reminder that early chemists recognized the crucial place occupied by amines in biological processes.

14.1   AMINE NOMENCLATURE

Unlike alcohols and alkyl halides, which are classified as primary, secondary, or tertiary according to the degree of substitution at the carbon that bears the functional group, amines are classified according to their *degree of substitution at nitrogen*. An amine with one carbon attached to nitrogen is a **primary amine,** an amine with two is a **secondary amine,** and an amine with three is a **tertiary amine.**

Notice that although
$(CH_3)_3CNH_2$ is a **primary**
amine, $(CH_3)_3COH$ is a **tertiary** alcohol.

$$R-\overset{H}{\underset{H}{N:}}$$

$$R-\overset{R'}{\underset{H}{N:}}$$

$$R-\overset{R'}{\underset{R''}{N:}}$$

Primary amine   Secondary amine   Tertiary amine

The groups attached to nitrogen may be any combination of alkyl or aryl groups.

Amines are named in two main ways in the IUPAC system: either as **alkylamines** or as **alkanamines.** When primary amines are named as alkylamines, the ending *-amine* is added to the name of the alkyl group that bears the nitrogen. When named as alkanamines, the alkyl group is named as an alkane and the *-e* ending replaced by *-amine*.

$$CH_3CH_2NH_2$$

$$\overset{NH_2}{\diagup\!\!\!\!\diagdown}$$

$$\underset{\overset{|}{NH_2}}{CH_3CHCH_2CH_2CH_3}$$

Ethylamine   Cyclohexylamine   1-Methylbutylamine
(ethanamine)   (cyclohexanamine)   (2-pentanamine)

PROBLEM 14.1   Give an acceptable alkylamine or alkanamine name for each of the following amines:

(a) $C_6H_5CH_2CH_2NH_2$

(b) $\underset{\overset{|}{CH_3}}{C_6H_5CHNH_2}$

Sample Solution   (a) The amino substituent is bonded to an ethyl group that bears a phenyl substituent at C-2. The compound $C_6H_5CH_2CH_2NH_2$ may be named as either 2-phenylethylamine or 2-phenylethanamine.

Aniline was first isolated in 1826 as a degradation product of indigo, a dark blue dye obtained from the West Indian plant *Indigofera anil*, from which the name "aniline" is derived.

Aniline is the parent IUPAC name for amino-substituted derivatives of benzene. Substituted derivatives of aniline are numbered beginning at the carbon that bears the amino group. Substituents are listed in alphabetical order, and the direction of numbering is governed by the usual "first point of difference" rule.

$$F-\overset{4}{\diagup\!\!\!\!\!\diagdown}-\overset{1}{}NH_2$$

$$Br-\overset{5}{}\overset{1}{\overset{NH_2}{\diagup\!\!\!\!\diagdown}}\overset{2}{}-CH_2CH_3$$

p-Fluoroaniline   5-Bromo-2-ethylaniline

Secondary and tertiary amines are named as *N*-substituted derivatives of primary amines. The parent primary amine is taken to be the one with the longest carbon chain. The prefix *N*- is added as a locant to identify substituents on the amino nitrogen as needed.

CH₃NHCH₂CH₃

N-Methylethylamine

(a secondary amine)

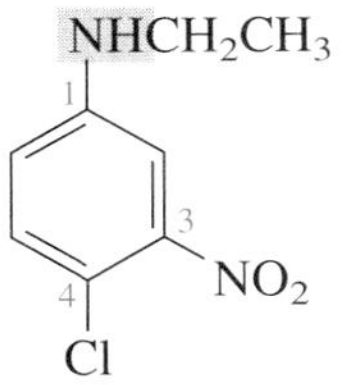

NHCH₂CH₃

NO₂

Cl

4-Chloro-N-ethyl-3-
nitroaniline

(a secondary amine)

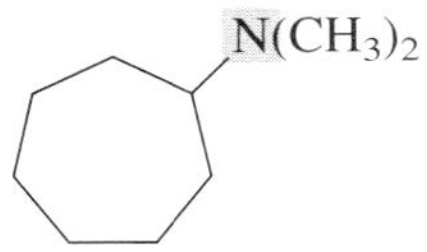

N(CH₃)₂

N,N-Dimethylcyclo-
heptylamine

(a tertiary amine)

PROBLEM 14.2 Assign alkanamine names to N-methylethylamine and to N,N-dimethylcycloheptylamine.

Sample Solution N-Methylethylamine (given as CH₃NHCH₂CH₃ in the preceding example) is an N-substituted derivative of ethanamine; it is N-methylethanamine.

PROBLEM 14.3 Classify the following amine as primary, secondary, or tertiary, and give it an acceptable IUPAC name.

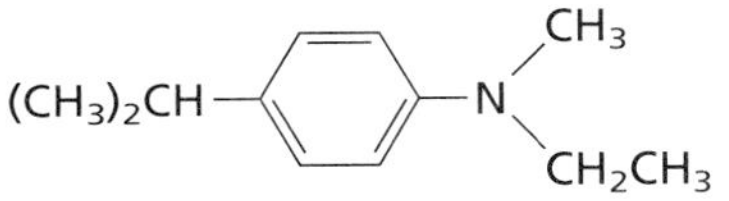

$$(CH_3)_2CH \underset{\overset{\displaystyle}{CH_2CH_3}}{\overset{\displaystyle CH_3}{N}}$$

A nitrogen that bears four substituents is positively charged and is named as an **ammonium** ion. The anion that is associated with it is also identified in the name.

$$\overset{+}{CH_3NH_3} \ Cl^-$$

Methylammonium
chloride

$$C_6H_5CH_2\overset{+}{N}(CH_3)_3 \ I^-$$

Benzyltrimethyl-
ammonium iodide

(a quaternary ammonium
salt)

Ammonium salts that have four alkyl groups bonded to nitrogen are called **quaternary ammonium salts.**

14.2 STRUCTURE AND BONDING

Alkylamines

Alkylamines have a pyramidal arrangement of bonds to nitrogen, similar to those of ammonia (Section 1.10). An orbital hybridization description of bonding in methylamine is shown in Figure 14.1. Nitrogen and carbon are both sp^3-hybridized and are joined by a σ bond. The unshared electron pair on nitrogen occupies an sp^3-hybridized orbital. This lone pair is involved in reactions in which amines act as bases or nucleophiles. The graphic that opened this chapter is an electrostatic potential map that clearly shows the concentration of electron density at nitrogen in methylamine.

Arylamines

Aniline, like alkylamines, has a pyramidal arrangement of bonds around nitrogen, but its pyramid is somewhat shallower.

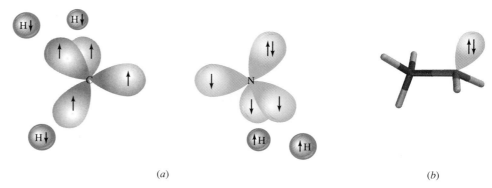

(a)   (b)

FIGURE 14.1   Orbital hybridization description of bonding in methylamine. (a) Carbon has four valence electrons; each of four equivalent sp^3-hybridized orbitals contains one electron. Nitrogen has five valence electrons. Three of its sp^3 hybrid orbitals contain one electron each; the fourth sp^3 hybrid orbital contains two electrons. (b) Nitrogen and carbon are connected by a σ bond in methylamine. This σ bond is formed by overlap of an sp^3 hybrid orbital on each atom. The five hydrogen atoms of methylamine are joined to carbon and nitrogen by σ bonds. The two remaining electrons of nitrogen occupy an sp^3-hybridized orbital.

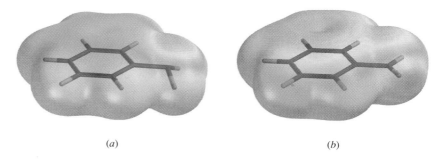

(a)   (b)

FIGURE 14.2   Electrostatic potential maps of the aniline in which the geometry at nitrogen is (a) nonplanar and (b) planar. In the nonplanar geometry, the unshared pair occupies an sp^3 hybrid orbital of nitrogen. The region of highest electron density in part (a) is associated with nitrogen. In the planar geometry, nitrogen is sp^2-hybridized and the electron pair is delocalized between a p orbital of nitrogen and the π system of the ring. The region of highest electron density in part (b) encompasses both the ring and nitrogen. The actual structure combines features of both; nitrogen adopts a hybridization state between sp^3 and sp^2.

The structure of aniline reflects a compromise between two modes of binding the nitrogen lone pair (Figure 14.2). The electrons are more strongly attracted to nitrogen when they are in an orbital with some s character—an sp^3-hybridized orbital, for example—than when they are in a p orbital. On the other hand, delocalization of these electrons into the aromatic π system is better achieved if they occupy a p orbital. A p orbital of nitrogen is better aligned for overlap with the p orbitals of the benzene ring to form an extended π system than is an sp^3-hybridized orbital. As a result of these two opposing forces, nitrogen adopts an orbital hybridization that is between sp^3 and sp^2.

The corresponding resonance description shows the delocalization of the nitrogen lone-pair electrons in terms of contributions from dipolar structures.

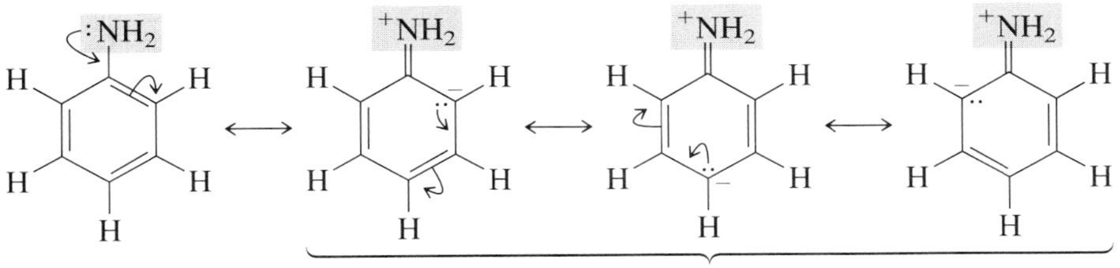

Most stable
Lewis structure
for aniline

Dipolar resonance forms of aniline

14.3 PHYSICAL PROPERTIES

We have often seen that the polar nature of a substance can affect physical properties such as boiling point. This is true for amines, which are more polar than alkanes but less polar than alcohols. For similarly constituted compounds, alkylamines have boiling points higher than those of alkanes but lower than those of alcohols.

$$CH_3CH_2CH_3 \qquad CH_3CH_2NH_2 \qquad CH_3CH_2OH$$

Propane   Ethylamine   Ethanol
bp −42°C   bp 17°C   bp 78°C

Dipole–dipole interactions, especially hydrogen bonding, are present in amines but absent in alkanes. Nitrogen is less electronegative than oxygen and the less polar nature of amines relative to alcohols makes these intermolecular forces weaker in amines than in alcohols.

Among isomeric amines, primary amines have the highest boiling points, and tertiary amines the lowest.

$$CH_3CH_2CH_2NH_2 \qquad CH_3CH_2NHCH_3 \qquad (CH_3)_3N$$

Propylamine   N-Methylethylamine   Trimethylamine
(a primary amine)   (a secondary amine)   (a tertiary amine)
bp 50°C   bp 34°C   bp 3°C

Primary and secondary amines can participate in intermolecular hydrogen bonding, but tertiary amines cannot.

Amines that have fewer than six or seven carbon atoms are soluble in water. All amines, even tertiary amines, can act as proton acceptors in hydrogen bonding to water molecules.

The simplest arylamine, aniline, is a liquid at room temperature and has a boiling point of 184°C. Almost all other arylamines have higher boiling points. Aniline is only slightly soluble in water (3 g/100 mL). Substituted derivatives of aniline tend to be even less water-soluble.

14.4 BASICITY OF AMINES

Two conventions are used to measure the basicity of amines. One of them defines a **basicity constant K_b** for the amine acting as a proton acceptor from water:

$$R_3N: + H-\overset{..}{\underset{..}{O}}H \rightleftharpoons R_3\overset{+}{N}-H + :\overset{..}{\underset{..}{O}}H^-$$

$$K_b = \frac{[R_3NH^+][HO^-]}{[R_3N]} \qquad \text{and} \qquad pK_b = -\log K_b$$

The ease with which amines are extracted into aqueous acid, combined with their regeneration on treatment with base, makes it a simple matter to separate amines from other plant materials, and nitrogen-containing natural products were among the earliest organic compounds to be studied.

Their basic properties led amines obtained from plants to be called **alkaloids**. The number of known alkaloids exceeds 5000. They are of special interest because most are characterized by a high level of biological activity. Some examples include cocaine, coniine, and morphine.

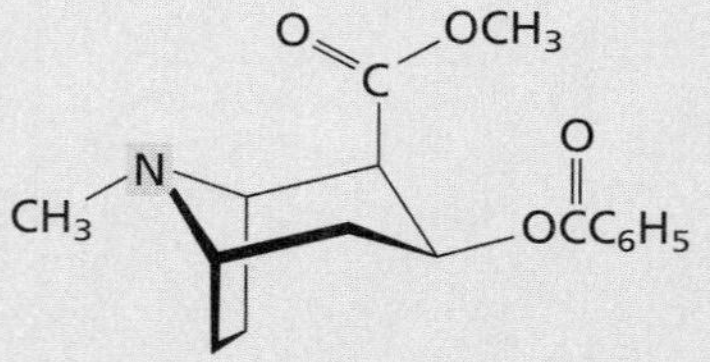

Cocaine

(A central nervous system stimulant obtained from the leaves of the coca plant.)

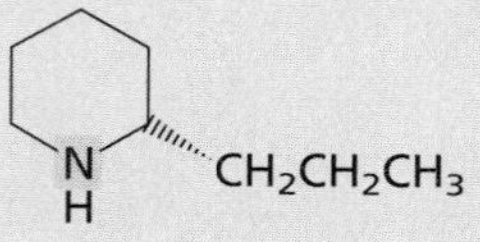

Coniine

(Present along with other alkaloids in the hemlock extract used to poison Socrates.)

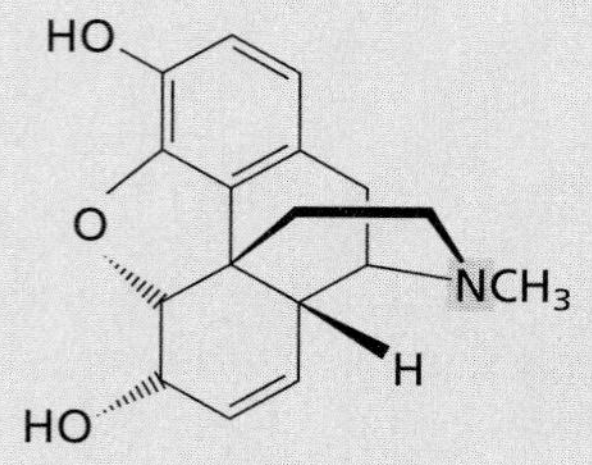

Morphine

(An opium alkaloid. Although it is an excellent analgesic, its use is restricted because of the potential for addiction. Heroin is the diacetate ester of morphine.)

Many alkaloids, such as nicotine and quinine, contain two (or more) nitrogen atoms. The nitrogens highlighted in yellow in quinine and nicotine are part of a substituted quinoline and pyridine ring, respectively.

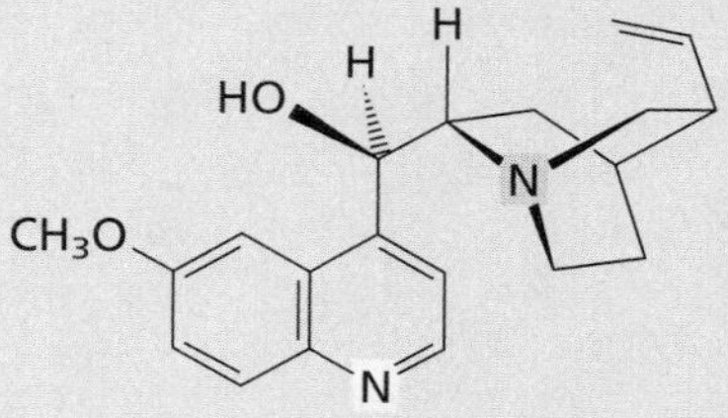

Quinine

(Alkaloid of cinchona bark used to treat malaria)

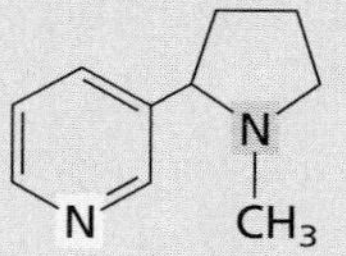

Nicotine

(An alkaloid present in tobacco; a very toxic compound sometimes used as an insecticide)

Several naturally occurring amines mediate the transmission of nerve impulses and are referred to as **neurotransmitters**. Two examples are epinephrine and serotonin. (Strictly speaking, these compounds are not classified as alkaloids, because they are not isolated from plants.)

Continued

For ammonia, $K_b = 1.8 \times 10^{-5}$ ($pK_b = 4.7$). A typical amine such as methylamine (CH_3NH_2) is a slightly stronger base than ammonia and has $K_b = 4.4 \times 10^{-4}$ ($pK_b = 3.3$).

The other convention relates the basicity of an amine (R_3N) to the acid dissociation constant K_a of its conjugate acid (R_3NH^+):

$$R_3\overset{+}{N}-H \Longrightarrow H^+ + R_3N:$$

where K_a and pK_a have their usual meaning:

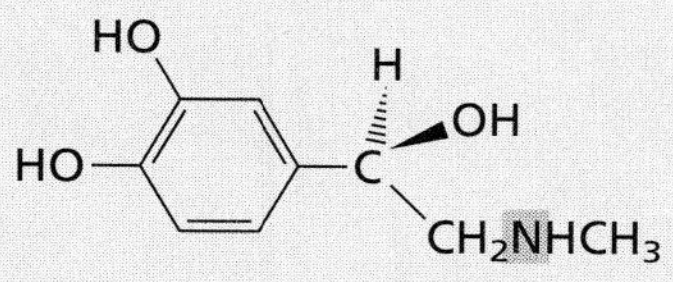

HO
HO
H
OH
C
CH$_2$NHCH$_3$

Epinephrine

(Also called adrenaline; a hormone secreted by the adrenal gland that prepares the organism for "flight or fight.")

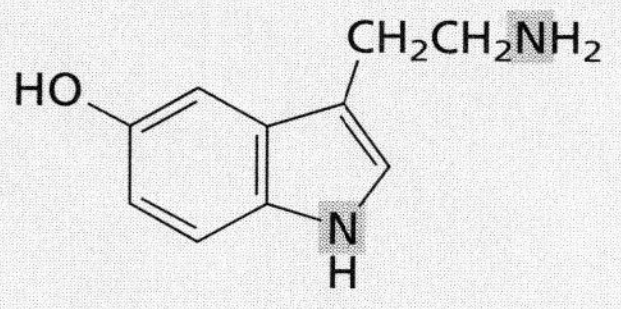

HO
CH$_2$CH$_2$NH$_2$
N
H

Serotonin

(A hormone synthesized in the pineal gland. Certain mental disorders are believed to be related to serotonin levels in the brain.)

Bioactive amines are also widespread in animals. A variety of structures and properties have been found in substances isolated from frogs, for example. One, called epibatidine, is a naturally occurring painkiller isolated from the skin of an Ecuadoran frog. Another family of frogs produces a toxic mixture of several stereoisomeric amines, called dendrobines, on their skin that protects them from attack.

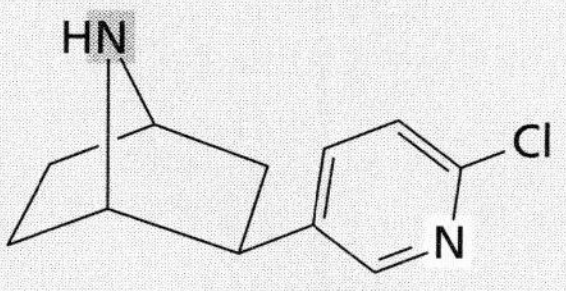

HN
Cl
N

Epibatidine

(Once used as an arrow poison, it is hundreds of times more powerful than morphine in relieving pain. It is too toxic to be used as a drug, however.)

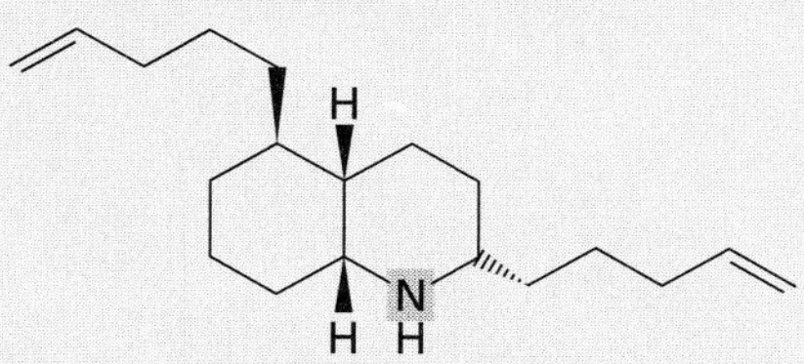

H
H
N
H H

Dendrobine

(Isolated from frogs of the Dendrobatidae family. Related compounds have also been isolated from certain ants.)

Among the more important amine derivatives found in the body are a group of compounds known as **polyamines**, which contain two to four nitrogen atoms separated by several methylene units:

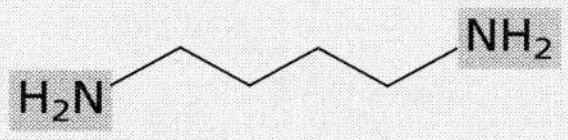

H$_2$N
NH$_2$

Putrescine

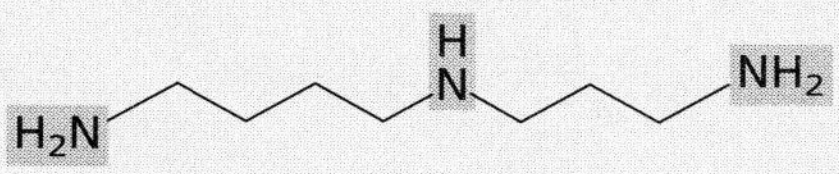

H$_2$N
H
N
NH$_2$

Spermidine

H$_2$N
H
N
N
H
NH$_2$

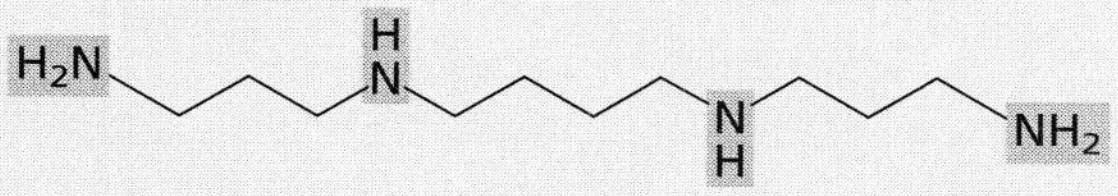

Spermine

These compounds are present in almost all mammalian cells, where they are believed to be involved in cell differentiation and proliferation. Because each nitrogen of a polyamine is protonated at physiological pH (7.4), putrescine, spermidine, and spermine exist as cations with a charge of + 2, + 3, and + 4, respectively, in body fluids. Structural studies suggest that these polyammonium ions affect the conformation of biological macromolecules by electrostatic binding to specific anionic sites—the negatively charged phosphate groups of DNA, for example.

$$K_a = \frac{[\text{H}^+][\text{R}_3\text{N}]}{[\text{R}_3\text{NH}^+]} \quad \text{and} \quad pK_a = -\log K_a$$

The conjugate acid of ammonia is ammonium ion (NH$_4^+$), which has $K_a = 5.6 \times 10^{-10}$ (p$K_a = 9.3$). The conjugate acid of methylamine is methylammonium ion (CH$_3$NH$_3^+$), which has $K_a = 2 \times 10^{-11}$ (p$K_a = 10.7$).

The more basic the amine, the weaker is its conjugate acid.

Methylamine is a stronger base than ammonia; methylammonium ion is a weaker acid than ammonium ion.

The relationship between the equilibrium constant K_b for an amine (R_3N) and K_a for its conjugate acid (R_3NH^+) is:

$$K_aK_b = 10^{-14} \quad \text{and} \quad pK_a + pK_b = 14$$

> **PROBLEM 14.4** A chemistry handbook lists K_b for quinine as 1×10^{-6}. What is pK_b for quinine? What are the values of K_a and pK_a for the conjugate acid of quinine?

Amines are weak bases, but as a class:

Amines are the strongest bases of all neutral molecules.

Table 14.1 lists basicity data for a number of amines. The most important relationships to be drawn from the data are

1. Alkylamines are slightly stronger bases than ammonia.
2. Alkylamines differ very little among themselves in basicity. Their basicities cover a range of less than 10 in equilibrium constant (1 pK unit).
3. Arylamines are much weaker bases than ammonia and alkylamines. Their basicity constants are on the order of 10^6 smaller than those of alkylamines (6 pK units).

TABLE 14.1	Base Strength of Amines As Measured by Their Basicity Constants and the Dissociation Constants of Their Conjugate Acids*				
		Basicity		**Acidity of Conjugate Acid**	
Compound	**Structure**	K_b	pK_b	K_a	pK_a
Ammonia	NH_3	1.8×10^{-5}	4.7	5.5×10^{-10}	9.3
Primary amines					
Methylamine	CH_3NH_2	4.4×10^{-4}	3.4	2.3×10^{-11}	10.6
Ethylamine	$CH_3CH_2NH_2$	5.6×10^{-4}	3.2	1.8×10^{-11}	10.8
Isopropylamine	$(CH_3)_2CHNH_2$	4.3×10^{-4}	3.4	2.3×10^{-11}	10.6
tert-Butylamine	$(CH_3)_3CNH_2$	2.8×10^{-4}	3.6	3.6×10^{-11}	10.4
Aniline	$C_6H_5NH_2$	3.8×10^{-10}	9.4	2.6×10^{-5}	4.6
Secondary amines					
Dimethylamine	$(CH_3)_2NH$	5.1×10^{-4}	3.3	2.0×10^{-11}	10.7
Diethylamine	$(CH_3CH_2)_2NH$	1.3×10^{-3}	2.9	7.7×10^{-12}	11.1
N-Methylaniline	$C_6H_5NHCH_3$	6.1×10^{-10}	9.2	1.6×10^{-5}	4.8
Tertiary amines					
Trimethylamine	$(CH_3)_3N$	5.3×10^{-5}	4.3	1.9×10^{-10}	9.7
Triethylamine	$(CH_3CH_2)_3N$	5.6×10^{-4}	3.2	1.8×10^{-11}	10.8
N,N-Dimethylaniline	$C_6H_5N(CH_3)_2$	1.2×10^{-9}	8.9	8.3×10^{-6}	5.1

* In water at 25°C.

The differences in basicity between ammonia, and primary, secondary, and tertiary alkylamines result from the interplay between steric and electronic effects on the molecules themselves and on the solvation of their conjugate acids. In total, the effects are small, and most alkylamines are very similar in basicity.

Arylamines are a different story, however; most are about a million times weaker as bases than ammonia and alkylamines.

As unfavorable as the equilibrium is for cyclohexylamine acting as a base in water,

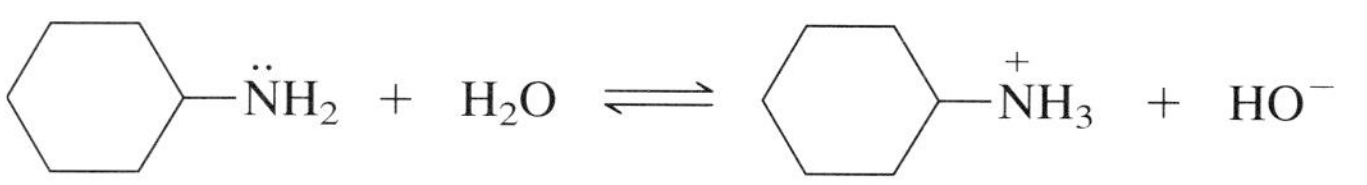

Cyclohexylamine	Water	Cyclohexylammonium ion	Hydroxide ion

it is far less favorable for aniline.

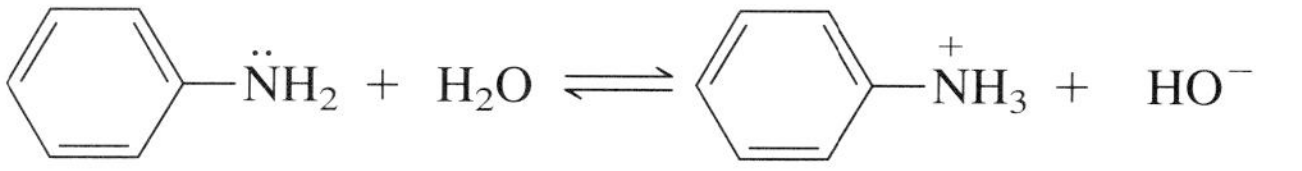

Aniline	Water	Anilinium ion	Hydroxide ion

Aniline is a much weaker base because its delocalized lone pair is more strongly held than the nitrogen lone pair in cyclohexylamine. The more strongly held the electron pair, the less able it is to abstract a proton.

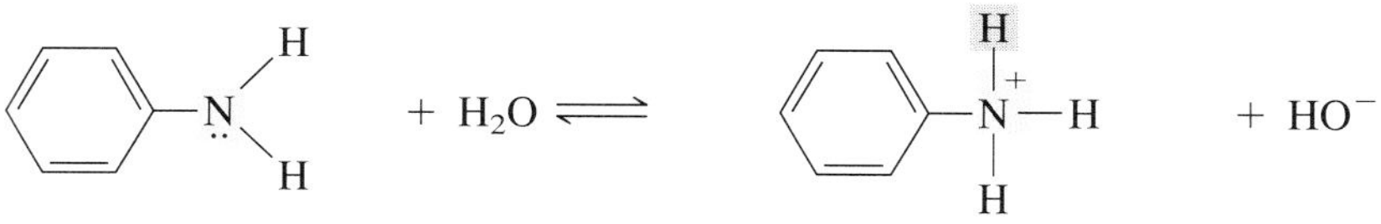

Aniline is stabilized by delocalization of lone pair into π system of ring, decreasing the electron density at nitrogen.

PROBLEM 14.5 The two amines shown differ by a factor of 40,000 in their K_b values. Which is the stronger base? Why?

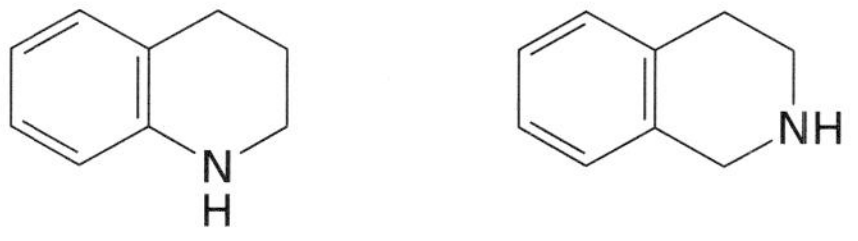

Tetrahydroquinoline   Tetrahydroisoquinoline

14.5 PREPARATION OF AMINES BY ALKYLATION OF AMMONIA

Alkylamines are, in principle, capable of being prepared by nucleophilic substitution reactions of alkyl halides with ammonia.

$$RX \ + \ 2NH_3 \ \longrightarrow \ RNH_2 \ + \ \overset{+}{N}H_4 \ X^-$$

Alkyl halide	Ammonia	Primary amine	Ammonium halide salt

This reaction, however, is not a general method for the synthesis of amines. Its major limitation is that the expected primary amine product is itself a nucleophile and competes with ammonia for the alkyl halide.

$$RX \; + \; RNH_2 \; + \; NH_3 \; \longrightarrow \; RNHR \; + \; \overset{+}{N}H_4 \; X^-$$

| Alkyl halide | Primary amine | Ammonia | Secondary amine | Ammonium halide salt |

When 1-bromooctane, for example, is allowed to react with ammonia, both the primary amine and the secondary amine are isolated in comparable amounts.

$$CH_3(CH_2)_6CH_2Br \xrightarrow{NH_3 \; (2 \; mol)} CH_3(CH_2)_6CH_2NH_2 \; + \; [CH_3(CH_2)_6CH_2]_2NH$$

| 1-Bromooctane (1 mol) | Octylamine (45%) | N,N-Dioctylamine (43%) |

In a similar manner, competitive alkylation may continue, resulting in formation of a trialkylamine or even a quaternary ammonium salt.

> **PROBLEM 14.6** Write equations that show the reaction of N,N-dioctylamine (from the preceding equation) with 1-bromooctane to form tertiary amine and quaternary ammonium salt by-products.

14.6 PREPARATION OF AMINES BY REDUCTION

Almost any nitrogen-containing organic compound can be reduced to an amine. The synthesis of amines then becomes a question of the availability of suitable precursors and the choice of an appropriate reducing agent.

Alkyl **azides,** prepared by nucleophilic substitution of alkyl halides by sodium azide (Section 8.1), are reduced to alkylamines by a variety of reagents, including lithium aluminum hydride. Catalytic hydrogenation is also effective.

$$R\!-\!\overset{..}{N}\!=\!\overset{+}{N}\!=\!\overset{..}{N}{:}^- \xrightarrow{reduce} R\overset{..}{N}H_2$$

| Alkyl azide | Primary amine |

$$C_6H_5CH_2CH_2N_3 \xrightarrow[\text{2. } H_2O]{\text{1. LiAlH}_4 \atop \text{diethyl ether}} C_6H_5CH_2CH_2NH_2$$

| 2-Phenylethyl azide | 2-Phenylethylamine (89%) |

The same reduction methods may be applied to the conversion of **nitriles** to primary amines.

$$RC\!\equiv\!N \xrightarrow[\text{H}_2, \text{ catalyst}]{\text{LiAlH}_4 \text{ or}} RCH_2NH_2$$

| Nitrile | Primary amine |

$$F_3C\!-\!\langle\!\rangle\!-\!CH_2CN \xrightarrow[\text{2. } H_2O]{\text{1. LiAlH}_4, \atop \text{diethyl ether}} F_3C\!-\!\langle\!\rangle\!-\!CH_2CH_2NH_2$$

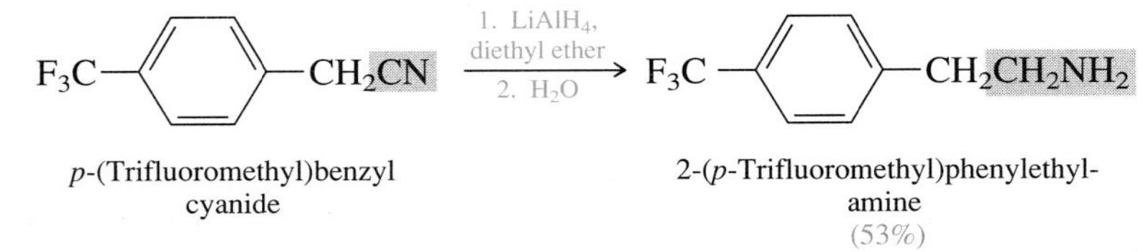

| p-(Trifluoromethyl)benzyl cyanide | 2-(p-Trifluoromethyl)phenylethyl-amine (53%) |

Because nitriles can be prepared from alkyl halides by nucleophilic substitution with cyanide ion, the overall process RX $\longrightarrow$ RC$\equiv$N $\longrightarrow$ RCH$_2$NH$_2$ leads to primary amines that have one more carbon atom than the starting alkyl halide.

Cyano groups in cyanohydrins (Section 11.7) are reduced under the same reaction conditions.

Nitro groups are readily reduced to primary amines by a variety of methods. Catalytic hydrogenation over platinum, palladium, or nickel is often used, as is reduction by iron or tin in hydrochloric acid. The ease with which nitro groups are reduced is especially useful in the preparation of arylamines, where the sequence ArH $\longrightarrow$ ArNO$_2$ $\longrightarrow$ ArNH$_2$ is the standard route to these compounds.

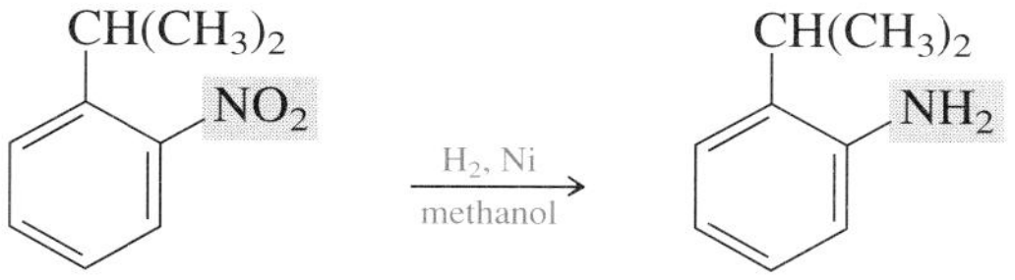

o-Isopropylnitrobenzene

o-Isopropylaniline
(92%)

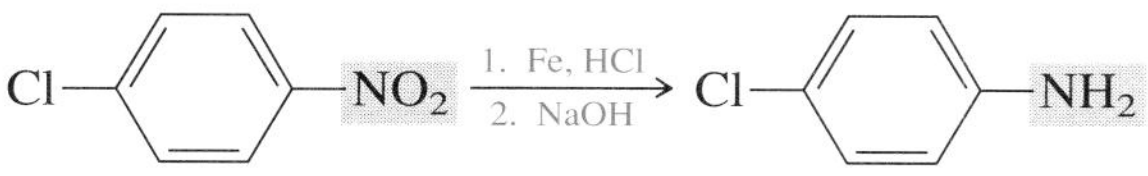

p-Chloronitrobenzene

p-Chloroaniline
(95%)

For reductions carried out in acidic media, a pH adjustment with sodium hydroxide is required in the last step in order to convert ArNH$_3{}^+$ to ArNH$_2$.

PROBLEM 14.7 Outline syntheses of each of the following arylamines from benzene:

(a) *o*-Isopropylaniline
(b) *p*-Isopropylaniline
(c) *p*-Chloroaniline

Sample Solution (a) The last step in the synthesis of *o*-isopropylaniline, the reduction of the corresponding nitro compound by catalytic hydrogenation, is given as one of the three preceding examples. The necessary nitroarene is obtained by fractional distillation of the ortho–para mixture formed during nitration of isopropylbenzene.

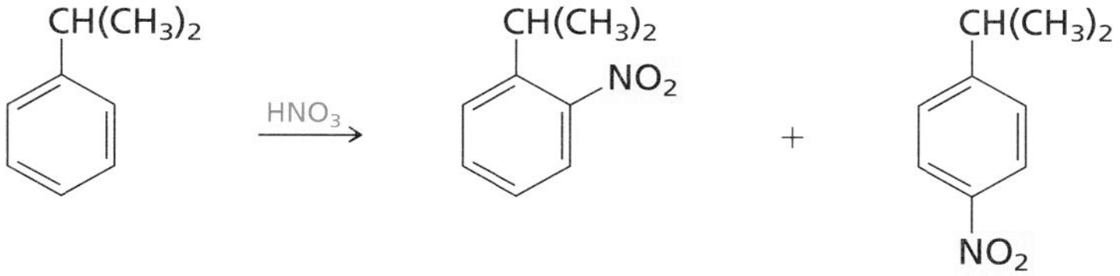

Isopropylbenzene

o-Isopropylnitrobenzene
(bp 110°C)

p-Isopropylnitrobenzene
(bp 131°C)

As actually performed, a 62% yield of a mixture of ortho and para nitration products has been obtained with an ortho–para ratio of about 1:3.

Isopropylbenzene is prepared by the Friedel–Crafts alkylation of benzene using isopropyl chloride and aluminum chloride (Section 6.6).

Reduction of an azide, a nitrile, or a nitro compound furnishes a primary amine. A method that provides access to primary, secondary, or tertiary amines is reduction of the carbonyl group of an amide by lithium aluminum hydride.

$$\underset{\text{Amide}}{RCNR'_2} \xrightarrow[\text{2. } H_2O]{\text{1. LiAlH}_4} \underset{\text{Amine}}{RCH_2NR'_2}$$

In this general equation, R and R' may be either alkyl or aryl groups. Because amides can be prepared readily (Section 13.11), this is a versatile method for the preparation of amines.

Acetanilide is an acceptable IUPAC synonym for N-phenylethanamide.

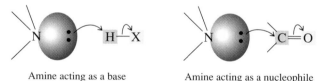

Acetanilide   N-Ethylaniline
                                                                           (92%)

> **PROBLEM 14.8** Outline a synthesis of N-ethylaniline, shown in the previous equation, beginning with acetic acid and any other necessary organic or inorganic reagents.

14.7 REACTIONS OF AMINES: A REVIEW AND A PREVIEW

The noteworthy properties of amines are their **basicity** and their **nucleophilicity.** The basicity of amines has been discussed in Section 14.4. Several reactions in which amines act as nucleophiles have already been encountered in earlier chapters. These are summarized in Table 14.2.

Both the basicity and the nucleophilicity of amines originate in the unshared electron pair of nitrogen. When an amine acts as a base, this electron pair abstracts a proton from a Brønsted acid. When an amine undergoes the reactions summarized in Table 14.2, the first step in each case is the attack of the unshared electron pair on the positively polarized carbon of a carbonyl group.

Amine acting as a base   Amine acting as a nucleophile

In addition to being more basic than arylamines, alkylamines are also more nucleophilic. All the reactions in Table 14.2 take place faster with alkylamines than with arylamines.

The sections that follow introduce some additional reactions of amines. In all cases our understanding of how these reactions take place starts with a consideration of the role of the unshared electron pair of nitrogen.

We will begin with an examination of the reactivity of amines as nucleophiles in S_N2 reactions.

14.8 REACTION OF AMINES WITH ALKYL HALIDES

Nucleophilic substitution results when primary alkyl halides are treated with amines.

TABLE 14.2	Reactions of Amines Discussed in Previous Chapters*

Reaction (section) and Comments	General Equation and Specific Example
Reaction of primary amines with aldehydes and ketones (Section 11.9) Imines are formed by nucleophilic addition of a primary amine to the carbonyl group of an aldehyde or a ketone. The key step is formation of a carbinolamine intermediate, which then dehydrates to the imine.	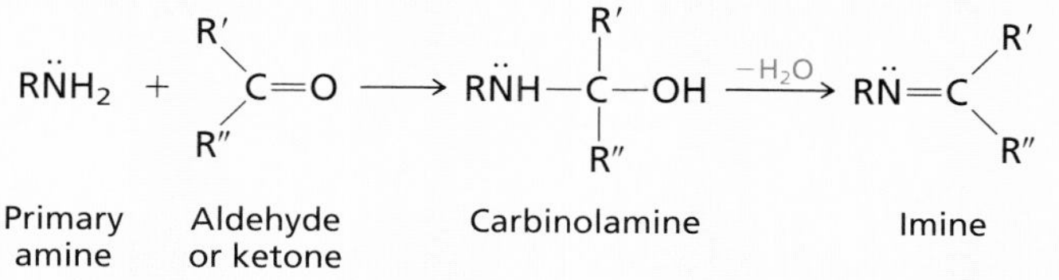 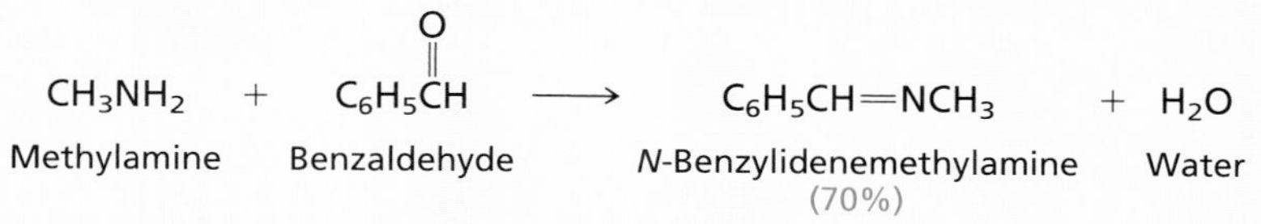 $\ddot{R}NH_2$ + $\overset{R'}{\underset{R''}{C}}{=}O$ ⟶ $RNH{-}\overset{R'}{\underset{R''}{C}}{-}OH$ $\xrightarrow{-H_2O}$ $R\ddot{N}{=}\overset{R'}{\underset{R''}{C}}$ Primary amine · Aldehyde or ketone · Carbinolamine · Imine
	CH_3NH_2 + $C_6H_5\overset{O}{\overset{\|}{C}}H$ ⟶ $C_6H_5CH{=}NCH_3$ + H_2O Methylamine · Benzaldehyde · N-Benzylidenemethylamine (70%) · Water
Reaction of amines with acyl chlorides (Section 13.11) Amines are converted to amides on reaction with acyl chlorides. Other acylating agents, such as carboxylic acid anhydrides and esters, may also be used.	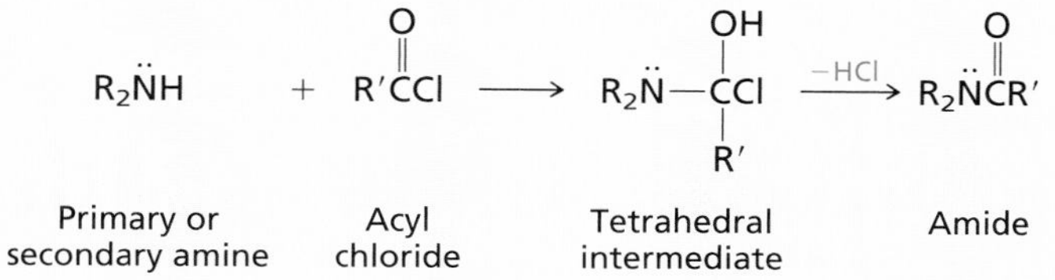 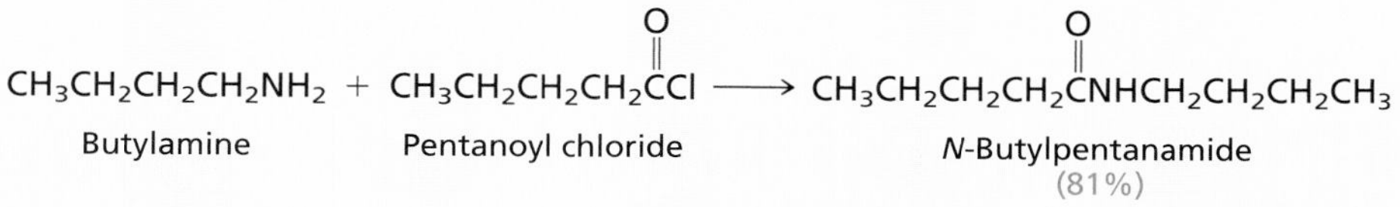 $R_2\ddot{N}H$ + $R'\overset{O}{\overset{\|}{C}}Cl$ ⟶ $R_2\ddot{N}{-}\overset{OH}{\underset{R'}{C}}{-}Cl$ $\xrightarrow{-HCl}$ $R_2\ddot{N}\overset{O}{\overset{\|}{C}}R'$ Primary or secondary amine · Acyl chloride · Tetrahedral intermediate · Amide
	$CH_3CH_2CH_2CH_2NH_2$ + $CH_3CH_2CH_2CH_2\overset{O}{\overset{\|}{C}}Cl$ ⟶ $CH_3CH_2CH_2CH_2\overset{O}{\overset{\|}{C}}NHCH_2CH_2CH_2CH_3$ Butylamine · Pentanoyl chloride · N-Butylpentanamide (81%)

* Both alkylamines and arylamines undergo these reactions.

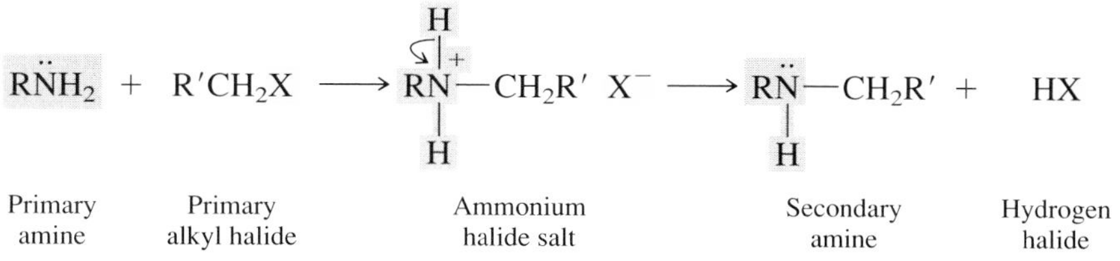

$\ddot{R}NH_2$ + $R'CH_2X$ ⟶ $\overset{H}{\underset{H}{R\overset{+}{N}{-}CH_2R'}}$ X^- ⟶ $\overset{\,}{\underset{H}{R\ddot{N}{-}CH_2R'}}$ + HX

Primary amine · Primary alkyl halide · Ammonium halide salt · Secondary amine · Hydrogen halide

$C_6H_5NH_2$ + $C_6H_5CH_2Cl$ $\xrightarrow[90\,°C]{NaHCO_3}$ $C_6H_5NHCH_2C_6H_5$

Aniline (4 mol) · Benzyl chloride (1 mol) · N-Benzylaniline (85–87%)

A second alkylation may follow, converting the secondary amine to a tertiary amine. Alkylation need not stop there; the tertiary amine may itself be alkylated, giving a quaternary ammonium salt.

$$\overset{..}{R}\overset{..}{N}H_2 \xrightarrow{R'CH_2X} R\overset{..}{N}HCH_2R' \xrightarrow{R'CH_2X} \overset{..}{R}\overset{..}{N}(CH_2R')_2 \xrightarrow{R'CH_2X} R\overset{+}{N}(CH_2R')_3 \; X^-$$

Primary amine	Secondary amine	Tertiary amine	Quaternary ammonium salt

Because of its high reactivity toward nucleophilic substitution, methyl iodide is the alkyl halide most often used to prepare quaternary ammonium salts.

$$\text{(Cyclohexylmethyl)-CH}_2\text{NH}_2 \; + \; 3\text{CH}_3\text{I} \xrightarrow[\text{heat}]{\text{methanol}} \text{(Cyclohexylmethyl)-CH}_2\overset{+}{\text{N}}(\text{CH}_3)_3 \; \text{I}^-$$

(Cyclohexylmethyl)- amine	Methyl iodide	(Cyclohexylmethyl)trimethyl- ammonium iodide (99%)

> **PROBLEM 14.9** Choline is an intermediate in formation of acetylcholine, a key substance involved in the transmission of nerve impulses in animals. Synthetic choline is added as a supplement to animal feed. Choline can be prepared by the reaction of trimethylamine with ethylene oxide in water. Choline is a hydroxide salt having the formula $(C_5H_{14}NO)^+(OH)^-$. What is its structure?
>
> $$(CH_3)_3N \; + \; H_2C\overset{O}{\triangle}CH_2 \xrightarrow{H_2O} \text{Choline}$$
>
> Trimethylamine   Ethylene oxide

14.9 NITROSATION OF AMINES

When solutions of sodium nitrite (NaNO$_2$) are acidified, a number of species are formed that act as **nitrosating agents.** That is, they react as sources of nitrosyl cation, $:\overset{+}{N}=\overset{..}{O}:$. To simplify discussion, organic chemists group all these species together and speak of the chemistry of one of them, nitrous acid, as a generalized precursor to nitrosyl cation.

$$:\overset{..}{\underset{..}{O}}-\overset{..}{N}=\overset{..}{O}: \xrightarrow{H^+} H-\overset{..}{\underset{..}{O}}-\overset{..}{N}=\overset{..}{O}: \xrightarrow{H^+} H-\overset{+}{\underset{H}{O}}-\overset{..}{N}=\overset{..}{O}: \xrightarrow{-H_2O} :\overset{+}{N}=\overset{..}{O}:$$

Nitrite ion (from sodium nitrite)	Nitrous acid	Nitrosyl cation

Nitrosation of amines is best illustrated by examining what happens when a secondary amine "reacts with nitrous acid." The amine acts as a nucleophile, attacking the nitrogen of nitrosyl cation.

$$R_2\overset{..}{N}\underset{H}{\overset{\frown}{\;}} \; + \; :\overset{+}{N}=\overset{..}{O}: \longrightarrow R_2\overset{+}{\underset{H}{N}}-\overset{..}{N}=\overset{..}{O}: \xrightarrow{-H^+} R_2\overset{..}{N}-\overset{..}{N}=\overset{..}{O}:$$

Secondary alkylamine	Nitrosyl cation		N-Nitroso amine

The intermediate that is formed in the first step loses a proton to give an *N*-nitroso amine as the isolated product.

$$(CH_3)_2\overset{..}{N}H \xrightarrow[\text{H}_2\text{O}]{\text{NaNO}_2, \text{HCl}} (CH_3)_2\overset{..}{N}-\overset{..}{N}=\overset{..}{O}:$$

Dimethylamine
N-Nitrosodimethylamine (88–90%)

> **PROBLEM 14.10** *N*-Nitroso amines are stabilized by electron delocalization. Write the two most stable resonance forms of *N*-nitrosodimethylamine, (CH₃)₂NNO.

N-Nitroso amines are more often called **nitrosamines,** and because many of them are potent carcinogens, they have been the object of much recent investigation. We encounter nitrosamines in the environment on a daily basis. A few of these, all of which are known carcinogens, are

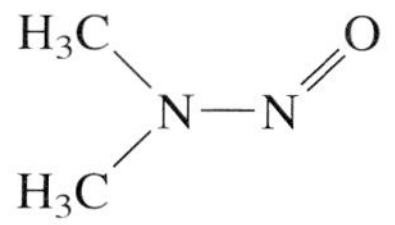

N-Nitrosodimethylamine
(formed during
tanning of leather;
also found in beer
and herbicides)

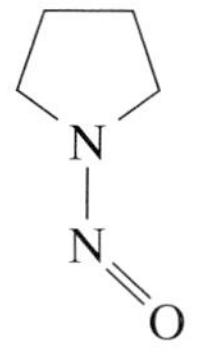

N-Nitrosopyrrolidine
(formed when bacon
that has been cured
with sodium nitrite
is fried)

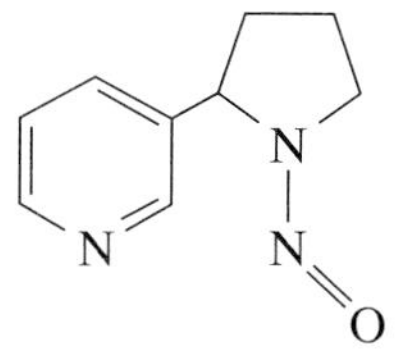

N-Nitrosonornicotine
(present in tobacco
smoke)

Nitrosamines are formed whenever nitrosating agents come in contact with secondary amines. Indeed, more nitrosamines are probably synthesized within our body than enter it by environmental contamination. Enzyme-catalyzed reduction of nitrate (NO_3^-) produces nitrite (NO_2^-), which combines with amines present in the body to form *N*-nitroso amines.

When primary amines are nitrosated, the *N*-nitroso compounds produced react further to form **diazonium ions.**

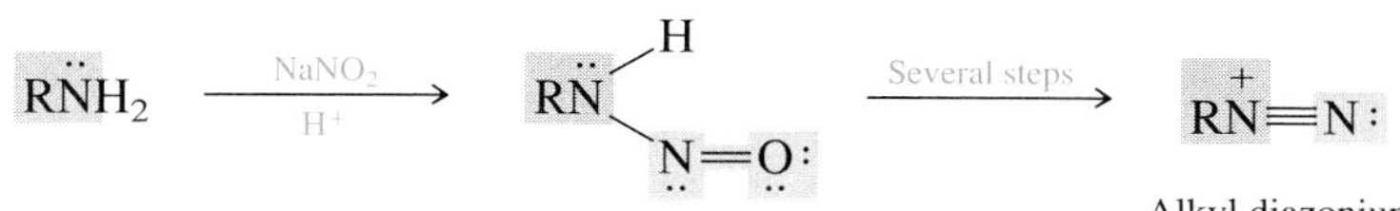

Primary
alkylamine

(Not isolable)

Alkyl diazonium
ion

Alkyl diazonium ions readily dissociate to carbocations and nitrogen (N_2) and are too unstable to be useful synthetically. Aryl diazonium ions, on the other hand, are more stable, and undergo a variety of reactions that make them versatile intermediates for the preparation of a host of ring-substituted aromatic compounds.

Aniline

Benzenediazonium chloride

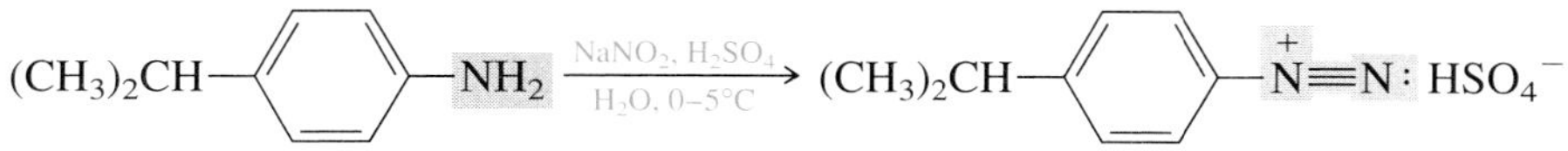

p-Isopropylaniline

p-Isopropylbenzenediazonium
hydrogen sulfate

14.10 SYNTHESIS USING ARYL DIAZONIUM SALTS

Figure 14.3 summarizes the important reactions of aryl diazonium salts that will be discussed in this section. An important method for the preparation of phenols, for example, is the hydrolysis of the corresponding aryl diazonium ion.

FIGURE 14.3 Flowchart showing the synthetic origin of aryl diazonium ions and their most useful transformations.

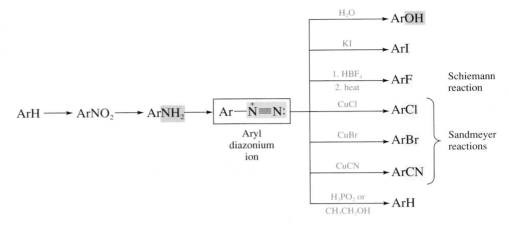

$$ArN\overset{+}{\equiv}N: \ + \ H_2O \ \longrightarrow \ ArOH \ + \ H^+ \ + \ :N\equiv N:$$

Aryl diazonium ion   Water   A phenol   Nitrogen

The aqueous acid solution used to prepare the diazonium ion is heated to give the phenol.

$(CH_3)_2CH$—⟨benzene⟩—NH_2 $\xrightarrow[\text{2. } H_2O, \text{ heat}]{\text{1. } NaNO_2, H_2SO_4, H_2O}$ $(CH_3)_2CH$—⟨benzene⟩—OH

p-Isopropylaniline   *p*-Isopropylphenol
(73%)

[**PROBLEM 14.11**   Design a synthesis of *m*-bromophenol from benzene.]

The reaction of an aryl diazonium salt with potassium iodide is the standard method for the preparation of aryl iodides. Addition of a solution of potassium iodide to the diazonium salt brings about the reaction.

$$Ar\overset{+}{\text—}N\equiv N: \ + \ I^- \ \longrightarrow \ ArI \ + \ :N\equiv N:$$

Aryl diazonium   Iodide   Aryl   Nitrogen
ion   ion   iodide

⟨benzene⟩—NH_2 with Br $\xrightarrow[\text{KI, room temperature}]{NaNO_2, HCl, H_2O, 0-5°C}$ ⟨benzene⟩—I with Br

o-Bromoaniline   *o*-Bromoiodobenzene
(72–83%)

[**PROBLEM 14.12**   Show by a series of equations how you could prepare *m*-bromoiodobenzene from benzene.]

Diazonium salt chemistry provides the principal synthetic method for the preparation of aryl fluorides through a process known as the **Schiemann reaction.** In this procedure the aryl diazonium ion is isolated as its fluoroborate salt, which then yields the desired aryl fluoride on being heated.

$$Ar\overset{+}{-}N\equiv N\colon \ \bar{B}F_4 \xrightarrow{\text{heat}} ArF \ + \ BF_3 \ + \ \colon N\equiv N\colon$$

Aryl diazonium   Aryl   Boron   Nitrogen
fluoroborate   fluoride   trifluoride

A standard way to form the aryl diazonium fluoroborate salt is to add fluoroboric acid (HBF_4) or a fluoroborate salt to the diazotization medium.

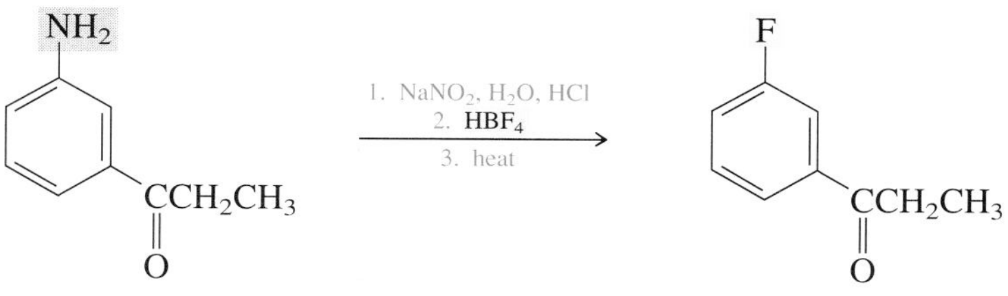

m-Aminophenyl ethyl ketone   Ethyl m-fluorophenyl ketone
(68%)

Reaction conditions:
1. $NaNO_2$, H_2O, HCl
2. HBF_4
3. heat

> **PROBLEM 14.13**   Show the proper sequence of synthetic transformations in the conversion of benzene to ethyl m-fluorophenyl ketone.

Although it is possible to prepare aryl chlorides and aryl bromides by electrophilic aromatic substitution, it is often necessary to prepare these compounds from an aromatic amine. The amine is converted to the corresponding diazonium salt and then treated with copper(I) chloride or copper(I) bromide as appropriate.

$$Ar\overset{+}{-}N\equiv N\colon \xrightarrow{\text{CuX}} ArX \ + \ \colon N\equiv N\colon$$

Aryl diazonium   Aryl chloride   Nitrogen
ion   or bromide

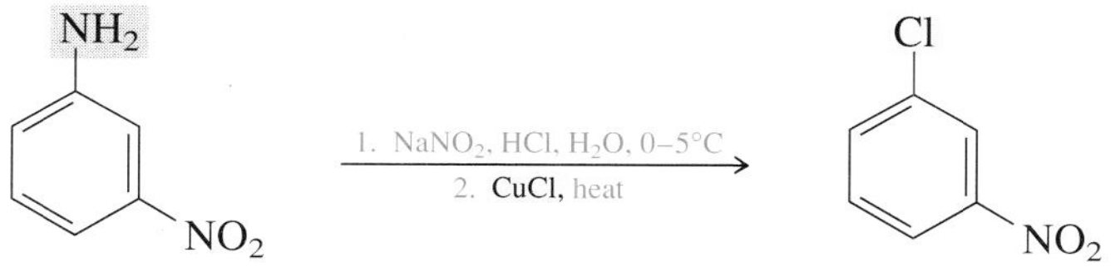

m-Nitroaniline   m-Chloronitrobenzene
(68–71%)

Reaction conditions:
1. $NaNO_2$, HCl, H_2O, 0–5°C
2. CuCl, heat

Reactions that employ copper(I) salts as reagents for replacement of nitrogen in diazonium salts are called **Sandmeyer reactions.** The Sandmeyer reaction using copper(I) cyanide is a good method for the preparation of aromatic nitriles:

$$Ar\overset{+}{-}N\equiv N\colon \xrightarrow{\text{CuCN}} ArCN \ + \ \colon N\equiv N\colon$$

Aryl diazonium   Aryl   Nitrogen
ion   nitrile

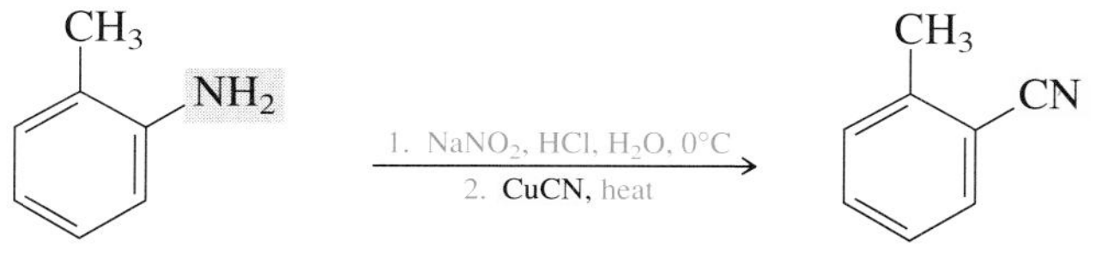

o-Toluidine   o-Methylbenzonitrile
(64–70%)

Reaction conditions:
1. $NaNO_2$, HCl, H_2O, 0°C
2. CuCN, heat

Because cyano groups may be hydrolyzed to carboxylic acids (Section 12.10), the Sandmeyer preparation of aryl nitriles is a key step in the conversion of arylamines to substituted benzoic acids.

It is possible to replace amino substituents on an aromatic nucleus by hydrogen by reducing a diazonium salt with hypophosphorous acid (H_3PO_2) or with ethanol. These reductions are free-radical reactions in which ethanol or hypophosphorous acid acts as a hydrogen atom donor:

$$Ar\overset{+}{-}N\equiv N: \xrightarrow[CH_3CH_2OH]{H_3PO_2 \text{ or}} ArH \ + \ :N\equiv N:$$

Aryl diazonium   Arene   Nitrogen
ion

Reactions of this type are called **reductive deaminations.**

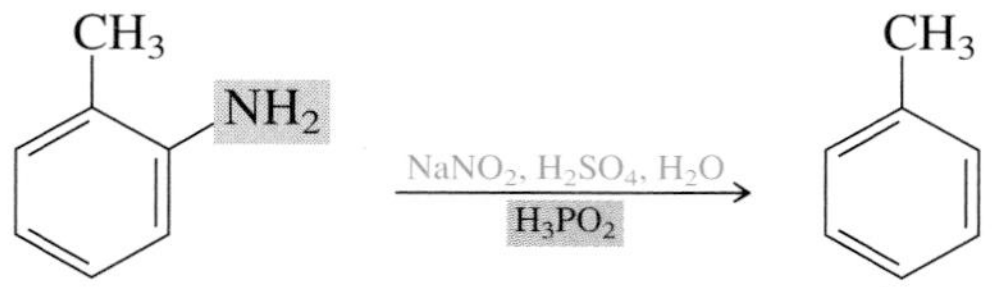

o-Toluidine   Toluene
                                                    (70–75%)

The value of diazonium salts in synthetic organic chemistry rests on two main points. Through the use of diazonium salt chemistry:

1. Substituents that are otherwise accessible only with difficulty, such as fluoro, iodo, cyano, and hydroxyl, may be introduced onto a benzene ring.

2. Compounds that have substitution patterns not directly available by electrophilic aromatic substitution can be prepared.

The first of these two features is readily apparent and is illustrated by Problems 14.11 to 14.13. If you have not done these problems yet, you are strongly encouraged to attempt them now.

The second point is somewhat less obvious but is readily illustrated by the synthesis of 1,3,5-tribromobenzene. This particular substitution pattern cannot be obtained by direct bromination of benzene, because bromine is an ortho, para director. Instead, advantage is taken of the powerful activating and ortho, para-directing effects of the amino group in aniline. Bromination of aniline yields 2,4,6-tribromoaniline in quantitative yield. Diazotization of the resulting 2,4,6-tribromoaniline and reduction of the diazonium salt gives the desired 1,3,5-tribromobenzene.

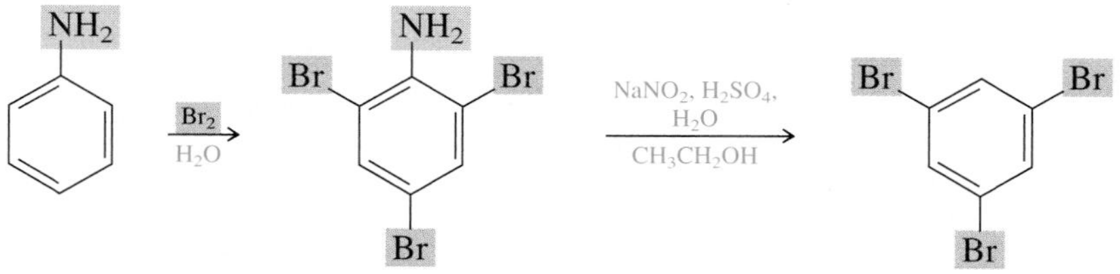

Aniline   2,4,6-Tribromoaniline   1,3,5-Tribromobenzene
                             (100%)   (74–77%)

14.11 AZO COUPLING

A reaction of aryl diazonium salts that does not involve loss of nitrogen takes place when they react with phenols and arylamines. Aryl diazonium ions are relatively weak electrophiles but have sufficient reactivity to attack strongly activated aromatic rings. The reaction is known as **azo coupling;** two aryl groups are joined together by an azo (—N=N—) function.

(ERG is a powerful electron-releasing group such as —OH or —NR$_2$)

Aryl diazonium ion

Intermediate in electrophilic aromatic substitution

Azo compound

The coupling of diazonium ions with phenols or other electron-rich aromatic compounds is a useful commercial reaction, as azo compounds are often highly colored and many of them are used as dyes. Two examples of azo dyes are shown: FD&C Orange No. 1 is used to color foods; para red is used to dye textiles.

A number of pH indicators—methyl red, for example—are azo compounds.

FD&C Orange No. 1

Para red

FROM DYES TO SULFA DRUGS

T he medicine cabinet was virtually bare of antibacterial agents until **sulfa drugs** burst on the scene in the 1930s. Before sulfa drugs became available, bacterial infection might transform a small cut or puncture wound to a life-threatening event. The story of how sulfa drugs were developed is an interesting example of being right for the wrong reasons. It was known that many bacteria absorbed dyes, and staining was a standard method for making bacteria more visible under the microscope. Might there not be some dye that is both absorbed by bacteria and toxic to them? Acting on this hypothesis, scientists at the German dyestuff manufacturer I. G. Farbenindustrie undertook a program to test the thousands of compounds in their collection for their antibacterial properties.

In general, *in vitro* testing of drugs precedes *in vivo* testing. The two terms mean, respectively, "in glass" and "in life." In vitro testing of antibiotics is carried out using bacterial cultures in test tubes or Petri dishes. Drugs that are found to be active in vitro progress to the stage of in vivo testing. In vivo testing is carried out in living organisms: laboratory animals or human volunteers. The I. G. Farben scientists found that some dyes did possess antibacterial properties, both in vitro and in vivo. Others were active in vitro but were converted to inactive substances in vivo and therefore of no use as drugs. Unexpectedly, an azo dye called Prontosil was inactive in vitro but active in vivo. In 1932, a member of the I. G. Farben research group, Gerhard Domagk, used Prontosil to treat a young child suffering from a

Continued

serious, potentially fatal staphylococcal infection. According to many accounts, the child was Domagk's own daughter; her infection was cured and her recovery was rapid and complete. Systematic testing followed and Domagk was awarded the 1939 Nobel Prize in medicine or physiology.

In spite of the rationale on which the testing of dyestuffs as antibiotics rested, subsequent research revealed that the antibacterial properties of Prontosil had nothing at all to do with its being a dye! In the body, Prontosil undergoes a reductive cleavage of its azo linkage to form **sulfanilamide,** which is the substance actually responsible for the observed biological activity. This is why Prontosil is active in vivo, but not in vitro.

Bacteria require *p*-aminobenzoic acid in order to biosynthesize **folic acid,** a growth factor. Struc-

Prontosil $\xrightarrow{\text{in vivo}}$ Sulfanilamide

turally, sulfanilamide resembles *p*-aminobenzoic acid and is mistaken for it by the bacteria. Folic acid biosynthesis is inhibited and bacterial growth is slowed sufficiently to allow the body's natural defenses to effect a cure. Because animals do not biosynthesize folic acid but obtain it in their food, sulfanilamide halts the growth of bacteria without harm to the host.

Identification of the mechanism by which Prontosil combats bacterial infections was an early triumph of **pharmacology,** a branch of science at the interface of physiology and biochemistry that studies the mechanism of drug action. By recognizing that sulfanilamide was the active agent, the task of preparing structurally modified analogs with potentially superior properties was considerably simplified. Instead of preparing Prontosil analogs, chemists synthesized sulfanilamide analogs. They did this with a vengeance; over 5000 compounds related to sulfanilamide were prepared during the period 1935–1946. Two of the most widely used sulfa drugs are sulfathiazole and sulfadiazine.

Sulfathiazole

Sulfadiazine

We tend to take the efficacy of modern drugs for granted. One comparison with the not-too-distant past might put this view into better perspective. Once sulfa drugs were introduced in the United States, the number of pneumonia deaths alone decreased by an estimated 25,000 per year.

The sulfa drugs are used less now than they were in the mid-twentieth century. Not only are more-effective, less-toxic antibiotics available, such as the penicillins and tetracyclines, but many bacteria that were once susceptible to sulfa drugs have become resistant.

LEARNING OBJECTIVES

This chapter focused on the chemical and physical properties of alkylamines and arylamines. The skills you have learned in this chapter should enable you to

- Give an acceptable IUPAC name for an alkylamine or arylamine.
- Draw the structure of a primary, secondary, or tertiary amine or a quaternary ammonium salt.
- Describe the bonding of amines by using an orbital hybridization model.

Continued

• Write a chemical equation for the reaction of an amine with an acid.
• Explain what is meant by K_b and pK_b for an amine and how they relate to the K_a and pK_a of the corresponding ammonium salt.
• Explain why arylamines are weaker bases than alkylamines.
• Write a chemical equation describing the preparation of an alkylamine by alkylation of ammonia.
• Write a chemical equation for the preparation of an alkylamine by reduction of an azo compound, nitrile, and an amide.
• Write a chemical equation describing the preparation of an arylamine from an aromatic nitro compound.
• Use a chemical equation to describe the reaction of an amine with an alkyl halide.
• Write a chemical equation describing the formation of a diazonium salt from a primary arylamine.
• Use chemical equations to describe the preparation of phenols, aryl halides, and aryl cyanides using diazonium salts.
• Write a chemical equation for the formation of an azo compound by a diazonium coupling reaction.

14.12 SUMMARY

Alkylamines are compounds of the type shown, where R, R′, and R″ are alkyl groups. One or more of these groups is an aryl group in arylamines.

Primary amine   Secondary amine   Tertiary amine

Alkylamines are named in two ways (Section 14.1). One method adds the ending *-amine* to the name of the alkyl group. The other applies the principles of substitutive nomenclature by replacing the *-e* ending of an alkane name by *-amine* and uses appropriate locants to identify the position of the amino group. Arylamines are named as derivatives of aniline.

Nitrogen's unshared electron pair is of major importance in understanding the structure and properties of amines (Section 14.2). Alkylamines have a pyramidal arrangement of bonds to nitrogen, and the unshared electron pair resides in an sp^3-hybridized orbital. The geometry at nitrogen in arylamines is somewhat flatter than in alkylamines, and the unshared electron pair is delocalized into the π system of the ring. Delocalization binds the electron pair more strongly in arylamines than in alkylamines. Arylamines are less basic and less nucleophilic than alkylamines.

Amines are less polar than alcohols (Section 14.3). Hydrogen bonding in amines is weaker than in alcohols because nitrogen is less electronegative than oxygen. Amines have lower boiling points than alcohols, but higher boiling points than alkanes. Primary amines have higher boiling points than isomeric secondary amines; tertiary amines, which cannot form intermolecular hydrogen bonds, have the lowest boiling points. Amines resemble alcohols in their solubility in water.

Basicity of amines is expressed either as a basicity constant K_b (pK_b) of the amine or as a dissociation constant K_a (pK_a) of its conjugate acid (Section 14.4).

$$R_3N\text{:} + H_2O \rightleftharpoons R_3\overset{+}{N}H + HO^- \qquad K_b = \frac{[R_3\overset{+}{N}H][HO^-]}{[R_3N]}$$

The basicity constants of alkylamines lie in the range 10^{-3}–10^{-5}. Arylamines are much weaker bases, with K_b values in the 10^{-9}–10^{-11} range.

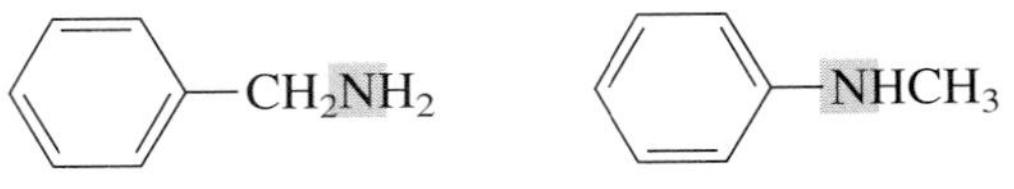

Benzylamine
(alkylamine: $pK_b = 4.7$)

N-Methylaniline
(arylamine: $pK_b = 11.8$)

Methods for the preparation of amines are summarized in Table 14.3 (Sections 14.5 and 14.6).

Amines are alkylated by primary alkyl halides (Section 14.8). The product may be a secondary or tertiary amine or quaternary ammonium salt.

$$RNH_2 \xrightarrow{R'CH_2X} RNHCH_2R' \xrightarrow{R'CH_2X} RN(CH_2R')_2 \xrightarrow{R'CH_2X} R\overset{+}{N}(CH_2R')_3\, X^-$$

| Primary amine | Secondary amine | Tertiary amine | Quaternary ammonium salt |

Nitrosation of amines occurs on acidification of a solution containing sodium nitrite in the presence of an amine (Section 14.9). Secondary amines react with nitrosating agents to give N-nitroso amines.

$$R_2NH \xrightarrow[H_2O]{NaNO_2,\ H^+} R_2N\text{—}N\text{=}O$$

Secondary amine   N-Nitroso amine

Primary arylamines yield aryl diazonium salts.

$$ArNH_2 \xrightarrow{NaNO_2,\ HCl} Ar\overset{+}{N}\text{≡}N\text{:}$$

Primary arylamine   Aryl diazonium ion

Aryl diazonium salts are useful synthetic intermediates (Section 14.10). Once formed, the diazonium salt may be treated with an appropriate reagent to give a phenol, aryl halide, or aryl cyanide (see Figure 14.3). Those reactions using copper(I) salts are called **Sandmeyer reactions.** The amino substituent of an arylamine can be replaced by hydrogen by treatment of its diazonium salt with ethanol or with hypophosphorous acid.

Aryl diazonium salts also react with strongly activated aromatic rings to form azo compounds, many of which are highly colored and are used as dyes (Section 14.11).

$$Ar\overset{+}{N}\text{≡}N\text{:} + Ar'H \longrightarrow ArN\text{=}NAr' + H^+$$

Aryl diazonium ion   Arylamine or phenol   Azo compound

TABLE14.3	Preparation of Amines

Reaction (section) and Comments	General Equation and Specific Example

Alkylation methods

Alkylation of ammonia (Section 14.5)
Ammonia can act as a nucleophile toward primary and some secondary alkyl halides to give primary alkylamines. Yields tend to be modest because the primary amine is itself a nucleophile and undergoes alkylation. Alkylation of ammonia can lead to a mixture containing a primary amine, a secondary amine, a tertiary amine, and a quaternary ammonium salt.

$$RX + 2NH_3 \longrightarrow RNH_2 + NH_4X$$

Alkyl halide   Ammonia   Alkylamine   Ammonium halide

$$C_6H_5CH_2Cl \xrightarrow{NH_3\ (8\ mol)} C_6H_5CH_2NH_2 + (C_6H_5CH_2)_2NH$$

Benzyl chloride (1 mol)   Benzylamine (53%)   Dibenzylamine (39%)

Reduction methods

Reduction of alkyl azides (Section 14.6)
Alkyl azides, prepared by nucleophilic substitution by azide ion in primary or secondary alkyl halides, are reduced to primary alkylamines by lithium aluminum hydride or by catalytic hydrogenation.

$$R\ddot{N}=\overset{+}{N}=\ddot{\underset{..}{N}}:^{-} \xrightarrow{reduce} R\ddot{N}H_2$$

Alkyl azide   Primary amine

$$CF_3CH_2CHCO_2CH_2CH_3 \xrightarrow{H_2,\ Pd} CF_3CH_2CHCO_2CH_2CH_3$$
$$\underset{N_3}{|}\qquad\qquad\qquad\qquad \underset{NH_2}{|}$$

Ethyl 2-azido-4,4,4-trifluorobutanoate   Ethyl 2-amino-4,4,4-trifluorobutanoate (96%)

Reduction of nitriles (Section 14.6)
Nitriles are reduced to primary amines by lithium aluminum hydride or by catalytic hydrogenation.

$$RC\equiv N \xrightarrow{reduce} RCH_2NH_2$$

Nitrile   Primary amine

$$\triangleright\!\!-CN \xrightarrow[2.\ H_2O]{1.\ LiAlH_4} \triangleright\!\!-CH_2NH_2$$

Cyclopropyl cyanide   Cyclopropylmethanamine (75%)

Reduction of aryl nitro compounds (Section 14.6) The standard method for the preparation of an arylamine is by nitration of an aromatic ring, followed by reduction of the nitro group. Typical reducing agents include iron or tin in hydrochloric acid or catalytic hydrogenation.

$$ArNO_2 \xrightarrow{reduce} ArNH_2$$

Nitroarene   Arylamine

$$C_6H_5NO_2 \xrightarrow[2.\ HO^-]{1.\ Fe,\ HCl} C_6H_5NH_2$$

Nitrobenzene   Aniline (97%)

Reduction of amides (Section 14.6)
Lithium aluminum hydride reduces the carbonyl group of an amide to a methylene group. Primary, secondary, or tertiary amines may be prepared by proper choice of the starting amide. R and R′ may be either alkyl or aryl.

$$\overset{O}{\overset{\|}{RCNR'_2}} \xrightarrow{reduce} RCH_2NR'_2$$

Amide   Amine

$$\overset{O}{\overset{\|}{CH_3CNHC(CH_3)_3}} \xrightarrow[2.\ H_2O]{1.\ LiAlH_4} CH_3CH_2NHC(CH_3)_3$$

N-tert-Butylacetamide   N-Ethyl-tert-butylamine (60%)

ADDITIONAL PROBLEMS

Structure, Nomenclature, and Properties

14.14 Write structural formulas for all the amines of molecular formula $C_4H_{11}N$. Give an acceptable name for each one, and classify it as a primary, secondary, or tertiary amine.

14.15 Provide a structural formula for each of the following compounds:

(a) Heptylamine
(b) 2-Ethyl-1-butanamine
(c) N-Ethylpentylamine
(d) Dibenzylamine
(e) Tetraethylammonium hydroxide
(f) N-Ethyl-4-methylaniline
(g) 2,4-Dichloroaniline
(h) N,N-Dimethylaniline

14.16 (a) Give the structures and an acceptable name for all the isomers of molecular formula C_7H_9N that contain a benzene ring.

(b) Which one of these isomers is the strongest base?

14.17 Name each of the following amines, indicating whether the amine is primary, secondary, or tertiary.

(a) $CH_3CH_2CH_2NHCH_2CH_2CH_3$
(b) $CH_3CH_2CH_2NCH_2CH_2CH_2CH_3$
$\qquad\qquad\quad |$
$\qquad\qquad\ \ CH_3$
(c) $(CH_3)_2CHCH_2CH_2NH_2$
(d) $(CH_3CH_2CH_2)_2NC_6H_5$
(e) $C_6H_5CH_2NHC_6H_5$

14.18 Many naturally occurring nitrogen compounds and many nitrogen-containing drugs are better known by their common names than by their systematic names. A few of these are given here. Write a structural formula for each one.

(a) 4-Aminobutanoic acid, better known as γ-aminobutyric acid or GABA, involved in metabolic processes occurring in the brain

(b) 2-(3,4,5-Trimethoxyphenyl)ethylamine, better known as mescaline, a hallucinogen obtained from the peyote cactus

(c) trans-2-Phenylcyclopropylamine, better known as tranylcypromine, an antidepressant drug

(d) 1-Phenyl-2-propanamine, better known as amphetamine, a stimulant

14.19 Arrange the following compounds in order of decreasing basicity.

$$C_6H_5NH_2 \qquad C_6H_5CH_2NH_2 \qquad \overset{\displaystyle O}{\overset{\displaystyle \|}{C_6H_5NHCCH_3}}$$

Aniline $\qquad$ Benzylamine $\qquad$ Acetanilide

Reactions

14.20 Give the structure of the expected product formed when benzylamine reacts with each of the following reagents:

(a) Hydrogen bromide
(b) Acetic acid
(c) Acetyl chloride
(d) Acetone
(e) Excess methyl iodide

14.21 Repeat Problem 14.20 for aniline.

14.22 Write the structure of the product formed on reaction of N-phenylpropanamide with each of the following:

(a) $LiAlH_4$, then H_2O
(b) HNO_3, H_2SO_4
(c) $(CH_3)_3CCl$, $AlCl_3$
(d) NaOH, H_2O, heat

14.23 Write the structure of the product formed from each of the following:

(a) *p*-Methylaniline and sodium nitrite, aqueous sulfuric acid, 0 to 5°C
(b) Product of part (a), heated in aqueous acid
(c) Product of part (a), treated with copper(I) chloride
(d) Product of part (a), treated with potassium iodide
(e) Product of part (a), treated with copper(I) bromide
(f) Product of part (a), treated with copper(I) cyanide
(g) Product of part (a), treated with hypophosphorous acid (H_3PO_2)
(h) Product of part (a), treated with fluoroboric acid, then heated
(i) Product of part (a), treated with phenol (product is $C_{13}H_{12}N_2O$)

14.24 Identify the principal organic products of each of the following reactions:

(a) 1,2-Diethyl-4-nitrobenzene $\xrightarrow[\text{ethanol}]{H_2, Pt}$

(b) $C_6H_5NHCCH_2CH_2CH_3$ (with O double-bonded to C) $\xrightarrow[\text{2. } H_2O]{\text{1. LiAlH}_4}$

(c) $(CH_3)_2CHNHCH(CH_3)_2$ $\xrightarrow[\text{HCl, } H_2O]{\text{NaNO}_2}$

(d) 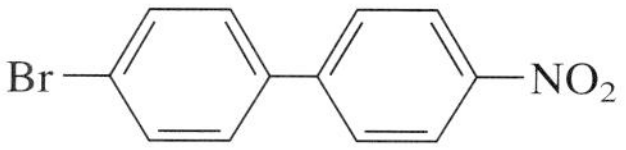 Br—⟨benzene⟩—⟨benzene⟩—NO_2 $\xrightarrow[\text{2. HO}^-]{\text{1. Fe, HCl}}$

(e) Product of part (d) $\xrightarrow[\text{2. } H_2O, \text{ heat}]{\text{1. NaNO}_2, H_2SO_4, H_2O}$

(f) 2,6-Dinitroaniline $\xrightarrow[\text{2. CuCl}]{\text{1. NaNO}_2, H_2SO_4, H_2O}$

(g) *m*-Bromoaniline $\xrightarrow[\text{2. CuBr}]{\text{1. NaNO}_2, HBr, H_2O}$

(h) *o*-Nitroaniline $\xrightarrow[\text{2. CuCN}]{\text{1. NaNO}_2, HCl, H_2O}$

(i) 2,6-Diiodo-4-nitroaniline $\xrightarrow[\text{2. KI}]{\text{1. NaNO}_2, H_2SO_4, H_2O}$

(j) Aniline $\xrightarrow[\text{2. 2,3,6-trimethylphenol}]{\text{1. NaNO}_2, H_2SO_4, H_2O}$

Synthesis

14.25 Outline a sequence of reactions that will convert

(a) Benzoic acid into *N*-methylbenzylamine
(b) Benzyl bromide ($C_6H_5CH_2Br$) into 2-phenylethylamine

14.26 Outline syntheses of each of the following aromatic compounds from benzene:

(a) *o*-Ethylaniline   (c) *m*-Chloroiodobenzene
(b) *p*-Bromoaniline   (d) *p*-Bromophenol

14.27 Outline a sequence of steps that would convert *p*-nitrotoluene into

(a) *p*-Methylaniline   (d) *p*-Iodotoluene
(b) *p*-Chlorotoluene   (e) *p*-Methylphenol
(c) *p*-Cyanotoluene   (f) 3,5-Dibromotoluene

14.28 Describe procedures for preparing each of the following compounds, using ethanol as the source of all their carbon atoms. Once you prepare a compound, you need not repeat its synthesis in a subsequent part of this problem.

(a) Ethylamine

(b) *N*-Ethylacetamide

(c) Diethylamine

(d) *N,N*-Diethylacetamide

(e) Triethylamine

(f) Tetraethylammonium bromide

14.29 Show by writing the appropriate sequence of equations how you could carry out each of the following transformations:

(a) 1-Butanol to 1-pentanamine

(b) *tert*-Butyl chloride to 2,2-dimethyl-1-propanamine

14.30 The following compounds have been prepared from *p*-nitroaniline. Outline a reasonable series of steps leading to each one.

(a) *p*-Nitrobenzonitrile (p-$O_2NC_6H_4CN$)

(b) *p*-Acetamidophenol

14.31 The following compounds have been prepared from *o*-anisidine (*o*-methoxyaniline). Outline a series of steps leading to each one.

(a) *o*-Bromoanisole

(b) *o*-Fluoroanisole

Miscellaneous Problems

14.32 *Physostigmine,* an alkaloid obtained from a West African plant, is used in the treatment of glaucoma. Treatment of physostigmine with methyl iodide gives a quaternary ammonium salt. What is the structure of this salt?

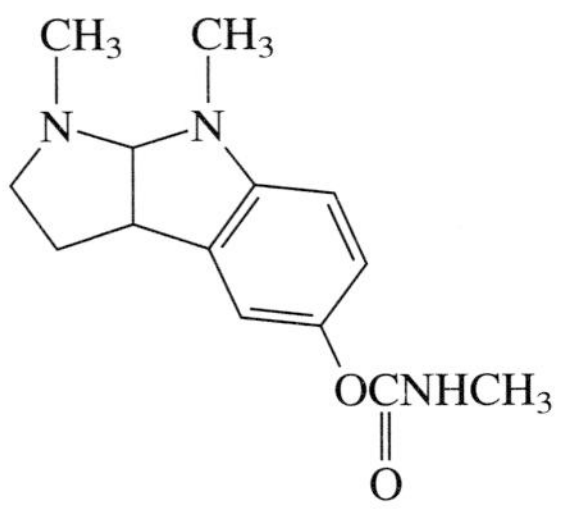

Physostigmine

14.33 Provide a reasonable explanation for each of the following observations:

(a) 4-Methylpiperidine has a higher boiling point than *N*-methylpiperidine.

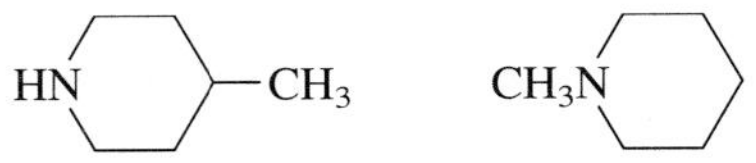

4-Methylpiperidine
(bp 129°C)

N-Methylpiperidine
(bp 106°C)

(b) Two isomeric quaternary ammonium salts are formed in comparable amounts when 4-*tert*-butyl-*N*-methylpiperidine is treated with benzyl chloride. (*Hint:* Building a molecular model will help.)

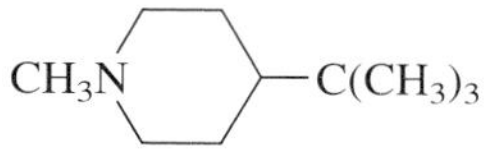

$$CH_3N \diagdown \hexagon \diagup C(CH_3)_3$$

4-*tert*-Butyl-*N*-methylpiperidine

14.34 Several amine-containing compounds are used medicinally as local anesthetics. The structure of one of these, lidocaine (also known as Xylocaine), is shown. To increase the water solubility of these compounds when given by injection, they are used as the hydrochloride salt. Draw the structure of lidocaine hydrochloride.

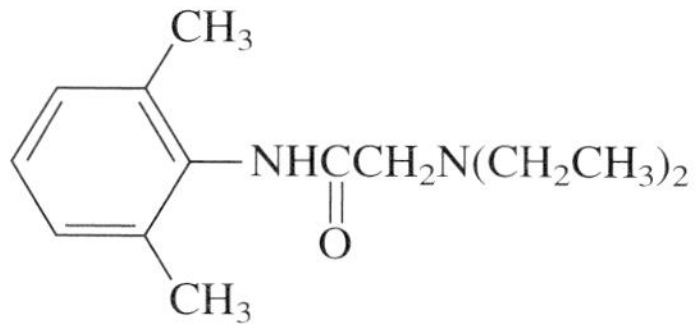

$$\begin{array}{c} CH_3 \\ \bigcirc \!\!-\! NHCCH_2N(CH_2CH_3)_2 \\ \underset{O}{\overset{\|}{}} \\ CH_3 \end{array}$$

Lidocaine

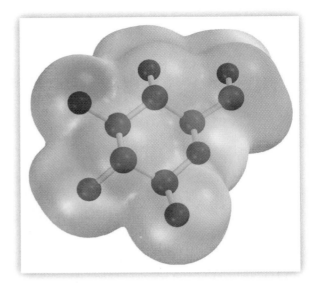

CHAPTER 15
CARBOHYDRATES

The major classes of organic compounds common to living systems are lipids, proteins, nucleic acids, and carbohydrates. Carbohydrates are very familiar to us—we call many of them "sugars." They make up a substantial portion of the food we eat and provide most of the energy that keeps the human engine running. Carbohydrates are structural components of the walls of plant cells and the wood of trees. Genetic information is stored and transferred by way of nucleic acids, specialized derivatives of carbohydrates, which we'll examine in more detail in Chapter 18.

Historically, carbohydrates were once considered to be "hydrates of carbon" because their molecular formulas in many (but not all) cases correspond to $C_n(H_2O)_m$. It is more realistic to define a carbohydrate as a polyhydroxy aldehyde or polyhydroxy ketone, a point of view closer to structural reality and more suggestive of chemical reactivity.

15.1 CLASSIFICATION OF CARBOHYDRATES

"Sugar" is a combination of the Sanskrit words *su* (sweet) and *gar* (sand). Thus, its literal meaning is "sweet sand."

The Latin word for "sugar" is *saccharum,* and the derived term "saccharide" is the basis of a system of carbohydrate classification. A **monosaccharide** is a simple carbohydrate, one that on attempted hydrolysis is not cleaved to smaller carbohydrates. Glucose ($C_6H_{12}O_6$), for example, is a monosaccharide. A **disaccharide** is cleaved on hydrolysis to two monosaccharides, which may be the same or different. Sucrose—common table sugar—is a disaccharide that yields one molecule of glucose and one of fructose on hydrolysis.

Sucrose ($C_{12}H_{22}O_{11}$) + H_2O $\longrightarrow$ glucose ($C_6H_{12}O_6$) + fructose ($C_6H_{12}O_6$)

TABLE 15.1	Some Classes of Monosaccharides	
Number of Carbon Atoms	**Aldose**	**Ketose**
Four	Aldotetrose	Ketotetrose
Five	Aldopentose	Ketopentose
Six	Aldohexose	Ketohexose
Seven	Aldoheptose	Ketoheptose
Eight	Aldooctose	Ketooctose

An **oligosaccharide** (*oligos* is a Greek word that in its plural form means "few") yields 3–10 monosaccharide units on hydrolysis. **Polysaccharides** are hydrolyzed to more than 10 monosaccharide units. Cellulose is a polysaccharide molecule that gives thousands of glucose molecules when completely hydrolyzed.

Over 200 different monosaccharides are known. They can be grouped according to the number of carbon atoms they contain and whether they are polyhydroxy aldehydes or polyhydroxy ketones. Monosaccharides that are polyhydroxy aldehydes are called **aldoses;** those that are polyhydroxy ketones are **ketoses.** Aldoses and ketoses are further classified according to the number of carbon atoms in the main chain. Table 15.1 lists the terms applied to monosaccharides having four to eight carbon atoms.

15.2   FISCHER PROJECTIONS AND D–L Notation

Stereochemistry is the key to understanding carbohydrate structure, a fact that was clearly appreciated by the German chemist Emil Fischer. The projection formulas used by Fischer to represent stereochemistry in chiral molecules are particularly well suited to studying carbohydrates. Figure 15.1 illustrates their application to the enantiomers of glyceraldehyde (2,3-dihydroxypropanal), a fundamental molecule in carbohydrate stereochemistry. When the Fischer projection is oriented as shown in the figure, with the carbon chain vertical and the aldehyde carbon at the top, the C-2 hydroxyl group points to the right in (+)-glyceraldehyde and to the left in (−)-glyceraldehyde.

Fischer determined the structure of glucose in 1900 and won the Nobel Prize in chemistry in 1902.

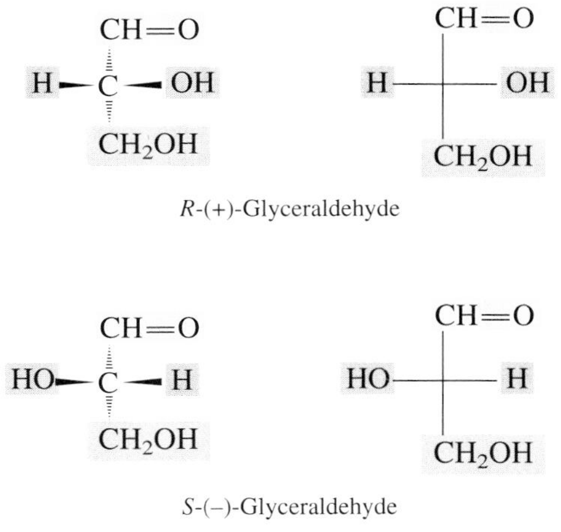

R-(+)-Glyceraldehyde

S-(−)-Glyceraldehyde

FIGURE 15.1 Three-dimensional representations and Fischer projections of the enantiomers of glyceraldehyde.

Techniques for determining the absolute configuration of chiral molecules were not developed until the 1950s, and so it was not possible for Fischer and his contemporaries to relate the sign of rotation of any substance to its absolute configuration. A system evolved based on the arbitrary assumption, later shown to be correct, that the enantiomers of glyceraldehyde have the signs of rotation and absolute configurations shown in Figure 15.1. Two stereochemical descriptors were defined: D and L. The absolute configuration of (+)-glyceraldehyde, as depicted in the figure, was said to be D and that of its enantiomer, (−)-glyceraldehyde, L. Compounds that had a spatial arrangement of substituents analogous to D-(+)- and L-(−)-glyceraldehyde were said to have the D and L configurations, respectively.

> Adopting the enantiomers of glyceraldehyde as stereochemical reference compounds originated with proposals made in 1906 by M. A. Rosanoff, a chemist at New York University.

PROBLEM 15.1   Identify each of the following as either D- or L-glyceraldehyde:

(a) 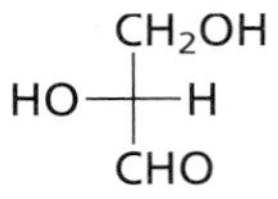   CH₂OH   (c) 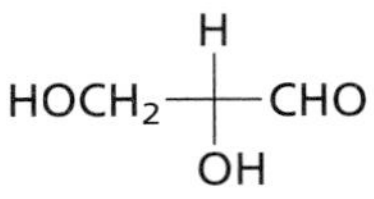   H   (b) 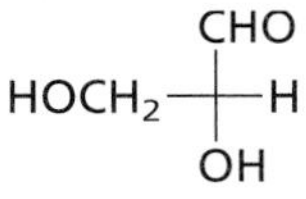   CHO

     HO———H   HOCH₂———CHO   HOCH₂———H

            CHO   OH   OH

Sample Solution   (a) Redraw the Fischer projection so as to more clearly show the true spatial orientation of the groups. Next, reorient the molecule so that its relationship to the glyceraldehyde enantiomers in Figure 15.1 is apparent.

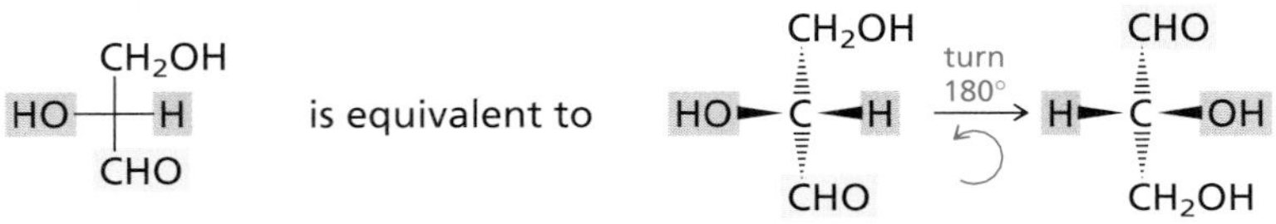

The structure is the same as that of (+)-glyceraldehyde in the figure. It is D-glyceraldehyde.

Fischer projections and D–L notation have proved to be so helpful in representing carbohydrate stereochemistry that the chemical and biochemical literature is replete with their use. To read that literature you need to be acquainted with these devices, as well as the more modern Cahn–Ingold–Prelog system.

15.3   ALDOTETROSES

Glyceraldehyde can be considered the simplest chiral carbohydrate. It is an **aldotriose** and because it contains one stereogenic center, it exists in two stereoisomeric forms: the D and L enantiomers. Moving up the scale in complexity, next come the **aldotetroses.** Examination of their structures illustrates the application of the Fischer system to compounds that contain more than one stereogenic center.

The aldotetroses are the four stereoisomers of 2,3,4-trihydroxybutanal. Fischer projections are constructed by orienting the molecule in an eclipsed conformation with the aldehyde group at what will be the top. The four carbon atoms define the main chain of the Fischer projection and are arranged vertically. Horizontal bonds are directed outward, vertical bonds back.

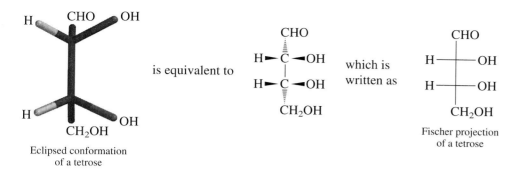

Eclipsed conformation
of a tetrose

is equivalent to

which is
written as

Fischer projection
of a tetrose

The particular aldotetrose just shown is called D-erythrose. The prefix D tells us that the configuration at the highest numbered stereogenic center is analogous to that of D-(+)-glyceraldehyde. Its mirror image is L-erythrose.

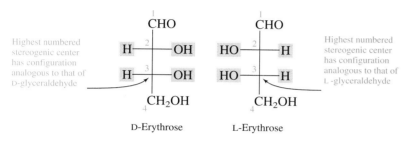

Highest numbered
stereogenic center
has configuration
analogous to that of
D-glyceraldehyde

Highest numbered
stereogenic center
has configuration
analogous to that of
L-glyceraldehyde

D-Erythrose   L-Erythrose

Relative to each other, both hydroxyl groups are on the same side in Fischer projections of the erythrose enantiomers. The remaining two stereoisomers have hydroxyl groups on opposite sides in their Fischer projections. They are diastereomers of D- and L-erythrose and are called D- and L-threose. The D and L prefixes again specify the configuration of the highest numbered stereogenic center. D-Threose and L-threose are enantiomers of each other:

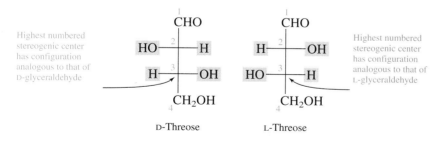

Highest numbered
stereogenic center
has configuration
analogous to that of
D-glyceraldehyde

Highest numbered
stereogenic center
has configuration
analogous to that of
L-glyceraldehyde

D-Threose   L-Threose

PROBLEM 15.2 Which aldotetrose is the structure shown? Is it D-erythrose, D-threose, L-erythrose, or L-threose? (Be careful! The conformation given is not the same as that used to generate a Fischer projection.)

As shown for the aldotetroses, an aldose belongs to the D or the L series according to the configuration of the stereogenic center farthest removed from the aldehyde function. Individual names, such as erythrose and threose, specify the particular arrangement of stereogenic centers within the molecule relative to each other. Optical activities cannot be determined directly from the D and L prefixes. As it turns out, both D-erythrose and D-threose are levorotatory, but D-glyceraldehyde is dextrorotatory.

15.4   ALDOPENTOSES AND ALDOHEXOSES

Aldopentoses have three stereogenic centers. The eight stereoisomers are divided into a set of four D-aldopentoses and an enantiomeric set of four L-aldopentoses. The aldopentoses are named ribose, arabinose, xylose, and lyxose. Fischer projections of the D stereoisomers of the aldopentoses are given in Figure 15.2. Notice that all these diastereomers have the same configuration at C-4 and that this configuration is analogous to that of D-(+)-glyceraldehyde.

> **PROBLEM 15.3** L-(+)-Arabinose is a naturally occurring L sugar. It is obtained by acid hydrolysis of the polysaccharide present in mesquite gum. Write a Fischer projection for L-(+)-arabinose.

Among the aldopentoses, D-ribose is a component of many biologically important substances, most notably the ribonucleic acids. D-Xylose is also very abundant and is isolated by hydrolysis of the polysaccharides present in corncobs and the wood of trees.

The aldohexoses include some of the most familiar of the monosaccharides, as well as one of the most abundant organic compounds on earth, D-(+)-glucose. With four stereogenic centers, 16 stereoisomeric aldohexoses are possible; 8 belong to the D series and 8 to the L series. All are known, either as naturally occurring substances or as the products of synthesis. The eight D-aldohexoses are given in Figure 15.2; it is the spatial arrangement at C-5, hydrogen to the left in a Fischer projection and hydroxyl to the right, that identifies them as carbohydrates of the D series.

Cellulose is more abundant than glucose, but each cellulose molecule is a polysaccharide composed of thousands of glucose units (Section 15.12). Methane may also be more abundant, but most of the methane comes from glucose.

> **PROBLEM 15.4**   Name the following sugar:
>
> ```
>                    CHO
>              H ——————— OH
>              H ——————— OH
>              H ——————— OH
>             HO ——————— H
>                   CH2OH
> ```

Of all the monosaccharides, D-(+)-glucose is the best known, most important, and most abundant. Its formation from carbon dioxide, water, and sunlight is the central theme of photosynthesis. Carbohydrate formation by photosynthesis is estimated to be on the order of 10^{11} tons per year, a source of stored energy used, directly or indirectly, by all higher forms of life on the planet. Glucose was isolated from raisins in 1747 and by hydrolysis of starch in 1811. Its structure was determined, in work culminating in 1900, by Emil Fischer.

D-(+)-Galactose is a constituent of numerous polysaccharides. It is best obtained by acid hydrolysis of lactose (milk sugar), a disaccharide of D-glucose and D-galactose. L-(−)-Galactose also occurs naturally and can be prepared by hydrolysis of flaxseed gum

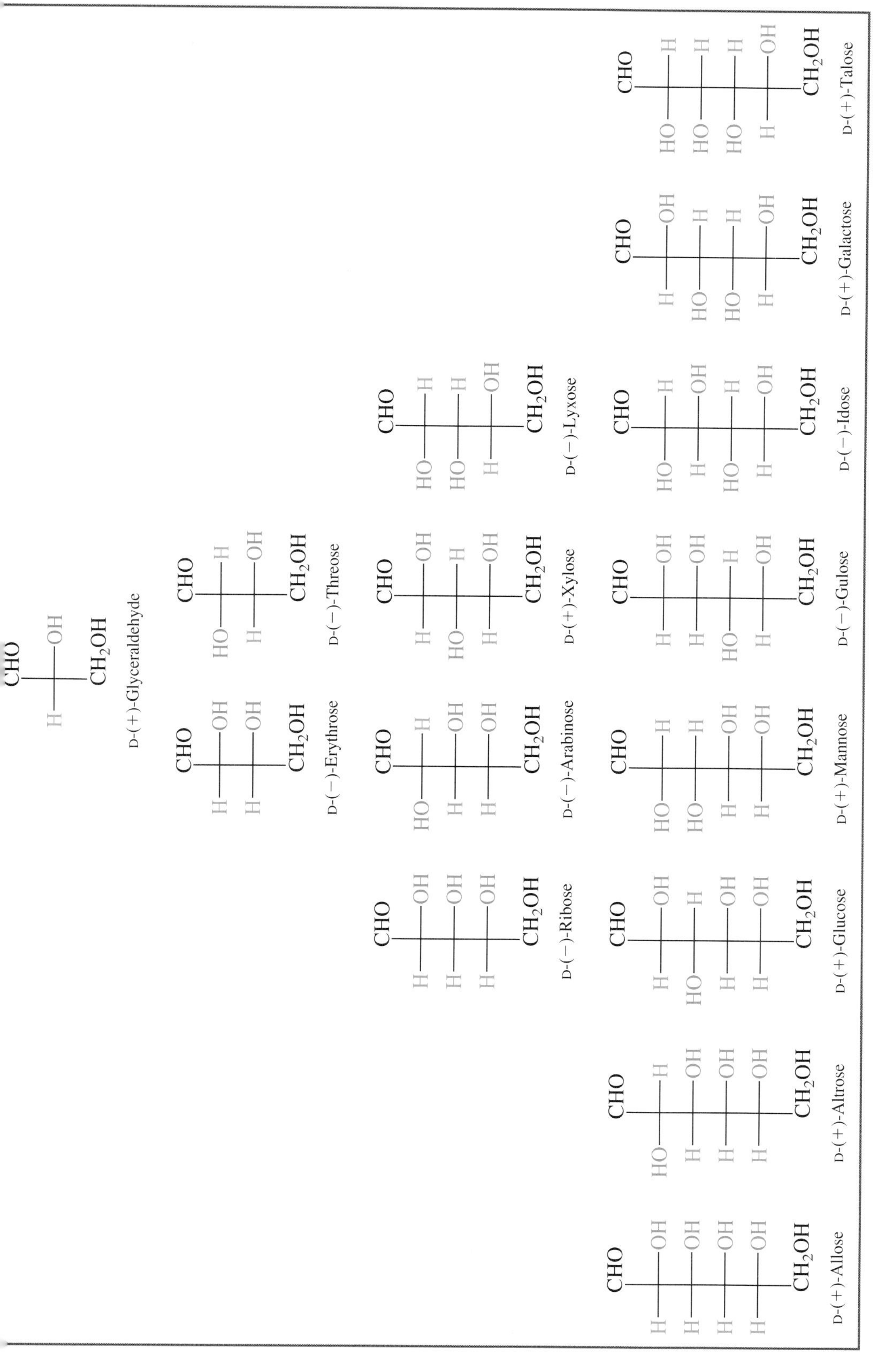

FIGURE 15.2 Configurations of the D series of aldoses containing three through six carbon atoms.

and agar. The principal source of D-(+)-mannose is hydrolysis of the polysaccharide of the ivory nut, a large, nut-like seed obtained from a South American palm.

15.5   CYCLIC FORMS OF CARBOHYDRATES: FURANOSE FORMS

Aldoses incorporate two functional groups, C=O and OH, which are capable of reacting with each other. We saw in Section 11.8 that nucleophilic addition of an alcohol function to a carbonyl group gives a hemiacetal. When the hydroxyl and carbonyl groups are part of the same molecule, a **cyclic hemiacetal** results, as illustrated in Figure 15.3.

Cyclic hemiacetal formation is most common when the ring that results is five- or six-membered. Five-membered cyclic hemiacetals of carbohydrates are called **furanose** forms; six-membered ones are called **pyranose** forms. The ring carbon that is derived from the carbonyl group, the one that bears two oxygen substituents, is called the **anomeric carbon.**

Aldoses exist almost exclusively as their cyclic hemiacetals; very little of the open-chain form is present at equilibrium. To understand their structures and chemical reactions, we need to be able to translate Fischer projections of carbohydrates into their cyclic hemiacetal forms.

Consider, for example, the furanose form of D-ribose, the carbohydrate that forms the backbone of ribonucleic acids (Chapter 18). Hemiacetal formation occurs between the carbonyl group and the hydroxyl at C-4. Furanose ring formation can be visualized by redrawing the Fischer projection in a form more suited to cyclization, being careful to maintain the stereochemistry at each stereogenic center.

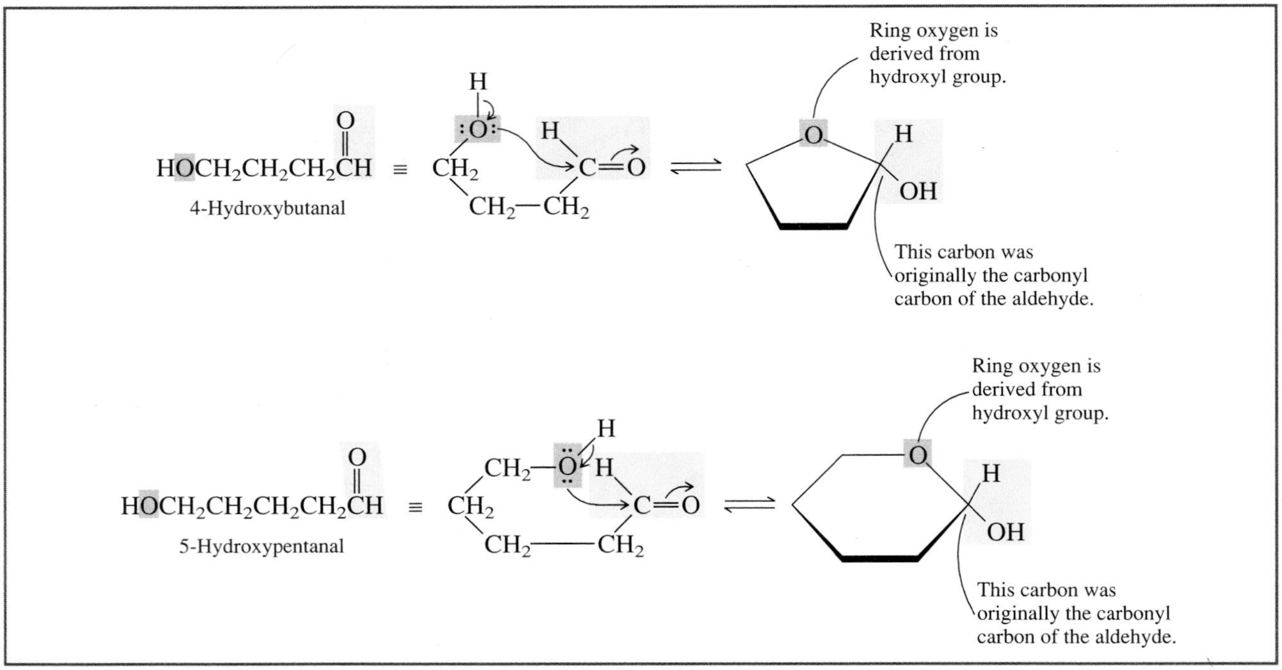

FIGURE 15.3   Cyclic hemiacetal formation in 4-hydroxybutanal and 5-hydroxypentanal.

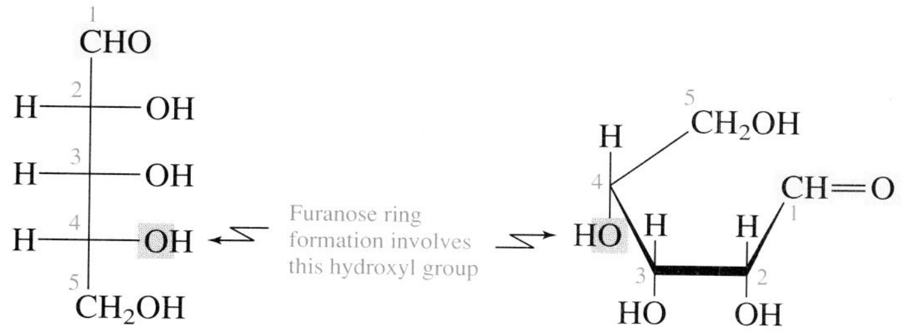

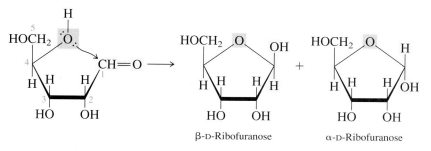

D-Ribose   Eclipsed conformation of D-ribose

Notice that the eclipsed conformation of D-ribose derived directly from the Fischer projection does not have its C-4 hydroxyl group properly oriented for furanose ring formation. We must redraw it in a conformation that permits the five-membered cyclic hemiacetal to form. This is accomplished by rotation about the C(3)–C(4) bond, taking care that the configuration at C-4 is not changed.

Conformation of D-ribose
suitable for furanose ring
formation

> Try using a molecular model to see this.

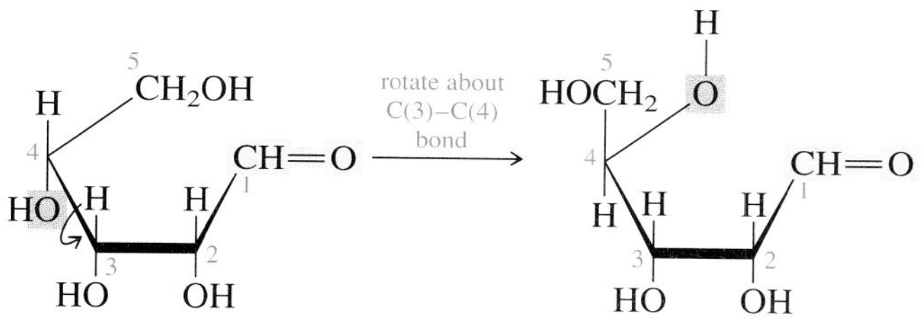

As viewed in the drawing, a 120° counterclockwise rotation of C-4 places its hydroxyl group in the proper position. At the same time, this rotation moves the CH$_2$OH group to a position such that it will become a substituent that is "up" on the five-membered ring. The hydrogen at C-4 then will be "down" in the furanose form.

β-D-Ribofuranose   α-D-Ribofuranose

> Structural drawings of carbohydrates of this type are called **Haworth formulas** after the British carbohydrate chemist Sir Walter Norman Haworth. Haworth was a corecipient of the 1937 Nobel Prize in chemistry.

PROBLEM 15.5 Write Haworth formulas corresponding to the furanose forms of each of the following carbohydrates:

(a) D-Xylose

(b) D-Arabinose

(c) L-Arabinose

(d) D-Threose

Sample Solution (a) The Fischer projection of D-xylose is given in Figure 15.2.

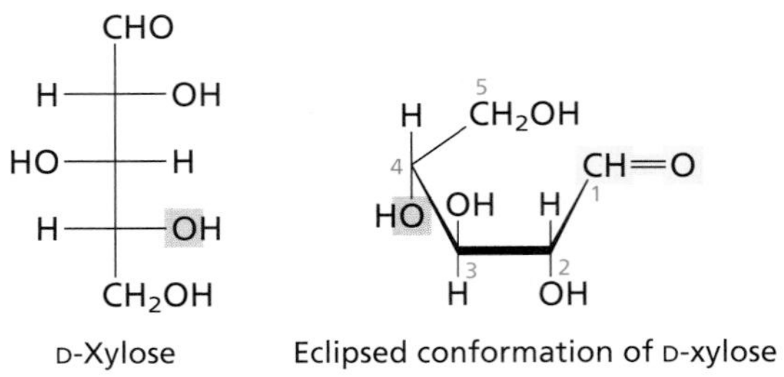

CHO
H——OH
HO——H
H——OH
CH₂OH

D-Xylose

Eclipsed conformation of D-xylose

Carbon-4 of D-xylose must be rotated in a counterclockwise sense to bring its hydroxyl group into the proper orientation to form a furanose ring.

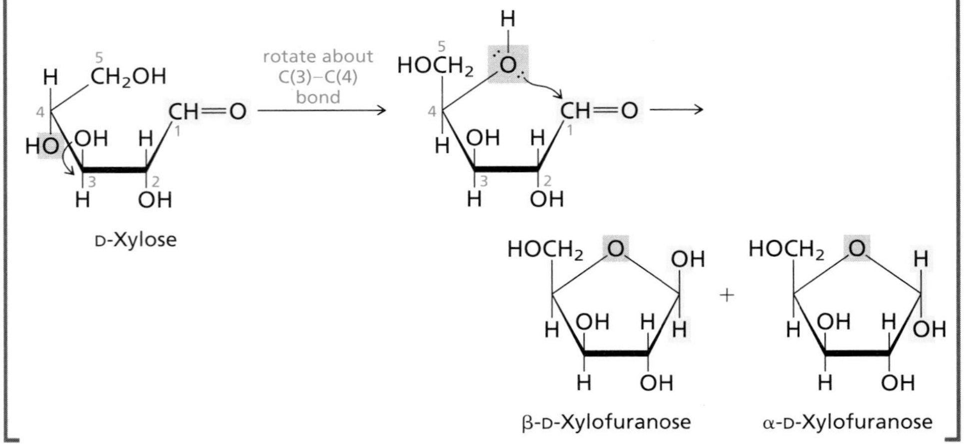

D-Xylose

rotate about C(3)—C(4) bond →

→

β-D-Xylofuranose   +   α-D-Xylofuranose

15.6   CYCLIC FORMS OF CARBOHYDRATES: PYRANOSE FORMS

During the discussion of hemiacetal formation in D-ribose in the preceding section, you may have noticed that aldopentoses have the potential of forming a six-membered cyclic hemiacetal via addition of the C-5 hydroxyl to the carbonyl group. This mode of ring closure leads to α- and β-pyranose forms:

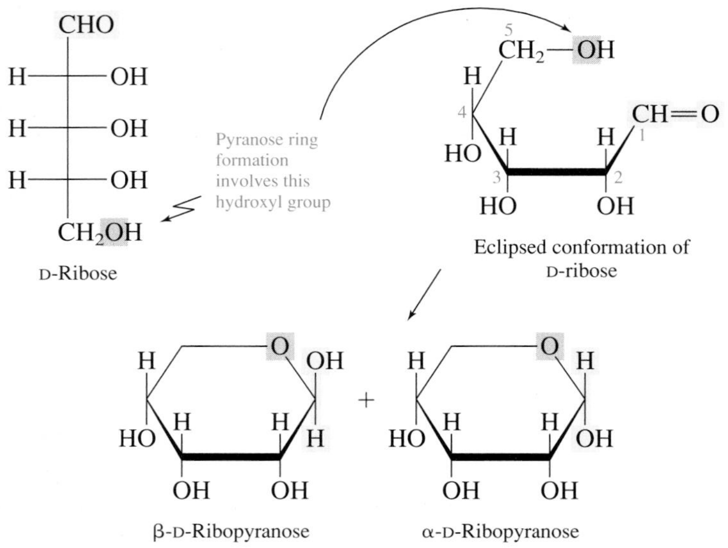

CHO
H——OH
H——OH
H——OH
CH₂OH

D-Ribose

Pyranose ring formation involves this hydroxyl group

Eclipsed conformation of D-ribose

β-D-Ribopyranose   +   α-D-Ribopyranose

Like aldopentoses, aldohexoses such as D-glucose are capable of forming two furanose forms (α and β) and two pyranose forms (α and β). The Haworth representations of the pyranose forms of D-glucose are constructed as shown in Figure 15.4; each has a CH₂OH group as a substituent on the six-membered ring.

Haworth formulas are satisfactory for representing *configurational* relationships in pyranose forms but are uninformative as to carbohydrate *conformations*. X-ray crystallographic studies of a large number of carbohydrates reveal that the six-membered pyranose ring of D-glucose adopts a chair conformation:

β-D-Glucopyranose

α-D-Glucopyranose

> Make a molecular model of the chair conformation of β-D-glucopyranose.

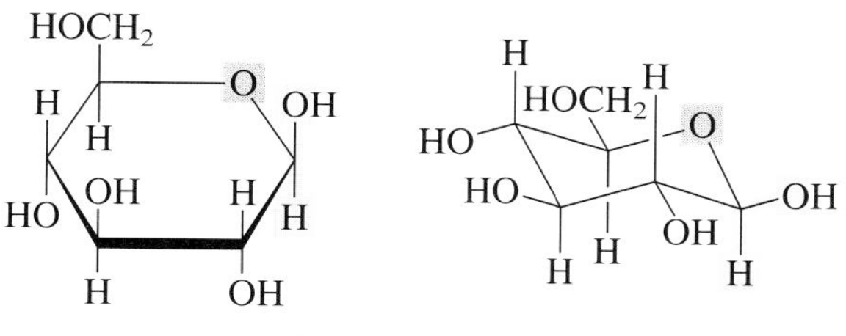

D-Glucose (hydroxyl group at C-5 is involved in pyranose ring formation)

Eclipsed conformation of D-Glucose; hydroxyl at C-5 is not properly oriented for ring formation

rotate about C-4–C-5 bond in counterclockwise direction

β-D-Glucopyranose 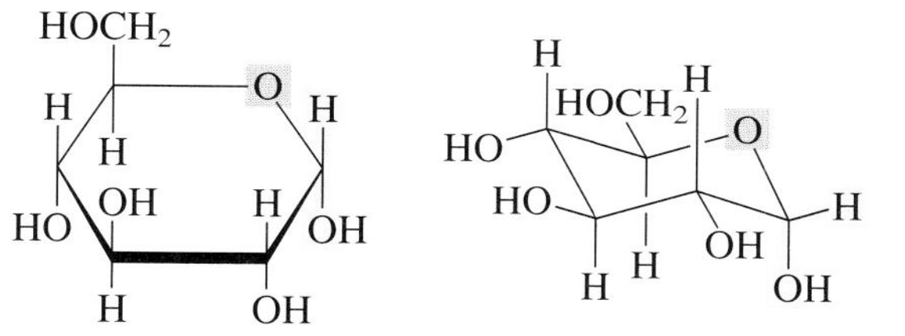   +   α-D-Glucopyranose   Eclipsed conformation of D-glucose in proper orientation for pyranose ring formation

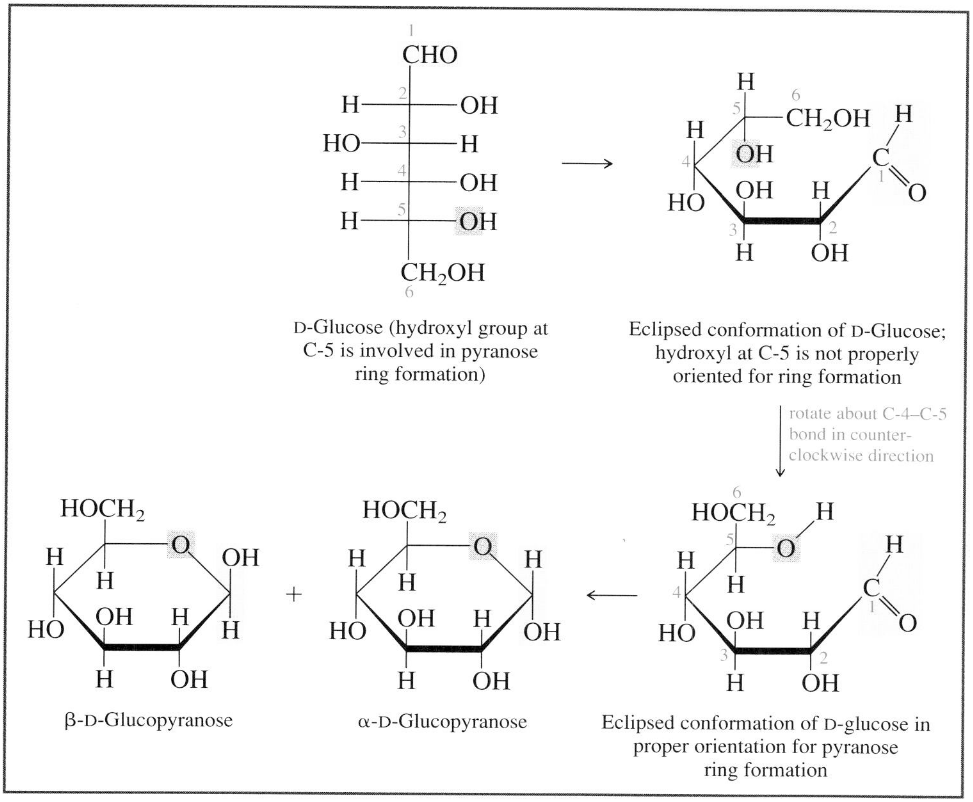

FIGURE 15.4 Haworth formulas for α- and β-pyranose forms of D-glucose.

All the ring substituents other than hydrogen in β-D-glucopyranose are equatorial in the most stable chair conformation. Only the anomeric hydroxyl group is axial in the α isomer; all the other substituents are equatorial.

Other aldohexoses behave similarly in adopting chair conformations that permit the CH_2OH substituent to occupy an equatorial orientation. Normally the CH_2OH group is the bulkiest, most conformationally demanding substituent in the pyranose form of a hexose.

PROBLEM 15.6 Clearly represent the most stable conformation of the β-pyranose form of each of the following sugars:

(a) D-Galactose

(c) L-Mannose

(b) D-Mannose

(d) L-Ribose

Sample Solution (a) By analogy with the procedure outlined for D-glucose in Figure 15.4, first generate a Haworth formula for β-D-galactopyranose:

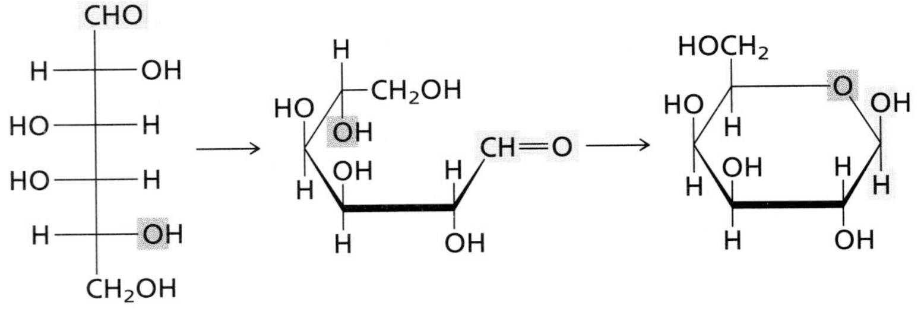

D-Galactose

β-D-Galactopyranose
(Haworth formula)

Next, redraw the planar Haworth formula more realistically as a chair conformation, choosing the one that has the CH_2OH group equatorial.

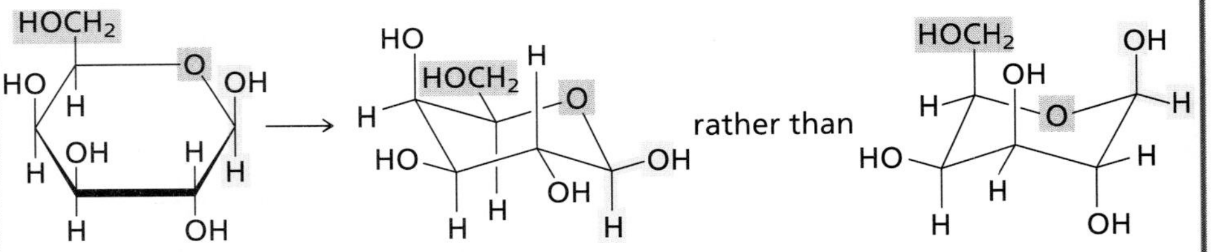

rather than

Most stable chair conformation of β-D-galactopyranose

Less stable chair; CH_2OH group is axial

Galactose differs from glucose in configuration at C-4. The C-4 hydroxyl is axial in β-D-galactopyranose, but it is equatorial in β-D-glucopyranose.

Because six-membered rings are normally less strained than five-membered ones, pyranose forms are usually present in greater amounts than furanose forms at equilibrium, and the concentration of the open-chain form is quite small.

15.7 HEMIACETAL EQUILIBRIUM

The reactions that give the cyclic hemiacetal forms of carbohydrates are reversible. Thus the furanose and pyranose forms of a carbohydrate in solution are in equilibrium with each other and with the open-chain form. The equilibrium for the pyranose forms of D-glucose is

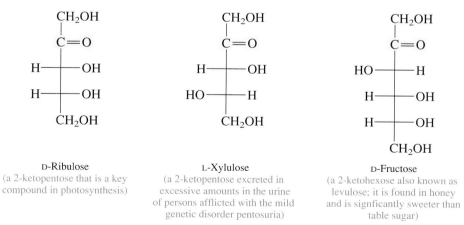

α-D-Glucopyranose
(mp 146°C;
$[\alpha]_D +112.2°$)

Open-chain form
of D-glucose

β-D-Glucopyranose
(mp 148–155°C;
$[\alpha]_D +18.7°$)

The optical rotations cited for each isomer are those measured immediately after each one is dissolved in water. On standing, the rotation of the solution containing the α isomer decreases from +112.2° to +52.5°; the rotation of the solution of the β isomer increases from +18.7° to the same value of +52.5°. The change in optical rotation resulting from equilibration of the anomeric forms is called **mutarotation.**

The β isomer of glucose is more stable and constitutes 64% of the equilibrium mixture. The less stable α isomer is 36% of the mixture, and the open-chain form constitutes less than 0.01%.

15.8 KETOSES

Up to this point all our attention has been directed toward aldoses, carbohydrates having an aldehyde function in their open-chain form. Aldoses are more common than ketoses, and their role in biological processes has been more thoroughly studied. Nevertheless, a large number of ketoses are known, and several of them are pivotal intermediates in carbohydrate biosynthesis and metabolism. Examples of some ketoses include D-ribulose, L-xylulose, and D-fructose:

D-Ribulose
(a 2-ketopentose that is a key
compound in photosynthesis)

L-Xylulose
(a 2-ketopentose excreted in
excessive amounts in the urine
of persons afflicted with the mild
genetic disorder pentosuria)

D-Fructose
(a 2-ketohexose also known as
levulose; it is found in honey
and is signficantly sweeter than
table sugar)

In these three examples the carbonyl group is located at C-2, which is the most common location for the carbonyl function in naturally occurring ketoses.

PROBLEM 15.7 How many ketotetroses are possible? Write Fischer projections for each.

How sweet is it?

There is no shortage of compounds, natural or synthetic, that taste sweet. The most familiar are naturally occurring sugars, especially sucrose, glucose, and fructose. All occur naturally, with worldwide production of sucrose from sugarcane and sugar beets exceeding 100 million tons per year. Glucose is prepared by the enzymatic hydrolysis of starch, and fructose is made by the isomerization of glucose.

D-(+)-Glucose   D-(−)-Fructose

Among sucrose, glucose, and fructose, fructose is the sweetest. Honey is sweeter than table sugar because it contains fructose formed by the isomerization of glucose as shown in the equation.

You may have noticed that most soft drinks contain "high-fructose corn syrup." Corn starch is hydrolyzed to glucose, which is then treated with glucose isomerase to produce a fructose-rich mixture. The enhanced sweetness permits less to be used, reducing the cost of production. Using less of a carbohydrate-based sweetener also reduces the number of calories.

Artificial sweeteners are a billion-dollar-per-year industry. The primary goal is, of course, to maximize sweetness and minimize calories. We'll look at the following three sweeteners to give us an overview of the field.

Saccharin   Sucralose   Aspartame

All three of these are hundreds of times sweeter than sucrose and variously described as "low-calorie" or "nonnutritive" sweeteners.

Saccharin was discovered at Johns Hopkins University in 1879 in the course of research on coal-tar derivatives and is the oldest artificial sweetener. In spite of its name, which comes from the Latin word for sugar, saccharin bears no structural relationship to any sugar. Nor is saccharin itself very soluble in water. The proton bonded to nitrogen, however, is fairly acidic and saccharin is normally marketed as its water-soluble sodium or calcium salt. Its earliest applications were not in weight control, but as a replacement for sugar in the diet of diabetics before insulin became widely available.

Sucralose has the structure most similar to sucrose. Galactose replaces the glucose unit of sucrose, and chlorines replace three of the hydroxyl groups. Sucralose is the newest artificial sweetener, having been approved by the U.S. Food and Drug Administration in 1998. The three chlorine substituents do not diminish sweetness, but do interfere with the ability of the body to metabolize sucralose. It, therefore, has no food value and is "noncaloric."

Aspartame is the market leader among artificial sweeteners. It is a methyl ester of a dipeptide, unrelated to any carbohydrate. It was discovered in the course of research directed toward developing drugs to relieve indigestion.

Saccharin, sucralose, and aspartame illustrate the diversity of structural types that taste sweet, and the vitality and continuing development of the industry of which they are a part.

15.9 STRUCTURAL VARIATIONS IN CARBOHYDRATES

Most sugars, including those discussed thus far in this chapter, possess a straight-chain carbon backbone and an oxygen atom on each carbon. Several variations of the general pattern of carbohydrate structure occur in nature. These variations usually involve replacement of one or more of the hydroxyl substituents on the main chain by some other atom or group. Examples of these variations include the following.

Deoxy Sugars

In **deoxy sugars** a hydroxyl group is replaced by hydrogen. Two examples of deoxy sugars are 2-deoxy-D-ribose and L-rhamnose:

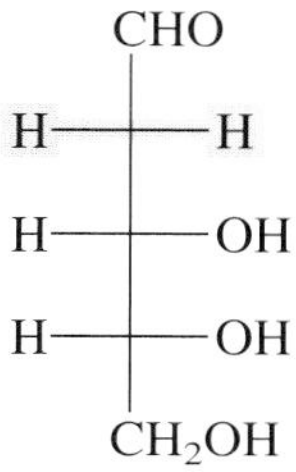

CHO
H
H
H
OH
H
OH
CH₂OH

2-Deoxy-D-ribose

CHO
H
OH
H
OH
HO
H
HO
H
CH₃

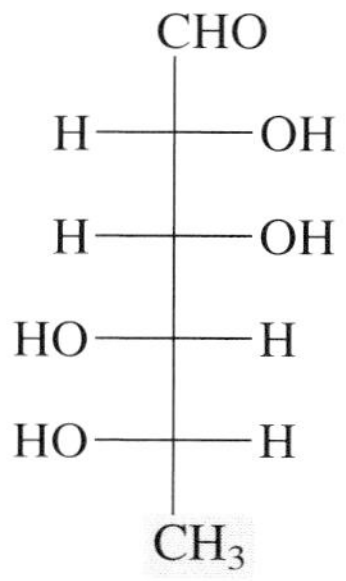

L-Rhamnose
(6-deoxy-L-mannose)

The hydroxyl at C-2 in D-ribose is absent in 2-deoxy-D-ribose. In Chapter 18 we shall see how derivatives of 2-deoxy-D-ribose, called **deoxyribonucleotides,** are the fundamental building blocks of deoxyribonucleic acid (DNA), the material responsible for storing genetic information. L-Rhamnose is a compound isolated from a number of plants. Its carbon chain terminates in a methyl rather than a CH_2OH group.

PROBLEM 15.8 Write Fischer projections of

(a) Cordycepose (3-deoxy-D-ribose): a deoxy sugar isolated by hydrolysis of the antibiotic substance cordycepin

(b) L-Fucose (6-deoxy-L-galactose): obtained from seaweed

Sample Solution (a) The hydroxyl group at C-3 in D-ribose is replaced by hydrogen in 3-deoxy-D-ribose.

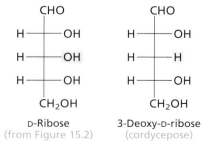

CHO
H
OH
H
OH
H
OH
CH₂OH

D-Ribose
(from Figure 15.2)

CHO
H
OH
H
H
H
OH
CH₂OH

3-Deoxy-D-ribose
(cordycepose)

Amino Sugars

Another structural variation is the replacement of a hydroxyl group in a carbohydrate by an amine group. Such compounds are called **amino sugars.** More than 60 amino sugars are presently known, many of them having been isolated and identified only recently as components of antibiotic substances. The anticancer drug doxorubicin hydrochloride (Adriamycin), for example, contains the amino sugar L-daunosamine as one of its structural units.

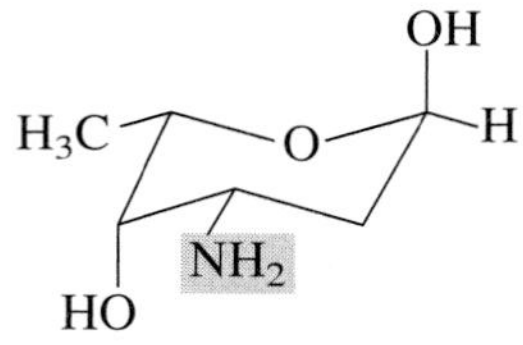

L-Daunosamine 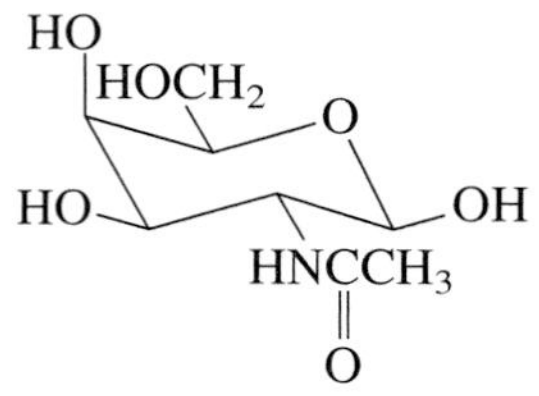   N-Acetyl-D-galactosamine

A second amino sugar, N-acetyl-D-galactosamine, is a component of chondroitin, a polysaccharide in cartilage.

Branched-Chain Carbohydrates

Carbohydrates that have a carbon substituent attached to the main chain are said to have a **branched chain.** D-Apiose and L-vancosamine are representative branched-chain carbohydrates:

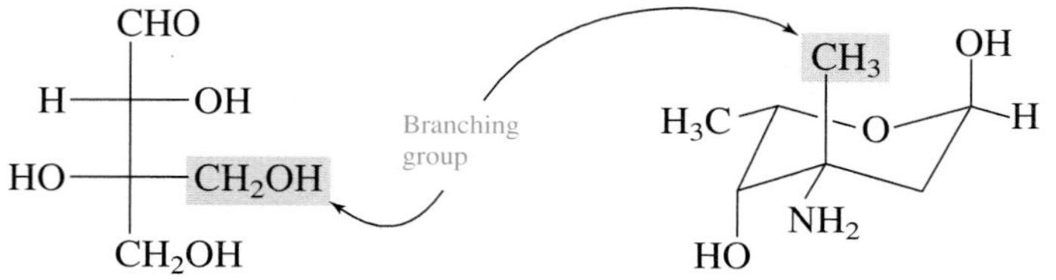

D-Apiose   L-Vancosamine

D-Apiose can be isolated from parsley and is a component of the cell wall polysaccharide of various marine plants. Among its novel structural features is the presence of only a single stereogenic center. L-Vancosamine is but one portion of vancomycin, a powerful antibiotic that is reserved for treating only the most stubborn infections. L-Vancosamine is not only a branched-chain carbohydrate, it is a deoxy sugar and an amino sugar as well.

15.10 GLYCOSIDES

Glycosides are a large and very important class of carbohydrate derivatives characterized by the replacement of the anomeric hydroxyl group by some other substituent. Glycosides are termed *O*-glycosides, *N*-glycosides, *S*-glycosides, and so on, according to the atom attached to the anomeric carbon.

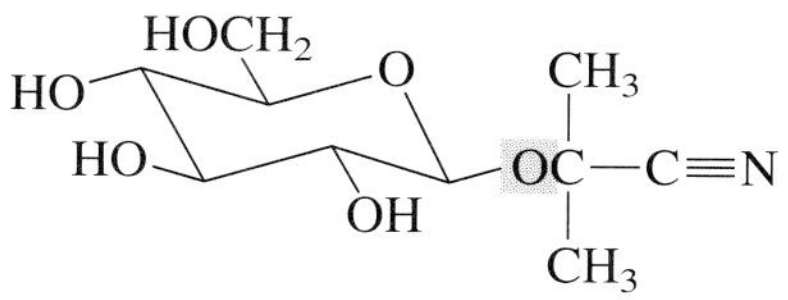

Linamarin
(an *O*-glycoside:
obtained from manioc,
a type of yam widely
distributed in
southeast Asia)

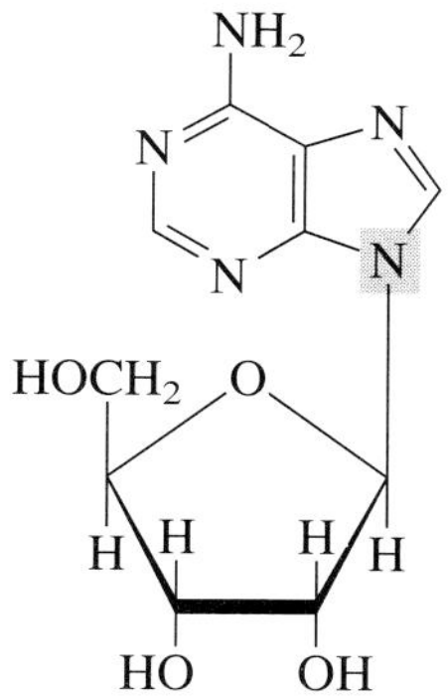

Adenosine
(an *N*-glycoside: also
known as a nucleoside;
adenosine is one of the
fundamental molecules
of biochemistry)

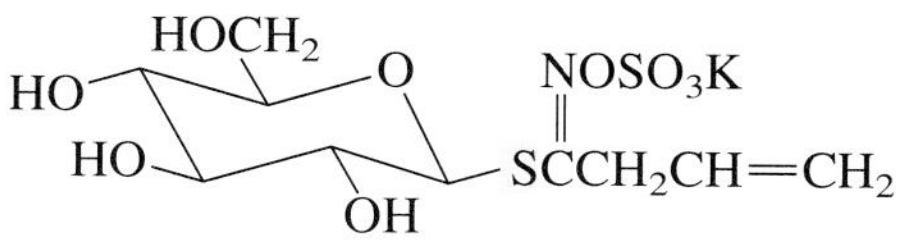

Sinigrin
(an *S*-glycoside:
contributes to the
characteristic flavor
of mustard and
horseradish)

Usually, the term "glycoside" without a prefix is taken to mean an *O*-glycoside and will be used that way in this chapter. Glycosides are classified as α or β in the customary way, according to the configuration at the anomeric carbon. All three of the glycosides just shown are β-glycosides.

PROBLEM 15.9   Identify by name the carbohydrate component of each of the three preceding glycosides.

Sample Solution   Replacement of the group attached to the anomeric carbon with —OH reveals the structure of the carbohydrate. Linamarin is a glycoside formed from the β-pyranose form of D-glucose.

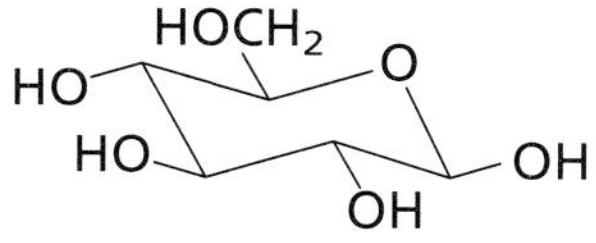

β-D-Glucopyranose

Structurally, *O*-glycosides are mixed acetals; they are ethers formed at the anomeric position of furanose or pyranose forms of carbohydrates. The laboratory preparation of glycosides can be illustrated by the reaction of D-glucose with methanol in the presence of an acid catalyst.

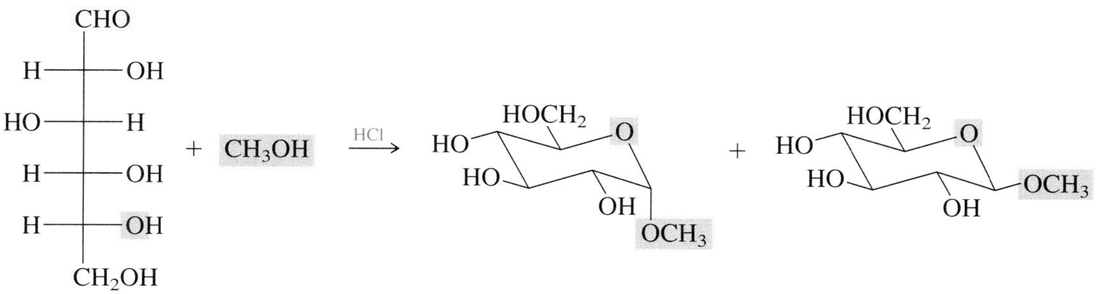

D-Glucose   Methanol   Methyl   Methyl
                                  α-D-glucopyranoside   β-D-glucopyranoside
                                  (major product; isolated   (minor product)
                                  in 49% yield)

Under neutral or basic conditions glycosides are configurationally stable; unlike the free sugars from which they are derived, glycosides do not exhibit mutarotation. Converting the anomeric hydroxyl group to an ether function (hemiacetal ⟶ acetal) prevents its reversion to the open-chain form in neutral or basic media. In aqueous acid, acetal formation can be reversed and the glycoside hydrolyzed to an alcohol and the free sugar.

15.11 DISACCHARIDES

Disaccharides are carbohydrates that yield two monosaccharide molecules on hydrolysis. Structurally, disaccharides are glycosides in which the alkoxy group attached to the anomeric carbon is derived from a second sugar molecule.

Maltose, obtained by the hydrolysis of starch, and **cellobiose,** by the hydrolysis of cellulose, are isomeric disaccharides. In both maltose and cellobiose two D-glucopyranose units are joined by a glycosidic bond between C-1 of one unit and C-4 of the other. The two are diastereomers, differing only in the stereochemistry at the anomeric carbon of the glycoside bond; maltose is an α-glycoside, cellobiose is a β-glycoside.

The stereochemistry and points of connection of glycosidic bonds are commonly designated by symbols such as α(1,4) for maltose and β(1,4) for cellobiose; α and β designate the stereochemistry at the anomeric position; the numerals specify the ring carbons involved.

Both maltose and cellobiose have a free anomeric hydroxyl group that is not involved in a glycoside bond. The configuration at the free anomeric center is variable and may be either α or β. Indeed, two stereoisomeric forms of maltose have been isolated. One has its anomeric hydroxyl group in an equatorial orientation; the other has an axial anomeric hydroxyl.

> The free anomeric hydroxyl group is the one shown at the far right of the preceding structural formula. The symbol ∿∿∿ is used to represent a bond of variable stereochemistry.

> **PROBLEM 15.10** The two stereoisomeric forms of maltose just mentioned undergo mutarotation when dissolved in water. What is the structure of the key intermediate in this process?

The single difference in their structures, the stereochemistry of the glycosidic bond, causes maltose and cellobiose to differ significantly in their three-dimensional shape, as the molecular models of Figure 15.5 illustrate. This difference in shape affects the way in which maltose and cellobiose interact with other chiral molecules such as proteins, and they behave much differently toward enzyme-catalyzed hydrolysis. An enzyme known as maltase catalyzes the hydrolytic cleavage of the α-glycosidic bond of maltose but is without effect in promoting the hydrolysis of the β-glycosidic bond of cellobiose. A different enzyme, emulsin, produces the opposite result. Emulsin catalyzes the hydrolysis of cellobiose but not of maltose. The behavior of each enzyme is general for glucosides (glycosides of glucose). Maltase catalyzes the hydrolysis of α-glucosides and is also known as α-glucosidase, whereas emulsin catalyzes the hydrolysis of β-glucosides and is known as β-glucosidase. The specificity of these enzymes offers a useful tool for

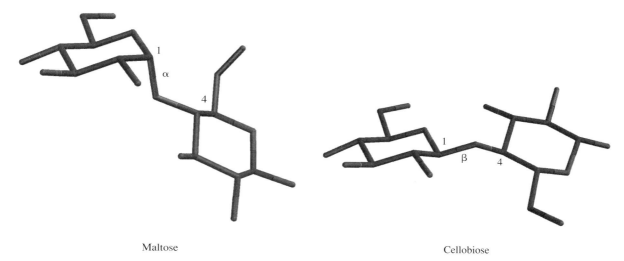

Maltose   Cellobiose

FIGURE 15.5   Molecular models of the disaccharides maltose and cellobiose. Two D-glucopyranose units are connected by a glycoside linkage between C-1 and C-4. The glycosidic bond has the α orientation in maltose and is β in cellobiose. Maltose and cellobiose are diastereomers.

structure determination because it allows the stereochemistry of glycosidic linkages to be assigned.

Lactose is a disaccharide constituting 2–6% of milk and is known as milk sugar. It differs from maltose and cellobiose in that only one of its monosaccharide units is D-glucose. The other monosaccharide unit, the one that contributes its anomeric carbon to the glycoside bond, is D-galactose. Like cellobiose, lactose is a β-glycoside.

$$
\begin{array}{c}
\text{Cellobiose: } \cdots\text{\tiny{I}}\text{I}\!\!\!| \\
\text{Lactose: } \blacktriangleright
\end{array}
$$

Digestion of lactose is facilitated by the β-glycosidase lactase. A deficiency of this enzyme makes it difficult to digest lactose and causes abdominal discomfort. Lactose intolerance is a genetic trait; it is treatable through over-the-counter formulations of lactase and by limiting the amount of milk in the diet.

The most familiar of all the carbohydrates is sucrose—common table sugar. **Sucrose** is a disaccharide in which D-glucose and D-fructose are joined at their anomeric carbons by a glycosidic bond (Figure 15.6). Its chemical composition is the same irrespective of its source; sucrose from sugarcane and sucrose from sugar beets are chemically identical. Because sucrose does not have a free anomeric hydroxyl group, it does not undergo mutarotation.

15.12 POLYSACCHARIDES

Cellulose is the principal structural component of vegetable matter. Wood is 30–40% cellulose, cotton over 90%. Photosynthesis in plants is responsible for the formation of 10^9 tons per year of cellulose. Structurally, cellulose is a polysaccharide composed

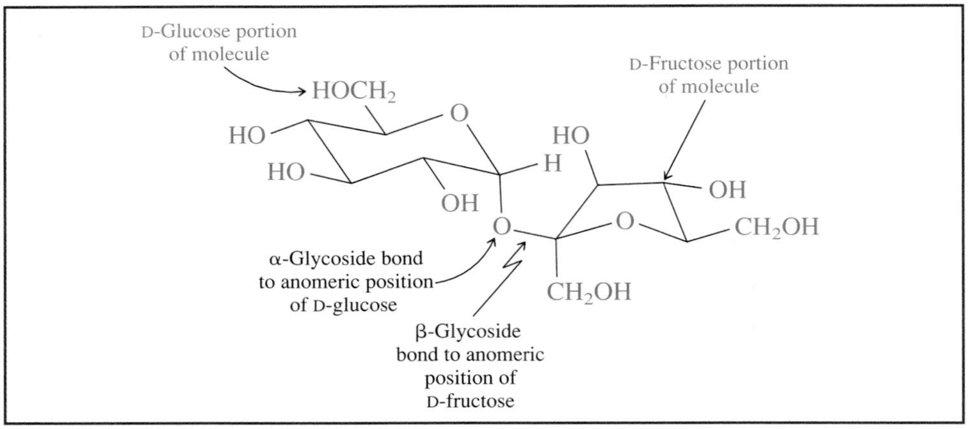

FIGURE 15.6 The structure of sucrose.

D-Glucose portion
of molecule

D-Fructose portion
of molecule

α-Glycoside bond
to anomeric position
of D-glucose

β-Glycoside
bond to anomeric
position of
D-fructose

of several thousand D-glucose units joined by β(1,4)-glycosidic linkages (Figure 15.7). Complete hydrolysis of all the glycosidic bonds of cellulose yields D-glucose. The disaccharide fraction that results from partial hydrolysis is cellobiose.

Animals lack the enzymes necessary to catalyze the hydrolysis of cellulose and so can't digest it. Cattle and other ruminants use cellulose as a food source in an indirect way. Colonies of microorganisms that live in their digestive tract consume cellulose and in the process convert it to other substances that the animal can digest.

A more direct source of energy for animals is provided by the starches found in many foods. **Starch** is a mixture of a water-dispersible fraction called amylose and a second component, amylopectin. Amylose is a polysaccharide made up of about 100 to several thousand D-glucose units joined by α(1,4)-glycosidic bonds.

Like amylose, amylopectin is a polysaccharide of α(1,4)-linked D-glucose units. Instead of being a continuous length of α(1,4) units, however, amylopectin is branched. Attached to C-6 at various points on the main chain are short polysaccharide branches of 24–30 glucose units joined by α(1,4)-glycosidic bonds (Figure 15.8).

Starch is a plant's way of storing glucose to meet its energy needs. Animals can tap that source by eating starchy foods and, with the aid of their α-glycosidase enzymes, hydrolyze the starch to glucose. When more glucose is available than is needed as fuel, animals store it as glycogen. Glycogen is similar to amylopectin in that it is a branched polysaccharide of α(1,4)-linked D-glucose units with subunits connected to C-6 of the main chain.

FIGURE 15.7 Cellulose is a polysaccharide in which D-glucose units are connected by β(1,4)-glycoside linkages analogous to cellobiose. Hydrogen bonding, especially between the C-2 and C-6 hydroxyl groups, causes adjacent glucose units to be turned at an angle of 180° with each other.

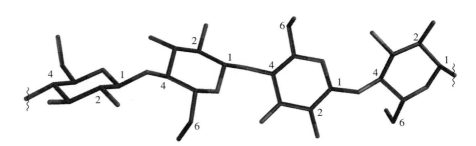

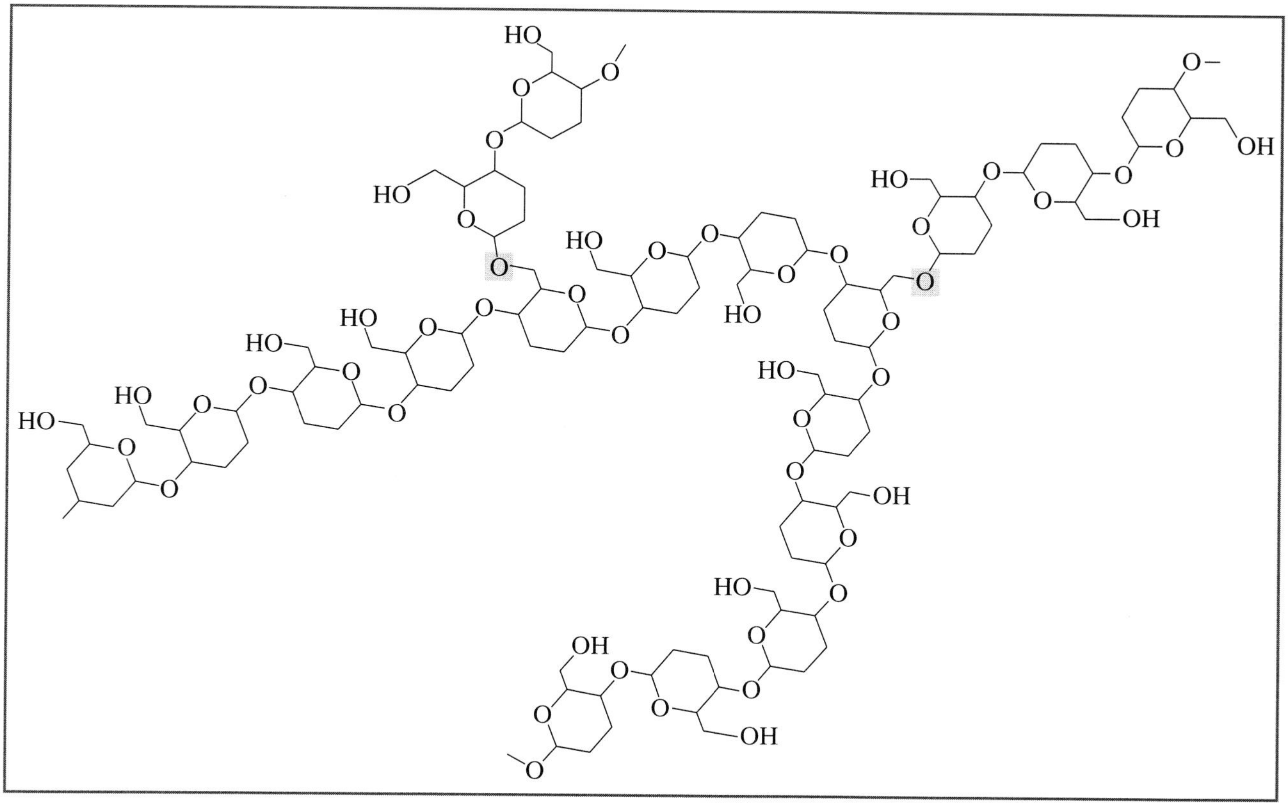

FIGURE 15.8   A portion of the amylopectin structure. α-Glycosidic bonds link C-1 and C-4 of adjacent glucose units. Glycoside bonds from C-6 to the polysaccharide chains are designated by the shaded oxygen atoms. Hydroxyl groups attached to individual rings have been omitted for clarity.

15.13 OXIDATION OF CARBOHYDRATES

A characteristic property of an aldehyde function is its sensitivity to oxidation (Section 11.14). A solution of copper(II) sulfate as its citrate complex (**Benedict's reagent**) is capable of oxidizing aliphatic aldehydes to the corresponding carboxylic acid.

| Aldehyde | From copper(II) sulfate | Hydroxide ion | | Carboxylate anion | Copper(I) oxide | Water |

The formation of a red precipitate of copper(I) oxide by reduction of Cu(II) is taken as a positive test for an aldehyde. Carbohydrates that give positive tests with Benedict's reagent are termed **reducing sugars.**

   Aldoses are reducing sugars because they possess an aldehyde function in their open-chain form. Ketoses are also reducing sugars. Under the conditions of the test, ketoses equilibrate with aldoses by way of enediol intermediates, and the aldoses are oxidized by the reagent.

Benedict's reagent is the key material in a test kit available from drugstores that permits individuals to monitor the glucose levels in their urine.

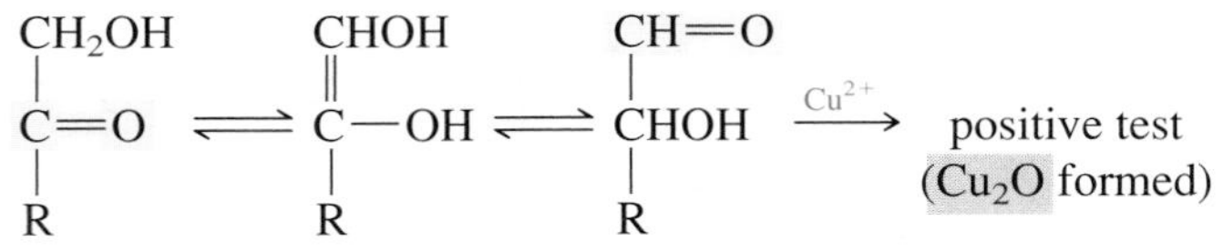

$$
\underset{\text{Ketose}}{\overset{\displaystyle \begin{array}{c} CH_2OH \\ | \\ C=O \\ | \\ R \end{array}}{}} \quad \rightleftharpoons \quad \underset{\text{Enediol}}{\overset{\displaystyle \begin{array}{c} CHOH \\ \| \\ C-OH \\ | \\ R \end{array}}{}} \quad \rightleftharpoons \quad \underset{\text{Aldose}}{\overset{\displaystyle \begin{array}{c} CH=O \\ | \\ CHOH \\ | \\ R \end{array}}{}} \quad \xrightarrow{Cu^{2+}} \quad \begin{array}{c} \text{positive test} \\ (Cu_2O \text{ formed}) \end{array}
$$

The same kind of equilibrium is available to α-hydroxy ketones generally; such compounds give a positive test with Benedict's reagent. Any carbohydrate that contains a free hemiacetal function is a reducing sugar. The free hemiacetal is in equilibrium with the open-chain form and through it is susceptible to oxidation. Maltose, for example, gives a positive test with Benedict's reagent.

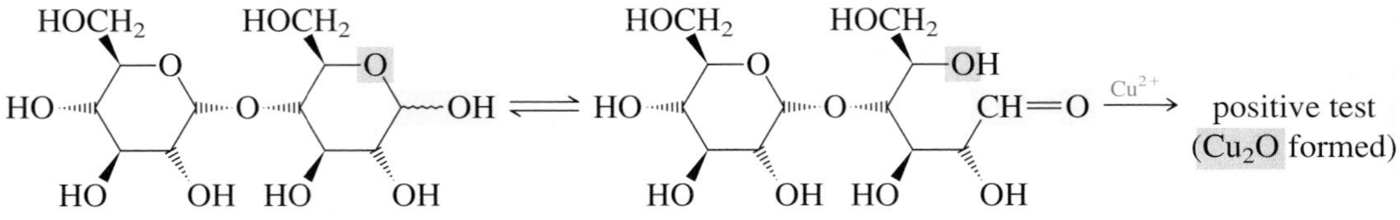

Maltose $\rightleftharpoons$ Open-chain form of maltose $\xrightarrow{Cu^{2+}}$ positive test (Cu_2O formed)

Glycosides, in which the anomeric carbon is part of an acetal function, are not reducing sugars and do not give a positive test.

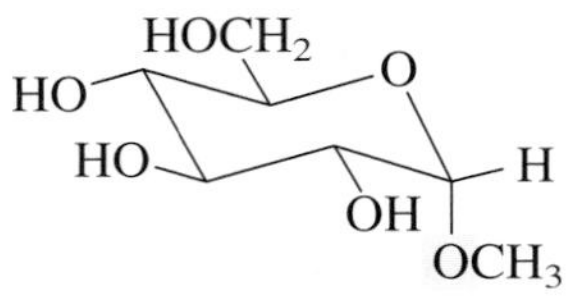

Methyl α-D-glucopyranoside
(not a reducing sugar)

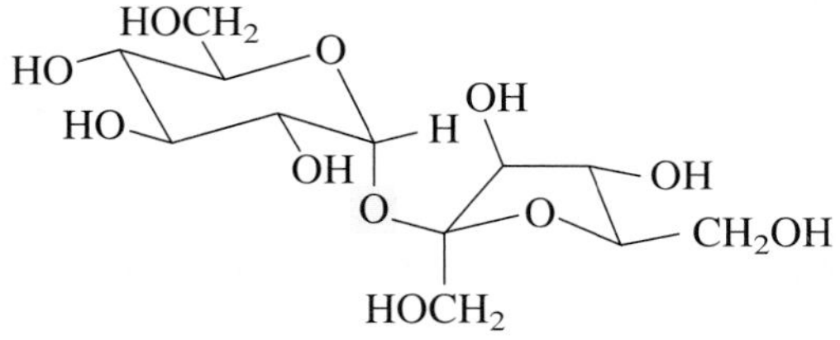

Sucrose
(not a reducing sugar)

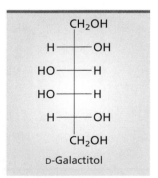

CH₂OH
H——OH
HO——H
HO——H
H——OH
CH₂OH

D-Galactitol

PROBLEM 15.11 Which of the following would be expected to give a positive test with Benedict's reagent? Why?

(a) D-Galactitol (see structure in margin)   (d) D-Fructose

(b) L-Arabinose   (e) Lactose

(c) 1,3-Dihydroxyacetone   (f) Amylose

Sample Solution (a) D-Galactitol lacks an aldehyde, an α-hydroxy ketone, or a hemiacetal function, so cannot be oxidized by Cu^{2+} and will not give a positive test with Benedict's reagent.

Fehling's solution, a tartrate complex of copper(II) sulfate, has also been used as a test for reducing sugars.

LEARNING OBJECTIVES

This chapter focused on one of the major classes of substances in living systems, carbohydrates. The skills you have learned in this chapter should enable you to

- Classify carbohydrates according to their structure.
- Represent the structure of a carbohydrate using a Fischer projection.
- Classify a carbohydrate as having the D or L configuration from its Fischer projection.
- Construct a Haworth projection of the furanose form of a pentose.
- Construct a Haworth projection of the pyranose form of a hexose.
- Explain how the α and β configurations of a carbohydrate differ.
- Use chemical equations to describe the equilibrium process associated with mutarotation.
- Describe the structural features characteristic of glycosides.
- Write the chemical structure of a disaccharide.
- Explain the structural difference between cellulose and starch.
- Describe the structural features that characterize reducing sugars.
- Explain the Benedict's test used to identify reducing sugars.

15.14 SUMMARY

Carbohydrates are classified according to their structure (Section 15.1). A **monosaccharide** cannot be hydrolyzed to a smaller carbohydrate; a **disaccharide** gives two monosaccharide molecules on hydrolysis; and so on. Carbohydrates containing an aldehyde group are called **aldoses;** carbohydrates having a ketone group are **ketoses.**

Fischer projection formulas offer a convenient way to represent configurational relationships in carbohydrates. When the hydroxyl group at the highest numbered stereogenic center is to the right, the carbohydrate belongs to the D series; when this hydroxyl is to the left, the carbohydrate belongs to the L series (Sections 15.2, 15.3).

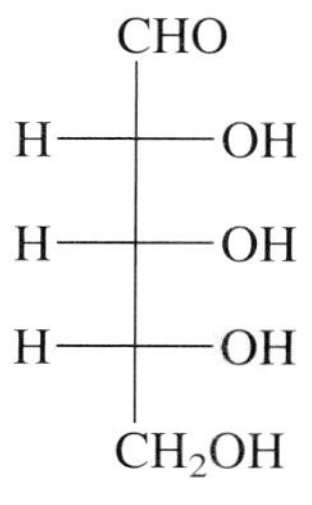

$$
\begin{array}{cc}
\text{CHO} & \text{CHO} \\
\text{H}\!-\!\!\!-\!\text{OH} & \text{H}\!-\!\!\!-\!\text{OH} \\
\text{H}\!-\!\!\!-\!\text{OH} & \text{HO}\!-\!\!\!-\!\text{H} \\
\text{H}\!-\!\!\!-\!\text{OH} & \text{HO}\!-\!\!\!-\!\text{H} \\
\text{CH}_2\text{OH} & \text{CH}_2\text{OH} \\
\text{D-Ribose} & \text{L-Arabinose}
\end{array}
$$

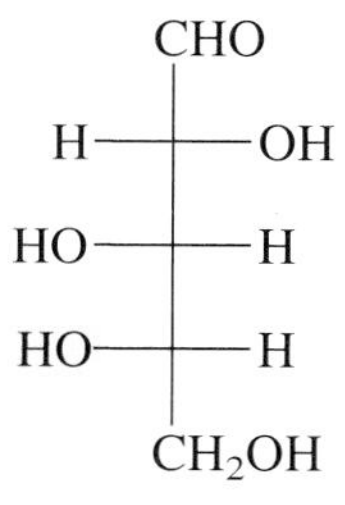

Sugars having five carbon atoms are called **pentoses;** those having six carbon atoms are called **hexoses** (Section 15.4). Most carbohydrates exist as cyclic hemiacetals. Cyclic acetals with five-membered rings are called **furanose** forms (Section 15.5); those with six-membered rings are called **pyranose** forms (Section 15.6).

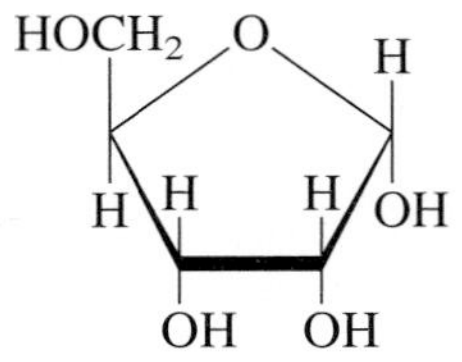

α-D-Ribofuranose 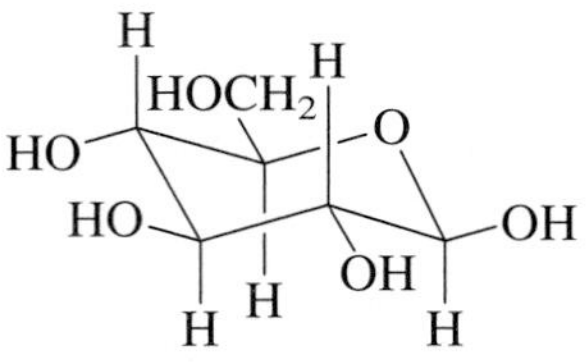   β-D-Glucopyranose

The **anomeric carbon** in a cyclic acetal is the one attached to *two* oxygens. It is the carbon that corresponds to the carbonyl carbon in the open-chain form. The symbols α and β refer to the configuration at the anomeric carbon.

A particular carbohydrate can interconvert between furanose and pyranose forms and between the α and β configuration of each form (Section 15.7). The change from one form to an equilibrium mixture of all the possible hemiacetals causes a change in optical rotation called **mutarotation.**

Structurally modified carbohydrates include **deoxy sugars, amino sugars,** and **branched-chain carbohydrates** (Section 15.9).

Glycosides are acetals, compounds in which the anomeric hydroxyl group has been replaced by an alkoxy group (Section 15.10). Glycosides are easily prepared by allowing a carbohydrate and an alcohol to stand in the presence of an acid catalyst.

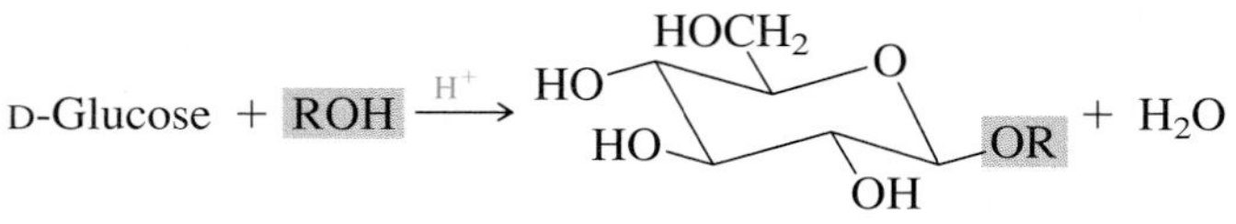

A glycoside

Disaccharides are carbohydrates in which two monosaccharides are joined by a glycoside bond (Section 15.11). **Polysaccharides** have many monosaccharide units connected through glycosidic linkages (Section 15.12). Complete hydrolysis of disaccharides and polysaccharides cleaves the glycoside bonds, yielding the free monosaccharide components.

Carbohydrates undergo chemical reactions characteristic of aldehydes and ketones. Aldoses or ketoses that contain a free hemiacetal function give positive tests with Benedict's reagent and are called **reducing sugars** (Section 15.13).

ADDITIONAL PROBLEMS

Classification and Structure

15.12 Using Fischer projections, draw the structure of

(a) A constitutional isomer of glyceraldehyde

(b) An L-ketotetrose

(c) The enantiomer of the compound in part (b)

15.13 Xylose is a pentose found in many varieties of apples and other fruits. It is also found in both wines and beers, in which it is not fermented by yeast. Refer to the Fischer projection formula of D-(+)-xylose in Figure 15.2 and give structural formulas for

(a) (−)-Xylose (Fischer projection)

(b) β-D-Xylopyranose

(c) Methyl β-D-xylopyranoside (a glycoside)

15.14 (a) Draw the most stable conformation of the β-pyranose form of D-galactose and
D-mannose. In what way does each of these differ from β-D-glucopyranose?
(b) Draw the α-pyranose forms of D-galactose and D-mannose, using Haworth formulas.

15.15 Write the Fischer projection of the open-chain form of each of the following:

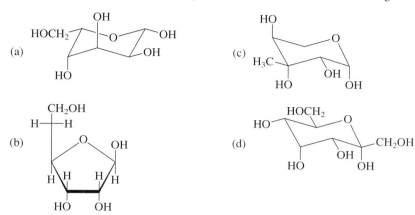

15.16 From among the carbohydrates shown in Problem 15.15 choose the one(s) that
(a) Belong to the L series
(b) Are deoxy sugars
(c) Are branched-chain sugars
(d) Are ketoses
(e) Are furanose forms
(f) Have the α configuration at their anomeric carbon

15.17 What are the *R*,*S* configurations of the three stereogenic centers in D-ribose? (A molecular model will be helpful here.)

15.18 How many ketopentoses are possible? Write their Fischer projections.

15.19 The Fischer projection of the branched-chain carbohydrate D-apiose has been presented in Section 15.9.
(a) How many stereogenic centers are in the open-chain form of D-apiose?
(b) How many stereogenic centers are in the furanose forms of D-apiose?
(c) How many stereoisomeric furanose forms of D-apiose are possible? Write their Haworth formulas.

Reactions

15.20 The aldehyde function of an aldose is reduced to —CH_2OH by sodium borohydride ($NaBH_4$) by way of the open-chain form. Give the Fischer projection formula of the product formed by sodium borohydride reduction of each of the following.
(a) D-Erythrose   (c) D-Glucose
(b) D-Arabinose   (d) D-Galactose

15.21 The products formed by reduction of aldoses are called alditols. Which of the alditols formed in Problem 15.20 are optically active?
(a) Erythritol   (c) Glucitol
(b) Arabinitol   (d) Galactitol

15.22 What two alditols are formed by reduction of D-fructose?

15.23 Treatment of D-mannose with methanol in the presence of an acid catalyst yields four isomeric methyl glycosides having the molecular formula $C_7H_{14}O_6$. Draw the structures of these four products.

Miscellaneous Problems

15.24 In 1987 a careful study of D-glucose established its composition in aqueous solution. The α- and β-pyranose forms were 99.7% of the mixture, with the α- and β-furanose forms accounting for another 0.29%. The free aldehyde was absent; however, 0.0045% of the aldehyde hydrate was present.

    (a) Write a Fischer projection formula for the aldehyde hydrate.
    (b) What portion of the glucose exists in a form containing a five-membered ring?
    (c) Is the anomeric hydroxyl group cis or trans to the —CH$_2$OH group in the β-pyranose form?

15.25 The specific optical rotations of pure α- and β-D-mannopyranose are +29.3° and −17.0°, respectively. When either form is dissolved in water, mutarotation occurs and the observed rotation of the solution changes until a final rotation of +14.2° is observed. Assuming that only α- and β-pyranose forms are present, calculate the percentage of each isomer at equilibrium.

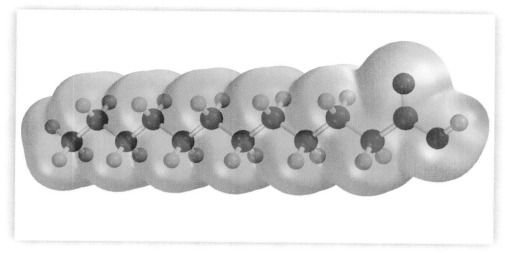

CHAPTER 16
LIPIDS

A long with proteins, carbohydrates, and nucleic acids, lipids form one of the major classes of organic compounds in biochemistry. Lipids play important roles in living systems, from their presence in cell membranes to their action as sex hormones. This chapter will describe various types of lipids, as well as examine how simple starting materials are converted by living organisms into complex terpenes.

16.1 CLASSIFICATION OF LIPIDS

Lipids are naturally occurring substances that are soluble in nonpolar solvents. This is an operational, rather than a structural, distinction. Samples of material from a natural source are shaken with a polar solvent (water or an alcohol–water mixture) and a relatively nonpolar one (diethyl ether, hexane, or dichloromethane). Carbohydrates, proteins, nucleic acids, and related compounds are polar and do not dissolve in the nonpolar solvent—they either dissolve in the polar aqueous phase or remain behind undissolved. The portion of the natural material that dissolves in the nonpolar solvent is the **lipid fraction.**

The various classes of compounds that make up the lipid fraction will be discussed in the sections which follow. One class of lipids, the prostaglandins, was the subject of the boxed essay in Chapter 12 (Section 12.3).

16.2 FATS, OILS, AND FATTY ACIDS

Fats are one type of lipid. They have a number of functions in living systems, including that of energy storage. Although carbohydrates serve as a source of readily available energy, an equal weight of fat delivers over twice the amount of energy. It is more efficient for an organism to store energy in the form of fat because it requires less mass than storing the same amount of energy in carbohydrates or proteins.

Fats and oils are naturally occurring mixtures of **triacylglycerols,** also called **triglycerides.** They differ in that fats are solids at room temperature and oils are liquids. We generally ignore this distinction and refer to both groups as fats.

Triacylglycerols are built on a glycerol framework.

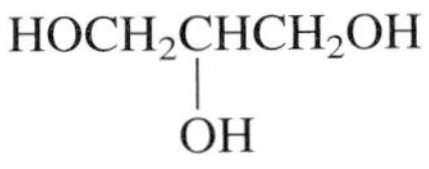

$$HOCH_2CHCH_2OH \qquad\qquad RCOCH_2CHCH_2OCR'$$

$$\underset{\text{OH}}{|} \qquad\qquad\qquad \underset{\underset{O}{\overset{\|}{OCR''}}}{|}$$

Glycerol 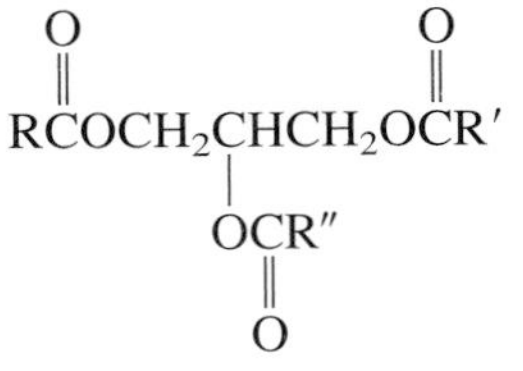   A triacylglycerol

All three acyl groups in a triacylglycerol may be the same, all three may be different, or one may be different from the other two.

Figure 16.1 shows the structures of two typical triacylglycerols, 2-oleyl-1,3-distearylglycerol (Figure 16.1a) and tristearin (Figure 16.1b). Both occur naturally—in cocoa butter, for example. All three acyl groups in tristearin are stearyl (octadecanoyl) groups. In 2-oleyl-1,3-distearylglycerol, two of the acyl groups are stearyl, but the one in the middle is oleyl (*cis*-9-octadecenoyl). As the figure shows, tristearin can be prepared by catalytic hydrogenation of the carbon–carbon double bond of 2-oleyl-1,3-distearylglycerol. Hydrogenation raises the melting point from 43°C in 2-oleyl-1,3-

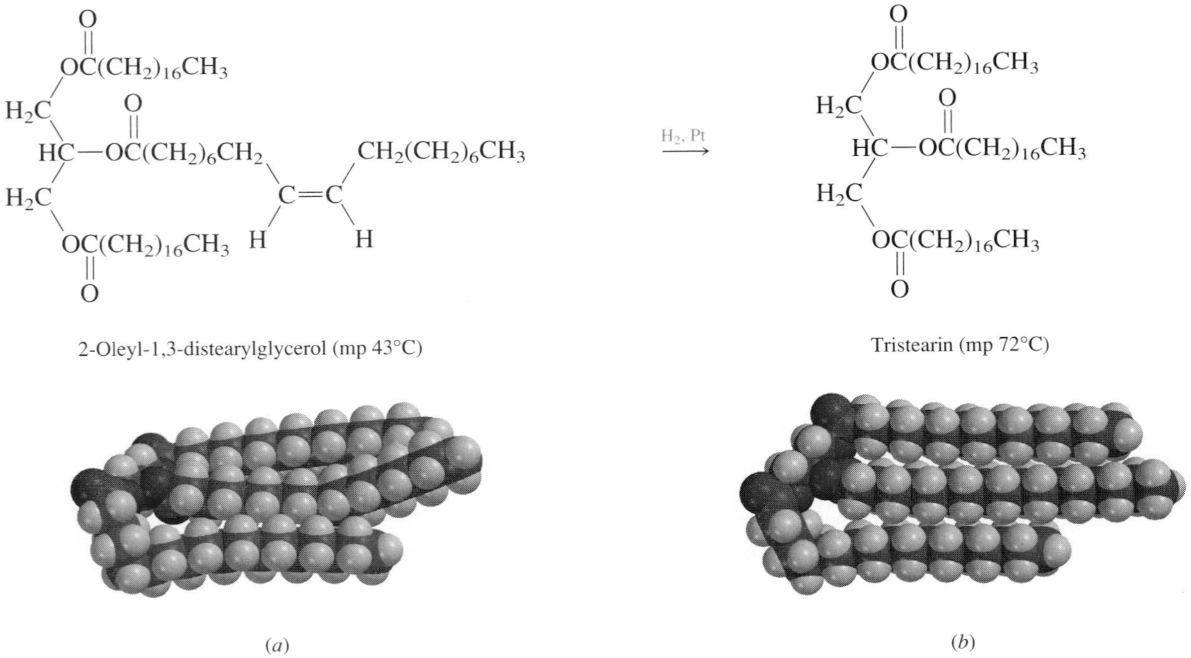

2-Oleyl-1,3-distearylglycerol (mp 43°C)

Tristearin (mp 72°C)

(a)   (b)

FIGURE 16.1   The structures of two typical triacylglycerols. (a) 2-Oleyl-1,3-distearylglycerol is a naturally occurring triacylglycerol found in cocoa butter. The cis double bond of its oleyl group gives the molecule a shape that interferes with efficient crystal packing. (b) Catalytic hydrogenation converts 2-oleyl-1,3-distearylglycerol to tristearin. Tristearin has a higher melting point than 2-oleyl-1,3-distearylglycerol.

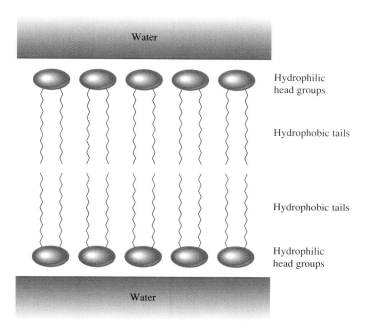

FIGURE 16.2 Cross section of a phospholipid bilayer.

Under certain conditions, such as at the interface of two aqueous phases, phosphatidyl-choline forms what is called a **lipid bilayer,** as shown in Figure 16.2. Because two long-chain acyl groups occur in each molecule, the most stable assembly has the hydrophilic polar groups solvated by water molecules at the top and bottom surfaces and the hydrophobic (nonpolar) acyl groups directed toward the interior of the bilayer.

Phosphatidylcholine is one of the principal components of cell membranes. These membranes are composed of lipid bilayers analogous to those of Figure 16.2. Nonpolar materials can diffuse through the bilayer from one side to the other relatively easily; polar materials, particularly metal ions such as Na^+, K^+, and Ca^{2+}, cannot. The transport of metal ions through a membrane is usually assisted by certain proteins present in the lipid bilayer, which contain a metal ion-binding site surrounded by a hydrophobic exterior. The metal ion is picked up at one side of the lipid bilayer and delivered at the other, surrounded at all times by a polar environment on its passage through the hydrocarbon-like interior of the membrane. Ionophore antibiotics such as monensin (Figure 10.2) disrupt the normal functioning of cells by facilitating metal ion transport across cell membranes.

> **PROBLEM 16.2** Lecithin is used in commercial food products such as mayonnaise to keep fat and water from separating into two layers. Explain what structural features of lecithin enable it to serve this function.

16.4   WAXES

Waxes are water-repelling solids that are part of the protective coatings of a number of living things, including the leaves of plants, the fur of animals, and the feathers of birds. They are usually mixtures of esters in which both the alkyl and acyl group are unbranched and contain a dozen or more carbon atoms. Beeswax, for example, contains the ester triacontyl hexadecanoate as one component of a complex mixture of hydrocarbons, alcohols, and esters.

$$CH_3(CH_2)_{14}\overset{\overset{\displaystyle O}{\|}}{C}OCH_2(CH_2)_{28}CH_3$$

Triacontyl hexadecanoate

> **PROBLEM 16.3**   Spermaceti is a wax obtained from the sperm whale. It contains, among other materials, an ester known as cetyl palmitate, which is used as an emollient in a number of soaps and cosmetics. The systematic name for cetyl palmitate is hexadecyl hexadecanoate. Write a structural formula for this substance.

16.5 STEROIDS: CHOLESTEROL

The name "cholesterol" is a combination of the Greek words for "bile" (*chole*) and "solid" (*stereos*) preceding the alcohol suffix -*ol*.

Steroids are a type of lipid that have a variety of functions in living systems. All steroids are characterized by the tetracyclic ring system shown in Figure 16.3a. **Cholesterol** (Figure 16.3b) is the most abundant steroid and the precursor for all other steroids in animals. An average adult human contains more than 200 g of cholesterol; it is found in almost all body tissues, with relatively large amounts present in the brain and spinal cord. Gallstones are almost entirely cholesterol. Cholesterol is also the principal constituent of the plaque that builds up on the walls of arteries and restricts the flow of blood in atherosclerosis (see the following boxed essay). Cholesterol is found in the body as both the free alcohol and esterified at its hydroxyl group by various fatty acids.

16.6 VITAMIN D

A steroid very closely related structurally to cholesterol is its 7-dehydro derivative. 7-Dehydrocholesterol is formed by enzymic oxidation of cholesterol and has a conjugated diene unit in its B ring. 7-Dehydrocholesterol is present in the tissues of the skin, where it is transformed to vitamin D_3 by a sunlight-induced photochemical reaction.

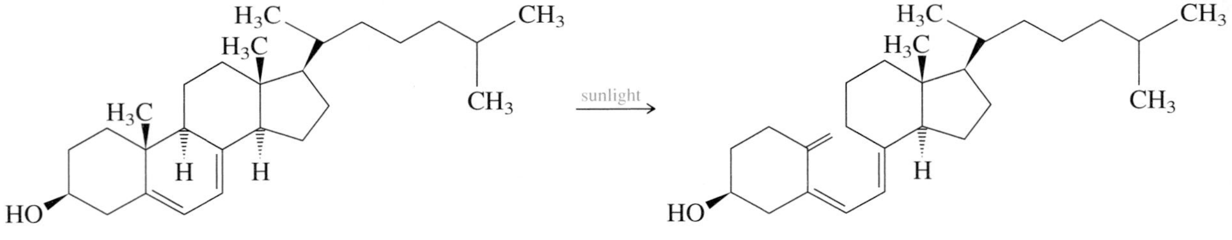

7-Dehydrocholesterol   sunlight →   Vitamin D_3

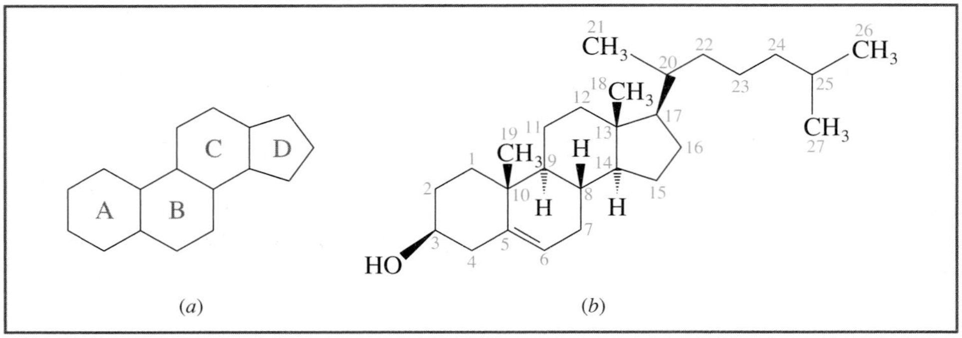

FIGURE 16.3   (a) The tetracyclic ring system characteristic of steroids. The rings are designated A, B, C, and D as shown. (b) The structure of cholesterol. A unique numbering system is used for steroids and is indicated in the structural formula.

(a)   (b)

"GOOD" CHOLESTEROL? "BAD" CHOLESTEROL? WHAT'S THE DIFFERENCE?

Cholesterol is biosynthesized in the liver, transported throughout the body to be used in a variety of ways, and returned to the liver where it serves as the biosynthetic precursor to other steroids. But cholesterol is a lipid and isn't soluble in water. How can it move through the blood if it doesn't dissolve in it? The answer is that it doesn't dissolve, but is instead carried through the blood and tissues as part of a **lipoprotein** (lipid + protein = lipoprotein).

The proteins that carry cholesterol from the liver are called **low-density lipoproteins**, or **LDLs;** those that return it to the liver are the **high-density lipoproteins**, or **HDLs.** If too much cholesterol is being transported by LDL, or too little by HDL, the extra cholesterol builds up on the walls of the arteries, causing atherosclerosis. A thorough physical examination nowadays measures not only total cholesterol concentration but also the distribution between LDL and HDL cholesterol. An elevated level of LDL cholesterol is a risk factor for heart disease. LDL cholesterol is sometimes called "bad cholesterol." HDLs, on the other hand, remove excess cholesterol and are protective. HDL cholesterol is considered "good cholesterol."

The distribution between LDL and HDL cholesterol depends mainly on genetic factors, but can be altered. Regular exercise increases HDL and reduces LDL cholesterol, as does limiting the amount of saturated fat in the diet. Much progress has been made in developing new drugs to lower cholesterol. The statin class, beginning with lovastatin in 1988 followed by simvastatin in 1991, have proven especially effective.

Simvastatin

The statins lower cholesterol by inhibiting the enzyme 3-hydroxy-3-methylglutaryl coenzyme A reductase, which is required for the biosynthesis of mevalonic acid (see Section 16.13). Mevalonic acid is an obligatory precursor to cholesterol, so less mevalonic acid translates into less cholesterol.

Vitamin D_3 is a key compound in the process by which Ca^{2+} is absorbed from the intestine. Low levels of vitamin D_3 lead to Ca^{2+} concentrations in the body that are insufficient to support proper bone growth, resulting in the bone disease called rickets.

Rickets was once more widespread than it is now. It was thought to be a dietary deficiency disease because it could be prevented in children by feeding them fish liver oil. Actually, rickets is an environmental disease brought about by a deficiency of sunlight. Where the winter sun is weak, children may not be exposed to enough of its light to convert the 7-dehydrocholesterol in their skin to vitamin D_3 at levels sufficient to promote the growth of strong bones. Fish have adapted to an environment that screens them from sunlight, and so they are not directly dependent on photochemistry for their vitamin D_3 and accumulate it by a different process. Although fish liver oil is a good source of vitamin D_3, it is not very palatable. Synthetic vitamin D_3, prepared from cholesterol, is often added to milk and other foods to ensure that children receive enough of the vitamin for their bones to develop properly. Irradiated ergosterol is another dietary supplement added to milk and other foods for the same purpose. Ergosterol, a steroid obtained from yeast, is structurally similar to 7-dehydrocholesterol and, on irradiation with sunlight or artificial light, is converted to vitamin D_2, a substance analogous to vitamin D_3 and comparable to it in antirachitic activity.

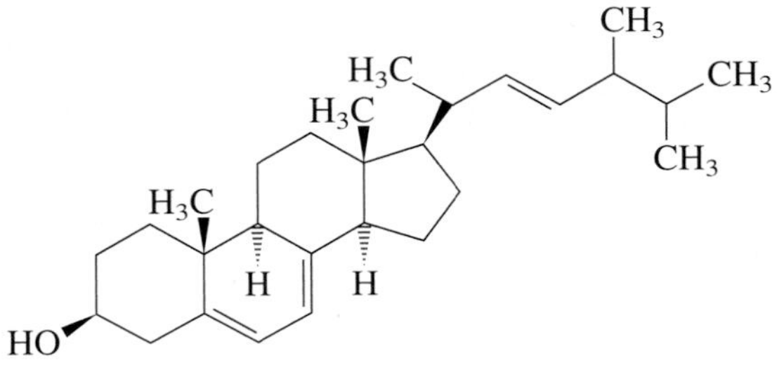

Ergosterol

[**PROBLEM 16.4** Suggest a reasonable structure for vitamin D$_2$.]

16.7 BILE ACIDS

A significant fraction of the body's cholesterol is used to form **bile acids.** Oxidation in the liver removes a portion of the C_8H_{17} side chain, and additional hydroxyl groups are introduced at various positions on the steroid nucleus. Cholic acid is the most abundant of the bile acids. In the form of certain amide derivatives called **bile salts,** of which sodium taurocholate is one example, bile acids act as emulsifying agents to aid the digestion of fats. Bile salts have detergent properties similar to those of salts of long-chain fatty acids and promote the transport of lipids through aqueous media.

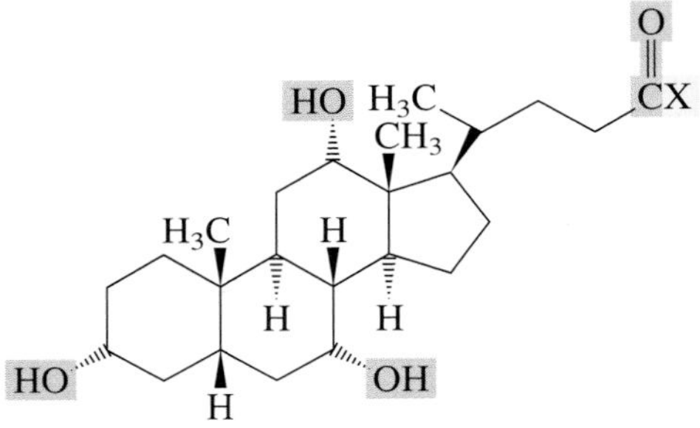

X = OH: cholic acid
X = NHCH$_2$CH$_2$SO$_3$Na: sodium taurocholate

16.8 CORTICOSTEROIDS

The outer layer, or **cortex,** of the adrenal gland is the source of a large group of substances known as **corticosteroids.** Like the bile acids, they are derived from cholesterol by oxidation, with cleavage of a portion of the alkyl substituent on the D ring. Cortisol is the most abundant of the corticosteroids, but cortisone is probably the best known. Cortisone is commonly prescribed as an antiinflammatory drug, especially in the treatment of rheumatoid arthritis.

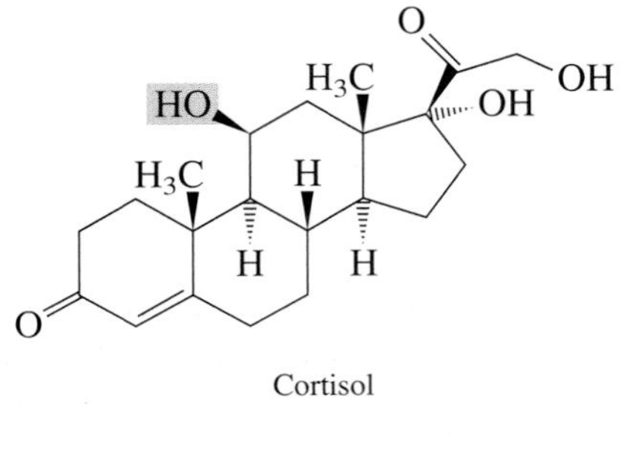

Cortisol

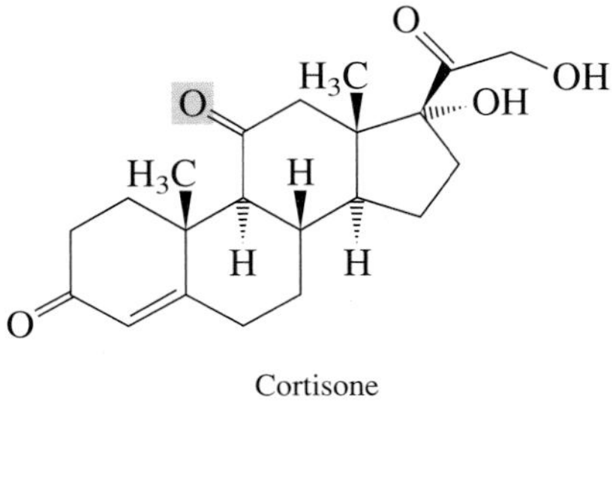

Cortisone

Corticosteroids exhibit a wide range of physiological effects. One important function is to assist in maintaining the proper electrolyte balance in body fluids. They also
play a vital regulatory role in the metabolism of carbohydrates and in mediating the allergic response.

Many antiitch remedies contain dihydrocortisone.

16.9   SEX HORMONES

Hormones are the chemical messengers of the body; they are secreted by the endocrine
glands and regulate biological processes. Corticosteroids, described in the preceding section,
are hormones produced by the adrenal glands. The sex glands—testes in males, ovaries in
females—secrete a number of hormones that are involved in sexual development and reproduction. Testosterone is the principal male sex hormone; it is an **androgen.** Testosterone
promotes muscle growth, deepening of the voice, the growth of body hair, and other male
secondary sex characteristics. Testosterone is formed from cholesterol and is the biosynthetic precursor of estradiol, the principal female sex hormone, or **estrogen.** Estradiol is a
key substance in the regulation of the menstrual cycle and the reproductive process. It is
the hormone most responsible for the development of female secondary sex characteristics.

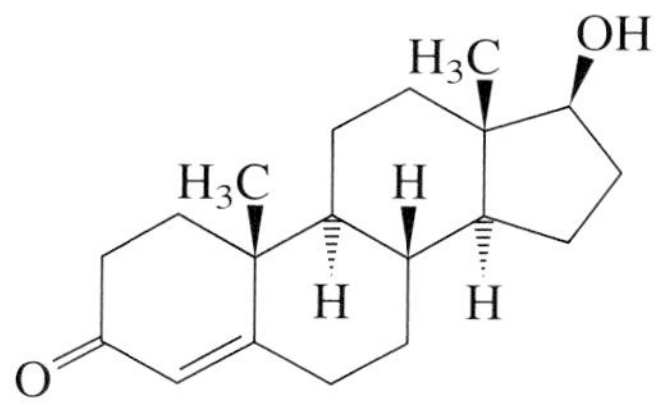

Testosterone 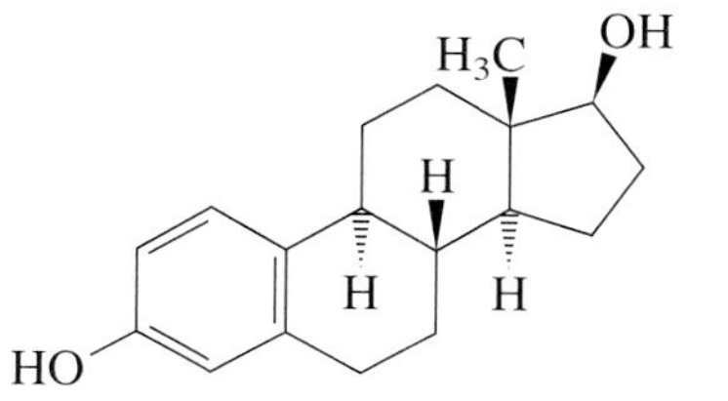   Estradiol

Testosterone and estradiol are present in the body in only minute amounts, and
their isolation and identification required heroic efforts. To obtain 0.012 g of estradiol
for study, for example, 4 tons of sow ovaries had to be extracted!

A separate biosynthetic pathway leads from cholesterol to progesterone, a female sex
hormone. One function of progesterone is to suppress ovulation at certain stages of the menstrual cycle and during pregnancy. Synthetic substances, such as norethindrone, have been
developed that are superior to progesterone when taken orally to "turn off" ovulation. By
inducing temporary infertility, they form the basis of most oral contraceptive agents.

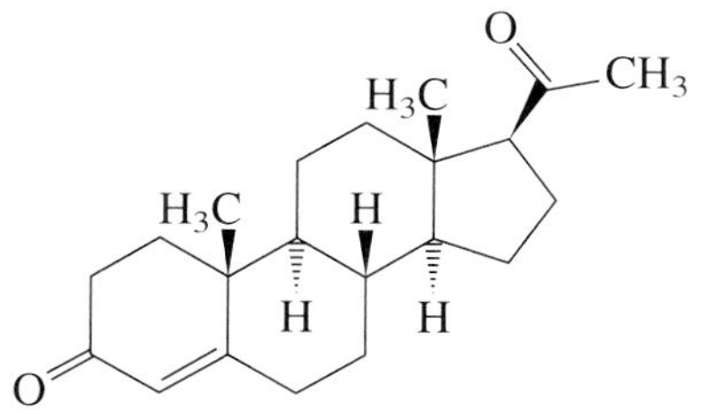

Progesterone 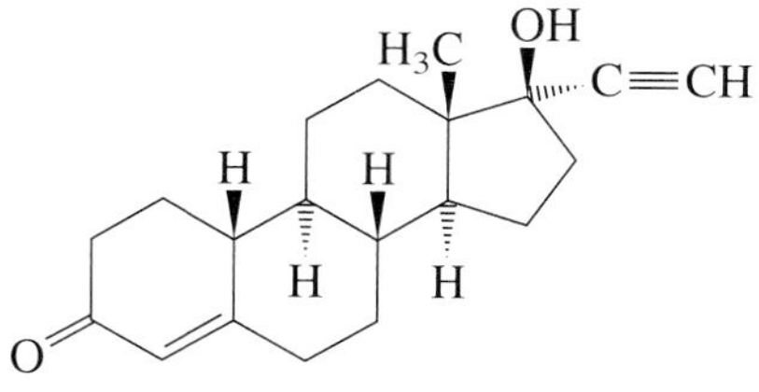   Norethindrone

16.10 CAROTENOIDS

Carotenoids are natural pigments characterized by a tail-to-tail linkage between two C_{20}
units and an extended conjugated system of double bonds. They are the most widely distributed of the substances that give color to our world and occur in flowers, fruits, plants,

ANABOLIC STEROIDS

As we have seen in this chapter, steroids have a number of functions in human physiology. Cholesterol is a component part of cell membranes and is found in large amounts in the brain. Derivatives of cholic acid assist the digestion of fats in the small intestine. Cortisone and its derivatives are involved in maintaining the electrolyte balance in body fluids. The sex hormones responsible for masculine and feminine characteristics as well as numerous aspects of pregnancy from conception to birth are steroids.

In addition to being an androgen, the principal male sex hormone testosterone promotes muscle growth and is classified as an **anabolic** steroid hormone. Biological chemists distinguish between two major classes of metabolism: catabolic and anabolic processes. **Catabolic processes** are degradative pathways in which larger molecules are broken down to smaller ones. **Anabolic processes** are the reverse; larger molecules are synthesized from smaller ones. Although the body mainly stores energy from food in the form of fat, a portion of that energy goes toward producing muscle from protein. An increase in the amount of testosterone, accompanied by an increase in the amount of food consumed, will cause an increase in the body's muscle mass.

Androstenedione, a close relative of testosterone, reached the public's attention in connection with Mark McGwire's successful bid to break Roger Maris's home run record in the summer of 1998. Androstenedione differs from testosterone in having a carbonyl group in the D ring where testosterone has a hydroxyl group. McGwire admitted to taking androstenedione, which is available as a nutritional supplement in health food stores and doesn't violate any of the rules of Major League Baseball. A controversy ensued as to the wisdom of androstenedione being sold without a prescription and the fairness of its use by athletes. Although the effectiveness of androstenedione as an anabolic steroid has not been established, it is clearly not nearly as potent as some others.

The pharmaceutical industry has developed and studied a number of anabolic steroids for use in veterinary medicine and in rehabilitation from injuries that are accompanied by deterioration of muscles. The ideal agent would be one that possessed the anabolic properties of testosterone without its androgenic (masculinizing) effects. Methandrostenolone (Dianabol) and stanozolol are among the many synthetic anabolic steroids that require a prescription.

Methandrostenolone
(Dianabol)

Stanozolol

Some scientific studies indicate that the gain in performance obtained through the use of anabolic steroids is small. This may be a case, though, in which the anecdotal evidence of the athletes may be closer to the mark than the scientific studies. The scientific studies are done under ethical conditions in which patients are treated with "prescription-level" doses of steroids. A 240-lb offensive tackle ("too small" by today's standards) may take several anabolic steroids at a time at 10–20 times their prescribed doses in order to weigh the 280 lb he (or his coach) feels is necessary. The price athletes pay for gains in size and strength can be enormous. This price includes emotional costs (friendships lost because of heightened aggressiveness), sterility, testicular atrophy (the testes cease to function once the body starts to obtain a sufficient supply of testosterone-like steroids from outside), and increased risk of premature death from liver cancer or heart disease.

Androstenedione

insects, and animals. It has been estimated that biosynthesis produces approximately a hundred million tons of carotenoids per year. The most familiar carotenoids are lycopene and β-carotene, pigments found in numerous plants and easily isolable from ripe tomatoes and carrots, respectively.

Lycopene

β-Carotene

Carotenoids absorb visible light (Section 19.11) and dissipate its energy as heat, thereby protecting the organism from any potentially harmful effects associated with sunlight-induced photochemistry. They are also indirectly involved in the chemistry of vision, owing to the fact that β-carotene is the biosynthetic precursor of vitamin A, also known as retinol, a key substance in the visual process.

> The structural chemistry of the visual process, beginning with β-carotene, was described in the boxed essay entitled "Imines in Biological Chemistry" in Chapter 11.

16.11 BIOSYNTHESIS: ACETYL COENZYME A

Many of the complex molecules in living systems are formed by **biosynthesis;** the molecule is prepared by the organism by a series of chemical reactions from simple starting materials. Most lipids, for example, are acetate-derived natural products. The most prevalent form of acetate in biosynthesis is a **thioester** known as acetyl coenzyme A (Figure 16.4a). Acetyl coenzyme A is itself formed from pyruvic acid; the sequence of steps is summarized in the following equation:

$$\underset{\substack{\text{Pyruvic} \\ \text{acid}}}{\text{CH}_3\text{CCOH}} + \underset{\text{Coenzyme A}}{\text{CoASH}} + \underset{\substack{\text{Oxidized} \\ \text{form of} \\ \text{nicotinamide} \\ \text{adenine} \\ \text{dinucleotide}}}{\text{NAD}^+} \longrightarrow \underset{\substack{\text{Acetyl} \\ \text{coenzyme A}}}{\text{CH}_3\text{CSCoA}} + \underset{\substack{\text{Reduced} \\ \text{form of} \\ \text{nicotinamide} \\ \text{adenine} \\ \text{dinucleotide}}}{\text{NADH}} + \underset{\substack{\text{Carbon} \\ \text{dioxide}}}{\text{CO}_2} + \underset{\text{Proton}}{\text{H}^+}$$

All the individual steps are catalyzed by enzymes. The cofactor NAD^+ is required as an oxidizing agent, and coenzyme A (Figure 16.4b) is the acetyl group acceptor. Coenzyme A is a **thiol;** its chain terminates in a **sulfhydryl** (—SH) **group.** Acetylation of the sulfhydryl group of coenzyme A gives acetyl coenzyme A.

Because sulfur does not stabilize a carbonyl group by electron donation as well as oxygen does, compounds of the type RCSR' (with C=O) are better acyl transfer agents than is RCOR' (with C=O). They also contain a greater proportion of enol ($-\overset{\text{OH}}{\underset{\text{}}{\text{C}}}=\overset{}{\underset{}{\text{C}}}-$) at equilibrium.

> Enzymes are biological catalysts. **Coenzymes** are molecules that act with enzymes to bring about a reaction. Because coenzymes undergo chemical change, they cannot be considered catalysts.

FIGURE 16.4 Structures of (*a*) acetyl coenzyme A and (*b*) coenzyme A.

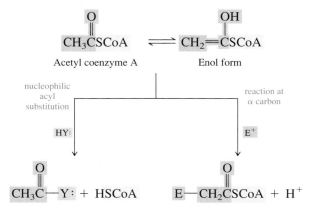

(*a*) $R = \overset{O}{\overset{\|}{C}}CH_3$ Acetyl coenzyme A (abbreviation: $CH_3\overset{O}{\overset{\|}{C}}SCoA$)

(*b*) $R = H$ Coenzyme A (abbreviation: CoASH)

Both properties are apparent in the chemistry of acetyl coenzyme A. In some of its reactions acetyl coenzyme A acts as an acyl transfer agent, whereas in others the α-carbon atom of the acetyl group is the reactive site.

$$CH_3\overset{O}{\overset{\|}{C}}SCoA \;\rightleftharpoons\; CH_2{=}\overset{OH}{\overset{|}{C}}SCoA$$

Acetyl coenzyme A   Enol form

nucleophilic acyl substitution   reaction at α carbon

HY: E⁺

$$CH_3\overset{O}{\overset{\|}{C}}{-}Y: + HSCoA \qquad E{-}CH_2\overset{O}{\overset{\|}{C}}SCoA + H^+$$

Acetyl coenzyme A is the biosynthetic starting material for the fatty acids found in triacylglycerols, phospholipids, and waxes. Because these compounds are built up by coupling acetate units together, fatty acids have an even number of carbon atoms in an unbranched chain. Cholesterol, and hence all the steroids, are formed through a complex series of transformations beginning with acetate. Most of these reactions have been studied and are understood in some detail. We will look briefly in the following sections at the biosynthesis of one class of lipids called terpenes.

16.12 TERPENE BIOSYNTHESIS: THE ISOPRENE RULE

The word "essential" as applied to naturally occurring organic substances can have two different meanings. For example, certain amino acids are essential, meaning they are necessary and must be present in our diet because humans lack the ability to biosynthesize them directly.

Essential is also used as the adjective form of the noun "essence." The mixtures of substances that create the fragrant material of plants are called essential oils because they contain the essence, or the odor, of the plant. The study of the composition of essential oils is one of the oldest areas of chemical research. Very often, the principal volatile component of an essential oil belongs to the class of chemical substances known as **terpenes.**

Myrcene, a hydrocarbon isolated from bayberry oil, is a typical terpene:

$$(CH_3)_2C=CHCH_2CH_2CCH=CH_2 \equiv$$

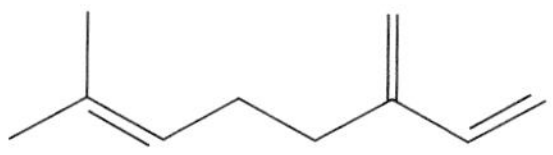

Myrcene

The structural feature that distinguishes terpenes from other natural products is the **isoprene unit.** The carbon skeleton of myrcene (exclusive of its double bonds) corresponds to the head-to-tail union of two isoprene units.

$$CH_2=C-CH=CH_2 \equiv$$

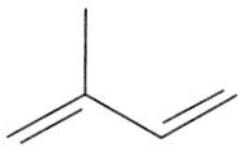

Isoprene
(2-methyl-1,3-butadiene)

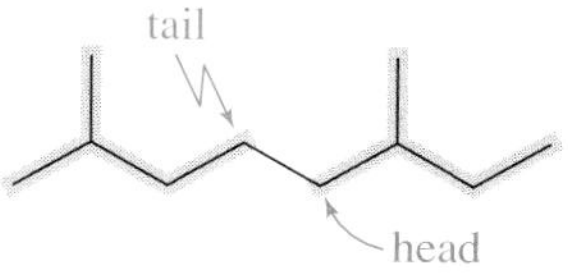

Two isoprene units
linked head to tail

Terpenes are often referred to as isoprenoid compounds. They are classified according to the number of carbon atoms they contain, as summarized in Table 16.2.

There are more than 23,000 known isoprenoid compounds.

Although the term "terpene" once referred only to hydrocarbons, current usage includes functionally substituted derivatives as well. Figure 16.5 presents the structural formulas for a number of representative terpenes. The isoprene units in some of these are relatively easy to identify. The three isoprene units in the sesquiterpene farnesol, for example, are indicated as follows in color. They are joined in a head-to-tail fashion.

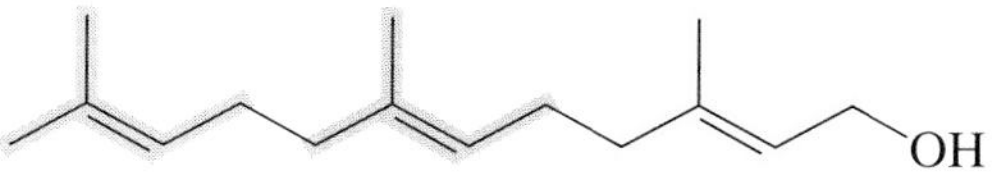

OH

Isoprene units in farnesol

Many terpenes contain one or more rings, but these also can be viewed as collections of isoprene units. An example is α-selinene. Like farnesol, it is made up of three isoprene units linked head to tail.

TABLE 16.2	Classification of Terpenes

Class	Number of Carbon Atoms
Monoterpene	10
Sesquiterpene	15
Diterpene	20
Sesterpene	25
Triterpene	30
Tetraterpene	40

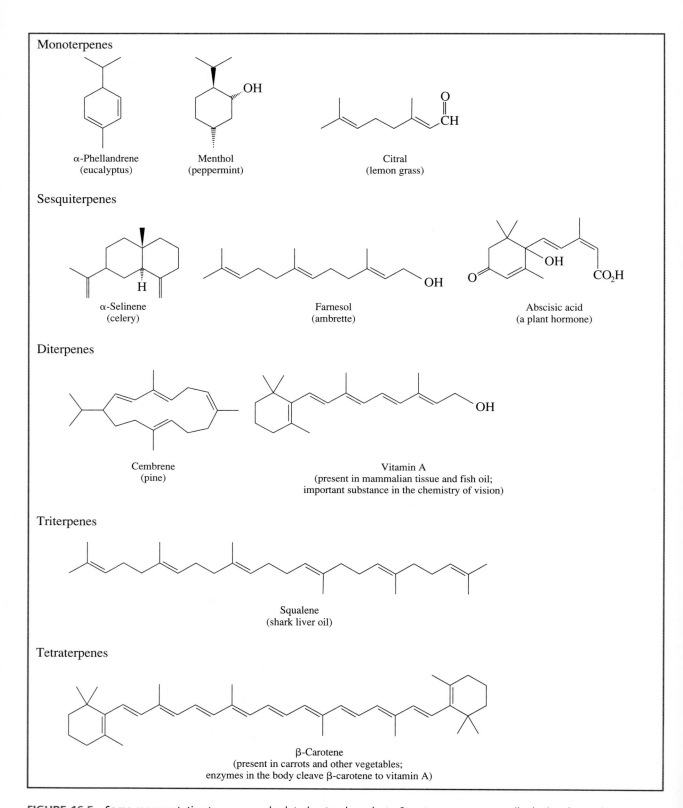

FIGURE 16.5 Some representative terpenes and related natural products. Structures are customarily depicted as carbon skeleton formulas when describing compounds of isoprenoid origin.

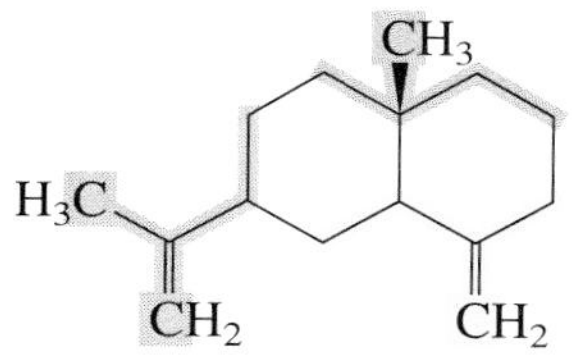

Isoprene units in α-selinene

PROBLEM 16.5   Locate the isoprene units in each of the monoterpenes, sesquiterpenes, and diterpenes shown in Figure 16.5. (In some cases there are two equally correct arrangements.)

Sample Solution   Isoprene units are 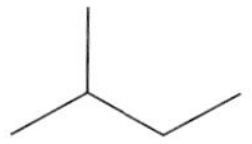 fragments in the carbon skeleton. Functional groups and multiple bonds are ignored when structures are examined for the presence of isoprene units. The monoterpene α-phellandrene has ten carbons and thus contains two C_5 isoprene units. There are two equally correct answers:

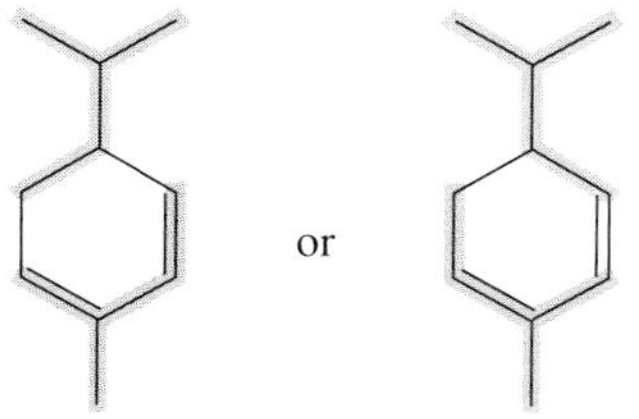

or

Tail-to-tail linkages of isoprene units sometimes occur, especially in the higher terpenes. The C(12)–C(13) bond of squalene unites two C_{15} units in a tail-to-tail manner. Notice, however, that isoprene units are joined head to tail within each C_{15} unit of squalene.

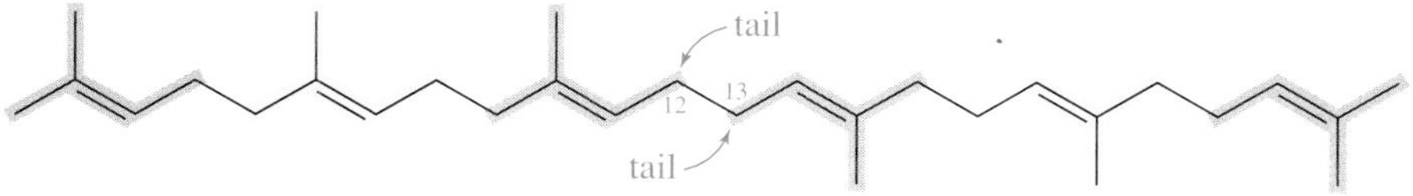

Isoprene units in squalene

PROBLEM 16.6   Identify the isoprene units in β-carotene (see Figure 16.5). Which carbons are joined by a tail-to-tail link between isoprene units?

The German chemist Otto Wallach (Nobel Prize in chemistry, 1910) established the structures of many monoterpenes and is credited with recognizing that they can be viewed as collections of isoprene units. Leopold Ruzicka of the Swiss Federal Institute of Technology (Zürich), in his studies of sesquiterpenes and higher terpenes, extended and refined what we now know as the **isoprene rule.** He was a corecipient of the Nobel Prize in chemistry in 1939. Although exceptions to it are known, the isoprene rule is a useful guide to terpene structures and has stimulated research in the biosynthetic origin of these compounds. It is a curious fact that terpenes contain isoprene units but isoprene does not occur naturally. What is the biological isoprene unit, how is it biosynthesized, and how do individual isoprene units combine to give terpenes?

16.13 ISOPENTENYL PYROPHOSPHATE: THE BIOLOGICAL ISOPRENE UNIT

Isoprenoid compounds are biosynthesized from acetate by a process that involves several stages. The first stage is the formation of mevalonic acid from three molecules of acetic acid:

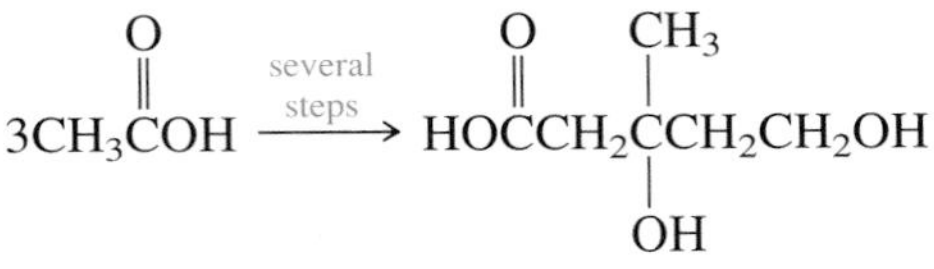

$$3CH_3\overset{\overset{\displaystyle O}{\parallel}}{C}OH \xrightarrow[\text{steps}]{\text{several}} HO\overset{\overset{\displaystyle O}{\parallel}}{C}CH_2\underset{\underset{\displaystyle OH}{|}}{\overset{\overset{\displaystyle CH_3}{|}}{C}}CH_2CH_2OH$$

Acetic acid   Mevalonic acid

In the second stage, mevalonic acid is converted to 3-methyl-3-butenyl pyrophosphate (isopentenyl pyrophosphate):

It is convenient to use the symbol —OPP to represent the pyrophosphate group.

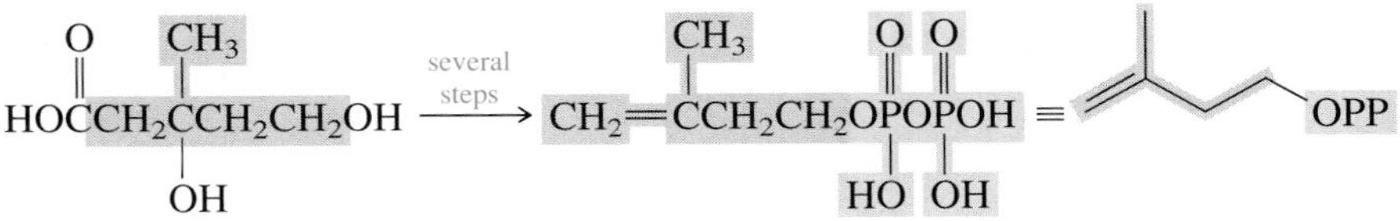

Mevalonic acid   Isopentenyl pyrophosphate

Isopentenyl pyrophosphate is the biological isoprene unit; it contains five carbon atoms connected in the same order as in isoprene.

Isopentenyl pyrophosphate undergoes an enzyme-catalyzed reaction that converts it, in an equilibrium process, to 3-methyl-2-butenyl pyrophosphate (dimethylallyl pyrophosphate):

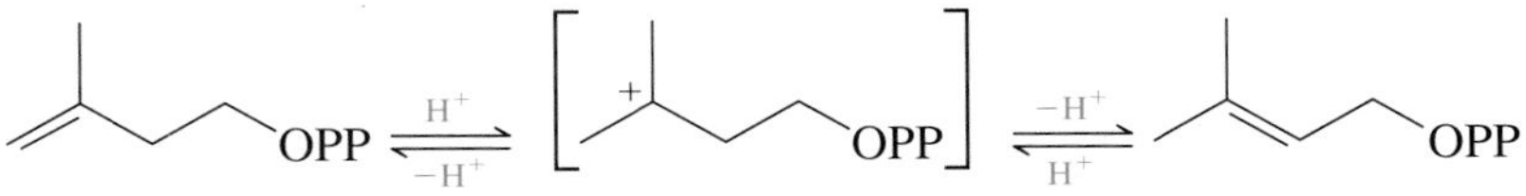

Isopentenyl   Carbocation intermediate   Dimethylallyl
pyrophosphate   pyrophosphate

Both isopentenyl pyrophosphate and dimethylallyl pyrophosphate are used in the biosynthesis of terpenes. This can be illustrated by outlining the formation of the monoterpene geraniol.

16.14 CARBON–CARBON BOND FORMATION IN TERPENE BIOSYNTHESIS

The chemical properties of isopentenyl pyrophosphate and dimethylallyl pyrophosphate are complementary in a way that permits them to react with each other to form a carbon–carbon bond that unites two isoprene units. Using the π electrons of its double bond, isopentenyl pyrophosphate acts as a nucleophile and displaces pyrophosphate from dimethylallyl pyrophosphate.

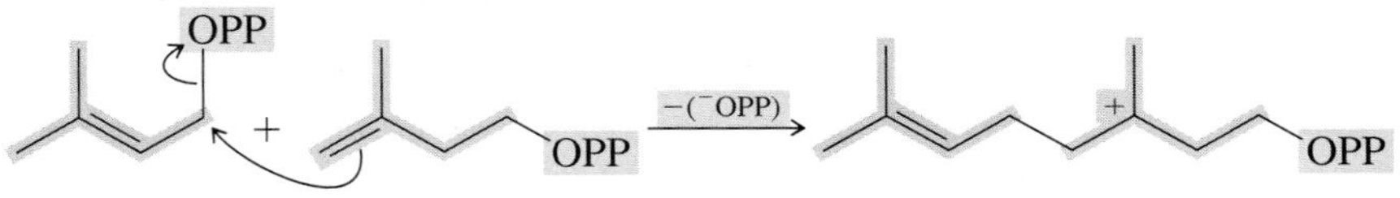

Dimethylallyl   Isopentenyl   Ten-carbon carbocation
pyrophosphate   pyrophosphate

The tertiary carbocation formed in this step can react according to any of the various reaction pathways available to carbocations. One of these is loss of a proton to give a double bond.

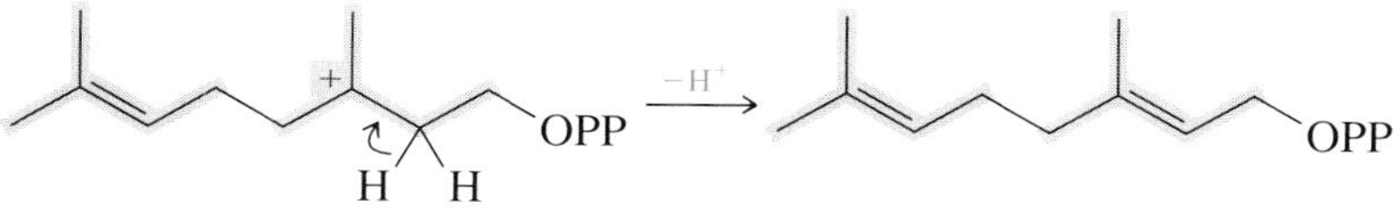

Geranyl pyrophosphate

The product of this reaction is geranyl pyrophosphate. Hydrolysis of the pyrophosphate ester group gives geraniol, a naturally occurring monoterpene found in rose oil.

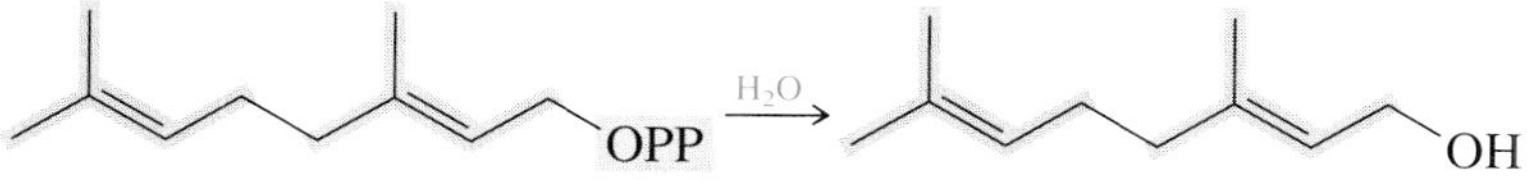

Geranyl pyrophosphate   Geraniol

> **PROBLEM 16.7**   Reaction of geranyl pyrophosphate with isopentenyl pyrophosphate leads to the sesquiterpene farnesol (see Figure 16.5). Outline a series of steps for this process.

LEARNING OBJECTIVES

This chapter focused on lipids, a major class of substances found in living organisms. The skills you have learned in this chapter should enable you to

- Explain what physical property defines a substance as a lipid.
- Write the chemical structure of a fat or oil and explain the difference between them.
- Describe the structural characteristics of a fatty acid.
- Write the structure of a phospholipid and explain the structural features that enable phospholipid molecules to form cell membranes.
- Write a structural formula typical of a wax.
- Explain the significance of cholesterol in the formation of steroids.
- Describe the function of vitamin D and the bile acids in the body.
- Describe the role of acetyl coenzyme A in the biosynthesis of terpenes.
- Identify the isoprene units in a terpene structure.

16.15   SUMMARY

Lipids are naturally occurring substances that are soluble in nonpolar solvents (Section 16.1). **Fats** and **oils** are glycerol esters of long-chain carboxylic acids (Section 16.2). Typically, the carbon chains of the fatty acids are unbranched and contain an even number of carbon atoms.

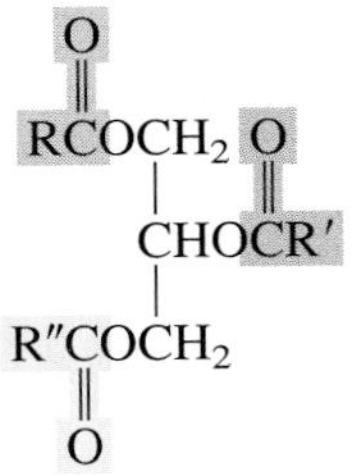

$$
\begin{array}{c}
O \\
\parallel \\
\text{RCOCH}_2 \quad O \\
\mid \qquad \parallel \\
\text{CHOCR}' \\
\mid \\
\text{R}''\text{COCH}_2 \\
\parallel \\
O
\end{array}
$$

Triacylglycerol
(R, R', and R'' may be the same or different)

Oils are usually of vegetable origin and tend to have one or more cis double bonds in their fatty acid chains. **Phospholipids** such as lecithin are structurally similar to triacylglycerols and are a principal component of cell membranes (Section 16.3). **Waxes** are mixtures of substances that usually contain esters of fatty acids and long-chain alcohols (Section 16.4).

Cholesterol is the most familiar of the **steroids** (Section 16.5). It is found in almost all body tissue and is the biological precursor of the other steroids.

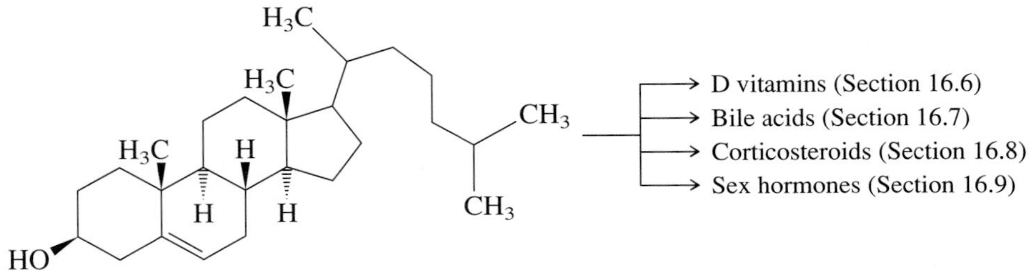

Cholesterol

→ D vitamins (Section 16.6)
→ Bile acids (Section 16.7)
→ Corticosteroids (Section 16.8)
→ Sex hormones (Section 16.9)

Most lipids share a common biosynthetic origin in that all their carbons are derived from acetic acid (acetate) (Section 16.11).

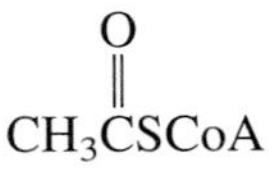

$$
\begin{array}{c}
O \\
\parallel \\
\text{CH}_3\text{CSCoA}
\end{array}
$$

Abbreviation for acetyl coenzyme A
(for complete structure, see Figure 16.4)

Terpenes and related isoprenoid compounds are biosynthesized from isopentenyl pyrophosphate (Sections 16.12 and 16.13).

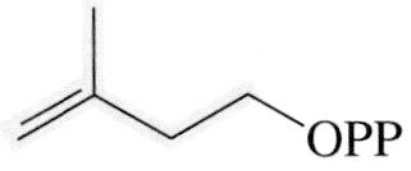

Isopentenyl pyrophosphate is
the "biological isoprene unit."

Carbon–carbon bond formation between isoprene units can be understood on the basis of nucleophilic attack of the π electrons of a double bond on a carbocation or an allylic carbon that bears a pyrophosphate leaving group (Section 16.14).

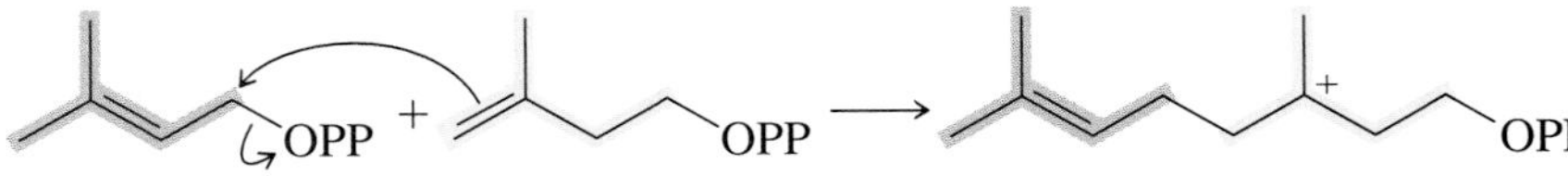

ADDITIONAL PROBLEMS

Fatty Acids and Esters

16.8 What structural features differentiate each of the following?

    (a) Animal fat and vegetable oil

    (b) A fat and a wax

    (c) A phospholipid and a triacylglycerol

16.9 Phospholipids are more soluble in water than triacylglycerols. Explain.

16.10 By means of chemical equations, show

    (a) The hydrolysis of trioleylglycerol

    (b) The catalytic hydrogenation of trioleylglycerol

16.11 Linolein is a lipid constituent of linseed oil and produces glycerol and linoleic acid on hydrolysis. Draw the structure of linolein.

16.12 (a) Elaidic acid is a stereoisomer of oleic acid. Write a structural formula for elaidic acid.

    (b) Ricinoleic acid, abundantly present as its glycerol triester in castor oil, is 12-hydroxy-oleic acid. Write a structural formula for ricinoleic acid.

16.13 One of the esters in the waxes responsible for the shiny coating on holly and rhododendron leaves is oleyl oleate. Write the structural formula and give the systematic name of this substance.

Steroids

16.14 Vitamin D is an example of a fat-soluble vitamin. Fat-soluble vitamins can be stored in moderate amounts in our bodies, and thus we are not dependent on a daily supply of them in our diet. Explain what structural characteristics make vitamin D fat-soluble.

16.15 How many stereogenic centers are in cholesterol? How many stereoisomers can have the cholesterol skeleton?

16.16 In spite of its complexity, cholesterol undergoes reactions with each of the following reagents that are typical of the functional groups it contains. Give the structure of the product formed in each case.

    (a) Br_2

    (b) H_2 (1 atm), Pt

    (c) Acetic anhydride

    (d) Pyridinium chlorochromate (PCC) in CH_2Cl_2

Terpenes and Isoprenoid Biosynthesis

16.17 Identify the isoprene units in each of the following naturally occurring substances:

    (a) Ascaridole, a naturally occurring peroxide present in chenopodium oil:

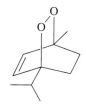

    (b) Dendrolasin, a constituent of the defense secretion of a species of ant:

(c) γ-Bisabolene, a sesquiterpene found in the essential oils of a large number of plants:

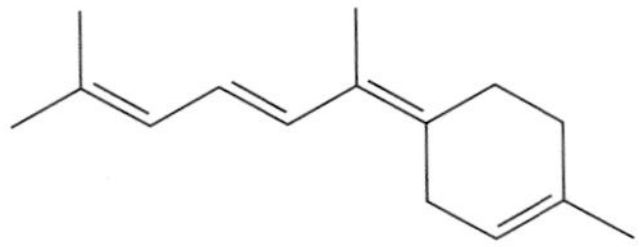

(d) α-Santonin, an anthelmintic (systemic drug for killing parasites) isolated from artemisia flowers:

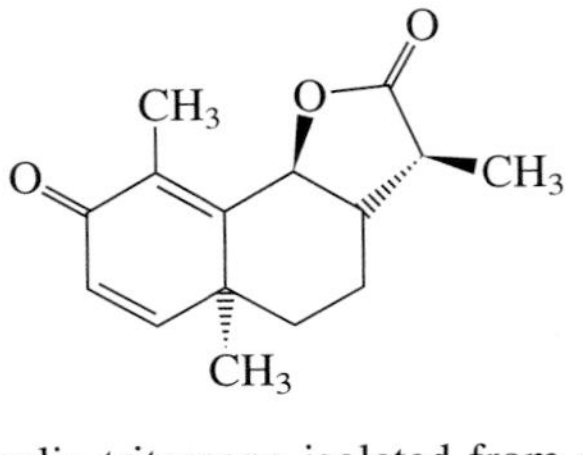

(e) Tetrahymanol, a pentacyclic triterpene isolated from a species of protozoans:

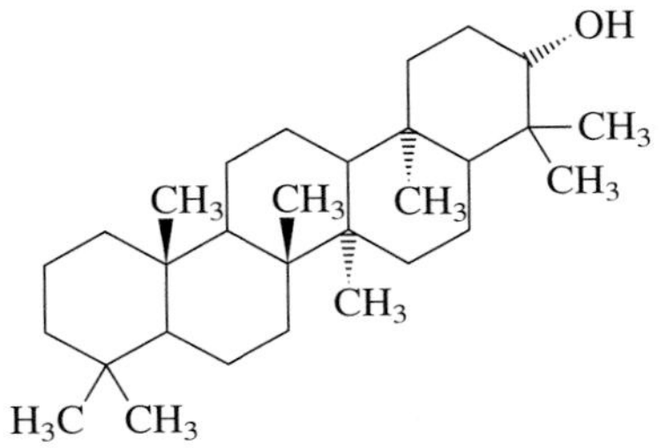

16.18 Cubitene is a diterpene present in the defense secretion of a species of African termite. What unusual feature characterizes the joining of isoprene units in cubitene?

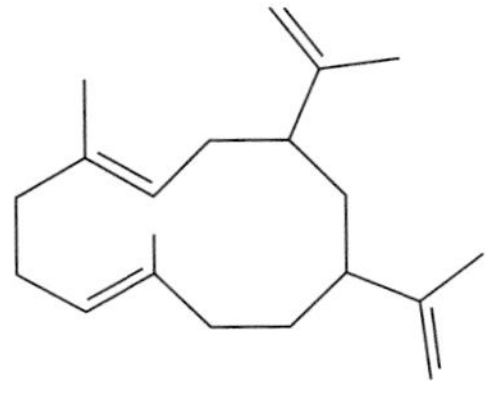

Miscellaneous

16.19 The Fischer projection of mevalonic acid (Section 16.13) is shown. Is the configuration of the stereogenic center *R* or *S*?

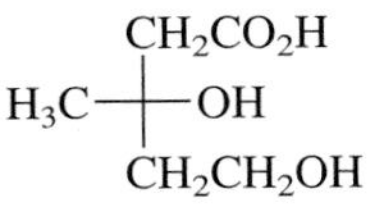

$$H_3C \overset{\displaystyle CH_2CO_2H}{\underset{\displaystyle CH_2CH_2OH}{\rule{0pt}{0pt}\Big|\!\!-\!\!OH}}$$

16.20 The isoprenoid compound shown is a scent marker present in the urine of the red fox. Suggest a reasonable synthesis for this substance from 4-bromo-2-methyl-1-butene and any necessary organic or inorganic reagents.

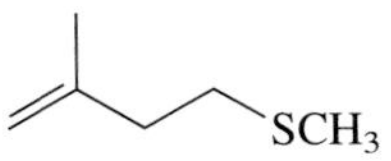

16.21 The monoterpenes α- and β-pinene are obtained from pine trees and are principal constituents of turpentine. Biosynthetically, both are derived from the same carbocation. Write a structural formula for this carbocation.

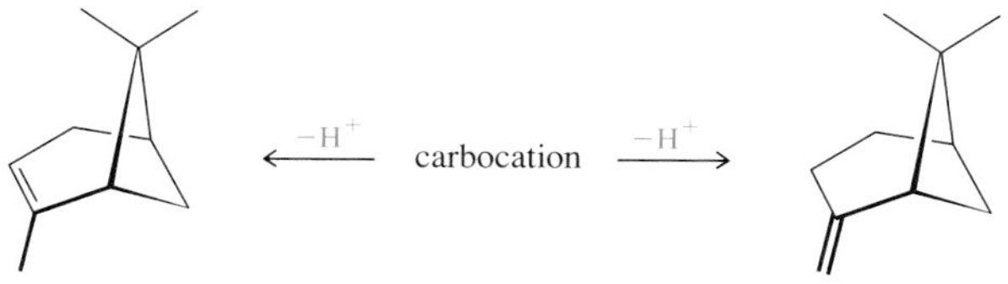

α-Pinene   β-Pinene

16.22 The ionones are fragrant substances present in the scent of iris and are used in perfume. A mixture of α- and β-ionone can be prepared by treatment of pseudoionone with sulfuric acid.

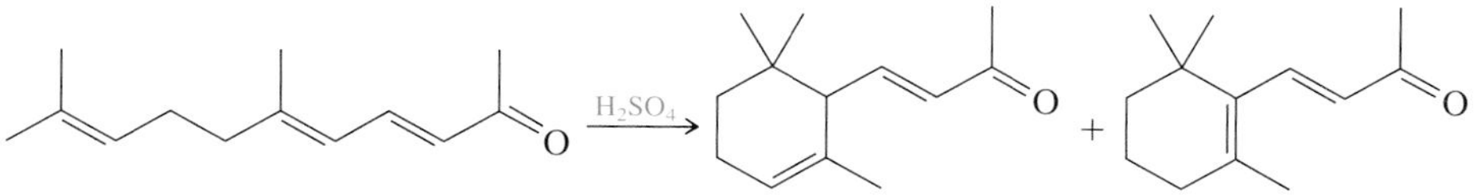

Pseudoionone   α-Ionone   β-Ionone

Write a stepwise mechanism for this reaction.

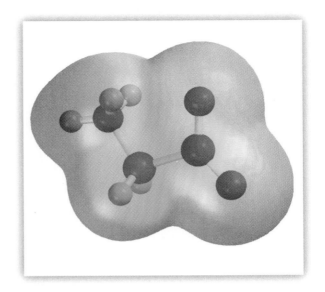

CHAPTER 17
AMINO ACIDS, PEPTIDES, AND PROTEINS

This chapter is devoted to amino acids and the polymers they form, called peptides and proteins. Proteins are striking in the diversity of roles they play in living systems: silk, hair, skin, muscle, and connective tissue are proteins. Most biological reactions are catalyzed by proteins called enzymes.

The relationship between structure and function reaches its ultimate expression in the chemistry of amino acids, peptides, and proteins. This chapter will explore the facets of protein structure by first concentrating on their fundamental building blocks, the α-amino acids. Then, after developing the principles of peptide structure and conformation, you will see how the insights gained from these smaller molecules aid our understanding of proteins.

17.1 STRUCTURE OF AMINO ACIDS

Amino acids are carboxylic acids that contain an amine functional group. Although more than 700 different amino acids are known to occur naturally, a group of 20 of them commands special attention. These 20 are the amino acids that are normally present in proteins and are shown in Figure 17.1 and in Table 17.1. All the amino acids from which proteins are derived are α-amino acids, and all but one of these contain a primary amino function and conform to the general structure

The one exception is proline, a secondary amine in which the amino nitrogen is incorporated into a five-membered ring.

Amino acids with nonpolar side chains

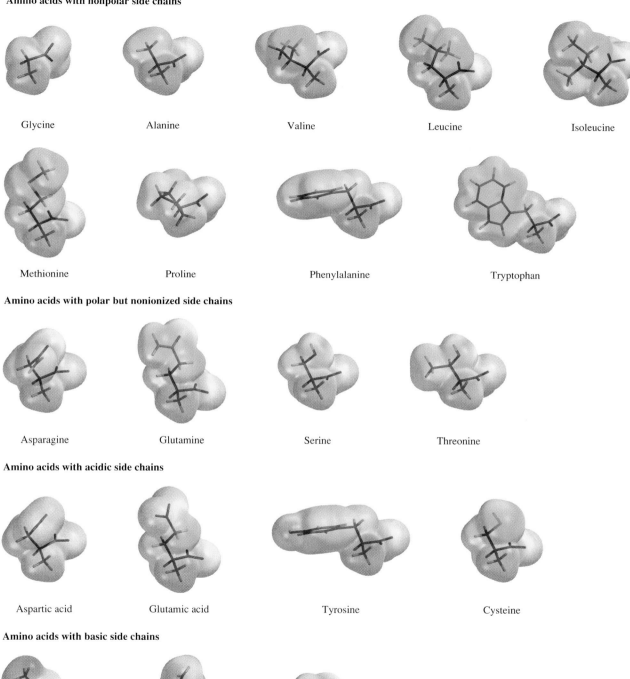

Glycine   Alanine   Valine   Leucine   Isoleucine

Methionine   Proline   Phenylalanine   Tryptophan

Amino acids with polar but nonionized side chains

Asparagine   Glutamine   Serine   Threonine

Amino acids with acidic side chains

Aspartic acid   Glutamic acid   Tyrosine   Cysteine

Amino acids with basic side chains

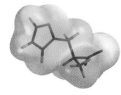

Lysine   Arginine   Histidine

FIGURE 17.1 Electrostatic potential maps of the 20 common amino acids listed in Table 17.1. Each amino acid is oriented so that its side chain is in the upper left corner. The side chains affect the shape and properties of the amino acids.

TABLE 17.1   α-Amino Acids Found in Proteins

Name	Abbreviation	Structural Formula*

Amino acids with nonpolar side chains

Name	Abbreviation	Structural Formula*
Glycine	Gly (G)	$\overset{\overset{+}{N}H_3}{H-CHCO_2^-}$
Alanine	Ala (A)	$\overset{\overset{+}{N}H_3}{CH_3-CHCO_2^-}$
Valine[†]	Val (V)	$\overset{\overset{+}{N}H_3}{(CH_3)_2CH-CHCO_2^-}$
Leucine[†]	Leu (L)	$\overset{\overset{+}{N}H_3}{(CH_3)_2CHCH_2-CHCO_2^-}$
Isoleucine[†]	Ile (I)	$\overset{CH_3\ \ \overset{+}{N}H_3}{CH_3CH_2CH-CHCO_2^-}$
Methionine[†]	Met (M)	$\overset{\overset{+}{N}H_3}{CH_3SCH_2CH_2-CHCO_2^-}$
Proline	Pro (P)	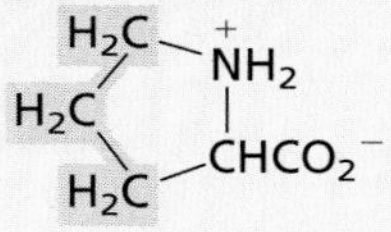 $\begin{array}{c} H_2C\diagdown \ \overset{+}{N}H_2 \\ H_2C \qquad \diagup \\ H_2C\diagup\ CHCO_2^- \end{array}$
Phenylalanine[†]	Phe (F)	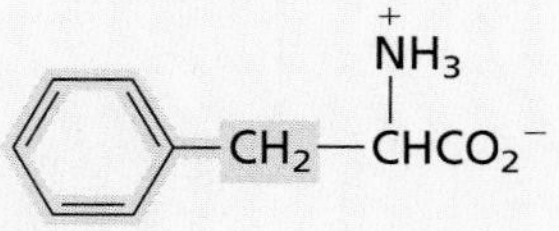 $C_6H_5-CH_2-\overset{\overset{+}{N}H_3}{CHCO_2^-}$
Tryptophan[†]	Trp (W)	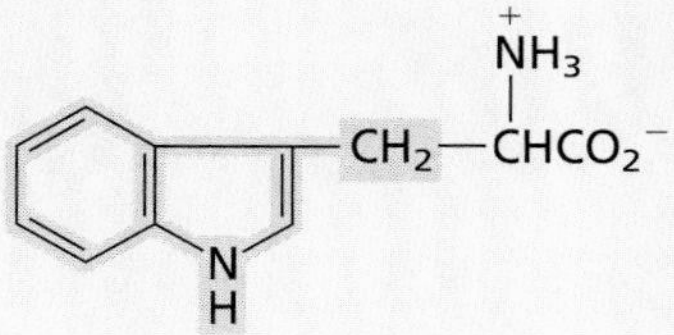 (indole)$-CH_2-\overset{\overset{+}{N}H_3}{CHCO_2^-}$

Amino acids with polar but nonionized side chains

Name	Abbreviation	Structural Formula*
Asparagine	Asn (N)	$\overset{O}{H_2NCCH_2}-\overset{\overset{+}{N}H_3}{CHCO_2^-}$

* All amino acids are shown in the form present in greatest concentration at pH 7.

[†] An essential amino acid, which must be present in the diet of animals to ensure normal growth.

(Continued)

TABLE 17.1	α-Amino Acids Found in Proteins *(Continued)*

Name	Abbreviation	Structural Formula*
Amino acids with polar but nonionized side chains		
Glutamine	Gln (Q)	$\overset{\overset{\displaystyle O}{\parallel}}{H_2NC}CH_2CH_2\overset{\overset{\displaystyle \overset{+}{N}H_3}{\mid}}{-CH}CO_2{}^-$
Serine	Ser (S)	$HOCH_2\overset{\overset{\displaystyle \overset{+}{N}H_3}{\mid}}{-CH}CO_2{}^-$
Threonine[†]	Thr (T)	$CH_3\overset{\overset{\displaystyle OH}{\mid}}{CH}\overset{\overset{\displaystyle \overset{+}{N}H_3}{\mid}}{-CH}CO_2{}^-$
Amino acids with acidic side chains		
Aspartic acid	Asp (D)	$\overset{\overset{\displaystyle O}{\parallel}}{{}^-OC}CH_2\overset{\overset{\displaystyle \overset{+}{N}H_3}{\mid}}{-CH}CO_2{}^-$
Glutamic acid	Glu (E)	$\overset{\overset{\displaystyle O}{\parallel}}{{}^-OC}CH_2CH_2\overset{\overset{\displaystyle \overset{+}{N}H_3}{\mid}}{-CH}CO_2{}^-$
Tyrosine	Tyr (Y)	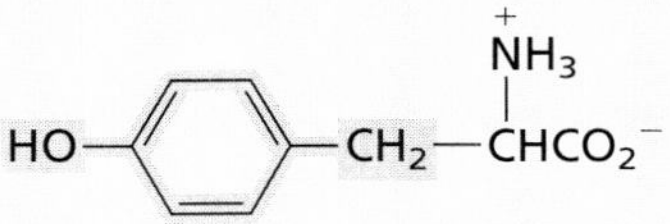 $HO-\underset{}{\bigcirc}-CH_2\overset{\overset{\displaystyle \overset{+}{N}H_3}{\mid}}{-CH}CO_2{}^-$
Cysteine	Cys (C)	$HSCH_2\overset{\overset{\displaystyle \overset{+}{N}H_3}{\mid}}{-CH}CO_2{}^-$
Amino acids with basic side chains		
Lysine[†]	Lys (K)	$\overset{+}{H_3}NCH_2CH_2CH_2CH_2\overset{\overset{\displaystyle \overset{+}{N}H_3}{\mid}}{-CH}CO_2{}^-$
Arginine[†]	Arg (R)	$H_2N\overset{\overset{\displaystyle \overset{+}{N}H_2}{\parallel}}{C}NHCH_2CH_2CH_2\overset{\overset{\displaystyle \overset{+}{N}H_3}{\mid}}{-CH}CO_2{}^-$
Histidine[†]	His (H)	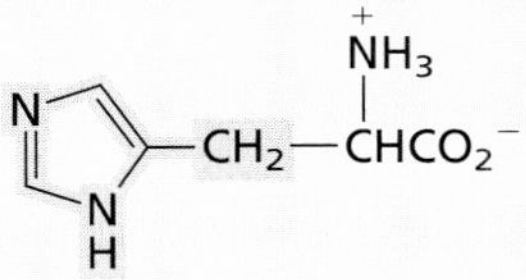 $\underset{\underset{\displaystyle H}{N}}{\overset{\displaystyle N}{\diagdown}}\!\!-CH_2\overset{\overset{\displaystyle \overset{+}{N}H_3}{\mid}}{-CH}CO_2{}^-$

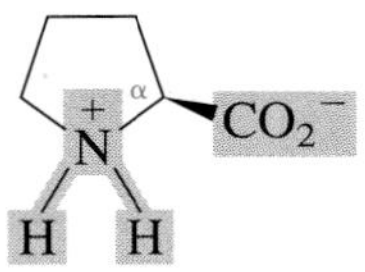

Table 17.1 includes three-letter and one-letter abbreviations for the amino acids. Both forms are widely used.

Our bodies can make some of the amino acids shown in the table. The others, which are called **essential amino acids,** we get from what we eat.

17.2 STEREOCHEMISTRY OF AMINO ACIDS

The graphic that opened this chapter is an electrostatic potential map of glycine.

Glycine is the simplest amino acid and the only one in Table 17.1 that is achiral. The α-carbon atom is a stereogenic center in all the others. Configurations in amino acids are normally specified by the D–L notational system. All the chiral amino acids obtained from proteins have the L configuration at their α-carbon atom.

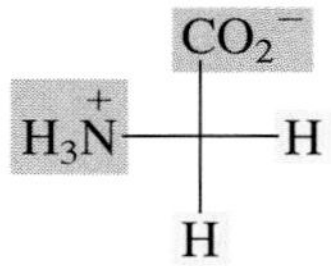

Glycine
(achiral)

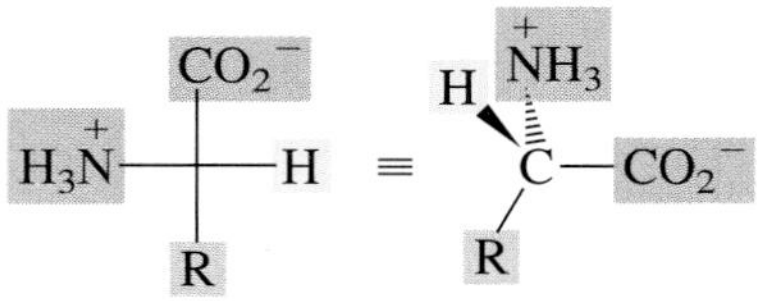

Fischer projection
of an L-amino acid

PROBLEM 17.1 What is the absolute configuration (R or S) at the α-carbon atom in each of the following L-amino acids?

(a)

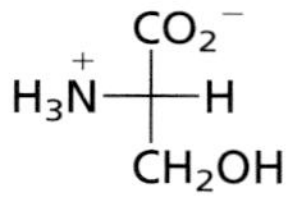

L-Serine

(c)

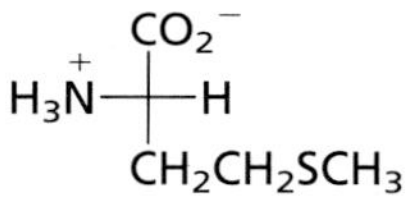

L-Methionine

(b)

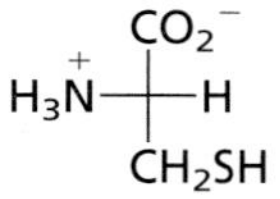

L-Cysteine

Sample Solution (a) First identify the four groups attached directly to the stereogenic center, and rank them in order of decreasing sequence rule precedence. For L-serine these groups are

$$H_3N- \quad > \quad -CO_2^- \quad > \quad -CH_2OH \quad > \quad H$$

Highest ranked   Lowest ranked

Next, translate the Fischer projection of L-serine to a three-dimensional representation, and orient it so that the lowest ranked substituent at the stereogenic center is directed away from you.

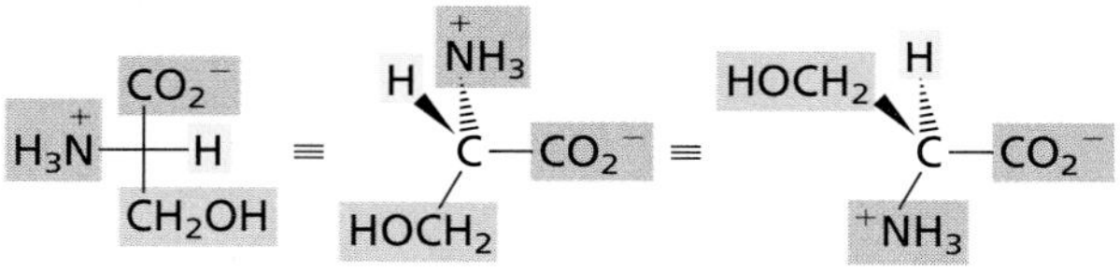

In order of decreasing precedence the three highest ranked groups trace a counterclockwise path.

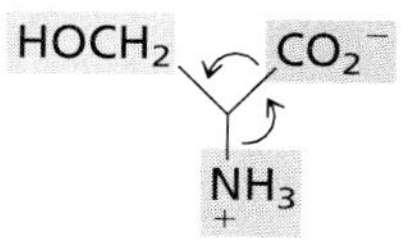

$$HOCH_2 \quad CO_2^-$$
$$NH_3^+$$

The absolute configuration of L-serine is *S*.

PROBLEM 17.2   Which of the amino acids in Table 17.1 have more than one stereogenic center?

Although all the chiral amino acids obtained from proteins have the L configuration at their α carbon, that should not be taken to mean that D-amino acids are unknown. In fact, quite a number of D-amino acids occur naturally. D-Alanine, for example, is a constituent of bacterial cell walls. The point is that D-amino acids are not constituents of proteins.

17.3   ACID–BASE BEHAVIOR OF AMINO ACIDS

The physical properties of a typical amino acid such as glycine suggest that it is a very polar substance, much more polar than would be expected on the basis of its formulation as $H_2NCH_2CO_2H$. Glycine is a crystalline solid; it does not melt, but on being heated it eventually decomposes at 233°C. It is very soluble in water but practically insoluble in nonpolar organic solvents. These properties are attributed to the fact that the stable form of glycine is a **zwitterion,** or **inner salt.**

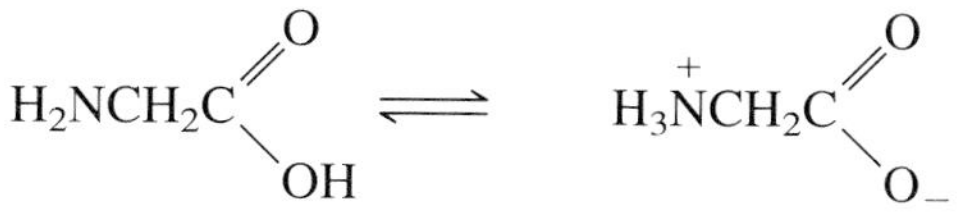

Zwitterionic form of glycine

> The zwitterion is also often referred to as a **dipolar ion.** Note, however, that it is not an ion, but a neutral molecule.

The equilibrium expressed by the preceding equation lies overwhelmingly to the side of the zwitterion.

Glycine and the other amino acids are **amphoteric;** they can act as both acids and bases. The acidic functional group is the ammonium ion, $\overset{+}{H_3}N\!-\!$; the basic functional group is the carboxylate ion, $-CO_2^-$. Consider glycine in aqueous solution as the pH changes. In a strongly acidic solution (low pH), the predominant species is the protonated form, an ammonium carboxylic acid. As the pH is raised, the carboxyl group is deprotonated, forming the zwitterion in solutions near neutrality. As the solution becomes more basic, the nitrogen of the zwitterion gives up a proton, forming an aminocarboxylate ion.

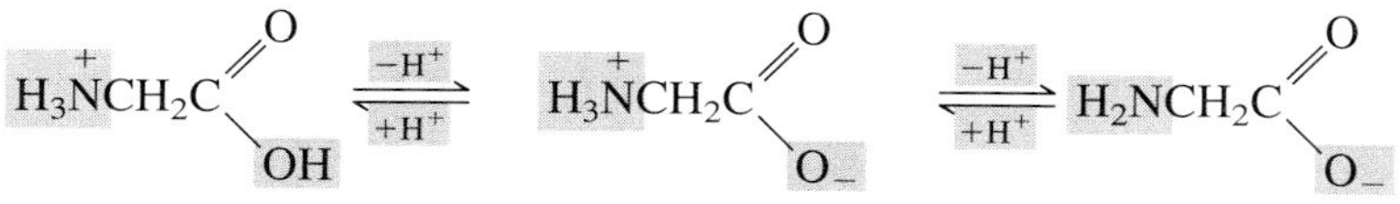

Species present in strong acid

Zwitterion; predominant species in solutions near neutrality

Species present in strong base

The pH of an aqueous solution at which the concentration of the zwitterion is a maximum is called the **isoelectric point,** or **pI,** of the amino acid. Each amino acid has a characteristic isoelectric point. Glycine, for example, has pI = 5.97.

PROBLEM 17.3 Write the structures of the predominant species present in a solution of phenylalanine (pI = 5.48) at pH:

(a) 2.0   (b) 5.5   (c) 9.0

Sample Solution (a) A pH of 2.0 represents an acidic solution; therefore the predominant species is an ammonium-carboxylic acid:

$$C_6H_5CH_2\underset{\overset{|}{{}^+NH_3}}{CH}\overset{\overset{O}{\|}}{C}OH$$

Several of the amino acids in Table 17.1 contain acidic or basic side chains. The presence of these ionizable groups has a dramatic effect on the value of the isoelectric point for that amino acid. Aspartic acid, for example, has a side chain containing a carboxylic acid group and an isoelectric point of 2.77. Lysine on the other hand, with a side chain containing a basic amine functional group, has an isoelectric point of 9.74.

PROBLEM 17.4 Write structural formulas for the principal species present when the pH of a solution containing lysine is raised from 1 to 9 and again to 13.

17.4   SYNTHESIS OF AMINO ACIDS

One of the oldest methods for the synthesis of amino acids dates back to the nineteenth century and is simply a nucleophilic substitution in which ammonia reacts with an α-halo carboxylic acid.

$$\underset{\overset{|}{Br}}{CH_3CH}CO_2H \quad + \quad 2NH_3 \quad \xrightarrow{H_2O} \quad \underset{\overset{|}{\underset{+}{NH_3}}}{CH_3CH}CO_2^- \quad + \quad NH_4Br$$

2-Bromopropanoic acid   Ammonia   Alanine   Ammonium bromide
                                                  (65–70%)

In the **Strecker synthesis** an aldehyde is converted to an α-amino acid with one more carbon atom by a two-stage procedure in which an α-amino nitrile is an intermediate. The α-amino nitrile is formed by reaction of the aldehyde with ammonia or an ammonium salt and a source of cyanide ion. Hydrolysis of the nitrile group to a carboxylic acid function (Section 12.10) completes the synthesis.

The synthesis is named after a nineteenth-century German chemist, Adolf Strecker, who first described the method.

$$\underset{}{CH_3}\overset{\overset{O}{\|}}{C}H \quad \xrightarrow[NaCN]{NH_4Cl} \quad \underset{\overset{|}{NH_2}}{CH_3CH}C\!\equiv\!N \quad \xrightarrow[\text{2. HO}^-]{\text{1. H}_2\text{O, HCl, heat}} \quad \underset{\overset{|}{\underset{+}{NH_3}}}{CH_3CH}CO_2^-$$

Acetaldehyde   2-Aminopropanenitrile   Alanine
                                                            (52–60%)

PROBLEM 17.5 Outline the steps in the preparation of valine by the Strecker synthesis.

ELECTROPHORESIS

lectrophoresis is a method for separation and purification that depends on the movement of charged particles in an electric field. Its principles can be introduced by considering the electrophoretic behavior of some representative amino acids. The medium is a cellulose acetate strip that is moistened with an aqueous solution buffered at a particular pH. The opposite ends of the strip are placed in separate compartments containing the buffer, and each compartment is connected to a source of direct electric current (Figure 17.2a). If the buffer solution is more acidic than the isoelectric point (pI) of the amino acid, the amino acid has a net positive charge and migrates toward the negatively charged electrode. Conversely, when the buffer is more basic than the pI of the amino acid, the amino acid has a net negative charge and mi-

grates toward the positively charged electrode. When the pH of the buffer corresponds to the pI, the amino acid has no net charge and does not migrate from the origin.

Thus if a mixture containing alanine, aspartic acid, and lysine is subjected to electrophoresis in a buffer that matches the isoelectric point of alanine (pH 6.0), aspartic acid (pI = 2.8) migrates toward the positive electrode, alanine remains at the origin, and lysine (pI = 9.7) migrates toward the negative electrode (Figure 17.2b).

$$^-O_2CCH_2CHCO_2^- \qquad CH_3CHCO_2^- \qquad \overset{+}{H_3}N(CH_2)_4CHCO_2^-$$
$$\underset{^+NH_3}{|} \qquad \underset{^+NH_3}{|} \qquad \underset{^+NH_3}{|}$$

Aspartic acid (monoanion)   Alanine (neutral)   Lysine (monocation)

A mixture of amino acids

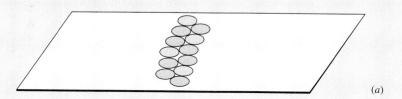

is placed at the center of a sheet of cellulose acetate. The sheet is soaked with an aqueous solution buffered at a pH of 6.0. At this pH aspartic acid ⬭ exists as its −1 ion, alanine ⬭ as its zwitterion, and lysine ⬭ as its +1 ion.

(a)

Application of an electric current causes the negatively charged ions to migrate to the + electrode, and the positively charged ions to migrate to the − electrode. The zwitterion, with a net charge of zero, remains at its original position.

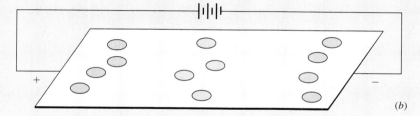

(b)

FIGURE 17.2 Application of electrophoresis to the separation of aspartic acid, alanine, and lysine according to their charge type at a pH corresponding to the isoelectric point (pI) of alanine.

Continued

Electrophoresis is used primarily to analyze mixtures of peptides and proteins, rather than individual amino acids, but analogous principles apply. Because they incorporate different numbers of amino acids and because their side chains are different, two peptides will have slightly different acid–base properties and slightly different net charges at a particular pH. Thus, their mobilities in an electric field will be different, and electrophoresis can be used to separate them. The medium used to separate peptides and proteins is typically a polyacrylamide gel, leading to the term **gel electrophoresis** for this technique.

A second factor that governs the rate of migration during electrophoresis is the size (length and shape) of the peptide or protein. Larger molecules move through the polyacrylamide gel more slowly than smaller ones. In current practice, the experiment is modified to exploit differences in size more than differences in net charge, especially in the **SDS gel electrophoresis** of proteins. Approxi-

mately 1.5 g of the detergent sodium dodecyl sulfate (SDS, Section 12.7) per gram of protein is added to the aqueous buffer. SDS binds to the protein, causing the protein to unfold so that it is roughly rod-shaped with the $CH_3(CH_2)_{10}CH_2-$ groups of SDS associated with the hydrophobic portions of the protein. The negatively charged sulfate groups are exposed to the water. The SDS molecules that they carry ensure that all the protein molecules are negatively charged and migrate toward the positive electrode. Furthermore, all the proteins in the mixture now have similar shapes and tend to travel at rates proportional to their chain length. Thus, when carried out on a preparative scale, SDS gel electrophoresis permits proteins in a mixture to be separated according to their molecular weight. On an analytical scale, it is used to estimate the molecular weight of a protein by comparing its electrophoretic mobility with that of proteins of known molecular weight.

Unless a resolution step is introduced into the reaction scheme (Section 7.13), the α-amino acids prepared by the synthetic methods just described are racemic. Optically active amino acids may be obtained by resolving a racemic mixture or by **enantioselective synthesis.** A synthesis is enantioselective if it produces one enantiomer of a chiral compound in an amount greater than its mirror image. Chemists have succeeded in preparing α-amino acids by techniques that are more than 95% enantioselective. Although this feat is impressive, we must not lose sight of the fact that the reactions that produce amino acids in living systems do so with 100% enantioselectivity.

17.5 PEPTIDES

A key biochemical reaction of amino acids is their conversion to peptides, polypeptides, and proteins. In all these substances amino acids are linked together by amide bonds. The amide bond between the amino group of one amino acid and the carboxyl of another is called a **peptide bond.** Alanylglycine is a representative dipeptide.

N-terminal amino acid $\overset{+}{H_3}NCHC$—$NHCH_2CO_2^-$ C-terminal amino acid

with $\overset{O}{\overset{\|}{C}}$ above and CH_3 below

Alanylglycine
(Ala-Gly)

By agreement, peptide structures are written so that the amino group (as $\overset{+}{H_3}N-$ or H_2N-) is at the left and the carboxyl group (as CO_2^- or CO_2H) is at the right. The left and right ends of the peptide are referred to as the **N terminus** (or amino terminus) and the **C terminus** (or carboxyl terminus), respectively. Alanine is the N-terminal amino acid in alanylglycine; glycine is the C-terminal amino acid. A dipeptide is named as an acyl derivative of the C-terminal amino acid. We call the precise order of bonding in a peptide its amino acid **sequence.** The amino acid sequence is conveniently specified by using the three-letter amino acid abbreviations for the respective amino acids and connecting them by hyphens or by using one-letter abbreviations. Individual amino acid components of peptides are often referred to as amino acid **residues.**

PROBLEM 17.6 Write structural formulas showing the constitution of each of the following dipeptides. Rewrite each sequence using one-letter abbreviations for the amino acids.

(a) Gly-Ala   (d) Gly-Glu
(b) Ala-Phe   (e) Lys-Gly
(c) Phe-Ala   (f) D-Ala-D-Ala

Sample Solution   (a) Gly-Ala is a constitutional isomer of Ala-Gly. Glycine is the N-terminal amino acid in Gly-Ala; alanine is the C-terminal amino acid.

$$\text{N-terminal amino acid} \qquad \overset{+}{H_3}NCH_2\overset{\overset{\displaystyle O}{\|}}{C}-NHCHCO_2^{-} \qquad \text{C-terminal amino acid}$$

$$\underset{CH_3}{}$$

Glycylalanine
(GA)

It is understood that α-amino acids occur as their L stereoisomers unless otherwise indicated. The D notation is explicitly shown when a D amino acid is present, and a racemic amino acid is identified by the prefix DL.

Figure 17.3 shows the structure of Ala-Gly as determined by X-ray crystallography. An important feature is the planar geometry associated with the peptide bond, and the most stable conformation with respect to this bond has the two α-carbon atoms anti to each other. Rotation about the amide linkage is slow because delocalization of the unshared electron pair of nitrogen into the carbonyl group gives partial double-bond character to the carbon–nitrogen bond.

A dipeptide derivative that is widely used as an artificial sweetener is the C-terminal methyl ester of Asp-Phe. Several years ago a chemist synthesized this compound in connection with research on digestive enzymes. The chemist noted that the new substance had a sweet taste; it is 200 times sweeter than table sugar. The artificial sweetener was given the common name aspartame and is marketed under the trade name Nutrasweet. Nutrasweet is found in a variety of diet soft drinks and diet foods (see boxed essay titled *How Sweet It Is* following Section 15.8).

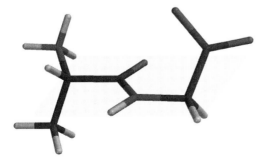

FIGURE   17.3 Structural features of the dipeptide L-alanylglycine as determined by X-ray crystallography.

> **PROBLEM 17.7** Draw the structural formula of Nutrasweet, the C-terminal methyl ester of Asp-Phe.

Up to this point we have dealt with dipeptides. A **tripeptide** contains three amino acid residues, and so on. **Polypeptides** contain many amino acid units. **Proteins** are polypeptides that usually contain 100–300 amino acids.

The number of amino acid residues need not be large for a peptide to have significant biological activity. Figure 17.4 shows the structure of leucine enkephalin, a pentapeptide. Enkephalins are components of **endorphins,** polypeptides present in the brain that act as the body's own painkillers. A second substance, known as methionine enkephalin, is also present in endorphins. Methionine enkephalin differs from leucine enkephalin only in having methionine instead of leucine as its C-terminal amino acid.

> **PROBLEM 17.8** What is the amino acid sequence (using three-letter abbreviations) of methionine enkephalin? Show it using one-letter abbreviations.

Peptides having structures slightly different from those described to this point are known. One such variation is seen in the nonapeptide oxytocin, shown in Figure 17.5. Oxytocin is a hormone secreted by the pituitary gland that stimulates uterine contractions during childbirth and milk letdown during nursing. Rather than terminating in a

FIGURE 17.4 The structure of the pentapeptide leucine enkephalin shown (a) as a structural drawing and (b) as a molecular model. The shape of the molecular model was determined by X-ray crystallography. Hydrogens have been omitted for clarity.

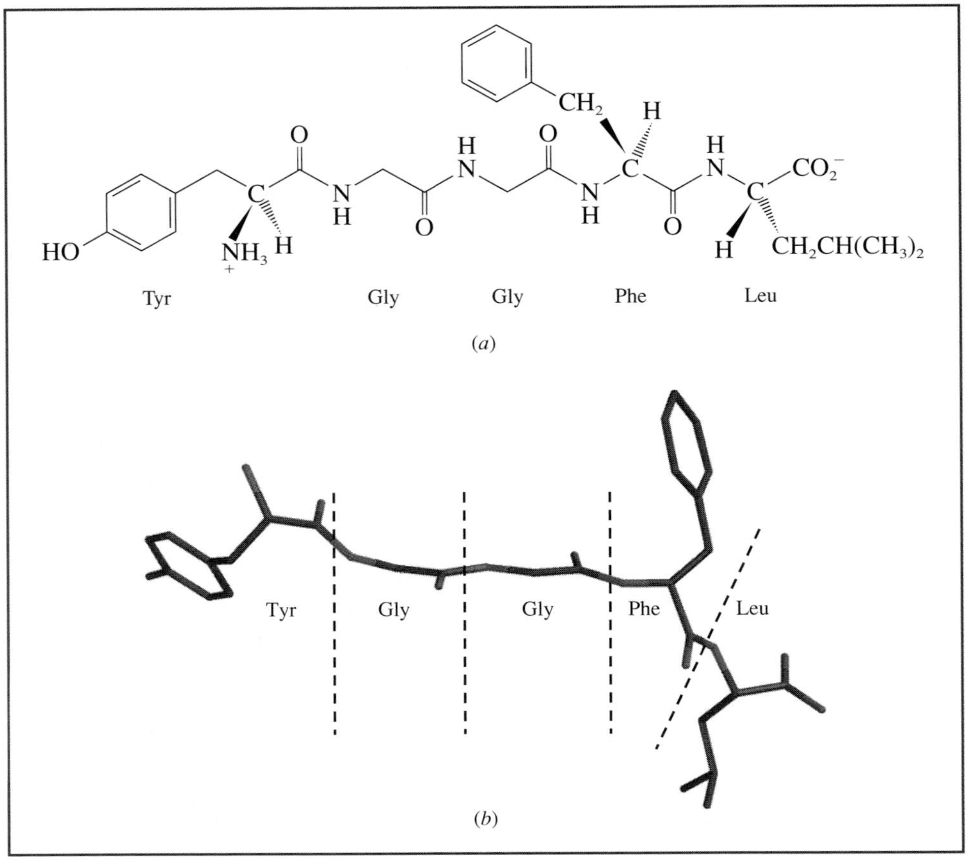

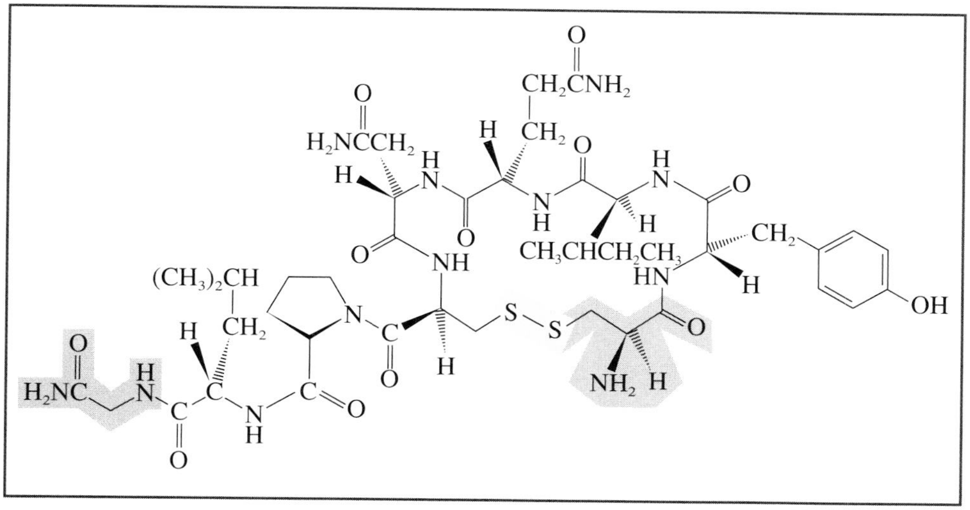

FIGURE 17.5   The structure of oxytocin, a nonapeptide containing a disulfide bond between two cysteine residues. One of these cysteines is the N-terminal amino acid and is highlighted in blue. The C-terminal amino acid is the amide of glycine and is highlighted in red. There are no free carboxyl groups in the molecule; all exist in the form of carboxamides.

carboxyl group, the terminal glycine residue in oxytocin has been modified so that it exists as the corresponding amide. Two cysteine units, one of them the N-terminal amino acid, are joined by the sulfur–sulfur bond of a large-ring cyclic disulfide unit. This is a common structural modification in polypeptides and proteins that contain cysteine residues. It provides a covalent bond between regions of peptide chains that may be many amino acid residues removed from each other.

Disulfide bridges are quite common in polypeptides and proteins, not only in ring formation, as in oxytocin, but in holding separate polypeptide chains together. For example, insulin has two polypeptide chains held together by two disulfide bridges, as depicted schematically in Figure 17.6.

A disulfide bridge forms when the thiol groups of two cysteines undergo oxidation.

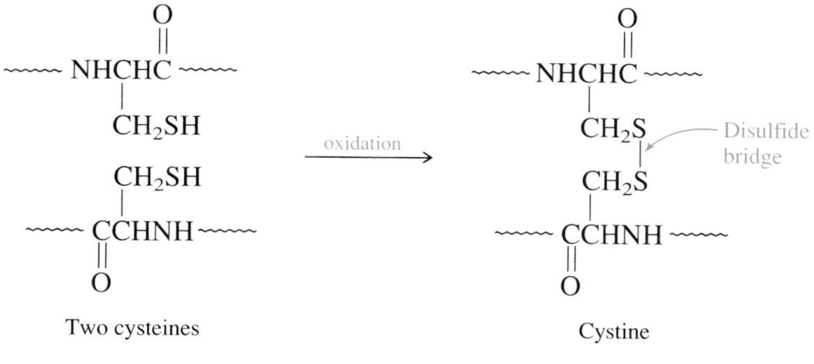

> In a hair permanent, the disulfide bridges of the hair are first broken and then re-formed, using an oxidizing agent, to hold the hair in the desired shape.

Two cysteines → Cystine

The bridged pair of cysteines is called a cystine residue.

17.6   PEPTIDE STRUCTURE DETERMINATION: AMINO ACID ANALYSIS

Chemists and biochemists distinguish between several levels of peptide and protein structures.

The **primary structure** of a peptide or proteins is the *sequence of amino acids plus any disulfide bridges.*

FIGURE 17.6 The amino acid sequence in bovine insulin. The A chain is shown in red and the B chain in blue. The A chain is joined to the B chain by two disulfide units (yellow). A disulfide bond also links cysteines 6 and 11 in the A chain. Human insulin has threonine and isoleucine at residues 8 and 10, respectively, in the A chain and threonine as the C-terminal amino acid in the B chain.

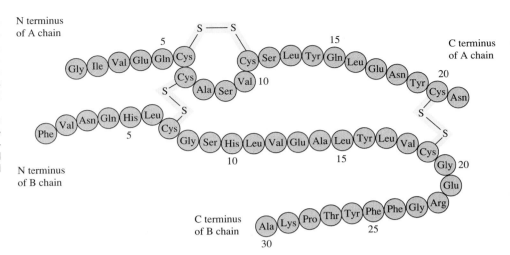

Determining the primary structure of a peptide can be a formidable task because the number of possible combinations of amino acids is staggering. Given a peptide of unknown structure, how is its amino acid sequence determined?

The first step is to determine which amino acids are present and the relative amounts of each. The unknown peptide is subjected to acid-catalyzed hydrolysis. Under these conditions each amide bond is cleaved, producing a solution that contains all the amino acids present in the original peptide. For example, hydrolysis of leucine enkephalin (see Figure 17.4) gives a solution containing 2 mol of glycine and 1 mol each of tyrosine, phenylalanine, and leucine.

$$\text{Leucine enkephalin} \quad \xrightarrow[\text{heat}]{\text{H}_3\text{O}^+} \quad 2\ \text{Gly} + \text{Tyr} + \text{Phe} + \text{Leu}$$

The hydrolysis mixture is then separated, and the amount of each amino acid present is determined by reaction with ninhydrin. Ninhydrin reacts with an amino acid to give a compound having a violet color. The relative number of moles of each amino acid present in the original peptide is proportional to the intensity of the violet color. The separation and identification of the amino acids present in a peptide is usually carried out automatically using an **amino acid analyzer.**

> **PROBLEM 17.9** Amino acid analysis of a certain tetrapeptide gave alanine, glycine, phenylalanine, and valine in equimolar amounts. What amino acid sequences are possible for this tetrapeptide?

17.7 PEPTIDE STRUCTURE DETERMINATION: PRINCIPLES OF SEQUENCE ANALYSIS

Determining the amino acid composition of a polypeptide is only the first step in a complete structure determination. The sequence of amino acids in the peptide chain must be determined for the structure to be complete. The number of possible arrangements of amino acids is astonishing. For example, the amino acids obtained from hydrolysis of leucine enkephalin (2 Gly, Tyr, Phe, Leu) could be used to construct 60 different polypeptides! Only one of these, of course, is the correct structure of leucine enkephalin.

Several techniques are used to determine the amino acid sequence in a peptide chain. These include enzyme-catalyzed partial hydrolysis of the peptide and end group analysis. These methods will be discussed in the sections that follow.

17.8 PARTIAL HYDROLYSIS OF PEPTIDES

Whereas acid-catalyzed hydrolysis of peptides cleaves amide bonds indiscriminately and eventually breaks all of them, enzymatic hydrolysis is much more selective and is the method used to convert a peptide into smaller fragments.

The enzymes that catalyze the hydrolysis of peptides are called **peptidases, proteases,** or **proteolytic enzymes.** One group of pancreatic enzymes, known as **carboxypeptidases,** catalyzes only the hydrolysis of the peptide bond to the C-terminal amino acid, for example. **Trypsin,** a digestive enzyme present in the intestine, catalyzes only the hydrolysis of peptide bonds involving the carboxyl group of a lysine or arginine residue. **Chymotrypsin,** another digestive enzyme, is selective for peptide bonds involving the carboxyl group of amino acids with aromatic side chains (phenylalanine, tryrosine, tryptophan). In addition to these, many other digestive enzymes are known and their selectivity exploited in the selective hydrolysis of peptides.

> Papain, the active component of most meat tenderizers, is a proteolytic enzyme.

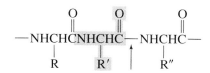

Site of chymotrypsin-catalyzed
hydrolysis when R′ is an
aromatic side chain

PROBLEM 17.10 Digestion of the tetrapeptide of Problem 17.9 with chymotrypsin gave a dipeptide that on amino acid analysis gave phenylalanine and valine in equimolar amounts. What amino acid sequences are possible for the tetrapeptide?

Partial hydrolysis, both enzymatic and nonenzymatic, is particularly useful in determining the sequence of a peptide. Fragments of the peptide having varying lengths are produced, and by identifying places where the fragments overlap, the sequence of amino acids in the original peptide can be deduced.

As an example, consider the following fragments obtained from hydrolysis of a pentapeptide: Ala-Gly-Phe + Gly-Phe-Leu + Phe-Leu-Gly. By aligning the fragments where there is overlap, the correct amino acid sequence is obtained.

$$
\begin{array}{ll}
\text{Hydrolysis} & \text{Ala-Gly-Phe} \\
\text{fragments} & \text{Gly-Phe-Leu} \\
& \text{Phe-Leu-Gly} \\
\hline
\text{Original peptide:} & \text{Ala-Gly-Phe-Leu-Gly}
\end{array}
$$

PROBLEM 17.11 Give the structure of the hexapeptide that gives the following fragments on partial hydrolysis.

Pro-Ala + Leu-Phe-Pro + Pro-Val + Val-Leu-Phe

17.9 END GROUP ANALYSIS

An amino acid sequence is ambiguous unless we know the direction in which to read it—left to right, or right to left. We need to know which end is the N terminus and which is the C terminus. As we saw in the preceding section, carboxypeptidase-catalyzed hydrolysis cleaves the C-terminal amino acid and so can be used to identify it. What about the N terminus?

Several chemical methods have been devised for identifying the N-terminal amino acid. They all take advantage of the fact that the N-terminal amino group is free and can act as a nucleophile. The α-amino groups of all the other amino acids are part of amide linkages, are not free, and are much less nucleophilic.

The **Edman degradation,** developed by Pehr Edman (University of Lund, Sweden), is the most frequently used method for determining the N-terminal amino acid. The N-terminal amino acid reacts with phenyl isothiocyanate ($C_6H_5N\!=\!C\!=\!S$) to give a **phenylthiohydantoin (PTH) derivative.** The PTH derivative is isolated and identified by comparison with PTH derivatives of known amino acids. The most significant feature of the Edman degradation is that following release of the N-terminal amino acid as the PTH derivative, the remainder of the peptide chain remains unchanged.

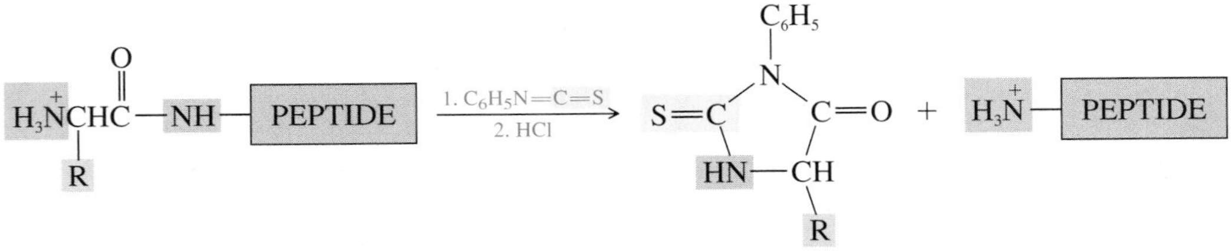

PTH derivative   Remainder of peptide

The new peptide has one less amino acid residue than the original and a new N-terminal amino acid, the second in the chain of the original peptide. Repeating the Edman procedure allows this newly exposed N terminus to be identified, and so on.

PROBLEM 17.12 Write the structures of the PTH derivatives obtained after the first three cycles of the Edman degradation of leucine enkephalin (see Figure 17.4).

Sample Solution The first Edman degradation cycle gives the PTH derivative of the N-terminal amino acid of the original peptide. In the case of leucine enkephalin, this amino acid is tyrosine.

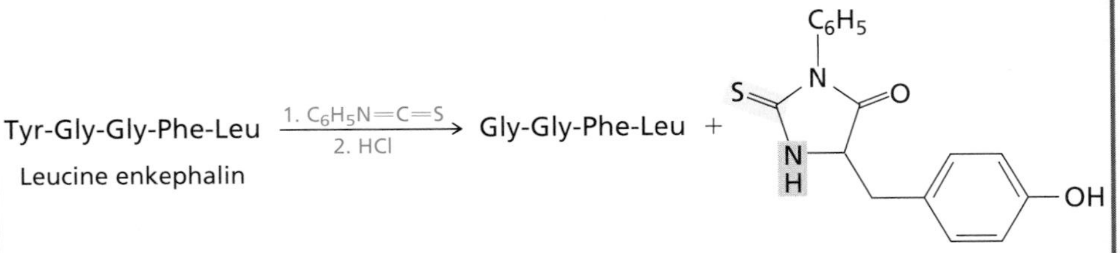

Tyr-Gly-Gly-Phe-Leu $\xrightarrow[\text{2. HCl}]{\text{1. } C_6H_5N=C=S}$ Gly-Gly-Phe-Leu +

Leucine enkephalin

PTH derivative
of tyrosine

It is a fairly routine matter to sequence the first 20 amino acids from the N terminus by repetitive Edman cycles, and even 60 residues have been determined on a single sample of the protein myoglobin. The entire procedure has been automated and incorporated into a device called an **Edman sequenator,** which carries out all the operations under computer control.

17.10 THE STRATEGY OF PEPTIDE SYNTHESIS

One way to confirm the structure proposed for a peptide is to synthesize a peptide having a specific sequence of amino acids and compare the two. This was done, for example, in the case of bradykinin, a peptide present in blood that acts to lower blood pressure. Excess bradykinin, formed as a response to the sting of wasps and other insects containing substances in their venom that stimulate bradykinin release, causes severe local pain. Bradykinin was originally believed to be an octapeptide containing two proline residues; however, a nonapeptide containing three prolines in the following sequence was synthesized and determined to be identical with natural bradykinin in every respect, including biological activity:

Arg-Pro-Pro-Gly-Phe-Ser-Pro-Phe-Arg

Bradykinin

A reevaluation of the original sequence data established that natural bradykinin was indeed the nonapeptide shown. Here the synthesis of a peptide did more than confirm structure; synthesis was instrumental in determining structure.

Chemists and biochemists also synthesize peptides to better understand how they act. By systematically altering the sequence, it's sometimes possible to find out which amino acids are intimately involved in the reactions that involve a particular peptide. Many synthetic peptides have been prepared in searching for new drugs.

The objective in peptide synthesis may be simply stated: to connect amino acids in a prescribed sequence by amide bond formation between them. A number of very effective methods and reagents have been designed for peptide bond formation, so that the joining together of amino acids by amide linkages is not difficult. The real difficulty lies in ensuring that the correct sequence is obtained. This can be illustrated by considering the synthesis of a representative dipeptide, Phe-Gly. Random peptide bond formation in a mixture containing phenylalanine and glycine would be expected to lead to four dipeptides:

$$\overset{+}{H_3}NCHCO_2^- \;+\; \overset{+}{H_3}NCH_2CO_2^- \longrightarrow \text{Phe-Gly} \;+\; \text{Phe-Phe} \;+\; \text{Gly-Phe} \;+\; \text{Gly-Gly}$$
$$\underset{\displaystyle CH_2C_6H_5}{|}$$

Phenylalanine   Glycine

To direct the synthesis so that only Phe-Gly is formed, the amino group of phenylalanine and the carboxyl group of glycine must be protected so that they cannot react under the conditions of peptide bond formation. We can represent the peptide bond formation step by the following equation, where X and Y are amine- and carboxyl-protecting groups, respectively:

$$X-NHCHCOH \;+\; H_2NCH_2C-Y \xrightarrow{\text{couple}} X-NHCHC-NHCH_2C-Y \xrightarrow{\text{deprotect}} \overset{+}{H_3}NCHC-NHCH_2CO^-$$

| N-Protected phenylalanine | C-Protected glycine | Protected Phe-Gly | Phe-Gly |

Thus, the synthesis of a dipeptide of prescribed sequence requires at least three operations:

1. *Protect* the amino group of the N-terminal amino acid and the carboxyl group of the C-terminal amino acid.

2. *Couple* the two protected amino acids by amide bond formation between them.

3. *Deprotect* the amino group at the N terminus and the carboxyl group at the C terminus.

Higher peptides are prepared in an analogous way by a direct extension of the logic just outlined for the synthesis of dipeptides.

17.11 PROTECTING GROUPS AND PEPTIDE BOND FORMATION

The reactivity of an amino group is suppressed by converting it to an amide, and amino groups are most often protected by acylation. The benzyloxycarbonyl group

$$
\overset{O}{\underset{\|}{(C_6H_5CH_2OC-)}}
$$

is one of the most often used amino-protecting groups. It is attached by acylation of an amino acid with benzyloxycarbonyl chloride.

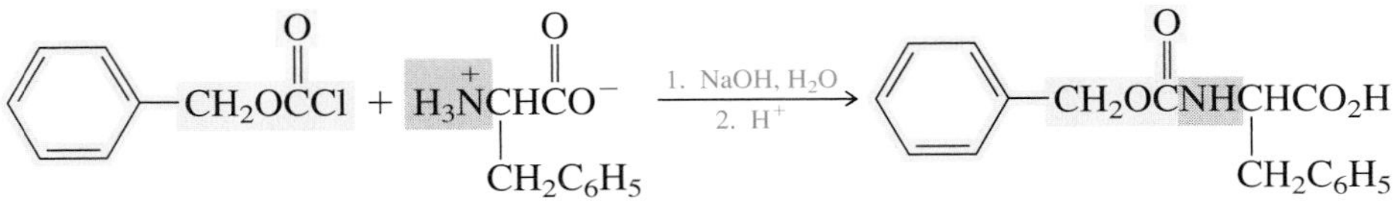

> Another name for the benzyl-oxycarbonyl group is carbobenzoxy. This name, and its abbreviation Cbz, are often found in the older literature, but are no longer a part of IUPAC nomenclature.

$$
\underset{\substack{\text{Benzyloxycarbonyl}\\\text{chloride}}}{C_6H_5-CH_2O\overset{O}{\overset{\|}{C}}Cl} + \underset{\substack{\text{Phenylalanine}}}{\overset{+}{H_3}N\overset{}{\underset{\underset{CH_2C_6H_5}{|}}{CH}}\overset{O}{\overset{\|}{C}}O^-} \xrightarrow[\text{2. H}^+]{\text{1. NaOH, H}_2\text{O}} \underset{\substack{N\text{-Benzyloxycarbonylphenylalanine}\\(82-87\%)}}{C_6H_5-CH_2O\overset{O}{\overset{\|}{C}}NH\overset{}{\underset{\underset{CH_2C_6H_5}{|}}{CH}}CO_2H}
$$

Just as it is customary to identify individual amino acids by abbreviations, so too with protected amino acids. The approved abbreviation for a benzyloxycarbonyl group is the letter "Z." Thus, *N*-benzyloxycarbonylphenylalanine is represented as

$$
\underset{\underset{CH_2C_6H_5}{|}}{ZNHCHCO_2H} \qquad \text{or more simply as} \qquad \text{Z-Phe}
$$

Carboxyl groups are normally protected as esters; benzyl esters are often used.

$$
\underset{\text{Glycine}}{\overset{+}{H_3}NCH_2CO_2^-} + \underset{\substack{\text{Benzyl}\\\text{alcohol}}}{C_6H_5CH_2OH} \xrightarrow[\text{2. HO}^-]{\text{1. H}^+} \underset{\text{Glycine benzyl ester}}{H_2NCH_2CO_2CH_2C_6H_5}
$$

The successful formation of a peptide bond between an N-protected amino acid and a C-protected one is often brought about by use of an activating agent, DCCI (*N,N'*-dicyclohexylcarbodiimide).

$$
\underset{\substack{\text{Z-Protected}\\\text{phenylalanine}}}{ZNH\overset{}{\underset{\underset{CH_2C_6H_5}{|}}{CH}}\overset{O}{\overset{\|}{C}}OH} + \underset{\substack{\text{Glycine}\\\text{benzyl ester}}}{H_2NCH_2\overset{O}{\overset{\|}{C}}OCH_2C_6H_5} \xrightarrow[\text{chloroform}]{\text{DCCI}} \underset{\substack{\text{Z-Protected Phe-Gly}\\\text{benzyl ester (83\%)}}}{ZNH\overset{}{\underset{\underset{CH_2C_6H_5}{|}}{CH}}\overset{O}{\overset{\|}{C}}-NHCH_2\overset{O}{\overset{\|}{C}}OCH_2C_6H_5}
$$

N,N'-Dicyclohexylcarbodiimide has the following structure:

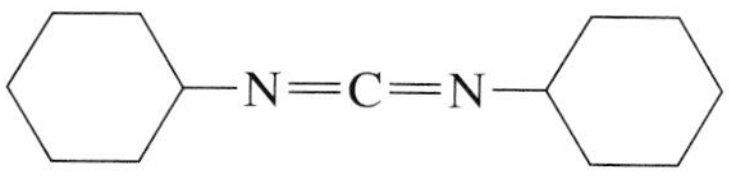

N,N′-Dicyclohexylcarbodiimide
(DCCI)

The value of using the Z group and the benzyl ester as protecting groups is that they can easily be removed (deprotected) by reactions other than hydrolysis. Treating the Z-protected peptide benzyl ester with hydrogen in the presence of a palladium catalyst removes both protecting groups by hydrogenolysis and gives the desired dipeptide.

Hydrogenolysis refers to the cleavage of a molecule under conditions of catalytic hydrogenation.

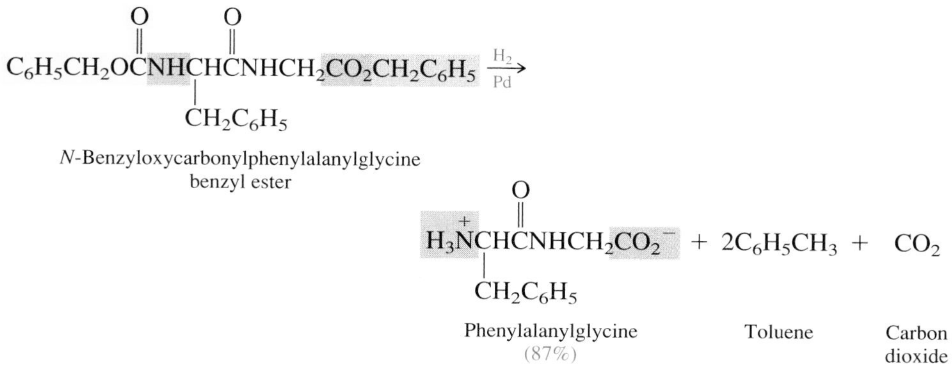

$$C_6H_5CH_2OCNHCHCNHCH_2CO_2CH_2C_6H_5 \xrightarrow[\text{Pd}]{H_2}$$

with substituent $CH_2C_6H_5$

N-Benzyloxycarbonylphenylalanylglycine
benzyl ester

$$\overset{+}{H_3}NCHCNHCH_2CO_2^- + 2C_6H_5CH_3 + CO_2$$

with substituent $CH_2C_6H_5$

| Phenylalanylglycine (87%) | Toluene | Carbon dioxide |

PROBLEM 17.13 Show the steps involved in the synthesis of Ala-Leu from alanine and leucine using benzyloxycarbonyl and benzyl ester protecting groups and DCCI-promoted peptide bond formation.

Higher peptides are prepared either by stepwise extension of peptide chains, one amino acid at a time, or by coupling of fragments containing several residues (the **fragment condensation** approach). Human pituitary adrenocorticotropic hormone (ACTH), for example, has 39 amino acids and was synthesized by coupling of smaller peptides containing residues 1–10, 11–16, 17–24, and 25–39. An attractive feature of this approach is that the various protected peptide fragments may be individually purified, which simplifies the purification of the final product. Among the substances that have been synthesized by fragment condensation are insulin (51 amino acids) and the protein ribonuclease A (124 amino acids).

17.12 SECONDARY STRUCTURES OF PEPTIDES AND PROTEINS

Recall from Section 17.6 that the primary structure of a peptide is its constitution, as dictated by the amino acid sequence. We also speak of the **secondary structure** of a peptide, that is, *the conformational relationship of nearest neighbor amino acids with respect to each other.* Hydrogen bonding between the N—H group of one amino acid residue and the C=O group of another is the interaction that plays the greatest role in the secondary structure of a peptide or protein.

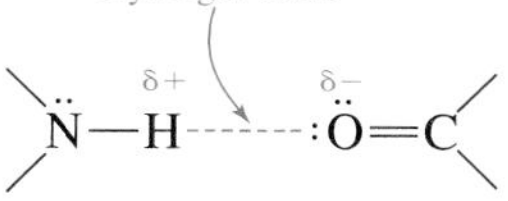

Two commonly encountered protein secondary structures are the **α helix** and the **pleated β sheet.** An example of a protein α helix is shown in Figure 17.8 (page 454). The helical structure is stabilized by hydrogen bonds within the chain. A right-handed helical conformation with about 3.6 amino acids per turn permits each carbonyl oxygen to be hydrogen-bonded to an amide proton.

The α helix is found in many proteins. The principal proteins in muscle (myosin) and wool (α-keratin), for example, contain high percentages of α helix. When wool fibers are stretched, these helical regions are elongated by breaking hydrogen bonds. After the stretching force is removed, the hydrogen bonds re-form spontaneously, and the wool fiber returns to its original shape.

The pleated β sheet secondary structure is quite different from the α helix. Hydrogen bonds form between carbonyl groups and amide protons of adjacent peptide chains, as shown in Figure 17.9 (page 454). The side-chain substituents of each chain alternate above and below the plane formed by the hydrogen-bonded amide groups, giving rise to the "pleating." Pleated-sheet structures are usually only stable in proteins which have a large percentage of small side chains such as H (glycine), CH_3 (alanine), or CH_2OH (serine). This allows the peptide chains to come close enough for hydrogen bonds to form.

17.13 TERTIARY STRUCTURE OF PEPTIDES AND PROTEINS

The **tertiary structure** of a peptide or protein refers to the folding of the chain. The way the chain is folded affects both the physical properties of a protein and its

SOLID-PHASE PEPTIDE SYNTHESIS: THE MERRIFIELD METHOD

n 1962, R. Bruce Merrifield of Rockefeller University (New York) reported the synthesis of the nonapeptide bradykinin by a novel method. The peptide-forming reactions were carried out on the surface of an insoluble polymer called a **solid support.**

The carboxyl end of the desired peptide is bonded to a modified polystyrene resin (Figure 17.7). The new peptide chain is extended toward the N-terminal end by adding amino acid residues one at a time. By anchoring the growing peptide to an insoluble polymer, excess reagents, impurities, and by-products can be removed by washing after each operation. When the peptide chain is complete, the C-terminal residue is cleaved from the polymer by treatment with hydrogen bromide in trifluoroacetic acid.

Merrifield was able to automate solid-phase peptide synthesis, and computer-controlled equipment is now available. Using an early version of his "peptide synthesizer," Merrifield and his coworkers were able to synthesize the enzyme ribonuclease in 1969. The 124 amino acid residues of ribonuclease were assembled by using 369 chemical reactions and 11,391 individual steps! In 1984 Merrifield was awarded the Nobel Prize in chemistry for the development of solid-phase peptide synthesis.

Merrifield's concept of a solid-phase method for peptide synthesis and his development of methods for carrying it out set the stage for an entirely new way to do chemical reactions. Solid-phase synthesis has been extended to include numerous other classes of compounds and has helped spawn a whole new field called **combinatorial chemistry.** Combinatorial synthesis allows a chemist, using solid-phase techniques, to prepare hundreds of related compounds (called **libraries**) at a time. It is one of the most active areas of organic synthesis, especially in the pharmaceutical industry.

Continued

Step 1: The Boc-protected* amino acid is anchored to the resin. Nucleophilic substitution of the benzylic chloride by the carboxylate anion gives an ester.

Step 2: The Boc protecting group is removed by treatment with hydrochloric acid in dilute acetic acid. After the resin has been washed, the C-terminal amino acid is ready for coupling.

Step 3: The resin-bound C-terminal amino acid is coupled to an N-protected amino acid by using N,N'-dicyclohexylcarbodiimide. Excess reagent and N,N'-dicyclohexylurea are washed away from the resin after coupling is complete.

Step 4: The Boc protecting group is removed as in step 2. If desired, steps 3 and 4 may be repeated to introduce as many amino acid residues as desired.

Step n: When the peptide is completely assembled, it is removed from the resin by treatment with hydrogen bromide in trifluoroacetic acid.

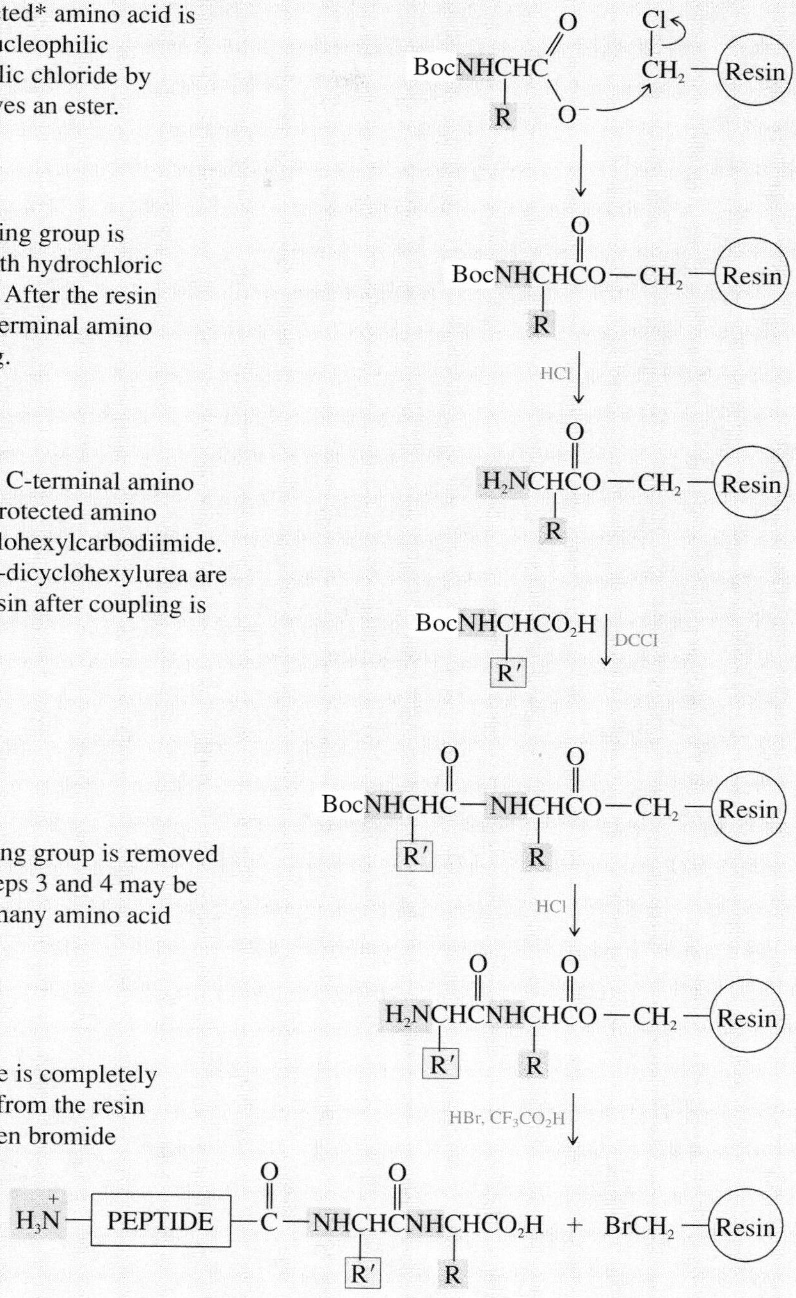

* The Boc protecting group is an abbreviation for *tert*-butoxycarbonyl, $(CH_3)_3COC-$.

FIGURE 17.7 Peptide synthesis by the solid-phase method of Merrifield. Amino acid residues are attached sequentially beginning at the C terminus.

FIGURE 17.8 An α helix of a portion of a protein in which all of the amino acids are alanine. The helix is stabilized by hydrogen bonds between the N—H proton of one amide group and the carbonyl oxygen of another. The methyl groups at the α carbon project away from the outer surface of the helix. When viewed along the helical axis, the chain turns in a clockwise direction (a right-handed helix).

biological function. Structural proteins, such as those present in skin, hair, tendons, wool, and silk, may have either helical or pleated-sheet secondary structures, but in general are elongated in shape, with a chain length many times the chain diameter. They are classed as **fibrous** proteins and, as befits their structural role, tend to be insoluble in water. Many other proteins, including most enzymes, operate in aqueous media; some are soluble, but most are dispersed as colloids. Proteins of this type are called **globular** proteins. Globular proteins are approximately spherical. Figure 17.10 shows carboxypeptidase A, a globular protein containing 307 amino acids. A typical protein such as carboxypeptidase A incorporates elements of a number of secondary structures: some segments are helical; others, pleated sheet; and still others correspond to no simple description.

The shape of a large protein is influenced by many factors, including, of course, its primary and secondary structure. The disulfide bond shown in Figure 17.10 links Cys-138 of carboxypeptidase A to Cys-161 and contributes to the tertiary structure. Carboxypeptidase A contains a Zn^{2+} ion, which is essential to the catalytic activity of the enzyme, and its presence influences the tertiary structure. The Zn^{2+} ion lies near the center of the enzyme, where it is coordinated to the imidazole nitrogens of two histidine residues (His-69, His-196) and to the carboxylate side chain of Glu-72.

Protein tertiary structure is also influenced by the environment. In water a globular protein usually adopts a shape that places its hydrophobic groups toward the interior, with its polar groups on the surface, where they are solvated by water molecules. About 65% of the mass of most cells is water, and the proteins present in cells are said to be in their **native state**—the tertiary structure in which they express their biological activity. When the tertiary structure of a protein is disrupted by adding substances that cause the protein chain to unfold, the protein becomes **denatured** and loses most, if not all, of its activity. Evidence that supports the view that the tertiary structure is dictated by the primary structure includes experiments in which proteins are denatured and allowed to stand, whereupon they are observed to spontaneously readopt their native-state conformation with full recovery of biological activity.

Most protein tertiary structures are determined by X-ray crystallography. The first, myoglobin, the oxygen storage protein of muscle, was determined in 1957. Since then thousands more have been determined. In the form of crystallographic coordinates, the data are

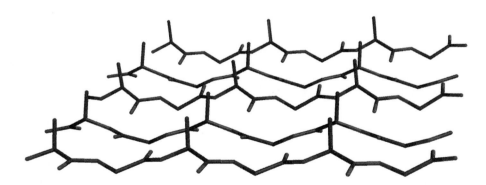

FIGURE 17.9 The β-sheet secondary structure of a protein, composed of alternating glycine and alanine residues. Hydrogen bonding occurs between the amide N—H of one chain and the carbonyl oxygen of another. Van der Waals repulsions between substituents at the α-carbon atoms, shown here as vertical methyl groups, introduce creases in the sheet.

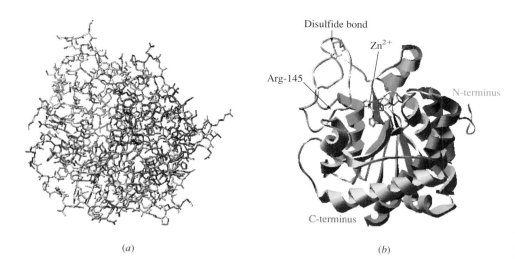

Disulfide bond

Zn^{2+}

Arg-145

N-terminus

C-terminus

(a)   (b)

FIGURE 17.10 The structure of carboxypeptidase A
displayed as (a) a tube model
and (b) a ribbon diagram.
The tube model shows all of
the amino acids and their
side chains. The most evident
feature illustrated by (a) is
the globular shape of the enzyme. The ribbon diagram
emphasizes the folding of
the chain and the helical regions. As can be seen in (b), a
substantial portion of the
protein, the sections colored
gray, is not helical but a random coil. The orientation of
the protein and the color-
coding are the same in both
views.

deposited in the **Protein Data Bank** and are freely available. The three-dimensional structure of carboxypeptidase in Figure 17.10, for example, was produced by downloading the
coordinates from the Protein Data Bank and converting them to a molecular model. At present, the Protein Data Bank averages about one new protein structure per day.

Knowing how the protein chain is folded is a key ingredient in understanding the
mechanism by which an enzyme catalyzes a reaction. Take carboxypeptidase for example. This enzyme catalyzes the hydrolysis of the peptide bond at the C terminus. It is
believed that an ionic bond between the positively charged side chain of an arginine
residue (Arg-145) of the enzyme and the negatively charged carboxylate group of the
substrate's terminal amino acid binds the peptide at the **active site,** the region of the
enzyme's interior where the catalytically important functional groups are located. There,
the Zn^{2+} ion acts as a Lewis acid toward the carbonyl oxygen of the peptide substrate,
increasing its susceptibility to attack by a water molecule (Figure 17.11).

Living systems contain thousands of different enzymes. As we have seen, all are structurally quite complex, and no sweeping generalizations can be made to include all aspects
of enzymic catalysis. The case of carboxypeptidase A illustrates one mode of enzyme action,
the bringing together of reactants and catalytically active functions at the active site.

Most, but not all, enzymes
are proteins. For identifying
certain RNA-catalyzed biological processes Sidney Alt-
man (Yale University) and
Thomas R. Cech (University
of Colorado) shared the 1989
Nobel Prize in chemistry.

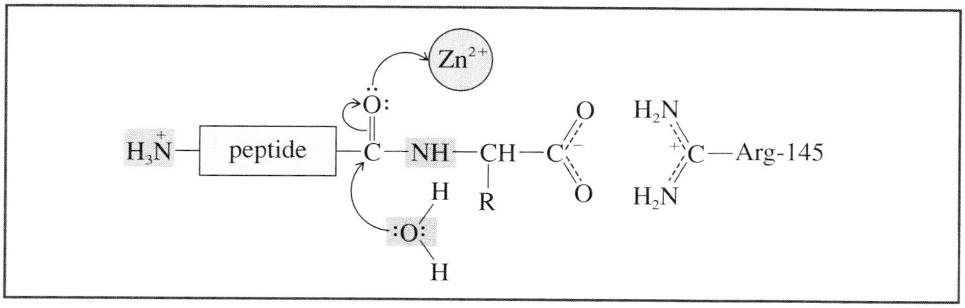

FIGURE 17.11 Proposed mechanism of hydrolysis of a peptide catalyzed by carboxypeptidase A.
The peptide is bound at the active site by an ionic bond between its C–terminal amino acid and the
positively charged side chain of arginine-145. Coordination of Zn^{2+} to oxygen makes the carbon of
the carbonyl group more positive and increases the rate of nucleophilic attack by water.

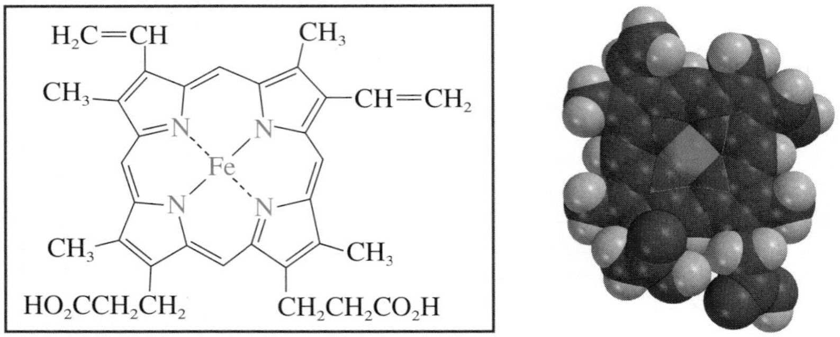

(a)   (b)

FIGURE 17.12   Heme shown as (a) a structural drawing and as (b) a space-filling model. The space-filling model shows the coplanar arrangement of the groups surrounding iron.

17.14 COENZYMES

Coenzyme A plays a crucial role in lipid biosynthesis and was discussed in Section 16.11.

The number of chemical processes that protein side chains can engage in is rather limited. Most prominent among them are proton donation, proton abstraction, and nucleophilic addition to carbonyl groups. In many biological processes a richer variety of reactivity is required, and proteins often act in combination with nonprotein organic molecules to bring about the necessary chemistry. These "helper molecules," referred to as **coenzymes, cofactors,** or **prosthetic groups,** interact with both the enzyme and the substrate to produce the necessary chemical change. Acting alone, for example, proteins lack the necessary functionality to be effective oxidizing or reducing agents. They can catalyze biological oxidations and reductions, however, in the presence of a suitable coenzyme.

Heme (Figure 17.12) is an important prosthetic group in which iron(II) is coordinated with the four nitrogen atoms of a type of tetracyclic aromatic substance known as a porphyrin. The oxygen-storing protein of muscle, myoglobin, represented schematically in Figure 17.13, consists of a heme group surrounded by a protein of 153 amino acids. Four of the six available coordination sites of Fe^{2+} are taken up by the nitrogens of the porphyrin, one by a histidine residue of the protein, and the last by a water molecule. Myoglobin stores oxygen obtained from the blood by formation of an $Fe-O_2$ complex. The oxygen displaces water as the sixth ligand on iron and is held there until needed. The protein serves as a container for the heme and prevents oxidation of Fe^{2+} to Fe^{3+}, an oxidation state in which iron lacks the ability to bind oxygen. Separately, neither heme nor the protein binds oxygen in aqueous solution; together, they do it very well.

17.15 PROTEIN QUATERNARY STRUCTURE: HEMOGLOBIN

Rather than existing as a single polypeptide chain, some proteins are assemblies of two or more chains. The manner in which these subunits are organized is called the **quaternary structure** of the protein.

For their work on myoglobin and hemoglobin, respectively, John C. Kendrew and Max F. Perutz were awarded the 1962 Nobel Prize in chemistry.

Hemoglobin is the oxygen-carrying protein of blood. It binds oxygen at the lungs and transports it to the muscles, where it is stored by myoglobin. Hemoglobin binds oxygen in very much the same way as myoglobin, using heme as the prosthetic group. Hemoglobin is much larger than myoglobin, however, having a molecular weight of 64,500, whereas that of myoglobin is 17,500; hemoglobin contains four heme units,

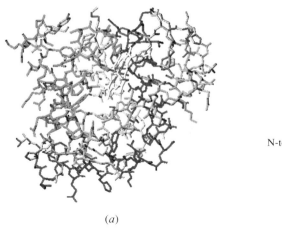

(a)

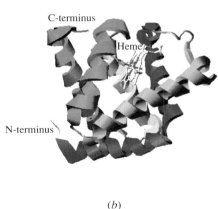

C-terminus

Heme

N-terminus

(b)

FIGURE 17.13 The structure of sperm-whale myoglobin displayed as (a) a tube model and (b) a ribbon diagram. The tube model shows all of the amino acids in the chain; the ribbon diagram shows the folding of the chain. The five separate regions of α helix in myoglobin are shown in different colors to show them more clearly. The heme portion is included in both drawings, but is easier to locate in the ribbon diagram, as is the histidine side chain that is attached to the iron of heme.

myoglobin only one. Hemoglobin is an assembly of four hemes and four protein chains, including two identical chains called the **alpha chains** and two identical chains called the **beta chains.**

Some substances, such as CO, form strong bonds to the iron of heme, strong enough to displace O_2 from it. Carbon monoxide binds 30–50 times more effectively than oxygen to myoglobin and hundreds of times better than oxygen to hemoglobin. Strong binding of CO at the active site interferes with the ability of heme to perform its biological task of transporting and storing oxygen, with potentially lethal results.

How function depends on structure can be seen in the case of the genetic disorder sickle cell anemia. This is a debilitating, sometimes fatal, disease in which red blood cells become distorted ("sickle-shaped") and interfere with the flow of blood through the capillaries. This condition results from the presence of an abnormal hemoglobin in affected people. The primary structures of the beta chain of normal and sickle cell hemoglobin differ by a single amino acid out of 149; sickle cell hemoglobin has valine in place of glutamic acid as the sixth residue from the N terminus. A tiny change in amino acid sequence can produce a life-threatening result! This modification is genetically controlled and probably became established in the gene pool because bearers of the trait have an increased resistance to malaria.

LEARNING OBJECTIVES

This chapter focused on the structure and properties of amino acids and the natural polymers they form, peptides and proteins. The skills you have learned in this chapter should enable you to

- Write the structure of an α-amino acid as a zwitterion.
- Portray the structure of naturally occurring amino acids in stereochemical detail.
- Write a chemical equilibrium which expresses the pH dependence of the ionization of an amino acid.

Continued

- Explain what is meant by the isoelectric point of an amino acid.
- Write a chemical equation describing the synthesis of an α-amino acid from an α-halo carboxylic acid.
- Write a chemical equation describing the synthesis of an α-amino acid from an aldehyde by the Strecker method.
- Write the chemical formula of a peptide and identify the N-terminal and C-terminal amino acid residues.
- Explain what is meant by the term *peptide bond.*
- Explain what is meant by the primary structure of a peptide.
- Write chemical equations which describe the analysis of the N terminus of a peptide using the Edman degradation.
- Explain how the C terminus of a peptide can be identified using carboxypeptidase-catalyzed hydrolysis.
- Deduce the amino acid sequence of a peptide from the fragments obtained by selective hydrolysis.
- Write a series of chemical equations describing the synthesis of a dipeptide using the appropriate protecting groups.
- Describe the predominant secondary structures of proteins and explain the role played by hydrogen bonds.
- Explain what is meant by the tertiary and quaternary structure of proteins.
- Explain what a coenzyme is and give an example.

17.16 SUMMARY

A group of 20 amino acids, listed in Table 17.1, regularly appears as the hydrolysis products of proteins. All are α-amino acids (Section 17.1).

Except for glycine, which is achiral, all of the α-amino acids present in proteins are chiral and have the L configuration at the α carbon (Section 17.2).

The most stable structure of a neutral amino acid is a **zwitterion.** The pH of an aqueous solution at which the concentration of the zwitterion is a maximum is called the isoelectric point (pI) (Section 17.3).

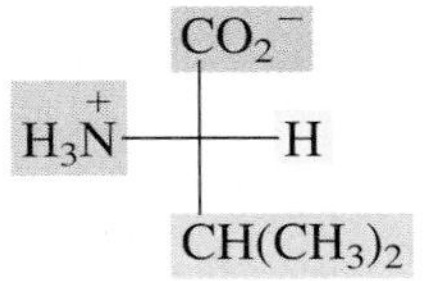

$$\overset{\displaystyle CO_2^-}{\underset{\displaystyle CH(CH_3)_2}{H_3\overset{+}{N}\text{———}H}}$$

Fischer projection of
L-valine in its zwitterionic form

The laboratory synthesis of α-amino acids may be accomplished by (Section 17.4):

1. Reaction of α-halo acids with ammonia:

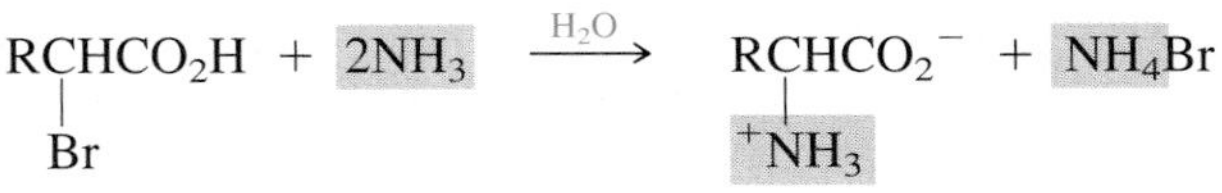

$$\underset{\underset{\displaystyle Br}{|}}{RCHCO_2H} + 2NH_3 \xrightarrow{H_2O} \underset{\underset{\displaystyle {}^+NH_3}{|}}{RCHCO_2^-} + NH_4Br$$

2. Reaction of aldehydes with ammonia and cyanide ion followed by hydrolysis of the intermediate α-aminonitrile (Strecker synthesis):

$$\underset{RCH}{\overset{O}{\parallel}} \xrightarrow[\text{NaCN}]{\text{NH}_4\text{Cl}} \underset{\text{NH}_2}{RCH\!\!\equiv\!\!N} \xrightarrow[\text{2. HO}^-]{\text{1. H}_2\text{O, HCl, heat}} \underset{^+\text{NH}_3}{RCHCO_2^-}$$

An amide linkage between two α-amino acids is called a **peptide bond** (Section 17.5). The **primary structure** of a peptide is given by its amino acid sequence plus any disulfide bonds between two cysteine residues. By convention, peptides are named and written beginning at the N terminus.

$$\underset{CH_3}{\overset{+}{H_3NCHC}}\overset{O}{\overset{\parallel}{-}}NH\underset{CH_2SH}{CHC}\overset{O}{\overset{\parallel}{-}}NHCH_2CO_2^- \qquad \text{Ala-Cys-Gly}$$

Alanylcysteinylglycine

Complete hydrolysis of a peptide gives a mixture of amino acids (Section 17.6). An **amino acid analyzer** identifies the individual amino acids and determines their molar ratios. Incomplete hydrolysis can be accomplished by using enzymes to catalyze cleavage at specific peptide bonds (Section 17.8). Carboxypeptidase-catalyzed hydrolysis can be used to identify the C-terminal amino acid. The N terminus is determined by chemical means. One method is the **Edman degradation** (Section 17.9).

Peptide synthesis requires that the number of possible reactions be limited through the use of protecting groups (Section 17.10). Amino-protecting groups include benzyloxycarbonyl (Z) (Section 17.11). Carboxyl groups can be protected as benzyl esters. Peptide bond formation between a protected amino acid having a free carboxyl group and a protected amino acid having a free amino group can be accomplished with the aid of *N,N'*-dicyclohexylcarbodiimide (DCCI). Protecting groups can be removed from the peptide product by using hydrogen and a palladium catalyst.

Two **secondary structures** of proteins are particularly prominent (Section 17.12). The β pleated sheet is stabilized by hydrogen bonds between N—H and C=O groups of adjacent chains. The α helix is stabilized by hydrogen bonds within a single polypeptide chain.

The folding of a peptide chain is its **tertiary structure** (Section 17.13). The tertiary structure has a tremendous influence on the properties of the peptide and the biological role it plays. The tertiary structure is normally determined by X-ray crystallography.

Many globular proteins are enzymes. They accelerate the rates of chemical reactions in biological systems, but the kinds of reactions that take place are the fundamental reactions of organic chemistry. One way in which enzymes accelerate these reactions is by bringing reactive functions together in the presence of catalytically active functions of the protein.

Often the catalytically active functions of an enzyme are nothing more than proton donors and proton acceptors. In many cases a protein acts in cooperation with a **coenzyme,** a small molecule having the proper functionality to carry out a chemical change not otherwise available to the protein itself (Section 17.14).

Many proteins consist of two or more chains, and the way in which the various units are assembled in the native state of the protein is called its **quaternary structure** (Section 17.15).

ADDITIONAL PROBLEMS

Amino Acids: Structure, Properties, and Synthesis

17.14 Ten of the amino acids present in proteins are designated as being "essential." Does this statement mean the other amino acids are less important in the construction of proteins? Explain.

17.15 L-Threonine has the D configuration at the carbon atom that bears the hydroxyl group. Write a Fischer projection formula for L-threonine.

17.16 Write structural formulas for the principal species present for each of the following amino acids at pH = 1, 13, and the isoelectric point (pI).
   (a) Isoleucine, pI = 6.02
   (b) Glutamic acid, pI = 3.22

17.17 The imidazole ring of the histidine side chain acts as a proton acceptor in certain enzyme-catalyzed reactions. Which is the more stable protonated form of the histidine residue, A or B? Why?

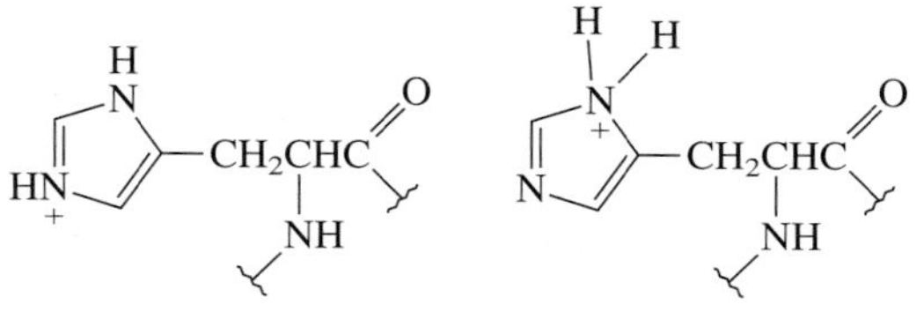

A   B

17.18 Outline a series of steps that would allow preparation of phenylalanine:
   (a) By the Strecker synthesis
   (b) From 2-bromo-3-phenylpropanoic acid

17.19 The penicillin and cephalosporin antibiotics are biosynthesized from amino acid precursors. What amino acids are incorporated into the structure of 6-aminopenicillanic acid?

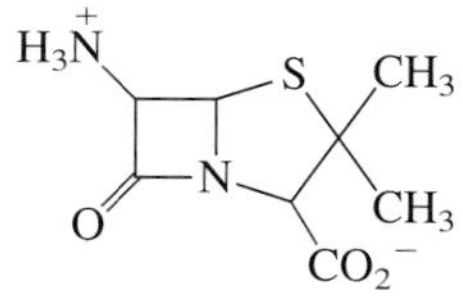

6-Aminopenicillanic acid

Primary Structure of Peptides

17.20 Write the structures and give acceptable names for the dipeptides which can be formed from one molecule each of L-valine and L-alanine.

17.21 If you set out to synthesize the dipeptide Val-Ala from racemic valine and alanine, how many stereoisomers might be obtained?

17.22 Expand your answer to Problem 17.6 (page 443) by showing the structural formula for each dipeptide in a manner that reveals the stereochemistry at the α-carbon atom.

17.23 (a) Using three-letter abbreviations for each amino acid residue, give the name of the following peptide.

(b) Which, if any, of the amino acid residues in this peptide has the D configuration?

17.24 Complete hydrolysis of a tripeptide gives a solution containing equimolar amounts of alanine, leucine, and proline. Using three-letter abbreviations, list all the possible peptides consistent with these data.

17.25 Treatment of the peptide from Problem 17.24 with carboxypeptidase releases proline. Reaction of the peptide with phenyl isothiocyanate followed by hydrochloric acid gives the PTH derivative of leucine. What is the structure of the peptide?

17.26 For the following, refer to the structure of leucine enkephalin, shown in Figure 17.4.
   (a) What derivative is obtained after four cycles of the Edman degradation?
   (b) What amino acid is released on treatment with carboxypeptidase?

17.27 Give the products of the hydrolysis of the peptide shown in the presence of the enzyme trypsin.

Ser-Ala-Arg-Phe-Gly-Ala

17.28 Complete hydrolysis of a tetrapeptide gives equimolar amounts of Leu, Gly, Phe, and Val. One cycle of the Edman degradation gives PTH-Phe. Partial hydrolysis yields a number of fragments, one of which is a tripeptide which contains Val, Gly, and Phe (not necessarily in that order). Also present in the hydrolysis mixture is a dipeptide containing Leu and Gly. What is the primary structure of the tetrapeptide?

17.29 Somatostatin is a tetradecapeptide of the hypothalamus that inhibits the release of pituitary growth hormone. Its amino acid sequence has been determined by a combination of Edman degradations and enzymic hydrolysis experiments. On the basis of the following data, deduce the primary structure of somatostatin:

1. Edman degradation gives PTH-Ala.
2. Selective hydrolysis gives peptides having the following indicated sequences:
   Phe-Trp
   Thr-Ser-Cys
   Lys-Thr-Phe
   Thr-Phe-Thr-Ser-Cys
   Asn-Phe-Phe-Trp-Lys
   Ala-Gly-Cys-Lys-Asn-Phe
3. Somatostatin has a disulfide bridge.

Peptide Synthesis

17.30 Lysine reacts with two equivalents of benzyloxycarbonyl chloride to give a derivative containing two benzyloxycarbonyl groups. What is the structure of this compound?

17.31 Show the steps involved in the synthesis of Val-Ala from valine and alanine using benzyloxycarbonyl and benzyl ester protecting groups and DCCI-promoted peptide bond formation.

17.32 Reaction with hydrogen in the presence of a palladium catalyst is commonly used to remove the benzyloxycarbonyl and benzyl ester protecting groups. Explain what difficulties might arise if hydrolysis were used to remove the protecting groups.

Miscellaneous Problems

17.33 What are the products of each of the following reactions? Your answer should account for all the amino acid residues in the starting peptides.

(a) Treatment of Ile-Glu-Phe with C_6H_5N=C=S, followed by hydrogen chloride

(b) Reaction of Val-Ser-Ala with benzyloxycarbonyl chloride

(c) Reaction of the product of part (b) with the benzyl ester of valine in the presence of DCCI

(d) Reaction of the product of part (c) with hydrogen and a palladium catalyst.

17.34 Explain the differences between the two commonly encountered types of protein secondary structures.

17.35 As disulfide bridges may be formed by oxidation, they may be cleaved by reduction. How many peptides would be obtained on reduction of the disulfide bridges of bovine insulin (Figure 17.6)? How many sulfhydryl (—SH) groups would each fragment contain?

17.36 Glutathione is the tripeptide Glu-Cys-Gly, and has the structure shown. What is unusual or different about the backbone of this peptide?

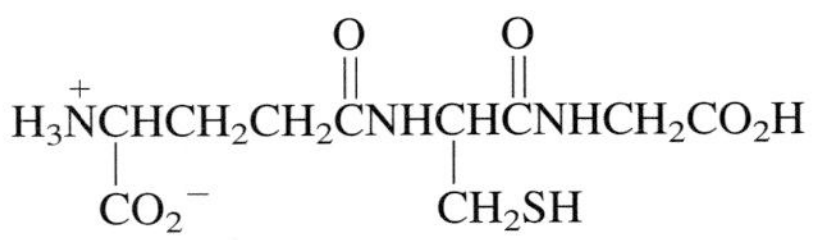

$$\overset{+}{H_3N}CHCH_2CH_2\overset{\overset{O}{\|}}{C}NHCHCH\overset{\overset{O}{\|}}{C}NHCH_2CO_2H$$

$$\underset{CO_2^-}{|}\qquad\qquad\underset{CH_2SH}{|}$$

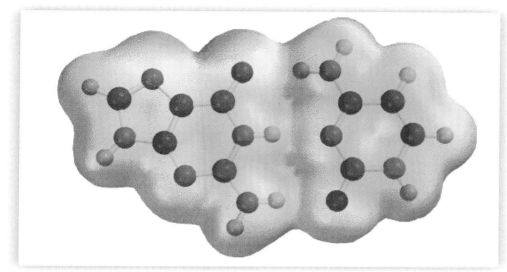

CHAPTER 18
NUCLEIC ACIDS

One of the major scientific achievements of the twentieth century was the identification, at the molecular level, of the chemical interactions involved in the transfer of genetic information and the control of protein biosynthesis. The substances involved are biological macromolecules called **nucleic acids.** We will conclude our look at biologically significant classes of organic compounds with a brief overview of these important molecules.

18.1 PYRIMIDINES AND PURINES

Nucleic acids were isolated over 100 years ago, and, as their name implies, they are acidic substances present in the nuclei of cells. There are two major kinds of nucleic acids: ribonucleic acid (RNA) and deoxyribonucleic acid (DNA). To understand the complex structure of nucleic acids, we first need to examine some simpler substances, nitrogen-containing aromatic heterocycles called **pyrimidines** and **purines.** The parent substance of each class and the numbering system used are shown:

Recall that heterocyclic aromatic compounds were introduced in Section 6.15.

Pyrimidine   Purine

The pyrimidines that occur in DNA are cytosine and thymine. Cytosine is also a structural unit in RNA, which, however, contains uracil instead of thymine. Other pyrimidine derivatives are sometimes present but in small amounts.

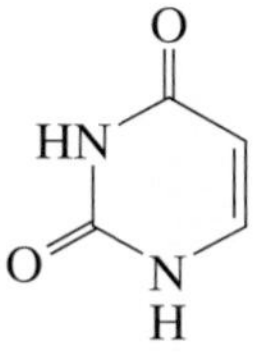

Uracil
(occurs in RNA)

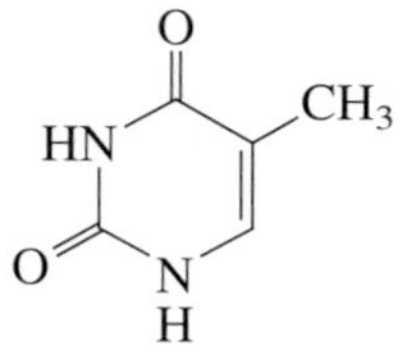

Thymine
(occurs in DNA)

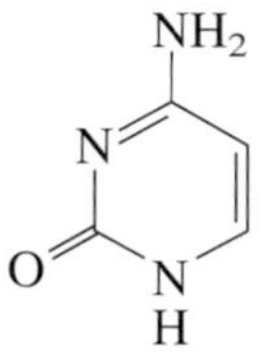

Cytosine
(occurs in both RNA and DNA)

PROBLEM 18.1 5-Fluorouracil (5-FU) is a drug used in cancer chemotherapy. What is its structure?

Adenine and guanine are the principal purines of both DNA and RNA.

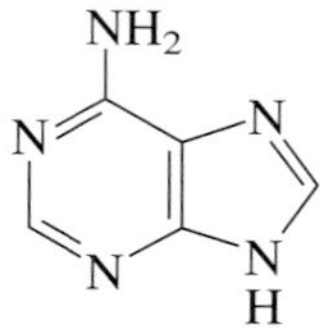

Adenine

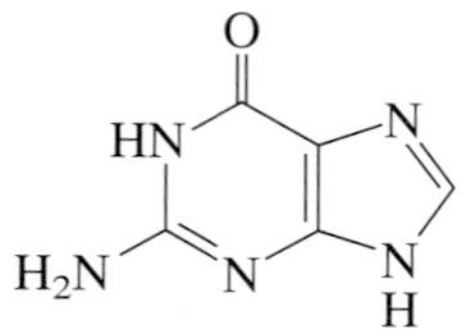

Guanine

The rings of purines and pyrimidines are aromatic and planar. You will see how important this flat shape is when we consider the structure of nucleic acids.

Pyrimidines and purines occur naturally in substances other than nucleic acids. Coffee, for example, is a familiar source of caffeine. Tea contains both caffeine and theobromine.

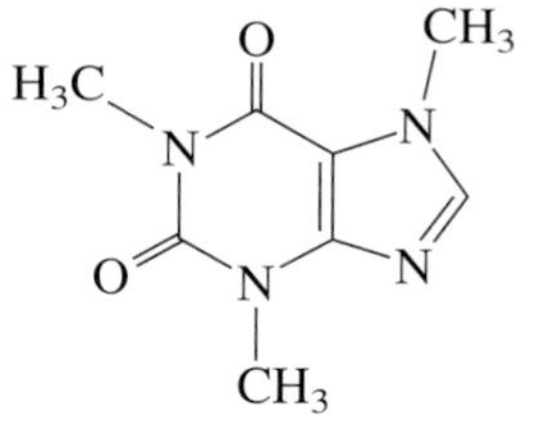

Caffeine

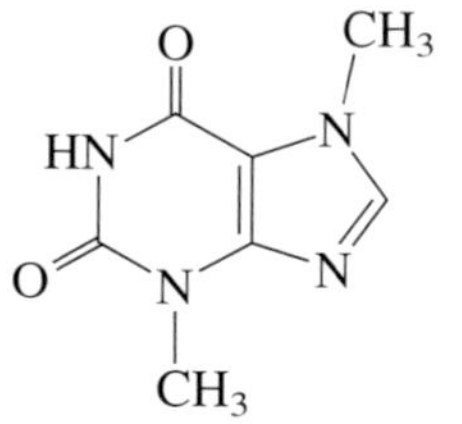

Theobromine

PROBLEM 18.2 Identify caffeine and theobromine as either purines or pyrimidines.

18.2 NUCLEOSIDES

The term **nucleoside** was once restricted to pyrimidine and purine *N*-glycosides of D-ribofuranose and 2-deoxy-D-ribofuranose, because these are the substances present in nucleic acids. The term is used more liberally now with respect to the carbohydrate portion, but is still usually limited to pyrimidine and purine substituents at the anomeric carbon. Uridine is a representative pyrimidine nucleoside; it bears a D-ribofuranose group at N-1. Adenosine is a representative purine nucleoside; its carbohydrate unit is attached at N-9.

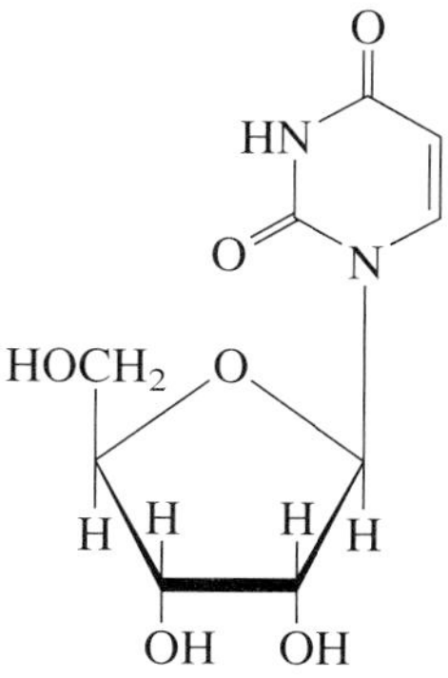

Uridine
(1-β-D-ribofuranosyluracil)

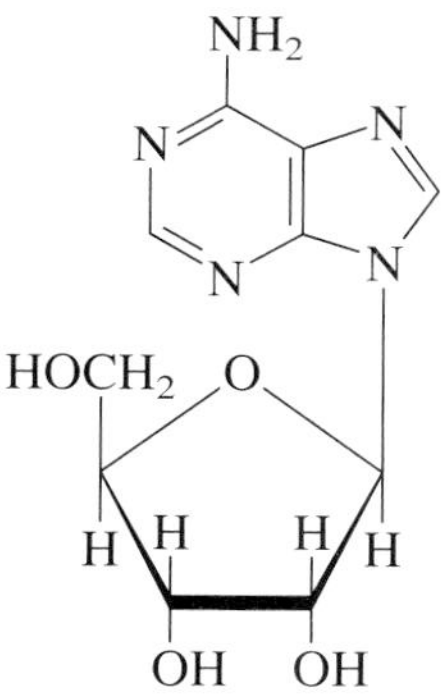

Adenosine
(9-β-D-ribofuranosyladenine)

It is customary to refer to the noncarbohydrate portion of a nucleoside as a purine or pyrimidine **base.**

PROBLEM 18.3 The names of the principal nucleosides obtained from RNA and DNA are listed. Write a structural formula for each one.

(a) Thymidine (thymine-derived nucleoside in DNA)

(b) Cytidine (cytosine-derived nucleoside in RNA)

(c) Guanosine (guanine-derived nucleoside in RNA)

Sample Solution (a) Thymine is a pyrimidine base present in DNA; its carbohydrate substituent is 2-deoxyribofuranose, which is attached to N-1 of thymine.

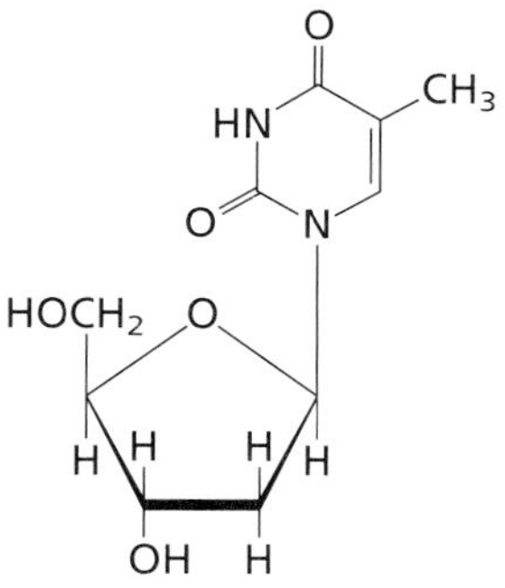

Thymidine

Nucleosides of 2-deoxyribose are named in the same way. Carbons in the carbohydrate portion of the molecule are identified as $1'$, $2'$, $3'$, $4'$, and $5'$ to distinguish them from atoms in the purine or pyrimidine base. Thus, the adenine nucleoside of 2-deoxyribose is called $2'$-deoxyadenosine or 9-β-$2'$-deoxyribofuranosyladenine.

18.3 NUCLEOTIDES

Nucleotides are phosphoric acid esters of nucleosides. The $5'$-monophosphate of adenosine is called $5'$-adenylic acid or adenosine $5'$-monophosphate (AMP).

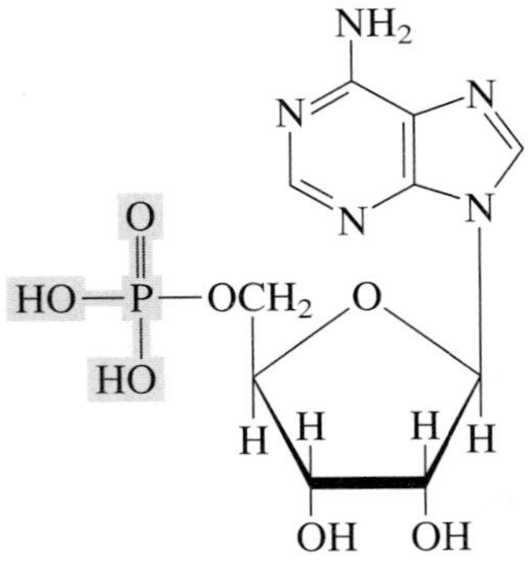

5'-Adenylic acid
(AMP)

As its name implies, 5'-adenylic acid is an acidic substance; it is a diprotic acid with pK_a's for ionization of 3.8 and 6.2, respectively. In aqueous solution at pH 7, both OH groups of the $P(O)(OH)_2$ unit are ionized.

Other important 5'-nucleotides of adenosine include adenosine diphosphate (ADP) and adenosine triphosphate (ATP):

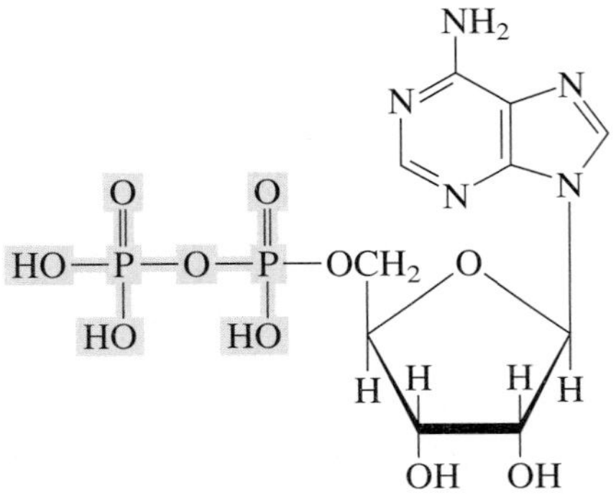

Adenosine diphosphate
(ADP)

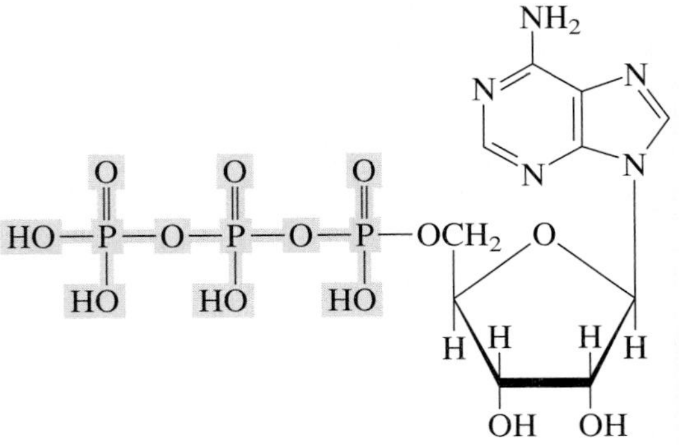

Adenosine triphosphate
(ATP)

Each phosphorylation step in the sequence shown is endothermic:

$$\text{Adenosine} \xrightarrow[\text{enzymes}]{PO_4^{3-}} \text{AMP} \xrightarrow[\text{enzymes}]{PO_4^{3-}} \text{ADP} \xrightarrow[\text{enzymes}]{PO_4^{3-}} \text{ATP}$$

The energy to drive each step comes from the breakdown of carbohydrates in the body. The body uses ATP as a storage vessel for energy released during the conversion of carbohydrates to carbon dioxide and water. ATP is often described as a "high-energy compound." The high energy is associated with the phosphoric anhydride linkages present.

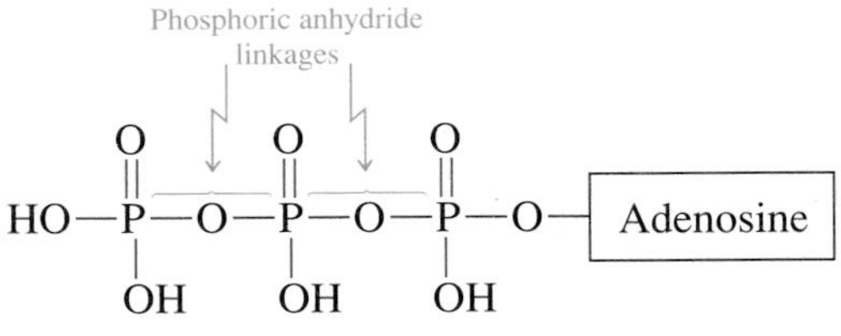

Adenosine triphosphate
(ATP)

That energy becomes available to cells when a phosphoric anhydride linkage of ATP undergoes hydrolysis. As with carboxylic acid anhydrides (Section 13.3), phosphoric acid anhydride hydrolysis is exothermic; conversion of 1 mol of ATP to ADP and phosphate releases 30 kJ (7.3 kcal) of energy.

Adenosine 3'-5'-cyclic monophosphate (cyclicAMP or cAMP) is an important regulator of a large number of biological processes. It is a cyclic ester of phosphoric acid and adenosine involving the hydroxyl groups at C-3' and C-5'.

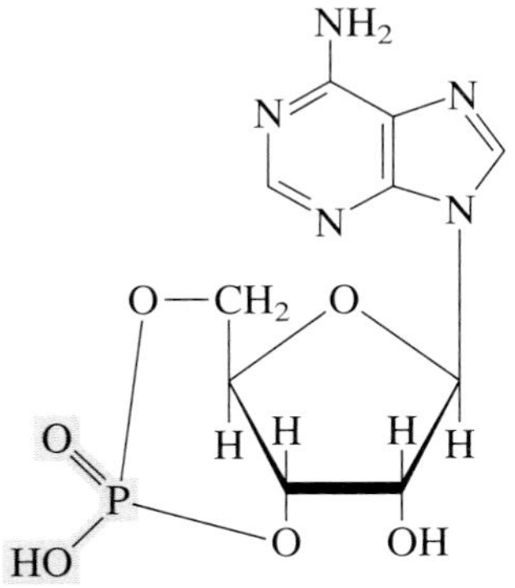

Adenosine 3'-5'-cyclic monophosphate
(cAMP)

18.4 NUCLEIC ACIDS

Nucleic acids are **polynucleotides** in which a phosphate ester unit links the 5' oxygen of one nucleotide to the 3' oxygen of another. Figure 18.1 is a generalized depiction of the structure of a nucleic acid. Nucleic acids are classified as ribonucleic acids (RNA) or deoxyribonucleic acids (DNA) depending on the carbohydrate present.

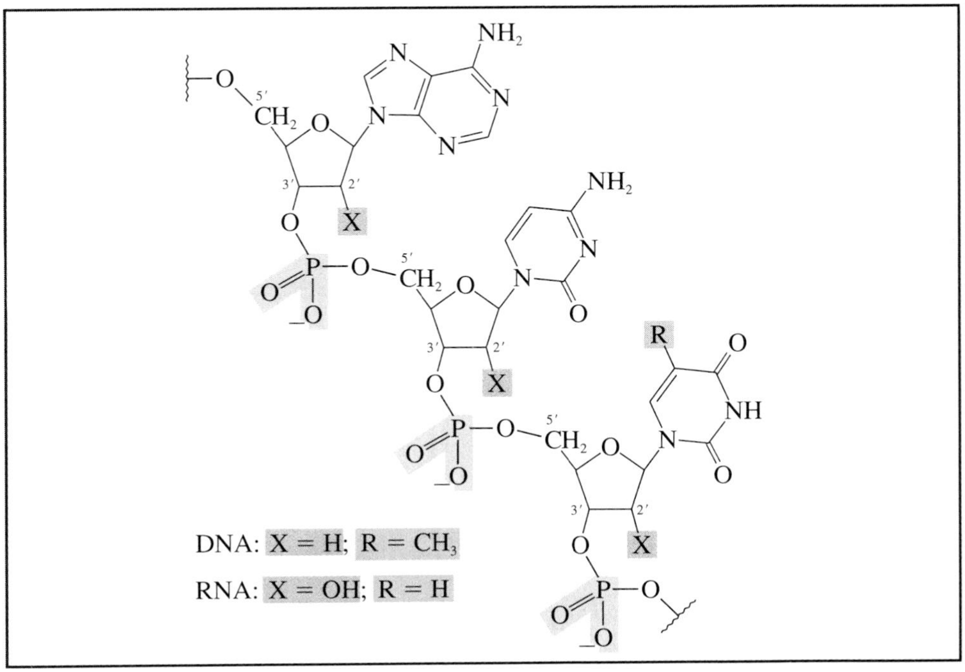

DNA: X = H; R = CH₃
RNA: X = OH; R = H

FIGURE 18.1 A portion of a polynucleotide chain.

Research on nucleic acids progressed slowly until it became evident during the 1940s that they played a role in the transfer of genetic information. It was known that the genetic information of an organism resides in the chromosomes present in each of its cells and that individual chromosomes are made up of smaller units called **genes.** When it became apparent that genes are DNA, interest in nucleic acids intensified. The feeling was that once the structure of DNA was established, the precise way in which it carried out its designated role would become more evident. In some respects the problems are similar to those of protein chemistry. Knowing that DNA is a polynucleotide is comparable to knowing that proteins are polyamides. What is the nucleotide sequence (primary structure)? What is the precise shape of the polynucleotide chain (secondary and tertiary structure)? Is the genetic material a single strand of DNA, or is it an assembly of two or more strands? The complexity of the problem can be indicated by noting that a typical strand of human DNA contains approximately 10^8 nucleotides; if uncoiled it would be several centimeters long, yet it and many others like it reside in cells too small to see with the naked eye.

In 1953 James D. Watson and Francis H. C. Crick pulled together data from biology, biochemistry, chemistry, and X-ray crystallography, along with the insight they gained from molecular models, to propose a structure for DNA and a mechanism for its replication. Their two brief papers paved the way for an explosive growth in our understanding of life processes at the molecular level, the field we now call molecular biology. Along with Maurice Wilkins, who was responsible for the X-ray crystallographic work, Watson and Crick shared the 1962 Nobel Prize in physiology or medicine.

> In the summer of 2000 an announcement was made in the United States that a "working draft" of the human genome had been achieved. Two research groups had identified overlapping fragments that account for 97% of the genome and had established the location and order of 85% of the base pairs.

> Watson and Crick have each written accounts of their work, and both are well worth reading. Watson's is entitled *The Double Helix.* Crick's is *What Mad Pursuit: A Personal View of Scientific Discovery.*

18.5 STRUCTURE AND REPLICATION OF DNA: THE DOUBLE HELIX

Watson and Crick were aided in their search for the structure of DNA by a discovery made by Erwin Chargaff (Columbia University), who found that the composition of DNAs from various sources followed a consistent pattern. Although the distribution of the bases among species varied widely, half the bases in all samples of DNA were purines and the other half were pyrimidines. Furthermore, the ratio of the purine adenine (A) to the pyrimidine thymine (T) was always close to 1:1. Likewise, the ratio of the purine guanine (G) to the pyrimidine cytosine (C) was also close to 1:1. Analysis of human DNA, for example, revealed it to have the following composition:

Purine	Pyrimidine	Base Ratio
Adenine (A) 30.3%	Thymine (T) 30.3%	A/T = 1.00
Guanine (G) 19.5%	Cytosine (C) 19.9%	G/C = 0.98
Total purines 49.8%	Total pyrimidines 50.1%	

Feeling that the constancy in the A/T and G/C ratios was no accident, Watson and Crick proposed that it resulted from a structural complementarity between A and T and between G and C. Consideration of various hydrogen bonding arrangements revealed that A and T could form the hydrogen-bonded **base pair** shown in Figure 18.2a and that G and C could associate as in Figure 18.2b. Specific base pairing of A to T and of G to C by hydrogen bonds is a key element in the Watson–Crick model for the structure of DNA. We shall see that it is also a key element in the replication of DNA.

Because each hydrogen-bonded base pair contains one purine and one pyrimidine, A----T and G----C are approximately the same size. Thus, two nucleic acid chains may be aligned side by side with their bases in the middle, as illustrated in Figure 18.3. The two chains are joined by the network of hydrogen bonds between the paired bases A----T and

> An electrostatic potential map of the guanine–cytosine base pair is pictured on the opening page of this chapter.

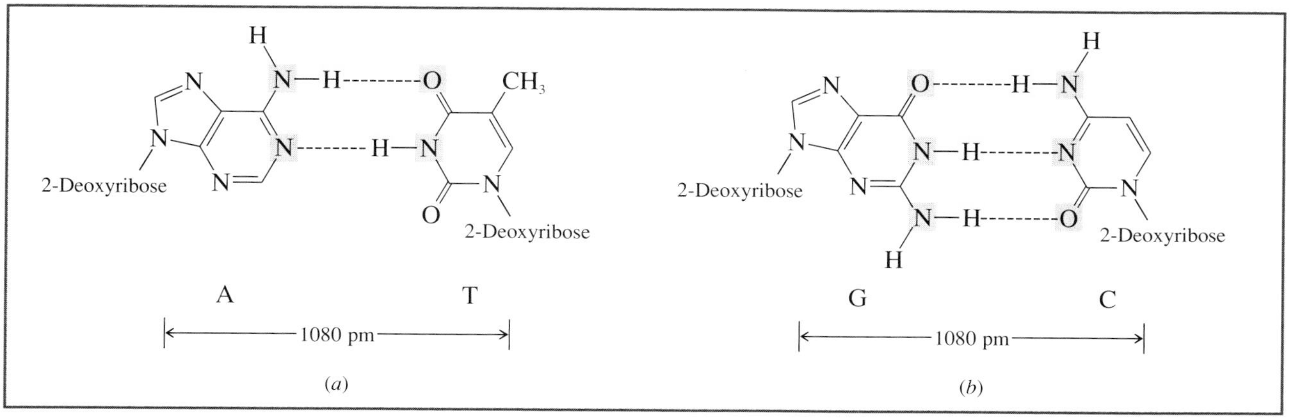

FIGURE 18.2 Base pairing between (*a*) adenine and thymine and (*b*) guanine and cytosine.

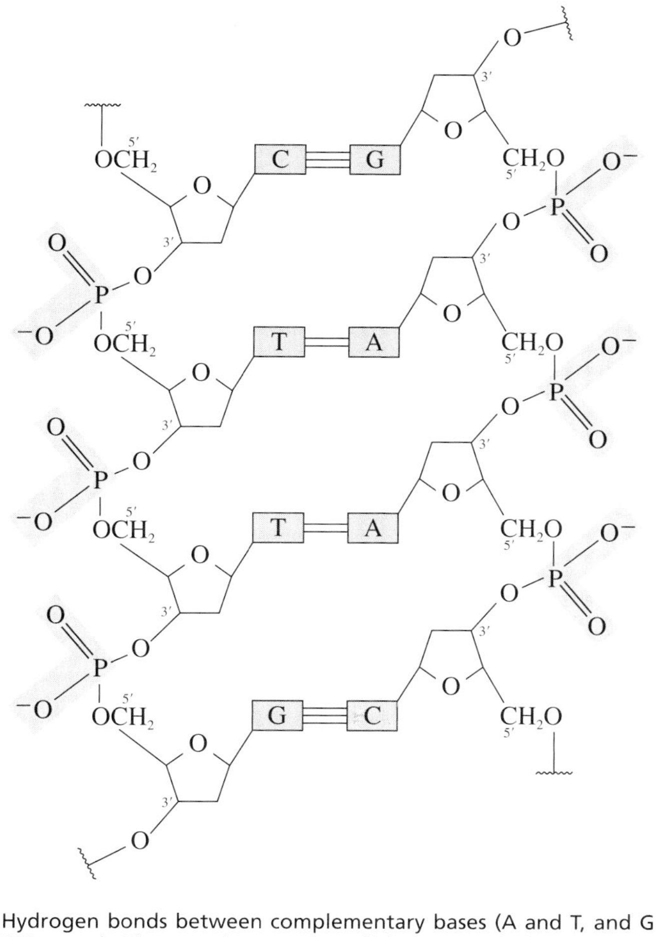

FIGURE 18.3 Hydrogen bonds between complementary bases (A and T, and G and C) permit pairing of two DNA strands. The strands are antiparallel; the 5′ end of the left strand is at the top, and the 5′ end of the right strand is at the bottom.

G----C. Because X-ray crystallographic data indicated a helical structure, Watson and Crick proposed that the two strands are intertwined as a **double helix** (Figure 18.4).

The Watson–Crick base pairing model for DNA structure holds the key to understanding the process of DNA **replication.** During cell division a cell's DNA is duplicated, so that the DNA in the new cell is identical with that in the original cell. At one stage of cell division the DNA double helix begins to unwind, separating the two chains. As portrayed in Figure 18.5, each strand serves as the template on which a new DNA strand is constructed. Each new strand is exactly like the original partner because the A----T, G----C base pairing requirement ensures that the new strand is the precise complement of the template, just as the old strand was. As the double helix unravels, each strand becomes one half of a new and identical DNA double helix.

The structural requirements for the pairing of nucleic acid bases are also critical for using genetic information, and in living systems this means protein biosynthesis.

18.6 DNA-DIRECTED PROTEIN BIOSYNTHESIS

Protein biosynthesis is directed by DNA through the agency of several types of ribonucleic acid called **messenger RNA (mRNA), transfer RNA (tRNA),** and **ribosomal RNA (rRNA).** The two main stages in protein biosynthesis are **transcription** and **translation.**

In the transcription stage a molecule of mRNA having a nucleotide sequence complementary to one of the strands of a DNA double helix is constructed. A diagram illustrating transcription is presented in Figure 18.6 (page 472). Transcription begins at the 5′ end of the DNA molecule, and ribonucleotides with bases complementary to the DNA bases are polymerized with the aid of the enzyme RNA polymerase. Thymine does not occur in RNA; the base that pairs with adenine in RNA is uracil. Unlike DNA, RNA is single-stranded.

In the translation stage, the nucleotide sequence of the mRNA is decoded and "read" as an amino acid sequence to be constructed. Because RNA has only four different bases and there are 20 amino acids to be coded for, codes using either one

FIGURE 18.4 Model of a DNA double helix. The carbohydrate–phosphate "backbone" is on the outside and can be roughly traced by the red oxygen atoms. The blue atoms belong to the purine and pyrimidine bases and lie on the inside.

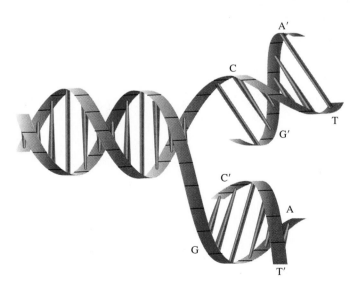

FIGURE 18.5 During DNA replication the double helix unwinds, and each of the original strands serves as a template for the synthesis of its complementary strand.

CANCER CHEMOTHERAPY

The treatment of diseases such as cancer with drugs is called **chemotherapy.** A characteristic common to all forms of cancer is the uncontrolled reproduction or growth of cells. The drugs used to control cancer attempt to stop this uncontrolled cell growth by interfering with the DNA replication process.

Anticancer drugs fall into one of several classes. An **antimetabolite** is a compound structurally similar to one normally used by an organism. The drug 5-FU (Problem 18.1), a fluorinated uracil derivative, is an example of an antimetabolite used to treat certain forms of gastrointestinal and breast cancer. The fluorine atom of 5-FU prevents conversion of this uracil derivative into thymine, a compound necessary for DNA synthesis.

A second antimetabolite used to treat uterine cancer and some forms of leukemia is methotrexate.

Methotrexate is an antimetabolite of folic acid. The nutritional requirements of rapidly dividing cancer cells are greater than those of healthy cells. Thus although folic acid is used by all cells, the growth of cancer cells is slowed to a greater extent with methotrexate therapy.

Methotrexate

A second important class of cancer chemotherapy agents is the **alkylating agents.** The first alkylating agents used as chemotherapy drugs were the same compounds found to be carcinogens (see the boxed essay titled *Nucleophilic Substitution and Cancer* following Section 8.2). Animal studies during World War II showed that nitrogen mustards destroyed lymphoid tissues. One nitrogen mustard derivative is cyclophosphamide, used to treat lymphomas such as Hodgkin's disease as well as breast, ovarian, and lung cancer.

The action of alkylating agents in killing cancer cells relies on the inhibition of DNA replication. The alkylating agent is difunctional; it has two groups that can be attacked by nucleophiles. Cyclophosphamide links the two strands of DNA together by a covalent bridge. Separation of the strands of DNA is a necessary step in replication (Section 18.5). Thus an alkylating agent drug prevents a cancer cell from reproducing. Of course even healthy, rapidly dividing cells are affected by the drug. This is a reason why cancer chemotherapy is associated with side effects such as hair loss and gastrointestinal upset. The hair follicles and the lining of the stomach and intestines have cells that normally reproduce rapidly, and these cells are affected by the anticancer drug to a greater extent than other healthy cells in the body.

Cyclophosphamide

nucleotide to one amino acid or two nucleotides to one amino acid are inadequate. If nucleotides are read in sets of three, however, the four mRNA bases (A, U, C, G) generate 64 possible "words," more than sufficient to code for 20 amino acids. It has been established that the **genetic code** is indeed made up of triplets of adjacent nucleotides called **codons.** The amino acids corresponding to each of the 64 possible codons of mRNA have been determined (Table 18.1).

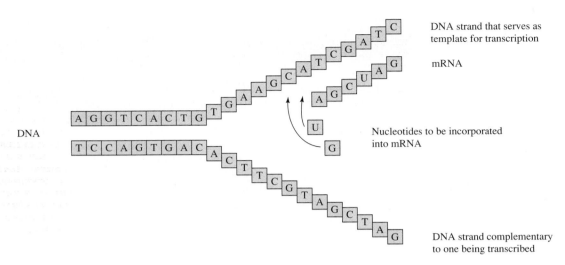

FIGURE 18.6 During transcription a molecule of mRNA is assembled by using DNA as a template.

> **PROBLEM 18.4** It was pointed out in Section 17.15 that sickle cell hemoglobin has valine in place of glutamic acid at one point in its protein chain. Compare the codons for valine and glutamic acid. How do they differ?

The mechanism of translation makes use of the same complementary base-pairing principle used in replication and transcription. Each amino acid is associated with a particular tRNA. Transfer RNA is much smaller than DNA and mRNA. It is single-stranded and contains 70–90 ribonucleotides arranged in a "cloverleaf" pattern (see Figure 18.7 on page 474). Its characteristic shape results from the presence of paired bases in some regions and their absence in others. All tRNAs have a CCA triplet at their 3′ terminus,

TABLE 18.1	The Genetic Code (Messenger RNA Codons)*			
Alanine GCU GCA GCC GCG	**Arginine** CGU CGA AGA CGC CGG AGG	**Asparagine** AAU AAC	**Aspartic acid** GAU GAC	**Cysteine** UGU UGC
Glutamic acid GAA GAG	**Glutamine** CAA CAG	**Glycine** GGU GGA GGC GGG	**Histidine** CAU CAC	**Isoleucine** AUU AUA AUC
Leucine UUA CUU CUA UUG CUC CUG	**Lysine** AAA AAG	**Methionine** AUG	**Phenylalanine** UUU UUC	**Proline** CCU CCA CCC CCG
Serine UCU UCA AGU UCC UCG AGC	**Threonine** ACU ACA ACC ACG	**Tryptophan** UGG	**Tyrosine** UAU UAC	**Valine** GUU GUA GUC GUG

* The first letter of each triplet corresponds to the nucleotide nearer the 5′ terminus, the last letter to the nucleotide nearer the 3′ terminus. UAA, UGA, and UAG are not included in the table; they are chain-terminating codons.

AIDS

The explosive growth of our knowledge of nucleic acid chemistry and its role in molecular biology in the 1980s happened to coincide with a challenge to human health that would have defied understanding a generation ago. That challenge is **acquired immune deficiency syndrome,** or **AIDS.** AIDS is a condition in which the body's immune system is devastated by a viral infection to the extent that it can no longer perform its vital function of identifying and destroying invading organisms. AIDS victims often die from "opportunistic" infections—diseases that are normally held in check by a healthy immune system but which can become deadly when the immune system is compromised. In the short time since its discovery, AIDS has claimed the lives of almost 22 million people worldwide, and the most recent estimates place the number of those infected at more than 36 million.

The virus responsible for almost all the AIDS cases in the United States was identified by scientists at the Louis Pasteur Institute in Paris in 1983 and is known as **human immunodeficiency virus 1 (HIV-1).** HIV-1 is believed to have originated in Africa, where a related virus, HIV-2, was discovered in 1986 by the Pasteur Institute group. Both HIV-1 and HIV-2 are classed as **retroviruses** because their genetic material is RNA rather than DNA. HIVs require a host cell to reproduce, and the hosts in humans are the so-called T4 lymphocytes, which are the cells primarily responsible for inducing the immune system to respond when provoked. The HIV penetrates the cell wall of a T4 lymphocyte and deposits both its RNA and an enzyme called **reverse transcriptase** inside the T4 cell, where the reverse transcriptase catalyzes the formation of a DNA strand that is complementary to the viral RNA. The transcribed DNA then serves as the template for formation of double-helical DNA, which, with the information it carries for reproduction of the HIV, becomes incorporated into the T4 cell's own genetic material. The viral DNA induces the host lymphocyte to begin producing copies of the virus, which then leave the host to infect other T4 cells. In the course of HIV reproduction, the ability of the T4 lymphocyte to reproduce itself is hampered. As the number of T4 cells decrease, so does the body's ability to combat infections.

At this time, there is no known cure for AIDS, but progress is being made in delaying the onset of symptoms and prolonging the lives of those infected with HIV. The first advance in treatment came with drugs such as zidovudine, also known as azidothymine, or AZT. AZT interferes with the ability of HIV to reproduce by blocking the action of reverse transcriptase. As seen by its structure

Zidovudine
(AZT)

AZT is a nucleoside. Several other nucleosides that are also reverse transcriptase inhibitors are in clinical use as well, sometimes in combination with AZT as "drug cocktails." A mixture makes it more difficult for a virus to develop resistance than a single drug does.

The most recent advance has been to simultaneously attack HIV on a second front using a protease inhibitor. Recall from Section 17.8 that proteases are enzymes that catalyze the hydrolysis of proteins at specific points. When HIV uses a cell's DNA to synthesize its own proteins, those proteins are in a form that must be modified by protease-catalyzed hydrolysis to become useful. Protease inhibitors prevent this modification and, in combination with reverse transcriptase inhibitors, slow the reproduction of HIV and have been found to dramatically reduce the "viral load" in HIV-infected patients.

The AIDS outbreak has been and continues to be a tragedy on a massive scale. Until a cure is discovered, or a vaccine developed, sustained efforts at preventing its transmission offer our best weapon against the spread of AIDS.

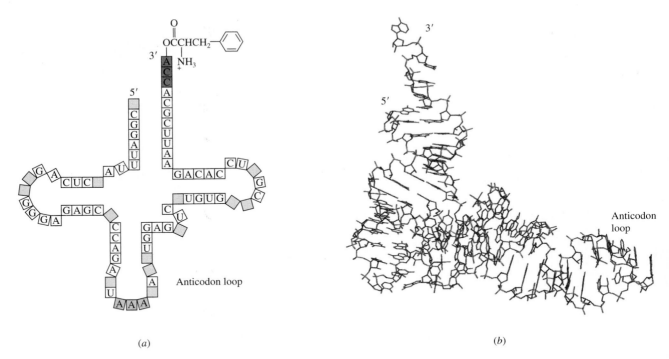

FIGURE 18.7 Phenylalanine tRNA. (*a*) A schematic drawing showing the sequence of bases. RNAs usually contain modified bases (green boxes), slightly different from those in other RNAs. The anticodon for phenylalanine is shown in red, and the CCA triplet which bears the phenylalanine is in blue. (*b*) The experimentally determined structure for yeast phenylalanine tRNA. Complementary base-pairing is present in some regions, but not in others.

to which is attached, by an ester linkage, to an amino acid unique to that particular tRNA. At one of the loops of the tRNA a nucleotide triplet called the **anticodon** occurs, which is complementary to a codon of mRNA. The codons of mRNA are read by the anticodons of tRNA, and the proper amino acids are transferred in sequence to the growing protein.

According to Crick, the so-called central dogma of molecular biology is "DNA makes RNA makes protein."

LEARNING OBJECTIVES

This chapter presented a brief look at nucleic acids, the macromolecules responsible for protein biosynthesis and the transfer of genetic information. The skills you have learned in this chapter should enable you to

- Distinguish between a purine and a pyrimidine.
- Explain the differences among a nucleoside, a nucleotide, and a nucleic acid.
- Describe how the base sequence of a DNA strand determines the sequence of bases in the complementary strand.
- Explain the roles played by messenger RNA, transfer RNA, and ribosomal RNA in protein biosynthesis.
- Explain what is meant by the term "codon."

18.7 SUMMARY

Nucleic acids are responsible for replication of genetic information and they direct protein biosynthesis. **Nucleic acids** are polymeric nucleotides (Section 18.4), **nucleosides** are purine and pyrimidine *N*-glycosides of D-ribose and 2-deoxy-D-ribose (Section 18.2), and **nucleotides** are phosphate esters of nucleosides (Section 18.3). **Deoxyribonucleic acid (DNA)** exists as a double helix (Section 18.5), in which hydrogen bonds are responsible for complementary base pairing between adenine and thymine and between guanine and cytosine. During cell division the strands unwind and are duplicated. Each strand acts as a template on which a complementary strand is constructed.

In the **transcription** stage of protein biosynthesis (Section 18.6), a molecule of **messenger RNA (mRNA)** having a nucleotide sequence complementary to that of DNA is assembled. Transcription is followed by **translation,** in which triplets of nucleotides of mRNA called **codons** are recognized by the **transfer RNA (tRNA)** for a particular amino acid and that amino acid is added to the growing peptide chain.

ADDITIONAL PROBLEMS

Structure

18.5 What is the structural distinction between a nucleotide and a nucleoside?

18.6 Nebularine is a toxic nucleoside isolated from a species of mushroom. Its systematic name is 9-β-D-ribofuranosylpurine. Write a structural formula for nebularine.

18.7 The compound vidarabine shows promise as an antiviral agent. Its structure is identical to that of adenosine (Section 18.2) except that D-arabinose (see Figure 15.2) replaces D-ribose as the carbohydrate component. Write a structural formula for this substance.

DNA Replication and Protein Biosynthesis

Problems 18.8 to 18.12 refer to the following nucleotide segments of a DNA strand:
- (a) A-A-A-G-G-T-C-C-C-G-T-A
- (b) T-A-C-T-C-G-C-G-G-A-T-G

18.8 Write the nucleotide sequence for the complementary DNA strand of each segment.

18.9 Write the mRNA nucleotide sequence that would be produced by transcription of each DNA segment.

18.10 List the codons present in each mRNA segment.

18.11 Determine the amino acid sequence that would be formed from each mRNA nucleotide segment.

18.12 What are the anticodons corresponding to each nucleotide segment in Problem 18.10?

18.13 Using Table 18.1, write a different mRNA nucleotide sequence that would code for each of the amino acid sequences in Problem 18.11.

18.14 Write the DNA nucleotide sequences that would serve as templates for the codons in Problem 18.13.

18.15 The structure of the pentapeptide leucine enkephalin was shown in Figure 17.4. What set of mRNA codons could produce this substance? Can more than one set of codons give the same compound?

18.16 What sequence of DNA bases would serve as a template for the mRNA codons in Problem 18.15?

Miscellaneous Problems

18.17 The generally accepted theory of how carcinogens act is that they interfere with DNA replication. One class of carcinogenic compounds acts as alkylating agents that react to change the structure of the purine and pyrimidine bases in the DNA molecule. An example is *O*-methylguanine. Explain how this change could lead to a base-pairing error in DNA replication.

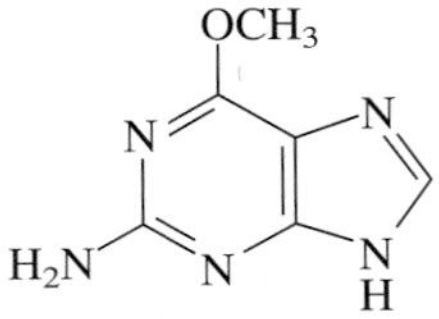

O-Methylguanine

18.18 Treatment of adenosine with nitrous acid gives a nucleoside known as inosine:

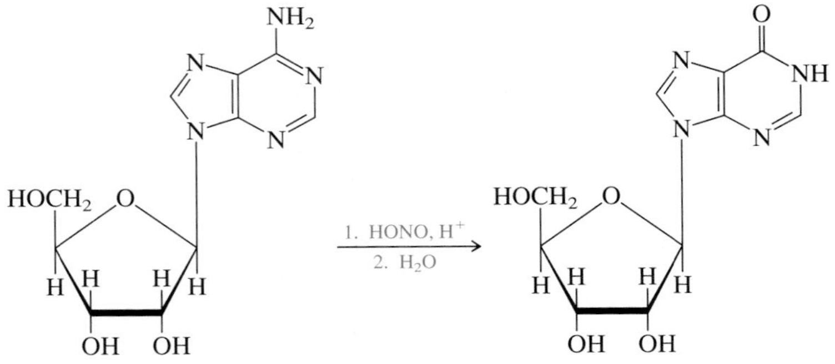

Adenosine   Inosine

Suggest a reasonable mechanism for this reaction. *Hint:* The reaction of nitrous acid with arylamines was described in Sections 14.9 and 14.10.

18.19 (a) The 5′-nucleotide of inosine, inosinic acid ($C_{10}H_{13}N_4O_8P$), is added to foods as a flavor enhancer. What is the structure of inosinic acid? (The structure of inosine is given in Problem 18.18.)

(b) The compound 2′,3′-dideoxyinosine (DDI) is used as a drug for the treatment of AIDS. What is the structure of DDI?

18.20 In one of the early experiments designed to elucidate the genetic code, Marshall Nirenberg of the U.S. National Institutes of Health (Nobel Prize in physiology or medicine, 1968) prepared a synthetic mRNA in which all the bases were uracil. He added this poly(U) to a cell-free system containing all the necessary materials for protein biosynthesis. A polymer of a single amino acid was obtained. What amino acid was polymerized?

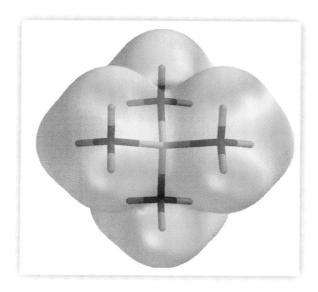

CHAPTER 19
SPECTROSCOPY

Until the second half of the twentieth century, the structure of a substance—a newly discovered natural product, for example—was determined using information obtained from chemical reactions. This information included the identification of functional groups by chemical tests, along with the results of experiments in which the substance was broken down into smaller, more readily identifiable fragments. After considering all the available chemical evidence, the chemist proposed a candidate structure (or structures) consistent with the observations. Proof of structure was provided either by converting the substance to some already known compound or by an independent synthesis.

Qualitative tests and chemical degradation have been supplemented and to a large degree replaced by instrumental methods of structure determination. The following are the most prominent methods and the structural clues they provide:

- **Nuclear magnetic resonance (NMR) spectroscopy** tells us about the carbon skeleton and the environments of the hydrogens attached to it.
- **Infrared (IR) spectroscopy** reveals the presence or absence of key functional groups.
- **Ultraviolet-visible (UV-Vis) spectroscopy** probes the electron distribution, especially in molecules that have conjugated π electron systems.
- **Mass spectrometry (MS)** gives the molecular weight and formula, both of the molecule itself and various structural units within it.

As diverse as these techniques are, all of them are based on the absorption of energy by a molecule, and all measure how a molecule responds to that absorption. In describing these techniques our emphasis will be on their application to structure determination.

We'll start with a brief discussion of electromagnetic radiation, which is the source of the energy that a molecule absorbs in NMR, IR, and UV-Vis spectroscopy.

19.1 PRINCIPLES OF MOLECULAR SPECTROSCOPY

Electromagnetic radiation, of which visible light is but one example, has the properties of both particles and waves. The particles are called **photons,** and each possesses an amount of energy referred to as a **quantum.** In 1900, the German physicist Max Planck proposed that the energy of a photon (E) is directly proportional to its frequency (ν).

$$E = h\nu$$

The SI units of frequency are reciprocal seconds (s^{-1}), given the name **hertz** and the symbol Hz in honor of the nineteenth-century physicist Heinrich R. Hertz. The constant of proportionality h is called **Planck's constant.**

Electromagnetic radiation travels at the speed of light ($c = 3.0 \times 10^8$ m/s), which is equal to the product of its frequency ν and its wavelength λ:

$$c = \nu\lambda$$

The range of photon energies is called the **electromagnetic spectrum** and is shown in Figure 19.1. Visible light occupies a very small region of the electromagnetic spectrum. It is characterized by wavelengths of 4×10^{-7} m (violet) to 8×10^{-7} m (red). When examining Figure 19.1 be sure to keep the following two relationships in mind:

1. *Frequency is inversely proportional to wavelength;* the greater the frequency, the shorter the wavelength.

2. *Energy is directly proportional to frequency;* electromagnetic radiation of higher frequency possesses more energy than radiation of lower frequency.

> "Modern" physics dates from Planck's proposal that energy is quantized, which set the stage for the development of quantum mechanics. Planck received the 1918 Nobel Prize in physics.

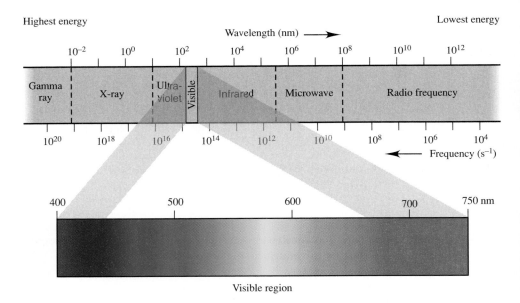

FIGURE 19.1 The electromagnetic spectrum.

(Reprinted, with permission, from M. Silberberg, *Chemistry,* 2nd ed., WCB/McGraw-Hill, New York, 2000, p. 260.)

PROBLEM 19.1 (a) Violet light has a wavelength of 4×10^{-7} m. Calculate the frequency of this light. (b) Red light has a wavelength of 8×10^{-7} m. Is the frequency of red light greater or less than violet light? Calculate the frequency of red light. (c) Which is of higher energy, a photon of violet light or a photon of red light?

Sample Solution (a) By rearranging the equation that relates the wavelength and frequency of radiation, frequency is given by

$$\nu = \frac{c}{\lambda}$$

Substituting the value for the wavelength of violet light,

$$\nu = \frac{3.0 \times 10^8 \text{ m/s}}{4 \times 10^{-7} \text{ m}}$$

$$= 7.5 \times 10^{14} \text{ s}^{-1}$$

The frequency of violet light is 7.5×10^{14} Hz.

Depending on its source, a photon can have a vast amount of energy; gamma rays and X-rays are streams of very high energy photons. Radio waves are of relatively low energy. Ultraviolet radiation is of higher energy than the violet end of visible light. Infrared radiation is of lower energy than the red end of visible light. When a molecule is exposed to electromagnetic radiation, it may absorb a photon, increasing its energy by an amount equal to the energy of the photon. Molecules are highly selective with respect to the frequencies they absorb. Only photons of certain specific frequencies are absorbed by a molecule. The particular photon energies absorbed by a molecule depend on molecular structure and can be measured with instruments called **spectrometers.** The data obtained are very sensitive indicators of molecular structure and have revolutionized the practice of chemical analysis.

The graphical representation of the molecule's absorption pattern is called a **spectrum.** A spectrum consists of a series of peaks at particular frequencies; its interpretation can provide information about the molecular structure of the sample.

With this general background, let us examine briefly three important spectroscopic methods: NMR, IR, and UV-Vis spectroscopy. Each method provides complementary information, and all are useful for the determination of molecular structures. We will discuss NMR spectroscopy at greater length than IR or UV-Vis, as it normally provides more structural information to an organic chemist than the other two methods.

19.2 NUCLEAR MAGNETIC RESONANCE SPECTROSCOPY

To the organic chemist, NMR spectroscopy is one of the most valuable tools available for determining molecular structures. To begin, let us look at what molecular process is involved in nuclear magnetic resonance (NMR) spectroscopy.

NMR spectroscopy is based on transitions between **nuclear spin states.** To understand how this can provide structural information, we must first discuss briefly the concept of nuclear spin. The nuclei of certain atoms possess a property called **spin;** those atoms of greatest interest to organic chemists are protium (^{1}H) and carbon-13 (^{13}C). The protium isotope is 99.9% of the natural abundance of hydrogen, and carbon-13 constitutes only about 1% of natural carbon. Thus proton NMR (^{1}H NMR) spectroscopy has been the type of NMR spectroscopy most widely used by organic chemists. Although

FIGURE 19.2 An external magnetic field causes the two nuclear spin states to have different energies. The difference in energy ΔE is proportional to the strength of the applied field.

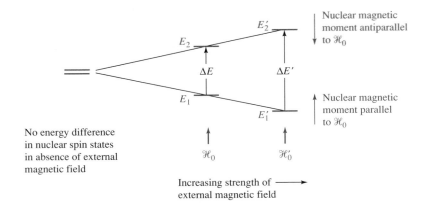

our discussion of NMR spectroscopy will focus on ^{1}H NMR, a brief description of ^{13}C NMR spectroscopy is included in Sections 19.8 and 9.

Like an electron, a proton has two spin states ($+\frac{1}{2}$ and $-\frac{1}{2}$), which are of equal energy. When a proton is placed in a strong magnetic field, the nuclear spin states are no longer of equal energy. As shown in Figure 19.2, the energy difference between the two states is directly proportional to the strength of the applied magnetic field. A magnetic field of 4.7 T, which is about 100,000 times stronger than earth's magnetic field, for example, separates the two spin states of ^{1}H by only 8×10^{-5} kJ/mol (1.9×10^{-5} kcal/mol). From Planck's equation $\Delta E = h\nu$, this energy gap corresponds to radiation having a frequency of 2×10^8 Hz (200 MHz), which lies in the radio frequency (rf) region of the electromagnetic spectrum (see Figure 19.1).

> The SI unit for magnetic field strength is the tesla (T), named after Nikola Tesla, a contemporary of Thomas Edison who, like Edison, was an inventor of electrical devices.

| Frequency of electromagnetic radiation (s^{-1} or Hz) | $\xrightarrow{\text{is proportional to}}$ | Energy difference between nuclear spin states (kJ/mol or kcal/mol) | $\xrightarrow{\text{is proportional to}}$ | Magnetic field (T) |

A proton will absorb electromagnetic radiation in the radio frequency region when it is placed in a strong magnetic field, causing the nuclear spin to "flip" from the low-energy state to the high-energy one.

> **PROBLEM 19.2** Most of the NMR spectra in this text were recorded on a spectrometer having a field strength of 4.7 T (200 MHz for ^{1}H). The first generation of widely used NMR spectrometers were 60-MHz instruments. What was the magnetic field strength of these earlier spectrometers?

19.3   NUCLEAR SHIELDING AND ^{1}H CHEMICAL SHIFTS

Protons in a molecule are connected to other atoms—carbon, oxygen, nitrogen, and so on—by covalent bonds. The electrons in these bonds, indeed all the electrons in a molecule, affect the magnetic environment of the protons. Alone, a proton would feel the full strength of the external field, but a proton in an organic molecule responds to both the external field plus any local fields within the molecule. The net field felt by a proton in a molecule will always be less than the applied field, and the proton is said to be **shielded.** All of the protons of a molecule are shielded from the applied field by the electrons, but some are less shielded than others. Sometimes the term "deshielded" is used to describe this decreased shielding of one proton relative to another.

The more shielded a proton, the greater the strength of the applied field must be to achieve resonance and produce a signal. A more shielded proton absorbs rf radiation at higher field strength (**upfield**) compared with one at lower field strength (**downfield**). Different protons give signals at different field strengths.

The dependence of the resonance position of a nucleus that results from its molecular environment is called its **chemical shift.**

This is where the real power of NMR lies. The chemical shifts of various protons in a molecule can be different and are characteristic of particular structural features.

Figure 19.3 shows the ^1H NMR spectrum of chloroform ($CHCl_3$) to illustrate how the terminology just developed applies to a real spectrum.

Instead of measuring chemical shifts in absolute terms, we measure them with respect to a standard—**tetramethylsilane** ($CH_3)_4Si$, abbreviated **TMS.** The protons of TMS are more shielded than those of most organic compounds, so all of the signals in a sample ordinarily appear at lower field than those of the TMS reference. When measured using a 100-MHz instrument, the signal for the proton in chloroform ($CHCl_3$), for example, appears 728 Hz downfield from the TMS signal. But because frequency is proportional to magnetic field strength, the same signal would appear 1456 Hz downfield from TMS on a 200-MHz instrument. We simplify the reporting of chemical shifts by converting them to parts per million (ppm) from TMS, which is assigned a value of 0. The TMS need not actually be present in the sample, or even appear in the spectrum, in order to serve as a reference.

$$\text{Chemical shift } (\delta) = \frac{\text{position of signal } - \text{ position of TMS peak}}{\text{spectrometer frequency}} \times 10^6$$

Thus, the chemical shift for the proton in chloroform is

$$\delta = \frac{1456 \text{ Hz} - 0 \text{ Hz}}{200 \times 10^6 \text{ Hz}} \times 10^6 = 7.28 \text{ ppm}$$

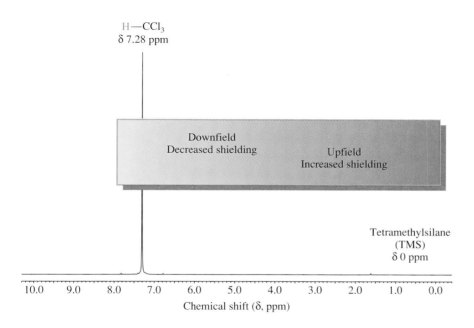

H—CCl$_3$
δ 7.28 ppm

Downfield
Decreased shielding

Upfield
Increased shielding

Tetramethylsilane
(TMS)
δ 0 ppm

| 10.0 | 9.0 | 8.0 | 7.0 | 6.0 | 5.0 | 4.0 | 3.0 | 2.0 | 1.0 | 0.0 |

Chemical shift (δ, ppm)

FIGURE 19.3 The 200-MHz ^1H NMR spectrum of chloroform (HCCl$_3$). Chemical shifts are measured along the *x*-axis in parts per million (ppm) from tetramethylsilane as the reference, which is assigned a value of zero.

When chemical shifts are reported this way, they are identified by the symbol δ and are independent of the field strength.

> **PROBLEM 19.3** The ^1H NMR signal for bromoform ($CHBr_3$) appears at 2065 Hz when recorded on a 300-MHz NMR spectrometer. (a) What is the chemical shift of this proton? (b) Is the proton in $CHBr_3$ more shielded or less shielded than the proton in $CHCl_3$?

NMR spectra are usually run in solution and, although chloroform is a good solvent for most organic compounds, it's rarely used because its own signal at δ 7.28 ppm would be so intense that it would obscure signals in the sample. Because the magnetic properties of deuterium (D = ^2H) are different from those of ^1H, $CDCl_3$ gives no signals at all in an ^1H NMR spectrum and is used instead. Indeed, $CDCl_3$ is the most commonly used solvent in ^1H NMR spectroscopy. Likewise, D_2O is used instead of H_2O for water-soluble substances such as carbohydrates.

19.4 EFFECTS OF MOLECULAR STRUCTURE ON ^1H CHEMICAL SHIFTS

Nuclear magnetic resonance spectroscopy is such a powerful tool for structure determination because

> *Protons in different environments experience different degrees of shielding and have different chemical shifts.*

In compounds of the type CH_3X, for example, the shielding of the methyl protons increases as X becomes less electronegative. Inasmuch as the shielding is due to the electrons, it isn't surprising to find that the chemical shift depends on the degree to which X draws electrons away from the methyl group.

> Problem 19.3 in the preceding section was based on the chemical shift difference between the proton in $CHCl_3$ and the proton in $CHBr_3$ and its relation to shielding.

Increased shielding of methyl protons
Decreasing electronegativity of attached atom

	CH_3F	CH_3OCH_3	$(CH_3)_3N$	CH_3CH_3
	Methyl fluoride	Dimethyl ether	Trimethylamine	Ethane
Chemical shift of methyl protons (δ), ppm:	4.3	3.2	2.2	0.9

A similar trend is seen in the methyl halides, in which the protons in CH_3F are the least shielded (δ 4.3 ppm) and those of CH_3I (δ 2.2 ppm) are the most.

The deshielding effects of electronegative substituents are cumulative, as the chemical shifts for various chlorinated derivatives of methane indicate:

	$CHCl_3$	CH_2Cl_2	CH_3Cl
	Chloroform (trichloromethane)	Methylene chloride (dichloromethane)	Methyl chloride (chloromethane)
Chemical shift (δ), ppm:	7.3	5.3	3.1

> **PROBLEM 19.4** The difference between the ^1H chemical shifts of $CHCl_3$ and CH_3CCl_3 is 4.6 ppm. What is the chemical shift for the protons in CH_3CCl_3? Explain your reasoning.

Table 19.1 collects chemical-shift information for protons of various types. Within each type, methyl (CH_3) protons are more shielded than methylene (CH_2) protons, and methylene protons are more shielded than methine (CH) protons. These differences are small—only about 0.7 ppm separates a methyl proton from a methine proton of the same type. Overall, proton chemical shifts among common organic compounds encompass a range of about 12 ppm. The protons in alkanes are the most shielded, whereas O—H protons of carboxylic acids are the least shielded.

The ability of an NMR spectrometer to separate signals that have similar chemical shifts is termed its **resolving power** and is directly related to the magnetic field strength of the instrument. Two closely spaced signals at 60 MHz become well separated if a 300-MHz instrument is used. (Remember, though, that the chemical shift δ, cited in parts per million, is independent of the field strength.)

TABLE 19.1	Chemical Shifts of Representative Types of Protons		
Type of Proton	**Chemical Shift (δ), ppm***	**Type of Proton**	**Chemical Shift (δ), ppm***
H—C—R	0.9–1.8	H—C—NR	2.2–2.9
H—C—C=C	1.6–2.6	H—C—Cl	3.1–4.1
H—C—C(=O)—	2.1–2.5	H—C—Br	2.7–4.1
H—C—C≡N	2.1–3	H—C—O	3.3–3.7
H—C≡C—	2.5		
H—C—Ar	2.3–2.8	H—NR	1–3[†]
H—C=C<	4.5–6.5	H—OR	0.5–5[†]
H—Ar	6.5–8.5	H—OAr	6–8[†]
H—C(=O)—	9–10	H—OC(=O)—	10–13[†]

* Approximate values relative to tetramethylsilane; other groups within the molecule can cause a proton signal to appear outside of the range cited.

[†] The chemical shifts of protons bonded to nitrogen and oxygen are temperature- and concentration-dependent.

19.5   INTERPRETING PROTON NMR SPECTRA

Analyzing an NMR spectrum in terms of a unique molecular structure begins with the information contained in Table 19.1. By knowing the chemical shifts characteristic of various proton environments, the presence of a particular structural unit in an unknown compound may be inferred. An NMR spectrum also provides other useful information, including:

1. *The number of signals,* which tells us how many different kinds of protons occur.
2. *The intensity of the signals* as measured by the area under each peak, which tells us the relative ratios of the different kinds of protons.
3. *The multiplicity, or splitting, of each signal,* which tells us how many protons are vicinal to the one giving the signal.

Protons that have different chemical shifts are said to be **chemical-shift-nonequivalent** (or **chemically nonequivalent**). A separate NMR signal is given for each chemical-shift-nonequivalent proton in a substance. Figure 19.4 shows the 200-MHz ^{1}H NMR spectrum of methoxyacetonitrile (CH_3OCH_2CN), a molecule with protons in two different environments. The three protons in the CH_3O group constitute one set, the two protons in the OCH_2CN group the other. These two sets of protons give rise to the peaks that we see in the NMR spectrum and can be assigned on the basis of their chemical shifts. The protons in the OCH_2CN group are connected to a carbon that bears two electronegative substituents (O and C≡N) and are less shielded than those of the CH_3O group, which are attached to a carbon that bears only one electronegative atom (O). The signal for the protons in the OCH_2CN group appears at δ 4.1 ppm; the signal corresponding to the CH_3O protons is at δ 3.3 ppm.

PROBLEM 19.5   Predict the number of signals and their approximate chemical shifts that would be expected in the ^{1}H NMR spectrum of each of the following compounds:

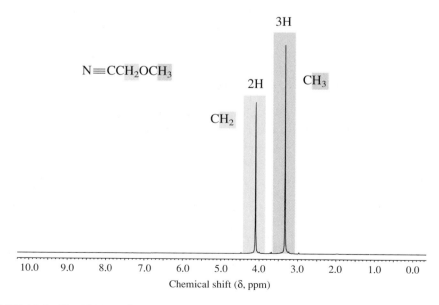

FIGURE 19.4   The 200-MHz ^{1}H NMR spectrum of methoxyacetonitrile (CH_3OCH_2CN).

(a) Ethanol, CH_3CH_2OH

(b) Diethyl ether, $CH_3CH_2OCH_2CH_3$

(c) *tert*-Butyl bromide, $(CH_3)_3CBr$

(d) 2-Chloropropane, CH_3CHCH_3
$\quad\quad\quad\quad\quad\quad\quad\quad\quad$ |
$\quad\quad\quad\quad\quad\quad\quad\quad\quad$ Cl

Sample Solution (a) Ethanol has three "types" of protons: the protons of the methyl (CH_3) group, the protons of the methylene (CH_2) group, and the hydroxyl proton. The NMR spectrum of ethanol has three signals. The approximate chemical shifts can be determined by looking at the appropriate entries in Table 19.1. Thus we could expect to find the CH_3 protons in the range δ 0.9–1.8 ppm, the CH_2 protons in the range δ 3.3–3.7 ppm, and the OH proton in the range δ 0.5–5 ppm.

Another way to assign the peaks is by comparing their intensities. The three equivalent protons of the CH_3O group give rise to a more intense peak than the two equivalent protons of the OCH_2CN group. This is clear by simply comparing the heights of the peaks in the spectrum. It is better, though, to compare peak *areas* by a process called **integration.** This is done electronically at the time the NMR spectrum is recorded, and the integrated areas are displayed on the computer screen or printed out. Peak areas are proportional to the number of equivalent protons responsible for that signal.

PROBLEM 19.6 Predict the relative intensities of the signals for each of the compounds given in Problem 19.5.

Sample Solution (a) The intensity of each signal is determined by the number of equivalent protons giving rise to that signal. Thus the three signals of ethanol have intensities in the ratio 3:2:1 from the CH_3, CH_2, and OH peaks, respectively.

It is important to remember that integration of peak areas gives relative, not absolute, proton counts. Thus, a 3:2 ratio of areas can, as in the case of CH_3OCH_2CN, correspond to a 3:2 ratio of protons. But in some other compound a 3:2 ratio of areas might correspond to a 6:4 or 9:6 ratio of protons.

PROBLEM 19.7 The 200-MHz ^1H NMR spectrum of 1,4-dimethylbenzene looks exactly like that of CH_3OCH_2CN (see Figure 19.4) except the chemical shifts of the two peaks are δ 2.2 ppm and δ 7.0 ppm. Assign the peaks to the appropriate protons of 1,4-dimethylbenzene.

19.6 SPIN–SPIN SPLITTING IN NMR SPECTROSCOPY

The ^1H NMR spectrum of CH_3OCH_2CN (see Figure 19.4) discussed in the preceding section is relatively simple because both signals are **singlets;** that is, each one consists of a single peak. It is quite common though to see a signal for a particular proton appear not as a singlet, but as a collection of peaks. The signal may be split into two peaks (a **doublet**), three peaks (a **triplet**), four peaks (a **quartet**), or even more. Figure 19.5 shows the ^1H NMR spectrum of 1,1-dichloroethane (CH_3CHCl_2), which is characterized by a doublet centered at δ 2.1 ppm for the methyl protons and a quartet at δ 5.9 ppm for the methine proton.

The number of peaks into which the signal for a particular proton is split is called its **multiplicity.** For simple cases the rule that allows us to predict splitting in ^1H NMR spectroscopy is

$$\text{Multiplicity of signal for } H_a = n + 1$$

FIGURE 19.5 The 200-MHz
^{1}H NMR spectrum of 1,1-
dichloroethane, showing the
methine proton as a quartet
and the methyl protons as a
doublet. The peak multiplici-
ties are seen more clearly in
the scale-expanded insets.

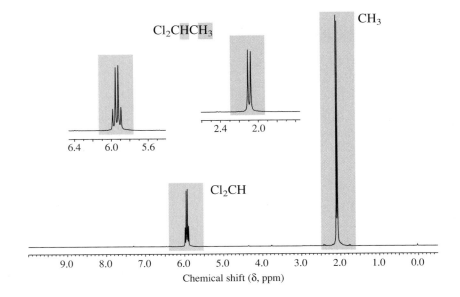

where n is equal to the number of equivalent protons that are vicinal to H_a. Two protons are vicinal to each other when they are bonded to adjacent atoms. Protons vicinal
to H_a are separated from H_a by three bonds. The three methyl protons of 1,1-
dichloroethane are vicinal to the methine proton and split its signal into a quartet. The
single methine proton, in turn, splits the methyl protons' signal into a doublet.

This proton splits the signal for the
methyl protons into a doublet.

$$H{-}\underset{\underset{CH_3}{|}}{\overset{\overset{Cl}{|}}{C}}{-}Cl$$

These three protons split the signal
for the methine proton into a quartet.

The physical basis for peak splitting involves interactions between the nuclear spins
of the vicinal protons, hence the phrase "spin–spin splitting." We will not discuss the
physical phenomenon further, however, but rather focus on the use of splitting in the
interpretation of ^{1}H NMR spectra.

The splitting typically observed in ^{1}H NMR spectra is between vicinal hydrogens,
as we saw for 1,1-dichloroethane. The hydrogens are separated by three bonds; hydrogens separated by four or more bonds do not normally give rise to observable splitting.
Also, splitting is generally only observed for protons bound to carbon. Thus hydroxyl
protons (O—H) and amine protons (N—H) are usually observed as singlets in ^{1}H NMR
spectra.

Lastly, we must note that only nonequivalent hydrogens will split each other. Thus
ethane, for example, shows only a single sharp peak in its ^{1}H NMR spectrum. Even
though the protons of one methyl group are vicinal to those of the other, they do not
split each other because they are equivalent.

PROBLEM 19.8 Describe the appearance of the ^{1}H NMR spectrum of each of the
following compounds. How many signals would you expect to find, and into how
many peaks will each signal be split?

(a) 1,2-Dichloroethane

(b) 1,1,1-Trichloroethane

(c) 1,1,2-Trichloroethane

(d) 1,2,2-Trichloropropane

(e) 1,1,1,2-Tetrachloropropane

Sample Solution (a) All the protons of 1,2-dichloroethane ($ClCH_2CH_2Cl$) are chemically equivalent and have the same chemical shift. Protons that have the same chemical shift do not split each other's signal, and so the NMR spectrum of 1,2-dichloroethane consists of a single sharp peak.

19.7 PATTERNS OF SPIN–SPIN SPLITTING

At first glance, splitting may seem to complicate the interpretation of NMR spectra. In fact, it makes structure determination easier because it provides additional information. It tells us how many protons are vicinal to a proton responsible for a particular signal. With practice, we learn to pick out characteristic patterns of peaks, associating them with particular structural types. One of the most common of these patterns is that of the ethyl group, represented in the NMR spectrum of ethyl bromide in Figure 19.6.

In compounds of the type CH_3CH_2X, especially where X is an electronegative atom or group, such as bromine in ethyl bromide, the ethyl group appears as a **triplet-quartet pattern.** The methylene proton signal is split into a quartet by coupling with the methyl protons. The signal for the methyl protons is a triplet because of vicinal coupling to the two protons of the adjacent methylene group.

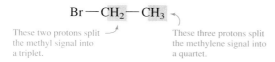

$$Br—CH_2—CH_3$$

These two protons split the methyl signal into a triplet.

These three protons split the methylene signal into a quartet.

The NMR spectrum of isopropyl chloride (Figure 19.7) illustrates the appearance of an isopropyl group. The signal for the six equivalent methyl protons at δ 1.5 ppm is split into a doublet by the proton of the H—C—Cl unit. In turn, the H—C—Cl proton

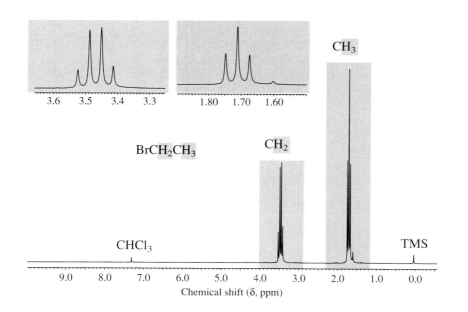

FIGURE 19.6 The 200-MHz ^{1}H NMR spectrum of ethyl bromide, showing the characteristic triplet–quartet pattern of an ethyl group.

FIGURE 19.7 The 200-MHz ^{1}H NMR spectrum of isopropyl chloride, showing the doublet–septet pattern of an isopropyl group.

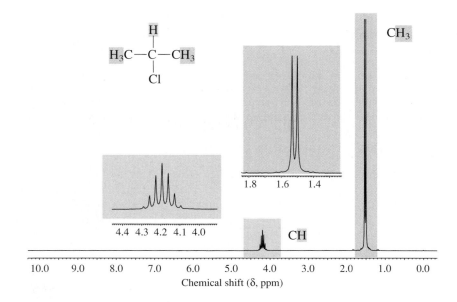

signal at δ 4.2 ppm is split into a septet by the six methyl protons. A **doublet-septet** pattern is characteristic of an isopropyl group.

This proton splits the signal for the methyl protons into a doublet.

These six protons split the methine signal into a septet.

PROBLEM 19.9 Describe the appearance of the ^{1}H NMR spectrum of each of the following compounds. How many signals would you expect to find, and into how many peaks will each signal be split?

(a) $ClCH_2OCH_2CH_3$ (d) *p*-Diethylbenzene

(b) $CH_3CH_2OCH_3$ (e) $ClCH_2CH_2OCH_2CH_3$

(c) $CH_3CH_2OCH_2CH_3$

Sample Solution (a) Along with the triplet–quartet pattern of the ethyl group, the NMR spectrum of this compound will contain a singlet for the two protons of the chloromethyl group.

$$ClCH_2-O-CH_2-CH_3$$

Singlet; no protons vicinal to these; therefore, no splitting

Split into quartet by three protons of methyl group

Split into triplet by two protons of adjacent methylene group

19.8 ^{13}C NMR SPECTROSCOPY

We pointed out in Section 19.2 that both ^{1}H and ^{13}C are nuclei that possess nuclear spin and thus can potentially provide useful structural information when studied by NMR. Although a ^{1}H NMR spectrum helps us infer much about the carbon skeleton of a molecule, a ^{13}C NMR spectrum has the obvious advantage of probing the carbon skeleton directly. ^{13}C NMR spectroscopy is analogous to ^{1}H NMR in that the number of signals informs us about the number of different kinds of carbons, and their chemical shifts are related to particular chemical environments.

However, unlike ^{1}H, which is the most abundant of the hydrogen isotopes (99.9%), only 1.1% of the carbon atoms in a sample are ^{13}C. Moreover, the intensity of the signal produced by ^{13}C nuclei is far weaker than the signal produced by the same number of ^{1}H nuclei. For ^{13}C NMR to be a useful technique in structure determination, a vast increase in the signal-to-noise ratio is required. Pulsed Fourier transform (FT)-NMR provides for this, and its development was the critical breakthrough that led to ^{13}C NMR becoming the routine tool that it is today.

To orient ourselves in the information that ^{13}C NMR provides, let's compare the ^{1}H and ^{13}C NMR spectra of 1-chloropentane (Figures 19.8a and 19.8b, respectively).

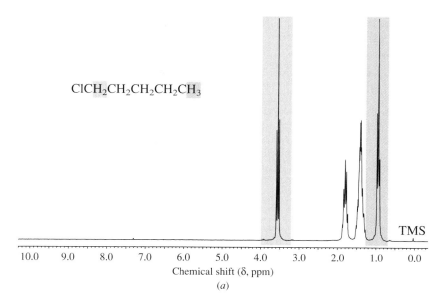

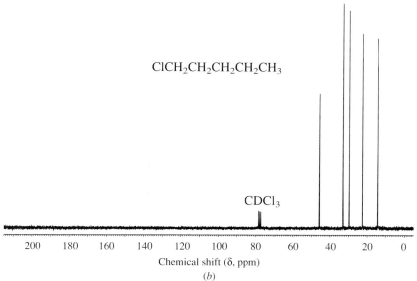

FIGURE 19.8 (a) The 200-MHz ^{1}H NMR spectrum and (b) the ^{13}C NMR spectrum of 1-chloropentane.

The ^1H NMR spectrum shows reasonably well defined triplets for the protons of the CH_3 and CH_2Cl groups (δ 0.9 and 3.55 ppm, respectively). The signals for the six CH_2 protons at C-2, C-3, and C-4 of $CH_3CH_2CH_2CH_2CH_2Cl$, however, appear as two unresolved multiplets at δ 1.4 and 1.8 ppm.

The ^{13}C NMR spectrum, on the other hand, is very simple:

A separate, distinct peak is observed for each carbon.

Notice, too, how well-separated these ^{13}C signals are: They cover a range of over 30 ppm, compared with less than 3 ppm for the proton signals of the same compound. In general, the window for proton signals in organic molecules is about 12 ppm; ^{13}C chemical shifts span a range of over 200 ppm. The greater spread of ^{13}C chemical shifts makes it easier to interpret the spectra.

PROBLEM 19.10 How many signals would you expect to see in the ^{13}C NMR spectrum of each of the following compounds?

(a) Propylbenzene

(d) 1,2,4-Trimethylbenzene

(b) Isopropylbenzene

(e) 1,3,5-Trimethylbenzene

(c) 1,2,3-Trimethylbenzene

Sample Solution (a) The two ring carbons that are ortho to the propyl substituent are equivalent and so must have the same chemical shift. Similarly, the two ring carbons that are meta to the propyl group are equivalent to each other. The carbon atom para to the substituent is unique, as is the carbon that bears the substituent. Thus, the ring carbons will have four signals, designated *w, x, y,* and *z* in the structural formula. These four signals for the ring carbons added to those for the three nonequivalent carbons of the propyl group yield a total of *seven* signals.

Propylbenzene

19.9 ^{13}C CHEMICAL SHIFTS

Just as chemical shifts in ^1H NMR are measured relative to the *protons* of tetramethylsilane, chemical shifts in ^{13}C NMR are measured relative to the *carbons* of tetramethylsilane as the zero point of the chemical-shift scale. Table 19.2 lists typical chemical-shift ranges for some representative types of carbon atoms.

In general, the factors that most affect ^{13}C chemical shifts are:

1. The hybridization of carbon

2. The electronegativity of the groups attached to carbon

Both can be illustrated by comparing the chemical shifts of the designated carbon in the compounds shown. (The numbers are the chemical shift of the indicated carbon in parts per million.)

23	138	61	202
Pentane	1-Pentene	1-Butanol	Butanal

TABLE 19.2	Chemical Shifts of Representative Carbons		
Type of Carbon	Chemical Shift (δ) ppm*	Type of Carbon	Chemical Shift (δ) ppm*
Hydrocarbons		**Functionally Substituted Carbons**	
RCH_3	0–35	RCH_2Br	20–40
R_2CH_2	15–40	RCH_2Cl	25–50
R_3CH	25–50	RCH_2NH_2	35–50
R_4C	30–40	RCH_2OH and RCH_2OR	50–65
$RC{\equiv}CR$	65–90	$RC{\equiv}N$	110–125
$R_2C{=}CR_2$	100–150	$\underset{\text{RCOH}}{\overset{O}{\parallel}}$ and $\underset{\text{RCOR}}{\overset{O}{\parallel}}$	160–185
(benzene ring)	110–175	$\underset{\text{RCH}}{\overset{O}{\parallel}}$ and $\underset{\text{RCR}}{\overset{O}{\parallel}}$	190–220

* Approximate values relative to tetramethylsilane.

sp^3-Hybridized carbons are more shielded than sp^2 as the chemical shifts for C-2 in pentane versus 1-pentene and C-1 in 1-butanol versus butanal demonstrate. The effect of substituent electronegativity is evident when comparing pentane with 1-butanol and 1-pentene with butanal. Replacing the methyl group in pentane by the more electronegative oxygen deshields the carbon in 1-butanol. Likewise, replacing C-1 in 1-pentene by oxygen deshields the carbonyl carbon in butanal.

> **PROBLEM 19.11** Consider carbons x, y, and z in p-methylanisole. One has a chemical shift of δ 20 ppm, another has δ 55 ppm, and the third δ 157 ppm. Match the chemical shifts with the appropriate carbons.
>
> $$\overset{x}{H_3C}-\langle \text{ring} \rangle\overset{y}{-}\overset{z}{O}\overset{}{C}H_3$$

sp-Hybridized carbons are a special case; they are less shielded than sp^3- but more shielded than sp^2-hybridized carbons.

19.10 INFRARED SPECTROSCOPY

Prior to the introduction of NMR spectroscopy, infrared (IR) spectroscopy was the instrumental method most often used for structure determination of organic compounds. IR still retains an important place in the chemist's inventory because of its usefulness in *identifying the presence of certain functional groups within a molecule.*

Infrared radiation comprises the portion of the electromagnetic spectrum (see Figure 19.1) between microwaves and visible light. The fraction of the infrared region of most use for structure determination lies between 2.5×10^{-6} m and 16×10^{-6} m in wavelength. Two derived units commonly employed in IR spectroscopy are the micrometer and the wave number. One **micrometer** (μm) is 10^{-6} m, and IR spectra record the region from 2.5 μm to 16 μm. **Wave numbers** are reciprocal centimeters (cm^{-1}), so that the

MAGNETIC RESONANCE IMAGING

Like all photographs, a chest X-ray is a two-dimensional projection of a three-dimensional object. It is literally a collection of shadows produced by all the bones and organs that lie between the source of the X-rays and the photographic plate. The clearest images in a chest X-ray are not the lungs (the customary reason for taking the X-ray in the first place) but rather the ribs and backbone. It would be desirable if we could limit X-ray absorption to two dimensions at a time rather than three. This is, in fact, what is accomplished by a technique known as **computerized axial tomography,** which yields its information in a form called a **CT** (or **CAT**) scan. With the aid of a computer, a CT scanner controls the movement of an X-ray source and detector with respect to the patient and to each other, stores the X-ray absorption pattern, and converts it to an image that is equivalent to an X-ray photograph of a thin section of tissue. It is a noninvasive diagnostic method, meaning that surgery is not involved nor are probes inserted into the patient's body.

As useful as the CT scan is, it has some drawbacks. Prolonged exposure to X-rays is harmful, and CT scans often require contrast agents to make certain organs more opaque to X-rays. Some patients are allergic to these contrast agents. An alternative technique was introduced in the 1980s that is not only safer but more versatile than X-ray tomography. This technique is **magnetic resonance imaging,** or **MRI.** MRI is an application of nuclear magnetic resonance spectroscopy that makes it possible to examine the inside of the human body using radio frequency radiation, which is lower in energy (see Figure 19.1) and less damaging than X-rays and requires no imaging or contrast agents. By all rights MRI should be called NMRI, but the word "nuclear" was dropped from the name so as to avoid confusion with nuclear medicine, which involves radioactive isotopes.

Although the technology of an MRI scanner is rather sophisticated, it does what we have seen other NMR spectrometers do; it detects protons. Thus, MRI is especially sensitive to biological materials such as water and lipids that are rich in hydrogen. Figure 19.9 shows an example of the use of MRI to detect a brain tumor. Regions of the image are lighter or darker according to the relative concentration of protons and to their environments.

Using MRI as a substitute for X-ray tomography is only the first of what are many medical applications. More lie on the horizon. If, for example, the rate of data acquisition could be increased, then it would become possible to make the leap from the equivalent of still photographs to motion pictures. One could watch the inside of the body as it works—see the heart beat, see the lungs expand and contract—rather than merely examine the structure of an organ.

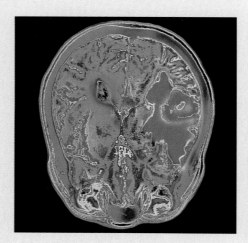

FIGURE 19.9   An MRI of a section of a brain that has a tumor in the left hemisphere. The image has been computer-enhanced to show the tumor and the surrounding liquid in different shades of red, fatty tissues in green, the normal part of the brain in blue, and the eyeballs in yellow.

(Photograph courtesy of Simon Fraser Science Photo Library, Newcastle upon Tyne.)

region 2.5–16 μm corresponds to 4000–625 cm^{-1}. An advantage to using wave numbers is that they are directly proportional to energy. Thus, 4000 cm^{-1} is the high-energy end of the scale and 625 cm^{-1} is the low-energy end.

When a molecule absorbs a photon of infrared radiation, its energy increases, and this increased energy is associated with bond vibrations, especially vibrations involving the *stretching and bending of bonds.* The vibrations for hydrogen cyanide, HCN, are shown in Figure 19.10.

How can the absorption of infrared radiation provide information about the structure of an unknown substance? Most common functional groups absorb infrared energy at characteristic frequencies in the infrared spectrum. The presence of a peak at a particular frequency can be used to determine whether an unknown molecule contains a certain functional group.

Stretching vibrations of many functional groups, including $C—H$, $O—H$, and $C=O$, are found in the region from 4000 to 1600 cm^{-1}. Table 19.3 lists the frequencies (in wave numbers) of the characteristic absorptions of several functional groups commonly found in organic compounds.

A typical IR spectrum, that of the ketone 2-hexanone, is shown in Figure 19.11. The spectrum appears as a series of absorption peaks (pointing down on the page) of varying shape and intensity. The data in Table 19.3 reveal that the absorption by a $\left(\diagdown C=O \right)$ group of a ketone appears in the range of 1710 to 1750 cm^{-1}, and we therefore assign the strong absorption at 1720 cm^{-1} to the carbonyl group. Carbonyl groups are among the structural units most readily identified by IR spectroscopy.

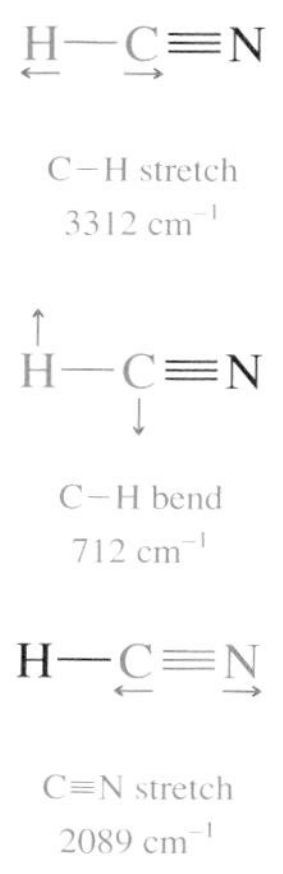

H—C≡N

C—H stretch
3312 cm^{-1}

H—C≡N

C—H bend
712 cm^{-1}

H—C≡N

C≡N stretch
2089 cm^{-1}

FIGURE 19.10 Stretching and bending vibrations of hydrogen cyanide, HCN.

TABLE 19.3   Infrared Absorption Frequencies of Some Common Structural Units

Structural Unit	Frequency, cm^{-1}	Structural Unit	Frequency, cm^{-1}
		Stretching Vibrations	
Single Bonds		**Double Bonds**	
—O—H (alcohols)	3200–3600	$\diagdown C=C \diagup$	1620–1680
—O—H (carboxylic acids)	2500–3600		
$\diagdown N—H \diagup$	3350–3500	$\diagdown C=O \diagup$	
		Aldehydes and ketones	1710–1750
sp C—H	3310–3320	Carboxylic acids	1700–1725
sp^2 C—H	3000–3100	Acid anhydrides	1800–1850 and 1740–1790
sp^3 C—H	2850–2950	Acyl halides	1770–1815
sp^2 C—O	1200	Esters	1730–1750
sp^3 C—O	1025–1200	Amides	1680–1700
		Triple Bonds	
		—C≡C—	2100–2200
		—C≡N	2240–2280
	Bending Vibrations of Diagnostic Value		
Alkenes:		*Substituted Derivatives of Benzene:*	
RCH=CH$_2$	910, 990	Monosubstituted	730–770 and 690–710
R$_2$C=CH$_2$	890	Ortho-disubstituted	735–770
cis-RCH=CHR′	665–730	Meta-disubstituted	750–810 and 680–730
trans-RCH=CHR′	960–980	Para-disubstituted	790–840
R$_2$C=CHR′	790–840		

FIGURE 19.11 The IR spectrum of 2-hexanone.

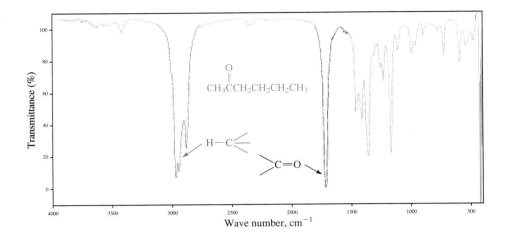

Although the IR spectrum of an organic compound usually contains a number of peaks, frequently only the peak associated with the functional group is assigned. One other area of the spectrum of 2-hexanone is worth noting, however. The group of peaks around 3000 cm^{-1} is seen in most organic molecules, because it is in this region that absorption due to carbon–hydrogen stretching occurs.

The region of an IR spectrum from 1400 to 625 cm^{-1} is where the variations from compound to compound are the greatest. This part of the spectrum is called the **fingerprint region,** because no two chemical compounds have the same exact pattern of peaks, just as no two persons have identical fingerprint patterns. The peaks in the fingerprint region are often difficult to assign, but the pattern they form can be useful when comparing the IR spectrum of a sample to that of a reference compound in the same way that criminals are identified by fingerprint comparisons.

Figure 19.12 is the IR spectrum of a second compound, 2-hexanol. The most noteworthy feature of this spectrum is the broad band centered at 3400 cm^{-1} due to the hydroxyl (O—H) functional group.

The IR spectrum of a carboxylic acid exhibits characteristic absorption patterns that allow identification of this functional group. The hydroxyl absorption is a broad band

FIGURE 19.12 The IR spectrum of 2-hexanol.

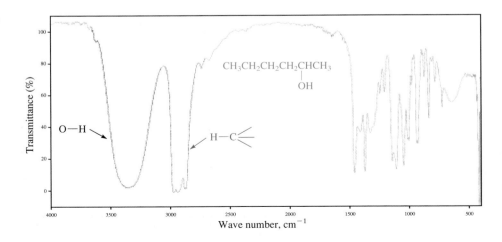

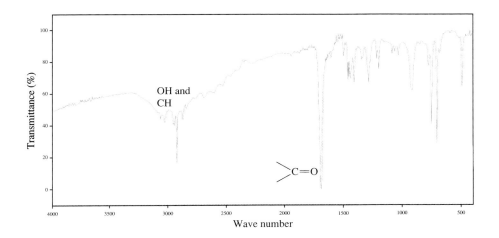

FIGURE 19.13 The IR spectrum of 4-phenylbutanoic acid.

in the 3500–2500 cm^{-1} region. The carbonyl group gives rise to a strong band near 1700 cm^{-1}. These features are apparent in the IR spectrum of 4-phenylbutanoic acid, shown in Figure 19.13.

PROBLEM 19.12   Which one of the following compounds is most consistent with the IR spectrum given in Figure 19.14? Explain your reasoning.

Acetophenone   Benzoic acid   Benzyl alcohol

19.11 ULTRAVIOLET-VISIBLE (UV-VIS) SPECTROSCOPY

The portion of the electromagnetic spectrum (see Figure 19.1) that lies just beyond the infrared region is visible light. The frequency range of visible light is between 12,500 and 25,000 cm^{-1}. Because wave numbers are directly proportional to energy, visible light is approximately ten times more energetic than infrared radiation. Red

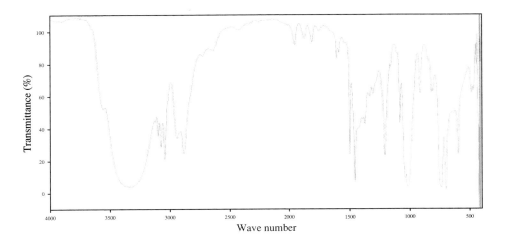

FIGURE 19.14 The IR spectrum of the unknown compound in Problem 19.12.

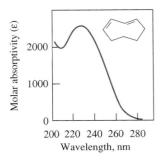

FIGURE 19.15 The UV spectrum of *cis,trans*-1,3-cyclooctadiene.

Molar absorptivity used to be called the **molar extinction coefficient.**

light is the low-energy end of the visible region; violet light is the high-energy end. Ultraviolet (UV) radiation lies beyond the violet end of visible light; it encompasses the region from 25,000 to 50,000 cm^{-1}. Positions of absorption in ultraviolet-visible (UV-Vis) spectroscopy are usually expressed as wavelengths in **nanometers, nm** (1 nm = 10^{-9} m). Thus, the visible region corresponds to 800–400 nm, and the UV region to 400–200 nm.

Electron excitation occurs when a molecule absorbs a photon having energy in the visible or UV region. An electron in the molecule undergoes a transition from its most stable state to the next higher energy level, an excited state. Thus UV-Vis spectroscopy probes the electron distribution in a molecule and is particularly useful when conjugated π electron systems are present.

Figure 19.15 shows the UV spectrum of a conjugated diene, *cis,trans*-1,3-cyclooctadiene. Note that there are fewer features than in an IR spectrum. The peak corresponds to the excitation of a π electron of the conjugated diene system. As is typical of most UV spectra, the absorption peak is rather broad. The wavelength at which absorption is a maximum is referred to as the λ_{max} of the sample. In this case λ_{max} is 230 nm.

Information about the π electron framework of a molecule can be learned from both the λ_{max} and the intensity of the absorption peak. The peak intensity is expressed by a unit called **molar absorptivity,** abbreviated ϵ_{max}. Table 19.4 lists the λ_{max} and ϵ_{max} for ethylene, a conjugated diene, a conjugated triene, and a conjugated polyene. The data show that as the number of double bonds in conjugation with one another increases, λ_{max} shifts to a longer wavelength and ϵ_{max} increases.

PROBLEM 19.13 Which one of the C$_5$H$_8$ isomers shown has its λ_{max} at the longest wavelength?

Chromophores are the structural units of conjugated compounds that absorb electromagnetic radiation in the ultraviolet or visible region. The chromophore is responsible

TABLE 19.4	The Effect of Conjugation on λ_{max} and Peak Intensity (ϵ_{max})	
	λ_{max}, nm	ϵ_{max}
Ethylene, CH$_2$=CH$_2$	175	15,000
1,3-Butadiene, CH$_2$=CH—CH=CH$_2$	217	21,000
1,3,5-Hexatriene, CH$_2$=CH—CH=CH—CH=CH$_2$	258	35,000
β-Carotene (11 double bonds)	465	125,000

β-Carotene

for the color in certain compounds. Lycopene, for example, is a carotenoid (a type of terpene, see Section 16.10) that contributes to the red color of tomatoes and paprika. The chromophore in lycopene is a conjugated system of 11 double bonds. Lycopene absorbs the blue-green fraction of visible light (λ_{max} = 505 nm) and appears red.

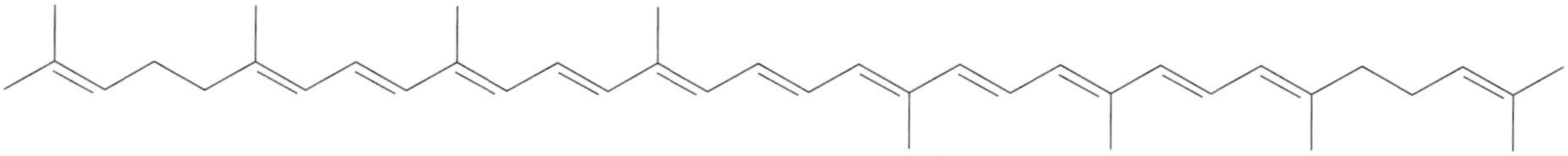

Lycopene

The maximum absorption of β-carotene (see Table 19.4), on the other hand, is at 465 nm, and β-carotene is yellow.

The concept of chromophores is at the heart of the chemistry of dyes. Dyes isolated from plant materials have been known for thousands of years. The blue dye indigo was used in India as long as 4000 years ago. For many centuries clothing has been dyed red with alizarin. In this century synthetic azo dyes have largely replaced natural ones in clothing and in processed foods (Section 14.11).

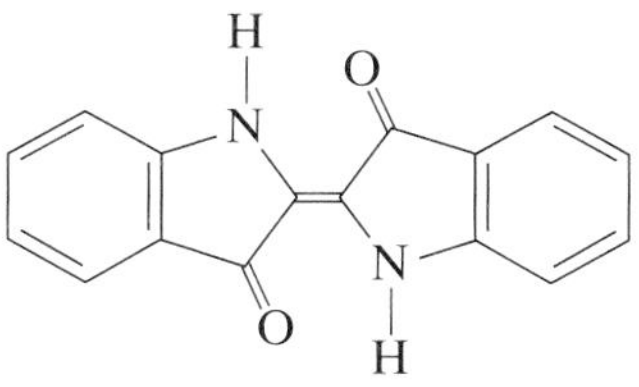

Indigo 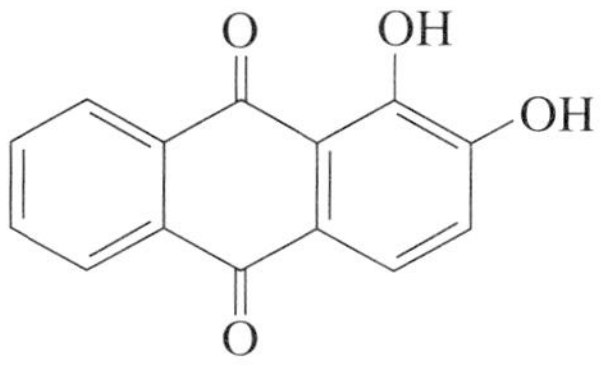   Alizarin

19.12 CONNECTING SPECTROSCOPY AND STRUCTURAL TYPE

The methods of spectroscopic analysis discussed thus far, in particular IR spectroscopy and NMR spectroscopy, are widely used by organic chemists in the determination of molecular structure. The functional group present in an organic molecule exhibits characteristic absorptions that aid in identification of that substance. In this section spectroscopic features that are characteristic of several common classes of organic compounds are discussed.

Alcohols

Infrared. As noted earlier (Section 19.10), alcohols exhibit a band in the IR spectrum characteristic of the O—H group, appearing in the 3200–3650 cm^{-1} region. In addition, the C—O absorption of an alcohol appears in the region between 1025 and 1200 cm^{-1}. These bands can be seen in the IR spectrum of 2-hexanol (see Figure 19.12).

^{1}H NMR. The most helpful signals in the NMR spectrum of alcohols result from the hydroxyl proton and the proton in the H—C—O unit of primary and secondary alcohols.

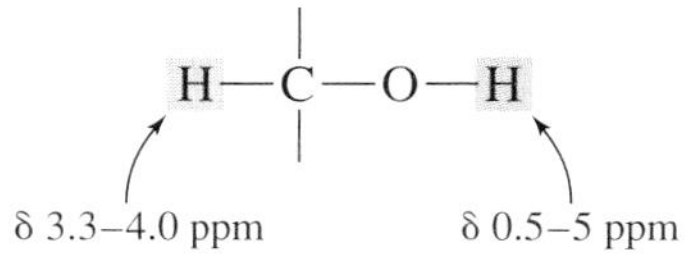

δ 3.3–4.0 ppm   δ 0.5–5 ppm

The chemical shift of the hydroxyl proton signal is variable, depending on solvent, temperature, and concentration. Its precise position is not particularly significant in structure determination. Because the signals due to hydroxyl protons are not usually split by other protons in the molecule and are often rather broad, they are often fairly easy to identify. For example, in the ^{1}H NMR spectrum of 2-phenylethanol, the hydroxyl proton signal appears as a singlet at δ 4.5 ppm. There are two triplets in this spectrum; the one at lower field strength (δ 4.0 ppm) corresponds to the protons of the CH_2O unit. The higher-field strength triplet at δ 3.1 ppm arises from the benzylic CH_2 group. The assignment of a particular signal to the hydroxyl proton can be confirmed by adding D_2O. The hydroxyl proton is replaced by deuterium, and its ^{1}H NMR signal disappears.

^{13}C NMR.   The electronegative oxygen of an alcohol decreases the shielding of the carbon to which it is attached. The chemical shift for the carbon of the C—OH unit is 60–75 ppm for most alcohols. Compared with an attached H, an attached OH causes a downfield shift of 35–50 ppm in the carbon signal.

$$CH_3CH_2CH_2CH_3 \qquad CH_3CH_2CH_2CH_2OH$$

δ 13.0 ppm δ 61.4 ppm

Butane   1-Butanol

Aldehydes and Ketones

Infrared.   Carbonyl groups are among the easiest functional groups to detect by IR spectroscopy. The C=O stretching vibration of aldehydes and ketones gives rise to strong absorption in the region 1710–1750 cm^{-1} as illustrated for butanal in Figure 19.16. In addition to a peak for C=O stretching, the CH=O group of an aldehyde exhibits two weak bands for C—H stretching near 2720 and 2820 cm^{-1}.

^{1}H NMR.   Aldehydes are readily identified by the presence of a signal for the hydrogen of CH=O at δ 9–10 ppm. This is a region where very few other protons ever appear. Figure 19.17 shows the ^{1}H NMR spectrum of 2-methylpropanal [$(CH_3)_2CHCH=O)$], where the large chemical shift difference between the aldehyde proton and the other protons in the molecule is clearly evident. As seen in the expanded-scale inset, the aldehyde proton is a doublet, split by the proton as C-2. Coupling between the protons in HC—CH=O is much smaller than typical vicinal couplings, making the multiplicity of the aldehyde peak difficult to see without expanding the scale.

FIGURE 19.16 Infrared spectrum of butanal showing peaks characteristic of the CH=O unit at 2720 and 2820 cm^{-1} (C—H) and at 1720 cm^{-1} (C=O).

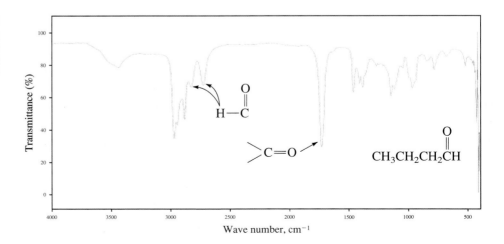

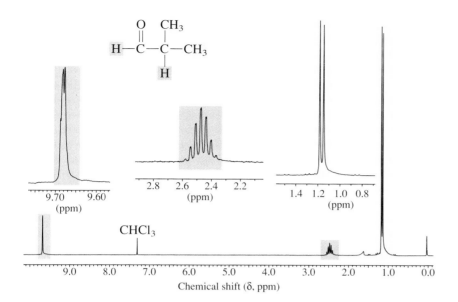

FIGURE 19.17 The 200-MHz ^1H NMR spectrum of 2-methylpropanal, showing the aldehyde proton as a doublet at low field strength (9.7 ppm).

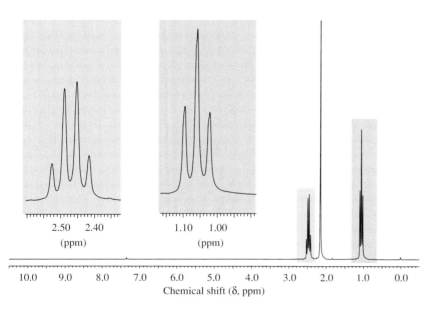

FIGURE 19.18 The 200-MHz ^1H NMR spectrum of 2-butanone. The triplet–quartet pattern of the ethyl group is more clearly seen in the scale-expanded insets.

Methyl ketones, such as 2-butanone in Figure 19.18, are characterized by sharp singlets near δ 2 ppm for the protons of $CH_3C{=}O$. Similarly, the deshielding effect of the carbonyl causes the protons of $CH_2C{=}O$ to appear at lower field (δ 2.4 ppm) than in a CH_2 group of an alkane.

^{13}C **NMR.**   The signal for the carbon of $C{=}O$ in aldehydes and ketones appears at very low field, some 190–220 ppm downfield from tetramethylsilane.

Carboxylic Acids

Infrared.   The most characteristic peaks in the IR spectra of carboxylic acids are those of the hydroxyl and carbonyl groups. As shown in the IR spectrum of 4-phenylbutanoic acid (see Figure 19.13) the O—H and C—H stretching frequencies overlap to produce a broad absorption in the 3500–2500 cm^{-1} region. The carbonyl group gives a strong band for C=O stretching at 1700 cm^{-1}.

^{1}H NMR.   The hydroxyl proton of a CO_2H group is normally the least shielded of all the protons in an NMR spectrum, appearing 10–12 ppm downfield from tetramethylsilane, often as a broad peak. As with other hydroxyl protons, the proton of a carboxyl group can be identified by adding D_2O to the sample. Hydrogen–deuterium exchange converts —CO_2H to —CO_2D, and the signal corresponding to the carboxyl group disappears.

^{13}C NMR.   Like other carbonyl groups, the carbon of the —CO_2H group of a carboxylic acid is strongly deshielded (δ 160–185 ppm), but not as much as that of an aldehyde or ketone.

Carboxylic Acid Derivatives

Infrared.   IR spectroscopy has been quite useful in determining structures of carboxylic acid derivatives. The carbonyl band is strong, and its position is sensitive to the substituents bound to the carbonyl group. As these examples illustrate, the characteristic absorption frequencies of the various acid derivatives vary considerably. Anhydrides have two carbonyl bands from interaction of the vibrations of the two carbonyl groups.

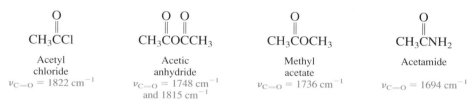

^{1}H NMR.   Chemical-shift differences in their ^{1}H NMR spectra aid the structure determination of esters. Consider two isomeric esters, ethyl acetate and methyl propanoate. Both have a methyl singlet and a triplet–quartet pattern for their ethyl group. Notice, however, the significant difference in the chemical shifts of the corresponding signals in each spectrum. The methyl is more shielded when it is bonded directly to the carbonyl group (δ 2.0 ppm), as in ethyl acetate, than when it is bonded to the oxygen of methyl propanoate (δ 3.6 ppm). Likewise, the methylene group is more shielded in methyl propanoate when it is bonded to the carbonyl group (δ 2.3 ppm) than when bonded to oxygen in ethyl acetate (δ 4.1 ppm).

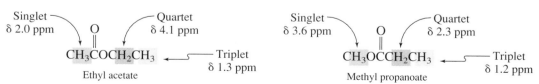

The chemical shift of the N—H proton of amides appears in the range δ 5–8 ppm. It is often a very broad peak; sometimes it is so broad that it does not rise much over the baseline and can be lost in the background noise.

¹³C NMR. The ¹³C NMR spectra of carboxylic acid derivatives, like the spectra of carboxylic acids themselves, are characterized by a low-field resonance for the carbonyl carbon in the range δ 160–180 ppm. The carbonyl carbons of carboxylic acid derivatives are more shielded than those of aldehydes and ketones, but less shielded than the sp^2-hybridized carbons of alkenes and arenes.

Amines

Infrared. The absorptions of interest in the IR spectra of amines are those associated with N—H vibrations. Primary alkyl- and arylamines exhibit two peaks in the range 3000–3500 cm⁻¹, which are due to symmetric and antisymmetric N—H stretching modes.

Symmetric N—H stretching of a primary amine

Antisymmetric N—H stretching of a primary amine

These two vibrations are clearly visible at 3270 and 3380 cm⁻¹ in the IR spectrum of butylamine, shown in Figure 19.19*a*. Secondary amines such as diethylamine, shown in Figure 19.19*b*, exhibit only one peak, which is due to N—H stretching, at 3280 cm⁻¹. Tertiary amines, of course, are transparent in this region, because they have no N—H bonds.

¹H NMR. Characteristics of the ¹H NMR spectra of amines may be illustrated by comparing 4-methylbenzylamine with 4-methylbenzyl alcohol. Nitrogen is less electronegative than oxygen and so shields neighboring nuclei to a greater extent. The benzylic methylene group attached to nitrogen in 4-methylbenzylamine appears at higher field (δ 3.7 ppm) than the benzylic methylene of 4-methylbenzyl alcohol (δ 4.6 ppm). The N—H protons are somewhat more shielded than the O—H protons of an alcohol. In 4-methylbenzylamine the protons of the amino group correspond to the signal at δ 1.4 ppm, and the hydroxyl proton signal of 4-methylbenzyl alcohol is found at δ 1.9 ppm. The chemical shifts of amino group protons, like those of hydroxyl protons, are variable and are sensitive to solvent, concentration, and temperature.

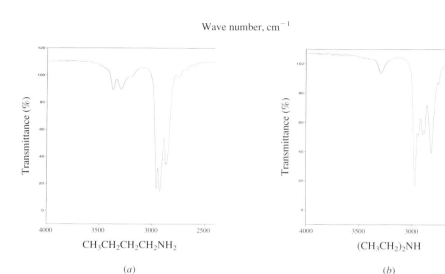

Wave number, cm⁻¹

$CH_3CH_2CH_2CH_2NH_2$

(*a*)

$(CH_3CH_2)_2NH$

(*b*)

FIGURE 19.19 Portions of the IR spectrum of (*a*) butylamine and (*b*) diethylamine. Primary amines exhibit two peaks due to N—H stretching, whereas secondary amines show only one.

δ 7.1 ppm

δ 2.3 ppm   δ 1.4 ppm   δ 2.4 ppm   δ 7.2 ppm   δ 1.9 ppm

H_3C —⟨benzene⟩— CH_2NH_2 H_3C —⟨benzene⟩— CH_2OH

δ 3.7 ppm   δ 4.6 ppm

4-Methylbenzylamine   4-Methylbenzyl alcohol

^{13}C NMR.   Similarly, carbons that are bonded to nitrogen are more shielded than those bonded to oxygen, as revealed by comparing the ^{13}C chemical shifts of methylamine and methanol.

δ 26.9 ppm CH_3NH_2 δ 48.0 ppm CH_3OH

Methylamine   Methanol

19.13 MASS SPECTROMETRY

Mass spectrometry is the instrumental method of structure determination that is most different from the other methods discussed in this chapter. It does not depend on the selective absorption of particular frequencies of electromagnetic radiation, but rather examines what happens to a molecule when it is bombarded with high-energy electrons. If an electron having an energy of about 10 electron volts [10 eV = 964 kJ/mol (230 kcal/mol)] collides with an organic molecule, the energy transferred as a result of that collision is sufficient to dislodge one of the molecule's electrons.

$$A{:}B + e^- \longrightarrow A{\cdot}\overset{+}{B} + 2e^-$$

We say the molecule AB has been ionized by **electron impact.** The species that results is called the **molecular ion, M^+.** The molecular ion is positively charged and has an odd number of electrons—it is a **cation radical.**

> *The molecular ion has the same mass as the neutral molecule less the negligible mass of a single electron.*

Although energies of about 10 eV are required, energies of about 70 eV are used in most mass spectrometers. Electrons this energetic not only bring about the ionization of a molecule but impart a large amount of energy to the molecular ion. The molecular ion dissipates this excess energy by dissociating into smaller fragments. Dissociation of a cation radical produces a neutral fragment and a positively charged fragment.

$$A{\cdot}\overset{+}{B} \longrightarrow A^+ + B{\cdot}$$

Cation radical   Cation   Radical

Ionization and fragmentation produce a mixture of particles, some neutral and some positively charged. The mass-to-charge ratio (*m/z*) of each of the positive ions is measured. As the charge is almost always 1, the value of *m/z* is equal to the mass of the ion. A **mass spectrum** is the distribution of the positive ions formed by fragmentation of the molecular ion. The **fragmentation pattern** is characteristic of a particular compound.

The most common format for displaying mass spectra is a bar graph on which relative ion intensity is plotted versus *m/z*. Figure 19.20 shows the mass spectrum of benzene in bar graph form. The most intense peak in the mass spectrum is called the **base peak** and is assigned a relative intensity of 100. Peak intensities are reported relative to

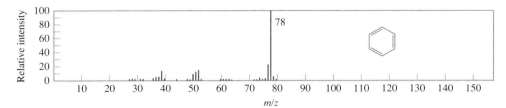

FIGURE 19.20 The mass spectrum of benzene. The peak at $m/z = 78$ corresponds to the C_6H_6 molecular ion.

the base peak. The base peak in the mass spectrum of benzene corresponds to the molecular ion (M^+) at $m/z = 78$.

Benzene   Electron   Molecular ion   Two
                          of benzene   electrons

Benzene does not undergo extensive fragmentation; none of the fragment ions in its mass spectrum are as abundant as the molecular ion.

Many of the fragmentation processes in mass spectrometry proceed to form a stable carbocation, and the principles developed in earlier chapters (Sections 3.9 and 6.5) are applicable. Alkylbenzenes of the type $C_6H_5CH_2R$ undergo cleavage of the bond to the benzylic carbon to give $m/z = 91$ ($C_7H_7^+$) as the base peak. The mass spectrum in Figure 19.21 and the following fragmentation diagram illustrate this for propylbenzene.

$$CH_2-CH_2-CH_3 \qquad M^+ \ 120$$

91

PROBLEM 19.14 The base peak appears at m/z 105 for one of the following compounds and at m/z 119 for the other two. Match the compounds with the appropriate m/z values for their base peaks.

CH$_2$CH$_3$ CH$_2$CH$_2$CH$_3$ CH$_3$CHCH$_3$

CH$_3$ CH$_3$ CH$_3$ CH$_3$

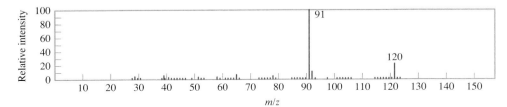

FIGURE 19.21 The mass spectrum of propylbenzene. The most intense peak is $C_7H_7^+$.

GAS CHROMATOGRAPHY, GC/MS, AND MS/MS

All of the spectra in this chapter (^{1}H NMR, ^{13}C NMR, IR, UV-Vis, and MS) were obtained using pure substances. It is much more common, however, to encounter an organic substance, either formed as the product of a chemical reaction or isolated from natural sources, as but one component of a mixture. Just as the last half of the twentieth century saw a revolution in the methods available for the *identification* of organic compounds, so too has it seen remarkable advances in methods for their *separation* and *purification*.

Classical methods for separation and purification include fractional distillation of liquids and recrystallization of solids, and these two methods are routinely included in the early portions of laboratory courses in organic chemistry. Because they are capable of being adapted to work on a large scale, fractional distillation and recrystallization are the preferred methods for purifying organic substances in the pharmaceutical and chemical industries.

Some other methods are more appropriate when separating small amounts of material in laboratory-scale work and are most often encountered there. Indeed, it is their capacity to deal with exceedingly small quantities that is the strength of a number of methods that together encompass the various forms of **chromatography.** The first step in all types of chromatography involves absorbing the sample onto some material called the **stationary phase.** Next, a second phase (the **mobile phase**) is allowed to move across the stationary phase. Depending on the properties of the two phases and the components of the mixture, the mixture is separated into its components according to the rate at which each is removed from the stationary phase by the mobile phase.

In **gas chromatography (GC),** the stationary phase consists of beads of an inert solid support coated with a high-boiling liquid, and the mobile phase is a gas, usually helium. Figure 19.22 shows a typical gas chromatograph. The sample is injected by

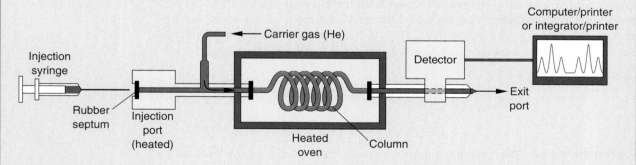

FIGURE 19.22 Schematic drawing of a gas chromatograph. When connected to a mass spectrometer, as in GC/MS, the effluent is split into two streams as it leaves the column.

(Reprinted, with permission, from A.M. Schoffstall, B.A. Gaddis, and M.L. Druelinger, *Microscale and Miniscale Organic Chemistry Laboratory Experiments,* McGraw-Hill, New York, 2000.)

syringe onto a heated block where a stream of helium carries it onto a coiled column packed with the stationary phase. The components of the mixture move through the column at different rates. They are said to have different **retention times.** Gas chromatography is also referred to as **gas–liquid partition chromatography,** because the technique depends on how different substances partition themselves between the gas phase (dispersed in the helium carrier

gas) and the liquid phase (dissolved in the coating on the beads of solid support).

Typically the effluent from a gas chromatograph is passed through a detector, which feeds a signal to a recorder whenever a substance different from pure carrier gas leaves the column. Thus, one determines the number of components in a mixture by counting the number of peaks on a strip chart. It is good practice to carry out the analysis under different

Continued

conditions by varying the liquid phase, the tempera-ture, and the flow rate of the carrier gas so as to en-sure that two substances have not eluted together and given a single peak under the original conditions. Gas chromatography can also be used to identify the components of a mixture by comparing their reten-tion times with those of authentic samples.

In **gas chromatography/mass spectrometry (GC/MS)**, the effluent from a gas chromatograph is passed into a mass spectrometer and a mass spectrum is taken every few milliseconds. Thus gas chromatogra-phy is used to separate a mixture, and mass spectrome-try used to analyze it. GC/MS is a very powerful analyti-cal technique. One of its more visible applications involves the testing of athletes for steroids, stimulants, and other performance-enhancing drugs. These drugs are converted in the body to derivatives called **metabo-lites,** which are then excreted in the urine. When the urine is subjected to GC/MS analysis, the mass spectra of

its organic components are identified by comparison with the mass spectra of known metabolites stored in the instrument's computer. Using a similar procedure, the urine of newborn infants is monitored by GC/MS for metabolite markers of genetic disorders that can be treated if detected early in life. GC/MS is also used to detect and measure the concentration of halogenated hydrocarbons in drinking water.

Although GC/MS is the most widely used analyt-ical method that combines a chromatographic separa-tion with the identification power of mass spectrom-etry, it is not the only one. Chemists have coupled mass spectrometers to most of the instruments that are used to separate mixtures. Perhaps the ultimate is **mass spectrometry/mass spectrometry (MS/MS),** in which one mass spectrometer generates and sepa-rates the molecular ions of the components of a mix-ture and a second mass spectrometer examines their fragmentation patterns!

Understanding how molecules fragment on electron impact permits a mass spec-trum to be analyzed in sufficient detail to deduce the structure of an unknown compound. Thousands of compounds of known structure have been examined by mass spectrome-try, and the fragmentation patterns that characterize different classes are well docu-mented.

19.14 MOLECULAR FORMULA AS A CLUE TO STRUCTURE

Chemists are often confronted with the problem of identifying the structure of an unknown compound. Sometimes the unknown can be shown to be a sample of a sub-stance previously reported in the chemical literature; on other occasions, the unknown may be truly that—a compound never before encountered by anyone. An arsenal of pow-erful techniques is available to simplify the task of structure determination. Even the molecular formula provides more information than might be apparent at first glance.

Consider, for example, a substance with the molecular formula C_7H_{16}. We know immediately that the compound is an alkane because its molecular formula corresponds to the general formula for that class of compounds, C_nH_{2n+2}, where $n = 7$.

What about a substance with the molecular formula C_7H_{14}? This compound can-not be an alkane but may be either a cycloalkane or an alkene because both these classes of hydrocarbons correspond to the general molecular formula C_nH_{2n}.

Any time a ring or a double bond is present in an organic molecule, its molecular formula has two fewer hydrogen atoms than that of an alkane with the same number of carbons.

The relationship between molecular formulas, multiple bonds, and rings is referred to as the **index of hydrogen deficiency** and can be expressed by the equation:

$$\text{Index of hydrogen deficiency} = \tfrac{1}{2}(C_nH_{2n+2} - C_nH_x)$$

where C_nH_x is the molecular formula of the compound.

Other terms that mean the same thing as the index of hydrogen deficiency include **elements of unsaturation, sites of unsaturation,** and **the sum of double bonds and rings.**

A molecule that has a molecular formula of C_7H_{14} has an index of hydrogen deficiency of 1:

$$\text{Index of hydrogen deficiency} = \tfrac{1}{2}(C_7H_{16} - C_7H_{14})$$

$$\text{Index of hydrogen deficiency} = \tfrac{1}{2}(2) = 1$$

Thus, the compound has one ring or one double bond. It can't have a triple bond.

A molecule of molecular formula C_7H_{12} has four fewer hydrogens than the corresponding alkane. It has an index of hydrogen deficiency of 2 and can have two rings, two double bonds, one ring and one double bond, or one triple bond.

What about substances other than hydrocarbons, 1-heptanol $[CH_3(CH_2)_5CH_2OH]$, for example? Its molecular formula $(C_7H_{16}O)$ contains the same carbon-to-hydrogen ratio as heptane and, like heptane, it has no double bonds or rings. Cyclopropyl acetate $(C_5H_8O_2)$, whose structure is shown at the left, has one ring and one double bond and an index of hydrogen deficiency of 2. Oxygen atoms have no effect on the index of hydrogen deficiency.

A halogen substituent, like hydrogen, is monovalent, and when present in a molecular formula is treated as if it were hydrogen for counting purposes.

How does one distinguish between rings and double bonds? This additional piece of information comes from catalytic hydrogenation experiments in which the amount of hydrogen consumed is measured exactly. Each of a molecule's double bonds consumes one molar equivalent of hydrogen, but rings are unaffected. For example, a substance with a hydrogen deficiency of 5 that takes up 3 moles of hydrogen must have two rings.

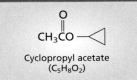

Cyclopropyl acetate
$(C_5H_8O_2)$

PROBLEM 19.15 How many rings are present in each of the following compounds? Each consumes 2 moles of hydrogen on catalytic hydrogenation.
(a) $C_{10}H_{18}$ (d) C_8H_8O
(b) C_8H_8 (e) $C_8H_{10}O_2$
(c) $C_8H_8Cl_2$ (f) C_8H_9ClO

Sample Solution (a) The molecular formula $C_{10}H_{18}$ contains four fewer hydrogens than the alkane having the same number of carbon atoms $(C_{10}H_{22})$. Therefore, the index of hydrogen deficiency of this compound is 2. Because it consumes two molar equivalents of hydrogen on catalytic hydrogenation, it must have two double bonds and no rings.

LEARNING OBJECTIVES

This chapter focused on the role of spectroscopy in the determination of chemical structures. The skills you have learned in this chapter should enable you to

- Explain how the frequency and wavelength of electromagnetic radiation are related.
- Explain what is meant by chemical shift in a proton nuclear magnetic resonance (^1H NMR) spectrum.
- Explain how spin–spin splitting affects the appearance of a ^1H NMR spectrum.

Continued

• Interpret a ^1H NMR spectrum in terms of the

Number of signals
Chemical shift of each signal
Relative intensity of each signal
Multiplicity of each signal

• Predict the number of signals in the ^{13}C NMR spectrum of a specific compound.

• Identify the functional group present in a molecule on the basis of its IR spectrum.

• Explain how conjugation affects the absorption of UV or visible light by a molecule.

• Explain the origin of the molecular ion in a mass spectrum.

• Calculate the index of hydrogen deficiency of a molecule from its molecular formula.

19.15 SUMMARY

Structure determination in modern-day organic chemistry relies heavily on instrumental methods. Several of the most widely used ones depend on the absorption of electromagnetic radiation (Section 19.1).

The spectroscopic methods most commonly used by organic chemists include:

^1H NMR spectroscopy (Sections 19.2 to 19.7). In the presence of an external magnetic field, the nuclear spin states of a proton have slightly different energies. Protons absorb electromagnetic radiation in the radio frequency region, causing the nuclear spin to "flip" from the low-energy state to the high-energy state. The energy depends on the extent to which a nucleus is shielded from the external magnetic field, called the **chemical shift** (Section 19.3). Protons in different environments within a molecule have different chemical shifts. The **number of signals** (Section 19.4) in a ^1H NMR spectrum reveals the number of chemical-shift-nonequivalent protons in a molecule; the **integrated areas,** or **peak intensities,** tell us their relative ratios; their chemical shifts indicate the kind of environment surrounding the proton; and the **splitting pattern** (Sections 19.6 and 19.7) is related to the number of protons on adjacent carbons.

^{13}C NMR spectroscopy (Sections 19.8 and 19.9). Using special techniques for signal enhancement, high-quality ^{13}C NMR spectra may be obtained, and these provide a useful complement to proton spectra. In many substances a separate signal is observed for each carbon atom.

IR spectroscopy(Section 19.10). This method probes molecular structure by examining transitions between vibrational energy levels using electromagnetic radiation in the $625–4000\text{-cm}^{-1}$ range. It is useful for determining the presence of certain **functional groups** based on their characteristic absorption frequencies.

UV-Vis spectroscopy (Section 19.11). Transitions between electronic energy levels involving electromagnetic radiation in the 200–800-nm range form the basis of UV-Vis spectroscopy. The absorption peaks tend to be broad but are often useful in indicating the presence of particular **conjugated π electron** systems within a molecule.

Mass spectrometry (Section 19.13). Mass spectrometry exploits the information obtained when a molecule is ionized by electron impact and then dissociates into smaller fragments. Positive ions are separated and detected according to their mass-to-charge ratio.

ADDITIONAL PROBLEMS

^{1}H NMR Spectra

19.16 How many signals would you expect to find in the ^{1}H NMR spectrum of each of the following compounds? Ignore splitting effects for this problem.

(a) 1-Bromobutane
(b) 2,2-Dibromobutane
(c) 1,4-Dibromobutane
(d) 1,1,4-Tribromobutane
(e) 1,1,1-Tribromobutane

19.17 Describe the appearance of the ^{1}H NMR spectrum of each of the following compounds. How many signals would you expect to find, and into how many peaks will each signal be split?

(a) $ClCH_2CH_2OCH_3$
(b) $ClCH_2C(CH_3)_2$
        |
        Cl
(c) $ClCH_2CH_2CH_2Cl$

(d) $Cl_2CHCH_2CCH_2CH_3$
              ‖
              O
(e) $CH_3OCH_2CH_2OCH_3$

19.18 Using the data in Table 19.1, predict the *approximate* chemical shift expected for each of the protons in Problem 19.17.

19.19 Each of the following compounds is characterized by an ^{1}H NMR spectrum that consists of only a single peak having the chemical shift indicated. Identify each compound.

(a) C_8H_{18}; δ 0.9 ppm
(b) C_5H_{10}; δ 1.5 ppm
(c) C_8H_8; δ 5.8 ppm
(d) C_4H_9Br; δ 1.8 ppm
(e) $C_2H_4Cl_2$; δ 3.7 ppm
(f) $C_2H_3Cl_3$; δ 2.7 ppm

19.20 Each of the following compounds is characterized by an ^{1}H NMR spectrum that consists of two peaks, both singlets, having the chemical shifts indicated. Identify each compound.

(a) C_6H_8; δ 2.7 ppm (4H) and 5.6 ppm (4H)
(b) $C_5H_{11}Br$; δ 1.1 ppm (9H) and 3.3 ppm (2H)
(c) $C_6H_{12}O$; δ 1.1 ppm (9H) and 2.1 ppm (3H)
(d) $C_6H_{10}O_2$; δ 2.2 ppm (6H) and 2.7 ppm (4H)

19.21 Deduce the structure of each of the following compounds on the basis of their ^{1}H NMR spectra and molecular formulas:

(a) C_8H_{10}; δ 1.2 ppm (triplet, 3H)
                      δ 2.6 ppm (quartet, 2H)
                      δ 7.1 ppm (broad singlet, 5H)

(b) $C_{10}H_{14}$; δ 1.3 ppm (singlet, 9H)
                        δ 7.0–7.5 ppm (multiplet, 5H)

(c) C_6H_{14}; δ 0.8 ppm (doublet, 12H)
                      δ 1.4 ppm (heptet, 2H)

(d) $C_4H_6Cl_4$; δ 3.9 ppm (doublet, 4H)
                       δ 4.6 ppm (triplet, 2H)

(e) $C_4H_6Cl_2$;   δ 2.2 ppm (singlet, 3H)
δ 4.1 ppm (doublet, 2H)
δ 5.7 ppm (triplet, 1H)

(f) C_3H_7ClO;   δ 2.0 ppm (pentet, 2H)
δ 2.8 ppm (singlet, 1H)
δ 3.7 ppm (triplet, 2H)
δ 3.8 ppm (triplet, 2H)

(g) $C_{14}H_{14}$;   δ 2.9 ppm (singlet, 4H)
δ 7.1 ppm (broad singlet, 10H)

19.22 From among the isomeric compounds of molecular formula C_4H_9Cl, choose the one having an ^1H NMR spectrum that

(a) Contains only a single peak
(b) Has several peaks including a doublet at δ 3.4 ppm
(c) Has several peaks including a triplet at δ 3.5 ppm
(d) Has several peaks including two distinct three-proton signals, one of them a triplet at δ 1.0 ppm and the other a doublet at δ 1.5 ppm

19.23 All the following questions pertain to ^1H NMR spectra of isomeric ethers having the molecular formula $C_5H_{12}O$.

(a) Which one has only singlets in its ^1H NMR spectrum?
(b) Along with other signals, this ether has a coupled doublet–septet pattern. None of the protons responsible for this pattern are coupled to protons anywhere else in the molecule. Identify this ether.
(c) In addition to other signals in its ^1H NMR spectrum, this ether exhibits two signals at relatively low field. One is a singlet; the other is a doublet. What is the structure of this ether?
(d) In addition to other signals in its ^1H NMR spectrum, this ether exhibits two signals at relatively low field. One is a triplet; the other is a quartet. Which ether is this?

^{13}C NMR Spectra

19.24 Identify each of the $C_4H_{10}O$ isomers on the basis of their ^{13}C NMR spectra:

(a) δ 18.9 ppm (CH_3) (two carbons)
δ 30.8 ppm (CH) (one carbon)
δ 69.4 ppm (CH_2) (one carbon)

(c) δ 31.2 ppm (CH_3) (three carbons)
δ 68.9 ppm (C) (one carbon)

(b) δ 10.0 ppm (CH_3)
δ 22.7 ppm (CH_3)
δ 32.0 ppm (CH_2)
δ 69.2 ppm (CH)

19.25 Identify the C_6H_{14} isomers on the basis of their ^{13}C NMR spectra:

(a) δ 19.1 ppm (CH_3)
δ 33.9 ppm (CH)

(b) δ 13.7 ppm (CH_3)
δ 22.8 ppm (CH_2)
δ 31.9 ppm (CH_2)

(c) δ 11.1 ppm (CH_3)
δ 18.4 ppm (CH_3)
δ 29.1 ppm (CH_2)
δ 36.4 ppm (CH)

(d) δ 8.5 ppm (CH_3)
δ 28.7 ppm (CH_3)
δ 30.2 ppm (C)
δ 36.5 ppm (CH_2)

(e) δ 14.0 ppm (CH_3)
δ 20.5 ppm (CH_2)
δ 22.4 ppm (CH_3)
δ 27.6 ppm (CH)
δ 41.6 ppm (CH_2)

IR and UV Spectra

19.26 An IR spectrum exhibits a broad band in the region between 3000 and 3500 cm^{-1} and a strong peak at 1710 cm^{-1}. Which of the following substances best fits this data? Explain your reasoning.

$$C_6H_5CH_2CH_2OH \qquad C_6H_5CH_2\overset{\overset{O}{\|}}{C}OH \qquad C_6H_5CH_2\overset{\overset{O}{\|}}{C}CH_3$$

19.27 What features would be present in the IR spectra of the following compounds that would allow each one to be distinguished from the others?

2-Pentanone   Pentanal   1-Pentanol

19.28 Explain how IR spectroscopy could be used to distinguish among the following three amines.

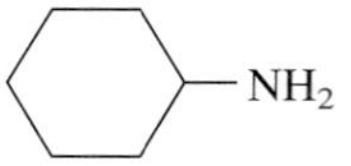

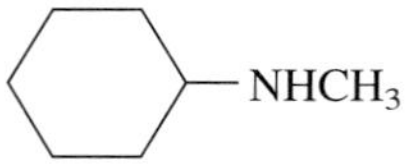

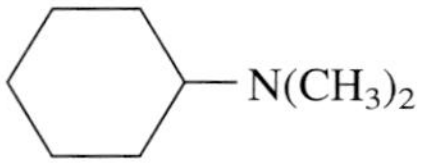

$$\text{C}_6\text{H}_{11}-\text{NH}_2 \qquad \text{C}_6\text{H}_{11}-\text{NHCH}_3 \qquad \text{C}_6\text{H}_{11}-\text{N(CH}_3)_2$$

19.29 Which of the following compounds has the longest wavelength UV absorption (λ_{max})? Explain your reasoning.

$$\text{CH}_3\text{CH}=\text{CHCH}_2\text{CH}_3 \qquad \text{CH}_2=\text{CHCH}=\text{CHCH}_3$$

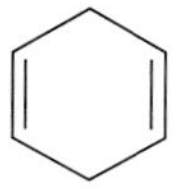

Combined Spectral Problems

19.30 Identify each of the following compounds based on the IR and ^{1}H NMR information provided:

(a) $\text{C}_{10}\text{H}_{12}\text{O}$: IR: 1710 cm^{-1}
^{1}H NMR: δ 1.0 ppm (triplet, 3H)
δ 2.4 ppm (quartet, 2H)
δ 3.6 ppm (singlet, 2H)
δ 7.2 ppm (singlet, 5H)

(b) $\text{C}_6\text{H}_{14}\text{O}_2$: IR: 3400 cm^{-1}
^{1}H NMR: δ 1.2 ppm (singlet, 12H)
δ 2.0 ppm (singlet, 2H)

(c) $\text{C}_4\text{H}_7\text{NO}$: IR: 2240 cm^{-1}
3400 cm^{-1} (broad)
^{1}H NMR: δ 1.65 ppm (singlet, 6H)
δ 3.7 ppm (singlet, 1H)

19.31 What structure of the compound having a molecular formula $\text{C}_5\text{H}_{12}\text{O}$ best fits the following spectroscopic information?

IR: Strong peak at 1150 cm^{-1}
No peak between 3000 and 3500 cm^{-1}
^{1}H NMR: 3H singlet (δ 3.4 ppm); 6H doublet (δ 1.0 ppm); plus other peaks
^{13}C NMR: Four signals

19.32 A compound ($\text{C}_8\text{H}_{10}\text{O}$) has the IR and ^{1}H NMR spectra presented in Figure 19.23. What is its structure?

19.33 Deduce the structure of a compound having the mass spectrum and ^{1}H NMR spectrum presented in Figure 19.24.

19.34 Figure 19.25 presents several types of spectroscopic data (IR, ^{1}H NMR, ^{13}C NMR, and mass spectra) for a particular compound. What is it?

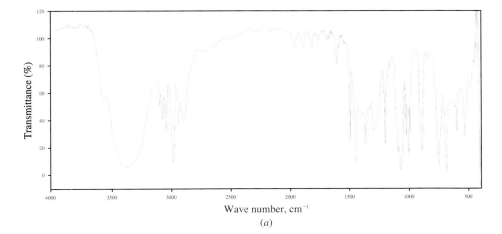

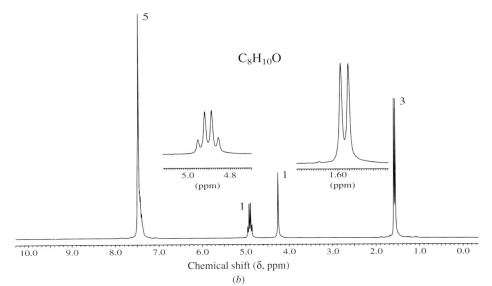

FIGURE 19.23 (a) IR and (b) 200-MHz ^{1}H NMR spectra of a compound $C_8H_{10}O$ (Problem 19.32).

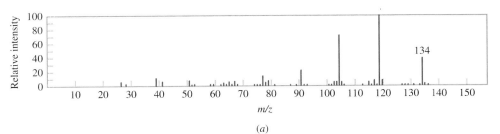

FIGURE 19.24 (a) Mass spectrum and (b) 200-MHz ^{1}H NMR spectrum of an unknown compound (Problem 19.33). (Figure continued on next page.)

FIGURE 19.24 Continued

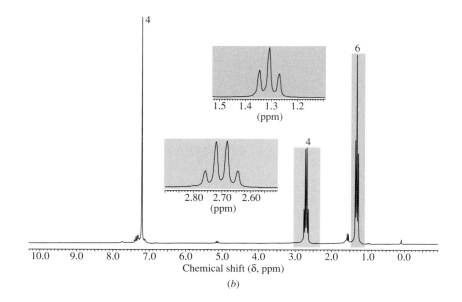

(b)

FIGURE 19.25 (a) Mass, (b) IR, (c) 200-MHz ^{1}H NMR, and (d) ^{13}C NMR spectra for the compound of Problem 19.34. (Figure continued on next page.)

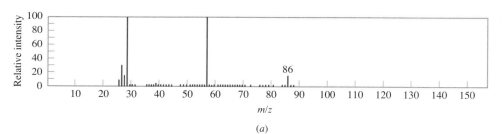

(a)

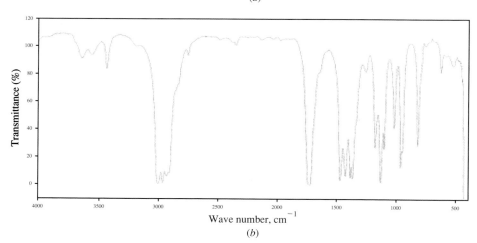

(b)

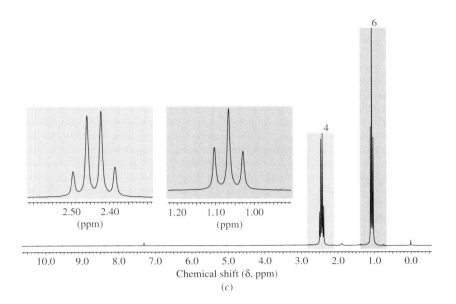

FIGURE 19.25 Continued

(c)

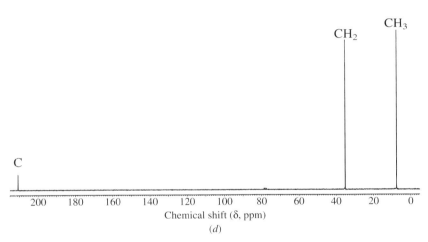

(d)

ANSWERS TO
IN-TEXT PROBLEMS

Problems are of two types: in-text problems that appear within the body of each chapter, and end-of-chapter problems. This appendix gives brief answers to all the in-text problems. More detailed discussions of in-text problems as well as detailed solutions to all the end-of-chapter problems are provided in a separate *Student Solutions Manual*. Answers to part (a) of those in-text problems with multiple parts have been provided in the form of a sample solution within each chapter and are not repeated here.

CHAPTER 1

1.1   4

1.2   All the third-row elements have a neon core containing 10 electrons ($1s^2 2s^2 2p^6$). The elements in the third row, their atomic numbers Z, and their electron configurations beyond the neon core are Na ($Z = 11$) $3s^1$; Mg ($Z = 12$) $3s^2$; Al ($Z = 13$) $3s^2 3p_x^1$; Si ($Z = 14$) $3s^2 3p_x^1 3p_y^1$; P ($Z = 15$) $3s^2 3p_x^1 3p_y^1 3p_z^1$; S ($Z = 16$) $3s^2 3p_x^2 3p_y^1 3p_z^1$; Cl ($Z = 17$) $3s^2 3p_x^2 3p_y^2 3p_z^1$; Ar ($Z = 18$) $3s^2 3p_x^2 3p_y^2 3p_z^2$.

1.3   Those ions that possess a noble gas electron configuration are (a) K^+; (b) H^-; (d) F^-; and (e) Ca^{2+}.

1.4   H:F̈:

1.5
```
     H H
H:C̈:C̈:H
     H H
```

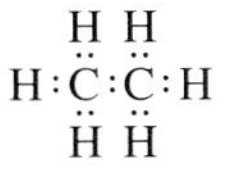

1.6   (b) 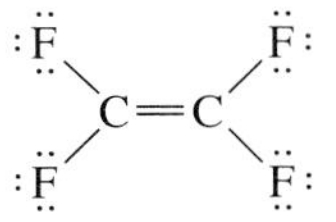 and (c) 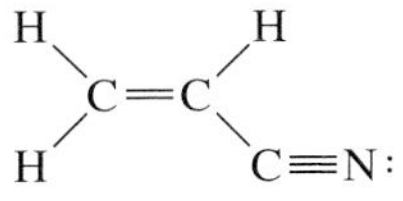 structures

1.7   Carbon bears a partial positive charge in CH_3Cl. It is partially negative in both CH_4 and CH_3Li, but the degree of negative charge is greater in CH_3Li.

1.8   (b) H: 0; C: -1; net: -1 (c) H: 0; C: 0; net: 0 (d) H: 0; C: $+1$; net: $+1$ (e) N: 0; C: -1; net: -1

1.9   (b), 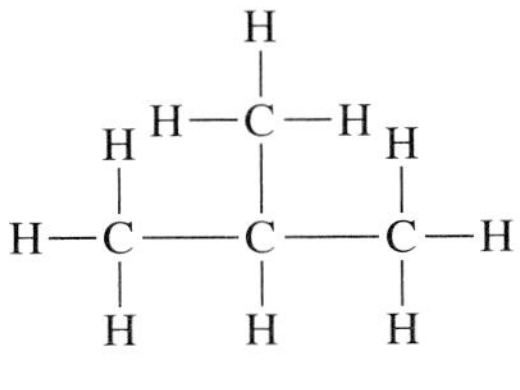 (c), 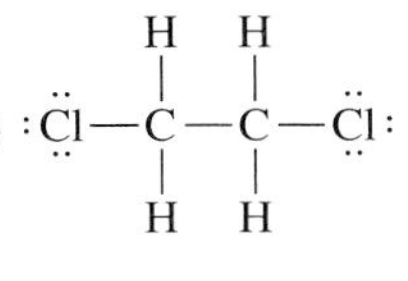 (d), 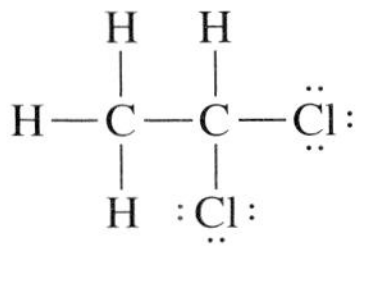 (e), 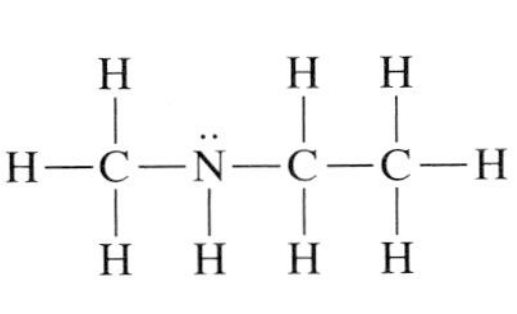 (f) 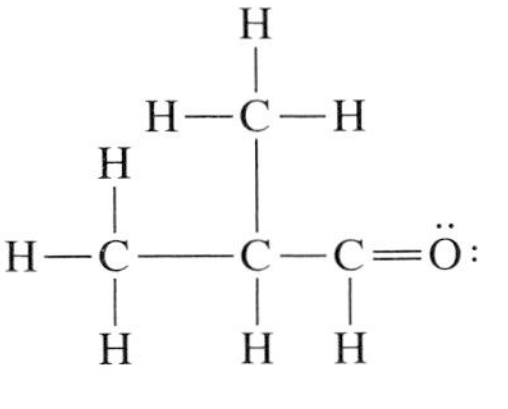 structures

1.10 (b) $(CH_3)_2CHCH(CH_3)_2$ (c) $HOCH_2CHCH(CH_3)_2$ with CH_3 substituent (d)

$$\begin{array}{ccc} & CH_2-CH_2 & \\ CH_2 & & CH-C(CH_3)_3 \\ & CH_2-CH_2 & \end{array}$$

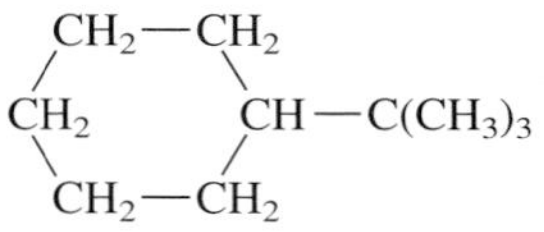

1.11 (b) $CH_3CH_2CH_2OH$, $(CH_3)_2CHOH$, and $CH_3CH_2OCH_3$. (c) There are seven isomers of $C_4H_{10}O$. Four have —OH groups: $CH_3CH_2CH_2CH_2OH$, $(CH_3)_2CHCH_2OH$, $(CH_3)_3COH$, and $CH_3CHCH_2CH_3$. Three have C—O—C units: $CH_3OCH_2CH_2CH_3$, $CH_3CH_2OCH_2CH_3$, and
|
OH
$(CH_3)_2CHOCH_3$

1.12 (b)

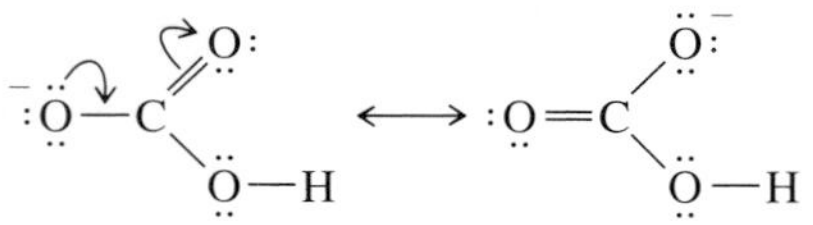

(c)

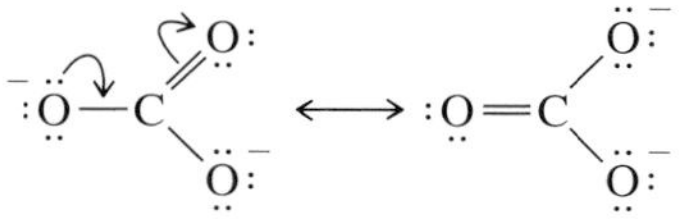

and

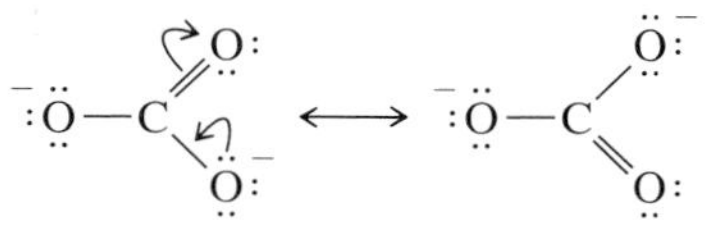

(d)

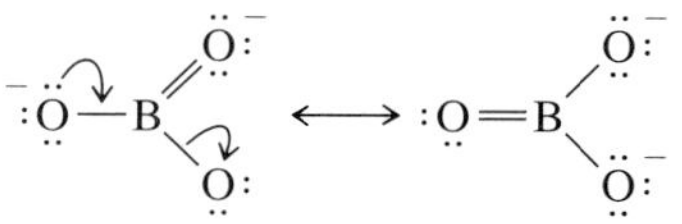

and

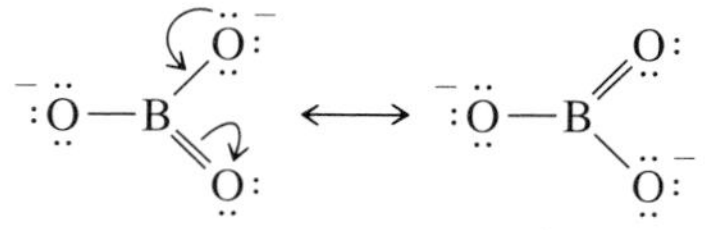

1.13 The H—B—H angles in BH_4^- are 109.5° (tetrahedral).

1.14 (b) Tetrahedral; (c) linear; (d) trigonal planar

1.15 (b) Oxygen is negative end of dipole moment directed along bisector of H—O—H angle; (c) no dipole moment; (d) dipole moment directed along axis of C—Cl bond, with chlorine at negative end, and carbon and hydrogens partially positive; (e) dipole moment aligned with axis of linear molecule, with nitrogen at negative end.

1.16 (b) sp^2; (c) carbon of CH_2 group is sp^2, and carbon of C=O is sp; (d) two doubly bonded carbons are each sp^2, whereas carbon of CH_3 group is sp^3; (e) carbon of C=O is sp^2, and carbons of CH_3 groups are sp^3; (f) two doubly bonded carbons are each sp^2, and carbon bonded to nitrogen is sp.

CHAPTER 2

2.1

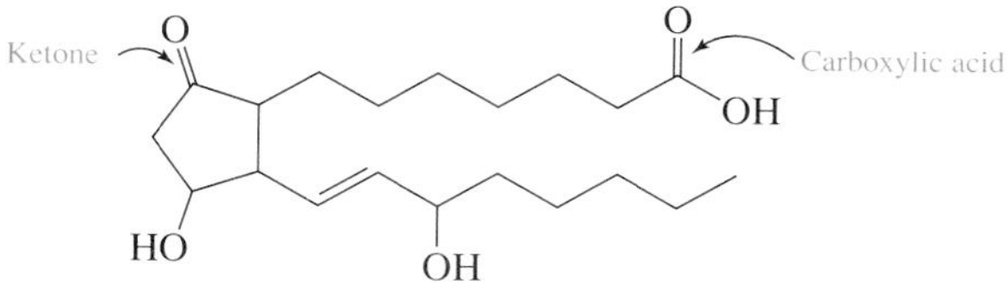

2.2 All three carbons; 10 σ bonds; each C—H σ bond is C sp^3–H $1s$; each C—C bond is C sp^3–C sp^3.

2.3 Propane has CH_3—H repulsions not present in ethane.

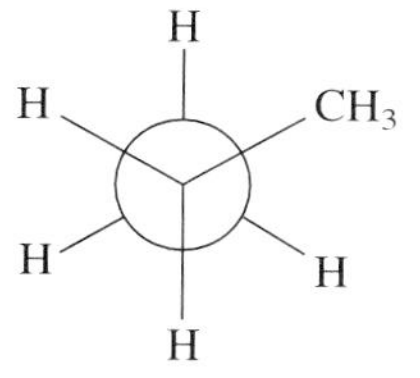

Staggered propane 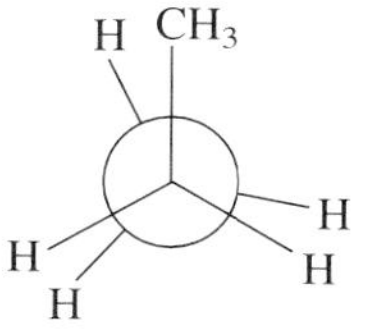   Eclipsed propane

2.4

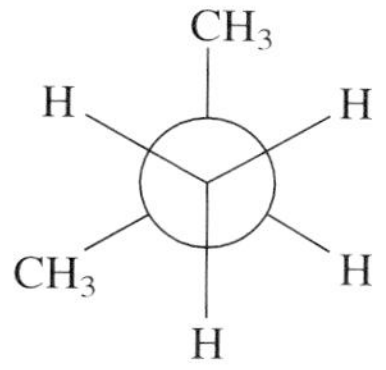

2.5 $CH_3(CH_2)_{26}CH_3$

2.6 The molecular formula is $C_{11}H_{24}$; the condensed structural formula is $CH_3(CH_2)_9CH_3$.

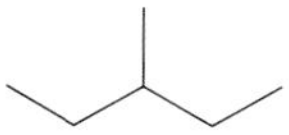

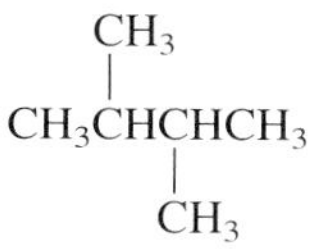

2.7

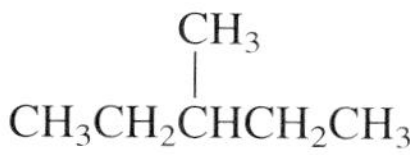

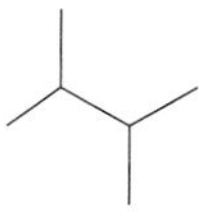

2.8 Undecane

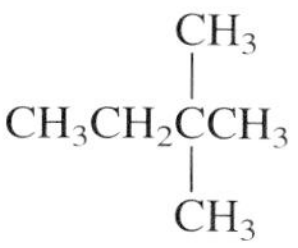

2.9

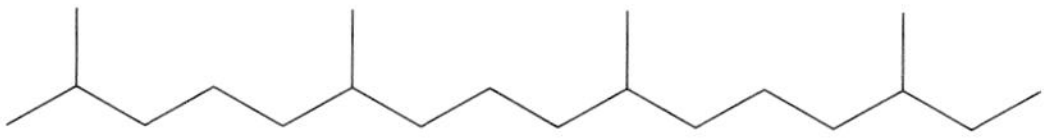

2.10 (b) $CH_3CH_2CH_2CH_2CH_3$ (pentane), $(CH_3)_2CHCH_2CH_3$ (2-methylbutane), $(CH_3)_4C$ (2,2-dimethylpropane); (c) 2,2,4-trimethylpentane; (d) 2,2,3,3-tetramethylbutane

2.11 (b) 4-Ethyl-2-methylhexane; (c) 8-ethyl-4-isopropyl-2,6-dimethyldecane

2.12 (b) 4-Isopropyl-1,1-dimethylcyclodecane; (c) cyclohexylcyclohexane

2.13 (b) 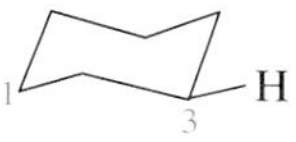 (c) 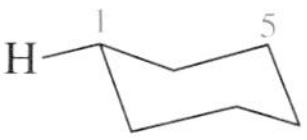 (d)

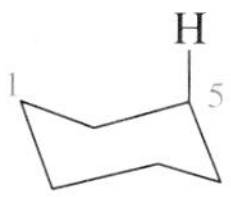

2.14

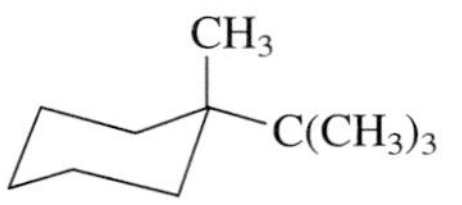

$$CH_3$$
$$C(CH_3)_3$$

2.15   1,1-Dimethylcyclopropane, ethylcyclopropane, methylcyclobutane, and cyclopentane

2.16   cis: $C(CH_3)_3$ trans: $C(CH_3)_3$
CH_3 CH_3

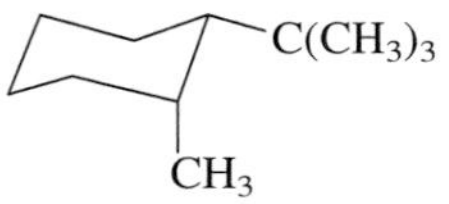

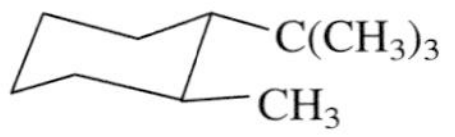

2.17   Octane (126°C); 2-methylheptane (116°C); 2,2,3,3-tetramethylbutane (106°C).

CHAPTER 3

3.1 $CH_3CH_2CH_2CH_2Cl$ $CH_3CHCH_2CH_3$
                                                       |
                                                       Cl

1-Chlorobutane   2-Chlorobutane
n-Butyl chloride   *sec*-Butyl chloride
or butyl chloride   or 1-methylpropyl chloride

$(CH_3)_2CHCH_2Cl$ $(CH_3)_3CCl$

1-Chloro-2-methylpropane   2-Chloro-2-methylpropane
Isobutyl chloride   *tert*-Butyl chloride
or 2-methylpropyl chloride   or 1,1-dimethylethyl chloride

3.2 $CH_3CH_2CH_2CH_2OH$ $CH_3CHCH_2CH_3$
                                                       |
                                                       OH

1-Butanol   2-Butanol
n-Butyl alcohol   *sec*-Butyl alcohol
or butyl alcohol   or 1-methylpropyl alcohol

$(CH_3)_2CHCH_2OH$ $(CH_3)_3COH$

2-Methyl-1-propanol   2-Methyl-2-propanol
Isobutyl alcohol   *tert*-Butyl alcohol
or 2-methylpropyl alcohol   or 1,1-dimethylethyl alcohol

3.3 $CH_3CH_2CH_2CH_2OH$ $CH_3CHCH_2CH_3$ $(CH_3)_2CHCH_2OH$ $(CH_3)_3COH$
                                              |
                                              OH

Primary   Secondary   Primary   Tertiary

3.4 $CH_3CH_2CHCHCH_2CH_2CH_3$
                   | |
                   HO CH_3

Secondary alcohol

3.5   Hydrogen bonding in ethanol (CH_3CH_2OH) makes its boiling point higher than that of dimethyl ether (CH_3OCH_3), in which hydrogen bonding is absent.

3.6 $H_3N\overset{\frown}{:}\, +\, H\overset{\frown}{—}\overset{..}{\underset{..}{Cl}}: \rightleftharpoons H_3\overset{+}{N}—H\, +\, :\overset{..}{\underset{..}{Cl}}:^{-}$

Base   Acid   Conjugate acid   Conjugate base

3.7   Formic acid: $K_a = 1.8 \times 10^{-4}$; Oxalic acid: $K_a = 6.5 \times 10^{-2}$. Oxalic acid is stronger.

3.8   Formate is the stronger base.

3.9 $(CH_3)_3C-\ddot{O}: \; + \; H-\ddot{C}l: \; \rightleftharpoons \; (CH_3)_3C-\overset{+}{\ddot{O}}-H \; + \; :\ddot{C}l:^-$
$\qquad\qquad\;\; |$ $\qquad\qquad\qquad\qquad\qquad\qquad\qquad\quad |$
$\qquad\qquad\;\; H$ $\qquad\qquad\qquad\qquad\qquad\qquad\qquad\quad H$

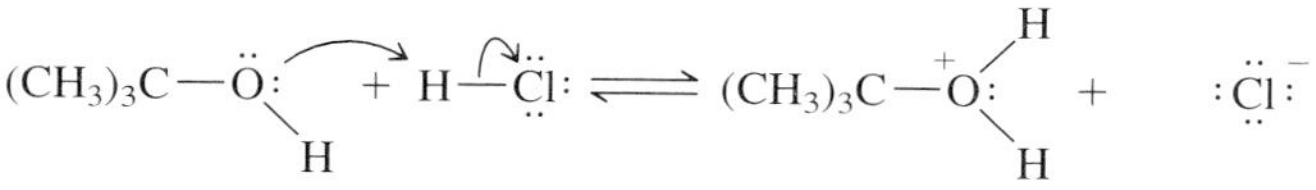

   Base   Acid   Conjugate   Conjugate base
                                      acid

3.10 $(CH_3)_3C-\overset{\delta+}{\underset{|}{O}}---H---\overset{\delta-}{Cl}$
$\qquad\qquad\qquad H$

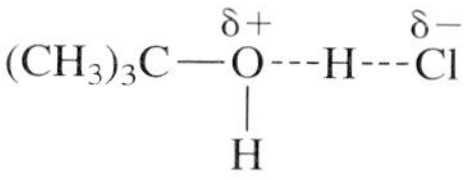

3.11   (b) $(CH_3CH_2)_3COH + HCl \longrightarrow (CH_3CH_2)_3CCl + H_2O$
           (c) $CH_3(CH_2)_{12}CH_2OH + HBr \longrightarrow CH_3(CH_2)_{12}CH_2Br + H_2O$

3.12 $CH_3CH_2CH_2\overset{+}{C}H_2 \qquad CH_3CH_2\overset{+}{C}HCH_3 \qquad CH_3\overset{+}{C}HCH_2 \qquad CH_3\overset{+}{C}CH_3$
$\qquad\qquad\qquad\qquad\qquad\qquad\qquad\qquad\qquad\quad |$ $\qquad\qquad |$
$\qquad\qquad\qquad\qquad\qquad\qquad\qquad\qquad\qquad CH_3$ $\qquad\quad CH_3$

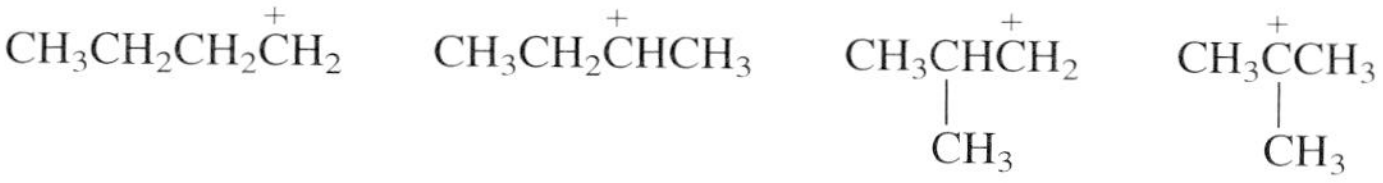

           (primary)   (secondary)   (primary)   (tertiary)

3.13 $CH_3\overset{+}{C}CH_2CH_3$
$\qquad\quad |$
$\qquad\quad CH_3$

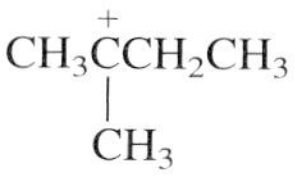

3.14 $CH_3\overset{\overset{\displaystyle OH}{|}}{C}CH_2CH_3$
$\qquad\quad |$
$\qquad\quad CH_3$

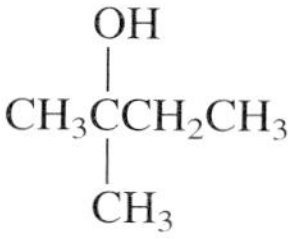

3.15   1-Butanol:

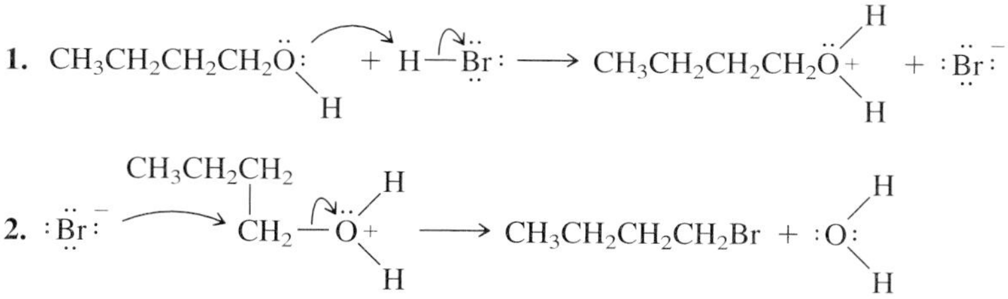

1. $CH_3CH_2CH_2CH_2\ddot{O}: \; + \; H-\ddot{B}r: \; \longrightarrow \; CH_3CH_2CH_2CH_2\overset{+}{O}-H \; + \; :\ddot{B}r:^-$
$\qquad\qquad\qquad\qquad |$ $\qquad\qquad\qquad\qquad\qquad\qquad\qquad\qquad\qquad\quad |$
$\qquad\qquad\qquad\qquad H$ $\qquad\qquad\qquad\qquad\qquad\qquad\qquad\qquad\qquad\quad H$

2. $:\ddot{B}r:^- \; \longrightarrow \; CH_2-\overset{+}{\ddot{O}}-H \; \longrightarrow \; CH_3CH_2CH_2CH_2Br \; + \; :\overset{\displaystyle H}{\underset{\displaystyle H}{O}}:$
$\qquad\qquad\qquad\;\; |$
$\qquad\qquad CH_3CH_2CH_2$

2-Butanol:

1. $CH_3CH_2CHCH_3 \; + \; H-\ddot{B}r: \; \longrightarrow \; CH_3CH_2CHCH_3 \; + \; :\ddot{B}r:^-$
$\qquad\qquad |$ $\qquad\qquad\qquad\qquad\qquad\qquad\qquad\qquad |$
$\qquad\qquad :\ddot{O}:$ $\qquad\qquad\qquad\qquad\qquad\qquad\qquad :\overset{+}{O}$
$\qquad\qquad\;\; |$ $\qquad\qquad\qquad\qquad\qquad\qquad\quad\; /\;\;\backslash$
$\qquad\qquad\;\; H$ $\qquad\qquad\qquad\qquad\qquad\qquad\; H\quad H$

2. $CH_3CH_2CHCH_3 \; \longrightarrow \; CH_3CH_2\overset{+}{C}HCH_3 \; + \; :\overset{\displaystyle H}{\underset{\displaystyle H}{O}}:$
$\qquad\qquad |$
$\qquad\qquad :\overset{+}{O}:$
$\qquad\quad /\;\;\backslash$
$\qquad H\quad H$

3. $:\ddot{B}r:^- \; + \; CHCH_3 \; \longrightarrow \; CH_3CH_2CHCH_3$
$\qquad\qquad\qquad\quad |\quad\overset{+}{}$ $\qquad\qquad\qquad\qquad\qquad |$
$\qquad\qquad CH_3CH_2$ $\qquad\qquad\qquad\qquad\qquad Br$

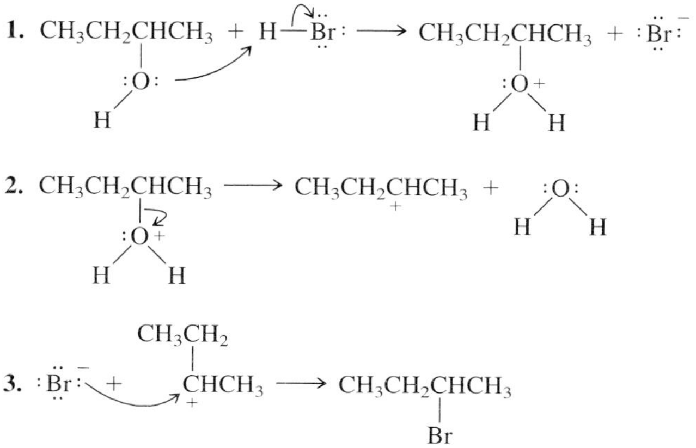

CHAPTER 4

4.1　(b) 3,3-Dimethyl-1-butene; (c) 2-methyl-2-hexene; (d) 4-chloro-1-pentene; (e) 4-penten-2-ol

4.2

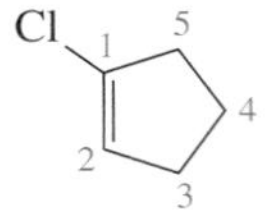

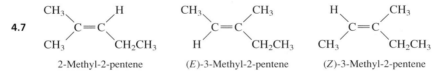

　　　　1-Chlorocyclopentene 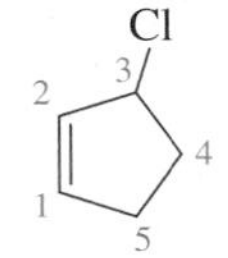　　　3-Chlorocyclopentene 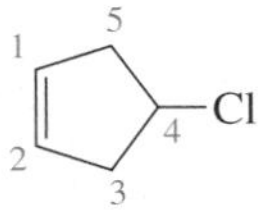　　　4-Chlorocyclopentene

4.3　(b) 3-Ethyl-3-hexene; (c) two carbons are sp^2-hybridized, six are sp^3-hybridized; (d) there are three sp^2–sp^3 σ bonds and three sp^3–sp^3 σ bonds.

4.4

　　　　1-Pentene 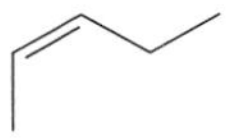　　　　　*cis*-2-Pentene 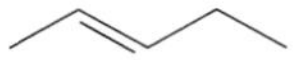　　　*trans*-2-Pentene

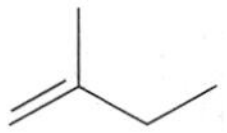

　　　　2-Methyl-1-butene 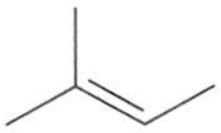　　　2-Methyl-2-butene 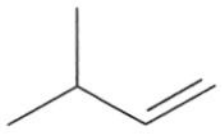　　　3-Methyl-1-butene

4.5　$CH_3(CH_2)_9$ 　 $(CH_2)_4CH(CH_3)_2$
　　　　　　　　C=C
　　　　　H　　　　　　H

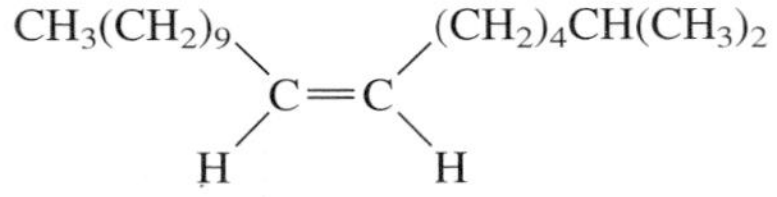

4.6　(b) Z; (c) E; (d) E

4.7

　　　　2-Methyl-2-pentene　　　(E)-3-Methyl-2-pentene　　　(Z)-3-Methyl-2-pentene

4.8　$(CH_3)_2C{=}C(CH_3)_2$

4.9　2-Methyl-2-butene > (E)-2-pentene > (Z)-2-pentene > 1-pentene
　　　(most stable)　　　　　　　　　　　　　　　　　(least stable)

4.10　(b) Propene; (c) propene; (d) 2,3,3-trimethyl-1-butene

4.11　(b)

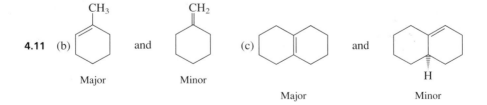

　　　　Major　　　　　　　Minor　　　　　　　　　　Major　　　　　　　Minor

4.12 (b)

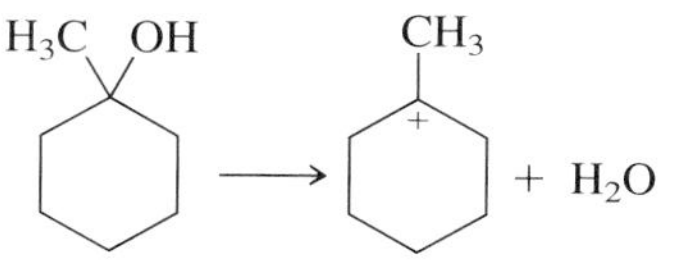

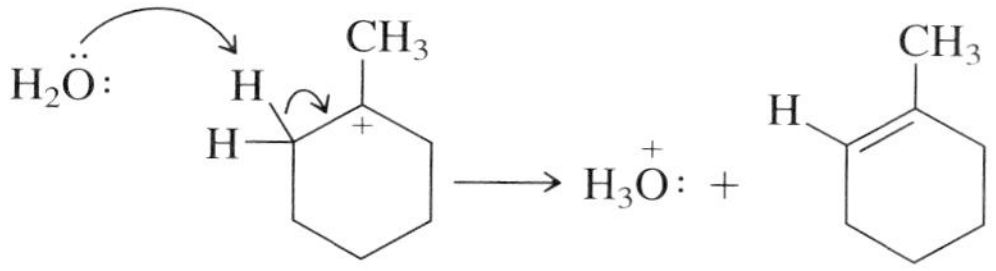

and

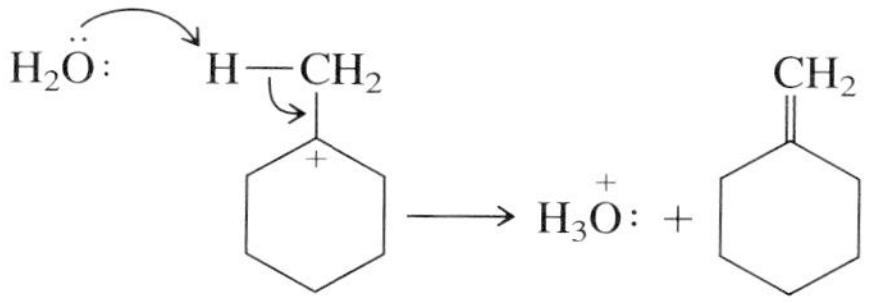

(c)

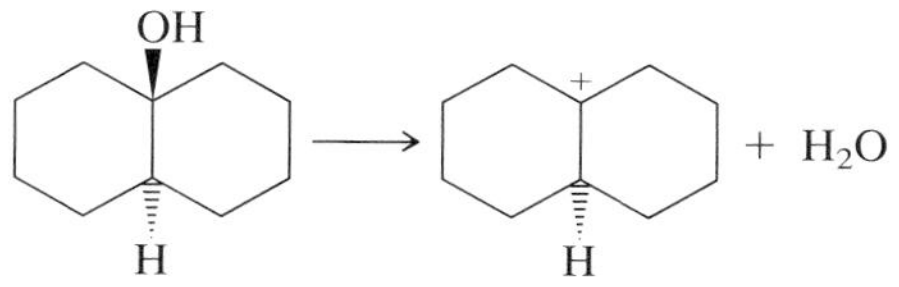

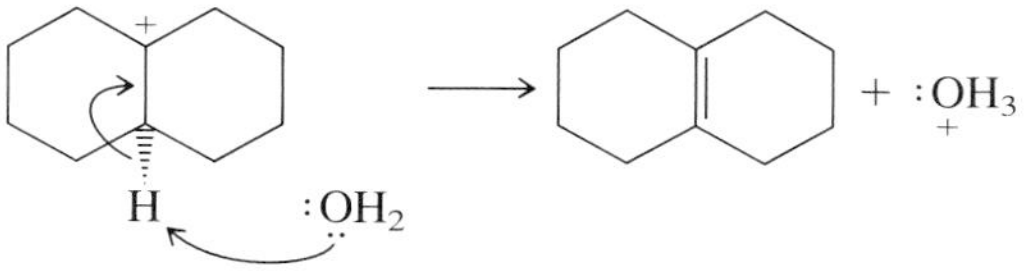

and

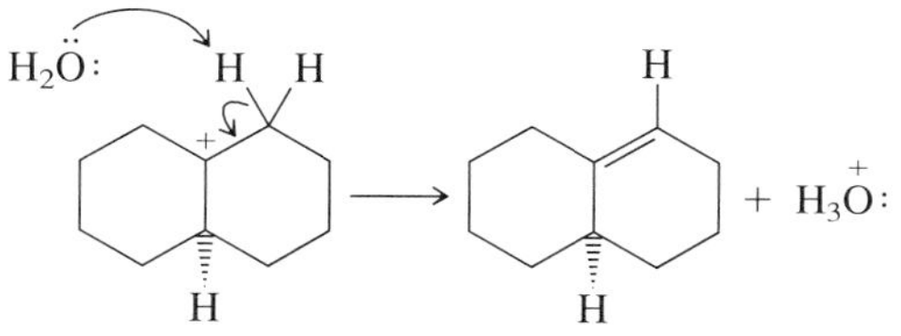

4.13 (b) $(CH_3)_2C=CH_2$; (c) $CH_3CH=C(CH_2CH_3)_2$; (d) $CH_3CH=C(CH_3)_2$ (major) and $CH_2=CHCH(CH_3)_2$ (minor)

4.14 $CH_2=CHCH_2CH_3$, $cis\text{-}CH_3CH=CHCH_3$, and $trans\text{-}CH_3CH=CHCH_3$

4.15

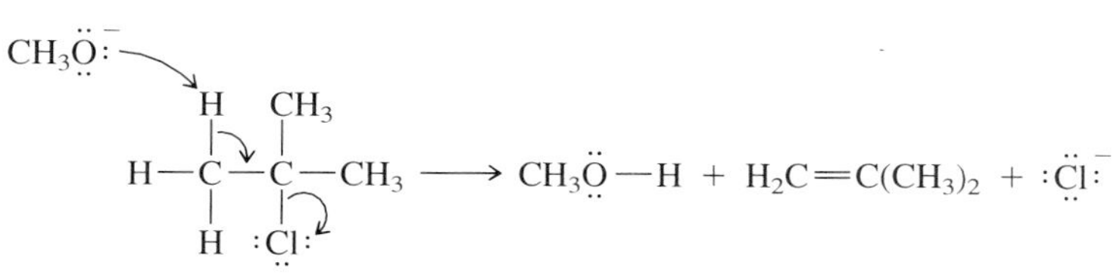

4.16 $CH_3CH_2CH_2C\equiv CH$ (1-pentyne), $CH_3CH_2C\equiv CCH_3$ (2-pentyne), $(CH_3)_2CHC\equiv CH$ (3-methyl-1-butyne)

4.17 $(CH_3)_3CCCH_3$ or $(CH_3)_3CCH_2CHBr_2$ or $(CH_3)_3CCHCH_2Br$
with Br substituents as shown

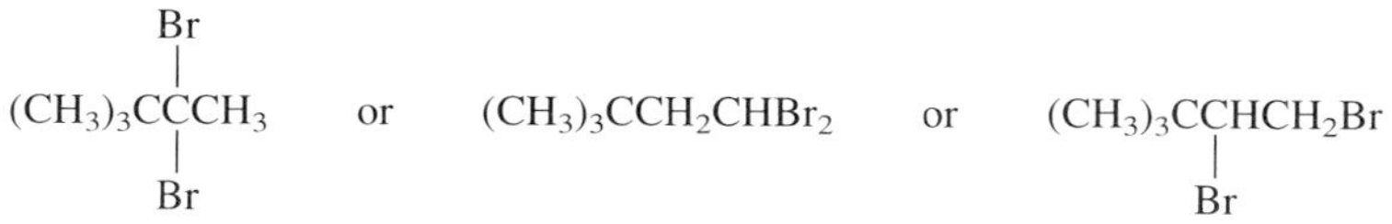

CHAPTER 5

5.1   2-Methyl-1-butene, 2-methyl-2-butene, and 3-methyl-1-butene

5.2   (b) $(CH_3)_2CCH_2CH_3$ (c) $CH_3CHCH_2CH_3$ (d) 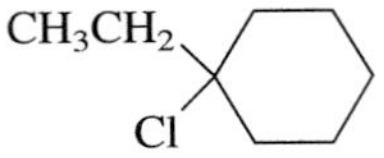 CH_3CH_2
                  | |
                  Cl   Cl

5.3   (b) $(CH_3)_2\overset{+}{C}CH_2CH_3$ (c) $CH_3\overset{+}{C}HCH_2CH_3$ (d) 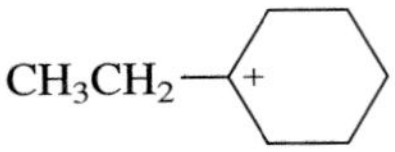 CH_3CH_2—(+)

5.4   The concentration of hydroxide ion is too small in acid solution to be chemically significant.

5.5 $CH_3\underset{\underset{CH_3}{|}}{\overset{\overset{Cl}{|}}{C}}CH_2Cl$

5.6   (b) $(CH_3)_2\underset{OH}{C}—\underset{Br}{C}HCH_3$ (c) $BrCH_2\underset{OH}{C}HCH(CH_3)_2$ (d)

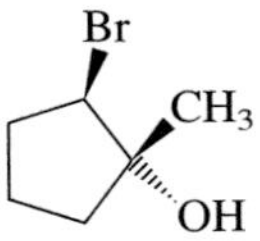

5.7 $(CH_3)_3CBr \xrightarrow[\text{heat}]{NaOCH_2CH_3} (CH_3)_2C{=}CH_2 \xrightarrow[H_2O]{Br_2} (CH_3)_2\underset{OH}{C}—CH_2Br$

5.8   (b) $(CH_3)_2\underset{Br}{C}—\underset{CH_3}{C}{=}CH_2$ $(CH_3)_2C{=}\underset{CH_3}{C}CH_2Br$

               1,2-addition   1,4-addition

        (c)

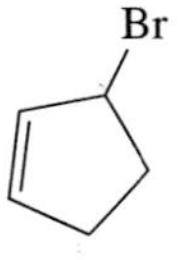

        Same product from 1,2-addition
        and 1,4-addition

5.9   Three: 3,4-dibromo-3-methyl-1-butene; 3,4-dibromo-2-methyl-1-butene; and 1,4-dibromo-2-methyl-2-butene

5.10   (b) $HC{\equiv}C{-}H + {:}\overset{-}{C}H_2CH_3 \underset{}{\overset{K \gg 1}{\rightleftharpoons}} HC{\equiv}\overset{-}{C}{:} + CH_3CH_3$

        Acetylene   Ethyl anion   Acetylide ion   Ethane
    (stronger acid)   (stronger base)   (weaker base)   (weaker acid)

  (c) $CH_3C{\equiv}CCH_2\overset{..}{\underset{..}{O}}{-}H + {:}\overset{-}{N}H_2 \overset{K \gg 1}{\rightleftharpoons} CH_3C{\equiv}CCH_2\overset{..}{\underset{..}{O}}{:}^- + {:}NH_3$

    2-Butyn-1-ol   Amide ion   2-Butyn-1-olate anion   Ammonia
   (stronger acid)   (stronger base)   (weaker base)   (weaker acid)

5.11   (b) $HC{\equiv}CH \xrightarrow[\text{2. } CH_3Br]{\text{1. } NaNH_2,\, NH_3} CH_3C{\equiv}CH \xrightarrow[\text{2. } CH_3CH_2CH_2CH_2Br]{\text{1. } NaNH_2,\, NH_3} CH_3C{\equiv}CCH_2CH_2CH_2CH_3$

  (c) $HC{\equiv}CH \xrightarrow[\text{2. } CH_3CH_2CH_2Br]{\text{1. } NaNH_2,\, NH_3} CH_3CH_2CH_2C{\equiv}CH \xrightarrow[\text{2. } CH_3CH_2Br]{\text{1. } NaNH_2,\, NH_3} CH_3CH_2CH_2C{\equiv}CCH_2CH_3$

5.12 $HC\equiv CCH_2CH_3$ $\xrightarrow[\text{2. } CH_3Br]{\text{1. } NaNH_2,\ NH_3}$ $CH_3C\equiv CCH_2CH_3$ $\xrightarrow[\text{Lindlar Pd}]{H_2}$

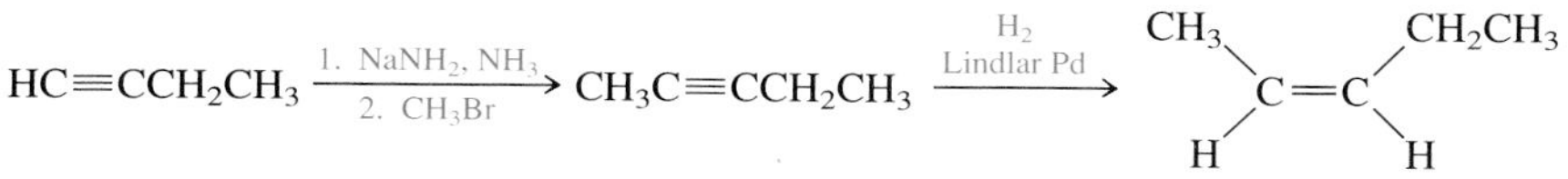

    CH_3 CH_2CH_3 on C=C (with H, H below)

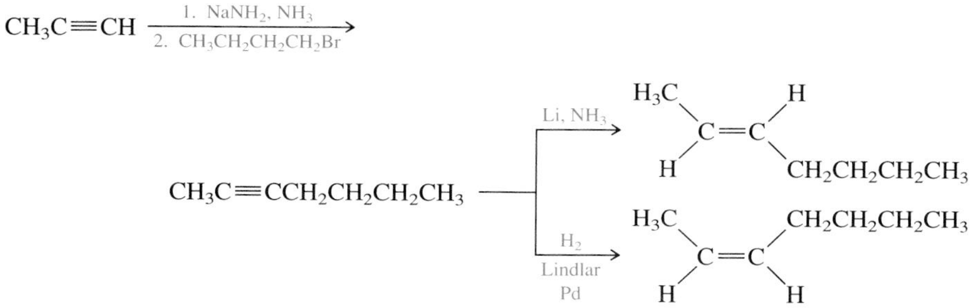

5.13 $CH_3C\equiv CH$ $\xrightarrow[\text{2. } CH_3CH_2CH_2CH_2Br]{\text{1. } NaNH_2,\ NH_3}$

$CH_3C\equiv CCH_2CH_2CH_2CH_3$

— Li, NH₃ → H_3C, H (top) C=C $CH_2CH_2CH_2CH_3$ (trans)

— $\xrightarrow[\text{Lindlar Pd}]{H_2}$ → H_3C, $CH_2CH_2CH_2CH_3$ (top) C=C H, H (cis)

5.14 $CH_3\overset{O}{\overset{\|}{C}}CH_2CH_3$ $CH_3\overset{OH}{\overset{|}{C}}=CHCH_3$

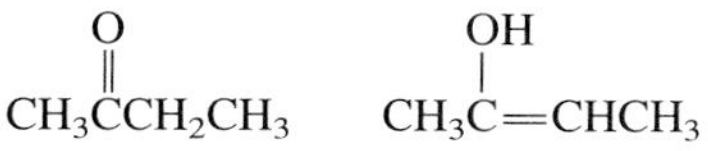

       Ketone   Enol

5.15 $H_2C=\overset{OH}{\overset{|}{C}}CH_2CH_2CH_2CH_2CH_2CH_3$

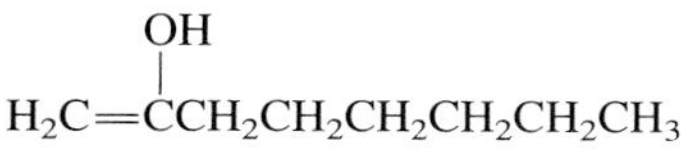

CHAPTER 6

6.1   (a) 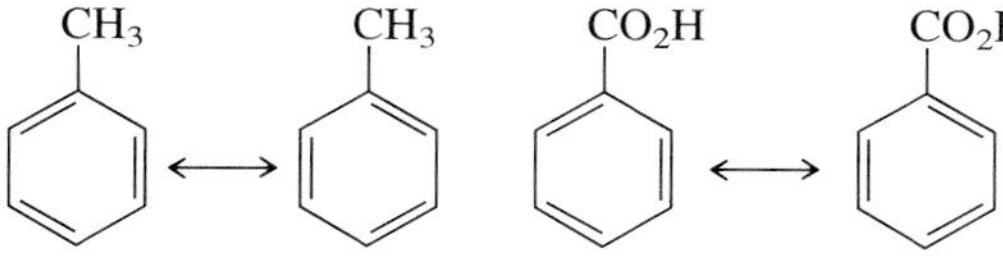 [toluene (CH_3)] ⟷ [toluene (CH_3)]   [benzoic acid (CO_2H)] ⟷ [benzoic acid (CO_2H)]   (b) 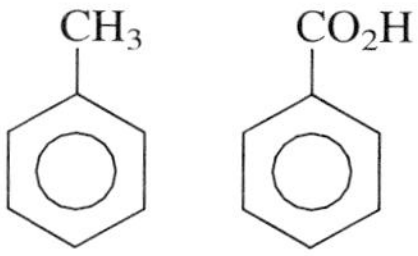 [toluene (CH_3)]   [benzoic acid (CO_2H)]

6.2   (b) 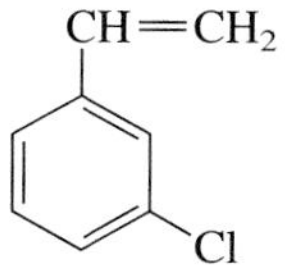 [3-chlorostyrene: ring with $CH=CH_2$ and Cl]   (c)  [ring with NH_2 top and NO_2 bottom]

6.3 $C_6H_5\overset{CH_3}{\underset{CH_3}{\overset{|}{\underset{|}{C}}}}Cl$

6.4 $(CH_3)_3C$—[ring with CO_2H and CO_2H]

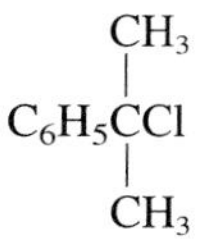

6.5   [ring with CH_3 top, NO_2, CH_3 bottom]

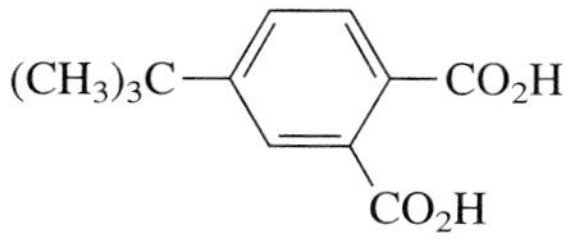

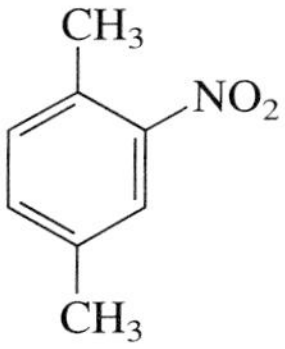

6.6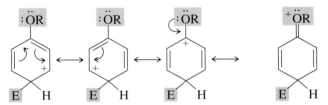

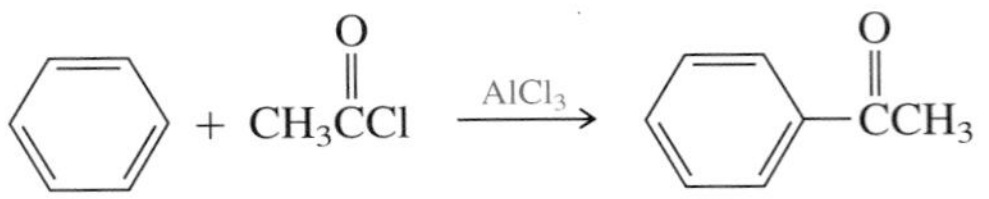

6.7 (b) Meta, slower than benzene; (c) ortho and para, slower than benzene

6.8 **Para attack:**

Most stable resonance
form; oxygen and all
carbons have octets of
electrons

6.9 (b) 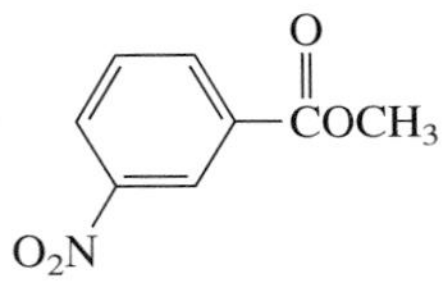   (c)

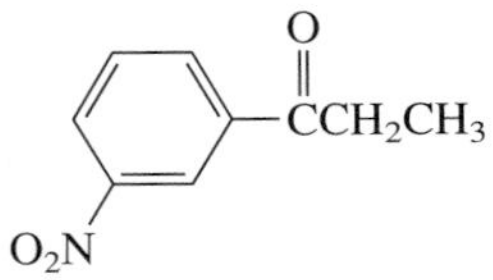

6.10 The group $-\overset{+}{N}(CH_3)_3$ is strongly deactivating and meta-directing. Its positively charged nitrogen makes it a powerful electron-withdrawing substituent. It resembles a nitro group.

6.11 *m*-**Bromonitrobenzene:**

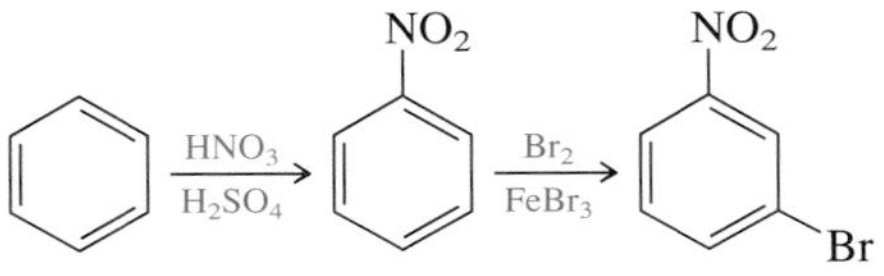

p-Bromonitrobenzene:

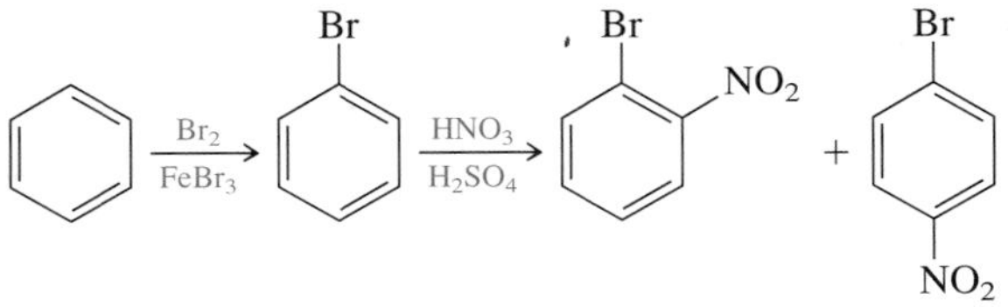

6.12 (b) Aromatic (6 π electrons in ring); (c) not aromatic (sp^3 carbon in ring); (d) aromatic (10 π electrons in ring)

CHAPTER 7

7.1 (c) C-2 is a stereogenic center; (d) no stereogenic center

7.2 $[\alpha]_D - 39°$

7.3 (+)-2-Butanol

7.4 (b) *R*; (c) *S*; (d) *S*

7.5

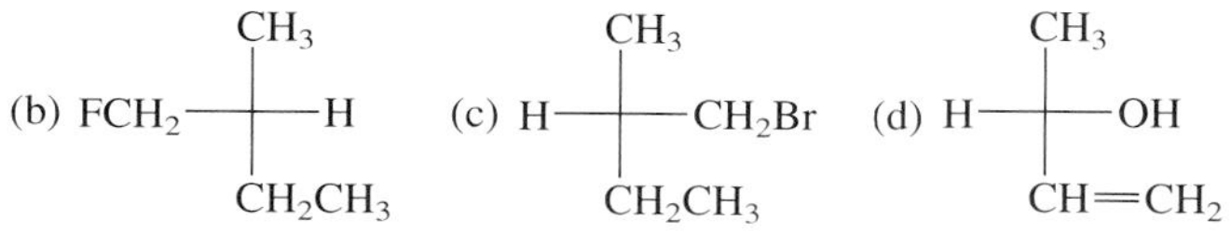

(b) FCH$_2$—[C with CH$_3$ top, H right, CH$_2$CH$_3$ bottom]   (c) H—[C with CH$_3$ top, CH$_2$Br right, CH$_2$CH$_3$ bottom]   (d) H—[C with CH$_3$ top, OH right, CH=CH$_2$ bottom]

7.6   *S*

7.7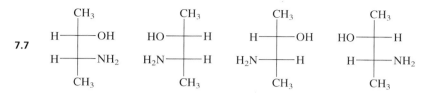

7.8   2*S*,3*R*

7.9   *RRR RRS RSR SRR SSS SSR SRS RSS*

7.10

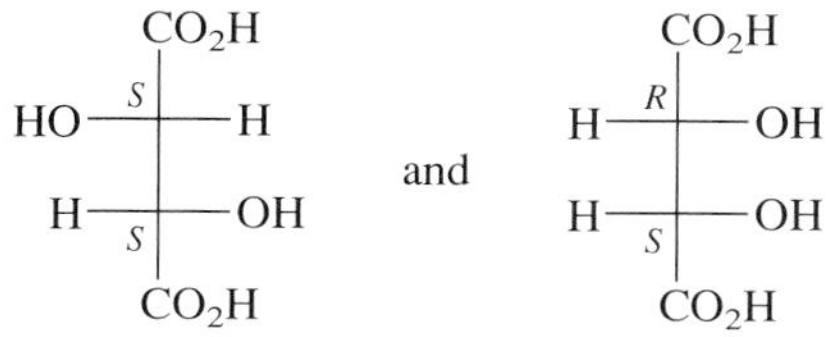

HO—[*S*]—H with CO$_2$H top, H—[*S*]—OH with CO$_2$H bottom   and   H—[*R*]—OH with CO$_2$H top, H—[*S*]—OH with CO$_2$H bottom

7.11   No

CHAPTER 8

8.1   (b) CH$_3$OCH$_2$CH$_3$ (c) CH$_3$N=$\overset{+}{N}$=$\overset{-}{\ddot{N}}$:

(d) CH$_3$C≡N   (e) CH$_3$SH   (f) CH$_3$I

8.2   No

8.3   HO—[C with CH$_3$ top, H right, CH$_2$(CH$_2$)$_4$CH$_3$ bottom]

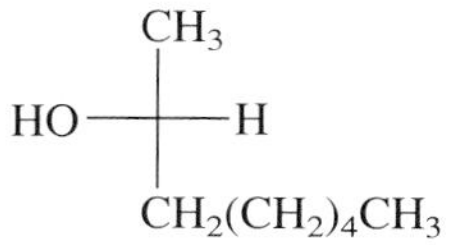

8.4   (b) 1-Bromopentane; (c) 2-bromo-5-methylhexane; (d) 1-bromodecane

8.5   Product is (CH$_3$)$_3$COCH$_3$. The mechanism of solvolysis is S$_N$1.

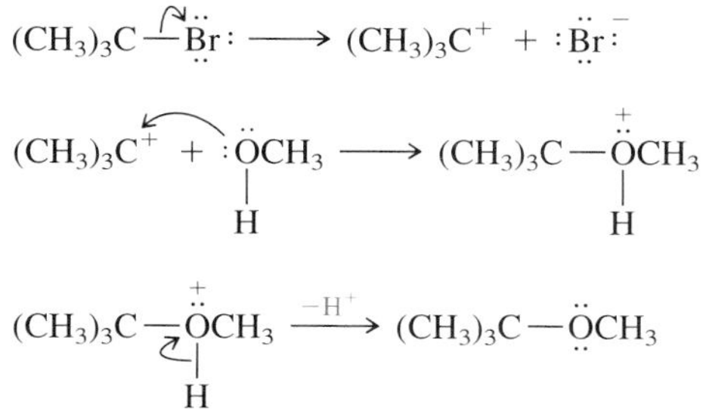

(CH$_3$)$_3$C—$\overset{..}{\underset{..}{Br}}$: ⟶ (CH$_3$)$_3C^+$ + :$\overset{..}{\underset{..}{Br}}$:$^-$

(CH$_3$)$_3$C$^+$ + :$\overset{..}{O}$CH$_3$ ⟶ (CH$_3$)$_3$C—$\overset{+}{\underset{|}{O}}CH_3$
                          | |
                          H   H

(CH$_3$)$_3$C—$\overset{+}{\underset{|}{O}}CH_3$ $\xrightarrow{-H^+}$ (CH$_3$)$_3$C—$\overset{..}{\underset{..}{O}}CH_3$
                  |
                  H

8.6   (b) 1-Methylcyclopentyl iodide; (c) cyclopentyl bromide

8.7   (CH$_3$)$_2$C=CHCH$_2$Cl

8.8 (b) 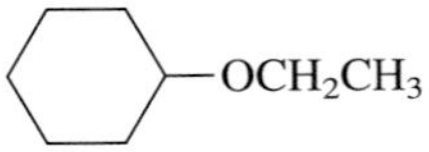 —OCH$_2$CH$_3$ (c) CH$_3$CHCH$_2$CH$_3$
                                                |
                                               OCH$_3$

(d) *cis*- and *trans*-CH$_3$CH=CHCH$_3$ and CH$_2$=CHCH$_2$CH$_3$

CHAPTER 9

9.1 H formal charge: 0; C formal charge: 0; Net: 0

9.2 (CH$_3$)$_2$ĊCH$_2$CH$_3$

9.3 (b) The carbon–carbon bond dissociation energy is lower for 2-methylpropane because it yields a more stable (secondary) radical; propane yields a primary radical. (c) The carbon–carbon bond dissociation energy is lower for 2,2-dimethylpropane because it yields a still more stable tertiary radical.

9.4   CH$_2$Cl$_2$ +   Cl$_2$ ⟶   CHCl$_3$ +   HCl

    Dichloromethane   Chlorine   Trichloromethane   Hydrogen chloride

        CHCl$_3$ +   Cl$_2$ ⟶   CCl$_4$ +   HCl

    Trichloromethane   Chlorine   Tetrachloromethane   Hydrogen chloride

9.5 **Initiation:**

$$:\ddot{C}l—\ddot{C}l: \longrightarrow :\ddot{C}l\cdot + \cdot\ddot{C}l:$$

    Chlorine   2 Chlorine atoms

Propagation:

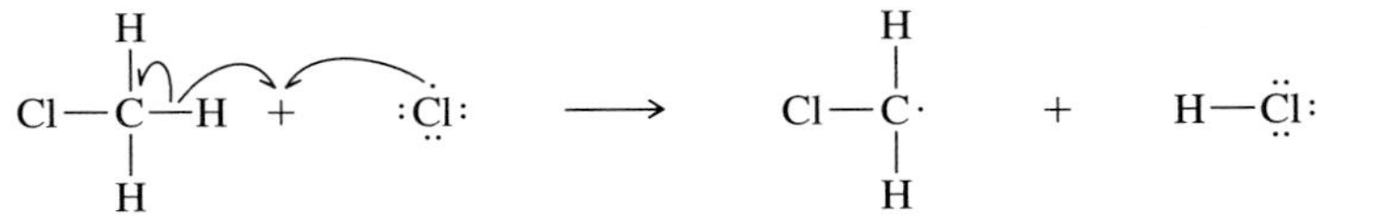

    Chloromethane   Chlorine atom   Chloromethyl radical   Hydrogen chloride

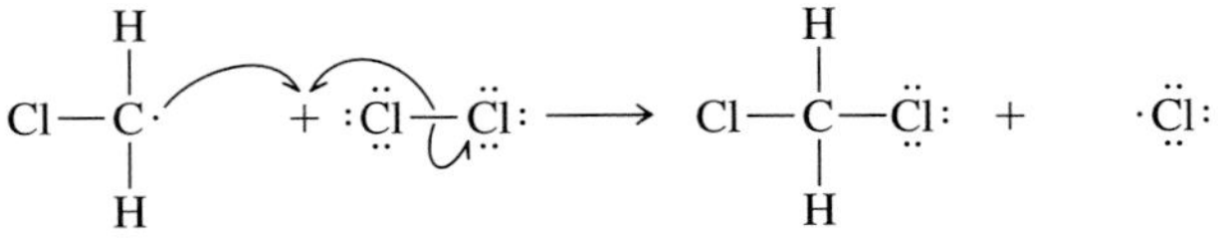

    Chloromethyl radical   Chlorine   Dichloromethane   Chlorine atom

9.6 CH$_3$CHCl$_2$ and ClCH$_2$CH$_2$Cl

9.7 (b) 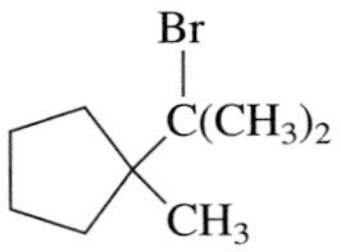   (c)

9.8 (CH$_3$)$_2$C=CHCH$_3$ + HBr $\xrightarrow{\text{peroxides}}$ (CH$_3$)$_2$CHCHCH$_3$
                                                                            |
                                                                           Br

9.9 $(CH_3)_2C{=}CHCH_3 \; + \; Br\cdot \longrightarrow \; (CH_3)_2\overset{\cdot}{C}CHCH_3$
$$\underset{\displaystyle Br}{\big|}$$

$(CH_3)_2\overset{\cdot}{C}CHCH_3 \; + \; HBr \longrightarrow \; (CH_3)_2CHCHCH_3 + Br\cdot$
$$\underset{\displaystyle Br}{\big|}$$

CHAPTER 10

10.1 Ethanol is oxidized, the chromium reagent is reduced.

10.2 The primary alcohols $CH_3CH_2CH_2CH_2OH$ and $(CH_3)_2CHCH_2OH$ can each be prepared by hydrogenation of an aldehyde. The secondary alcohol $CH_3CHCH_2CH_3$ can be prepared by hydro-

                                                                   $\overset{\displaystyle |}{OH}$

genation of a ketone. The tertiary alcohol $(CH_3)_3COH$ cannot be prepared by hydrogenation.

10.3 (b) $(CH_3)_2CHCH_2CH_2OH$ (c) 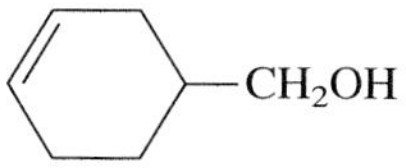 —CH_2OH (d) 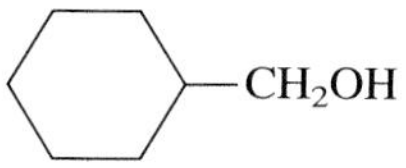 —CH_2OH

10.4 (b) $CH_3\overset{\displaystyle O}{\overset{\|}{C}}(CH_2)_5CH_3$ (c) $CH_3(CH_2)_5\overset{\displaystyle O}{\overset{\|}{C}}H$

10.5 $CH_3CHCH_2CH_2SH$
        $\overset{\displaystyle |}{CH_3}$

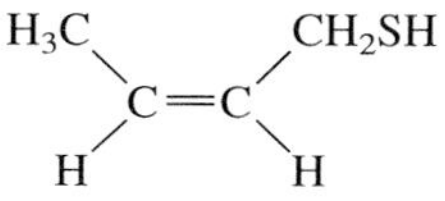

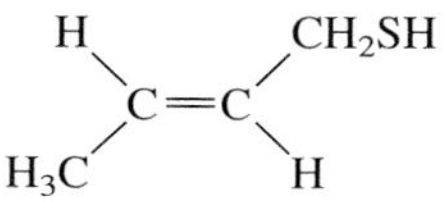

    3-Methyl-1-butanethiol   *cis*-2-Butene-1-thiol   *trans*-2-Butene-1-thiol

10.6 (b)  $H_2C{-}CHCH{=}CH_2$
          $\underset{\displaystyle O}{\diagdown\diagup}$

10.7 $C_6H_5CH_2ONa + CH_3CH_2Br \longrightarrow C_6H_5CH_2OCH_2CH_3 \; + \; NaBr$
    and $CH_2CH_2ONa + C_6H_5CH_2Br \longrightarrow C_6H_5CH_2OCH_2CH_3 + NaBr$

10.8 (b) $(CH_3)_2CHONa + CH_2{=}CHCH_2Br \longrightarrow CH_2{=}CHCH_2OCH(CH_3)_2 + NaBr$
    (c) $(CH_3)_3COK + C_6H_5CH_2Br \longrightarrow (CH_3)_3COCH_2C_6H_5 + KBr$

10.9 $(CH_3)_2CHCH_2(CH_2)_2CH_2$ $CH_2(CH_2)_8CH_3$
                        $\diagdown$ $\diagup$
                             $C{=}C$
                      $\diagup$ $\diagdown$
                    H H

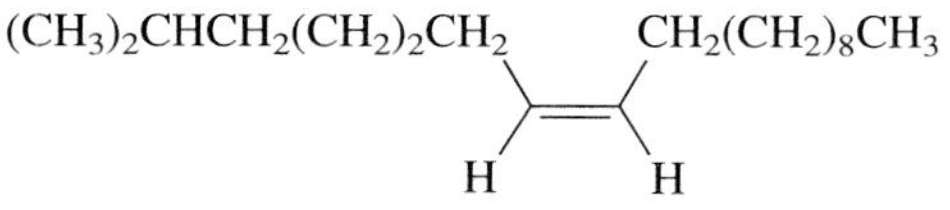

                Z-configuration

10.10 $CH_3CH_2CH_2CH_2OCH_2CH_2OH$

10.11 (b) 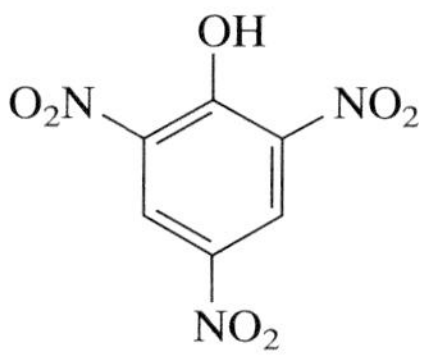   (c)

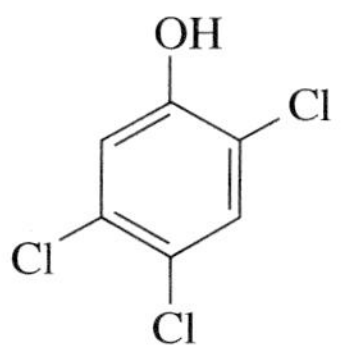

10.12 2,4,6-Trinitrophenol is more acidic. All three nitro groups are directly conjugated to the phenoxide oxygen and thus stabilize the phenoxide ion.

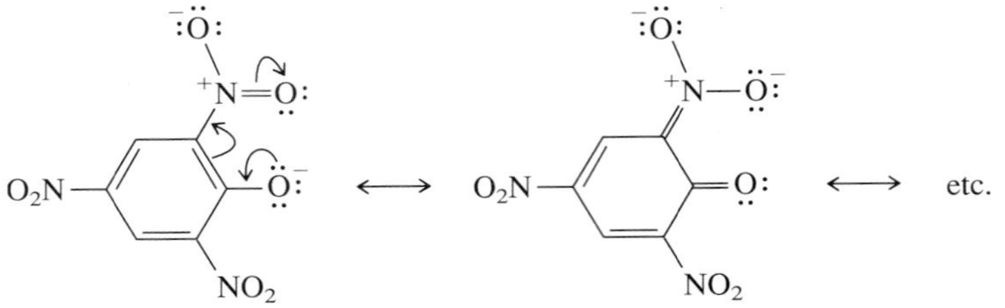

CHAPTER 11

11.1 (b) Trichloroethanal (or trichloroacetaldehyde); (c) 3-phenyl-2-propenal; (d) 4-hydroxy-3-methoxybenzaldehyde

11.2 (b) 1-Phenylethanone (methyl phenyl ketone); (c) 3,3-dimethyl-2-butanone (*tert*-butyl methyl ketone)

11.3 $Cl_3CCH(OH)_2$

11.4 A catalyst does not change the concentrations of species present at equilibrium, only the rate at which equilibrium is achieved. The percentage of hydrate remains unchanged under conditions of acid or base catalysis.

11.5 $CH_2{=}CC{\equiv}N$ with CH_3 substituent

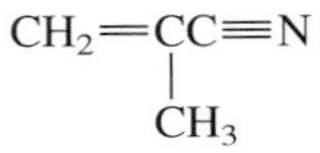

11.6

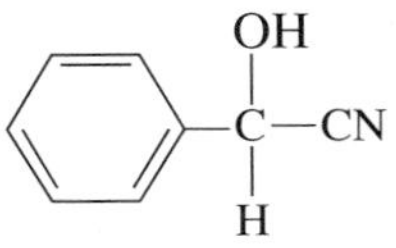

11.7 Step 1:

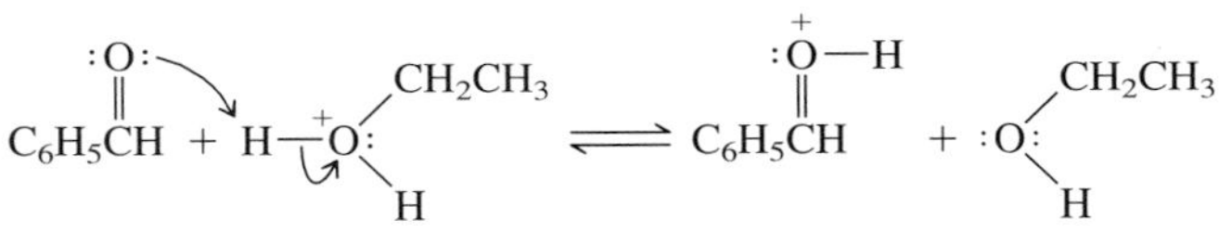

Step 2:

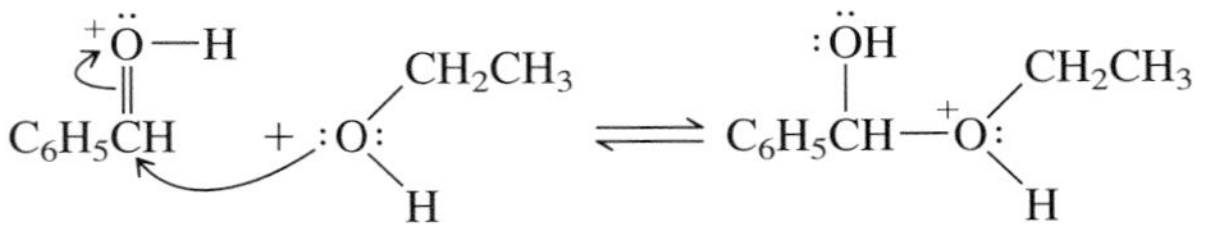

Step 3:

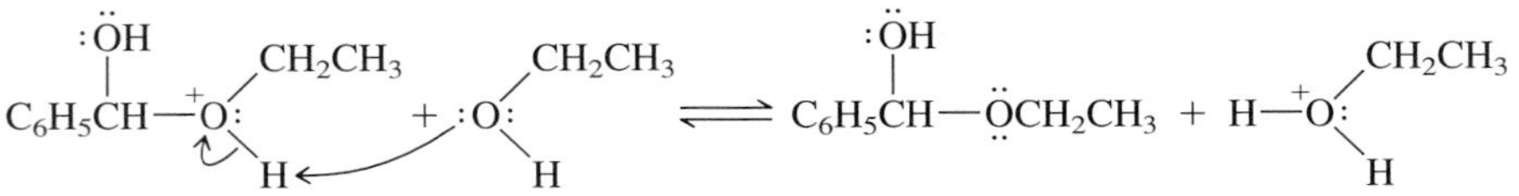

Formation of the hemiacetal is followed by loss of water to give a carbocation.

Step 4:

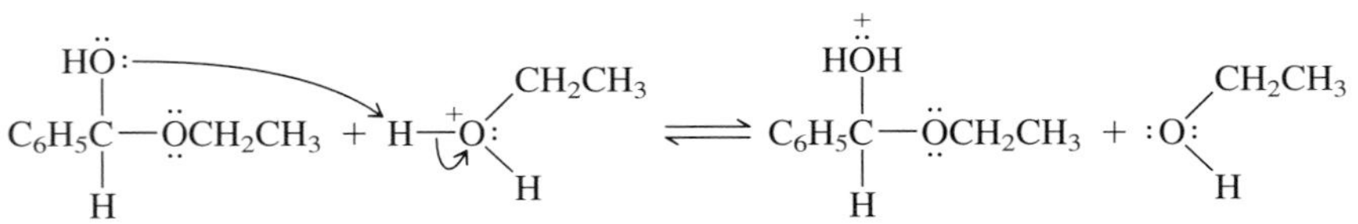

Step 5: 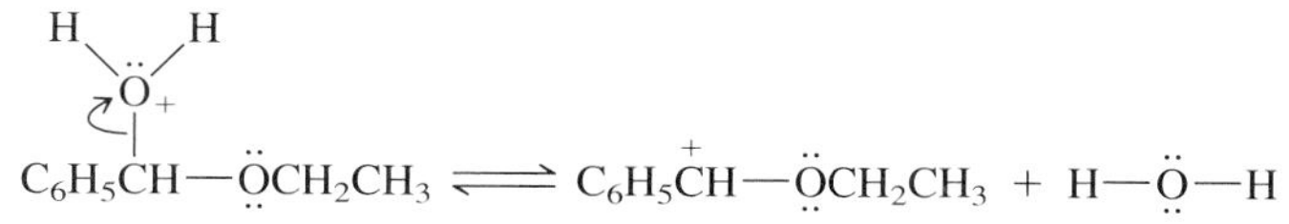

Step 6: 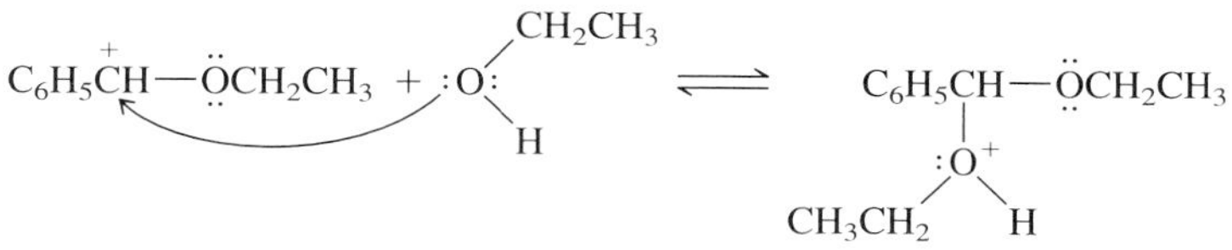

Step 7:

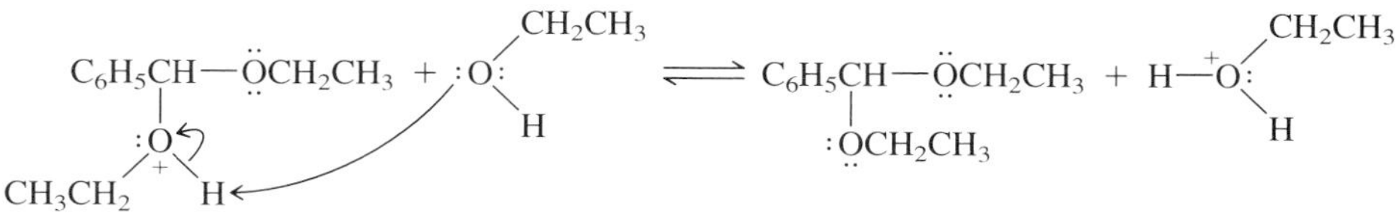

11.8 (b) $C_6H_5CH(OH)NHCH_2CH_2CH_2CH_3 \longrightarrow C_6H_5CH{=}NCH_2CH_2CH_2CH_3$

(c) 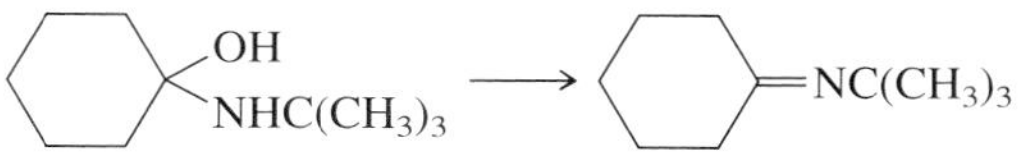

11.9 (b) $CH_2{=}CHCH_2MgCl$ (c) 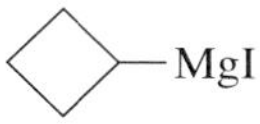 (d) 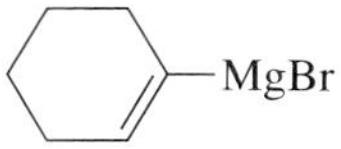

11.10 (b) $C_6H_5CH(OH)CH_2CH_2CH_3$ (c) 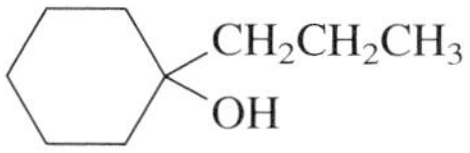 (d) $CH_3CH_2CH_2C(CH_3)(CH_2CH_3)OH$

11.11 (b) $CH_3MgBr + HC(=O)CH_2CH_2CH_2CH_3 \xrightarrow[\text{2. } H_3O^+]{\text{1. diethyl ether}} CH_3CH(OH)CH_2CH_2CH_2CH_3$

$CH_3C(=O)H + BrMgCH_2CH_2CH_2CH_3 \xrightarrow[\text{2. } H_3O^+]{\text{1. diethyl ether}} CH_3CH(OH)CH_2CH_2CH_2CH_3$

(c) $C_6H_5MgBr + CH_3CH_2C(=O)H \xrightarrow[\text{2. } H_3O^+]{\text{1. diethyl ether}} C_6H_5CH(OH)CH_2CH_3$

$CH_3CH_2MgBr + C_6H_5C(=O)H \xrightarrow[\text{2. } H_3O^+]{\text{1. diethyl ether}} C_6H_5CH(OH)CH_2CH_3$

11.12 CH_3MgI + [ketone: phenyl–CO–CH₂CH₃] $\xrightarrow[\text{2. } H_3O^+]{\text{1. diethyl ether}}$ [product: phenyl–C(OH)(CH₃)CH₂CH₃]

CH_3CH_2MgBr + [ketone: phenyl–CO–CH₃] $\xrightarrow[\text{2. } H_3O^+]{\text{1. diethyl ether}}$ [product: phenyl–C(OH)(CH₃)CH₂CH₃]

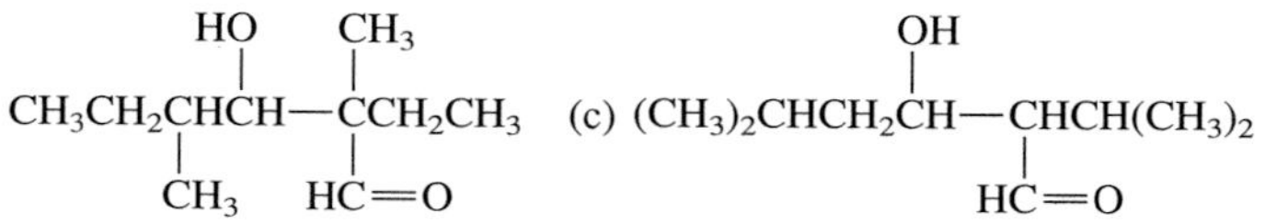

[phenyl]–MgBr + $CH_3CCH_2CH_3$ (ketone) $\xrightarrow[\text{2. } H_3O^+]{\text{1. diethyl ether}}$ [product: phenyl–C(OH)(CH₂CH₃)CH₃]

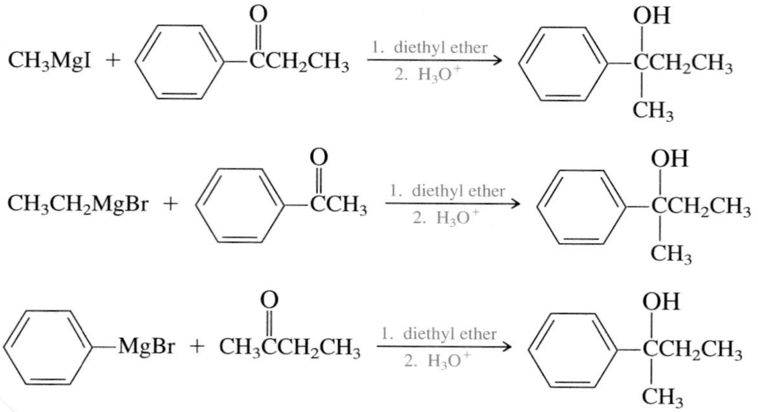

11.13 (b) Zero; (c) five; (d) four

11.14 (b) $C_6H_5\overset{\text{OH}}{\underset{}{C}}{=}CH_2$ (c) 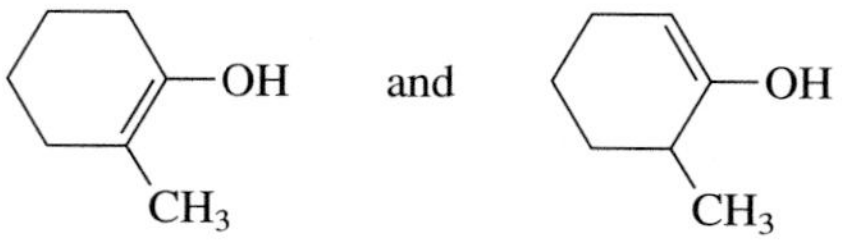 [cyclohexene ring with –OH and –CH₃ substituents]   and   [cyclohexene ring with –OH and –CH₃ substituents]

11.15 (b) $CH_3CH_2\overset{\text{HO}}{\underset{\text{CH}_3}{C}}H\!-\!\overset{\text{CH}_3}{\underset{\text{HC}=O}{C}}CH_2CH_3$ (c) $(CH_3)_2CHCH_2\overset{\text{OH}}{\underset{\text{HC}=O}{C}}H\!-\!CHCH(CH_3)_2$

11.16 (b) $CH_3CH_2\overset{\text{HO}}{\underset{\text{CH}_3}{C}}H\!-\!\overset{\alpha}{\underset{\text{HC}=O}{\overset{\text{CH}_3}{C}}}CH_2CH_3$ (c) $(CH_3)_2CHCH_2CH{=}\underset{\text{HC}=O}{C}CH(CH_3)_2$

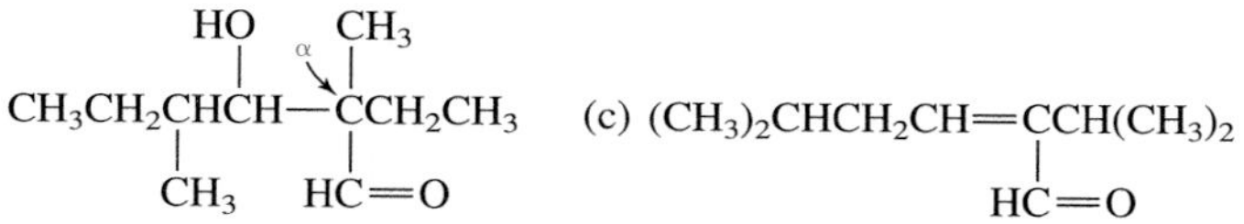

Cannot dehydrate; no protons
on α-carbon atom

CHAPTER 12

12.1 (b) (*E*)-2-butenoic acid; (c) *p*-methylbenzoic acid or 4-methylbenzoic acid

12.2 [structure showing acetic acid dimer with hydrogen bonding: CH_3C with $\overset{\delta-}{O}$⋯$\overset{\delta+}{H}$—O—H and O—$\overset{\delta+}{H}$⋯$\overset{\delta-}{O}$—H bonds to water molecules]

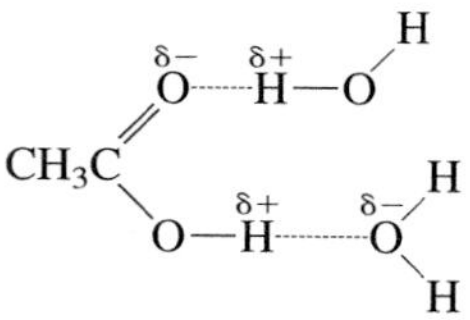

12.3 pK_a = 3.48. Acetylsalicylic acid is a stronger acid than benzoic acid.

12.4 (b) $CH_3\overset{}{\underset{\text{OH}}{C}}HCO_2H$ (c) $CH_3\overset{O}{\overset{\|}{C}}CO_2H$

12.5 (b) CH_3CO_2H + $(CH_3)_3CO^-$ $\rightleftharpoons$ $CH_3CO_2^-$ + $(CH_3)_3COH$
(The position of equilibrium lies to the right.)
(c) CH_3CO_2H + Br^- $\rightleftharpoons$ $CH_3CO_2^-$ + HBr
(The position of equilibrium lies to the left.)
(d) CH_3CO_2H + $HC{\equiv}C:^-$ $\rightleftharpoons$ $CH_3CO_2^-$ + $HC{\equiv}CH$
(The position of equilibrium lies to the right.)

(e) CH_3CO_2H + NO_3^- $\rightleftharpoons$ $CH_3CO_2^-$ + HNO_3
(The position of equilibrium lies to the left.)
(f) CH_3CO_2H + H_2N^- $\rightleftharpoons$ $CH_3CO_2^-$ + NH_3
(The position of equilibrium lies to the right.)

12.6 (b) $(CH_3)_2CHBr$ $\xrightarrow[\substack{2.\ CO_2 \\ 3.\ H_3O^+}]{1.\ \text{Mg, diethyl ether}}$ $(CH_3)_2CHCO_2H$

(c) —Br $\xrightarrow[\substack{2.\ CO_2 \\ 3.\ H_3O^+}]{1.\ \text{Mg, diethyl ether}}$ —CO_2H

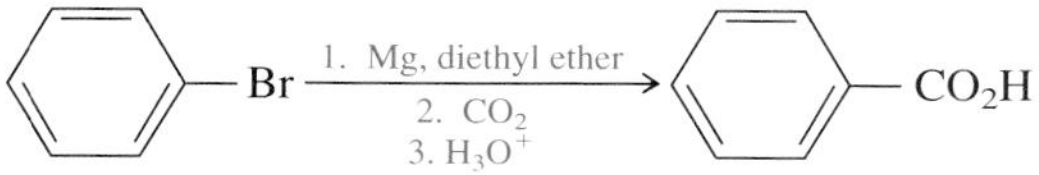

12.7 (b) $(CH_3)_2CHBr$ $\xrightarrow{\text{NaCN}}$ $(CH_3)_2CHCN$ $\xrightarrow[\text{heat}]{H_2O,\ H_2SO_4}$ $(CH_3)_2CHCO_2H$

(c) Sequence not feasible because first step (nucleophilic substitution by cyanide on bromobenzene) fails.

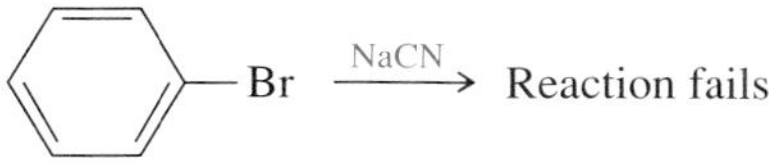

—Br $\xrightarrow{\text{NaCN}}$ Reaction fails

12.8 (b) $C_6H_5CH_2\overset{\overset{\displaystyle O}{\|}}{C}Cl$ (c) $C_6H_5CH_2CH_2OH$ (d) $C_6H_5CH_2\overset{\overset{\displaystyle O}{\|}}{C}OCH_2CH_3$

CHAPTER 13

13.1 (b) $CH_3CH_2\underset{\underset{\displaystyle C_6H_5}{|}}{CH}\overset{\overset{\displaystyle O}{\|}}{C}O\overset{\overset{\displaystyle O}{\|}}{C}\underset{\underset{\displaystyle C_6H_5}{|}}{CH}CH_2CH_3$ (c) $CH_3CH_2\underset{\underset{\displaystyle C_6H_5}{|}}{CH}\overset{\overset{\displaystyle O}{\|}}{C}OCH_2CH_2CH_3$

(d) $CH_3CH_2CH_2\overset{\overset{\displaystyle O}{\|}}{C}OCH_2\underset{\underset{\displaystyle C_6H_5}{|}}{CH}CH_2CH_3$ (e) $CH_3CH_2\underset{\underset{\displaystyle C_6H_5}{|}}{CH}\overset{\overset{\displaystyle O}{\|}}{C}NH_2$

(f) $CH_3CH_2\underset{\underset{\displaystyle C_6H_5}{|}}{CH}\overset{\overset{\displaystyle O}{\|}}{C}NHCH_2CH_3$

13.2 $C_6H_5\overset{\overset{\displaystyle O}{\|}}{C}Cl$ + H_2O $\longrightarrow$ $C_6H_5\overset{\overset{\displaystyle O}{\|}}{C}OH$ + HCl

Nucleophilic addition of water to give the tetrahedral intermediate

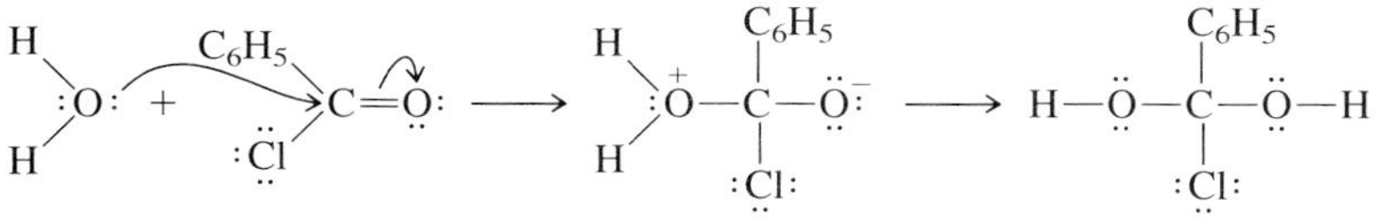

Dissociation of the tetrahedral intermediate by dehydrohalogenation

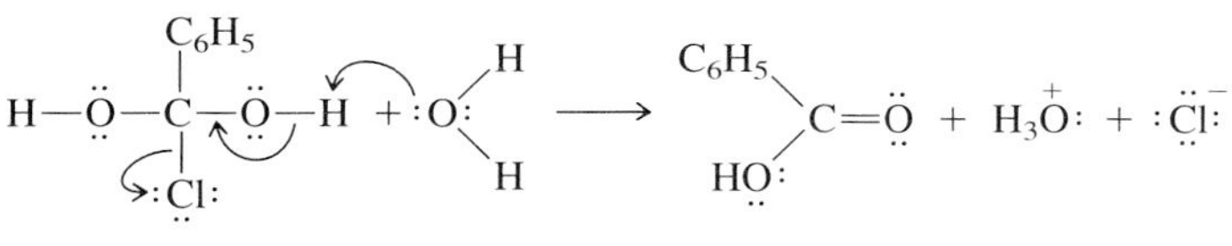

13.3 $CH_3CH_2CH_2CH_2OH$ + $CH_3CH_2\overset{\overset{\text{O}}{\|}}{C}OH$ $\xrightarrow[\text{heat}]{H_2SO_4}$ $CH_3CH_2\overset{\overset{\text{O}}{\|}}{C}OCH_2CH_2CH_2CH_3$ + H_2O

　　　　 1-Butanol　　　　　　Propanoic　　　　　　　　　　　Butyl propanoate　　　　　　　　Water
　　　　　　　　　　　　　　　 acid

$$CH_3CH_2\overset{\overset{\ddot{\text{O}}:}{\|}}{C}-\underset{\ddot{\text{O}}H}{}H \quad \underset{\longleftarrow}{\overset{H^+}{\rightleftharpoons}} \quad CH_3CH_2\overset{\overset{+\ddot{\text{O}}H}{\|}}{C}-\underset{\ddot{\text{O}}H}{}H$$

$$CH_3CH_2\overset{\overset{+}{\ddot{\text{O}}H}}{C} \quad + \quad H\ddot{\text{O}}CH_2CH_2CH_2CH_3 \quad \overset{-H^+}{\rightleftharpoons} \quad CH_3CH_2\underset{:\ddot{\text{O}}H}{\overset{:\ddot{\text{O}}H}{C}}-OCH_2CH_2CH_2CH_3$$

　　　　　　　　　　　　　　　　　　　　　　　　　　　　　　Tetrahedral intermediate

$$CH_3CH_2\underset{:\ddot{\text{O}}CH_2CH_2CH_2CH_3}{\overset{:\ddot{\text{O}}H}{C}}-\ddot{\text{O}}H \quad \overset{H^+}{\rightleftharpoons} \quad CH_3CH_2\underset{:\ddot{\text{O}}CH_2CH_2CH_2CH_3}{\overset{:\ddot{\text{O}}H}{C}}-\overset{+}{\text{O}}H_2$$

$$\Big\updownarrow$$

$$CH_3CH_2\overset{\overset{\ddot{\text{O}}:}{\|}}{C}-\underset{\ddot{\text{O}}CH_2CH_2CH_2CH_3}{} \quad \rightleftharpoons \quad CH_3CH_2\overset{\overset{+}{\ddot{\text{O}}H}}{C}-\underset{\ddot{\text{O}}CH_2CH_2CH_2CH_3}{}$$

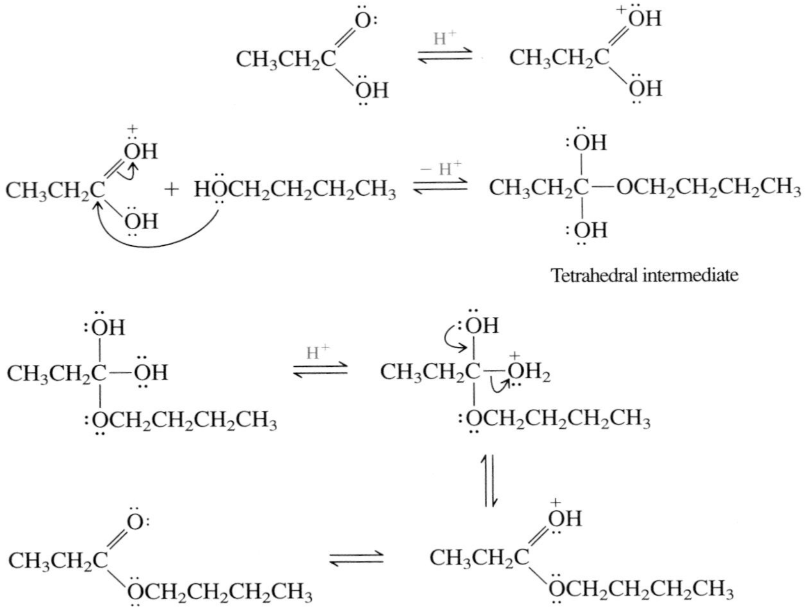

13.4 (b) $CH_3\overset{\overset{\text{O}}{\|}}{C}OH \xrightarrow{SOCl_2} CH_3\overset{\overset{\text{O}}{\|}}{C}Cl \xrightarrow[\text{pyridine}]{C_6H_5CH_2OH} CH_3\overset{\overset{\text{O}}{\|}}{C}OCH_2C_6H_5$

　　　　 (c) $(CH_3)_2CH\overset{\overset{\text{O}}{\|}}{C}OH \xrightarrow{SOCl_2} (CH_3)_2CH\overset{\overset{\text{O}}{\|}}{C}Cl \xrightarrow[\text{pyridine}]{(CH_3)_2CHOH} (CH_3)_2CH\overset{\overset{\text{O}}{\|}}{C}OCH(CH_3)_2$

13.5

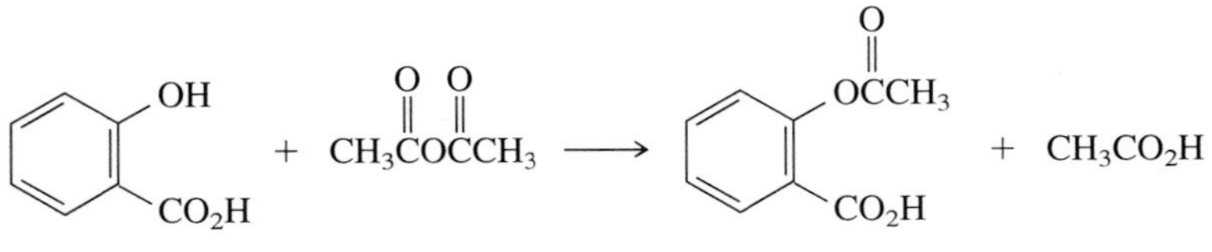

13.6 $(CH_3)_3C$ $\xrightarrow[\text{anhydride}]{\text{acetic}}$ $(CH_3)_3C$

$(CH_3)_3C$ $\xrightarrow[\text{anhydride}]{\text{acetic}}$ $(CH_3)_3C$

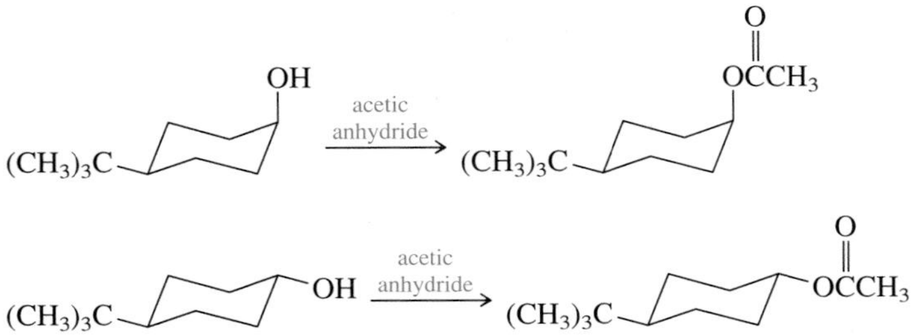

13.7 $CH_3(CH_2)_{14}\overset{\overset{\text{O}}{\|}}{C}OCH_2(CH_2)_{28}CH_3 + H_2O \xrightarrow{H^+} CH_3(CH_2)_{14}\overset{\overset{\text{O}}{\|}}{C}OH + HOCH_2(CH_2)_{28}CH_3$

13.8 $CH_3(CH_2)_{12}\overset{\overset{\textstyle O}{\|}}{C}O\,CH_2CH\,CH_2O\overset{\overset{\textstyle O}{\|}}{C}(CH_2)_{12}CH_3$

$$\overset{\overset{\textstyle O}{\|}}{OC}(CH_2)_{12}CH_3$$

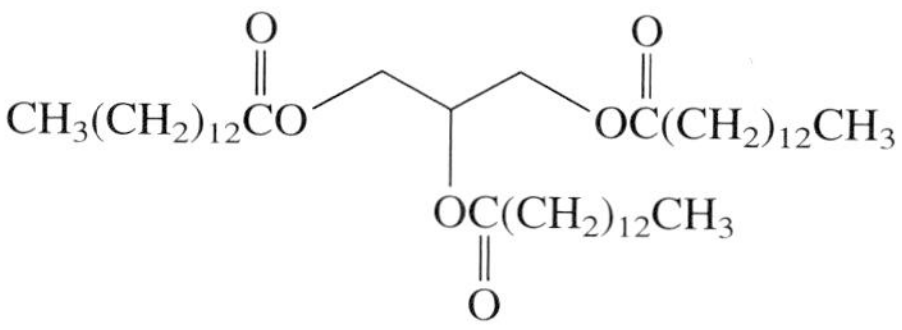

13.9   Step 1: Nucleophilic addition of hydroxide ion to the carbonyl group

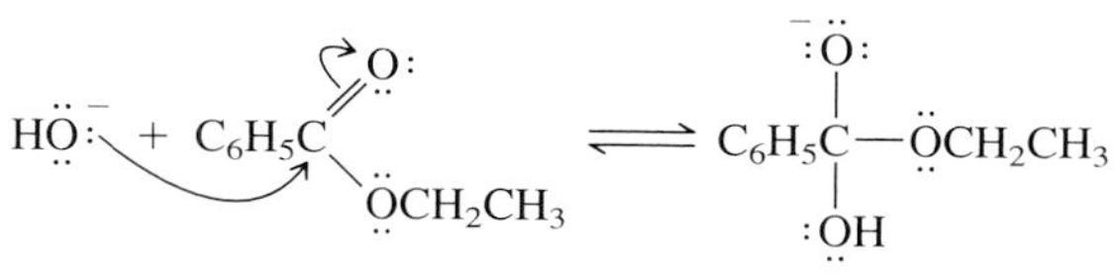

$$HO:^- + C_6H_5\overset{\overset{\textstyle O:}{\|}}{C}{\overset{\textstyle}{\underset{\textstyle \ddot{O}CH_2CH_3}{}}} \rightleftharpoons C_6H_5\overset{\overset{\textstyle :\ddot{O}:^-}{|}}{\underset{\textstyle :OH}{C}}{-}\ddot{O}CH_2CH_3$$

Step 2: Proton transfer from water to give neutral form of tetrahedral intermediate

$$C_6H_5\overset{\overset{\textstyle :\ddot{O}:^-}{|}}{\underset{\textstyle :OH}{C}}{-}\ddot{O}CH_2CH_3 + H{-}\ddot{O}H \rightleftharpoons C_6H_5\overset{\overset{\textstyle :\ddot{O}H}{|}}{\underset{\textstyle :OH}{C}}{-}\ddot{O}CH_2CH_3 + {}^-:\ddot{O}H$$

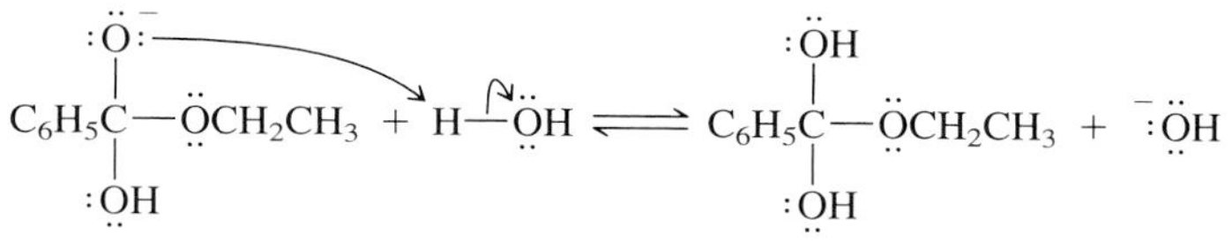

Step 3: Hydroxide ion-promoted dissociation of tetrahedral intermediate

$$HO:^- + C_6H_5\overset{\overset{\textstyle H{-}\ddot{O}:}{|}}{\underset{\textstyle :OH}{C}}{-}\ddot{O}CH_2CH_3 \rightleftharpoons H\ddot{O}H + C_6H_5\overset{\overset{\textstyle \ddot{O}:}{\|}}{\underset{\textstyle OH}{C}} + {}^-:\ddot{O}CH_2CH_3$$

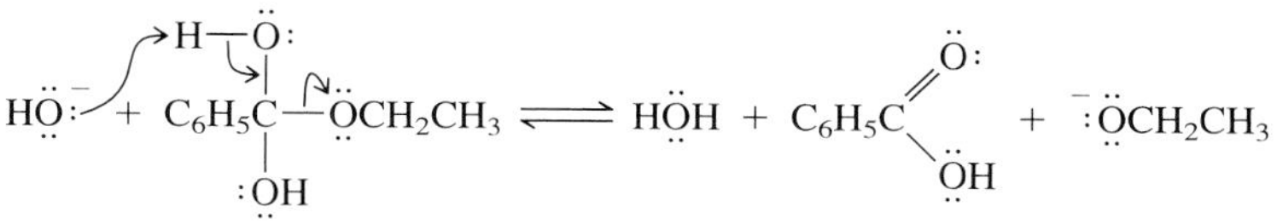

Step 4: Proton abstraction from benzoic acid

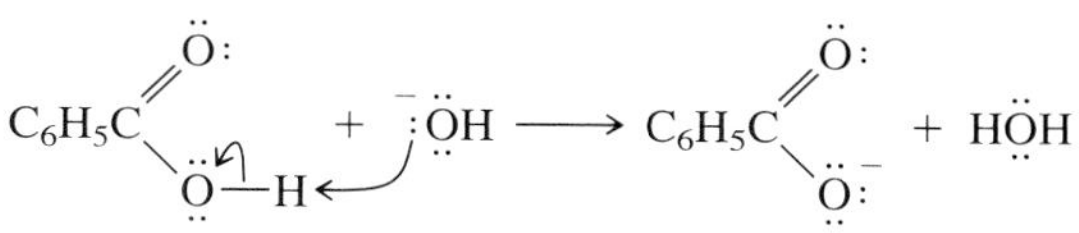

$$C_6H_5\overset{\overset{\textstyle \ddot{O}:}{\|}}{\underset{\textstyle \ddot{O}{-}H}{C}} + {}^-:\ddot{O}H \longrightarrow C_6H_5\overset{\overset{\textstyle \ddot{O}:}{\|}}{\underset{\textstyle \ddot{O}:^-}{C}} + H\ddot{O}H$$

13.10 (b) 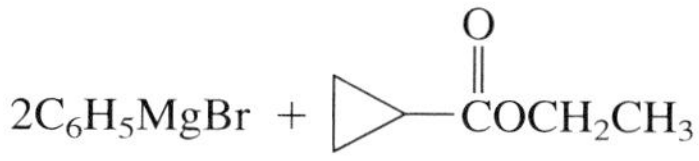 $2C_6H_5MgBr + {\triangleright}{-}\overset{\overset{\textstyle O}{\|}}{C}OCH_2CH_3$

13.11 $CH_3CH_2\overset{\overset{\textstyle O}{\|}}{C}OCH(CH_3)_2$

13.12 (b) $CH_3\overset{\overset{\textstyle O}{\|}}{C}O\overset{\overset{\textstyle O}{\|}}{C}CH_3 + 2CH_3NH_2 \longrightarrow CH_3\overset{\overset{\textstyle O}{\|}}{C}NHCH_3 + CH_3\overset{\overset{\textstyle O}{\|}}{C}O^- \; {}^+CH_3NH_3$

(c) $H\overset{\overset{\textstyle O}{\|}}{C}OCH_3 + HN(CH_3)_2 \longrightarrow H\overset{\overset{\textstyle O}{\|}}{C}N(CH_3)_2 + CH_3OH$

13.13 (b) $CH_3\overset{\overset{\textstyle O}{\|}}{C}NHCH_3 + H_2O \xrightarrow{\text{NaOH}} CH_3\overset{\overset{\textstyle O}{\|}}{C}ONa + CH_3NH_2$

13.14 Step 1: Nucleophilic addition of hydroxide ion to the carbonyl group

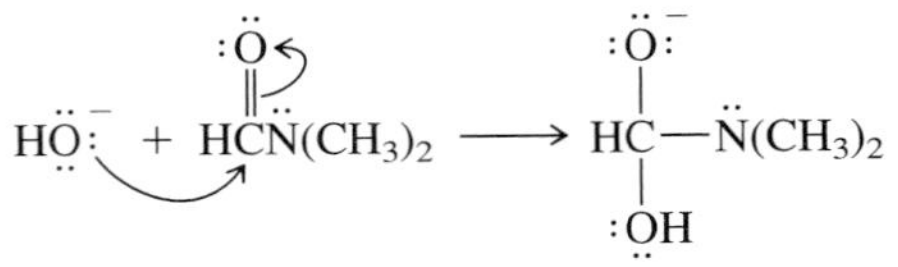

$$HO:^- + HCN(CH_3)_2 \longrightarrow HC\!-\!N(CH_3)_2$$

Step 2: Proton transfer to give neutral form of tetrahedral intermediate

$$HC\!-\!N(CH_3)_2 + H\!-\!OH \longrightarrow HC\!-\!N(CH_3)_2 + \,^-\!:OH$$

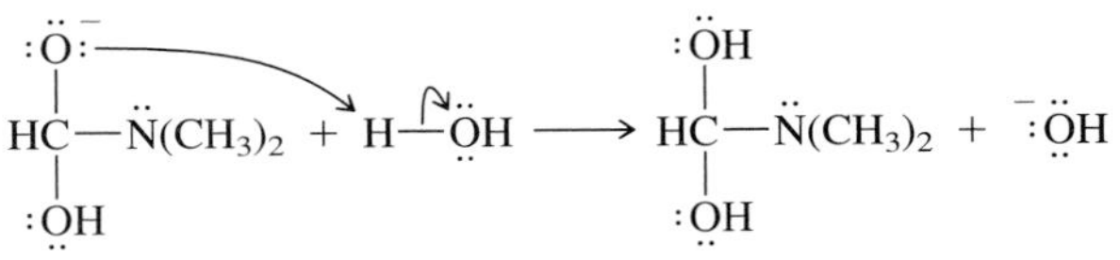

Step 3: Proton transfer from water to nitrogen of tetrahedral intermediate

$$HC\!-\!N(CH_3)_2 + H\!-\!OH \rightleftharpoons HC\!-\!\overset{+}{N}H(CH_3)_2 + \,^-\!:OH$$

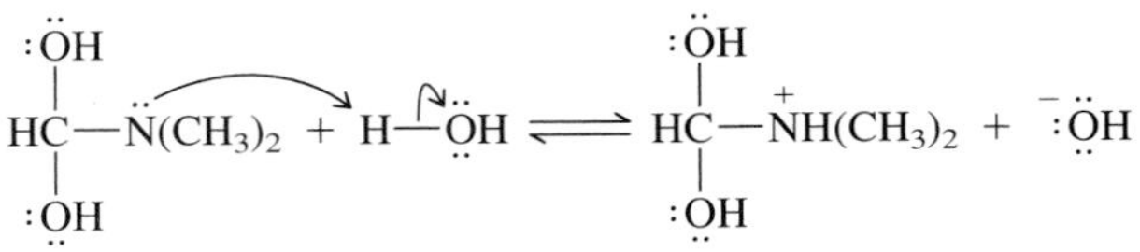

Step 4: Dissociation of N-protonated form of tetrahedral intermediate

$$HO:^- + HC\!-\!\overset{+}{N}H(CH_3)_2 \rightleftharpoons H_2O: + HC\overset{O:}{\underset{OH}{\diagdown}} + HN(CH_3)_2$$

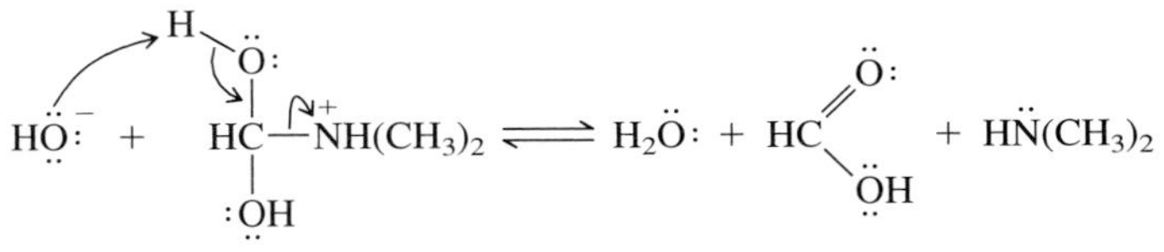

Step 5: Irreversible formation of formate ion

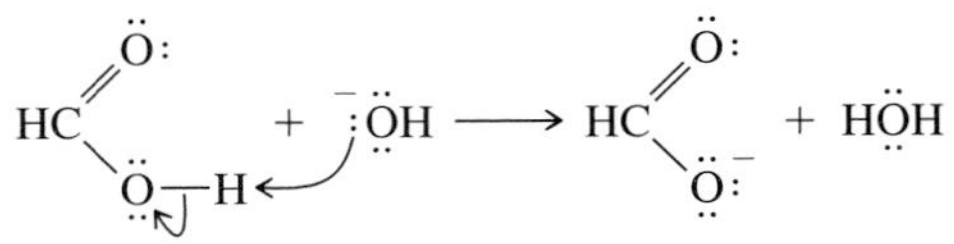

$$HC\overset{O:}{\underset{O\!-\!H}{\diagup}} + \,^-\!:OH \longrightarrow HC\overset{O:}{\underset{O:^-}{\diagdown}} + HOH$$

CHAPTER 14

14.1 (b) 1-Phenylethanamine or 1-phenylethylamine;

14.2 N,N-Dimethylcycloheptanamine

14.3 Tertiary amine; N-ethyl-4-isopropyl-N-methylaniline

14.4 $pK_b = 6$; K_a of conjugate acid $= 1 \times 10^{-8}$; pK_a of conjugate acid $= 8$

14.5 Tetrahydroisoquinoline is a stronger base than tetrahydroquinoline. The unshared electron pair of tetrahydroquinoline is delocalized into the aromatic ring, and this substance resembles aniline in its basicity, whereas tetrahydroisoquinoline resembles an alkylamine.

14.6 $2[CH_3(CH_2)_6CH_2]_2NH + CH_3(CH_2)_6CH_2Br \longrightarrow$
$$[CH_3(CH_2)_6CH_2]_3N + [CH_3(CH_2)_6CH_2]_2\overset{+}{N}H_2 \; Br^-$$

$$[CH_3(CH_2)_6CH_2]_3N + CH_3(CH_2)_6CH_2Br \longrightarrow [CH_3(CH_2)_6CH_2]_4\overset{+}{N} \; Br^-$$

14.7 (b) Prepare p-isopropylnitrobenzene as in part (a); then reduce with H_2, Ni (or Fe + HCl or Sn + HCl, followed by base). (c) Chlorinate benzene with Cl_2 + $FeCl_3$; then nitrate (HNO_3, H_2SO_4), separate the desired para isomer from the unwanted ortho isomer, and reduce.

14.8 CH_3CO_2H $\xrightarrow{SOCl_2}$ $CH_3\overset{O}{\underset{\|}{C}}Cl$ $\xrightarrow{-NH_2}$ $CH_3\overset{O}{\underset{\|}{C}}NH-\!\!\!\bigcirc$

$CH_3\overset{O}{\underset{\|}{C}}NH-\!\!\!\bigcirc$ $\xrightarrow[\text{2. H}_2\text{O}]{\text{1. LiAlH}_4}$ $CH_3CH_2NH-\!\!\!\bigcirc$

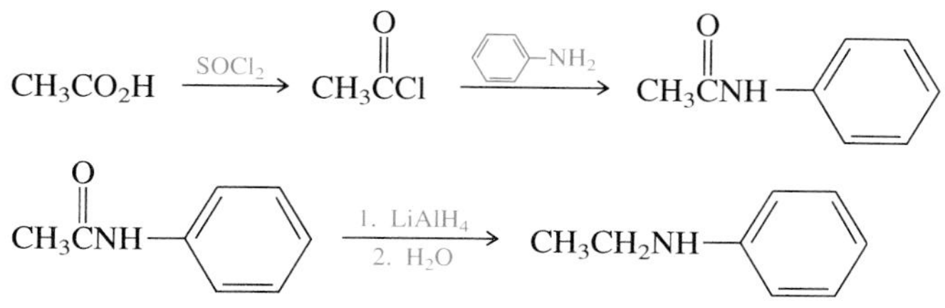

14.9 $(CH_3)_3\overset{+}{N}CH_2CH_2OH$ HO^-

14.10
$\underset{H_3C}{\overset{H_3C}{>}}\!\!N\!-\!\!N\!\!\overset{\curvearrowright}{}\!\!\overset{..}{\underset{..}{O}}:$ $\longleftrightarrow$ $\underset{H_3C}{\overset{H_3C}{>}}\!\!\overset{+}{N}\!\!=\!\!N\!\!\overset{..}{\underset{..}{O}}:^-$

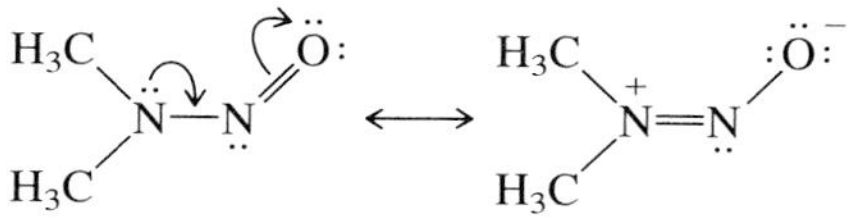

14.11 Intermediates: benzene to nitrobenzene to *m*-bromonitrobenzene to *m*-bromoaniline to *m*-bromophenol. Reagents: HNO_3, H_2SO_4; Br_2, $FeBr_3$; Fe, HCl, then HO^-; $NaNO_2$, H_2SO_4, H_2O, then heat in H_2O.

14.12 Prepare *m*-bromoaniline as in Problem 14.11; then $NaNO_2$, HCl, H_2O followed by KI.

14.13 Intermediates: benzene to ethyl phenyl ketone to ethyl *m*-nitrophenyl ketone to *m*-aminophenyl ethyl ketone to ethyl *m*-fluorophenyl ketone. Reagents: propanoyl chloride, $AlCl_3$; HNO_3, H_2SO_4; Fe, HCl, then HO^-; $NaNO_2$, H_2O, HCl, then HBF_4, then heat.

CHAPTER 15

15.1 (b) L-Glyceraldehyde; (c) D-glyceraldehyde

15.2 L-Erythrose

15.3

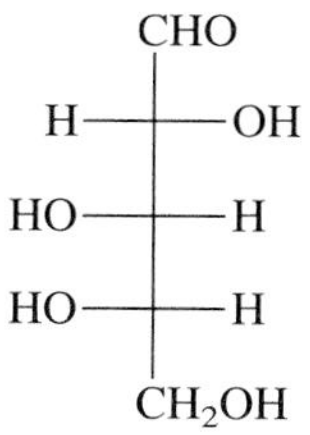

```
        CHO
    H ──┼── OH
   HO ──┼── H
   HO ──┼── H
        CH2OH
```

15.4 L-Talose

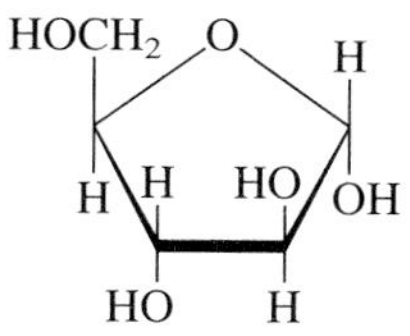

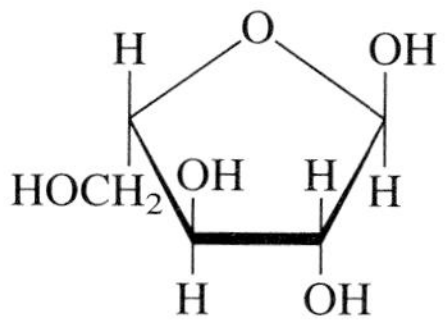

15.5 (b), 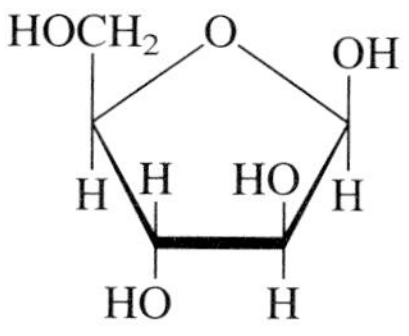 (c), (d)

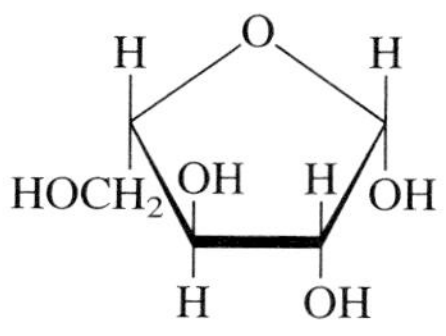

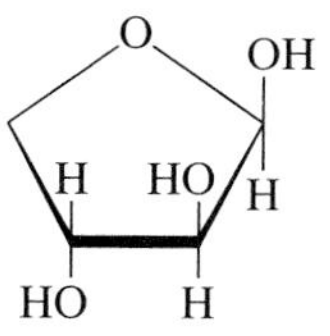

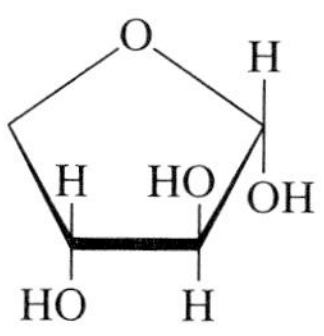

15.6 (b) 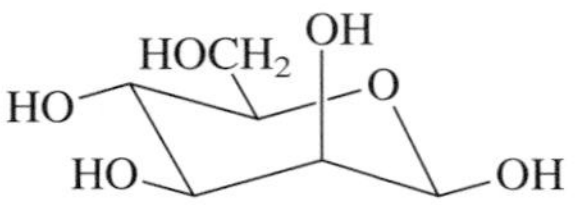 (c)

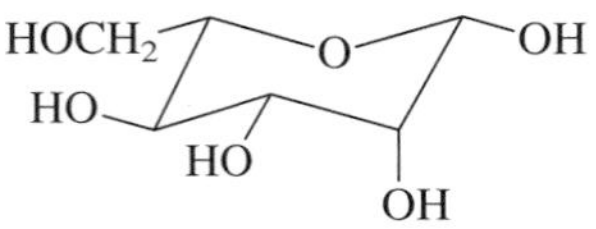

(d)

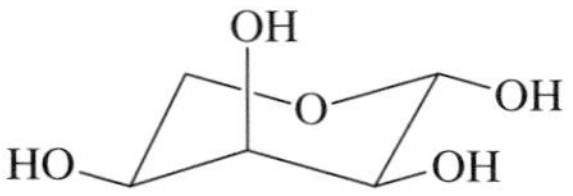

15.7

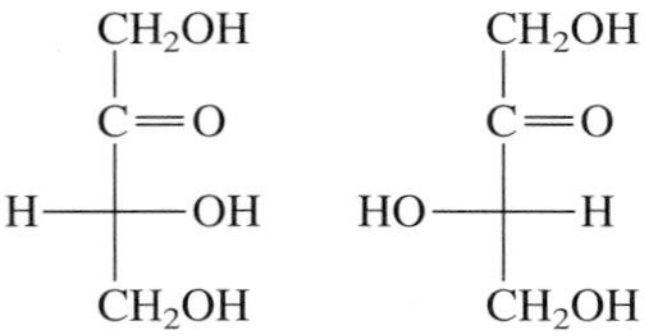

15.8 (b)

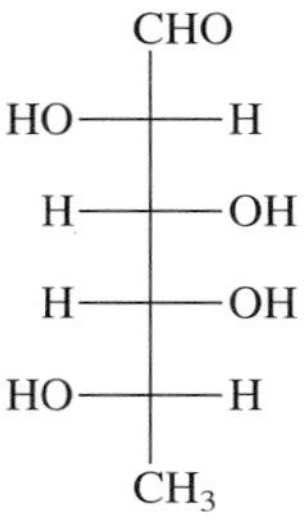

15.9 Adenosine: β-D-Ribofuranose; Sinigrin: β-D-glucopyranose.

15.10

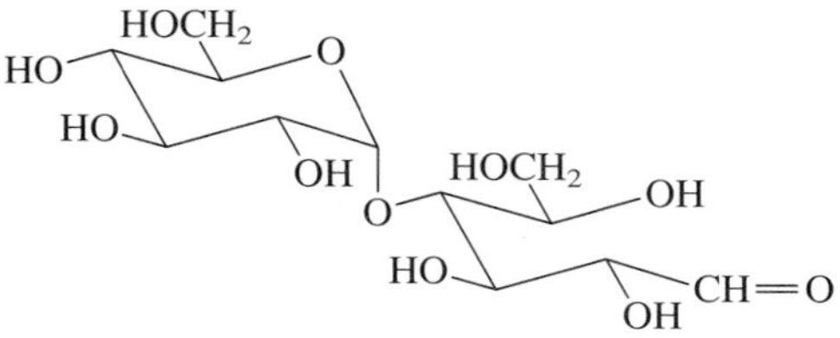

15.11 All [(b) through (f)] will give positive tests.

CHAPTER 16

16.1 Hydrolysis gives $CH_3(CH_2)_{16}CO_2H$ (2 mol) and (Z)-$CH_3(CH_2)_7CH=CH(CH_2)_7CO_2H$ (1 mol). The same mixture of products is formed from 1-oleyl-2,3-distearylglycerol.

16.2 Lecithin contains both hydrophilic and hydrophobic structural units. Fat-type molecules bind to the hydrophobic portion of lecithin and the fat + lecithin complexes are dispersed in the water.

16.3 $CH_3(CH_2)_{14}\overset{\displaystyle O}{\overset{\displaystyle \|}{C}}O(CH_2)_{15}CH_3$

16.4 The structure of vitamin D_2 is the same as that of vitamin D_3 except that vitamin D_2 has a double bond between C-22 and C-23 and a methyl substituent at C-24.

16.5

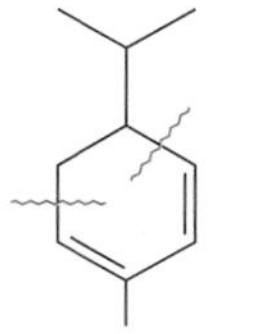

α-Phellandrene 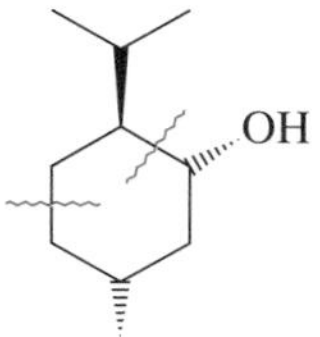   Menthol 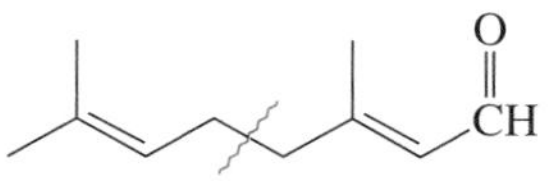   Citral

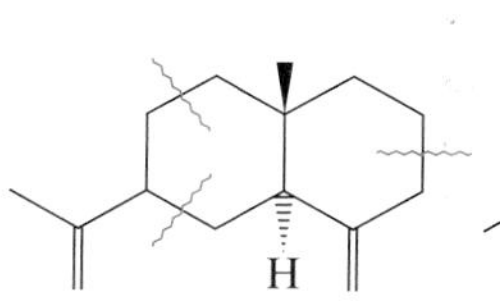
H

α-Selinene

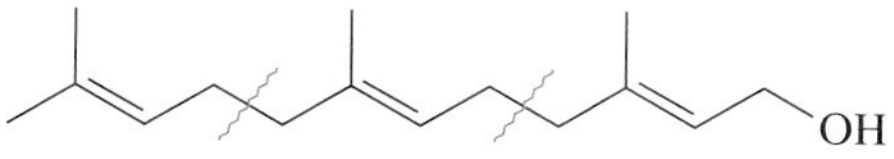
OH

Farnesol

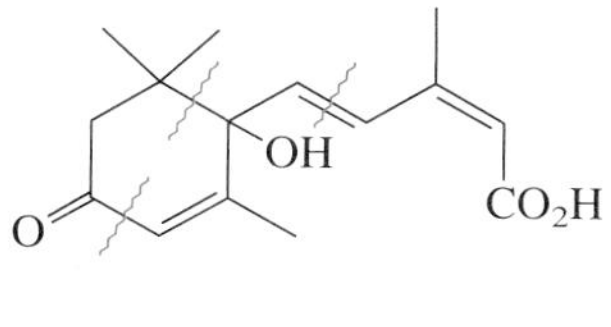
OH
O
CO2H

Abscisic acid

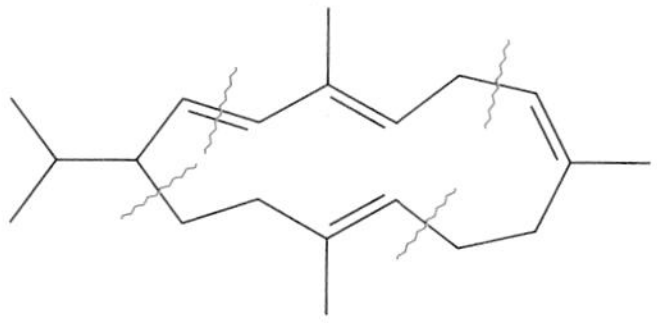

Cembrene

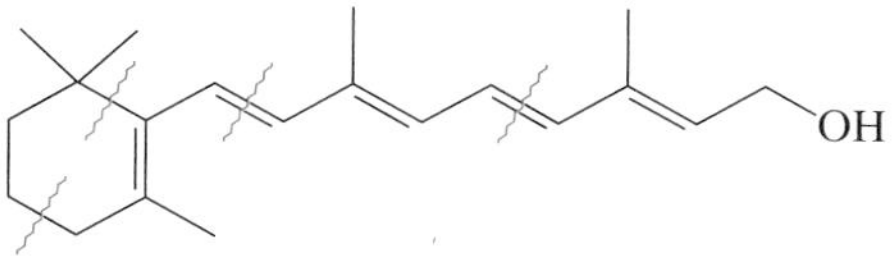
OH

Vitamin A

16.6

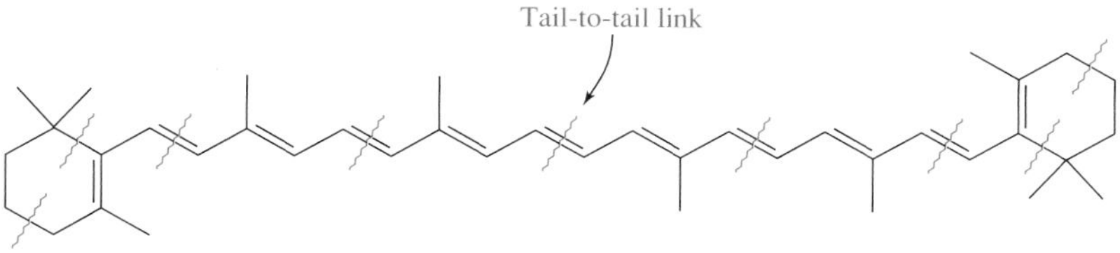
Tail-to-tail link

16.7

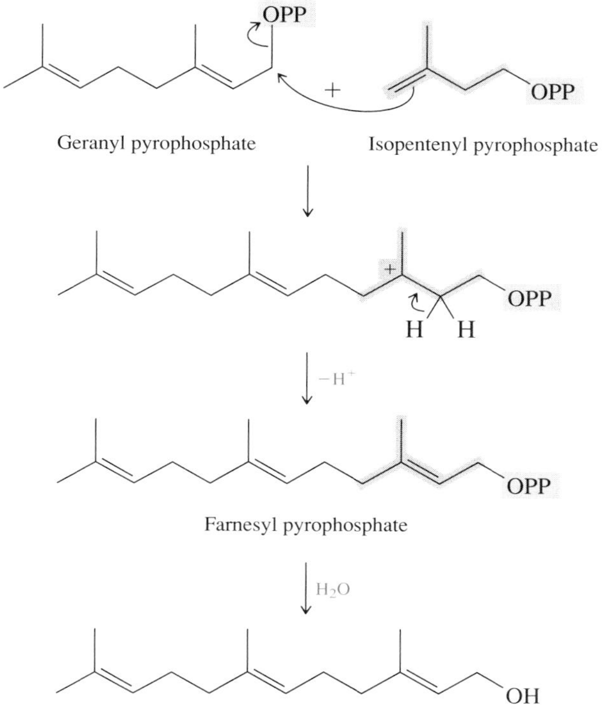
OPP
+
OPP
Geranyl pyrophosphate
Isopentenyl pyrophosphate
+
H
H
OPP
−H+
OPP
Farnesyl pyrophosphate
H2O
OH

Farnesol

CHAPTER 17

17.1 (b) *R*; (c) *S*

17.2 Isoleucine and threonine

17.3 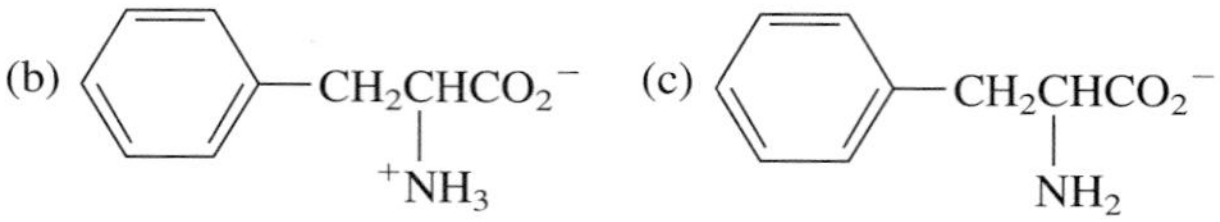 (b) C_6H_5—$CH_2CHCO_2^-$ with $^+NH_3$ (c) C_6H_5—$CH_2CHCO_2^-$ with NH_2

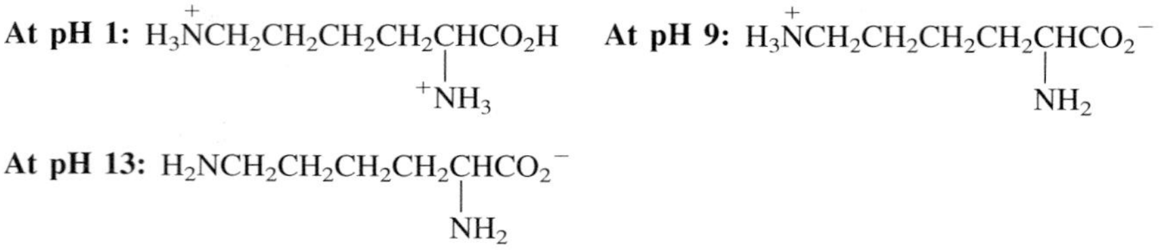

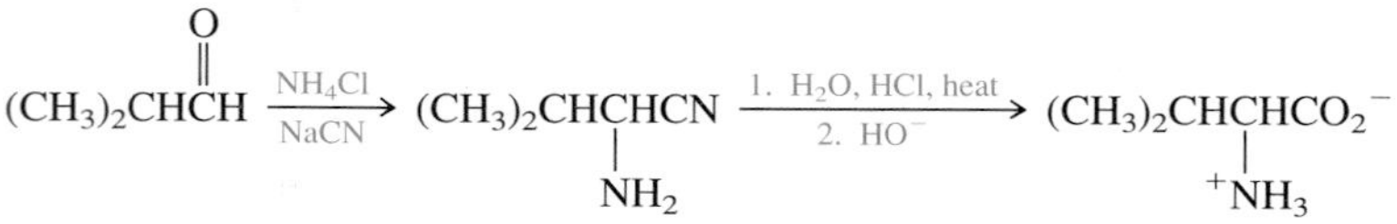

17.4 **At pH 1:** $H_3\overset{+}{N}CH_2CH_2CH_2CH_2CHCO_2H$ with $^+NH_3$ **At pH 9:** $H_3\overset{+}{N}CH_2CH_2CH_2CH_2CHCO_2^-$ with NH_2

At pH 13: $H_2NCH_2CH_2CH_2CH_2CHCO_2^-$ with NH_2

17.5 $(CH_3)_2CHCH\overset{O}{\overset{\|}{}}$ $\xrightarrow[\text{NaCN}]{\text{NH}_4\text{Cl}}$ $(CH_3)_2CHCHCN$ with NH_2 $\xrightarrow[\text{2. HO}^-]{\text{1. H}_2\text{O, HCl, heat}}$ $(CH_3)_2CHCHCO_2^-$ with $^+NH_3$

17.6 (b) $H_3\overset{+}{N}CHCNHCHCO_2^-$ with CH_3 and $CH_2C_6H_5$ (c) $H_3\overset{+}{N}CHCNHCHCO_2^-$ with $C_6H_5CH_2$ and CH_3

(d) $H_3\overset{+}{N}CH_2CNHCHCO_2^-$ with $CH_2CH_2CO_2^-$ (e) $H_3\overset{+}{N}CHCNHCH_2CO_2^-$ with $H_3\overset{+}{N}CH_2CH_2CH_2CH_2$

(f) $H_3\overset{+}{N}CHCNHCHCO_2^-$ with CH_3 and CH_3

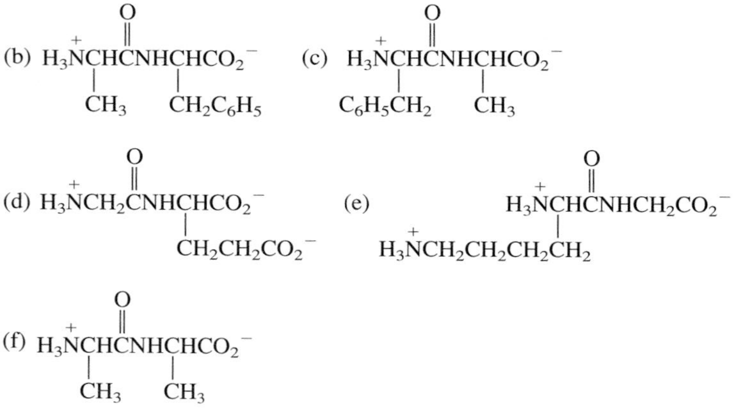

One-letter abbreviations: (b) AF; (c) FA; (d) GE; (e) KG; (f) D-A-D-A

17.7 $H_3\overset{+}{N}CHCNHCHCOCH_3$ with $^-O_2CCH_2$ and $CH_2C_6H_5$

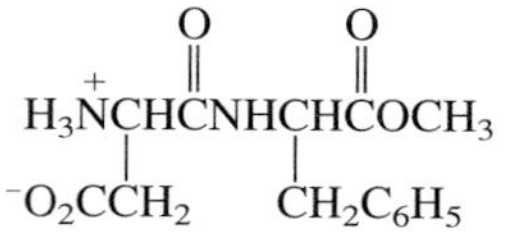

17.8 Tyr-Gly-Gly-Phe-Met; YGGFM

17.9

Ala-Gly-Phe-Val	Gly-Ala-Phe-Val	Phe-Gly-Ala-Val	Val-Gly-Phe-Ala
Ala-Gly-Val-Phe	Gly-Ala-Val-Phe	Phe-Gly-Val-Ala	Val-Gly-Ala-Phe
Ala-Phe-Gly-Val	Gly-Phe-Ala-Val	Phe-Ala-Gly-Val	Val-Phe-Gly-Ala
Ala-Phe-Val-Gly	Gly-Phe-Val-Ala	Phe-Ala-Val-Gly	Val-Phe-Ala-Gly
Ala-Val-Gly-Phe	Gly-Val-Ala-Phe	Phe-Val-Gly-Ala	Val-Ala-Gly-Phe
Ala-Val-Phe-Gly	Gly-Val-Phe-Ala	Phe-Val-Ala-Gly	Val-Ala-Phe-Gly

17.10 Val-Phe-Gly-Ala   Val-Phe-Ala-Gly

17.11 Pro-Val-Leu-Phe-Pro-Ala

17.12 After tyrosine, the next two amino acids are both glycine and give an unsubstituted PTH.

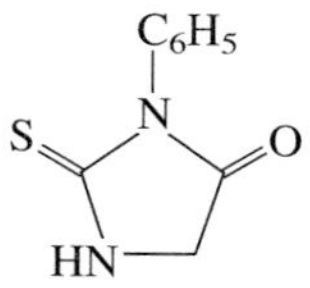

17.13 $H_3\overset{+}{N}CHCO_2^-$ + $C_6H_5CH_2O\overset{O}{\overset{\|}{C}}Cl$ $\longrightarrow$ $C_6H_5CH_2O\overset{O}{\overset{\|}{C}}NHCHCO_2H$

　　　　　　|CH_3　　　　　　　　　　　　　　　　　　　　　|CH_3

Alanine

$H_3\overset{+}{N}CHCO_2^-$ + $C_6H_5CH_2OH$ $\xrightarrow[\text{2. HO}^-]{\text{1. H}^+\text{, heat}}$ $H_2NCHCO_2CH_2C_6H_5$

(CH_3)_2CHCH_2　　　　　　　　　　　　　　　　　(CH_3)_2CHCH_2

Leucine

$C_6H_5CH_2O\overset{O}{\overset{\|}{C}}NHCHCO_2H$ + $H_2NCH\overset{O}{\overset{\|}{C}}OCH_2C_6H_5$ $\xrightarrow{\text{DCCI}}$

　　　　　　|CH_3　　　　　(CH_3)_2CHCH_2

$C_6H_5CH_2O\overset{O}{\overset{\|}{C}}NHCH\overset{O}{\overset{\|}{C}}NHCH\overset{O}{\overset{\|}{C}}OCH_2C_6H_5$

　　　　　　　　　|CH_3　　　|CH_2CH(CH_3)_2

$C_6H_5CH_2O\overset{O}{\overset{\|}{C}}NHCH\overset{O}{\overset{\|}{C}}NHCH\overset{O}{\overset{\|}{C}}OCH_2C_6H_5$ $\xrightarrow[\text{Pd}]{\text{H}_2}$ Ala-Leu

　　　　　　　　　|CH_3　　　|CH_2CH(CH_3)_2

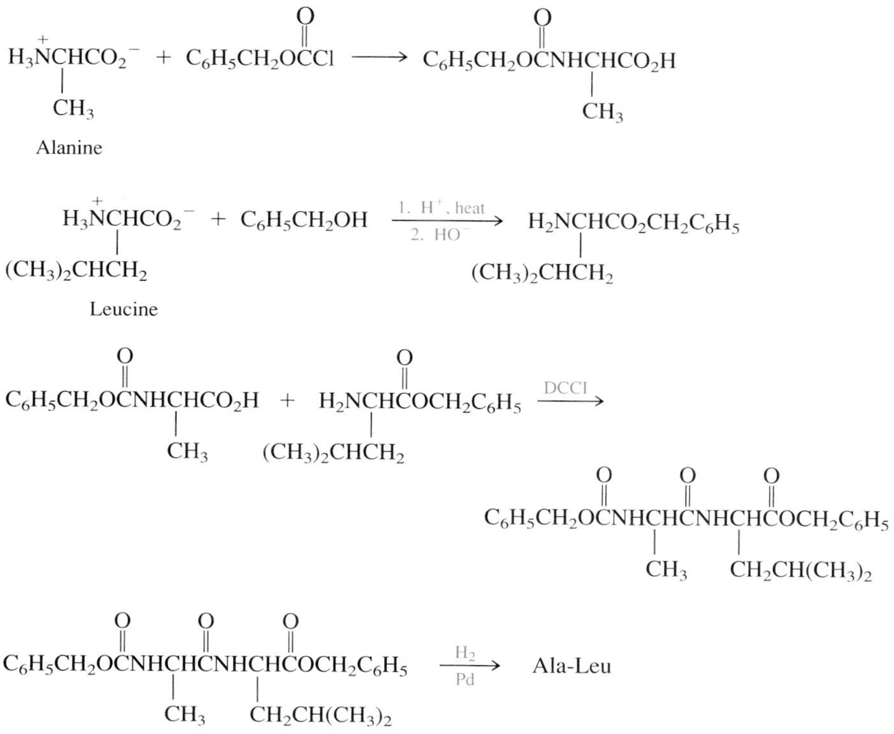

CHAPTER 18

18.1

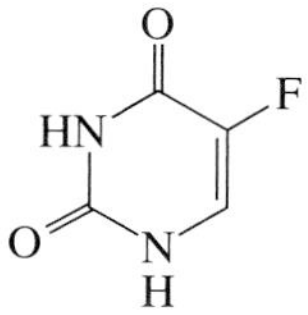

18.2 Caffeine and theobromine are both purines.

18.3 (b) Cytidine　　　　　　　　　　　　　(c) Guanosine

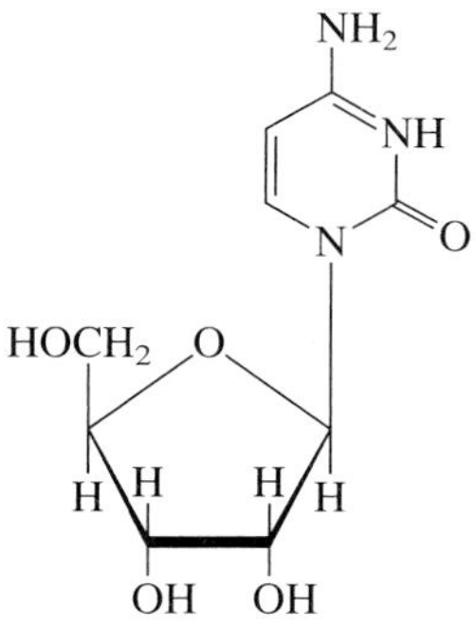

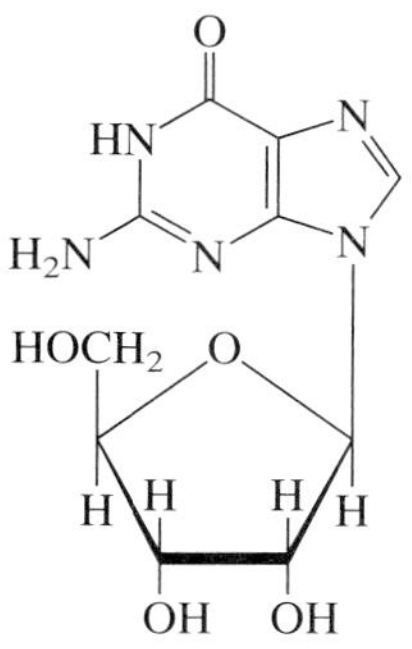

18.4 The codons for glutamic acid (GAA and GAG) differ by only one base from two of the codons for valine (GUA and GUG).

CHAPTER 19

19.1 (b) The frequency of red light ($\nu = 3.8 \times 10^{-14}s^{-1}$) is less than the frequency of violet light ($\nu = 7.5 \times 10^{14}s^{-1}$). (c) A photon of violet light is of higher energy.

19.2 1.41 T

19.3 (a) 6.88 ppm; (b) higher field; more shielded

19.4 H in CH_3CCl_3 is more shielded than H in $CHCl_3$. If H in $CHCl_3$ appears at δ 7.28 ppm, then H in CH_3CCl_3 appears 4.6 ppm upfield of 7.28 ppm. Its chemical shift is δ 2.7 ppm.

19.5 (b) Two: CH_3 in range δ 0.9–1.8 ppm, CH_2 in range δ 3.3–3.7 ppm; (c) one in range δ 0.9–1.8 ppm; (d) two: CH_3 in range δ 0.9–1.8 ppm, CH in range δ 3.1–4.1 ppm

19.6 (b) CH_3 (3), CH_2 (2); (c) CH_3 (only signal); (d) CH_3 (6), CH (1)

19.7 The chemical shift of the methyl protons is δ 2.2 ppm. The chemical shift of the protons attached to the aromatic ring is δ 7.0 ppm.

19.8 (b) One signal (singlet); (c) two signals (doublet and triplet); (d) two signals (both singlets); (e) two signals (doublet and quartet)

19.9 (b) Three signals (singlet, triplet, and quartet); (c) two signals (triplet and quartet); (d) three signals (singlet, triplet, and quartet); (e) four signals (three triplets and quartet)

19.10 (b) Six; (c) six; (d) nine; (e) three

19.11 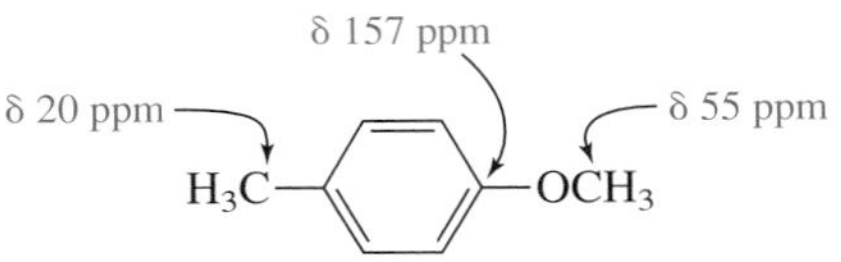
δ 157 ppm
δ 20 ppm
δ 55 ppm
H₃C
OCH₃

19.12 Benzyl alcohol. Infrared spectrum has peaks for O—H and sp^3 C—H; lacks peak for C=O.

19.13 2-Methyl-1,3-butadiene

19.14 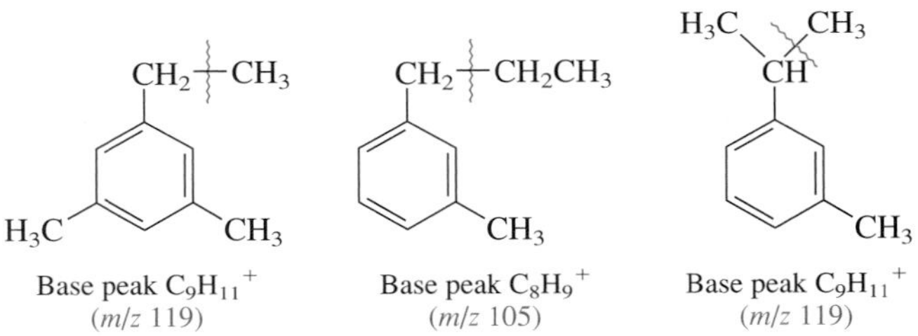
CH₂—CH₃
H₃C
CH₃
Base peak C₉H₁₁⁺
(m/z 119)
CH₂—CH₂CH₃
CH₃
Base peak C₈H₉⁺
(m/z 105)
H₃C
CH₃
CH
CH₃
Base peak C₉H₁₁⁺
(m/z 119)

19.15 (b) 3; (c) 2; (d) 3; (e) 2; (f) 2

G L O S S A R Y

Absolute configuration (Section 7.5): The three-dimensional arrangement of atoms or groups at a stereogenic center.

Acetal (Section 11.8): Product of the reaction of an aldehyde or a ketone with 2 moles of an alcohol according to the equation

$$\underset{\text{O}}{\overset{\text{O}}{\underset{\parallel}{\text{RCR}'}}} + 2\text{R}''\text{OH} \xrightarrow{\text{H}^+} \underset{\overset{|}{\text{OR}''}}{\overset{\text{OR}''}{\overset{|}{\text{RCR}'}}} + \text{H}_2\text{O}$$

Acetyl coenzyme A (Section 16.11): A thiol ester abbreviated as

$$\underset{\text{CH}_3\text{CSCoA}}{\overset{\text{O}}{\overset{\parallel}{}}}$$

that acts as the source of acetyl groups in biosynthetic processes involving acetate.

Acetylene (Sections 1.15 and 4.12): The simplest alkyne, $\text{HC}\equiv\text{CH}$.

Achiral (Section 7.1): Opposite of *chiral*. An achiral object is superimposable on its mirror image.

Acid (Section 3.5): According to the Arrhenius definition, a substance that ionizes in water to produce protons. According to the Brønsted–Lowry definition, a substance that donates a proton to some other substance. According to the Lewis definition, an electron-pair acceptor.

Acid anhydride (Sections 2.3 and 13.1): Compound of the type

$$\underset{\text{RCOCR}}{\overset{\text{O O}}{\overset{\parallel \ \parallel}{}}}$$

Both R groups are usually the same, although they need not always be.

Acid dissociation constant K_a (Section 3.5): Equilibrium constant for dissociation of an acid:

$$K_a = \frac{[\text{H}^+][\text{A}^-]}{[\text{HA}]}$$

Activating substituent (Sections 6.9 and 6.10): A group that when present in place of a hydrogen causes a particular reaction to occur faster. Term is most often applied to substituents that increase the rate of electrophilic aromatic substitution.

Active site (Section 17.13): The region of an enzyme at which the substrate is bound.

Acylation (Section 6.6 and Chapter 13): Reaction in which an acyl group becomes attached to some structural unit in a molecule. Examples include the Friedel–Crafts acylation and the conversion of amines to amides.

Acyl chloride (Sections 2.3 and 13.1): Compound of the type

$$\underset{\text{RCCl}}{\overset{\text{O}}{\overset{\parallel}{}}}$$

R may be alkyl or aryl.

Acyl group (Sections 6.8 and 13.1): The group

$$\underset{\text{RC}-}{\overset{\text{O}}{\overset{\parallel}{}}}$$

R may be alkyl or aryl.

Acylium ion (Section 6.8): The cation $\text{R}-\overset{+}{\text{C}}\equiv\overset{..}{\text{O}}:$.

Acyl transfer (Section 13.3): A nucleophilic acyl substitution. A reaction in which one type of carboxylic acid derivative is converted to another.

Addition (Section 5.1): Reaction in which a reagent X—Y adds to a multiple bond so that X becomes attached to one of the carbons of the multiple bond and Y to the other.

1,2 Addition (Section 5.8): Addition of reagents of the type X—Y to conjugated dienes in which X and Y add to adjacent doubly bonded carbons:

$$\text{R}_2\text{C}=\text{CH}-\text{CH}=\text{CR}_2 \xrightarrow{\text{X}-\text{Y}} \underset{\overset{|}{\text{X}}\ \overset{|}{\text{Y}}}{\text{R}_2\text{C}-\text{CH}-\text{CH}=\text{CR}_2}$$

1,4 Addition (Section 5.8): Addition of reagents of the type X—Y to conjugated dienes in which X and Y add to the termini of the diene system:

$$\text{R}_2\text{C}=\text{CH}-\text{CH}=\text{CR}_2 \xrightarrow{\text{X}-\text{Y}} \underset{\overset{|}{\text{X}}}{\text{R}_2\text{C}}-\text{CH}=\text{CH}-\underset{\overset{|}{\text{Y}}}{\text{CR}_2}$$

Alcohol (Section 3.1): Compound of the type ROH.

Alcohol dehydrogenase (Section 10.5): Enzyme in the liver that catalyzes the oxidation of alcohols to aldehydes and ketones.

Aldehyde (Sections 2.3 and 11.1): Compound of the type

$$\underset{\text{RCH}}{\overset{\displaystyle O}{\|}} \quad \text{or} \quad \underset{\text{ArCH}}{\overset{\displaystyle O}{\|}}$$

Aldol addition (Section 11.18): Nucleophilic addition of an aldehyde or ketone enolate to the carbonyl group of an aldehyde or a ketone. The most typical case involves two molecules of an aldehyde, and is usually catalyzed by bases.

$$2RCH_2CH \xrightarrow{\ HO^-\ } RCH_2\underset{\underset{CH=O}{|}}{CHCHR}$$

Aldol condensation (Section 11.18): When an aldol addition is carried out so that the β-hydroxy aldehyde or ketone dehydrates under the conditions of its formation, the product is described as arising by an aldol condensation.

$$2RCH_2CH \xrightarrow[\text{heat}]{HO^-} RCH_2CH{=}\underset{\underset{CH=O}{|}}{CR} + H_2O$$

Aldose (Section 15.1): Carbohydrate that contains an aldehyde carbonyl group in its open-chain form.

Alicyclic (Section 2.12): Term describing an *ali*phatic *cyclic* structural unit.

Aliphatic (Section 2.1): Term applied to compounds that do not contain benzene or benzene-like rings as structural units. (Historically, *aliphatic* was used to describe compounds derived from fats and oils.)

Alkadiene (Section 4.1): Hydrocarbon that contains two carbon–carbon double bonds; commonly referred to as a *diene*.

Alkaloid (Section 14.3): Amine that occurs naturally in plants. The name derives from the fact that such compounds are weak bases.

Alkane (Section 2.1): Hydrocarbon in which all the bonds are single. Alkanes have the general formula C_nH_{2n+2}.

Alkene (Section 2.1): Hydrocarbon that contains a carbon–carbon double bond (C=C); also known by the older name *olefin*.

Alkoxide ion (Section 10.10): Conjugate base of an alcohol; a species of the type $R{-}\ddot{O}:^-$.

Alkylamine (Section 14.1): Amine in which the organic groups attached to nitrogen are alkyl groups.

Alkylation (Section 5.10): Reaction in which an alkyl group is attached to some structural unit in a molecule.

Alkyl group (Section 2.10): Structural unit related to an alkane by replacing one of the hydrogens by a potential point of attachment to some other atom or group. The general symbol for an alkyl group is R—.

Alkyl halide (Section 3.1): Compound of the type RX, in which X is a halogen substituent (F, Cl, Br, I).

Alkyloxonium ion (Section 3.5): Positive ion of the type ROH_2^+.

Alkyne (Section 2.1): Hydrocarbon that contains a carbon–carbon triple bond.

Allyl cation (Section 5.8): The carbocation

$$CH_2{=}CHCH_2^+$$

The carbocation is stabilized by delocalization of the π electrons of the double bond, and the positive charge is shared by the two CH_2 groups. Substituted analogs of allyl cation are called *allylic carbocations*.

Amide (Sections 2.3 and 13.1): Compound of the type $R\overset{\displaystyle O}{\overset{\|}{C}}NR'_2$

Amine (Chapter 14): Molecule in which a nitrogen-containing group of the type $-NH_2$, $-NHR$, or $-NR_2$ is attached to an alkyl or aryl group.

α-Amino acid (Section 17.1): A carboxylic acid that contains an amino group at the α-carbon atom. α-Amino acids are the building blocks of peptides and proteins. An α-amino acid normally exists as a *zwitterion*.

$$\underset{\overset{|}{{}^+NH_3}}{RCHCO_2^-}$$

L-Amino acid (Section 17.2): A description of the stereochemistry at the α-carbon atom of a chiral amino acid. The Fischer projection of an α-amino acid has the amino group on the left when the carbon chain is vertical with the carboxyl group at the top.

$$H_3\overset{+}{N}\underset{R}{\overset{CO_2^-}{-\!\!\!\!|\!\!\!\!-}}H$$

Amino acid residues (Section 17.5): Individual amino acid components of a peptide or protein.

Amino sugar (Section 15.9): Carbohydrate in which one of the hydroxyl groups has been replaced by an amino group.

Amylopectin (Section 15.12): A polysaccharide present in starch. Amylopectin is a polymer of α(1,4)-linked glucose units, as is amylose (see *amylose*). Unlike amylose, amylopectin contains branches of 24–30 glucose units connected to the main chain by an α(1,6) linkage.

Amylose (Section 15.12): The water-dispersible component of starch. It is a polymer of α(1,4)-linked glucose units.

Anabolic steroid (Section 16.9): A steroid that promotes muscle growth.

Androgen (Section 16.9): A male sex hormone.

Angle strain (Section 2.13): The strain a molecule possesses because its bond angles are distorted from their normal values.

Anion (Section 1.2): Negatively charged ion.

Anomeric carbon (Section 15.5): The carbon atom in a furanose or pyranose form that is derived from the carbonyl carbon of the open-chain form. It is the ring carbon that is bonded to two oxygens.

Anti (Section 2.6): Term describing relative position of two substituents on adjacent atoms when the angle between their bonds is on the order of 180°. Atoms X and Y in the structure shown are anti to each other.

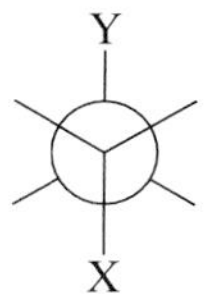

Anti addition (Section 5.1): Addition reaction in which the two portions of the attacking reagent X—Y add to opposite faces of the double bond.

Anticodon (Section 18.6): Sequence of three bases in a molecule of tRNA that is complementary to the codon of mRNA for a particular amino acid.

Ar— (Section 2.2): Symbol for an aryl group.

Arene (Section 2.1): Aromatic hydrocarbon. Often abbreviated ArH.

Arenium ion (Section 6.7): The carbocation intermediate formed by attack of an electrophile on an aromatic substrate in electrophilic aromatic substitution. See *cyclohexadienyl cation.*

Aromatic compound (Section 6.1): An electron-delocalized species that is much more stable than any structure written for it in which all the electrons are localized either in covalent bonds or as unshared electron pairs.

Aromaticity (Section 6.1): Special stability associated with aromatic compounds.

Arylamine (Section 14.1): An amine that has an aryl group attached to the amine nitrogen.

Asymmetric center (Section 7.2): Obsolete name for a *stereogenic center.*

Atomic number (Section 1.1): The number of protons in the nucleus of a particular atom. The symbol for atomic number is Z and each element has a unique atomic number.

Axial bond (Section 2.14): A bond to a carbon in the chair conformation of cyclohexane oriented like the six "up-and-down" bonds in the following:

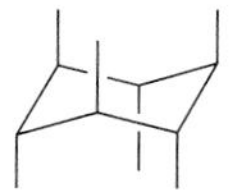

Azo coupling (Section 14.11): Formation of a compound of the type ArN=NAr' by reaction of an aryl diazonium salt with an arene. The arene must be strongly activated toward electrophilic aromatic substitution; that is, it must bear a powerful electron-releasing substituent such as —OH or —NR₂.

Ball-and-stick model (Section 1.10): Type of molecular model in which balls representing atoms are connected by sticks representing bonds.

Base (Section 3.5): According to the Arrhenius definition, a substance that ionizes in water to produce hydroxide ions. According to the Brønsted–Lowry definition, a substance that accepts a proton from some suitable donor. According to the Lewis definition, an electron-pair donor.

Base pair (Section 18.5): Term given to the purine of a nucleotide and its complementary pyrimidine. Adenine (A) is complementary to thymine (T), and guanine (G) is complementary to cytosine (C).

Base peak (Section 19.13): The most intense peak in a mass spectrum. The base peak is assigned a relative intensity of 100, and the intensities of all other peaks are cited as a percentage of the base peak.

Basicity constant K_b (Section 14.4): A measure of base strength, especially of amines.

$$K_b = \frac{[R_3NH^+][HO^-]}{[R_3N]}$$

Bending vibration (Section 19.10): The regular, repetitive motion of an atom or a group along an arc the radius of which is the bond connecting the atom or group to the rest of the molecule. Bending vibrations are one type of molecular motion that gives rise to a peak in the infrared spectrum.

Benedict's reagent (Section 15.13): A solution containing the citrate complex of $CuSO_4$. It is used to test for the presence of reducing sugars.

Benzene (Section 6.1): The most typical aromatic hydrocarbon:

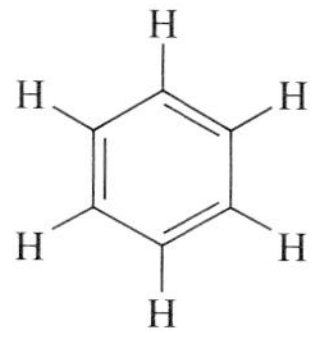

Benzylic carbon (Section 6.5): A carbon directly attached to a benzene ring. A hydrogen attached to a benzylic carbon is a benzylic hydrogen. A carbocation in which the benzylic carbon is positively charged is a benzylic carbocation. A free radical in which the benzylic carbon bears the unpaired electron is a benzylic radical.

Bile acids (Section 16.7): Steroid derivatives biosynthesized in the liver that aid digestion by emulsifying fats.

Bimolecular (Section 3.6): A process in which two particles react in the same elementary step.

Biological isoprene unit (Section 16.13): Isopentenyl pyrophosphate, the biological precursor to terpenes and steroids:

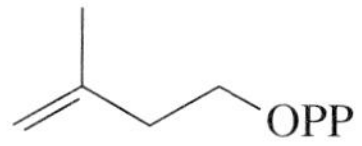

Boat conformation (Section 2.14): An unstable conformation of cyclohexane, depicted as

π bond (Section 1.14): In alkenes, a bond formed by overlap of *p* orbitals in a side-by-side manner. A π bond is weaker than a σ bond. The carbon–carbon double bond in alkenes consists of two *sp²*-hybridized carbons joined by a σ bond and a π bond.

σ bond (Section 1.12): A connection between two atoms in which the electron probability distribution has rotational symmetry along the internuclear axis. A cross section perpendicular to the internuclear axis is a circle.

Bond dissociation energy (Section 9.2): For a substance A:B, the energy required to break the bond between A and B so that each retains one of the electrons in the bond. Table 9.1 gives bond dissociation energies for some representative compounds.

Bond-line formula (Section 1.7): Formula in which connections between carbons are shown but individual carbons and hydrogens are not. The bond-line formula

represents the compound $(CH_3)_2CHCH_2CH_3$.

Branched-chain carbohydrate (Section 15.9): Carbohydrate in which the main carbon chain bears a carbon substituent in place of a hydrogen or hydroxyl group.

Bromohydrin (Section 5.6): A halohydrin in which the halogen is bromine (see *halohydrin*).

Bromonium ion (Section 5.6): A halonium ion in which the halogen is bromine (see *halonium ion*).

Brønsted acid: See *acid.*

Brønsted base: See *base.*

n-Butane (Section 2.6): Common name for butane $CH_3CH_2CH_2CH_3$.

n-Butyl group (Section 2.10): The group $CH_3CH_2CH_2CH_2$—.

sec-Butyl group (Section 2.10): The group

$$CH_3CH_2CHCH_3$$

tert-Butyl group (Section 2.10): The group $(CH_3)_3C$—.

Cahn–Ingold–Prelog notation (Section 7.6): System for specifying absolute configuration as *R* or *S* on the basis of the order in which atoms or groups are attached to a stereogenic center. Groups are ranked in order of precedence according to rules based on atomic number.

Carbanion (Section 5.9): Anion in which the negative charge is borne by carbon. An example is acetylide ion.

Carbinolamine (Section 11.9): Compound of the type

$$HO-\overset{\displaystyle |}{\underset{\displaystyle |}{C}}-NR_2$$

Carbinolamines are formed by nucleophilic addition of an amine to a carbonyl group and are intermediates in the formation of imines.

Carbocation (Section 3.8): Positive ion in which the charge resides on carbon. An example is *tert*-butyl cation, $(CH_3)_3C^+$. Carbocations are unstable species that, though they cannot normally be isolated, are believed to be intermediates in certain reactions.

Carboxylate ion (Section 12.7): The conjugate base of a carboxylic acid, an ion of the type RCO_2^-.

Carboxylation (Section 12.9): In the preparation of a carboxylic acid, the reaction of a carbanion with carbon dioxide. Typically, the carbanion source is a Grignard reagent.

$$RMgX \xrightarrow[\text{2. } H_3O^+]{\text{1. } CO_2} RCO_2H$$

Carboxylic acid (Sections 2.3 and 12.1): Compound of the type
$$\overset{\displaystyle O}{\overset{\displaystyle \|}{RCOH}}$$
, also written as RCO_2H.

Carboxylic acid derivative (Chapter 13): Compound that yields a carboxylic acid on hydrolysis. Carboxylic acid derivatives include acyl chlorides, anhydrides, esters, and amides.

Carotenoids (Section 16.10): Naturally occurring tetraterpenoid plant pigments.

Cation (Section 1.2): Positively charged ion.

Cellobiose (Section 15.11): A disaccharide in which two glucose units are joined by a β(1,4) linkage. Cellobiose is obtained by the hydrolysis of cellulose.

Cellulose (Section 15.12): A polysaccharide in which thousands of glucose units are joined by β(1,4) linkages.

Chain reaction (Section 9.4): Reaction mechanism in which a sequence of individual steps repeats itself many times, usually because a reactive intermediate consumed in one step is regenerated in a subsequent step. The halogenation of alkanes is a chain reaction proceeding via free-radical intermediates.

Chair conformation (Section 2.14): The most stable conformation of cyclohexane:

Chemical shift (Section 19.3): A measure of how shielded the nucleus of a particular atom is. Nuclei of different atoms have different chemical shifts, and nuclei of the same atom have chemical shifts that are sensitive to their molecular environment. In proton and carbon-13 NMR, chemical shifts are cited as δ, or parts per million (ppm), from the hydrogens or carbons, respectively, of tetramethylsilane.

Chiral (Section 7.1): Term describing an object that is not superimposable on its mirror image.

Chiral carbon atom (Section 7.2): A carbon that is bonded to four groups, all of which are different from one another. Also called an asymmetric carbon atom. A more modern term is *stereogenic center.*

Chiral center (Section 7.2): See *stereogenic center.*

Chlorohydrin (Section 5.6): A halohydrin in which the halogen is chlorine (see *halohydrin*).

Chloronium ion (Section 5.6): A halonium ion in which the halogen is chlorine (see *halonium ion*).

Cholesterol (Section 16.5): The most abundant steroid in animals and the biological precursor to other naturally occurring steroids, including the bile acids, sex hormones, and corticosteroids.

Chromatography (Section 19.13): A method for separation and analysis of mixtures based on the different rates at which different compounds are removed from a stationary phase by a moving phase.

Chromophore (Section 19.11): The structural unit of a molecule principally responsible for absorption of radiation of a particular frequency; a term usually applied to ultraviolet-visible spectroscopy.

Chymotrypsin (Section 17.8): A digestive enzyme that catalyzes the hydrolysis of proteins. Chymotrypsin selectively catalyzes the cleavage of the peptide bond between the carboxyl group of phenylalanine, tyrosine, or tryptophan and some other amino acid.

cis- (Sections 2.17 and 4.3): Stereochemical prefix indicating that two substituents are on the same side of a ring or double bond. (Contrast with the prefix *trans-*.)

Closed-shell electron configuration (Section 1.1): Stable electron configuration in which all the lowest energy orbitals of an atom (in the case of the noble gases), an ion (e.g., Na^+), or a molecule (e.g., benzene) are filled.

^{13}C NMR (Section 19.8): Nuclear magnetic resonance spectroscopy in which the environments of individual carbon atoms are examined via their mass 13 isotope.

Codon (Section 18.6): Set of three successive nucleotides in mRNA that is unique for a particular amino acid. The 64 codons possible from combinations of A, T, G, and C code for the 20 amino acids from which proteins are constructed.

Coenzyme (Section 17.14): Molecule that acts in combination with an enzyme to bring about a reaction.

Coenzyme Q (Section 10.17): Naturally occurring group of related quinones involved in the chemistry of cellular respiration. Also known as ubiquinone.

Combustion (Section 2.20): Burning of a substance in the presence of oxygen. All hydrocarbons yield carbon dioxide and water when they undergo combustion.

Common nomenclature (Section 2.8): Names given to compounds on some basis other than a comprehensive, systematic set of rules.

Concerted reaction (Section 3.6): Reaction that occurs in a single elementary step.

Condensation polymer (Section 13.12): Polymer in which the bonds that connect the monomers are formed by condensation reactions. Typical condensation polymers include polyesters and polyamides.

Condensed structural formula (Section 1.7): A standard way of representing structural formulas in which subscripts are used to indicate replicated atoms or groups, as in $(CH_3)_2CHCH_2CH_3$.

Conformational analysis (Section 2.5): Study of the conformations available to a molecule, their relative stability, and the role they play in defining the properties of the molecule.

Conformations (Section 2.5): Nonidentical representations of a molecule generated by rotation about single bonds.

Conformers (Section 2.5): Different conformations of a single molecule.

Conjugate acid (Section 3.5): The species formed from a Brønsted base after it has accepted a proton.

Conjugate addition (Section 5.8): Addition reaction in which the reagent adds to the termini of the conjugated system with migration of the double bond; synonymous with 1,4 addition.

Conjugate base (Section 3.5): The species formed from a Brønsted acid after it has donated a proton.

Conjugated diene (Section 4.1): System of the type $C=C-C=C$, in which two pairs of doubly bonded carbons are joined by a single bond. The π electrons are delocalized over the unit of four consecutive sp^2-hybridized carbons.

Connectivity (Section 1.7): Order in which a molecule's atoms are connected. Synonymous with *constitution.*

Constitution (Section 1.7): Order of atomic connections that defines a molecule.

Constitutional isomers (Section 1.8): Isomers that differ in respect to the order in which the atoms are connected. Butane $(CH_3CH_2CH_2CH_3)$ and isobutane $[(CH_3)_3CH]$ are constitutional isomers.

Covalent bond (Section 1.3): Chemical bond between two atoms that results from their sharing of two electrons.

C terminus (Section 17.5): The amino acid at the end of a peptide or protein chain that has its carboxyl group intact—that is, in which the carboxyl group is not part of a peptide bond.

Cumulated diene (Section 4.1): Diene of the type $C=C=C$, in which a single carbon atom participates in double bonds with two others.

Cyanohydrin (Section 11.7): Compound of the type

$$\begin{matrix} & OH & \\ & | & \\ R & C & R' \\ & | & \\ & C \equiv N & \end{matrix}$$

Cyanohydrins are formed by nucleophilic addition of HCN to the carbonyl group of an aldehyde or a ketone.

Cycloalkane (Section 2.12): An alkane in which a ring of carbon atoms is present.

Cycloalkene (Section 4.1): A cyclic hydrocarbon characterized by a double bond between two of the ring carbons.

Cyclohexadienyl cation (Section 6.7): The key intermediate in electrophilic aromatic substitution reactions. It is represented by the general structure

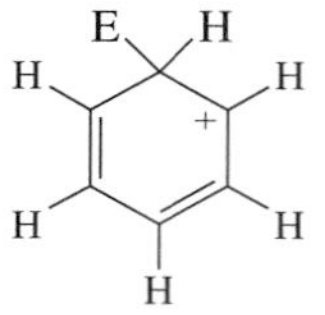

where E is derived from the electrophile that attacks the ring.

Deactivating substituent (Sections 6.9 and 6.11): A group that when present in place of a hydrogen substituent causes a particular reaction to occur more slowly. The term is most often applied to the effect of substituents on the rate of electrophilic aromatic substitution.

Dehydration (Section 4.7): Removal of H and OH from adjacent atoms. The term is most commonly employed in the preparation of alkenes by heating alcohols in the presence of an acid catalyst.

Dehydrogenation (Section 4.1): Removal of the elements of H_2 from adjacent atoms. The term is most commonly encountered in the industrial preparation of ethylene from ethane, propene from propane, 1,3-butadiene from butane, and styrene from ethylbenzene.

Dehydrohalogenation (Section 4.9): Reaction in which an alkyl halide, on being treated with a base such as sodium ethoxide, is converted to an alkene by loss of a proton from one carbon and the halogen from the adjacent carbon.

Delocalization (Section 1.9): Association of an electron with more than one atom. The simplest example is the shared electron pair (covalent) bond. Delocalization is important in conjugated π electron systems, where an electron may be associated with several carbon atoms.

Deoxy sugar (Section 15.9): A carbohydrate in which one of the hydroxyl groups has been replaced by a hydrogen.

Detergents (Section 12.7): Substances that clean by micellar action. Although the term usually refers to a synthetic detergent, soaps are also detergents.

Diastereomers (Section 7.10): Stereoisomers that are not enantiomers—stereoisomers that are not mirror images of each other.

1,3-Diaxial repulsion (Section 2.16): Repulsive forces between axial substituents on the same side of a cyclohexane ring.

Diazonium ion (Section 14.9): Ion of the type $R\overset{+}{-N}\equiv N:$. Aryl diazonium ions are formed by treatment of primary aromatic amines with nitrous acid. They are extremely useful in the preparation of aryl halides, phenols, and aryl cyanides.

Dimer (Section 12.3): Molecule formed by the combination of two identical molecules.

Dipeptide (Section 17.5): A compound in which two α-amino acids are linked by an amide bond between the amino group of one and the carboxyl group of the other:

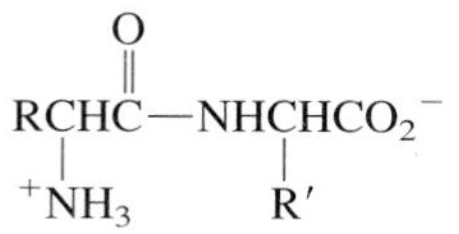

Dipole–dipole attraction (Section 3.4): A force of attraction between oppositely polarized atoms.

Dipole/induced-dipole attraction (Section 3.4): A force of attraction that results when a species with a permanent dipole induces a complementary dipole in a second species.

Dipole moment (Section 1.5): Product of the attractive force between two opposite charges and the distance between them.

Disaccharide (Section 15.11): A carbohydrate that yields two monosaccharide units (which may be the same or different) on hydrolysis.

Disubstituted alkene (Section 4.5): Alkene of the type $R_2C=CH_2$ or $RCH=CHR$. The groups R may be the same or different, they may be any length, and they may be branched or unbranched. The significant point is that two carbons are *directly* bonded to the carbons of the double bond.

Disulfide bridge (Section 17.5): An S—S bond between the sulfur atoms of two cysteine residues in a peptide or protein.

DNA (deoxyribonucleic acid) (Section 18.5): A polynucleotide of 2′-deoxyribose present in the nuclei of cells that serves to store and replicate genetic information. Genes are DNA.

Double bond (Section 1.4): Bond formed by the sharing of four electrons between two atoms.

Double dehydrohalogenation (Section 4.14): Reaction in which a geminal dihalide or vicinal dihalide, on being treated with a very strong base such as sodium amide, is converted to an alkyne by loss of two protons and the two halogen substituents.

Double helix (Section 18.5) The form in which DNA normally occurs in living systems. Two complementary strands of DNA are associated with each other by hydrogen bonds between their base pairs, and each DNA strand adopts a helical shape.

Downfield (Section 19.3): The low-field region of an NMR spectrum. A signal that is downfield with respect to another lies to its left on the spectrum.

Eclipsed conformation (Section 2.5): Conformation in which bonds on adjacent atoms are aligned with one another. For example, the C—H bonds indicated in the structure shown are eclipsed.

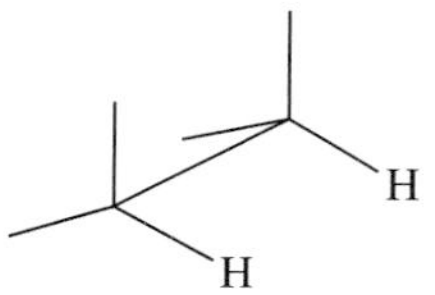

Edman degradation (Section 17.9): Method for determining the N-terminal amino acid of a peptide or protein. It involves treating the material with phenyl isothiocyanate (C_6H_5N=C=S), cleaving with acid, and then identifying the phenylthiohydantoin (PTH derivative) produced.

Electromagnetic radiation (Section 19.1): Various forms of radiation propagated at the speed of light. Electromagnetic radiation includes (among others) visible light; infrared, ultraviolet, and microwave radiation; and radio waves, cosmic rays, and X-rays.

Electronegativity (Section 1.5): A measure of the ability of an atom to attract the electrons in a covalent bond toward itself. Fluorine is the most electronegative element.

Electronic effect (Section 4.5): An effect on structure or reactivity that is attributed to the change in electron distribution that a substituent causes in a molecule.

Electron impact (Section 19.13): Method for producing positive ions in mass spectrometry whereby a molecule is bombarded by high-energy electrons.

Electrophile (Section 3.10): A species (ion or compound) that can act as a Lewis acid, or electron pair acceptor; an "electron seeker." Carbocations are one type of electrophile.

Electrophilic addition (Section 5.2): Mechanism of addition in which the species that first attacks the multiple bond is an electrophile ("electron seeker").

Electrophilic aromatic substitution (Section 6.6): Fundamental reaction type exhibited by aromatic compounds. An electrophilic species (E^+) attacks an aromatic ring and replaces one of the hydrogens.

$$Ar\text{—}H + E\text{—}Y \longrightarrow Ar\text{—}E + H\text{—}Y$$

Electrophoresis (Section 17.3): Method for separating substances on the basis of their tendency to migrate to a positively or negatively charged electrode at a particular pH.

Elementary step (Section 3.6): A step in a reaction mechanism in which each species shown in the equation for this step participates in the same transition state. An elementary step is characterized by a single transition state.

Elements of unsaturation: See *index of hydrogen deficiency.*

Elimination (Section 4.6): Reaction in which a double or triple bond is formed by loss of atoms or groups from adjacent atoms. (See *dehydration, dehydrogenation, dehydrohalogenation,* and *double dehydrohalogenation.*)

Elimination bimolecular (E2) mechanism (Section 4.10): Mechanism for elimination of alkyl halides characterized by a transition state in which the attacking base removes a proton at the same time that the bond to the halide leaving group is broken.

Elimination unimolecular (E1) mechanism (Section 4.11): Mechanism for elimination characterized by the slow formation of a carbocation intermediate followed by rapid loss of a proton from the carbocation to form the alkene.

Enantiomers (Section 7.1): Stereoisomers that are related as an object and its nonsuperimposable mirror image.

Enediyne antibiotics (Section 4.13): A family of tumor-inhibiting substances that is characterized by the presence of a C≡C—C=C—C≡C unit as part of a nine- or ten-membered ring.

Energy of activation (Section 3.6): Minimum energy that a reacting system must possess above its most stable state in order to undergo a chemical or structural change.

Enol (Section 11.16): Compound of the type

$$\underset{\underset{\displaystyle RC=CR_2}{|}}{OH}$$

Enols are in equilibrium with an isomeric aldehyde or ketone, but are normally much less stable than aldehydes and ketones.

Enolate ion (Section 11.17): The conjugate base of an enol. Enolate ions are stabilized by electron delocalization.

$$\underset{\underset{\displaystyle RC=CR_2}{|}}{O^-} \longleftrightarrow \underset{\underset{\displaystyle RC-\overset{..}{C}R_2}{||}}{O}$$

Enzyme (Section 17.13): A protein that catalyzes a chemical reaction in a living system.

Epoxidation (Section 10.11): Conversion of an alkene to an epoxide by treatment with a peroxy acid.

Epoxide (Section 10.9): Compound of the type

$$R_2C\underset{O}{\overset{\diagup\ \ \diagdown}{\text{———}}}CR_2$$

Equatorial bond (Section 2.14): A bond to a carbon in the chair conformation of cyclohexane oriented approximately along the equator of the molecule.

Essential amino acids (Section 17.1): Amino acids that must be present in the diet for normal growth and good health.

Essential oils (Section 16.12): Pleasant-smelling oils of plants, consisting of mixtures of terpenes, esters, alcohols, and other volatile organic substances.

Ester (Sections 2.3 and 13.1): Compound of the type

$$\underset{\displaystyle RCOR'}{\overset{\displaystyle O}{||}}$$

Estrogen (Section 16.9): A female sex hormone.

Ethene (Section 4.1): IUPAC name for CH_2=CH_2. The common name *ethylene,* however, is used far more often, and the IUPAC rules permit its use.

Ether (Sections 2.3 and 10.8): Molecule that contains a C—O—C unit such as ROR', ROAr, or ArOAr. When the two groups bonded to oxygen are the same, the ether is described as a symmetrical ether. When the groups are different, it is called a mixed ether.

Ethylene (Section 4.1): $CH_2\!=\!CH_2$, the simplest alkene and the most important industrial organic chemical.

Ethyl group (Section 2.10): The group $CH_3CH_2\!-\!$.

Exothermic (Section 5.1): Term describing a reaction or process that gives off heat.

Extinction coefficient: See *molar absorptivity*.

E–Z notation for alkenes (Section 4.4): System for specifying double-bond configuration that is an alternative to cis–trans notation. When higher ranked substituents are on the same side of the double bond, the configuration is Z. When higher ranked substituents are on opposite sides, the configuration is E. Rank is determined by the Cahn–Ingold–Prelog system.

Fats and oils (Section 16.2): Triesters of glycerol. Fats are solids at room temperature, oils are liquids.

Fatty acids (Section 16.2): Carboxylic acids obtained by hydrolysis of fats and oils. Fatty acids typically have unbranched chains and contain an even number of carbon atoms in the range of 12–20 carbons. They may include one or more double bonds.

Fingerprint region (Section 19.10): The region 1400–625 cm^{-1} of an infrared spectrum. This region is less characteristic of functional groups than others, but varies so much from one molecule to another that it can be used to determine whether two substances are identical or not.

Fischer esterification (Section 13.5): Acid-catalyzed ester formation between an alcohol and a carboxylic acid:

$$\underset{\text{RCOH}}{\overset{\overset{\textstyle O}{\|}}{}} + R'OH \xrightarrow{H^+} \underset{\text{RCOR}'}{\overset{\overset{\textstyle O}{\|}}{}} + H_2O$$

Fischer projection (Section 7.7): Method for representing stereochemical relationships. The four bonds to a stereogenic carbon are represented by a cross. The horizontal bonds are understood to project toward the viewer and the vertical bonds away from the viewer.

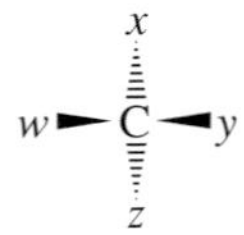

$$w \blacktriangleright \underset{z}{\overset{x}{C}} \blacktriangleleft y \qquad \text{is represented in a Fischer projection as} \qquad w \underset{z}{\overset{x}{-\!\!\!|\!\!\!-}} y$$

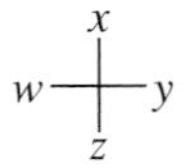

Formal charge (Section 1.6): The charge, either positive or negative, on an atom calculated by subtracting from the number of valence electrons in the neutral atom a number equal to the sum of its unshared electrons plus half the electrons in its covalent bonds.

Fragmentation pattern (Section 19.13): In mass spectrometry, the ions produced by dissociation of the molecular ion.

Free radical (Section 9.1): Neutral species in which one of the electrons in the valence shell of carbon is unpaired. An example is methyl radical, $\cdot CH_3$.

Frequency (Section 19.1): Number of waves per unit time. Although often expressed in hertz (Hz), or cycles per second, the SI unit for frequency is s^{-1}.

Friedel–Crafts acylation (Section 6.6): An electrophilic aromatic substitution in which an aromatic compound reacts with an acyl chloride or a carboxylic acid anhydride in the presence of aluminum chloride. An acyl group becomes bonded to the ring.

$$Ar\!-\!H + \underset{\text{RC}}{\overset{\overset{\textstyle O}{\|}}{}}\!-\!Cl \xrightarrow{AlCl_3} \underset{\text{Ar}}{}\!-\!\underset{\text{CR}}{\overset{\overset{\textstyle O}{\|}}{}}$$

Friedel–Crafts alkylation (Section 6.6): An electrophilic aromatic substitution in which an aromatic compound reacts with an alkyl halide in the presence of aluminum chloride. An alkyl group becomes bonded to the ring.

$$Ar\!-\!H + R\!-\!X \xrightarrow{AlCl_3} Ar\!-\!R$$

Functional group (Section 2.2): An atom or a group of atoms in a molecule responsible for its reactivity under a given set of conditions.

Furanose form (Section 15.5): Five-membered ring arising via cyclic hemiacetal formation between the carbonyl group and a hydroxyl group of a carbohydrate.

Gauche (Section 2.6): Term describing the position relative to each other of two substituents on adjacent atoms when the angle between their bonds is on the order of 60°. Atoms X and Y in the structure shown are gauche to each other.

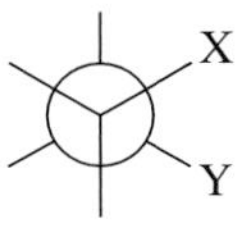

Geminal dihalide (Section 4.14): A dihalide of the form R_2CX_2, in which the two halogen substituents are located on the same carbon.

Geminal diol (Section 11.6): The hydrate $R_2C(OH)_2$ of an aldehyde or a ketone.

Genome (Section 18.4): The aggregate of all the genes that determine what an organism becomes.

Globular protein (Section 17.13): An approximately spherically shaped protein that forms a colloidal dispersion in water. Most enzymes are globular proteins.

Glycogen (Section 15.12): A polysaccharide present in animals that is derived from glucose. Similar in structure to amylopectin.

Glycoside (Section 15.10): A carbohydrate derivative in which the hydroxyl group at the anomeric position has been replaced by some other group. An *O*-glycoside is an ether of a carbohydrate in which the anomeric position bears an alkoxy group.

Grain alcohol (Section 3.1): A common name for ethanol (CH_3CH_2OH).

Grignard reagent (Section 11.11): An organomagnesium compound of the type RMgX formed by the reaction of magnesium with an alkyl or aryl halide.

Halogenation (Sections 6.6 and 9.3): Replacement of a hydrogen by a halogen. The most frequently encountered examples are the free-radical halogenation of alkanes and the halogenation of arenes by electrophilic aromatic substitution.

Halohydrin (Section 5.6): A compound that contains both a halogen atom and a hydroxyl group. The term is most often used for compounds in which the halogen and the hydroxyl group are on adjacent atoms (vicinal halohydrins). The most commonly encountered halohydrins are *chlorohydrins* and *bromohydrins.*

Halonium ion (Section 5.6): A species that incorporates a positively charged halogen. Bridged halonium ions are intermediates in the addition of halogens to the double bond of an alkene.

Haworth formulas (Section 15.5): Planar representations of furanose and pyranose forms of carbohydrates.

α-Helix (Section 17.12): One type of protein secondary structure. It is a right-handed helix characterized by hydrogen bonds between NH and C=O groups. It contains approximately 3.6 amino acids per turn.

Hemiacetal (Section 11.8): Product of nucleophilic addition of one molecule of an alcohol to an aldehyde or a ketone. Hemiacetals are compounds of the type

$$\begin{array}{c} \text{OH} \\ | \\ R_2C{-}OR' \end{array}$$

Hemiketal (Section 11.8): An old name for a hemiacetal derived from a ketone.

Heteroatom (Section 1.7): An atom in an organic molecule that is neither carbon nor hydrogen.

Heterocyclic compound (Section 6.15): Cyclic compound in which one or more of the atoms in the ring are elements other than carbon. Heterocyclic compounds may or may not be aromatic.

Heterolytic cleavage (Section 9.2): Dissociation of a two-electron covalent bond in such a way that both electrons are retained by one of the initially bonded atoms.

Hexose (Section 15.1): A carbohydrate with six carbon atoms.

High-density lipoprotein (HDL) (Section 16.5): A protein that carries cholesterol from the tissues to the liver where it is metabolized. HDL is often called "good cholesterol."

Homolytic cleavage (Section 9.2): Dissociation of a two-electron covalent bond in such a way that one electron is retained by each of the initially bonded atoms.

Hückel's rule (Section 6.14): Completely conjugated planar monocyclic hydrocarbons possess special stability when the number of their π electrons $=4n + 2$, where n is an integer.

Hund's rule (Section 1.1): When two orbitals are of equal energy, they are populated by electrons so that each is half-filled before either one is doubly occupied.

Hybrid orbital (Section 1.12): An atomic orbital represented as a mixture of various contributions of that atom's s, p, d, etc. orbitals.

Hydration (Section 5.5): Addition of the elements of water (H, OH) to a multiple bond.

Hydrocarbon (Section 2.1): A compound that contains only carbon and hydrogen.

Hydrogenation (Section 5.1): Addition of H_2 to a multiple bond.

Hydrogen bonding (Section 3.4): Type of dipole–dipole attractive force in which a positively polarized hydrogen of one molecule is weakly bonded to a negatively polarized atom of an adjacent molecule. Hydrogen bonds typically involve the hydrogen of one —OH group and the oxygen of another.

Hydrolysis (Sections 8.5 and 13.7): Water-induced cleavage of a bond.

Hydronium ion (Section 3.5): The species H_3O^+.

Hydrophilic (Section 12.7): Literally, "water-loving"; a term applied to substances that are soluble in water, usually because of their ability to form hydrogen bonds with water.

Hydrophobic (Section 12.7): Literally, "water-hating"; a term applied to substances that are not soluble in water, but are soluble in nonpolar, hydrocarbon-like media.

Imine (Section 11.9): Compound of the type $R_2C{=}NR'$ formed by the reaction of an aldehyde or a ketone with a primary amine ($R'NH_2$). Imines are sometimes called *Schiff's bases.*

Index of hydrogen deficiency (Section 19.14): A measure of the total double bonds and rings a molecule contains. It is determined by comparing the molecular formula C_nH_x of the compound to that of an alkane that has the same number of carbons according to the equation:

$$\text{Index of hydrogen deficiency} = \tfrac{1}{2}(C_nH_{2n+2} - C_nH_x)$$

Induced-dipole/induced-dipole attraction (Section 2.19): Force of attraction resulting from a mutual and complementary polarization of one molecule by another.

Inductive effect (Section 3.9): An electronic effect transmitted by successive polarization of the σ bonds within a molecule or an ion.

Infrared (IR) spectroscopy (Section 19.10): Analytical technique based on energy absorbed by a molecule as it vibrates by stretching and bending bonds. Infrared spectroscopy is useful for analyzing the functional groups in a molecule.

Initiation step (Section 9.4): A process that causes a reaction, usually a free-radical reaction, to begin but that by itself is not the principal source of products. The initiation step in the halogenation of an alkane is the dissociation of a halogen molecule to two halogen atoms.

Integrated area (Section 19.5): The relative area of a signal in an NMR spectrum. Areas are proportional to the number of equivalent protons responsible for the peak.

Intermediate (Section 3.8): Transient species formed during a chemical reaction. Typically, an intermediate is not stable under the conditions of its formation and proceeds further to form the product. Unlike a transition state, which corresponds to a maximum along a potential energy surface, an intermediate lies at a potential energy minimum.

Intermolecular forces (Section 2.19): Forces, either attractive or repulsive, between two atoms or groups in *separate* molecules.

Inversion of configuration (Section 8.3): Reversal of the three-dimensional arrangement of the four bonds to sp^3-hybridized carbon. The representation shown illustrates inversion of configuration in a nucleophilic substitution where LG is the leaving group and Nu is the nucleophile.

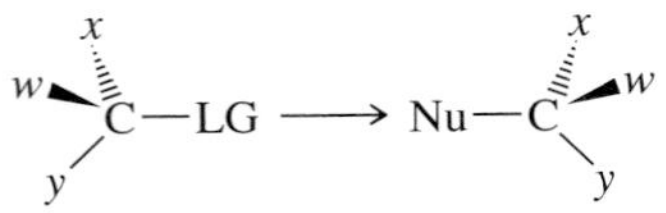

Ionic bond (Section 1.2): Chemical bond between oppositely charged particles that results from the electrostatic attraction between them.

Isobutane (Section 2.6): The common name for 2-methylpropane, $(CH_3)_3CH$.

Isobutyl group (Section 2.10): The group $(CH_3)_2CHCH_2—$.

Isoelectric point (Section 17.3): pH at which the concentration of the zwitterionic form of an amino acid is a maximum. At a pH below the isoelectric point the dominant species is a cation. At higher pH, an anion predominates. At the isoelectric point the amino acid has no net charge.

Isolated diene (Section 4.1): Diene of the type

$$C=C—(C)_x—C=C$$

in which the two double bonds are separated by one or more sp^3-hybridized carbons. Isolated dienes are slightly less stable than isomeric conjugated dienes.

Isomers (Section 1.8): Different compounds that have the same molecular formula. Isomers may be either constitutional isomers or stereoisomers.

Isoprene unit (Section 16.13): The characteristic five-carbon structural unit found in terpenes:

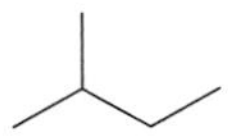

Isopropyl group (Section 2.10): The group $(CH_3)_2CH—$.

IUPAC nomenclature (Section 2.8): The most widely used method of naming organic compounds. It uses a set of rules proposed and periodically revised by the International Union of Pure and Applied Chemistry.

Kekulé structure (Section 6.1): Structural formula for an aromatic compound that satisfies the customary rules of bonding and is usually characterized by a pattern of alternating single and double bonds. There are two Kekulé formulations for benzene:

A single Kekulé structure does not completely describe the actual bonding in the molecule.

Ketal (Section 11.8): An old name for an acetal derived from a ketone.

Keto–enol tautomerism (Section 11.16): Process by which an aldehyde or a ketone and its enol equilibrate:

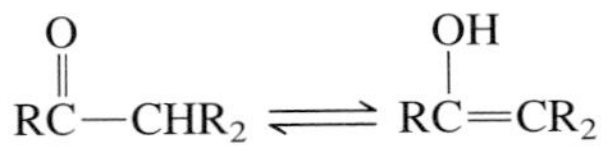

Ketone (Sections 2.3 and 11.1): A member of the family of compounds in which both atoms attached to a carbonyl group (C=O) are carbon, as in

Ketose (Section 15.8): A carbohydrate that contains a ketone carbonyl group in its open-chain form.

Lactose (Section 15.11): Milk sugar; a disaccharide formed by a β-glycosidic linkage between C-4 of glucose and C-1 of galactose.

Leaving group (Sections 4.10 and 8.1): The group, normally a halide ion, that is lost from carbon in a nucleophilic substitution or elimination.

Le Châtelier's principle (Section 5.5): A reaction at equilibrium responds to any stress imposed on it by shifting the equilibrium in the direction that minimizes the stress.

Lewis acid: See *acid*.

Lewis base: See *base*.

Lewis structure (Section 1.3): A chemical formula in which electrons are represented by dots. Two dots (or a line) between two atoms represent a covalent bond in a Lewis structure. Unshared electrons are explicitly shown, and stable Lewis structures are those in which the octet rule is satisfied.

Lindlar catalyst (Section 5.11): A catalyst for the hydrogenation of alkynes to *cis*-alkenes. It is composed of palladium, which has been "poisoned" with lead(II) acetate and quinoline, supported on calcium carbonate.

Lipid bilayer (Section 16.3): Arrangement of two layers of phospholipids that constitutes cell membranes. The polar termini are located at the inner and outer membrane–water interfaces, and the hydrophobic hydrocarbon tails cluster on the inside.

Lipids (Section 16.1): Biologically important natural products characterized by high solubility in nonpolar organic solvents.

Lipophilic (Section 12.7): Literally, "fat-loving;" synonymous in practice with *hydrophobic*.

Locant (Section 2.9): In IUPAC nomenclature, a prefix that designates the atom that is associated with a particular structural unit. The locant is most often a number, and the structural unit is usually an attached substituent as in 2-chlorobutane.

Low-density lipopropein (LDL) (Section 16.5): A protein that carries cholesterol from the liver through the blood to the tissues. Elevated LDL levels are a risk factor for heart disease; LDL is often called "bad cholesterol."

Magnetic resonance imaging (MRI) (Section 19.9): A diagnostic method in medicine in which tissues are examined by NMR.

Maltose (Section 15.11): A disaccharide obtained from starch in which two glucose units are joined by an $\alpha(1,4)$-glycosidic link.

Markovnikov's rule (Section 5.3): An unsymmetrical reagent adds to an unsymmetrical double bond in the direction that places the positive part of the reagent on the carbon of the double bond that has the greater number of hydrogens.

Mass spectrometry (Section 19.13): Analytical method in which a molecule is ionized and the various ions are examined on the basis of their mass-to-charge ratio.

Mechanism (Section 3.6): The sequence of steps that describes how a chemical reaction occurs; a description of the intermediates and transition states that are involved during the transformation of reactants to products.

Mercaptan (Section 10.6): An old name for the class of compounds now known as *thiols*.

Merrifield method: See *solid-phase peptide synthesis*.

Meso stereoisomer (Section 7.11): An achiral molecule that has stereogenic centers. The most common kind of meso compound is a molecule with two stereogenic centers and a plane of symmetry.

Messenger RNA (mRNA) (Section 18.6): A polynucleotide of ribose that "reads" the sequence of bases in DNA and interacts with tRNAs in the ribosomes to promote protein biosynthesis.

Meta (Section 6.3): Term describing a 1,3 relationship between substituents on a benzene ring.

Meta director (Section 6.9): A group that when present on a benzene ring directs an incoming electrophile to a position meta to itself.

Methanogen (Section 2.4): An organism that produces methane.

Methine group (Section 2.6): The group CH.

Methylene group (Section 2.6): The group $-CH_2-$.

Methyl group (Section 1.13): The group CH_3-.

Mevalonic acid (Section 16.13): An intermediate in the biosynthesis of steroids from acetyl coenzyme A.

Micelle (Section 12.7): A spherical aggregate of species such as carboxylate salts of fatty acids that contain a hydrophobic end and a hydrophilic end. Micelles containing 50–100 carboxylate salts of fatty acids are soaps.

Molar absorptivity (Section 19.11): A measure of the intensity of a peak, usually in UV-Vis spectroscopy.

Molecular formula (Section 1.7): Chemical formula in which subscripts are used to indicate the number of atoms of each element present in one molecule. In organic compounds, carbon is cited first, hydrogen second, and the remaining elements in alphabetical order.

Molecular ion (Section 19.13): In mass spectrometry, the species formed by loss of an electron from a molecule.

Monomer (Section 9.8): The simplest stable molecule from which a particular polymer may be prepared.

Monosaccharide (Section 15.1): A carbohydrate that cannot be hydrolyzed further to yield a simpler carbohydrate.

Monosubstituted alkene (Section 4.5): An alkene of the type $RCH=CH_2$, in which only one carbon is *directly* bonded to the carbons of the double bond.

Multiplicity (Section 19.6): The number of peaks into which a signal is split in nuclear magnetic resonance spectroscopy. Signals are described as singlets, doublets, triplets, and so on, according to the number of peaks into which they are split.

Mutarotation (Section 15.7): The change in optical rotation that occurs when a single form of a carbohydrate is allowed to equilibrate to a mixture of isomeric hemiacetals.

Neurotransmitter (Section 14.3): Substance, usually a naturally occurring amine, that mediates the transmission of nerve impulses.

Newman projection (Section 2.5): Method for depicting conformations in which one sights down a carbon–carbon bond and represents the front carbon by a point and the back carbon by a circle.

Nitration (Section 6.6): Replacement of a hydrogen by an $-NO_2$ group. The term is usually used in connection with electrophilic aromatic substitution.

$$Ar-H \xrightarrow[H_2SO_4]{HNO_3} Ar-NO_2$$

Nitrile (Section 12.10): A compound of the type $RC\equiv N$. R may be alkyl or aryl. Also known as alkyl or aryl cyanides.

Nitrosamine See *N-nitroso amine*.

N-Nitroso amine (Section 14.9): A compound of the type $R_2N-N=O$. R may be alkyl or aryl groups, which may be the same or different. *N*-Nitroso amines are formed by nitrosation of secondary amines.

Nitrosation (Section 14.9): The reaction of a substance, usually an amine, with nitrous acid. Primary amines yield diazonium ions; secondary amines yield *N*-nitroso amines. Tertiary aromatic amines undergo nitrosation of their aromatic ring.

Noble gases (Section 1.1): The elements in group VIIIA of the periodic table (helium, neon, argon, krypton, xenon, radon). Also known as the rare gases, they are, with few exceptions, chemically inert.

N terminus (Section 17.5): The amino acid at the end of a peptide or protein chain that has its α-amino group intact; that is, the α-amino group is not part of a peptide bond.

Nuclear magnetic resonance (NMR) spectroscopy (Section 19.2): A method for structure determination based on the effect of molecular environment on the energy required to promote a given nucleus from a lower energy spin state to a higher energy state.

Nucleic acid (Section 18.4): A polynucleotide present in the nuclei of cells.

Nucleophile (Section 3.10): An atom that has an unshared electron pair which can be used to form a bond to carbon. Nucleophiles are Lewis bases.

Nucleophilic acyl substitution (Section 13.3): Nucleophilic substitution at the carbon atom of an acyl group.

Nucleophilic addition (Section 11.6): The characteristic reaction of an aldehyde or a ketone. An atom possessing an unshared electron pair bonds to the carbon of the C=O group, and some other species (normally hydrogen) bonds to the oxygen.

$$\underset{\text{RCR}'}{\overset{\text{O}}{\|}} + \text{H}-\text{Y}: \longrightarrow \underset{\underset{\text{R}'}{|}}{\overset{\text{OH}}{\underset{|}{\text{RC}}}}-\text{Y}:$$

Nucleophilic substitution (Chapter 8): Reaction in which a nucleophile replaces a leaving group, usually a halide ion, from sp^3-hybridized carbon. Nucleophilic substitution may proceed by either an S_N1 or an S_N2 mechanism.

Nucleoside (Section 18.2): The combination of a purine or pyrimidine base and a carbohydrate, usually ribose or 2-deoxyribose.

Nucleotide (Section 18.3): The phosphate ester of a nucleoside.

Octane rating (Section 2.20): The capacity of a sample of gasoline to resist "knocking," expressed as a number equal to the percentage of 2,2,4-trimethylpentane ("isooctane") in an isooctane–heptane mixture that has the same knocking characteristics.

Octet rule (Section 1.3): When forming compounds, atoms gain, lose, or share electrons so that the number of their valence electrons is the same as that of the nearest noble gas. For the elements carbon, nitrogen, oxygen, and the halogens, this number is 8.

Oligosaccharide (Section 15.1): A carbohydrate that gives three to ten monosaccharides on hydrolysis.

Optical activity (Section 7.4): Ability of a substance to rotate the plane of polarized light. To be optically active, a substance must be chiral, and one enantiomer must be present in excess of the other.

Orbital (Section 1.1): The region in space where the probability of finding an electron is high.

σ Orbital (Section 1.12): A bonding orbital characterized by rotational symmetry.

Organometallic compound (Section 11.10): A compound that contains a carbon-to-metal bond.

Ortho (Section 6.3): Term describing a 1,2 relationship between substituents on a benzene ring.

Ortho, para director (Section 6.9): A group that when present on a benzene ring directs an incoming electrophile to the positions ortho and para to itself.

Oxidation (Section 10.2): A decrease in the number of electrons associated with an atom. In organic chemistry, oxidation of carbon occurs when a bond between carbon and an atom that is less electronegative than carbon is replaced by a bond to an atom that is more electronegative than carbon.

Oxonium ion (Section 3.5): Specific name for the species H_3O^+ (also called *hydronium ion*). General name for species such as alkyloxonium ions ROH_2^+ analogous to H_3O^+.

Para (Section 6.3): Term describing a 1,4 relationship between substituents on a benzene ring.

Pauli exclusion principle (Section 1.1): No two electrons can have the same set of four quantum numbers. An equivalent expression is that only two electrons can occupy the same orbital, and then only when they have opposite spins.

PCC (Section 10.5): Abbreviation for pyridinium chlorochromate $C_5H_5NH^+$ $ClCrO_3^-$. When used in an anhydrous medium, PCC oxidizes primary alcohols to aldehydes and secondary alcohols to ketones.

PDC (Section 10.5): Abbreviation for pyridinium dichromate $(C_5H_5NH)_2^{2+}$ $Cr_2O_7^{2-}$. Used in same manner and for same purposes as PCC (see preceding entry).

***n*-Pentane** (Section 2.7): The common name for pentane, $CH_3CH_2CH_2CH_2CH_3$.

Pentose (Section 15.4): A carbohydrate with five carbon atoms.

Peptide (Section 17.5): Structurally, a molecule composed of two or more α-amino acids joined by peptide bonds.

Peptide bond (Section 17.5): An amide bond between the carboxyl group of one α-amino acid and the amino group of another.

$$-\underset{\underset{\text{R}'}{|}}{\text{NHCHC}}\overset{\text{O}}{\overset{\|}{}}-\underset{\underset{\text{R}}{|}}{\text{NHCHC}}\overset{\text{O}}{\overset{\|}{}}-$$

(The bond highlighted in red is the peptide bond.)

Period (Section 1.1): A horizontal row of the periodic table.

Peroxide (Section 9.7): A compound of the type ROOR.

Phenols (Section 10.13): Family of compounds characterized by a hydroxyl substituent on an aromatic ring as in ArOH. *Phenol* is also the name of the parent compound, C_6H_5OH.

Phenyl group (Section 6.3): The group

It is often abbreviated C_6H_5—.

Phospholipid (Section 16.3): A diacylglycerol bearing a choline-phosphate "head group." Also known as phosphatidylcholine.

Photon (Section 19.1): Term for an individual "bundle" of energy, or particle, of electromagnetic radiation.

pKₐ (Section 3.5): A measure of acid strength defined as $-\log K_a$. The stronger the acid, the smaller the value of pK_a.

Planck's constant (Section 19.1): Constant of proportionality (h) in the equation $E = h\nu$, which relates the energy (E) to the frequency (ν) of electromagnetic radiation.

Plane of symmetry (Section 7.3): A plane that bisects an object, such as a molecule, into two mirror-image halves; also called a mirror plane. When a line is drawn from any element in the object perpendicular to such a plane and extended an equal distance in the opposite direction, a duplicate of the element is encountered.

Pleated β sheet (Section 17.12): Type of protein secondary structure characterized by hydrogen bonds between NH and C=O groups of adjacent parallel peptide chains. The individual chains are in an extended zigzag conformation.

Polar covalent bond (Section 1.5): A shared electron pair bond in which the electrons are drawn more closely to one of the bonded atoms than the other.

Polarimeter (Section 7.4): An instrument used to measure optical activity.

Polarized light (Section 7.4): Light in which the electric field vectors vibrate in a single plane. Polarized light is used in measuring optical activity.

Polyamide (Section 13.12): A polymer in which individual structural units are joined by amide bonds. Nylon is a synthetic polyamide; proteins are naturally occurring polyamides.

Polycyclic aromatic hydrocarbon (Section 6.4): An aromatic hydrocarbon characterized by the presence of two or more benzene-like rings.

Polycyclic hydrocarbon (Section 2.18): A hydrocarbon in which two carbons are common to two or more rings.

Polyester (Section 13.12): A polymer in which individual structural units are joined by ester bonds.

Polyether (Section 10.9): A molecule that contains many ether linkages. Polyethers occur naturally in a number of antibiotic substances.

Polyethylene (Section 9.8): A polymer of ethylene.

Polymer (Section 9.8): Large molecule formed by the repetitive combination of many smaller molecules (monomers).

Polymerization (Section 9.8): Process by which a polymer is prepared. The principal processes include *free-radical, cationic,* coordination, and *condensation polymerization.*

Polypeptide (Section 17.5): A polymer made up of "many" (more than eight to ten) amino acid residues.

Polypropylene (Section 9.8): A polymer of propene.

Polysaccharide (Section 15.12): A carbohydrate that yields "many" monosaccharide units on hydrolysis.

Potential energy (Section 3.6): The energy a system has exclusive of its kinetic energy.

Potential energy diagram (Section 3.6): Plot of potential energy versus some arbitrary measure of the degree to which a reaction has proceeded (the reaction coordinate). The point of maximum potential energy is the transition state.

Primary alkyl group (Section 2.10): Structural unit of the type RCH_2—, in which the point of attachment is to a primary carbon.

Primary amine (Section 14.1): An amine with a single alkyl or aryl substituent and two hydrogens: an amine of the type RNH_2 (primary alkylamine) or $ArNH_2$ (primary arylamine).

Primary carbon (Section 2.10): A carbon that is directly attached to only one other carbon.

Primary structure (Section 17.6): The sequence of amino acids in a peptide or protein.

Principal quantum number (Section 1.1): The quantum number (n) of an electron that describes its energy level. An electron with $n = 1$ must be an s electron; one with $n = 2$ has s and p states available.

Propagation steps (Section 9.4): Elementary steps that repeat over and over again in a chain reaction. Almost all of the products in a chain reaction arise from the propagation steps.

Protecting group (Section 17.11): A temporary alteration in the nature of a functional group so that it is rendered inert under the conditions in which reaction occurs somewhere else in the molecule. To be synthetically useful, a protecting group must be stable under a prescribed set of reaction conditions, yet be easily introduced and removed.

Protein (Chapter 17): A naturally occurring polymer that typically contains 100–300 amino acid residues.

Protein Data Bank (Section 17.15): A central repository in which crystallographic coordinates for biological molecules, especially proteins, are stored. The data are accessible via the World-Wide Web and can be transformed into three-dimensional images with appropriate molecular-modeling software.

Protic solvent (Section 8.5): A solvent that has easily exchangeable protons, especially protons bonded to oxygen as in hydroxyl groups, as in water and alcohols.

Purine (Section 18.1): The heterocyclic aromatic compound.

Pyranose form (Section 15.6): Six-membered ring arising via cyclic hemiacetal formation between the carbonyl group and a hydroxyl group of a carbohydrate.

Pyrimidine (Section 18.1): The heterocyclic aromatic compound.

Quantum (Section 19.1): The energy associated with a photon.

Quaternary ammonium salt (Section 14.1): Salt of the type $R_4N^+ X^-$. The positively charged ion contains a nitrogen with a total of four organic substituents (any combination of alkyl and aryl groups).

Quaternary carbon (Section 2.10): A carbon that is directly attached to four other carbons.

Quaternary structure (Section 17.15): Description of the way in which two or more protein chains, not connected by chemical bonds, are organized in a larger protein.

Quinone (Section 10.17): The product of oxidation of an ortho or para dihydroxybenzene derivative. Examples of quinones include

and

R (Section 2.2): Symbol for an alkyl group.

Racemic mixture (Section 7.4): Mixture containing equal quantities of enantiomers.

Rate-determining step (Section 4.10): Slowest step of a multistep reaction mechanism. The overall rate of a reaction can be no faster than its slowest step.

Reducing sugar (Section 15.13): A carbohydrate that can be oxidized with substances such as Benedict's reagent. In general, a carbohydrate with a free hydroxyl group at the anomeric position.

Reduction (Section 10.2): Gain in the number of electrons associated with an atom. In organic chemistry, reduction of carbon occurs when a bond between carbon and an atom that is more electronegative than carbon is replaced by a bond to an atom that is less electronegative than carbon.

Refining (Section 2.20): Conversion of crude oil to useful materials, especially gasoline.

Reforming (Section 2.20): Step in oil refining in which the proportion of aromatic and branched-chain hydrocarbons in petroleum is increased so as to improve the octane rating of gasoline.

Regioselective (Section 4.7): Term describing a reaction that can produce two (or more) constitutional isomers but gives one of them in greater amounts than the other. A reaction that is 100% regioselective is termed regiospecific.

Relative configuration (Section 7.5): Stereochemical configuration on a comparative, rather than an absolute, basis. Terms such as D, L, α, and β describe relative configuration.

Resolution (Section 7.13): Separation of a racemic mixture into its enantiomers.

Resonance (Section 1.9): Method by which electron delocalization may be shown using Lewis structures. The true electron distribution in a molecule is regarded as a hybrid of the various Lewis structures that can be written for a molecule.

Resonance energy (Section 6.1): Extent to which a substance is stabilized by electron delocalization. It is the difference in energy between the substance and a hypothetical model in which the electrons are localized.

Retention of configuration (Section 8.7): Stereochemical pathway observed when a new bond is made that has the same spatial orientation as the bond that was broken.

Retrosynthetic analysis (Section 11.13): Technique for synthetic planning based on reasoning backward from the target molecule to appropriate starting materials. An arrow of the type ⟹ designates a retrosynthetic step.

Ring inversion (Section 2.15): Process by which a chair conformation of cyclohexane is converted to a mirror-image chair. All of the equatorial substituents become axial, and vice versa. Also called ring flipping, or chair – chair interconversion.

RNA (ribonucleic acid) (Section 8.6): A polynucleotide of ribose.

Sandmeyer reaction (Section 14.10): Reaction of an aryl diazonium ion with CuCl, CuBr, or CuCN to give, respectively, an aryl chloride, aryl bromide, or aryl cyanide (nitrile).

Saponification (Section 13.7): Hydrolysis of esters in basic solution. The products are an alcohol and a carboxylate salt. The term means "soap making" and derives from the process whereby animal fats were converted to soap by heating with wood ashes.

Saturated hydrocarbon (Section 5.1): A hydrocarbon in which no multiple bonds occur.

Sawhorse formula (Section 2.5): A representation of the three-dimensional arrangement of bonds in a molecule by a drawing of the type shown.

Schiff's base (Section 11.9): Another name for an imine; a compound of the type $R_2C{=}NR'$.

Secondary alkyl group (Section 2.10): Structural unit of the type $R_2CH—$, in which the point of attachment is to a secondary carbon.

Secondary amine (Section 14.1): An amine with any combination of two alkyl or aryl substituents and one hydrogen on nitrogen; an amine of the type

RNHR′   or   RNHAr   or   ArNHAr′

Secondary carbon (Section 2.10): A carbon that is directly attached to two other carbons.

Secondary structure (Section 17.12): The conformation with respect to nearest neighbor amino acids in a peptide or protein. The α helix and the β pleated sheet are examples of protein secondary structures.

Sequence rule (Section 7.6): Foundation of the Cahn–Ingold–Prelog system. It is a procedure for ranking substituents on the basis of atomic number.

Shielding (Section 19.3): Effect of a molecule's electrons that decreases the strength of an external magnetic field felt by a proton or another nucleus.

Soaps (Section 12.7): Cleansing substances obtained by the hydrolysis of fats in aqueous base. Soaps are sodium or potassium salts of unbranched carboxylic acids having 12–18 carbon atoms.

Solid-phase peptide synthesis (Section 17.11): Method for peptide synthesis in which the C-terminal amino acid is covalently attached to an inert solid support and successive amino acids are attached via peptide bond formation. At the completion of the synthesis the polypeptide is removed from the support.

Solvolysis reaction (Section 8.5): Nucleophilic substitution in a medium in which the only nucleophiles present are the solvent and its conjugate base.

Space-filling model (Section 1.10): A type of molecular model that attempts to represent the volume occupied by the atoms.

Specific rotation (Section 7.4): Optical activity of a substance per unit concentration per unit path length:

$$[\alpha] = \frac{100\alpha}{cl}$$

where α is the observed rotation in degrees, c is the concentration in g/100 mL, and l is the path length in decimeters.

Spectrometer (Section 19.1): Device designed to measure absorption of electromagnetic radiation by a sample.

Spectrum (Section 19.1): Output, usually in chart form, of a spectrometer. Analysis of a spectrum provides information about molecular structure.

sp **Hybridization** (Section 1.15): Hybridization state adopted by carbon when it bonds to two other atoms as, for example, in alkynes. The *s* orbital and one of the 2*p* orbitals mix to form two equivalent *sp*-hybridized orbitals. A linear geometry is characteristic of *sp* hybridization.

sp² **Hybridization** (Section 1.14): A model to describe the bonding of a carbon attached to three other atoms or groups. The carbon 2*s* orbital and the two 2*p* orbitals are combined to give a set of three equivalent *sp²* orbitals having 33.3% *s* character and 66.7% *p* character. One *p* orbital remains unhybridized. A trigonal planar geometry is characteristic of *sp²* hybridization.

sp³ **Hybridization** (Section 1.13): A model to describe the bonding of a carbon attached to four other atoms or groups. The carbon 2*s* orbital and the three 2*p* orbitals are combined to give a set of four equivalent orbitals having 25% *s* character and 75% *p* character. These orbitals are directed toward the corners of a tetrahedron.

Spin quantum number (Section 1.1): One of the four quantum numbers that describe an electron. An electron may have either of two different spin quantum numbers, $+\frac{1}{2}$ or $-\frac{1}{2}$.

Spin–spin coupling (Section 19.6): The communication of nuclear spin information between two nuclei.

Spin–spin splitting (Section 19.6): The splitting of NMR signals caused by the coupling of nuclear spins. Only nonequivalent nuclei (such as protons with different chemical shifts) can split one another's signals.

Squalene (Section 16.12): A naturally occurring triterpene from which steroids are biosynthesized.

Staggered conformation (Section 2.5): Conformation of the type shown, in which the bonds on adjacent carbons are as far away from one another as possible.

Stereochemistry (Chapter 7): Chemistry in three dimensions; the relationship of physical and chemical properties to the spatial arrangement of the atoms in a molecule.

Stereogenic center (Section 7.2): An atom that has four nonequivalent atoms or groups attached to it. At various times stereogenic centers have been called *asymmetrical centers* or *chiral centers.*

Stereoisomers (Section 2.17): Isomers that have the same constitution but that differ in respect to the arrangement of their atoms in space. Stereoisomers may be either *enantiomers* or *diastereomers.*

Stereoselective reaction (Section 4.7): Reaction in which a single starting material has the capacity to form two or more stereoisomeric products but forms one of them in greater amounts than any of its stereoisomers. Terms such as "addition to the less hindered side" describe stereoselectivity.

Steric hindrance (Section 8.4): An effect on structure or reactivity that depends on van der Waals repulsive forces.

Steric strain (Section 2.6): Destabilization of a molecule as a result of van der Waals repulsion, distorted bond distances, bond angles, or torsion angles.

Steroid (Section 16.5): Type of lipid present in both plants and animals characterized by a nucleus of four fused rings (three are six membered, one is five membered). Cholesterol is the most abundant steroid in animals.

Strecker synthesis (Section 17.4): Method for preparing amino acids in which the first step is reaction of an aldehyde with ammonia and hydrogen cyanide to give an amino nitrile, which is then hydrolyzed.

$$\underset{\text{RCH}}{\overset{\text{O}}{\|}} \xrightarrow[\text{HCN}]{\text{NH}_3} \underset{\underset{\text{NH}_2}{|}}{\text{RCHC}\equiv\text{N}} \xrightarrow{\text{hydrolysis}} \underset{\underset{^+\text{NH}_3}{|}}{\text{RCHCO}_2^-}$$

Stretching vibration (Section 19.10): A regular, repetitive motion of two atoms or groups along the bond that connects them.

Structural isomer (Section 1.8): Synonymous with *constitutional isomer.*

Substitution nucleophilic bimolecular (S$_N$2) mechanism (Section 8.2): Concerted mechanism for nucleophilic substitution in which the nucleophile attacks carbon from the side opposite the bond to the leaving group and assists the departure of the leaving group.

Substitution nucleophilic unimolecular (S$_N$1) mechanism (Section 8.5): Mechanism for nucleophilic substitution characterized by a two-step process. The first step is rate-determining and is the ionization of an alkyl halide to a carbocation and a halide ion.

Substitution reaction (Section 3.8): Chemical reaction in which an atom or a group of a molecule is replaced by a different atom or group.

Sucrose (Section 15.11): A disaccharide of glucose and fructose in which the two monosaccharides are joined at their anomeric positions.

Sulfonation (Section 6.6): Replacement of a hydrogen by an —SO$_3$H group. The term is usually used in connection with electrophilic aromatic substitution.

$$\text{Ar}-\text{H} \xrightarrow[\text{H}_2\text{SO}_4]{\text{SO}_3} \text{Ar}-\text{SO}_3\text{H}$$

Syn addition (Section 5.1): Addition reaction in which the two portions of the reagent that add to a multiple bond add from the same side.

Systematic nomenclature (Section 2.8): Names for chemical compounds that are developed on the basis of a prescribed set of rules. Usually the IUPAC system is meant when the term "systematic nomenclature" is used.

Tautomerism (Section 11.16): Process by which two isomers are interconverted by movement of an atom or a group. Enolization is a form of tautomerism.

$$\underset{\text{RC}-\text{CHR}_2}{\overset{\text{O}}{\|}} \rightleftharpoons \underset{\text{RC}=\text{CR}_2}{\overset{\text{OH}}{|}}$$

Terminal alkyne (Section 4.12): Alkyne of the type RC≡CH, in which the triple bond appears at the end of the chain.

Termination steps (Section 9.4): Reactions that halt a chain reaction. In a free-radical chain reaction, termination steps consume free radicals without generating new radicals to continue the chain.

Terpenes (Section 16.12): Compounds that can be analyzed as clusters of isoprene units. Terpenes with 10 carbons are classified as monoterpenes, those with 15 are sesquiterpenes, those with 20 are diterpenes, and those with 30 are triterpenes.

Tertiary alkyl group (Section 2.10): Structural unit of the type R$_3$C—, in which the point of attachment is to a tertiary carbon.

Tertiary amine (Section 14.1): Amine of the type R$_3$N with any combination of three alkyl or aryl substituents on nitrogen.

Tertiary carbon (Section 2.10): A carbon that is directly attached to three other carbons.

Tertiary structure (Section 17.13): A description of how a protein chain is folded.

Tesla (Section 19.2): SI unit for magnetic field strength.

Tetrahedral intermediate (Section 13.3): The key intermediate in nucleophilic acyl substitution. Formed by nucleophilic addition to the carbonyl group of a carboxylic acid derivative.

Tetramethylsilane (TMS) (Section 19.3): The molecule (CH$_3$)$_4$Si, used as a standard to calibrate proton and carbon-13 NMR spectra.

Tetrasubstituted alkene (Section 4.5): Alkene of the type R$_2$C=CR$_2$, in which four carbons are *directly* bonded to the carbons of the double bond. (The R groups may be the same or different.)

Tetrose (Section 15.3): A carbohydrate with four carbon atoms.

Thiol (Section 10.6): Compound of the type RSH or ArSH.

Torsional strain (Section 2.5): Decreased stability of a molecule that results from the eclipsing of bonds.

trans- (Section 2.17): Stereochemical prefix indicating that two substituents are on opposite sides of a ring or a double bond. (Contrast with the prefix *cis-*.)

Transcription (Section 18.6): Construction of a strand of mRNA complementary to a DNA template.

Transfer RNA (tRNA) (Section 18.6): A polynucleotide of ribose that is bound at one end to a unique amino acid. This amino acid is incorporated into a growing peptide chain.

Transition state (Section 3.6): The point of maximum energy in an elementary step of a reaction mechanism.

Translation (Section 18.6): The "reading" of mRNA by various tRNAs, each one of which is unique for a particular amino acid.

Triacylglycerol (Section 16.2): A derivative of glycerol (1,2,3-propanetriol) in which the three oxygens bear acyl groups derived from fatty acids.

Tripeptide (Section 17.5): A compound in which three α-amino acids are linked by peptide bonds.

Triple bond (Section 1.4): Bond formed by the sharing of six electrons between two atoms.

Trisubstituted alkene (Section 4.5): Alkene of the type R$_2$C=CHR, in which three carbons are *directly* bonded to the carbons of the double bond. (The R groups may be the same or different.)

Trivial nomenclature (Section 2.8): Term synonymous with *common nomenclature.*

Trypsin (Section 17.8): A digestive enzyme that catalyzes the hydrolysis of proteins. Trypsin selectively catalyzes the cleavage of the peptide bond between the carboxyl group of lysine or arginine and some other amino acid.

Ultraviolet-visible (UV-Vis) spectroscopy (Section 19.11): An analytical method based on transitions between electronic energy states in molecules. Useful in studying conjugated systems such as polyenes.

Unimolecular (Section 4.11): Describing a step in a reaction mechanism in which only one particle undergoes a chemical change at the transition state.

Unsaturated hydrocarbon (Section 5.1): A hydrocarbon that can undergo addition reactions; that is, one that contains multiple bonds.

Upfield (Section 19.3): The high-field region of an NMR spectrum. A signal that is upfield with respect to another lies to its right on the spectrum.

Valence bond theory (Section 1.12): Theory of chemical bonding based on overlap of half-filled atomic orbitals between two atoms. Orbital hybridization is an important element of valence bond theory.

Valence electrons (Section 1.1): The outermost electrons of an atom. For second-row elements these are the $2s$ and $2p$ electrons.

Valence shell electron-pair repulsion (VSEPR) model (Section 1.10): Method for predicting the shape of a molecule based on the notion that electron pairs surrounding a central atom repel each other. Four electron pairs will arrange themselves in a tetrahedral geometry, three will assume a trigonal planar geometry, and two electron pairs will adopt a linear arrangement.

Van der Waals forces (Section 2.19): Intermolecular forces that do not involve ions (dipole–dipole, dipole/induced-dipole, and induced-dipole/induced-dipole forces).

Van der Waals strain (Section 2.6): Destabilization that results when two atoms or groups approach each other too closely. Also known as van der Waals repulsion.

Vicinal (Section 4.14): Describing two substituents that are located on adjacent atoms.

Vicinal coupling (Section 19.6): Coupling of the nuclear spins of atoms X and Y when they are substituents on adjacent atoms as in X—A—B—Y. Vicinal coupling is the most common cause of spin–spin splitting in ^{1}H NMR spectroscopy.

Walden inversion (Section 8.3): Originally, a reaction sequence developed by Paul Walden whereby a chiral starting material was transformed to its enantiomer by a series of stereospecific reactions. Current usage is more general and refers to the inversion of configuration that attends any bimolecular nucleophilic substitution.

Wavelength (Section 19.1): Distance between two successive maxima (peaks) or two successive minima (troughs) of a wave.

Wave numbers (Section 19.10): Conventional units in infrared spectroscopy that are proportional to frequency. Wave numbers are in reciprocal centimeters (cm^{-1}).

Wax (Section 16.4): A mixture of water-repellent substances that form a protective coating on the leaves of plants, the fur of animals, and the feathers of birds, among other things. A principal component of a wax is often an ester in which both the acyl portion and the alkyl portion are characterized by long carbon chains.

Williamson ether synthesis (Section 10.10): Method for the preparation of ethers involving an S_N2 reaction between an alkoxide ion and a primary alkyl halide:

$$RONa \ + \ R'CH_2Br \ \longrightarrow \ R'CH_2OR \ + \ NaBr$$

Wood alcohol (Section 3.1): A common name for methanol, CH_3OH.

Zaitsev's rule (Section 4.7): When two or more alkenes are capable of being formed by an elimination reaction, the one with the more highly substituted double bond (the more stable alkene) is the major product.

Zwitterion (Section 17.3): The form in which neutral amino acids actually exist. The amino group is in its protonated form and the carboxyl group is present as a carboxylate

$$\underset{\overset{|}{\overset{+}{N}H_3}}{RCHCO_2^-}$$

INDEX